S0-BDO-881

IES LIGHTING HANDBOOK

1981 Application Volume

"Its object shall be . . . the advancement of the theory and practice of illuminating engineering and the dissemination of knowledge relating thereto."

IES LIGHTING HANDBOOK

1981 **Application Volume**

JOHN E. KAUFMAN, PE, FIES
Editor

HOWARD HAYNES
Associate Editor

Published by
ILLUMINATING ENGINEERING SOCIETY OF NORTH AMERICA
345 East 47th Street, New York, N. Y. 10017

ISBN 0-87995-008-0
Library of Congress Catalog Card Number: 80-84964

Composed and Printed by WAVERLY PRESS, INC., Baltimore, Maryland

PRINTED IN THE UNITED STATES OF AMERICA

Preface

1981 is the Society's 75th year of continuous dissemination of knowledge relating to the advancement of the art and science of illuminating engineering. Today, more than ever before, illuminating engineering involves a changing cross-section of individuals with varied interests in an expanding range of research and application activities. As a result, lighting technology is changing rapidly. To accommodate this rapid change the *IES Lighting Handbook* is being published in two volumes—a Reference Volume and an Application Volume—each to be updated more frequently than the single volume Handbook which was first published in 1949, with a second edition in 1952, a third in 1959, a fourth in 1966 and a fifth in 1972. Furthermore, it is planned to revise each volume on its own cycle to provide more up-to-date reference and application materials as needed. To help the user identify the most recent material, each volume has a date rather than an edition number.

This *IES Lighting Handbook—1981 Application Volume* is an expanded and up-to-date version of the application type materials in the 5th edition Handbook. Updated and expanded reference type material appears in the *1981 Reference Volume.* As an aid to the reader the Table of Contents for the *1981 Reference Volume* appears along with that of this volume and the locations of subject matter in both volumes are included in the Index at the end of this volume.

Major changes in content within the two volumes include:

1. The use of SI units (International System of Units) in which lux rather than footcandle is used as the unit of luminous flux density, candela per square meter rather than footlambert for luminance and meter and millimeter rather than foot and inch for length, but with the replaced values and units given in parentheses or brackets (the latter where conversions are approximate);
2. The use of the term "illuminance" to replace "illumination" when referring to luminous flux density;
3. New lighting terms and their definitions;
4. The addition of more energy management related materials, including a separate section on the subject;
5. A new method for selecting illuminance values for interior lighting design;
6. Current industry tabulations of lamp data;
7. New calculation and measurement procedures;
8. More color illustrations to explain color phenomena;
9. More discussion of lighting design with a new section on that subject;
10. A section on lighting system design considerations covering the application of lighting quality and quantity criteria;
11. Current quality and quantity criteria for offices, educational facilities, health care facilities, merchandising spaces, residences, roadways, sports facilities, airports, public conveyances and outdoor decorative applications and advertising;
12. Expanded coverage of the nonvisual applications and effects of ultraviolet, visible and infrared radiation.

All Handbook material has been reviewed for content and accuracy either by the Society's technical committees, under the counsel of the Handbook Editor as Technical Director, or where outside the scope of the Society's committee structure by other organizations and individuals, and for that we wish to acknowledge sincere appreciation to the Society's technical committees and to the individuals and organizations listed on the following page.

The Handbook Committee (listed below) developed policy and plans for operation, and helped establish the format of the 1981 Volumes.

Handbook Committee

D. H. Shapiro, *Chairman*	H. G. Jones
C. L. Amick	R. C. LeVere
S. W. Bruun, *Advisory*	J. J. Neidhart
G. F. Dean	M. N. Waterman
J. H. Jensen	R. S. Wissoker

In addition to those acknowledged above, appreciation is also given to the officers of the Illuminating Engineering Society during the Society's years 1976 to 1980 under whose administration and stimulation these volumes were prepared, and especially to Frank M. Coda, Executive Vice President, who was helpful in consultation and budgetary operations.

Editing, coordination, artwork and production of the Handbook were carried out through the Society's Technical Office with Howard Haynes as Associate Editor, and LaRee Di Stasio, Donna Brauer, Edward A. Campbell and Jack F. Christensen.

With the continued objective of providing essential information on light and lighting in a simple, condensed style and with the extent of revisions and additions of new subjects and material it is evident that errors or omissions will occur. Whenever such errors or omissions are found, your comments will be appreciated.

John E. Kaufman
Editor

Contributors

The numbers in parentheses following each name refer to the Sections where contributions were made. Credits for Illustrations and Tables are found on page C-1.

Contributing IES Technical Committees

Aviation (15)
Design Practice (2)
Emergency (2)
Energy Management (4)
Farm (IES/ASAE) (9)
Industrial (9)
Institutions (7)
Lighting Economics (3)
Lighting & Thermal Environment (2)
Merchandising (8)
Office (5)
Photobiology (19)
Residence (10)
Roadway (14)
School and College (6)
Sports and Recreational Areas (13)
Theatre, Television and Film (11)

Contributing Individuals and Organizations

M. A. Aimone (4)
E. W. Bickford (19)
W. C. Burkhardt (17)
F. Clark (4)
D. L. DiLaura (2)
C. A. Douglas (20)
J. R. Fairweather (12)
W. L. Fink (2)
J. E. Flynn (2)
R. H. Goodman (17)
P. H. Greenlee (15)
A. L. Hart (12) (17)
R. L. Henderson (16)
International Association of Lighting Designers (1)
IEEE Production and Application of Light (2)
J. H. Jensen (1)
J. S. Kinney (18)
P. D. Lee (18)
R. E. Levin (2)
W. B. Maley (16)
P. W. Mauer (16)
Outdoor Advertising Association of America (17)
W. F. Parrish (18)
Society of Motion Picture and Television Engineers (11)
W. K. Y. Tao (4)

1981 Application Volume

Contents

1981 Reference Volume

Contents

Lighting Design

SECTION

1

Lighting is an Art as well as a Science. Illumination produced from daylight and electric light sources gives one the freedom to see anywhere at any time. But lighting is not merely utilitarian—it is capable of providing more than just a high degree of visibility for seeing tasks and objects (see Fig. 1-1). It is increasingly evident that lighting is not only an engineering discipline: it also is an art of marvelous subtlety and resource.

Lighting as an art has always pervaded the theatre, the creative retail display and the photographer's set. In these applications lighting is or primary importance to a goal of enhancing reality; massive numbers of luminaires are often employed and hidden outside of the photographer's frame or conveniently behind prosceniums and valances.

Today, lighting as an art, combined with new kinds of light sources, luminaires and techniques, has brought a more subtle enhancement to the illumination of homes, restaurants, lobbies, offices, museums, arenas, gardens, terminals, vehicles and, in fact, every kind of space used by people.

To master the humanistic aspect of lighting, the lighting designer should learn to see light with a knowing eye. As a part of one's everyday life, one should constantly ask the question: "Where is the light coming from and what is it doing?" This question should be applied to whatever one sees—to mountains in sunlight, to faces in offices lighted with fluorescent sources, to tables in a restaurant romantically lighted with candles. This question is concerned with shadows and textures and colors and reflections, and it

Fig. 1-1. Today's lighting design approach is broadly concerned with providing light for the visual tasks to be performed, along with creating a balanced, comfortable and esthetically appealing environment, coordinated with the decorative and architectural scheme.

NOTE: References are listed at the end of each section.

should lead to more questions: "How does the lighting make me feel?" and "How would I feel if the lighting were changed in some particular way?"

The designer should look at the lighting in great paintings and photographs; go to the theatre; and watch what the sun does, indoors and out, and how it changes with the seasons. Above all, buildings of all kinds should be visited and their lighting observed with a knowing eye.

Vertical vs. Horizontal Illuminance. A thorough study of classical lighting installations will reveal that the most exciting lighting, the most interesting lighting, the "lighting that people like" is not usually the *horizontal* uniform illumination that the engineer has historically been trained to calculate and lay out. Rather, it is the interesting play of light and shadow in the *vertical* plane that causes the most favorable human response. Although the specification and calculation of horizontal illuminance is called for in most work areas, it must not be considered all-important.

An example of the inappropriateness of horizontal illuminance as a measure is a lobby with a very dark floor and little human traffic. If the floor has a 5 per cent reflectance, what is the use of specifying even 500 lux (50 footcandles) on the floor, for it will only yield a luminance of 8.6 candelas per square meter (2.5 footlamberts) on the floor. One must hope instead for a light wall, and concentrate most of the available watts of energy for lighting there.

But what if there is no light wall? Then it is incumbent on the lighting designer to tell the architect that the lighting will be a disaster! There *are* lighting impossibilities—a mirror cannot be lighted, nor can true stained-glass, except by lighting an equally large diffusing surface behind it (see page 7–4). Yet costly attempts have been made to achieve these impossibilities. In the long run it is always better to say "this can't be done" than to equivocate.

Lighting designers must be aware that light is not manifest as it passes through the air—it must first be intercepted by an object. In the case of the dark lobby, designers must go beyond providing illuminance—horizontal or vertical—they must authoritatively propose to the architect the inclusion of objects that can be lighted! This might mean murals, paintings, tapestries, plantings, fish tanks, sculpture, screens or furniture—anything that is light-colored and will reflect light in an interesting way, and is appropriate to the design.

THE DESIGN PROCESS

Any design process is rooted in analysis and synthesis: first there is a gathering of pertinent information; then the examining, organizing and distilling of the information (analysis); and finally the selecting and combining of elements to produce the "design" (synthesis). The difficulty in lighting design is that not only is there an enormous body of physical phenomena to consider, but also an equivalently large group of functional, formal and emotional factors which must be balanced and blended.

The lighting designer, architect, contractor, interior designer and consulting engineer are all concerned with the physical organization of human activities and with the problem of providing pleasant, protective shelter to house these activities. This problem includes the search for an understanding of optimum environmental conditions. And in this search, it has been recognized that a major criterion in building design for optimum environments is to properly solve the problem of bringing light into an enclosed, protected space.

In this respect, the contemporary design team has a greater degree of freedom and flexibility than its predecessors. Contemporary builders are free to use available daylight, as architects have done throughout history; or they can solve the lighting problem somewhat independent of structure by using electric light to achieve their objectives.

Light can influence an observer's unconscious interpretation of a space, for the observer's judgment is based not only on form, but on form as modified by light. There are both esthetic and psychological implications in this.

Analysis

The first step in the design process is analytic. Starting on the scale of the largest possible context of the project at hand, one should identify the influences that exist. A design problem is like a set of nested boxes in that any decision about how to build one of the boxes is influenced by the next larger box and the next smaller box. Lighting design differs from architectural design at this early stage because a proposed space or area is usually already defined when the lighting design process begins. That is to say, some of the largest boxes have already been built. Adjustments of those early decisions may be necessary when daylight factors are brought to the attention of the design team by the lighting consultant. But usually, many fundamental decisions have already been made in response to such questions as: "Which site should be chosen?", "How big is the space or area?", "Where is the entrance?", "What is the basic building envelope or area boundary?", "What is the surrounding space or area (the next larger box)?", etc.

The answers to these questions must include the usual environmental design factors and concerns, but must be particularly thorough in the area of visual matters: "What does one see?" and "What is the nature of what is seen—what are the brightness patterns and what are the adaptation factors?" This line of analysis is used for all levels of the project.

Once the contextual analysis and itinerary are complete, the questions are: "What do you want to see?", "What does the space or area or surround emphasize by virtue of its configuration?", "Should that emphasis be reinforced by the lighting?", "What is the over-all composition of the lighting?", "What is the day/night set of differences?", "Where are the highlights, where are the shadows?", "What should the lighting reinforce?" and "What is the character of the light in each area and sub-area (always the nested boxes!)?"

Synthesis

Synthesis is the act of putting together the given, the found and the rationalized elements from the previous sections. It is of utmost importance to keep the *whole* design problem in mind, not just lighting issues, in this synthesis. A pattern—or combination of decisions—is created which is the "design", in the largest sense: it incorporates the functional, formal and emotional findings and decisions. Later, this section discusses some of the aspects of lighting design that must be included in this pattern or master plan. They are not mutually exclusive: indeed they are strongly interrelated. Their synthesis is a simultaneous one, in which one part yields dominance to another, and each part is continually checked against the other for compatibility. Some of these general aspects will include: light and shadow, sparkle, diffusion, and the descriptive explanations of several design practice systems, etc.

As stated previously, in any project in which lighting designers are asked to participate, their work should be an integral part of the whole. If the building or other facility is conceived of as an organism, it becomes evident that the lighting system must grow and develop in coordination with all other systems to perfect its function. The design process is one of intercommunication between the disciplines of architecture, interior design and engineering.

APPROACHES TO LIGHTING DESIGN

Two approaches to the lighting design process are outlined in this section—from the luminous environment viewpoint and from the visual task viewpoint (both entirely compatible)—but before they are presented several general considerations are given (for specific lighting design considerations, see Sections 2 through 4).

Consideration of Space Function. The function of a space greatly influences the way in which lighting is applied. In an office, for example, the typical visual task may be considered as reading at desk top level. However, the same type of visual task (reading) may be encountered, regardless of location, in a factory, in a store, or in a home. But such factors as economics, appearance, continuity of effort, and quality of lighting results desired, influence the lighting design developed for the task. Thus application techniques generally designated as industrial lighting, store lighting, office lighting, and so on have developed based on lighting solutions for the types of visual tasks encountered generally in each type of occupancy. Each of these is a synthesis of engineering theory, application experience, and consumer acceptance and desire in a particular field. Because these include more than an objective assessment of engineering considerations, it is necessary to relate the design of a lighting installation to the particular occupancy of the space it is to serve.

Provision of Quality and Quantity of Illumination. The lighting designer needs to know and understand fully the visual sense, or *how we see* (Section 3 of the 1981 Reference Volume). Such knowledge is basic in selecting the actual luminance of the visual task and the relative luminance of the task, its immediate surroundings and anything else in the peripheral field of view. These luminances affect visual comfort and the performance of the task. Research indicates that desirable seeing conditions exist when the luminances of the surroundings and the visual task are relatively uniform, and veiling reflections are eliminated or effectively reduced and diminished. Since this condition is not always practical or easy to attain, especially in view of decorative and other considerations which are usually involved, luminance limitation recommendations have been developed. When followed, these recommendations will provide a generally satisfactory visual environment. See later application sections.

The lighting designer has considerable latitude in the selection of a lighting solution for a specific area. Good practice calls for provision of both quality and quantity of lighting, which are commensurate with the degree of severity of the seeing tasks which will exist in the area, and which will aid performance and minimize the fatigue resulting from visual effort. These provisions are in the form of recommended values (see Section 2 and later application sections).

As mentioned earlier, it is important to remember that not all visual tasks are in a horizontal plane. Much critical seeing in business and industry is involved with tasks in a vertical or other non-horizontal plane. Typical design methods do not always take this into account, so the designer must often make special provision for luminaire distributions and placements that provide task luminances in these non-horizontal planes.

Selection of Lighting Systems, Sources and Luminaires. Lighting designers have at their disposal a wide range of types and sizes of light sources, of luminaires, and of lighting equipment and lighting components. Here, they have an opportunity to exercise professional judgment, personal taste or choice based on economic analyses, and come up with any of several solutions, each of which may be justified on its own merit.

Lighting systems have been simply classified by illuminating engineers to reflect the general type of lighting produced, and the general layout of luminaires. They are generically described as general, local, localized general, supplementary and task-ambient. Also, luminaires have been divided into five classifications, based on light distribution characteristics, as follows: direct, semi-direct, general diffuse, semi-indirect, and indirect (see page 1-12). These classifications are useful primarily in that they simplify professional discussion relating to lighting techniques as employed for lighting any specific area.

The choice of light source, of luminaire characteristics, and of the system layout are closely interrelated in an application technique. A method easily applied with one type of light source may be equally applicable, or most impractical, with another. Frequently local conditions of vibration, ambient temperature, dust and dirt, continuity of service, and color influence light source selection, application, operation, output and maintenance and, indirectly therefore, the lighting application technique. See also Section 8 of the 1981 Reference Volume. Where various light sources are equally applicable, economics may be the deciding factor.

The lighting designer may have a choice in the method of installation of the lighting system; or, architectural design and structural conditions may dictate a particular method of installation. In either case, a competent knowledge of the principles of light control and of the lighting tools and devices which are available for such control, will be most helpful in the design of the most satisfactory and efficient lighting system possible for the problem involved. This is particularly true where the designer strives for energy efficiency in lighting design.

Selection of Lighting Control Systems. Today's interest in lighting energy management (see Section 4) dictates more consideration of lighting control. Lighting control systems range from simple photoelectric controls to turn an individual luminaire on and off to sophisticated microprocessor controllers that oversee all the lighting in a multi-building complex. Switches, dimmers, timers, photocells, and other controls can be simple or complex, but are a necessary part of lighting design as they provide flexibility to the lighting design and contribute significantly to energy conservation.

Consideration of Economics. The design of a lighting system is affected by both initial and operating costs. In this era of rapidly-escalating energy costs, considerations of operating costs often outweigh the initial cost. Life-cycle costing is often the best way to look at lighting costs. There is no sharp line of demarcation between excellent and good lighting, between good and average, or between average and poor. Also, there is no easy way to predict the exact value of commercial or industrial lighting in terms of production, safety, quality control, employee morale or employee health; or to weigh the importance of home lighting in dollars and cents. Nevertheless, lighting designers must balance costs against the attainable results in developing any lighting design, relying to a great extent upon experience gained in the solution of comparable problems if such is available. See Section 3.

Coordination with Mechanical and Acoustical Systems. The enjoyment of a space or activity involves all of the senses, and though lighting systems and the visual implications of light can be discussed separately as an abstraction, the final design should reflect the needs of the total environment. The development of air-conditioning systems and the evolution of acoustical control techniques have combined with electric light to provide an extensive and unprecedented capacity for environmental control.

It has become increasingly critical that electrical and mechanical components be compatible with other elements in the same building system so that the resulting assembly is an efficient and economical operating unit *and* an architecturally coordinated design. In this regard, an increasing number of luminaire types has been developed to supply both light and air to a space. See page 2-43. This equipment offers the initial advantage of a simplified ceiling appearance, due to the elimination of overlapping (and sometimes conflicting) mechanical and electrical ceiling patterns. These units may also offer considerable flexibility in space planning, where the "mechanically coordinated module" is employed. Within each standard dimensional module, there should be: (a) lighting, (b) air supply, (c) air return, (d) sound absorption and (e) sprinkler heads where necessary. Where this can be achieved, it means that modular floor and partition systems can be used (initially or in future modifications) to set off any module or any combination of modules as a separate room. Each room, then, would automatically include all of the basic elements of the mechanical-electrical environment. To achieve this degree of extreme flexibility with conventional equipment would require a luminaire, an air diffuser and an air return in each module.

Coordination with Furniture. In an effort to simplify the flow of work through an office space, interior designers and space planners have been utilizing the "open office plan" or "office landscape" approach. In these designs, work stations are situated close to room divider panels which can interfere with the distribution of light from ceiling-mounted luminaires. Such arrangements often call for *task lighting* built into the furniture and directing the light onto localized work areas.

Such lighting must be carefully designed to cover all of the key work areas and to minimize veiling reflections. Such a system of localized task lighting could create severe luminance ratios in the field of view unless supplemented by some form of general or *ambient* lighting.

Sometimes this ambient lighting is located in the furniture itself (or free-standing nearby) and directed to the ceiling to create an indirect general lighting system.

A great deal of coordination is required among the lighting designer, the space planner or interior designer, and the owner to assure that the lighting is located correctly and that ample provision has been made for future flexibility.

Interior-Exterior Relationships. With increasingly widespread circulation and activity at night, the problem of building esthetics extends beyond daytime architecture. Particularly where there is an extensive use of glass, the organization and design of the interior lighting system will affect the exterior impression of the building at night and on days that are relatively dull and overcast. The brightness of visible interior surfaces, the pattern of luminaires, and the color of the light source may exert an important influence on the exterior appearance of the building during these periods.

Luminous Environment Design Approach

It is suggested that lighting designers, when using the luminous environment design approach, include the following in their analysis and synthesis steps: (1) determine the desired visual composition in the space; (2) determine the desired appearance of visual tasks and objects in the space; (3) select luminaires that fit the concept of visual composition and implement the desired appearance of objects.

Step 1. Determining the Visual Composition of the Space. Through the design and placement of lighting elements, the designer can specify the combination of surfaces to be lighted, or left in darkness. In this sense, the designer can specify how the pattern of brightness is to merge with the structural pattern. Looking at it in another way, the designer can specify how the visual perception of space is to merge with the activity involved—the use of light to identify centers of interest and attention, and otherwise to complement the basic mood of the activity as this is understood and interpreted by the individual designer.

In practice this responsibility seldom lies entirely in the hands of a single individual. Usually, the design is the result of some combination of skills in architecture, interior design, illuminating engineering, and electrical engineering. The specific responsibility of lighting designers, then, is to recognize the importance of both spatial and task-oriented attributes so that their technical efforts will contribute to the total result. To do this, the following should be considered:

1. ***Focal Centers.*** Does the lighting system properly identify centers of primary and secondary attention? A display, a picture, planting or a featured wall are examples. (See Fig. 1-2.) Depending on the relative dominance desired, the luminance of these focal areas should range from 5 to 10 times that of other nearby surfaces.

2. ***The Overhead Zone.*** Since the ceiling is usually of secondary or subordinant interest relative to the activity, the appropriate influence of form, pattern, and brightness in this area should be considered. (See Figs. 1-3 and 1-4.)

3. ***The Perimeter Zone.*** In many cases (particularly where a sense of relaxation is desired), perimeter brightness should be greater than the brightness of the overhead zone. (See Fig. 1-5.) In general, simplicity is desirable in the perimeter zone. "Visual clutter" here may cause confusion in spatial comprehension and orientation. It may also complicate identification of meaningful focal centers in merchandise and display areas.

4. ***The Occupied Zone.*** The functional or activity emphasis is usually in this zone. For this reason, illumination objectives here are generally those listed in Fig. 2-2. The balance among this zone and the two previously mentioned determines over-all effectiveness. (See Figs. 1-6 and 1-7.)

Fig. 1-2. An example of a very strong focal center. While general illumination in the store is well shielded and low key, the silverware display, through spotlighting and color, is the eye-catching feature.

5. ***Transitional Considerations.*** When the luminance ratio between adjoining spaces and activities approximates 2 or 3 to 1, or less, the sense of visual change tends to be subliminal. When a sense of "change" is desired, rather than "continuity," luminance differences should substantially exceed this 3 to 1 ratio. In general, a clearly noticeable transition is approximately a 10 to 1 difference. A dominant or abrupt change is approximately 100 to 1 or greater.

Color continuity or change ("whiteness") should also be considered in the lighting of adjacent spaces.

Fig. 1-3. The spatial influence of the lighting system affects orientation and comprehension of the room. The system on the left is spatially confusing despite the fact that work surface illumination and visual comfort are both relatively good. By way of contrast, the lighting system on the right functions in a sense of unity with the basic rhythm of the space.

Fig. 1–4. All four photos illustrate methods of providing relatively uniform lighting for large work areas. In the upper left, the ceiling is a unifying factor, but completely dominates the space. The upper right photo shows a pattern introduced into the lighting system. Architectural interest is achieved along with a reduction in ceiling luminance. In the lower left photo, a high degree of brightness control has been achieved through the use of louvers with precision cut-off. The slight configuration of the ceiling provides relief from a flat plane. There is also coordination of the ceiling and window patterns. In the office at lower right, attention is removed from the overhead zone by a ceiling configuration which causes the luminaires to disappear from view as one looks across the room. Used in combination, the dark and light walls provide relief. If all walls had been dark or all had been light, the effect would have been less pleasing in this large space.

Fig. 1–5. Lighting of vertical surfaces is often more important to the impact of a space than horizontal illumination. Such lighting can help create a bold architectural statement, can play one plane against another and, by reflection, provide illumination for safe passage in public areas.

Fig. 1–6. For an effective appearance there should be a balance among the brightnesses in the overhead, perimeter and occupied zones. In the store at the upper left, the ceiling dominates, tending to compete with merchandise. In the upper right, attention has been removed from the ceiling by low brightness luminaires but space tends to be dingy because of low perimeter brightness. An example of proper combination of minimum attention on ceiling and accent on perimeter in a general merchandise operation is shown at the right.

Fig. 1–7. Two examples of lighting in a private office. In the one at left, the ceiling dominates and there is no focal emphasis for the occupant. In the one on the right, a carefully shielded and precisely located lighting panel focuses attention on the occupant, reduces ceiling brightness and minimizes veiling reflections. Balancing luminances are provided by accent spotlighting and wall lighting (to emphasize rough texture of paneling). A portable lamp provides accent and interest.

6. ***Level of Stimulation.*** There are few absolutes here; but some significant variables can be outlined—along with the direction of their influence.

General Luminance Level. Levels below 35 candelas per square meter [10 footlamberts] are associated with twilight conditions. These levels may appear "dingy" unless higher brightness accents are provided.

Color and Shade. Warm colors tend to stimulate; cool tones tend to soothe. Saturated colors are more stimulating than tints.

Areas of darkness tend to be subduing or relaxing, and may be dramatic if focal accents are provided.

Information Content. When a person is understimulated, he or she may become bored with the task or activity. In these situations, techniques involving spatial pattern (brightness, color, sparkle) often tend to stimulate interest and vitality.

Step II. Determining the Desired Appearance of Objects in the Space. In addition to the above spatial factors, parallel lighting criteria relate to the desired appearance of objects in the space.

1. ***Diffusion.*** Diffuse light tends to reduce the variations that relate to form (contour gradients), pattern, and texture. (Consider that a tennis ball on an overcast day may actually seem to disappear.) However, diffusion is desired in most work areas to prevent distracting shadows at the task itself.

Sculpture should generally be developed with some directionality in lighting (see page 7-30)—but with significant diffusion to relieve the harshness. The appearance of facial forms and expressions are best treated with a similar combination of directional and diffuse lighting.

2. ***Sparkle.*** Sparkle and highlight can enhance the sense of vitality in a space. For this reason, consideration should be given to the value of concentrated or "point" light sources interacting with polished and refractive surfaces.

3. ***Color Rendition.*** Color schemes (paint, fabrics, etc.) should be chosen under the lighting to be used in the space. In choosing light sources, consider the importance of appearance of people, merchandise, etc. Visual contact is nearly always important. Light sources should not be compared side-by-side because in a given space the eye tends to adapt over a wide range. The eye is extremely sensitive to color differences in transition from one space to another or from one display to another.

Step III. Selection of Luminaires to Fit the Concept of Visual Composition and to Implement the Desired Appearance of Objects. This section has, so far, stressed the visual implications of light in a space. But lighting is a design objective—an end result to be achieved through the careful selection and placement of lighting components.

The terms "luminaire" and "lighting element" imply the ingredient of light control. Such factors as brightness control (for glare) and appropriate beam distribution (for direction or diffusion of light) should be considered. It is in the design and placement of these devices and elements that the designer specifies the luminous environment.

A. *The Engineering Study.* An initial engineering study is needed to determine the alternative techniques available to achieve the design objectives. The background for this study may come from the experience of the designer or from a consultant in this field. From these alternative techniques, a preliminary selection can be made that reflects such considerations as initial and operating costs, maintenance, ruggedness, candlepower distribution and brightness control.

Once this preliminary selection of techniques is made, then the problem becomes primarily one of testing and design assimilation. In this regard, the following engineering-oriented criteria are important in guiding the final selection of lighting equipment from among the alternatives available:

1. ***Distribution Characteristics and Color of Light.*** Is the selection suitable to achieve the desired visual result (spatial illumination and illumination at the various task centers, as discussed under Steps I and II)?

2. ***Dimensional Characteristics and Form.*** Are the physical characteristics regarding shape and size generally appropriate to meet the needs of the illumination concept? Reflector size and finish, lamp-to-enclosure distance, shielding and cut-off angles, lamp ventilation, etc., should be considered as well as the dimensional intrusion into the space when pendant or bracket equipment is involved.

3. ***Lighted and Unlighted Appearance of Lighting Materials.*** Is the appearance and quality of detail compatible with the general quality of other materials in the building design?

4. ***Initial and Operating Costs.*** Is the cost structure compatible with the general quality of

other materials and systems in the building design?

5. *Maintenance.* Is the system accessible for lamp replacement and cleaning? Are the characteristics regarding dirt collection and deterioration compatible with the use of the space? What is the recommended maintenance interval? See Section 4.

6. *Energy.* Is the lighting system energy efficient? See Section 4.

B. *The Architectural Study.* In addition to the engineering-oriented considerations, it should be recognized that luminaires and other lighting components become potentially prominent factors in the architectural composition. In this sense, modern lighting techniques should be carefully assimilated into the basic architectural design concept, because this system assumes an inherent esthetic significance far beyond the normal connotations of electrical-mechanical design. For this reason, lighting devices and elements

Fig. 1–8. Styles and Lighting Effects of Architectural Periods

Period	Architectural Style	Natural Lighting Effects†
Greek 700–146 B.C. Orders: Doric, Ionic, Corinthian Important buildings: Temples	Column and lintel, with entablature. Harmony of design so as to obtain perfect balance between horizontal and vertical elements. Perfect proportion, simple decoration	Emphasis on the statue of the god or goddess to whom the temple was dedicated. Light was obtained from roof openings usually over the statue, or from clerestory openings, or from doorways. Temples were usually oriented so that the rising sun might light up the statue. Direction of incident light mainly from above, at oblique angles.
Roman 146 B.C.–365 A.D. Orders: Tuscan, Doric, Ionic, Corinthian, Composite Important buildings: Temples, basilicas, thermae (baths), palaces	Column and lintel, with entablature. Arch developed. Vault and dome evolved. Elaborate decoration	The Romans used windows extensively. They obtained light by means of clerestories, openings in the center of domes, or windows at the base of domes. Direction of incident light mainly from above, at oblique angles. Light used to enhance the elaborate decoration and majestic proportions of interiors
Early Christian 300–900 A.D. Important buildings: Basilican churches	Column and lintel, with a long interior perspective. Occasional domes and rotundas supported on arched colonnades	Oblique lighting from upper angles obtained through clerestories and window openings, usually small. Emphasis on altar obtained by columnar perspective as well as the convergent perspective of windows in clerestories. Glass mosaics reflecting light often used for the high altar
Byzantine 324 A.D. Important buildings: Churches	The dome on pendentives is the main feature of Byzantine architecture. In Roman architecture domes were used only over circular or polygonal buildings, but in Byzantine architecture domes were placed also over square structures. Here the earlier horizontal motif changes almost imperceptibly to a vertical motif	Lighting from upper angles obtained through windows at the base of domes. The dome being highly illuminated acted as a huge reflector. Small glass and translucent marble windows prevented glare and added color to the interior. Brilliant mosaics glowed with numerous subdued reflections. To relieve their flat wall decoration, the Byzantine builders obtained "depth" by means of arcades
Romanesque 800–1200 A.D. Important buildings: Churches, castles	Massive Roman walls coupled with the round arch	The effect of solemnity and vastness was produced by the contrast between great wall spaces and small windows. Such windows, single or grouped together, admitted rays of light through clerestories
Gothic 1200–1500 A.D. Important buildings: Churches, monasteries, castles, mansions, town halls	This aspiring style with its pointed arches definitely introduced the vertical motif. Solids prevailed in Roman architecture, but in Gothic architecture, voids prevailed instead, since slender buttresses were used instead of massive walls	In churches the mood of solemnity was produced by the lofty, dimly illuminated ceiling, while long rays of light penetrated stained glass windows. In castles and manor houses larger windows than ever had been used before in domestic architecture became the vogue
Renaissance 1400–present day Important buildings: Churches, castles, town halls, palaces, villas, chateaux, civic buildings	The rebirth of classical ideals brought the ideal of architectural harmony again into vogue. Buildings were so designed that the vertical and horizontal members obeyed the classical laws of proportion. For decoration Greek and Roman details were copied	Lighting effects became more numerous to suit different types of buildings. Domes were supported on "drums" which were pierced with large windows. The dome lighting of the Byzantine period was revived and improved. The direction of incident light was still mainly from above, though lower windows also were enlarged. Windows became more numerous, and more light was sought than before
Modern (twentieth century) All types of buildings	The twentieth-century style strives for structural logic. For skyscraper design the vertical motif is emphasized. For smaller buildings the supporting steel structure is not camouflaged but rather is indicated by simple "wall lines" and other decorative devices. Stone, glass, chromium and other metals are used without elaborate ornamentation	Electric illumination now is recognized as an architectural medium. Modern lighting systems vary from the layout with outlets located with mathematical symmetry to the decorative system with light sources in arcades, columns, recesses, panels, cornices, coves, wall pockets, urns, etc. Luminaires differ widely in design and in material

† Some use was made of flame sources (wooden torches, tapers, candles, and oil and gas lamps) even in very early periods. The design of luminaire in period interiors frequently follows the pattern established by the characteristics of these early lamps.

should be analyzed by architectural standards, in addition to their engineering function and performance.

1. ***Brightness, Color, Scale and Form.*** Does the equipment assume an appropriate textural or pattern role in the architectural composition? Does it contribute to a sense of unity?

2. ***Compatability with "Period" Designs.*** Is the detailing in appropriate harmony with the architectural period of the building? See Fig. 1-8. The use of wall lanterns, chandeliers, etc., are generally decorative, and require another unobtrusive system to produce comfortable illumination without glare.

3. ***Space Requirements and Architectural Detailing.*** Are the physical space allowances sufficient? Is the building cubage and design sufficient to provide for necessary lighting cavities and recesses? Is the detailing and use of materials compatible with the detailing of other elements and systems in the building?

4. ***Coordination with other Environmental Systems.*** Is the lighting system functionally and physically compatible with other environmental systems in the building design? Consider coordination with other ceiling elements and wall systems.

C. ***The Architectural Context of Luminaires.*** A study of architectural history reveals two basic alternatives in the approach to lighting systems and lighting design: (1) the visually-subordinant system, and (2) the visually-prominent system.

1. ***Visually-Subordinant Lighting Systems.*** Throughout the history of building, some designers have attempted to introduce light in a way that the observer will be conscious of the effect of the light, while the light source itself is played down in the architectural composition. For example, in some Byzantine churches, small unobtrusive windows were placed at the base of a dome to light this large structural element. The brilliant dome then became a major focal center; and serving as a huge reflector, the dome (not the windows) became the apparent primary light source for the interior space. Similarly, the windows of some Baroque interiors were placed so that they were somewhat concealed from the normal view of the observer, and the observer's attention was focused on a brightly lighted adjacent decorative wall. In both cases, the objective was to place emphasis on the surfaces to be lighted while minimizing any distracting influence from the lighting system itself.

This design attitude can be seen in the development of some electric lighting systems (see Figs. 1-4 (lower left and right) and 1-5). In this regard, indirect systems and concentrating direct lighting systems are useful design tools. With appropriate shielding and careful control of luminaire brightness, these devices direct light toward a specific surface, plane or object—emphasizing these objects or areas, with little distracting influence from the lighting device itself. Inherently, then, the space is visually organized as a composition of *reflected* brightness patterns (horizontal and vertical).

2. ***Visually-Prominent Lighting Systems.*** A light source or luminaire may itself compel attention, even to the extent that such elements become dominant factors in the visual environment. In architectural history, the large stained glass windows of the Gothic period are probably the most obvious examples of this approach. In contemporary building, transilluminated ceilings and walls are a similar dominant influence as seen in Figs. 1-4 and 1-6 (upper left).

Where light-transmitting (rather than opaque) materials are prominently involved in the lighting unit itself, the units become architectural forms and surfaces, as well as lighting elements. Such self-luminous elements help to visually define a space and are important to the general architectural organization of the room.

Visual Task Oriented Design Approach

In areas where the primary function of the lighting installation is to provide illumination for the quick, accurate performance of visual tasks, the task itself is the starting point in the lighting design. Some of the factors to be considered in the design approach are briefly listed below; however, more detailed information can be found in the later application sections.

1. ***Visual Task.***

a. What are the commonly found visual tasks?

b. How should the task be portrayed by the lighting? Should the lighting be diffuse or directional? Are shadows important for a three dimensional effect? Will the task be susceptible to veiling reflections? Is color important?

c. What illuminances should be provided?

2. ***Area in Which Task is Performed.***

a. What are the dimensions of the area and reflectances of surfaces?

b. What should the surface luminances be to minimize transient adaptation effects without creating a bland environment?
c. Might the surfaces produce reflected glare?
d. Is illumination uniformity desirable?

3. ***Luminaire Selection.***
a. What type of distribution, control medium and spectral quality is needed to properly portray the task (for diffusion, shadows or avoiding veiling reflections) and provide a comfortable environment (visually, thermally and sonically)?
b. What type is needed to illuminate the area surfaces (for transient adaptation, for avoiding reflected glare)?
c. What should the luminaire look like and how should it be mounted (see previous design approach)?
d. What is the area atmosphere and, therefore, the type of maintenance characteristics needed?
e. What are the economics of the lighting system?

4. ***Calculation, Layout and Evaluation.***
a. What layout of luminaires will portray the task best (illuminance, direction of illumination, veiling reflections, disability glare)?
b. What layout will be most comfortable (visually—direct and reflected glare—thermally)?
c. What layout will be most pleasing esthetically?
d. What energy management considerations apply?

Fig. 1-9. One form of task-ambient lighting is illustrated here. Fluorescent luminaires provide both the upward general illumination and the light directed to the work area. With the "task light" placed directly in front of the user, care must be exercised to keep veiling reflections to a minimum. In this installation, light from the center section of the luminaire has been blocked by an opaque shield. Light is directed from the two ends of the luminaire to the task area by special lenses.

TYPES OF LIGHTING SYSTEMS

Lighting systems are often classified in accordance with their layout or location with respect to the visual task or object lighted— general lighting, localized general lighting, local (supplementary) lighting and task-ambient lighting. They are also classified in accordance with the CIE* type of luminaire used—direct, semi-direct, general diffuse (direct-indirect), semi-indirect, and indirect. See Fig. 1-10.

Classification by Layout and Location

General Lighting. Lighting systems which provide an approximately uniform illuminance on the work-plane over the entire area are called general lighting systems. The luminaires are usually arranged in a symmetrical plan fitted into the physical characteristics of the area. See Fig. 1-4 (lower right). Such installations thus blend well with the room architecture. They are relatively simple to install and require no coordination with furniture or machinery that may not be in place at the time of the installation. Perhaps the greatest advantage of general lighting systems is that they permit complete flexibility in task location.

Localized General Lighting. A localized general lighting system consists of a functional arrangement of luminaires with respect to the visual task or work areas. See Fig. 1-7 (right) where the ceiling panel illuminates the desk. It also provides illumination for the entire room area. Such a lighting system requires special coordination in installation and careful consideration to ensure adequate general lighting for the room. This system has the advantages of better utilization of the light on the work area and the opportunity to locate the luminaires so that annoying shadows, direct glare and veiling reflections are prevented or minimized.

* Commission Internationale de l'Eclairage (International Commission on Illumination).

Local Lighting. A local lighting system provides lighting only over a relatively small area occupied by the task and its immediate surroundings. The illumination may be from luminaires mounted near the task or from remote spotlights (portable table lamp and wall washing units in Fig. 1-7, right). It is an economical means of providing higher illuminances over a small area, and it usually permits some adjustment of the lighting to suit the requirements of the individual. Improper adjustments may, however, cause annoying glare for nearby workers. Local lighting, by itself, is seldom desirable. To prevent excessive changes in adaptation, it should be used in conjunction with general lighting that is at least 20 to 30 per cent of the local lighting level; it then becomes *supplementary lighting.*

Task-Ambient Lighting. A relatively new type of lighting arrangement is called *task-ambient* lighting. It is most often built into the furniture in an open plan office layout. *Task* lights are located close to work areas (as in local lighting, above) and are supplemented by indirect *ambient* illumination from sources concealed in the furniture and directed to the ceiling (as in general lighting, above). See Fig. 1-9. Such systems are characterized by extensive ceiling areas almost totally devoid of lighting equipment. Electrical service to the lighting equipment is often through the floor.

Classification by CIE Luminaire Type

Direct Lighting. When luminaires direct 90 to 100 per cent of their output downward, they form a direct lighting system. The distribution may vary from wide spread to highly concentrating depending on the reflector material, finish and contour and on the shielding or control media employed.

Direct lighting units can have the highest utilization of all types, but this utilization may be reduced in varying degrees by brightness control media required to minimize direct glare. Direct glare may also be reduced by using area units with minimum lampings, *e.g.*, a two-foot-wide fluorescent unit with just two lamps.

Veiling reflections may be excessive unless distribution of light is designed to reduce the effect.

Reflected glare and shadows may be problems unless close spacings are employed. Large area units are also advantageous in this respect.

High reflectance room surfaces are particularly important with direct lighting to improve brightness relationships, and higher illuminances provided by controlled brightness equipment will also tend to improve the brightness relationships throughout the room. With very concentrating distributions, care should be taken to ensure adequate wall luminances and illuminances on vertical surfaces.

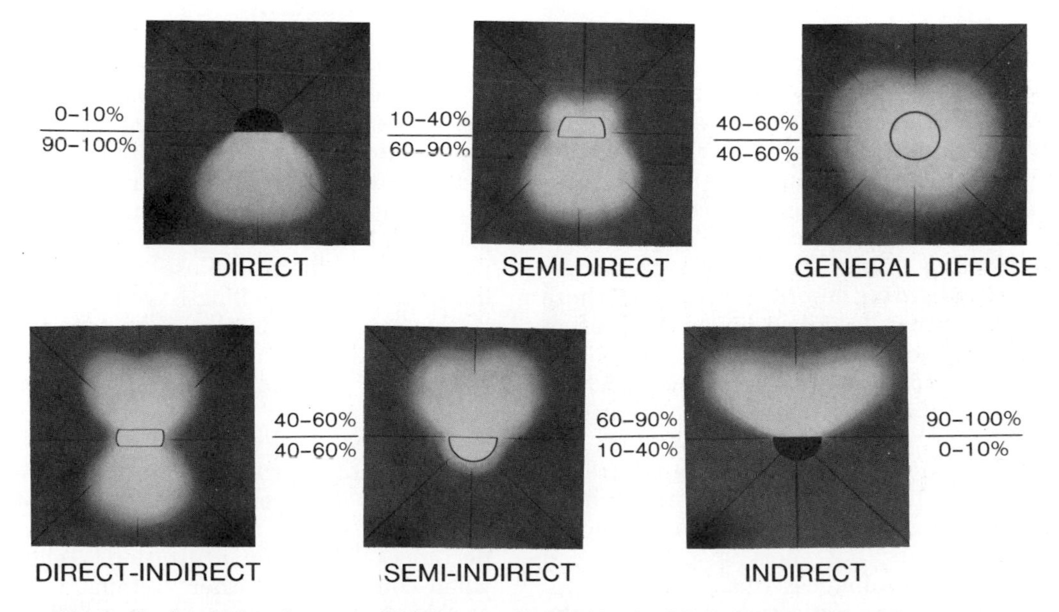

Fig. 1-10. Luminaires for general lighting are classified by the CIE in accordance with the percentages of total luminaire output emitted above and below horizontal. The light distribution curves may take many forms within the limits of upward and downward distribution, depending on the type of light source and the design of the luminaire.

Luminous ceilings, louvered ceilings, and large-area modular lighting elements are forms of direct lighting having characteristics similar to those of indirect lighting discussed in later paragraphs below. These forms of lighting are frequently used to obtain the higher illuminances, but care should be taken to limit the luminance of the shielding medium to 860 candelas per square meter [250 footlamberts] or less to prevent direct glare if critical, prolonged seeing is involved. Reflected glare may be a problem with systems employing cellular louvers as the shielding medium since the images of the light sources above the louvers may be reflected by shiny surfaces at the work-plane.

Semi-Direct Lighting. The distribution from semi-direct units is predominantly downward (60 to 90 per cent) but with a small upward component to illuminate the ceiling and upper walls. The characteristics are essentially the same as for direct lighting except that the upward component will tend to soften shadows and improve room brightness relationships. Care should be exercised with close-to-ceiling mounting of some types to prevent overly bright ceilings directly above the luminaire. Utilization can approach, or even sometimes exceed, that of well-shielded direct units.

General Diffuse Lighting. When downward and upward components of light from luminaires are about equal (each 40 to 60 per cent of total luminaire output), the system is classified as general diffuse. **Direct-Indirect** is a special (non-CIE) category within the classification for luminaires which emit very little light at angles near the horizontal. Since this characteristic results in lower luminances in the direct-glare zone, direct-indirect luminaires are usually more suitable than general diffuse luminaires which distribute the light about equally in all directions.

General diffuse units combine the characteristics of direct lighting described above and those of indirect lighting described below. Utilization is somewhat lower than for direct or semi-direct units, but it is still quite good in rooms with high reflectance surfaces. Brightness relationships throughout the room are generally good, and shadows from the direct component are softened by the upward light reflected from the ceiling. Excellent direct glare control can be provided by well-shielded units, but short suspensions can result in ceiling luminances that exceed the luminaire luminances. Reflected glare from the downward component can be a problem, but it is mitigated by the reflected upward light; close spacings or layouts that locate units so that they are not reflected in the task result in further reductions.

Luminaires designed to provide a general-diffuse or direct-indirect distribution when pendant mounted are frequently installed on or very close to the ceiling. It should be recognized that such mountings change the distribution to direct or semidirect since the ceiling acts as a top reflector redirecting the upward light back through the luminaire. Photometric data obtained with the luminaire equipped with top reflectors or installed on a simulated ceiling board should be employed to determine the luminaire characteristics for such application conditions.

Semi-Indirect Lighting. Lighting systems which emit 60 to 90 per cent of their output upward are defined as semi-indirect. The characteristics of semi-indirect lighting are similar to those of indirect systems discussed below except that the downward component usually produces a luminaire luminance that closely matches that of the ceiling. However, if the downward component becomes too great and is not properly controlled, direct or reflected glare may result. An increased downward component improves utilization of light somewhat over that of indirect lighting. This factor makes somewhat higher illuminances possible with fewer semi-indirect luminaires and without excessive ceiling luminance.

Indirect Lighting. Lighting systems classified as indirect are those which direct 90 to 100 per cent of the light upward to the ceiling and upper side walls. In a well-designed installation the entire ceiling becomes the primary source of illumination, and shadows will be virtually eliminated. Also, since the luminaires direct very little light downward, both direct and reflected glare will be minimized if the installation is well planned. Luminaires whose luminance approximates that of the ceiling have some advantages in this respect. It is also important to suspend the luminaires a sufficient distance below the ceiling to obtain reasonable uniformity of ceiling luminance without excessive luminance immediately above the luminaires.

Since with indirect lighting the ceiling and upper walls must reflect light to the work-plane, it is essential that these surfaces have high reflectances. Even then, utilization is relatively low when compared to other systems. Care should also be exercised to prevent over-all ceiling luminance from becoming too high and thus glaring.

THE TOTAL CONCEPT OF LIGHTING DESIGN

The purpose of this Application Volume is to discuss the general concept of *Lighting Design* as a method for implementing illuminating engineering technology and to present specific methods and solutions for a variety of application problems. While the engineering aspects of lighting design have been refined to a state of high precision, a clear operative technique for the application of these factors to reveal and enhance the entire built environment has proven much more elusive. This is due, in large measure, to the fact that most of the first hundred years of practical electric light have been devoted to the develoment of sources and to the invention of means for moving light from source to task in an efficient and predictable manner. The result of this has been an overwhelming emphasis on the calculation and evaluation of illuminance.

This emphasis has obscured the fact that the viewer does not "see" illuminance. Rather, the viewer sees luminance in the form of perceived brightness, and the world as an arrangement of objects and surfaces in varying brightness patterns presented in a three-dimensional array. Indeed, the fourth dimension of time is a partner in this arrangement. Since lighting systems are designed for the human viewer, the designer must constantly strive to understand what viewers really perceive, both as a function of what they see and how their intelligence and experience inform their sight.

With this is mind, it is appropriate to establish a definition of lighting design as a method for the determination and composition of all the perceived brightness relationships in three-dimensional space. Thus, beginning with the illuminated task, a lighting design could be developed which results in an orderly hierarchy of brightnesses determining what will (and will not) be seen (and comprehended) through a sequence of time. With this approach, a designer would systematically review all pertinent features of a problem and pursue an integrated solution. From this would come systems which address *all* the functions of good lighting: illumination of primary task, clarification of space and structure, guiding the user about, drawing attention to points of interest and hazards, acknowledging the characteristics of the human vision system, and creating a wide and diversified environment in an economical and responsible manner.

REFERENCES

1. Flynn, J. E. and Mills, S. M.: *Architectural Lighting Graphics,* Reinhold Publishing Corp., New York, 1962.
2. Harris, C. M.: *Dictionary of Architecture and Construction,* McGraw-Hill Book Company, New York, 1975.
3. Design Practice Committee of the IES: "An Interim Report Relating the Lighting Design Procedure to Effective Energy Utilization", EMS-2, *Light. Des. Appl.,* Vol. 5, p. 34, September, 1975.

Lighting System Design Considerations

SECTION
2

As an introduction to the subject of lighting design, Section 1 discusses lighting design and lighting systems in general terms. On the other hand, Sections 5 through 18 provide specific design and evaluation information for the most common interior and exterior facilities, and Sections 3 and 4 cover economics and energy management, in depth, as they relate to lighting. This section is presented as a bridge between the general design and the specific application discussions by providing material on design considerations that apply to each of the application sections that follow. The 1981 Reference Volume provides the basic data and procedures as reference for design purposes.

PSYCHOLOGICAL CONSIDERATIONS

In the planning of lighting systems, lighting designers should be sensitive to more than simple task requirements for reading, typing, sewing, shaving, cooking, etc. They should, of course, be sensitive to the adverse effects of glare. They should also become sensitive to the uses of silhouette, sparkle, focal emphasis, color, and other patterns of spatial light, and become sensitive to the fact that the correct use of these patterns is fundamental in satisfying some space-activity requirements—as when they need to reinforce attraction or attention, reinforce impressions of spaciousness, stimulate sensations of spatial intimacy or warmth, and reinforce impressions of cheerfulness or playfulness.

Categories of Environmental Lighting. In considering the space-activity requirements mentioned above, there are basically two kinds of environmental lighting systems as discussed below:

1. *Lighting systems that flood a space somewhat indiscriminately with permissive illumination from general overhead luminaires.* These systems tend to be behaviorally neutral, in the sense that they tend *not* to exert an intentional reinforcing or guiding influence on user impressions or behavior. Instead, these systems are usually intended to permit:

a. Easy perception of reading or manual tasks.
b. Random circulation and unguided attention.
c. Flexible relocation of furniture and work centers without changes in the room lighting.

This type of system may, therefore, offer many advantages in regard to utility, flexibility and general clarity. But for some types of activities, the resulting diffusion and uniformity are rather significant shortcomings due to the often bland psychological effect of the room when it is lighted in this manner. (It should be recognized that this blandness can often be relieved by a sensitive use of surface color.)

2. *Systems that develop specific patterns of light and shade to reinforce selected information or room cues.* This kind of lighting is much more active behaviorally, in the sense that such systems are generally intended to reinforce a specific pattern of user impression or behavior. As such, this approach requires more specific design intentions.

These designs are often based on the idea that light can be a vehicle that influences users' selective attention or alters the information content in the visual field. As such, these lighting designs should be carefully evaluated for their role in establishing *cues* that reinforce users' understanding of their environment and the activities around them.

NOTE: References are listed at the end of each section.

In this regard consideration should be given to the following categories of visual experience:

a. The effect of light on user orientation and room comprehension. Some lighting patterns seem to affect personal orientation and user understanding of the room and its artifacts. For example, spotlighting or shelf-lighting affects user attention and consciousness; wall-lighting or corner-lighting affects user understanding of room size and shape. Together, these lighting elements can establish or modify the users' sense of visual limits or enclosure.

b. The effect of light on impressions of activity setting or mood. Other lighting patterns seem to involve the communication of ideas or impressions—with the suggestion that light is, in part, a medium that assists communication of spatial ideas and moods.

In this sense, spatial lighting patterns are becoming recognized as part of a visual language that can assist the designer in implementing impressions such as "somberness," "playfulness," "pleasantness," "tension," etc. Similarly, the designer can use light patterns to affect psychosocial impressions such as "intimacy," "privacy" and "warmth." In other words, lighting can be used in one way to produce a carnival-like atmosphere and in another way to produce a somber place for quiet meditation. Lighting can be used to produce a cold, impersonal public place or conversely, a warm, intimate place where one feels a greater sense of privacy. More than esthetic amenities are to be considered here, because these impressions or moods are often fundamental in satisfying some experience and activity requirements in a designed space.

"Light Structure" Models. There is considerable evidence that light can (and does) make an identifiable contribution to the quality of a room—and this contribution clearly goes beyond simplistic concepts of task visibility.

Qualitative influences can be identified through the use of "light structure" models that are developed with contemporary research methods in the behavioral sciences.[1-4] This concept is based on the theory that the experience of room lighting is, in part, an experience of recognizing and assimilating communicative patterns. There is the suggestion that patterns of light can convey information; and that the brain constructs an impression of the phenomenal world from this information. This concept of information content and "meaning" further suggests that lighting should be considered not merely as a stimulus but also as a "structure."

An example of a "light structure" model is shown in Fig. 2-1. Models have been developed to serve as a partial guide for the use of lighting effects appropriate for various task and non-task applications.

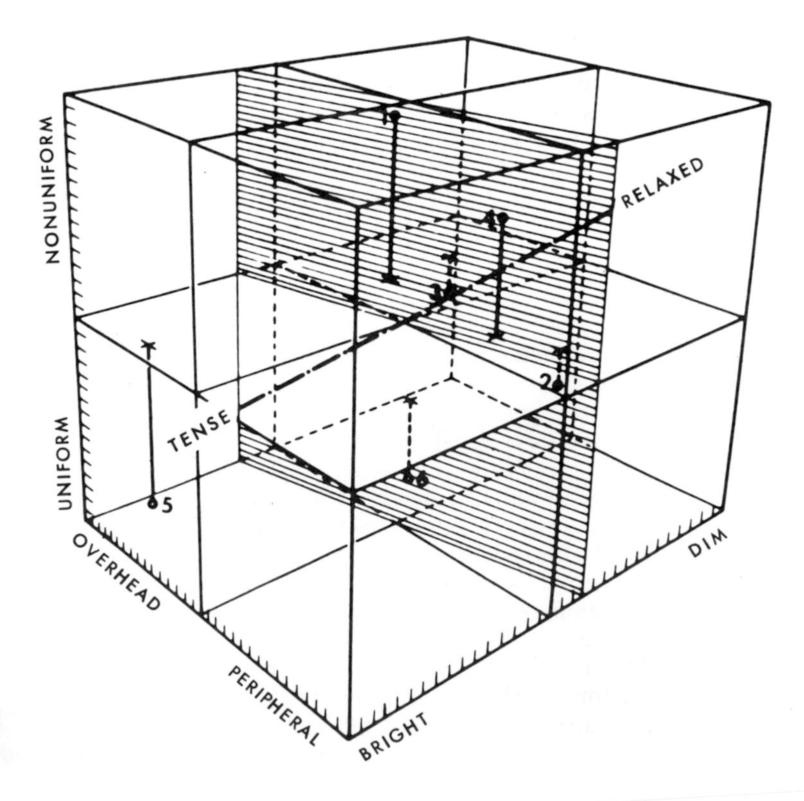

Fig. 2-1. Light structure model indicating lighting design decision for affecting impressions of relaxation (and tension).

Impressions of Relaxation. Impression of relaxation is an important subjective factor to be considered in the design of more casual areas, such as waiting rooms, lounges, some restaurants, etc. It is a subjective visual impression that appears to be reinforced by the following lighting influences (see Fig. 2-1):
1. Uniformity (*i.e.*, reinforced by nonuniform lighting).
2. Distribution (*i.e.*, reinforced more specifically by nonuniform wall lighting).
3. Color (*i.e.*, some studies indicate reinforcement by warm colors of white light).

Ongoing research in the behaviorial sciences will provide valuable knowledge about the influences of lighting quality and quantity on the human organism. Lighting designers should seek the results of these studies for guidance in their design.

Impressions of Perceptual Clarity. Impressions of perceptual clarity are an important subjective factor to be considered in the design of work spaces. It is a subjective visual impression that appears to be reinforced by four lighting influences:
1. Luminance (*i.e.*, reinforced by higher luminance on the horizontal plane).
2. Locational (*i.e.*, reinforced by luminance in the central part of the room).
3. Color (*i.e.*, reinforced by cool tone, continuous spectrum light sources).
4. Distribution (*i.e.*, reinforced by peripheral wall brightness).

Impressions of Spaciousness. Impression of spaciousness is an important subjective factor to be considered in the design of circulation and assembly spaces, such as corridors, lobbies, assembly halls, etc. It is a subjective visual impression that appears to be reinforced by uniform, peripheral lighting. Color of white light (warm or cool) appears to be a negligible subjective factor.

ILLUMINATION CONSIDERATIONS

In this section, illumination considerations include: illuminance, luminance ratios, visual comfort, reflected glare, disability glare, veiling reflections, color and shadows. Each is an important consideration and no emphasis is intended by order of presentation or length of the discussions that follow (some have additional coverage in other sections of this volume and in the 1981 Reference Volume).

Illuminance Selection and Application

A New Illuminance Selection Procedure. Since 1958 the Society has been publishing single-value illuminance recommendations based on a method established at that time.[5] In recent years it became apparent, through on-going research and design experience, that it was time to move away from the single-value recommendations to a range approach—illuminance ranges accompanied by a weighting-factor guidance system reflecting lighting-performance trends found in research. In 1979 the Society established such a new procedure.[6]

Since early 1979 the Society's committees have applied the new procedure in preparing new interior illuminance recommendations* for this Handbook and for the Society's recommended practices and committee reports.

It is intended that this new procedure will accommodate a need for flexibility in determining illuminance levels so that lighting designers can tailor lighting systems to specific needs, especially in an energy conscious era. Such flexibility requires that additional information be available to effectively use the new range approach—a *lighting task* must be considered to be composed of the following:
1. The visual display (details to be seen).
2. The age of the observers.
3. The importance of speed and/or accuracy for visual performance.
4. The reflectance of the task (background on which the details are seen).

The visual display is the object being viewed—it will present some inherent visual difficulty. The age of the observer is a predictor of the condition of the observer's visual system. The importance of speed and/or accuracy distinguishes between casual, important and critical seeing requirements. The reflectance will determine the adaptation luminance produced by the illuminance. These characteristics, considered in concert, determine the appropriate amount of light for the *lighting task.* All four must be considered as comprising the lighting task.

In applying the new procedure the first step is to determine a range of illuminances appropriate for the visual difficulty presented by the visual display, the first of the above characteristics; and then to determine a target value from that range on the basis of the remaining three characteristics. The Society's application committees, on a consensus basis, have established appropriate

* At this time illuminances for exteriors and for certain applications continue to be provided as single-value recommendations established on a consensus basis. See Fig. 2-2. See also page 2-4.

ranges of illuminances for various types of visual displays. Experience and judgment played an obvious, important role in the pairing of visual displays and these ranges. Nine ranges, called *Illuminance Categories*, have been established, patterned after those in the CIE Report No. 29.[7] These are designated "A" through "I," covering illuminance levels from 20 to 20,000 lux [2 to 2000 footcandles]. See Part I of Fig. 2-2. Further work of the application committees resulted in recommended ranges of illuminance for specific visual displays (tasks) and areas (see Fig. 2-2, Parts II and III).

Alternatively, if the inherent visual difficulty of a visual display has been measured in terms of its equivalent contrast, $\tilde{C}$, the Illuminance Category can be determined from Fig. 2-3.[6] This table of equivalent contrasts and Illuminance Categories was established on a consensus basis by the Society's Committee on Recommendations for Quality and Quantity of Illumination.

For a given visual display, a specific value of illuminance can be chosen from the recommended range only if the second, third, and fourth characteristics of the lighting task are known; *i.e.*, observers age, importance of speed and/or accuracy, and task reflectance. These should be determined at design time, by the designer in conjunction with the user. Specific target values of maintained illuminance *cannot* be determined before hand by the Society. Thus, the recommendations for most interior lighting tasks consist of an Illuminance Category determined by the visual display.

A guide for using the second, third and fourth characteristics of the lighting task, to determine a specific target value of illuminance, takes the form of a table of Weighting Factors (see Fig. 2-4). The designer or user determines the weight of each characteristic. A combined weighting factor then indicates whether the lower, middle or upper value of illuminance in the range is appropriate (see the procedure outlined below).

It can be seen that over-all design of this procedure makes it an illuminance *selection* procedure, where consensus-determined recommended ranges combine with user supplied information and judgment. The result is the determination of a specific target value of illuminance appropriate for the lighting task under consideration.

Limitations of the New Selection Procedure. This illuminance selection procedure is intended for use in interior environments where visual performance is an important consideration. It has been developed from a consideration of experience and research results from visual performance experiments. Its use is then limited to applications where this information can be applied directly. Thus, the illuminance selection procedure[6] is *not* used to determine the appropriate illuminances when:

1. Merchandising is the principal activity in the space and the advantageous display of goods is the purpose of lighting.
2. Advertising, sales promotion or attraction is the purpose of lighting.
3. Lighting is for sensors other than the eye, as in film and television applications.
4. The principle purpose of lighting is to achieve artistic effects.
5. Luminance ratios have a greater importance than adaptation luminance, as when it is desired to achieve a particular psychological or emotional setting rather than provide for visual performance.
6. Minimum illuminances are required for safety.
7. Maximum illuminances are established to prevent nonvisual effects, such as bleaching or deterioration due to ultraviolet and infrared radiation in a museum.
8. Illuminances are part of a test procedure for evaluating equipment, such as for surgical lighting systems.

Procedure for Selecting Illuminances. The procedure provides a method for determining a target maintained illuminance value for a single visual task, and as such will not assure an adequate illuminance level for a given space. This is especially true for those spaces in which a variety of visual tasks occurs. To help assure appropriate task illuminance as well as provide potential for increased energy savings, the designer should consider an illuminance target as the quantity of light required on the plane of the task.

The designer should be aware of, or assume, the potential visual tasks to be performed within the space. The illuminance level determined using this procedure is a function of the visual characteristics of that task. Therefore, the importance, duration and difficulty of each task in the space must be considered as each may dictate a different illuminance level. The importance of providing various illuminance levels can then be rated accordingly. Multiple level lighting systems, segregation of certain visual tasks, nonuniform lighting systems, or single level systems to meet the commonly occurring most critical visual task requirements, are options the designer must consider for system optimization.

The four step procedure described below requires the designer to select an Illuminance Cat-

Continued on page 2-20

Fig. 2-2. Currently Recommended Illuminance Categories and Illuminance Values for Lighting Design—Target Maintained Levels

The tabulation that follows is a consolidated listing of the Society's current illuminance recommendations. This listing is intended to guide the lighting designer in selecting an appropriate illuminance for design and evaluation of lighting systems.

Guidance is provided in two forms: (1), in Parts I, II and III as an *Illuminance Category*, representing a range of illuminances (see page 2-4 for a method of selecting a value within each illuminance range); and (2), in parts IV, V and VI as an *Illuminance Value.* Illuminance Categories are represented by letter designations A through I. Illuminance Values are given in *lux* with an approximate equivalence in *footcandles* and as such are intended as *target* (nominal) values with deviations expected. These target values also represent *maintained* values (see page 2-24).

This table has been divided into the six parts for ease of use. Part I provides a listing of both Illuminance Categories and Illuminance Values for generic types of interior activities and normally is to be used when Illuminance Categories for a specific Area/Activity cannot be found in parts II and III. Parts IV, V and VI provide target maintained Illuminance Values for outdoor facilities, sports and recreational areas, and transportation vehicles where special considerations apply as discussed on page 2-4.

In all cases the recommendations in this table are based on the assumption that the lighting will be properly designed to take into account the visual characteristics of the task. See the design information in the particular application sections in this Application Handbook for further recommendations.

I. Illuminance Categories and Illuminance Values for Generic Types of Activities in Interiors

Type of Activity	Illuminance Category	Ranges of Illuminances		Reference Work-Plane
		Lux	Footcandles	
Public spaces with dark surroundings	A	20-30-50	2-3-5	General lighting throughout spaces
Simple orientation for short temporary visits	B	50-75-100	5-7.5-10	General lighting throughout spaces
Working spaces where visual tasks are only occasionally performed	C	100-150-200	10-15-20	General lighting throughout spaces
Performance of visual tasks of high contrast or large size	D	200-300-500	20-30-50	Illuminance on task
Performance of visual tasks of medium contrast or small size	E	500-750-1000	50-75-100	Illuminance on task
Performance of visual tasks of low contrast or very small size	F	1000-1500-2000	100-150-200	Illuminance on task
Performance of visual tasks of low contrast and very small size over a prolonged period	G	2000-3000-5000	200-300-500	Illuminance on task, obtained by a combination of general and local (supplementary lighting)
Performance of very prolonged and exacting visual tasks	H	5000-7500-10000	500-750-1000	Illuminance on task, obtained by a combination of general and local (supplementary lighting)
Performance of very special visual tasks of extremely low contrast and small size	I	10000-15000-20000	1000-1500-2000	Illuminance on task, obtained by a combination of general and local (supplementary lighting)

II. Commercial, Institutional, Residential and Public Assembly Interiors

Area/Activity	Illuminance Category
Air terminals (see **Transportation terminals**)	
Armories	C[1]
Art galleries (see **Museums**)	
Auditoriums	
Assembly	C[1]
Social activity	B
Banks (also see **Reading**)	
Lobby	
General	C
Writing area	D
Tellers' stations	E[3]
Barber shops and beauty parlors	E
Churches and synagogues	(see page 7-2)[4]
Club and lodge rooms	
Lounge and reading	D
Conference rooms	
Conferring	D
Critical seeing (refer to individual task)	
Court rooms	
Seating area	C
Court activity area	E[3]
Dance halls and discotheques	B

For footnotes, see page 2-19.

Fig. 2–2. *Continued*

II. *Continued*

Area/Activity	Illuminance Category
Depots, terminals and stations (see **Transportation terminals**)	
Drafting	
Mylar	
High contrast media; India ink, plastic leads, soft graphite leads	E[3]
Low contrast media; hard graphite leads	F[3]
Vellum	
High contrast	E[3]
Low contrast	F[3]
Tracing paper	
High contrast	E[3]
Low contrast	F[3]
Overlays[5]	
Light table	C
Prints	
Blue line	E
Blueprints	E
Sepia prints	F
Educational facilities	
Classrooms	
General (see **Reading**)	
Drafting (see **Drafting**)	
Home economics (see **Residences**)	
Science laboratories	E
Lecture rooms	
Audience (see **Reading**)	
Demonstration	F
Music rooms (see **Reading**)	
Shops (see Part III, Industrial Group)	
Sight saving rooms	F
Study halls (see **Reading**)	
Typing (see **Reading**)	
Sports facilities (see Part V, Sports and Recreational Areas)	
Cafeterias (see **Food service facilities**)	
Dormitories (see **Residences**)	
Elevators, freight and passenger	C
Exhibition halls	C[1]
Fire halls (see **Municipal buildings**)	
Food service facilities	
Dining areas	
Cashier	D
Cleaning	C
Dining	B[6]
Food displays (see Merchandising spaces)	
Kitchen	E
Garages—parking	(see page 14–24)
Gasoline stations (see **Service stations**)	
Graphic design and material	
Color selection	F[11]
Charting and mapping	F
Graphs	E
Keylining	F
Layout and artwork	F
Photographs, moderate detail	E[13]
Health care facilities	
Ambulance (local)	E
Anesthetizing	E
Autopsy and morgue[17, 18]	
Autopsy, general	E
Autopsy table	G
Morgue, general	D
Museum	E
Cardiac function lab	E
Central sterile supply	
Inspection, general	E
Inspection	F
At sinks	E
Work areas, general	D
Processed storage	D
Corridors[17]	
Nursing areas—day	C
Nursing areas—night	B
Operating areas, delivery, recovery, and laboratory suites and service	E
Critical care areas[17]	
General	C
Examination	E
Surgical task lighting	H
Handwashing	F
Cystoscopy room[17, 18]	E
Dental suite[17]	
General	D
Instrument tray	E
Oral cavity	H
Prosthetic laboratory, general	D
Prosthetic laboratory, work bench	E
Prosthetic laboratory, local	F
Recovery room, general	C
Recovery room, emergency examination	E
Dialysis unit, medical[17]	F
Elevators	C
EKG and specimen room[17]	
General	B
On equipment	C
Emergency outpatient[17]	
General	E
Local	F
Endoscopy rooms[17, 18]	
General	E
Peritoneoscopy	D
Culdoscopy	D
Examination and treatment rooms[17]	
General	D
Local	E
Eye surgery[17, 18]	F
Fracture room[17]	
General	E
Local	F
Inhalation therapy	D
Laboratories[17]	
Specimen collecting	E
Tissue laboratories	F
Microscopic reading room	D
Gross specimen review	F

For footnotes, see page 2-19. For illuminance ranges for each Illuminance Category, see page 2-5.

Fig. 2-2. *Continued*

II. *Continued*

Area/Activity	Illuminance Category
Chemistry rooms	E
Bacteriology rooms	
General	E
Reading culture plates	F
Hematology	E
Linens	
Sorting soiled linen	D
Central (clean) linen room	D
Sewing room, general	D
Sewing room, work area	E
Linen closet	B
Lobby	C
Locker rooms	C
Medical illustration studio[17,18]	F
Medical records	E
Nurseries[17]	
General[18]	C
Observation and treatment	E
Nursing stations[17]	
General	D
Desk	E
Corridors, day	C
Corridors, night	A
Medication station	E
Obstetric delivery suite[17]	
Labor rooms	
General	C
Local	E
Birthing room	F[7]
Delivery area	
Scrub, general	G
General	G
Delivery table	(see page 7-15)
Resuscitation	G
Postdelivery recovery area	E
Substerilizing room	B
Occupational therapy[17]	
Work area, general	D
Work tables or benches	E
Patients' rooms[17]	
General[18]	B
Observation	A
Critical examination	E
Reading	D
Toilets	D
Pharmacy[17]	
General	E
Alcohol vault	D
Laminar flow bench	F
Night light	A
Parenteral solution room	D
Physical therapy departments	
Gymnasiums	D
Tank rooms	D
Treatment cubicles	D
Postanesthetic recovery room[17]	
General[18]	E
Local	H
Pulmonary function laboratories[17]	E
Radiological suite[17]	
Diagnostic section	
General[18]	A
Waiting area	A
Radiographic/fluoroscopic room	A
Film sorting	F
Barium kitchen	E
Radiation therapy section	
General[18]	B
Waiting area	B
Isotope kitchen, general	E
Isotope kitchen, benches	E
Computerized radiotomography section	
Scanning room	B
Equipment maintenance room	E
Solarium	
General	C
Local for reading	D
Stairways	C
Surgical suite[17]	
Operating room, general[18]	F
Operating table	(see page 7-12)
Scrub room[18]	E
Instruments and sterile supply room	D
Clean up room, instruments	E
Anesthesia storage	C
Substerilizing room	C
Surgical induction room[17,18]	E
Surgical holding area[17,18]	E
Toilets	C
Utility room	D
Waiting areas[17]	
General	C
Local for reading	D
Homes (see **Residences)**	
Hospitals (see **Health care facilities)**	
Hotels	
Bathrooms, for grooming	D
Bedrooms, for reading	D
Corridors, elevators and stairs	C
Front desk	E[3]
Linen room	
Sewing	F
General	C
Lobby	
General lighting	C
Reading and working areas	D
Canopy (see Part IV, Outdoor Facilities)	
Kitchens (see **Food service facilities** or **Residences)**	
Libraries	
Reading areas (see **Reading)**	
Book stacks (vertical 760 millimeters (30 inches) above floor)	
Active stacks	D
Inactive stacks	B
Book repair and binding	D

For footnotes, see page 2-19. For illuminance ranges for each Illuminance Category, see page 2-5.

Fig. 2-2. *Continued*

II. *Continued*

Area/Activity	Illuminance Category
Cataloging	D[3]
Card files	E
Carrels, individual study areas (see **Reading**)	
Circulation desks	D
Map, picture and print rooms (see **Graphic design and material**)	
Audiovisual areas	D
Audio listening areas	D
Microform areas (see **Reading**)	
Locker rooms	C
Merchandising spaces	
Alteration room	F
Fitting room	
Dressing areas	D
Fitting areas	F
Locker rooms	C
Stock rooms	D
Wrapping and packaging	D
Sales transaction area	E
Circulation	(see page 8-6)[8]
Merchandise	(see page 8-6)[8]
Feature display	(see page 8-6)[8]
Show windows	(see page 8-6)[8]
Motels (see **Hotels**)	
Municipal buildings—fire and police	
Police	
Identification records	F
Jail cells and interrogation rooms	D
Fire hall	D
Museums	
Displays of non-sensitive materials	D
Displays of sensitive materials	(see page 7-29)[2]
Lobbies, general gallery areas, corridors	C
Restoration or conservation shops and laboratories	E
Nursing homes (see **Health care facilities**)	
Offices	
Accounting (see **Reading**)	
Conference areas (see **Conference rooms**)	
Drafting (see **Drafting**)	
General and private offices (see **Reading**)	
Libraries (see **Libraries**)	
Lobbies, lounges and reception areas	C
Mail sorting	E
Off-set printing and duplicating area	D
Post offices (see **Offices**)	
Reading	
Copied tasks	
Ditto copy	E[3]
Micro-fiche reader	B[12, 13]
Mimeograph	D
Photographs, moderate detail	E[13]
Thermal copy, poor copy	F[3]
Xerograph	D
Xerography, 3rd generation and greater	E

Area/Activity	Illuminance Category
Electronic data processing tasks	
CRT screens	B[12, 13]
Impact printer	
good ribbon	D
poor ribbon	E
2nd carbon and greater	E
Ink jet printer	D
Keyboard reading	D
Machine rooms	
Active operations	D
Tape storage	D
Machine area	C
Equipment service	E[10]
Thermal print	E
Handwritten tasks	
#3 pencil and softer leads	E[3]
#4 pencil and harder leads	F[3]
Ball-point pen	D[3]
Felt-tip pen	D
Handwritten carbon copies	E
Non photographically reproducible colors	F
Chalkboards	E[3]
Printed tasks	
6 point type	E[3]
8 and 10 point type	D[3]
Glossy magazines	D[13]
Maps	E
Newsprint	D
Typed originals	D
Typed 2nd carbon and later	E
Telephone books	E
Residences	
General lighting	
Conversation, relaxation and entertainment	B
Passage areas	B
Specific visual tasks[20]	
Dining	C
Grooming	
Makeup and shaving	D
Full-length mirror	D
Handcrafts and hobbies	
Workbench hobbies	
Ordinary tasks	D
Difficult tasks	E
Critical tasks	F
Easel hobbies	E
Ironing	D
Kitchen duties	
Kitchen counter	
Critical seeing	E
Noncritical	D
Kitchen range	
Difficult seeing	E
Noncritical	D
Kitchen sink	
Difficult seeing	E
Noncritical	D
Laundry	
Preparation and tubs	D
Washer and dryer	D

For footnotes, see page 2-19. For illuminance ranges for each Illuminance Category, see page 2-5.

Fig. 2-2. *Continued*

II. *Continued*

Area/Activity	Illuminance Category
Music study (piano or organ)	
Simple scores	D
Advanced scores	E
Substand size scores	F
Reading	
In a chair	
Books, magazines and newspapers	D
Handwriting, reproductions and poor copies	E
In bed	
Normal	D
Prolonged serious or critical	E
Desk	
Primary task plane, casual	D
Primary task plane, study	E
Sewing	
Hand sewing	
Dark fabrics, low contrast	F
Light to medium fabrics	E
Occasional, high contrast	D
Machine sewing	
Dark fabrics, low contrast	F
Light to medium fabrics	E
Occasional, high contrast	D
Table games	D
Restaurants (see **Food service facilities**)	
Safety	(see page 2-45)
Schools (see **Educational facilities**)	
Service spaces (see also **Storage rooms**)	
Stairways, corridors	C
Elevators, freight and passenger	C
Toilets and wash rooms	C
Service stations	
Service bays (see Part III, Industrial Group)	
Sales room (see **Merchandising spaces**)	
Show windows	(see page 8-6)
Stairways (see **Service spaces**)	
Storage rooms (see Part III, Industrial Group)	
Stores (see **Merchandising spaces** and **Show windows**)	
Television	(see Section 11)
Theatre and motion picture houses (see Section 11)	
Toilets and washrooms	C
Transportation terminals	
Waiting room and lounge	C
Ticket counters	E
Baggage checking	D
Rest rooms	C
Concourse	B
Boarding area	C

III. Industrial Group

Area/Activity	Illuminance Category
Aircraft maintenance	(see page 9-12)[21]
Aircraft manufacturing	(see page 9-12)[21]
Assembly	
Simple	D
Moderately difficult	E
Difficult	F
Very difficult	G
Exacting	H
Automobile manufacturing	(see page 9-17)[21]
Bakeries	
Mixing room	D
Face of shelves	D
Inside of mixing bowl	D
Fermentation room	D
Make-up room	
Bread	D
Sweet yeast-raised products	D
Proofing room	D
Oven room	D
Fillings and other ingredients	D
Decorating and icing	
Mechanical	D
Hand	E
Scales and thermometers	D
Wrapping	D
Book binding	
Folding, assembling, pasting	D
Cutting, punching, stitching	E
Embossing and inspection	F
Breweries	
Brew house	D
Boiling and keg washing	D
Filling (bottles, cans, kegs)	D
Building construction (see Part IV, Outdoor Facilities)	
Building exteriors (see Part IV, Outdoor Facilities)	
Candy making	
Box department	D
Chocolate department	
Husking, winnowing, fat extraction, crushing and refining, feeding	D
Bean cleaning, sorting, dipping, packing, wrapping	D
Milling	E
Cream making	
Mixing, cooking, molding	D
Gum drops and jellied forms	D
Hand decorating	D
Hard candy	
Mixing, cooking, molding	D

For footnotes, see page 2-19. For illuminance ranges for each Illuminance Category, see page 2-5.

Fig. 2-2. *Continued*

III. *Continued*

Area/Activity	Illuminance Category
Die cutting and sorting	E
Kiss making and wrapping	E
Canning and preserving	
Initial grading raw material samples	D
Tomatoes	E
Color grading and cutting rooms	F
Preparation	
Preliminary sorting	
Apricots and peaches	D
Tomatoes	E
Olives	F
Cutting and pitting	E
Final sorting	E
Canning	
Continuous-belt canning	E
Sink canning	E
Hand packing	D
Olives	E
Examination of canned samples	F
Container handling	
Inspection	F
Can unscramblers	E
Labeling and cartoning	D
Casting (see **Foundries**)	
Central stations (see **Electric generating stations**)	
Chemical plants (see **Petroleum and chemical plants**)	
Clay and concrete products	
Grinding, filter presses, kiln rooms	C
Molding, pressing, cleaning, trimming	D
Enameling	E
Color and glazing—rough work	E
Color and glazing—fine work	F
Cleaning and pressing industry	
Checking and sorting	E
Dry and wet cleaning and steaming	E
Inspection and spotting	G
Pressing	F
Repair and alteration	F
Cloth products	
Cloth inspection	I
Cutting	G
Sewing	G
Pressing	F
Clothing manufacture (men's)	
Receiving, opening, storing, shipping	D
Examining (perching)	I
Sponging, decating, winding, measuring	D
Piling up and marking	E
Cutting	G
Pattern making, preparation of trimming, piping, canvas and shoulder pads	E
Fitting, bundling, shading, stitching	D
Shops	F
Inspection	G
Pressing	F
Sewing	G
Control rooms (see **Electric generating stations—interior**)	
Corridors (see **Service spaces**)	
Cotton gin industry	
Overhead equipment—separators, driers, grid cleaners, stick machines, conveyers, feeders and catwalks	D
Gin stand	D
Control console	D
Lint cleaner	D
Bale press	D
Dairy farms (see **Farms**)	
Dairy products	
Fluid milk industry	
Boiler room	D
Bottle storage	D
Bottle sorting	E
Bottle washers	[22]
Can washers	D
Cooling equipment	D
Filling: inspection	E
Gauges (on face)	E
Laboratories	E
Meter panels (on face)	E
Pasteurizers	D
Separators	D
Storage refrigerator	D
Tanks, vats	
Light interiors	C
Dark interiors	E
Thermometer (on face)	E
Weighing room	D
Scales	E
Dispatch boards (see **Electric generating stations—interior**)	
Dredging (see Part IV, Outdoor Facilities)	
Electrical equipment manufacturing	
Impregnating	D
Insulating: coil winding	E
Electric generating stations—interior (see also **Nuclear power plants**)	
Air-conditioning equipment, air preheater and fan floor, ash sluicing	B
Auxiliaries, pumps, tanks, compressors, gauge area	C
Battery rooms	D
Boiler platforms	B
Burner platforms	C
Cable room	B
Coal handling systems	B
Coal pulverizer	C
Condensers, deaerator floor, evaporator floor, heater floors	B
Control rooms	
Main control boards	D[23]
Auxiliary control panels	D[23]
Operator's station	E[23]

For footnotes, see page 2-19. For illuminance ranges for each Illuminance Category, see page 2-5.

Fig. 2-2. *Continued*

III. *Continued*

Area/Activity	Illuminance Category
Maintenance and wiring areas	D
Emergency operating lighting	C
Gauge reading	D
Hydrogen and carbon dioxide manifold area	C
Laboratory	E
Precipitators	B
Screen house	C
Soot or slag blower platform	C
Steam headers and throttles	B
Switchgear and motor control centers	D
Telephone and communication equipment rooms	D
Tunnels or galleries, piping and electrical	B
Turbine building	
Operating floor	D
Below operating floor	C
Visitor's gallery	C
Water treating area	D
Electric generating stations—exterior (see Part IV, Outdoor Facilities)	
Elevators (see **Service spaces**)	
Explosives manufacturing	
Hand furnaces, boiling tanks, stationary driers, stationary and gravity crystallizers	D
Mechanical furnace, generators and stills, mechanical driers, evaporators, filtration, mechanical crystallizers	D
Tanks for cooking, extractors, percolators, nitrators	D
Farms—dairy	
Milking operation area (milking parlor and stall barn)	
General	C
Cow's udder	D
Milk handling equipment and storage area (milk house or milk room)	
General	C
Washing area	E
Bulk tank interior	E
Loading platform	C
Feeding area (stall barn feed alley, pens, loose housing feed area)	C
Feed storage area—forage	
Haymow	A
Hay inspection area	C
Ladders and stairs	C
Silo	A
Silo room	C
Feed storage area—grain and concentrate	
Grain bin	A
Concentrate storage area	B
Feed processing area	B
Livestock housing area (community, maternity, individual calf pens, and loose housing holding and resting areas)	B
Machine storage area (garage and machine shed)	B
Farm shop area	
Active storage area	B
General shop area (machinery repair, rough sawing)	D
Rough bench and machine work (painting, fine storage, ordinary sheet metal work, welding, medium benchwork)	D
Medium bench and machine work (fine woodworking, drill press, metal lathe, grinder)	E
Miscellaneous areas	
Farm office (see **Reading**)	
Restrooms (see **Service spaces**)	
Pumphouse	C
Farms—poultry (see **Poultry industry**)	
Flour mills	
Rolling, sifting, purifying	E
Packing	D
Product control	F
Cleaning, screens, man lifts, aisleways and walkways, bin checking	D
Forge shops	E
Foundries	
Annealing (furnaces)	D
Cleaning	D
Core making	
Fine	F
Medium	E
Grinding and chipping	F
Inspection	
Fine	G
Medium	F
Molding	
Medium	F
Large	E
Pouring	E
Sorting	E
Cupola	C
Shakeout	D
Garages—service	
Repairs	E
Active traffic areas	C
Write-up	D
Glass works	
Mix and furnace rooms, pressing and lehr, glass-blowing machines	C
Grinding, cutting, silvering	D
Fine grinding, beveling, polishing	E
Inspection, etching and decorating	F
Glove manufacturing	
Pressing	G
Knitting	F
Sorting	F
Cutting	G
Sewing and inspection	G
Hangars (see **Aircraft manufacturing**)	
Hat manufacturing	
Dyeing, stiffening, braiding, cleaning, refining	E

For footnotes, see page 2-19. For illuminance ranges for each Illuminance Category, see page 2-5.

Fig. 2–2. *Continued*

III. *Continued*

Area/Activity	Illuminance Category
Forming, sizing, pouncing, flanging, finishing, ironing	F
Sewing	G
Inspection	
Simple	D
Moderately difficult	E
Difficult	F
Very difficult	G
Exacting	H
Iron and steel manufacturing	(see page 9–63)[21]
Jewelry and watch manufacturing	G
Laundries	
Washing	D
Flat work ironing, weighing, listing, marking	D
Machine and press finishing, sorting	E
Fine hand ironing	E
Leather manufacturing	
Cleaning, tanning and stretching, vats	D
Cutting, fleshing and stuffing	D
Finishing and scarfing	E
Leather working	
Pressing, winding, glazing	F
Grading, matching, cutting, scarfing, sewing	G
Loading and unloading platforms (see Part IV, Outdoor Facilities)	
Locker rooms	C
Logging (see Part IV, Outdoor Facilities)	
Lumber yards (see Part IV, Outdoor Facilities)	
Machine shops	
Rough bench or machine work	D
Medium bench or machine work, ordinary automatic machines, rough grinding, medium buffing and polishing	E
Fine bench or machine work, fine automatic machines, medium grinding, fine buffing and polishing	G
Extra-fine bench or machine work, grinding, fine work	H
Materials handling	
Wrapping, packing, labeling	D
Picking stock, classifying	D
Loading, inside truck bodies and freight cars	C
Meat packing	
Slaughtering	D
Cleaning, cutting, cooking, grinding, canning, packing	D
Nuclear power plants (see also **Electric generating stations)**	
Auxiliary building, uncontrolled access areas	C
Controlled access areas	
Count room	E[23]
Laboratory	E
Health physics office	F
Medical aid room	F
Hot laundry	D
Storage room	C
Engineered safety features equipment	D
Diesel generator building	D
Fuel handling building	
Operating floor	D
Below operating floor	C
Off gas building	C
Radwaste building	D
Reactor building	
Operating floor	D
Below operating floor	C
Packing and boxing (see **Materials handling)**	
Paint manufacturing	
Processing	D
Mix comparison	F
Paint shops	
Dipping, simple spraying, firing	D
Rubbing, ordinary hand painting and finishing art, stencil and special spraying	D
Fine hand painting and finishing	E
Extra-fine hand painting and finishing	G
Paper-box manufacturing	E
Paper manufacturing	
Beaters, grinding, calendering	D
Finishing, cutting, trimming, papermaking machines	E
Hand counting, wet end of paper machine	E
Paper machine reel, paper inspection, and laboratories	F
Rewinder	F
Parking areas	(see page 14–24)
Petroleum and chemical plants	(see page 9–51)[21]
Plating	D
Polishing and burnishing (see **Machine shops)**	
Power plants (see **Electric generating stations)**	
Poultry industry (see also **Farm—dairy)**	
Brooding, production, and laying houses	
Feeding, inspection, cleaning	C
Charts and records	D
Thermometers, thermostats, time clocks	D
Hatcheries	
General area and loading platform	C
Inside incubators	D
Dubbing station	F
Sexing	H
Egg handling, packing, and shipping	
General cleanliness	E
Egg quality inspection	E
Loading platform, egg storage area, etc.	C
Egg processing	
General lighting	E
Fowl processing plant	
General (excluding killing and unloading area)	E
Government inspection station and grading stations	E
Unloading and killing area	C

For footnotes, see page 2–19. For illuminance ranges for each Illuminance Category, see page 2–5.

Fig. 2-2. *Continued*

III. *Continued*

Area/Activity	Illuminance Category
Feed storage	
Grain, feed rations	C
Processing	C
Charts and records	D
Machine storage area (garage and machine shed)	B
Printing industries	
Type foundries	
Matrix making, dressing type	E
Font assembly—sorting	D
Casting	E
Printing plants	
Color inspection and appraisal	F
Machine composition	E
Composing room	E
Presses	E
Imposing stones	F
Proofreading	F
Electrotyping	
Molding, routing, finishing, leveling molds, trimming	E
Blocking, tinning	D
Electroplating, washing, backing	D
Photoengraving	
Etching, staging, blocking	D
Routing, finishing, proofing	E
Tint laying, masking	E
Receiving and shipping (see **Materials handling)**	
Railroad yards (see Part IV, Outdoor Facilities)	
Rubber goods—mechanical	(see page 9–56)[21]
Rubber tire manufacturing	(see page 9–56)[21]
Safety	(see page 2–45)
Sawmills	
Secondary log deck	B
Head saw (cutting area viewed by sawyer)	E
Head saw outfeed	B
Machine in-feeds (bull edger, resaws, edgers, trim, hula saws, planers)	B
Main mill floor (base lighting)	A
Sorting tables	D
Rough lumber grading	D
Finished lumber grading	F
Dry lumber warehouse (planer)	C
Dry kiln colling shed	B
Chipper infeed	B
Basement areas	
Active	A
Inactive	A
Filing room (work areas)	E
Service spaces (see also **Storage rooms)**	
Stairways, corridors	B
Elevators, freight and passenger	B
Toilets and wash rooms	C
Sheet metal works	
Miscellaneous machines, ordinary bench work	E
Presses, shears, stamps, spinning, medium bench work	E
Punches	E
Tin plate inspection, galvanized	F
Scribing	F
Shoe manufacturing—leather	
Cutting and stitching	
Cutting tables	G
Marking, buttonholing, skiving, sorting, vamping, counting	G
Stitching, dark materials	G
Making and finishing, nailers, sole layers, welt beaters and scarfers, trimmers, welters, lasters, edge setters, sluggers, randers, wheelers, treers, cleaning, spraying, buffing, polishing, embossing	F
Shoe manufacturing—rubber	
Washing, coating, mill run compounding	D
Varnishing, vulcanizing, calendering, upper and sole cutting	D
Sole rolling, lining, making and finishing processes	E
Soap manufacturing	
Kettle houses, cutting, soap chip and powder	D
Stamping, wrapping and packing, filling and packing soap powder	D
Stairways (see **Service spaces)**	
Steel (see **Iron and steel)**	
Storage battery manufacturing	D
Storage rooms or warehouses	
Inactive	B
Active	
Rough, bulky items	C
Small items	D
Storage yards (see Part IV, Outdoor Facilities)	
Structural steel fabrication	E
Sugar refining	
Grading	E
Color inspection	F
Testing	
General	D
Exacting tests, extra-fine instruments, scales, etc.	F
Textile mills	
Staple fiber preparation	
Stock dyeing, tinting	D
Sorting and grading (wool and cotton)	E[16]
Yarn manufacturing	
Opening and picking (chute feed)	D
Carding (nonwoven web formation)	D[24]
Drawing (gilling, pin drafting)	D
Combing	D[24]
Roving (slubbing, fly frame)	E
Spinning (cap spinning, twisting, texturing)	E
Yarn preparation	
Winding, quilling, twisting	E
Warping (beaming, sizing)	F[16]
Warp tie-in or drawing-in (automatic)	E

For footnotes, see page 2–19. For illuminance ranges for each Illuminance Category, see page 2–5.

Fig. 2-2. *Continued*

III. *Continued*

Area/Activity	Illuminance Category
Fabric production	
Weaving, knitting, tufting	F
Inspection	G[16]
Finishing	
Fabric preparation (desizing, scouring, bleaching, singeing, and mercerization)	D
Fabric dyeing (printing)	D
Fabric finishing (calendaring, sanforizing, sueding, chemical treatment)	E[16]
Inspection	G[16, 25]
Tobacco products	
Drying, stripping	D
Grading and sorting	F
Toilets and wash rooms (see **Service spaces)**	
Upholstering	F
Warehouse (see **Storage rooms)**	
Welding	
Orientation	D
Precision manual arc-welding	H
Woodworking	
Rough sawing and bench work	D
Sizing, planing, rough sanding, medium quality machine and bench work, gluing, veneering, cooperage	D
Fine bench and machine work, fine sanding and finishing	E

IV. Outdoor Facilities

Area/Activity	Lux	Footcandles
Building (construction)		
General construction	100	10
Excavation work	20	2
Building exteriors		
Entrances		
Active (pedestrian and/or conveyance)	50	5
Inactive (normally locked, infrequently used)	10	1
Vital locations or structures	50	5
Building surrounds	10	1
Buildings and monuments, floodlighted		
Bright surroundings		
Light surfaces	150	15
Medium light surfaces	200	20
Medium dark surfaces	300	30
Dark surfaces	500	50
Dark surroundings		
Light surfaces	50	5
Medium light surfaces	100	10
Medium dark surfaces	150	15
Dark surfaces	200	20
Bulletin and poster boards		
Bright surroundings		
Light surfaces	500	50
Dark surfaces	1000	100
Dark surroundings		
Light surfaces	200	20
Dark surfaces	500	50
Central station (see **Electric generating stations—exterior)**		
Coal yards (protective)	2	0.2
Dredging	20	2
Electric generating stations—exterior		
Boiler areas		
Catwalks, general areas	20	2
Stairs and platforms	50	5
Ground level areas including precipitators, FD and ID fans, bottom ash hoppers	50	5
Cooling towers		
Fan deck, platforms, stairs, valve areas	50	5
Pump areas	20	2
Fuel handling		
Barge unloading, car dumper, unloading hoppers, truck unloading, pumps, gas metering	50	5
Conveyors	20	2
Storage tanks	10	1
Coal storage piles, ash dumps	2	0.2
Hydroelectric		
Powerhouse roof, stairs, platform and intake decks	50	5
Inlet and discharge water area	2	0.2
Intake structures		
Deck and laydown area	50	5
Value pits	20	2
Inlet water area	2	0.2
Parking areas		
Main plant parking	20	2
Secondary parking	10	1
Substation		
Horizontal general area	20	2
Vertical tasks	50	5
Transformer yards		
Horizontal general area	20	2
Vertical tasks	50	5
Turbine areas		
Building surrounds	20	2
Turbine and heater decks, unloading bays	50	5

For footnotes, see page 2-19. For illuminance ranges for each Illuminance Category, see page 2-5.

Fig. 2-2. *Continued*

IV. *Continued*

Area/Activity	Lux	Footcandles
Entrances, stairs and platforms	50[9]	5[9]
Flags, floodlighted (see **Bulletin and poster boards)**		
Gardens[19]		
General lighting	5	0.5
Path, steps, away from house	10	1
Backgrounds—fences, walls, trees, shrubbery	20	2
Flower beds, rock gardens	50	5
Trees, shrubbery, when emphasized	50	5
Focal points, large	100	10
Focal points, small	200	20
Gasoline station (see **Service stations** in Part II)		
Highways (see page 14-8)		
Loading and unloading platforms	200	20
Freight car interiors	100	10
Logging (see also **Sawmills)**		
Yarding	30	3
Log loading and unloading	50	5
Log stowing (water)	5	0.5
Active log storage area (land)	5	0.5
Log booming area (water)—foot traffic	10	1
Active log handling area (water)	20	2
Log grading—water or land	50	5
Log bins (land)	20	2
Lumber yards	10	1
Parking areas (see page 14-24)		
Piers		
Freight	200	20
Passenger	200	20
Active shipping area surrounds	50	5
Prison yards	50	5
Quarries	50	5
Railroad yards		
Retarder classification yards		
Receiving yard		
Switch points	20	2
Body of yard	10	1
Hump area (vertical)	200	20
Control tower and retarder area (vertical)	100	10
Head end	50	5
Body	10	1
Pull-out end	20	2
Dispatch or forwarding yard	10	1
Hump and car rider classification yard		
Receiving yard		
Switch points	20	2
Body of yard	10	1
Hump area	50	5
Flat switching yards		
Side of cars (vertical)	50	5
Switch points	20	2
Trailer-on-flatcars		
Horizontal surface of flatcar	50	5
Hold-down points (vertical)	50	5
Container-on-flatcars	30	3
Roadways (see page 14-8)		
Sawmills (see also **Logging)**		
Cut-off saw	100	10
Log haul	20	2
Log hoist (side lift)	20	2
Primary log deck	100	10
Barker in-feed	300	30
Green chain	200 to 300[26]	20 to 30[26]
Lumber strapping	150 to 200[26]	15 to 20[26]
Lumber handling areas	20	2
Lumber loading areas	50	5
Wood chip storage piles	5	0.5
Service station (at grade)		
Dark surrounding		
Approach	15	1.5
Driveway	15	1.5
Pump island area	200	20
Building faces (exclusive of glass)	100[14]	10[14]
Service areas	30	3
Landscape highlights	20	2
Light surrounding		
Approach	30	3
Driveway	50	5
Pump island area	300	30
Building faces (exclusive of glass)	300[14]	30[14]
Service areas	70	7
Landscape highlights	50	5
Ship yards		
General	50	5
Ways	100	10
Fabrication areas	300	30
Smokestacks with advertising messages (see **Bulletin and poster boards)**		
Storage yards		
Active	200	20
Inactive	10	1
Streets (see page 14-8)		
Water tanks with advertising messages (see **Bulletin and poster boards)**		

For footnotes, see page 2-19.

Fig. 2–2. *Continued*

V. Sports and Recreational Areas

Area/Activity	Lux	Footcandles
Archery (indoor)		
Target, tournament	500[14]	50[14]
Target, recreational	300[14]	30[14]
Shooting line, tournament	200	20
Shooting line, recreational	100	10
Archery (outdoor)		
Target, tournament	100[14]	10[14]
Target, recreational	50[14]	5[14]
Shooting line, tournament	100	10
Shooting line, recreational	50	5
Badminton		
Tournament	300	30
Club	200	20
Recreational	100	10
Baseball		
Major league		
Infield	1500	150
Outfield	1000	100
AA and AAA league		
Infield	700	70
Outfield	500	50
A and B league		
Infield	500	50
Outfield	300	30
C and D league		
Infield	300	30
Outfield	200	20
Semi-pro and municipal league		
Infield	200	20
Outfield	150	15
Recreational		
Infield	150	15
Outfield	100	10
Junior league (Class I and Class II)		
Infield	300	30
Outfield	200	20
On seats during game	20	2
On seats before and after game	50	5
Basketball		
College and professional	500	50
College intramural and high school	300	30
Recreational (outdoor)	100	10
Bathing beaches		
On land	10	1
150 feet from shore	30[14]	3[14]
Billiards (on table)		
Tournament	500	50
Recreational	300	30
Bowling		
Tournament		
Approaches	100	10
Lanes	200	20
Pins	500[14]	50[14]
Recreational		
Approaches	100	10
Lanes	100	10
Pins	300[14]	30[14]
Bowling on the green		
Tournament	100	10
Recreational	50	5
Boxing or wrestling (ring)		
Championship	5000	500
Professional	2000	200
Amateur	1000	100
Seats during bout	20	2
Seats before and after bout	50	5
Casting—bait, dry-fly, wet-fly		
Pier or dock	100	10
Target (at 24 meters [80 feet] for bait casting and 15 meters [50 feet] for wet or dry-fly casting)	50[14]	5[14]
Combination (outdoor)		
Baseball/football		
Infield	200	20
Outfield and football	150	15
Industrial softball/football		
Infield	200	20
Outfield and football	150	15
Industrial softball/6-man football		
Infield	200	20
Outfield and football	150	15
Croquet or Roque		
Tournament	100	10
Recreational	50	5
Curling		
Tournament		
Tees	500	50
Rink	300	30
Recreational		
Tees	200	20
Rink	100	10
Fencing		
Exhibitions	500	50
Recreational	300	30
Football		
Distance from nearest sideline to the farthest row of spectators		
Class I Over 30 meters [100 feet]	1000	100
Class II 15 to 30 meters [50 to 100 feet]	500	50
Class III 9 to 15 meters [30 to 50 feet]	300	30
Class IV Under 9 meters [30 feet]	200	20
Class V No fixed seating facilities	100	10

It is generally conceded that the distance between the spectators and the play is the first consideration in determining the class and lighting requirements. However, the potential seating capacity of the stands should also be considered and the following ratio is suggested: Class I for

For footnotes, see page 2–19.

Fig. 2–2. *Continued*

V. *Continued*

Area/Activity	Lux	Footcandles
over 30,000 spectators; Class II for 10,000 to 30,000; Class III for 5000 to 10,000; and Class IV for under 5000 spectators.		
Football, Canadian—rugby (see **Football**)		
Football, six-man		
High school or college	200	20
Jr. high and recreational	100	10
Golf		
Tee	50	5
Fairway	10, 30[14]	1, 3[14]
Green	50	5
Driving range		
At 180 meters [200 yards]	50[14]	5[14]
Over tee area	100	10
Miniature	100	10
Practice putting green	100	10
Gymnasiums (refer to individual sports listed)		
General exercising and recreation	300	30
Handball		
Tournament	500	50
Club		
Indoor—four-wall or squash	300	30
Outdoor—two-court	200	20
Recreational		
Indoor—four-wall or squash	200	20
Outdoor—two-court	100	10
Hockey, field	200	20
Hockey, ice (indoor)		
College or professional	1000	100
Amateur	500	50
Recreational	200	20
Hockey, ice (outdoor)		
College or professional	500	50
Amateur	200	20
Recreational	100	10
Horse shoes		
Tournament	100	10
Recreational	50	5
Horse shows	200	20
Jai-alai		
Professional	1000	100
Amateur	700	70
Lacrosse	200	20
Playgrounds	50	5
Quoits	50	5
Racing (outdoor)		
Auto	200	20
Bicycle		
Tournament	300	30
Competitive	200	20
Recreational	100	10
Dog	300	30

Area/Activity	Lux	Footcandles
Dragstrip		
Staging area	100	10
Acceleration, 400 meters [1320 feet]	200	20
Deceleration, first 200 meters [660 feet]	150	15
Deceleration, second 200 meters [660 feet]	100	10
Shutdown, 250 meters [820 feet]	50	5
Horse	200	20
Motor (midget of motorcycle)	200	20
Racquetball (see **Handball**)		
Rifle 45 meters [50 yards]—outdoor)		
On targets	500[14]	50[14]
Firing point	100	10
Range	50	5
Rifle and pistol range (indoor)		
On targets	1000[14]	100[14]
Firing point	200	20
Range	100	10
Rodeo		
Arena		
Professional	500	50
Amateur	300	30
Recreational	100	10
Pens and chutes	50	5
Roque (see **Croquet**)		
Shuffleboard (indoor)		
Tournament	300	30
Recreational	200	20
Shuffleboard (outdoor)		
Tournament	100	10
Recreational	50	5
Skating		
Roller rink	100	10
Ice rink, indoor	100	10
Ice rink, outdoor	50	5
Lagoon, pond, or flooded area	10	1
Skeet		
Targets at 18 meters [60 feet]	300[14]	30[14]
Firing points	50	5
Skeet and trap (combination)		
Targets at 30 meters [100 feet] for trap, 18 meters [60 feet] for skeet	300[14]	30[14]
Firing points	50	5
Ski slope	10	1
Soccer (see **Football**)		
Softball		
Professional and championship		
Infield	500	50
Outfield	300	30
Semi-professional		
Infield	300	30
Outfield	200	20

For footnotes, see page 2–19.

Fig. 2–2. *Continued*

V. *Continued*

Area/Activity	Lux	Footcandles
Industrial league		
Infield	200	20
Outfield	150	15
Recreational (6-pole)		
Infield	100	10
Outfield	70	7
Slow pitch, tournament—see industrial league		
Slow pitch, recreational (6-pole)—see recreational (6-pole)		
Squash (see **Handball**)		
Swimming (indoor)		
Exhibitions	500	50
Recreational	300	30
Underwater—1000 [100] lamp lumens per square meter [foot] of surface area		
Swimming (outdoor)		
Exhibitions	200	20
Recreational	100	10
Underwater—600 [60] lamp lumens per square meter [foot] of surface area		
Tennis (indoor)		
Tournament	1000	100
Club	750	75
Recreational	500	50
Tennis (outdoor)		
Tournament	300	30
Club	200	20
Recreational	100	10
Tennis, platform	500	50
Tennis, table		
Tournament	500	50
Club	300	30
Recreational	200	20
Trap		
Targets at 30 meters [100 feet]	300[14]	30[14]
Firing points	50	5
Volley ball		
Tournament	200	20
Recreational	100	10

VI. Transportation Vehicles

Area/Activity	Lux	Footcandles
Aircraft		
Passenger compartment		
General	50	5
Reading (at seat)	200	20
Airports		
Hangar apron	10	1
Terminal building apron		
Parking area	5	0.5
Loading area	20[14]	2[14]
Rail conveyances		
Boarding or exiting	100	10
Fare box (rapid transit train)	150	15
Vestibule (commuter and inter-city trains)	100	10
Aisles	100	10
Advertising cards (rapid transit and commuter trains)	300	30
Back-lighted advertising cards (rapid transit and commuter trains)—860 cd/m^2 (250 fL) average maximum.		
Reading	300[3]	30[3]
Rest room (inter-city train)	200	20
Dining area (inter-city train)	500	50
Food preparation (inter-city train)	700	70
Lounge (inter-city train)		
General lighting	200	20
Table games	300	30
Sleeping car		
General lighting	100	10
Normal reading	300[3]	30[3]
Prolonged seeing	700[3]	70[3]
Road Conveyances		
Step well and adjacent ground area	100	10
Fare box	150	15
General lighting (for seat selection and movement)		
City and inter-city buses at city stop	100	10
Inter-city bus at country stop	20	2
School bus while moving	150	15
School bus at stops	300	30
Advertising cards	300	30
Back-lighted advertising cards (see Rail conveyances)		
Reading	300[3]	30[3]
Emergency exit (school bus)	50	5
Ships		
Living Areas		
Staterooms and Cabins		
General lighting	100	10
Reading and writing	300[15,3]	30[15,3]
Prolonged seeing	700[16,3]	70[16,3]
Baths (general lighting)	100	10
Mirrors (personal grooming)	500	50
Barber shop and beauty parlor	500	50
On subject	1000	100
Day rooms		
General lighting	200[15]	20[15]
Desks	500[16,3]	50[16,3]
Dining rooms and messrooms	200	20

For footnotes, see page 2–19.

Fig. 2-2. *Continued*

VI. *Continued*

Area/Activity	Lux	Footcandles
Enclosed promenades		
General lighting	100	10
Entrances and passageways		
General	100	10
Daytime embarkation	300	30
Gymnasiums		
General lighting	300	30
Hospital		
Dispensary (general lighting)	300[16]	30[16]
Operating room		
General lighting	500[16]	50[16]
Doctor's office	300[16]	30[16]
Operating table	20000	2000
Wards		
General lighting	100	10
Reading	300	30
Toilets	200	20
Libraries and lounges		
General lighting	200	20
Reading	300[16,3]	30[16,3]
Prolonged seeing	700[16,3]	70[16,3]
Purser's office	200[16]	20[16]
Shopping areas	200	20
Smoking rooms	150	15
Stairs and foyers	200	20
Recreation areas		
Ball rooms	150[15]	15[15]
Cocktail lounges	150[15]	15[15]
Swimming pools		
General	150[15]	15[15]
Underwater		
Outdoors—600 [60] lamp lumens/square meter [foot] of surface area		
Indoors—1000 [100] lamp lumens/square meter [foot] of surface area		
Theatre		
Auditorium		
General	100[15]	10[15]
During picture	1	0.1
Navigating Areas		
Chart room		
General	100	10
On chart table	500[16,3]	50[16,3]
Gyro room	200	20
Radar room	200	20
Radio room	100[16]	10[16]
Radio room, passenger foyer	100	10
Ship's offices		
General	200[16]	20[16]
On desks and work tables	500[16,3]	50[16,3]
Wheelhouse	100	10
Service Areas		
Food preparation		
General	200[16]	20[16]
Butcher shop	200[16]	20[16]
Galley	300[16]	30[16]
Pantry	200[16]	20[16]
Thaw room	200[16]	20[16]
Sculleries	200[16]	20[16]
Food storage (non-refrigerated)	100	10
Refrigerated spaces (ship's stores)	50	5
Laundries		
General	200[16]	20[16]
Machine and press finishing, sorting	500	50
Lockers	50	5
Offices		
General	200	20
Reading	500[16,3]	50[16,3]
Passenger counter	500[16,3]	50[16,3]
Storerooms	50	5
Telephone exchange	200	20
Operating Areas		
Access and casing	100	10
Battery room	100	10
Boiler rooms	200[16]	20[16]
Cargo handling (weather deck)	50[16]	5[16]
Control stations (except navigating areas)		
General		
Control consoles	200	20
Gauge and control boards	300	30
	300	30
Switchboards	300	30
Engine rooms	200[16]	20[16]
Generator and switchboard rooms	200[16]	20[16]
Fan rooms (ventilation & air conditioning)	100	10
Motor rooms	200	20
Motor generator rooms (cargo handling)	100	10
Pump room	100	10
Shaft alley	100	10
Shaft alley escape	30	3
Steering gear room	200	20
Windlass rooms	100	10
Workshops		
General	300[16]	30[16]
On top of work bench	500[16]	50[16]
Tailor shop	500[16]	50[16]
Cargo holds		
Permanent luminaires	30[16]	3[16]
Passageways and trunks	100	10

[1] Include provisions for higher levels for exhibitions.
[2] Specific limits are provided to minimize deterioration effects.
[3] Task subject to veiling reflections. Illuminance listed is not an ESI value. Currently, insufficient experience in the use of ESI target values precludes the direct use of Equivalent Sphere Illumination in the present consensus approach to recommend illuminance values. Equivalent Sphere Illumination may be used as a tool in determining the effectiveness of controlling veiling reflections and as a part of the evaluation of lighting systems.
[4] Illuminance values are listed based on experience and consensus. Values relate to needs during various religious ceremonies.
[5] Degradation factors: Overlays—add 1 weighting factor for each overlay; Used material—estimate additional factors.

Fig. 2-2. *Continued*

[6] Provide higher level over food service or selection areas.
[7] Supplying illumination as in delivery room must be available.
[8] Illuminance values developed for various degrees of store area activity.
[9] Or not less than ⅕ the level in the adjacent areas.
[10] Only when actual equipment service is in process. May be achieved by a general lighting system or by localized or portable equipment.
[11] For color matching, the spectral quality of the color of the light source is important.
[12] Veiling reflections may be produced on glass surfaces. It may be necessary to treat plus weighting factors as minus in order to obtain proper illuminance.
[13] Especially subject to veiling reflections. It may be necessary to shield the task or to reorient it.
[14] Vertical.
[15] Illuminance values may vary widely, depending upon the effect desired, the decorative scheme, and the use made of the room.
[16] Supplementary lighting should be provided in this space to produce the higher levels required for specific seeing tasks involved.
[17] Good to high color rendering capability should be considered in these areas. As lamps of higher luminous efficacy and higher color rendering capability become available and economically feasible, they should be applied in all areas of health care facilities.
[18] Variable (dimming or switching).
[19] Values based on a 25 per cent reflectance, which is average for vegetation and typical outdoor surfaces. These figures must be adjusted to specific reflectances of materials lighted for equivalent brightnesses. Levels give satisfactory brightness patterns when viewed from dimly lighted terraces or interiors. When viewed from dark areas they may be reduced by at least ½; or they may be doubled when a high key is desired.
[20] General lighting should not be less than ⅓ of visual task illuminance nor less than 200 lux [20 footcandles].
[21] Industry representatives have established a table of single illuminance values which, in their opinion, can be used in preference to employing reference 6. Illuminance values for specific operations can also be determined using illuminance categories of similar tasks and activities found in this table and the application of the appropriate weighting factors in Fig. 2-4.
[22] Special lighting such that (1) the luminous area is large enough to cover the surface which is being inspected and (2) the luminance is within the limits necessary to obtain comfortable contrast conditions. This involves the use of sources of large area and relatively low luminance in which the source luminance is the principal factor rather than the illuminance produced at a given point.
[23] Maximum levels—controlled system.
[24] Additional lighting needs to be provided for maintenance only.
[25] Color temperature of the light source is important for color matching.
[26] Select upper level for high speed conveyor systems. For grading redwood lumber 3000 lux [300 footcandles] is required.

egory based on types of visual tasks to be performed in the design space. Each category prescribes a range of illuminances permitting the designer to establish a target illuminance responsive to several task and observer characteristics, including the importance of speed and/or accuracy in performing the task, and the age of the observer.

Step 1. Define Visual Task. Determine the type of activity for which the level of lighting is to be selected (*e.g.*, reading typed originals). Also establish the plane of the visual task to which the illuminance level is to be applied.

Step 2. Select Illuminance Category. Select the appropriate Illuminance Category from one of the following:

a. Fig. 2-2, Parts II and III—when a review of typical tasks reveals specific task types.

b. Fig. 2-2, Part I—if specific tasks cannot be established, generic task descriptions must be used.

c. Fig. 2-3—if an equivalent contrast ($\tilde{C}$) has been determined.

Step 3. Determine Illuminance Range. Referring to Fig. 2-2, Part I, and using the Illuminance Category selected in Step 2, determine the recommended range of illuminances.

Because of the characteristics of the functions in Categories A through C, illuminances are required over the entire area of the interior space considered. For instance, in a lobby area, one visual task is walking to an elevator lobby. This visual task remains constant throughout time and space; therefore, a general level of illumination should be provided throughout the lobby.

Categories D through F, however, are for tasks which remain relatively fixed at one location for meaningful visual performance, although tasks may change considerably from one location to another within a given space. For example, an accounting office may have a secretarial pool where reading felt-tip-pen hand-written notes and proofreading typed originals are prominent tasks, while at the same time accountants may be reading computer printouts. Each task calls for a particular illuminance level for satisfactory visual performance, and so each task should be lighted accordingly. Therefore, Categories D through F should be applied to the appropriate task areas only.

Categories G through I are for extremely difficult visual tasks, and may be difficult to illuminate. For practical and economical reasons, lighting systems for these tasks may require a combination of general over-all illumination and task area illumination. Because of the unusual conditions associated with tasks in Categories G through I, very careful analysis is recommended.

Step 4. Establish Illuminance Target Value. From the range of illuminances determined in Step 3, a design illuminance is to be established based upon several factors. These factors vary depending upon the visual task. For Illuminance Categories A through C use step *a*, below, for establishing a design illuminance. Use step *b* for Categories D through I.

a. For Categories A Through C. To establish an appropriate illuminance target value, the designer should be familiar with the design space and intended occupants, to the extent that the following information can be determined:

(1) Occupants ages (*e.g.*, if the design space is an elevator lobby in a senior citizens' housing complex, then establish ages of the housing occupants).

(2) Surface reflectances (*e.g.*, if the design space is a building lobby, and the floor is to be slate, with walls of teak, their reflectances must be established).

After the above information has been established, the designer may determine an appropriate target value from the Illuminance Category by using Fig. 2-4a as follows:

(a) Review each of the two characteristics and determine the appropriate weighting factors (−1, 0, +1).

(b) Add the two factors algebraically taking into account the signs.

(c) If the total factor is −2, use the lowest of the three illuminances in the established range; if the total factor is +2, use the highest of the three illuminances; otherwise use the middle illuminance.

b. For Categories D Through I. At this point the designer should become thoroughly familiar with the anticipated task and anticipated space occupants to the extent that the following information can be established:

(1) The precise task considered (*e.g.*, if the task is reading computer printouts, obtain a

Fig. 2–4. Weighting Factors to be Considered in Selecting Specific Illuminance Within Ranges of Values for Each Category.

a. For Illuminance Categories A through C

Room and Occupant Characteristics	Weighting Factor		
	−1	0	+1
Occupants ages	Under 40	40–55	Over 55
Room surface reflectances*	Greater than 70 per cent	30 to 70 per cent	Less than 30 per cent

b. For Illuminance Categories D through I

Task and Worker Characteristics	Weighting Factor		
	−1	0	+1
Workers ages	Under 40	40–55	Over 55
Speed and/or accuracy**	Not important	Important	Critical
Reflectance of task background***	Greater than 70 per cent	30 to 70	Less than 30 per cent

* Average weighted surface reflectances, including wall, floor and ceiling reflectances, if they encompass a large portion of the task area or visual surround. For instance, in an elevator lobby, where the ceiling height is 7.6 meters (25 feet), neither the task nor the visual surround encompass the ceiling, so only the floor and wall reflectances would be considered.

** In determining whether speed and/or accuracy is not important, important or critical, the following questions need to be answered: What are the time limitations? How important is it to perform the task rapidly? Will errors produce an unsafe condition or product? Will errors reduce productivity and be costly? For example, in reading for leisure there are no time limitations and it is not important to read rapidly. Errors will not be costly and will not be related to safety. Thus, speed and/or accuracy is not important. If however, prescription notes are to be read by a pharmacist, accuracy is critical because errors could produce an unsafe condition and time is important for customer relations.

*** The task background is that portion of the task upon which the meaningful visual display is exhibited. For example, on this page the meaningful visual display includes each letter which combines with other letters to form words and phrases. The display medium, or task background, is the paper, which has a reflectance of approximately 85 per cent.

Fig. 2–3. Illuminance Categories of Fig. 2–2, Part I, for Measured Equivalent Contrast Values of Task Visual Displays.

Equivalent Contrast $\tilde{C}$†	Illuminance Category**
over 1.0	*
.75–1.0	D
.62– .75	E
.50– .62	F
.40– .50	G
.30– .40	H
under .30	I

* Use 200 lux [20 footcandles] and omit use of Fig. 2–4 and footnote (**) below.

** If task reflectance is between 5 and 20 per cent use next higher illuminance category; *i.e.*, D to E, E to F, etc. If less than 5 per cent use two categories higher.

† As determined using a visibility meter and the procedure outlined in Reference 8.

Note: Although specific equivalent contrasts are established scientifically, a consensus procedure has been used in establishing corresponding Illuminance Categories.

sample to determine the reflectance of the computer paper alone—this is the task background reflectance).
(2) Occupant ages (*e.g.*, if the task is writing payroll checks, and only the senior accountants perform this task, then establish the approximate ages of the senior accountants).
(3) Importance of speed (*e.g.*, if the occupants are under abnormal time constraints, as in a newscopy proofreading room, then speed might be considered critical).
(4) Importance of accuracy (*e.g.*, if accuracy could be a life-death matter as in prescription reading/filling, then accuracy is considered critical).

After the above information has been established, the designer may determine an appropriate target value from the Illuminance Category by using Fig. 2-4b as follows:
(a) Review each of the three characteristics and determine the appropriate weighting factors (−1, 0, +1).
(b) Add the three factors algebraically, taking into account the signs.
(c) If the total weighting factor is −2 or −3, use the lowest of the three illuminances in the established range; if the total factor is +2 or +3, use the highest of the three illuminances; otherwise use the middle illuminance.
(d) When designing spaces with tasks in Categories D through I, it is recommended that 200 lux [20 footcandles] be regarded as the minimum acceptable horizontal illuminance for the general, non-task area.

Proper determination of the weighting factors requires information and judgment on the part of the user. Guessed values are poor substitutes for information and can result in over or under design. Whenever possible, design information should be used in determining values of the weighting factors for each characteristic of the lighting task.

Simplification of Steps 3 and 4. Fig. 2–5 is provided as a means of combining the tables referred to in Steps 3 and 4 as a short cut method once Steps 3 and 4 are understood. In Fig. 2–5, the Illuminance Category from Step 2 and Weighting Factor information (age, speed and accuracy, and reflectance) are used to directly select the illuminances in lux (if footcandles are desired, divide by 10). For a rough estimate of reflectances a gray scale marked with per cent reflectances may prove helpful. Where surfaces are in color, the Munsell value scales for judging reflectance will be found to be helpful. See Section 5 of the 1981 Reference Volume.

Example of Illuminance Selection. A classroom in a high school is to be relighted. The designer in consultation with the teacher and school administrators has determined the following:
1. The task is reading mimeograph material with a reflectance of about 80 per cent.
2. The students are teenagers.
3. The students practice typing to improve speed and accuracy, thus speed and accuracy are considered to be important, but not critical.

Using the above step-by-step procedure:

Step 1. The visual task is defined above.

Step 2. Referring to Fig. 2–2 an Illuminance Category of D is found under Reading, Mimeograph, on page 2–8.

Step 3. Referring to Part I of Fig. 2–2, the illuminance range is found to be 200-300-500 lux [20-30-50 footcandles].

Step 4. Referring to Fig. 2–4b and the above information, the weighting factors selected are: −1 for workers' ages; 0 for speed and/or accuracy; and −1 for reflectance of task background. The algebraic sum is −1 +0 −1 = −2. Therefore, the illuminance to be selected is the lowest value, *i.e.*, 200 lux [20 footcandles].

If the task were reading #3 pencil handwriting on 80 per cent reflectance paper and the students were older (an adult education course), the Illuminance Category would change to E, the illuminance range would become 500-750-1000 lux [50-75-100 footcandles], and the weighting factor for age would be 0. The new algebraic sum of the weighting factors is 0 + 0 −1 = −1. Therefore, the illuminance to be selected is the mid value in the new range, *i.e.*, 750 lux [75 footcandles].

By referring to Fig. 2–5 after step 2, the illuminance can be selected without referring to Part I of Fig. 2–2 or to Fig. 2–4.

Application of Illuminance Values Selected. The use of selected illuminance values may be influenced by work areas involving many visual tasks. The designer, usually through client/occupant/designer interaction, must establish the task of prime importance, with the subsequent heirarchy of remaining tasks. Similarly, the time duration of each task, worker ages, expected task performance, and task characteristics must be determined. If all or many of the tasks require similar lighting qualities, then the designer might design the lighting system to meet one task, and will therefore meet the majority of the other tasks' requirements. If however, the tasks vary considerably in lighting requirements, then the designer should consider

Fig. 2–5. Illuminance Values, Maintained, in Lux, for a Combination of Illuminance Categories and User, Room and Task Characteristics (For Illuminance in Footcandles, Divide by 10).

a. General Lighting Throughout Room

Weighting Factors		Illuminance Categories		
Average of Occupants Ages	Average Room Surface Reflectance (per cent)	A	B	C
Under 40	Over 70	20	50	100
	30–70	20	50	100
	Under 30	20	50	100
40–55	Over 70	20	50	100
	30–70	30	75	150
	Under 30	50	100	200
Over 55	Over 70	30	75	150
	30–70	50	100	200
	Under 30	50	100	200

b. Illuminance on Task

Weighting Factors			Illuminance Categories					
Average of Workers Ages	Demand for Speed and/or Accuracy*	Task Background Reflectance (per cent)	D	E	F	G**	H**	I**
Under 40	NI	Over 70	200	500	1000	2000	5000	10000
		30–70	200	500	1000	2000	5000	10000
		Under 30	300	750	1500	3000	7500	15000
	I	Over 70	200	500	1000	2000	5000	10000
		30–70	300	750	1500	3000	7500	15000
		Under 30	300	750	1500	3000	7500	15000
	C	Over 70	300	750	1500	3000	7500	15000
		30–70	300	750	1500	3000	7500	15000
		Under 30	300	750	1500	3000	7500	15000
40–55	NI	Over 70	200	500	1000	2000	5000	10000
		30–70	300	750	1500	3000	7500	15000
		Under 30	300	750	1500	3000	7500	15000
	I	Over 70	300	750	1500	3000	7500	15000
		30–70	300	750	1500	3000	7500	15000
		Under 30	300	750	1500	3000	7500	15000
	C	Over 70	300	750	1500	3000	7500	15000
		30–70	300	750	1500	3000	7500	15000
		Under 30	500	1000	2000	5000	10000	20000
Over 55	NI	Over 70	300	750	1500	3000	7500	15000
		30–70	300	750	1500	3000	7500	15000
		Under 30	300	750	1500	3000	7500	15000
	I	Over 70	300	750	1500	3000	7500	15000
		30–70	300	750	1500	3000	7500	15000
		Under 30	500	1000	2000	5000	10000	20000
	C	Over 70	300	750	1500	3000	7500	15000
		30–70	500	1000	2000	5000	10000	20000
		Under 30	500	1000	2000	5000	10000	20000

* NI = not important, I – important, and C = critical
** Obtained by a combination of general and supplementary lighting.

multiple level systems, variable control systems or a combination of systems in order to accommodate a number of tasks of varying visual requirements in an energy-economic manner.

The target values obtained from this procedure are conventional illuminance values in lux or footcandles and are values to be maintained in service. For many visual displays this is a reasonable indicator of display visibility. Some displays, however, can exhibit veiling reflections and serious contrast loss as indicated by a superscript 3 in Fig. 2–2. In these cases, the illuminance alone is not a reliable indicator of visibility; the contrast of the display must also be taken into account. Equivalent sphere illumination (ESI) is a measure of visibility that takes both illuminance and contrast into account.

Currently, insufficient experience with the use of ESI target values precludes the direct use of ESI values as part of the consensus approach recommendation process. Thus, the recommendations are in conventional units of illuminance. However, ESI may be used as a tool in determining the effectiveness of controlling veiling reflection and as part of the evaluation of lighting systems.

The target values of illuminance for Illuminance Categories A to C are *average maintained illuminances*, and the lumen method, using zonal-cavity calculated coefficients of utilization for luminaires, or for daylighting, predicts such average illuminance values. The target values of illuminance obtained for visual displays in the last six categories (D through I) are localized values, that is, *maintained illuminance on the task* and point calculation methods are appropriate. In either case the procedure for determining light loss factors should be used in calculating maintained average or point illuminances. See page 9-1 of the 1981 Reference Volume.

Luminance Ratios

Luminances in the visual field which surrounds an object or a task can have different effects on visual ability depending upon the areas involved, their location with respect to the line of sight, and their actual luminances as compared with that of the task. These luminances may produce a decrement in visual ability, visual comfort, or both. For this reason, the luminances of the various surfaces in the visual field should be controlled and limited.

When there is a large difference in luminances between areas (a high luminance ratio), for example a large difference between the luminance of a task and that of a bright window during the day or a dark window at night, there may be losses in the ability to see the task display if one looks away from the task to the window and then back at the task. This is due to transient adaptation and the changes in sensitivity of the eye (see Section 3 of the 1981 Reference Volume). If the ratio is high, there also may be a reaction of discomfort.

As a guide for design purposes, luminance ratio limits have been recommended for various applications, such as offices, educational facilities, institutions, industrial areas and residences (see Sections 5 through 10). For additional guidance, recommended limits of reflectances (both upper and lower), of large area surfaces, are given for the same applications. The use of these reflectance limits, along with a selection of appropriate colors, should help to control luminances and keep within the ratio limits without creating a bland and uninteresting environment.

Visual Comfort

Visual discomfort may occur when excessively high luminances are within the visual field. High luminances also can distract and even reduce visibility (see above).

When luminances and their relationships in the field of view cause visual discomfort but do not necessarily interfere with seeing, the sensation experienced by an observer is termed *discomfort glare*. It usually is produced by direct glare from light sources or luminaires which are too bright, inadequately shielded, or of too great an area. Discomfort glare also can be caused by annoying reflection of bright areas in specular surfaces (known as reflected glare). The latter should not be confused with veiling reflections which impair visual performance rather than cause discomfort.

Both maximum and average luminances of a potential glare source are significant factors in evoking a glare sensation, but average luminances are recognized as being the more pertinent. A rating system based on the degree of freedom from discomfort glare in a lighting installation called Visual Comfort Probability (VCP), uses average luminances. This system evaluates lighting systems in terms of direct glare. It is an estimator of the fraction of the observer population which will accept the lighting system and its environment as being comfortable, using the perception of glare due to direct light from luminaires to the observer as a criterion.

This evaluation of comfort is based on the following factors which influence subjective judgments of visual comfort: room size and shape; room surface reflectances; illuminance levels; luminaire type, size, luminance, maximum luminance and light distribution; number and location of luminaires; luminance of the entire field of view; observer location and line of sight; and differences in individual glare sensitivity.

Extensive investigations and analyses (See page 3-12 in the 1981 Reference Volume) have resulted in a comprehensive standard discomfort glare evaluation procedure which takes these factors into account. The final product of the evaluation procedure is a Visual Comfort Probability (VCP) rating of the lighting system expressed as a per cent of people viewing along a specified line of sight who will be expected to find it acceptable.

VCP can be calculated for specific lighting systems and given observer lines of sight (see page 9-71 in the 1981 Reference Volume). However, in order to systematize the calculations, to aid in the development of VCP tables, and to permit comparison of luminaires, standard conditions have been adopted:[9]

1. An initial level of 1000 lux [100 footcandles].
2. Room surfaces of 80 per cent for the effective ceiling cavity reflectance, 50 per cent for the wall reflectance and 20 per cent for the effective floor cavity reflectance.
3. Mounting heights above the floor of 2.6, 3, 4 and 4.9 meters (8.5, 10, 13 and 16 feet).
4. A range of room dimensions to include square, long-narrow, and short-wide rooms.
5. A standard layout involving luminaires uniformly distributed throughout the space.
6. An observation point 1.2 meters (four feet) in front of the center of the rear wall and 1.2 meters (four feet) above the floor.
7. A horizontal line of sight directly forward.
8. A limit to the field of view corresponding to an angle of 53 degrees above and directly forward from the observer.

By consensus, direct glare will not be a problem in lighting installations if all three of the following conditions are satisfied.[9]

1. The VCP is 70 or more;
2. The ratio of maximum*-to-average luminaire luminance does not exceed five to one at 45, 55, 65, 75 and 85 degrees from nadir crosswise and lengthwise; and
3. Maximum luminaire luminances crosswise and lengthwise do not exceed the following values:

Angle Above Nadir (degrees)	Maximum Luminance (candelas per square meter)	(footlamberts)
45	7710	2250
55	5500	1605
65	3860	1125
75	2570	750
85	1695	495

The principal research used to establish the VCP system involved luminances of a magnitude compatible with fluorescent lamps. Further, the most extensive field validification utilized lighting systems typical of fluorescent luminaires. Although the mathematics can be applied to virtually any situation, extrapolation to significantly different visual fields has not been validified.

The VCP system is based on empirical relations derived from a variety of experiments. It has been concluded that differences of 5 percentage points or less are meaningless, *i.e.*, if two lighting systems do not differ by more than 5 in VCP, the VCP system provides no basis for judging a difference in visual comfort. Here, it is assumed that this entire difference is due to the lighting systems. Artifacts introduced by using different computational procedures for two lighting systems can further spread the VCP values for two systems that are not demonstrably different.[10]

An alternate simplified method of providing an acceptable degree of comfort has been derived from the formulas for discomfort glare. This simplified method is based on the premise that luminaire designers do not design different units for rooms of different sizes, but consider the probable range of room sizes and design for the "commonly found more difficult" potential glare situation. (In rooms less than 6 meters (20 feet) in length and width, the luminaires are largely out of the field of view.) This simplified method is only applicable to flat bottom luminaires. See page 9-73 in the 1981 Reference Volume.[11]

Reducing Discomfort Glare. Discomfort glare can be reduced by:

1. Decreasing the luminance of lighting equipment or other sources of objectionable glare, such as windows and overhead skylights, relative to the over-all luminance.
2. Diminishing the area of uncomfortable luminances (with level of constant luminance).
3. Increasing the angle between the source and line of sight.
4. Increasing the general luminance in the room (see the recommended luminance ratios).

* Brightest 6.54 square centimeters (one square inch) area.

Reflected Glare

Glare resulting from specular reflections of high luminance in polished or glossy surfaces is known as reflected glare and, as mentioned above, it may cause discomfort. If the reflections are of high luminance and of significant area, they also may produce disability glare if near the line of sight and may prove distracting to workers.[12] Bright reflections of small size often do not cause reflected glare but provide *enhancing reflections*, desired sparkle, such as in jewelry, glassware or tableware on display. When specular reflections of high or low luminance veil the visual task, this effect is known as veiling reflections. (See below.)

There is little in the way of research to quantify the visual impact of reflections in specular surfaces. Generally, the conditions which cause reflected glare are sufficiently obvious as to indicate the lighting solution. In general, large area low luminance sources are preferred where the task has even a small degree of specularity. Additional considerations in this regard are discussed in the various application sections. For example, finishes on office furniture and equipment should be matte.

Disability Glare

Disability glare is caused by stray light within the eyes, producing a veiling luminance upon the retinal image of an object to be seen, resulting in reduced object visibility (see page 3-12 in the 1981 Reference Volume). In most interior lighting situations, disability glare does not present a severe problem if discomfort glare and reflected glare are minimized and luminance ratios are controlled as discussed above. However, in the case of roadway lighting, vehicle headlights, and certain types of industrial lighting, the effects of stray light may have to be taken into consideration.

Veiling Reflections

For years lighting designers and those engaged in vision research have recognized that substantial losses in contrast, hence, in visibility and visual performance, can result when light sources are reflected in specular or semi-specular visual tasks.[13, 14] This has been known as the general subject of reflected glare. The effects vary from the reflection of an incandescent filament in a polished metal surface at one extreme, in which case the result is annoying, distracting and disabling, from the visual standpoint, to the other extreme where there is a reflection of a large luminous area in the surface of a magazine printed on dull or matte paper. With the latter, the effect may be undetectable by the naked eye and may be almost unmeasurable by instruments. The term "reflected glare" is reserved today for effects toward the first extreme and the term "veiling reflection" for effects near the second. Perhaps the greatest and certainly the most insidious problem is the reflection of luminaires or skylights in semi-specular and semi-matte surfaces such as a printed page and pencil writing on paper.

There are a great many factors that contribute to veiling reflections and each of them individually has long been known. The problem is to integrate the effects of these interrelated factors. Much has been learned, particularly the contributions made by Illuminating Engineering Research Institute (IERI) research.

Factors Causing Contrast Loss. In the study of contrast losses due to veiling reflections, the visual task (printing or handwriting on paper are usually considered as the visual task in the following discussions), the workers' orientation and viewing angle, and the lighting system all must be analyzed.

The Visual Task. The specularity of paper covers a wide range of degrees and modes of appearance. Most papers consist of rough fibers that have been matted together. Generally the fibers are somewhat shiny in themselves, but because of their random orientation they reflect light more or less equally in all directions. The harder the paper is pressed the more specular it becomes. Some papers are filled with clay or other coating so that the surface is very smooth. Some papers are actually glazed. The luminance of the paper depends both on the amount of light being diffusely reflected from it and whatever bright surface may be reflected in it. This reflection may be discernible as in the case of coated papers or glossy photographs, but frequently it is so indistinct as to go undetected even though serious losses in visibility occur.

The specularity of the graphic medium—pencil, pen, ink, carbon, etc.—again covers a very wide range. The degree of specularity depends on how the medium is deposited. For example, a very soft pencil brushed lightly across rough paper would leave a very diffuse mark. On the other hand, a hard pencil applied with pressure on a smooth surface can be very shiny. The

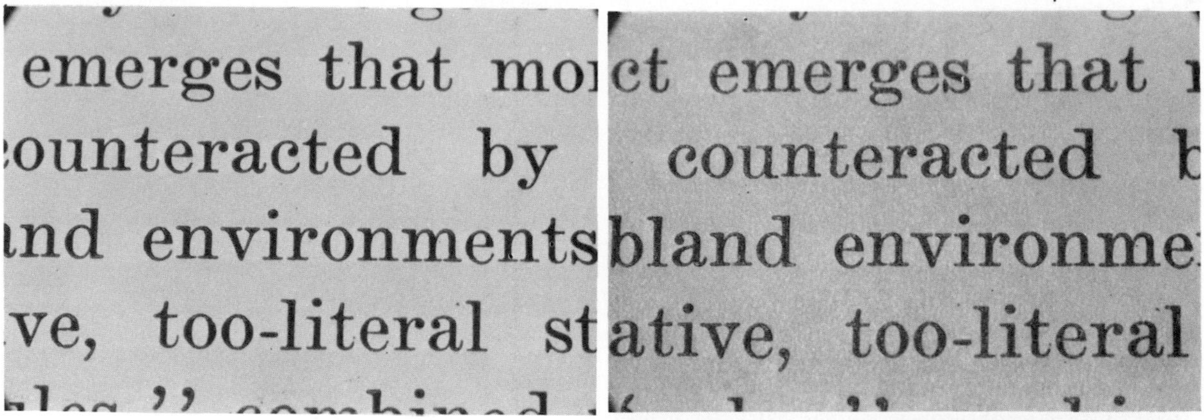

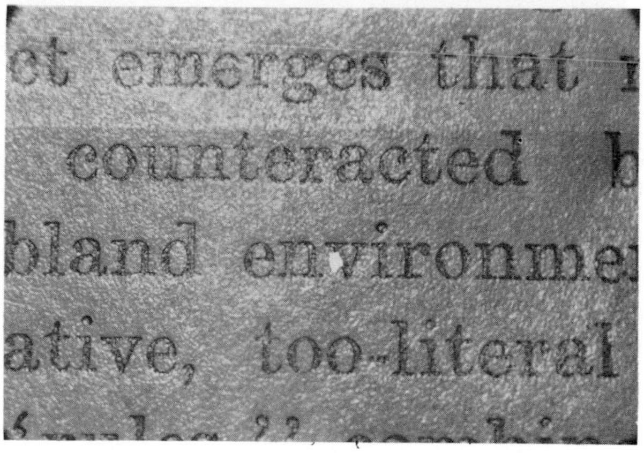

Fig. 2-6. An example of black printing on glossy paper. In the photograph above, a spotlight is located behind the camera in a position to cause minimum veiling reflections. The upper right photograph shows the same task lighted by a troffer that is positioned above and in front of the task in the offending zone. Note the subtle character of the specular reflections so that the loss of contrast is not immediately obvious. Note also that the specular reflections occur in different positions on different letters. The photograph at the right shows the same paper with a spotlight in the offending zone which creates the maximum veiling reflection effect. This condition might be called reflected glare. The image on the page is apparent as are the highlights in the letter stroke. Note how highlights tend to occur along the edges of the letter indicating that the type has embossed the paper.

luminance of the mark again depends on the amount of light being diffusely reflected from it and the reflection in it of luminous areas. Therefore, when considering task contrast, both diffuse and specular reflectance of both paper and graphic medium should be considered as well as the reflection of light sources in relation to the illuminance level. Figs. 2-6 and 2-7 illustrate a range of conditions of veiling reflections.

If the paper and the graphic medium could be considered as being perfect planes, the problem would be simpler than is the actual case. The pressure applied by the pencil, pen, typewriter key or printing type actually embosses the paper. The groove thus created causes the reflection of the light source to occur from positions on the ceiling other than the normal angle of reflection from the plane of the paper. Thus the part of the ceiling that is causing the problem may not be immediately obvious.

Consideration should be given to whether the task is lying in a horizontal plane or is slanted as is the case of a letter being hand-held, a book on a slanted school desk top, or a pencil drawing on a vertical drafting board. Chalkboards in classrooms, merchandise in stores, and signs are tasks of prime importance that lie in a vertical plane. The relative importance of various tasks and the planes in which they occur also should be considered. For example, office lighting should really be designed not just in terms of a task lying flat on a desk but also for one hand-held at about 45 degrees.

The Worker. The orientation of the worker with respect to the task greatly influences the magnitude of the effect of veiling reflections. First, for one eye position and one point of regard, a simplified relationship between the eye, the task, the perpendicular to the task, and an "offending zone" can be established. See Fig. 2-8. If the task were perfectly specular and flat, the offending zone would merely be a point. However, since the types of tasks involved here are more or less diffusing, the theoretical offending point becomes enlarged to an offending zone. Now if the eye is in such a position that the rays of light from the offending zone are reflected toward it, veiling reflections will occur. The angle of reflection is considered as the viewing angle. As the viewing angle increases, effects of the specular characteristics of the paper and the ink or pencil increase.

It has been found that people work throughout

Fig. 2-7. The upper left photograph shows pencil stenographic notes lighted by a spotlight located behind the camera. It can be seen that in this case the pencil stroke is relatively light because even in the darker parts of the stroke, areas of white paper show through. The photograph directly above shows the same task with the spotlight in the offending zone (above and in front of the task). As is frequently the case with this type of task, negative contrast or contrast reversal occurs where the pencil stroke can actually become brighter than the paper. In the photograph at the left the same task is lighted with an indirect lighting system.

a range of viewing angles[15, 16] with a peak at about 25 degrees as indicated by the approximate frequency distribution curves for office workers and for school children shown in Fig. 2-9a. Fig. 2-9b shows that 85 per cent of seeing occurs within 0 to 40-degree viewing angles, with higher angles used for only occasional glances. This is due to foreshortening of the task and to the increased viewing distance (see Fig. 2-9c) and to the resulting increase of task difficulty as shown in Fig. 2-9d. On this basis it seems reasonable to use 0 to 40 degrees as the practical range of viewing angles for design purposes.

Next the location and orientation of the worker and the task in the room must be evaluated. A worker with his back to a wall and facing out toward the center of the room has the maximum ceiling area as a potential offending zone. Furthermore, he has relatively little light coming from behind him which would cause little or no veiling reflections. A worker facing the wall would have minimum veiling reflections (not an appealing position psychologically). A person in the center of the room has light falling on the task from all directions. The ceiling may or may not constitute the offending zone, depending on the viewing angle.

A person sitting beside a window would have greatly reduced veiling reflections because the major source of illumination is outside the offending zone. It is also true that if a person were seated facing a window he would have potentially serious veiling reflection conditions.

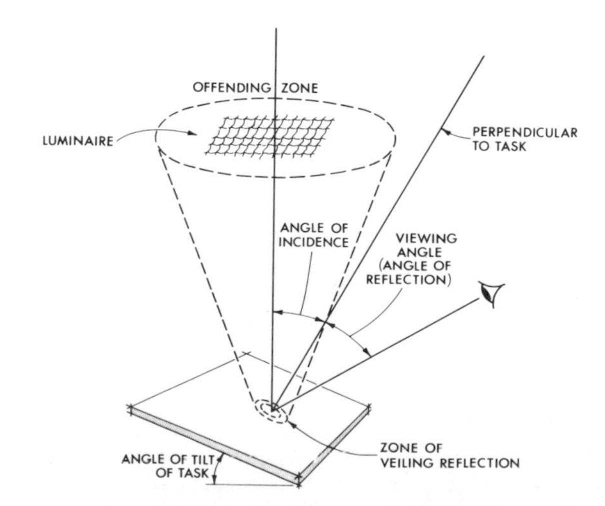

Fig. 2-8. Generalized description of angular relationships in analyzing veiling reflections.

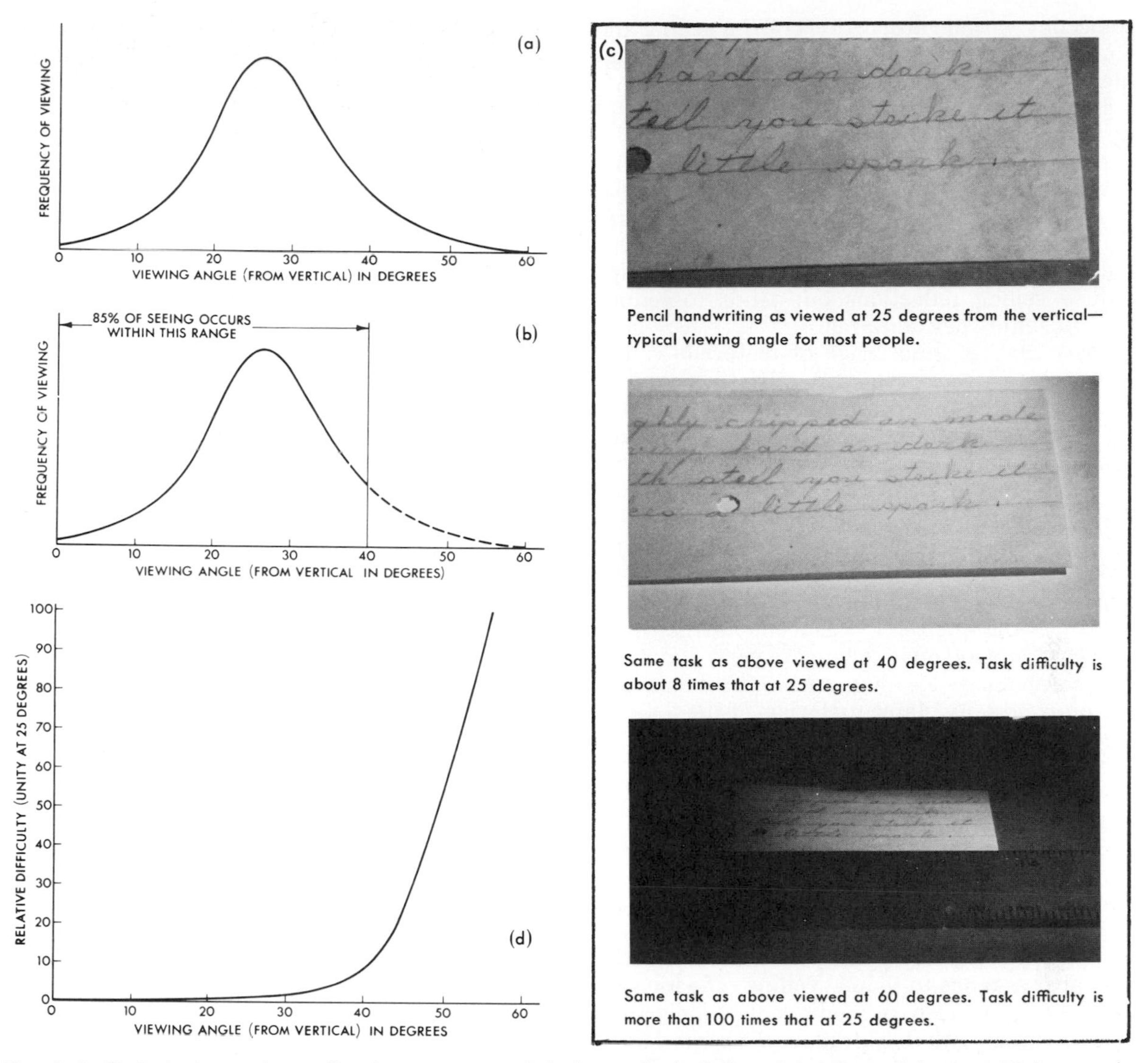

Fig. 2-9. Task viewing angles. *a*. People use a range of viewing angles in their work but the peak is about 25 degrees. *b*. Eighty-five per cent of seeing occurs within a range of 0 to 40 degrees, with seeing at large angles limited to occasional glances due to foreshortening and increased viewing distance. *c*. Photographs of an actual pencil handwriting sample as seen at viewing angles of 25, 40, and 60 degrees. *d*. Curve showing the relative difficulty of the pencil handwriting shown in *c* as measured with the Visual Task Evaluator.

The Lighting System. The worst condition is represented by a highly concentrated, very bright source with maximum candlepower directed toward the task. It is also likely to be uncomfortable as one looks around the room, and it may create shadows that interfere with writing, etc. Paradoxically, it is also the condition under which the worker can most easily escape veiling reflections by tilting the task so that the reflected rays do not reach his eye. When the task is truly a flat surface this is an effective solution. Because of embossing and curvature of many tasks such as books and magazines, reflections are not so easily eliminated.

Another condition would be a luminous dome placed over the worker and the task. Here the effect of veiling reflections would be reduced, but could not be escaped since there would always be a luminous area in the offending zone. Furthermore, through a rather wide range of illuminances the lighting system would generally be considered comfortable.

Between these two conditions lies the full range of luminaires of various sizes spaced so as to occupy various proportions of the ceiling and employing materials that produce varying candlepower distribution.

Lighting materials with reference to horizontal tasks may be compared as follows:

Diffusing. At small viewing angles (looking nearly straight down at the task) this material has the minimum luminance exposed to be reflected in the task. At large viewing angles it tends to have the highest luminance. This helps reduce veiling reflections but tends to reduce visual comfort per unit area.

Prismatic. A wide range of materials comes under this category, but generally they tend to expose greater flux toward the work for small viewing angles. Many of them have good light control at large viewing angles. Prismatic materials have been designed to produce special flux distributions referred to below.

Louvers. These materials expose the maximum luminance to the work at small viewing angles. They generally have lower luminance at large viewing angles. Translucent louvers have the least control; opaque louvers next; and specular parabolic wedge louvers have the maximum.

Polarizing. Available materials of the flake or layer type have a degree of diffusion that exposes less luminance to the work than prismatic or louver materials. They have less luminance at large angles than diffusing, about the same as prismatic, and more than opaque louvers.

Polarization can reduce veiling reflections. The effect is greatest at large viewing angles and least at small viewing angles. For any single ray of light, polarization in a plane perpendicular to the task always tends to reduce veiling reflections.

These have been termed "radial" polarizers because in azimuth they produce the same degree of polarization in all directions. "Linear" and "dichroic" polarizers can also be useful—particularly for specialized application.

Special. New optical designs of luminaires and materials have been produced with candlepower distributions that reduce the flux coming from the offending zone and minimize the luminance directed to the eye of the worker. While there are distinct variations in effective illumination and visual comfort for various orientations and positions of the worker, very significant improvements are provided in controlling veiling reflections.

Guides for Reducing Veiling Reflections.

The Task. Where possible the written or printed task should be on matte paper using non-glossy inks. The use of glossy paper stock and hard pencils should be minimized.

The Worker. The orientation and the position of the worker is very important. The desirable position only can be determined by actually determining CRF values (see page 9-63 in the 1981 Reference Volume). It is also true that various orientations will produce varying degrees of visual comfort.

The Lighting System. In smaller spaces such as offices where desk positions can be determined, substantial gains can be made by not positioning lighting equipment in the general area above and forward of the desk. Positions on either side and behind the worker are preferred. Where desk positions are random, as in large general offices, it is desirable to have as much light as possible reach the task from sources outside the offending zone. Taken to the extreme this will suggest the utilization of over-all ceiling treatments.

Any decision on a lighting installation should be made on an over-all basis rather than on any one factor. Thus in addition to considering the illuminance and the effect of veiling reflections produced by a lighting system and the material, the efficiency of the system and the visual comfort in the space should be considered as prime factors.

Methods of Evaluation. Those tasks which are subject to veiling reflections are subject to the visibility criteria known as equivalent sphere illumination (ESI). ESI is best used as a tool in determining the effectiveness of controlling veiling reflections and as part of the evaluation of lighting systems. The concept of ESI can best be understood by reviewing some basic principles behind ESI.

The concept begins by the establishment of a reference lighting condition. Sphere lighting (perfectly diffuse lighting) is used as the reference, since spheres are relatively easy to construct and have repeatable illumination characteristics. Sphere illumination is not said to be the best lighting condition and does not necessarily dictate the use of diffuse light in the real environment. It is an arbitrary benchmark type of lighting used to measure relative visibility potential. This reference lighting condition is such that the same amount of (sphere) illumination will always produce the same amount of visibility.

Equivalent sphere illumination is the same concept, except taken one step further—to the real lighting environment. The ESI of a visual task in a real environment is the equivalent illuminance produced by a sphere which makes the task as visible in the sphere as it is in the real environment. That is, the visibility of the task in the real environment is *equivalent* to that produced by a certain amount of sphere illumination. Thus, the term Equivalent Sphere Illumination. ESI is analogous to a measure of visibility in the real environment and as such can be used as a tool in evaluating lighting equipment or alternate lighting schemes.

Erroneous evaluations can result if ESI is not completely understood. The visual task, the observers orientation in the space, the lighting system, and the detrimental effects of veiling reflections all play a part in the determination of ESI values. See Sections 3 and 9 of the 1981 Reference Volume for discussions of visibility, veiling reflections and ESI.

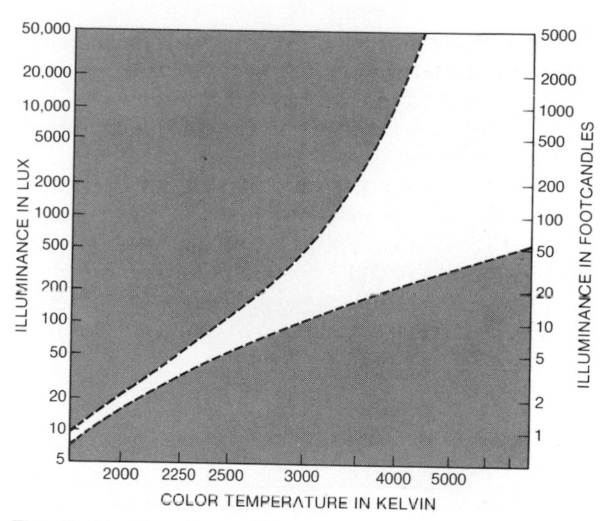

Fig. 2-11. Experiments[18] have shown the preferred color temperature of light sources at various illuminance levels (the unshaded area). Color temperatures—illuminance combinations in the lower shaded area produce cold, drab environments, while those in the upper shaded area can produce overly colorful and unnatural appearances.

Color

The importance of color in illuminating engineering and particularly lighting design should not be underestimated. An entire section (Section 5) in the 1981 Reference Volume is devoted to the basic concepts of color, chromaticity, color rendering and the use of color. The lighting designer is referred to that material, especially the portion on the use of color on page 5-18.

There often is confusion between two color characteristics of light sources—chromaticity and color rendering—and especially their use. In simple terms, chromaticity refers to color appearance of a light source, or its color temperature. (Fig. 2-10 diagrammatically shows the approximate color temperatures in kelvins of several electric light sources and daylight.) Color rendering refers to the ability of a light source, with its particular chromaticity, to render colors of objects as one would expect them to appear at the same color temperature. A useful tool in selecting an appropriate color temperature is shown in Fig. 2-11, which has been developed from experiments by Kruithof.[18] If in the design process the designer has selected an illuminance level, Fig. 2-11 can be referred to in selecting a light source color temperature that should be acceptable, in that it produces neither a cold, drab nor too warm, overly colorful environment.

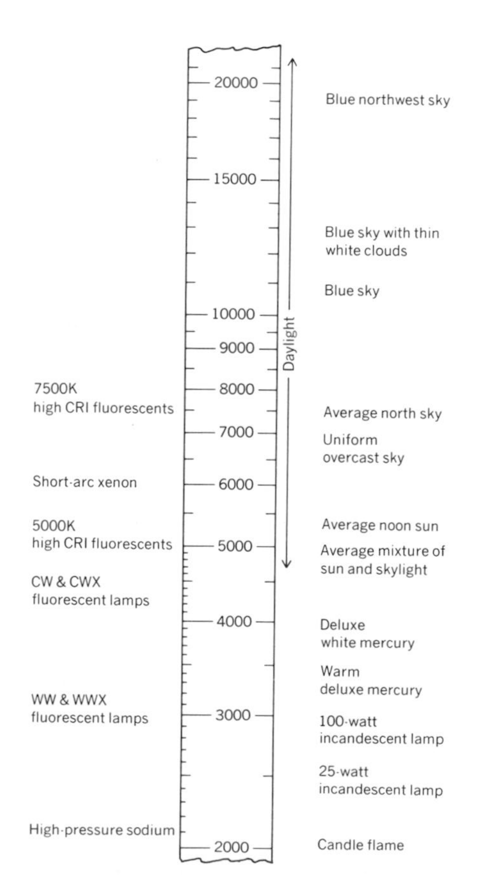

Fig. 2-10. Correlated color temperature in kelvins, of several electric light and daylight sources.

Shadows

The direction of illumination is especially important when viewing three dimensional objects. As is illustrated in Figs. 2-12 and 2-13, shadows can aid or hinder the seeing of details. In the

Fig. 2–12. Harsh shadows produced by unidirectional illumination (left) and soft shadows produced by diffuse illumination (right).

case of curved and faceted surfaces which are polished or semi-polished, the direction of the lighting is important in controlling highlights. Some shadow contributes to the identification of form.

PHYSICAL CONSIDERATIONS

Luminaire Spacing

At one time in the past, "good lighting" was specified in terms of horizontal illuminance with the criterion that the point-to-point variations did not exceed plus or minus one-sixth of the average value. Even if this were an acceptable basis, an installation which provides uniform horizontal illuminance may still represent a poor lighting installation. For example, if there is insufficient overlapping of luminaire distributions as may occur with sharp cut-off luminaires near the maximum calculated spacings, the illuminance at many points can be unidirectional with the consequential strong shadows, dark areas with single lamp failures, etc.

Today, with our broader understanding of the requirements for good lighting, it is quite possible that the old concept of an average horizontal illuminance level plus or minus one-sixth will be violated in the process of designing good lighting. However, usually there still will be a requirement that the illuminance meet some degree of uniformity.

Many factors contribute to the total quality of a lighting system. Horizontal work-plane illuminance represents only one aspect. Horizontal illuminance at other levels throughout the space cannot be ignored. The directional qualities of the illuminance at specific points within the room can be considered in terms of cast shadows,[19] vector and scalar illuminance,[20] vertical components of illuminance,[21] and the like. Often, varying visual tasks occur at different locations within the space. These and similar factors must be considered for lighting system evaluation. In specific cases, some factors may be extremely important while others can be safely omitted from

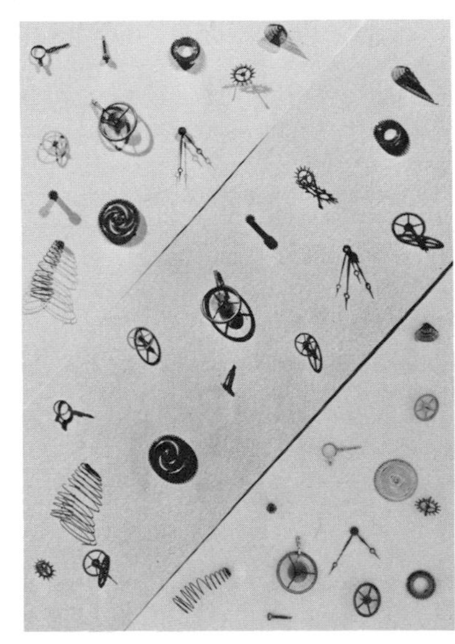

Fig. 2–13. Multiple shadows (upper left) are confusing; single shadows (center) may confuse but can help; diffused light (lower right) erases the shadows.

consideration. Nevertheless, the lighting designer must deliberately make the choice.

In those situations for which a relatively uniform illuminance is an accepted performance criterion, certain guidelines can be an important aid to the lighting designer in determining the luminaire layout. To achieve acceptable uniformity, luminaires should not be spaced too far apart or too far from the walls. Spacing limitations between luminaires are related to the intensity distribution of the luminaires, placement of the luminaires within the room, and the reflectances of the room surfaces. The principal factor for direct, semi-direct and general diffuse luminaires is the mounting height above the work-plane; for semi-indirect and indirect luminaires, it is the ceiling height above the work-plane.

The lighting designer often is faced with the problem of selecting luminaires that are potential candidates for a particular design, and the Luminaire Spacing Criterion (SC) is a parameter to assist in this decision. This standard method of classifying luminaire spread (SC) is given numerically. (See the 1981 Reference Volume, Section 9.) It is a spacing of luminaires expressed as a fraction or multiple of the luminaire mounting height above the work-plane.

The SC is a measure of the beam spread or coverage of the direct component of illuminance from direct and semi-direct illuminaires. It is a guide to permit the designer to evaluate the potential suitability of a luminaire before executing a complete design analysis. The SC permits a designer to separate luminaires into two categories for a particular lighting layout: those which are likely to produce reasonably uniform horizontal illuminance and those which will not produce uniform horizontal illuminance in a specific layout pattern. It does this by evaluating one aspect of uniformity at a few select points. This limited evaluation can be based on a luminaire characteristic alone (luminaire intensity distribution) and consequently is a property attributable to the luminaire.

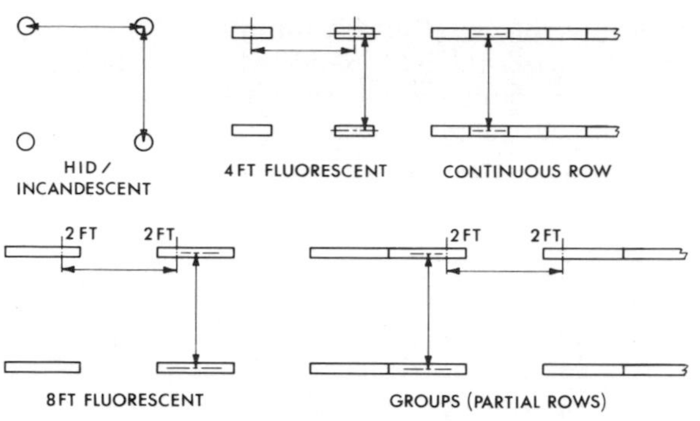

The immediate use of such SC values is to estimate the relative direct coverage of luminaires. Further, there is a general trend that the uniformity of horizontal illuminance decreases as the spacing between luminaires is increased. Fig. 2-14 illustrates the spacing between luminaires. The product of SC times the mounting height (MH) gives a spacing which generally is in the vicinity of the dividing point between a reasonably acceptable uniformity of horizontal illuminance and a noticeably poorer uniformity of horizontal illuminance.

As a general trend, horizontal illuminance will tend to be relatively uniform for luminaire spacing less than that given by the SC, and uniformity will tend to decrease as the spacing exceeds that

Fig. 2-14. Spacing dimensions (in feet) to be used in relation to Spacing Criterion (SC). Mounting height is from luminaire to work-plane for direct, semi-direct and general-diffuse luminaires and from ceiling to work-plane for semi-indirect and indirect luminaires.

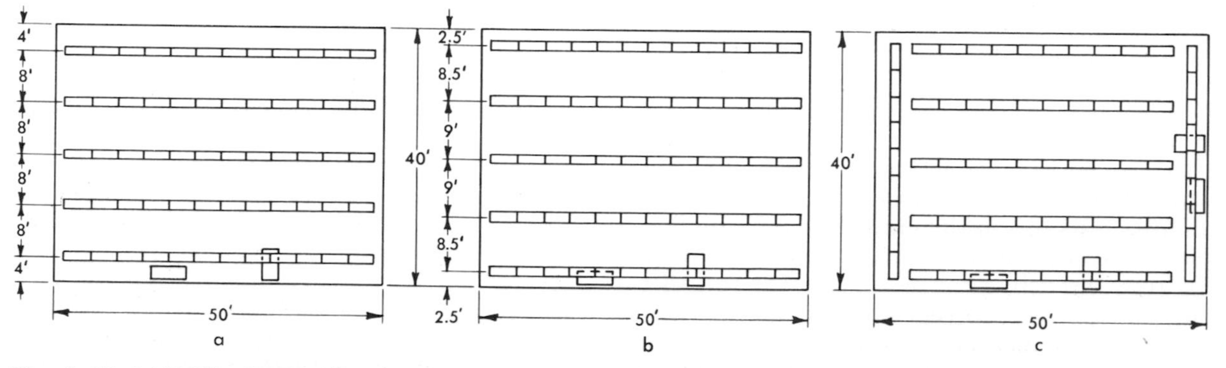

Fig. 2-15. (a) Lighting layout using equal spacing between continuous rows of luminaires. (b) Layout is changed to provide higher illuminance near side walls. (c) By adding four more units on each end, layout (b) can be modified to provide 80 per cent more light near the end walls and prevent possible scallop effects. Dimensions are in feet.

given by the SC. Certain luminaires* may produce a horizontal illuminance where the uniformity is acceptable over a specific limited range of spacings. In such cases the SC is not an applicable measure for the luminaire. Here, the manufacturer states specific spacing instructions rather than giving an SC value.

The commonly used practice of letting the distance from the luminaires to the wall equal one-half the distance between rows (see Fig. 2–15a) results in inadequate illuminance near the walls. Since desks and benches are frequently located along the walls, a distance of 760 millimeters (2½ feet) from the wall to the center of the luminaire should be employed to avoid excessive drop-off in illuminance. This will locate the luminaires over the edge of desks facing the wall or over the center of desks that are perpendicular to the wall (see Fig. 2–15b). To further improve illuminance uniformity across the room, it is often desirable to use somewhat closer spacings between outer rows of luminaires than between central rows, taking care to be sure that no spacing exceeds the maximum permissible spacing.

To prevent excessive reduction in illuminance at the ends of the room, the ends of fluorescent luminaire rows should preferably be 150 to 300 millimeters (6 to 12 inches) from the walls, or in no case more than 610 millimeters (2 feet) from the walls. Even the 150- to 300-millimeter (6- to 12-inch) spacing leaves much to be desired from the standpoint of uniformity and, where practicable, the arrangement shown in Fig. 2–15c is much more satisfactory. With this arrangement, the units at each end of the row are replaced by a continuous row parallel to, and 760 millimeters (2½ feet) from, the end wall. In the example shown, 5 units were replaced by 9 units providing a potential increase in the illuminance at the end of the room of 80 per cent over what it would be with the layout shown by Fig. 2–15b. This technique not only improves uniformity but also eliminates scallops of light on the end walls and provides a uniform wash of light on all four walls.

Another excellent method of compensating for the normal reduction in illuminance that may be expected at the ends of rows is to use a greater number of lamps in the end units. Still another technique is to provide additional units between the rows at each end. The units could be either parallel or at right angles to the rows.

Spacings closer than the maximum permissible are often highly desirable to reduce harsh shadows and veiling reflections in the task as well as further improve uniformity. This is particularly true for direct and semi-direct equipment, and spacings that are substantially less than the maximum permissible spacing should be seriously considered.

* For example, some sharp cut-off batwing distribution luminaires.

The following formulas can be used to calculate the values:

1. For individually mounted luminaires, the wall-to-luminaire spacing should be:

$$\text{Wall-to-Luminaire Spacing} = \frac{\text{Luminaire-to-Luminaire Spacing}}{3}$$

2. For individual units or crosswise spacing of continuous rows:

$$\text{Minimum Number of Rows} = \frac{\text{Room Width}}{\text{Maximum Spacing Allowed}}$$

3. For lengthwise spacing in continuous rows:

$$\text{Maximum Number of Units Per Row} = \frac{\text{Room Length} - 0.3}{\text{Luminaire Length}}$$

(allows 0.15 meters end spacing when lengths are in meters)

or

$$\frac{\text{Room Length} - 1}{\text{Luminaire Length}}$$

(allows one-half-foot end spacing when lengths are in feet)

$$\text{Minimum Number of Units Per Row} = \frac{\text{Room Length} - 1.22}{\text{Luminaire Length}}$$

(allows 0.61-meter end spacing when lengths are in meters)

or

$$\frac{\text{Room Length} - 4}{\text{Luminaire Length}}$$

(allows two-foot end spacing when lengths are in feet)

Dimming Devices

The most practical method of controlling light output for nearly every purpose is to control the electrical input to the light source. There are several methods by which this may be accomplished, based on one of two principles: the input current can be varied by changing the amplitude of the current, or it can be varied by changing the amount of time during a cycle that it is permitted to flow.

Change in amplitude can be accomplished by resistance dimmers. The metallic rheostatic dimmer, one of the earliest controls developed, is still made and offers features which prescribe its use under certain circumstances. A resistance dimmer is connected in series in the circuit and the voltage which appears across the lamp is equal to the line voltage less the voltage drop (IR) across the resistance dimmer. To vary the amount of light, adjustment to the rheostat is made by moving a contact which adds or subtracts resistance to the circuit. The dimmer capacity must be selected to match the lamp load fairly closely if dimming control to blackout is to be achieved. The resistance loss in the dimmer is an appreciable percentage of the lamp load at low luminance levels, liberating large amounts of heat which must be disposed of and consuming considerable power.

Change in amplitude can also be accomplished by continuously adjustable autotransformers. A single layer of copper magnet wire is wound over an iron core to form a toroid. Each wire turn is "bared" to form a commutator for the carbon brush. When a 120-volt ac line is applied to this winding, sliding the brush from turn to turn allows tapping off the desired ac output voltage. Wire size, number of turns, and brush dimensions have been carefully designed so brush contact is made with the next turn before leaving the previous conductor, assuring perfectly smooth, flickerless type of light control of almost infinite fineness from zero to full luminance. A dimmer of such design has excellent regulation—there is no visual change in light intensity as lamps are added or removed from the circuit, and it will also dim equally well all sizes of lamps at the same time. Autotransformer dimmers provide fairly good efficiency up to a maximum of 95 per cent. Although solid-state dimmers have tended to replace autotransformer dimmers, 1000- to 2000-watt units are still available. Larger capacity autotransformers (above 2000 watts) are available with motor drives.

The saturable reactor was the first of the electronic type or phase angle control dimmers. It consists of a magnetic core associated with dc and ac windings. By adjusting the current of the dc winding, the inductive reactance of the ac winding may be varied smoothly from approximately zero to its maximum value. Saturable reactors are not in much use today in lighting applications because of their weight, size and slowness of response; however, because they are good in the control of very heavy loads, they do fill some requirements.

Thyratron tube control is another of the earlier electronic methods. A gating type of control is obtained from a thyratron tube when ac voltage is applied to its plate and an ac voltage of adjustable phase is applied to its grid. As the phase angle between the voltage on the plates and grids is adjusted, the tube can be made to conduct various length portions of the cycle. Two thyratrons are connected back to back so that each one may conduct during each half cycle. While response is very rapid, all filaments in this tube system must be preheated approximately ½ minute or more. Should a power failure occur during use, the need to preheat becomes a great inconvenience.

Magnetic amplifiers or self saturating reactors became popular as high power dimmers for a short period. Although the performance of this type of system was generally good, it was replaced by thyristor controlled dimmers because of their size and weight advantages.

Solid-state electronics has virtually taken over the field of present day dimming equipment, from the 600-watt wallbox dimmer to the 100-kilowatt or higher theatrical or architectural system. The thyristor has become the main component in most incandescent and fluorescent dimmer manufacturing. The original solid-state equipment designs utilized dual back-to-back silicon controlled rectifiers (SCR). This arrangement has given way to the equally reliable but more cost-efficient triac devices in most dimmers rated at less than 6 kilowatts. However, many of the larger theatrical units rated at 6 and 12 kilowatts still employ dual SCRs because of the lack of high current triacs.

Because thyristors are fast switching devices, some radio frequency interference will be generated unless suitable filtering is designed into the dimmer. Extensive filtering is generally provided in larger architectural or theatrical systems to completely eliminate any possible interaction with sensitive equipment. While solid-state dimmers are typically 98 to 99 per cent efficient, they still have to dissipate heat, and cool operation is the key to long dimmer lifetime.

Thyristor controlled fluorescent dimmers are being applied in many commercial applications, such as conference rooms, offices, restaurants, schools and churches.

Recent advances in device technology and circuit design have made it possible to dim high intensity discharge lamps (see Section 8 of the 1981 Reference Volume). It has also become practical to convert low frequency ac to high frequency ac in order to improve the efficacy of standard fluorescent lamps. Further advances have been made in energy management of lighting through integrated circuit technology. Automatic dimming system control by means of photoelectric feedback, time programming, and other computer controlled demand limiting are all available. Photoelectric devices can adjust the lighting to allow for minimum power consumption when daylight is available. Time programming can automatically adjust light levels to suit task changes, such as from work to leisure to cleaning or security modes. Computer operated controls can be used to reduce light levels to reduce the load in peak demand periods.

Solid-state dimming offers the following:

For incandescent sources

1. Full range adjustability of light for various tasks or moods, down to zero light output.
2. Energy savings—only 1 to 2 per cent of the connected load is dissipated in the dimmer.
3. Increase in lamp life—lamps will last up to 20 times the rated life at 50 per cent light output (70 per cent power).

For fluorescent sources

1. Continuously variable light output down to a fraction of a per cent of full output, with no discernible color shift.
2. Power control—per cent energy savings equal to the per cent light reduction down to about 50 per cent light output (see Fig. 8-46 in the 1981 Reference Volume).

For high intensity discharge sources

1. Range of light output control from 2 per cent minimum for mercury to 40 per cent minimum for metal halide and high pressure sodium lamps.
2. Power control—energy savings nearly the same as shown in the fluorescent power curve, down to 50 to 60 per cent light output.

Energy management devices as previously mentioned can be effectively applied to all sources listed above.

Wiring for Lighting

Every electric lighting system, regardless of its size, scope, simplicity or complexity should have a well designed, trouble free electrical wiring system. Its size and capacity, its electrical characteristics (voltages, frequencies, phases, etc.), feeders, branch circuit layouts, and switch and dimmer controls, must all be specifically selected and designed to conform to the layout and design of the lighting systems which it is to operate and control.

Lighting designers should know the basic fundamentals of electrical wiring system design to insure that they can obtain the maximum flexibility and efficiency from the lighting system. Quite often lighting designers are the consulting electrical engineers on the project. In this case, they usually are qualified to design and specify the electrical wiring system for the lighting. However, if the lighting designers are not qualified to design the electrical wiring system, they should seek the services of a qualified electrical consultant or electrical contractor when designing the lighting system, to insure that the wiring system and controls will provide all the lighting variations and flexibility that is desired and intended. All electrical systems must be designed and installed in accordance with the provisions and requirements of the *National Electrical Code* and other local or state code requirements, and the electrical consultant is qualified to include these provisions and requirements in the design and specifications.

The first step in the design of the electrical wiring system is to determine the total electrical load for the lighting system. On large projects, it may be desirable to break down the lighting load into logical sub-loads, for serving individually from separate load centers. These sub-loads can then be further broken down for individual panelboard control. On smaller projects, this usually resolves itself into selection and location of one or more lighting panels, each conveniently located near the center of the lighting load it serves, compatible with the character, use, and structural configuration of the building.

When lighting loads have been determined, the characteristics of the electric power supply, such as voltage, phase and frequency, must be considered and evaluated in order to select optimum locations for load centers and panelboards.

For relatively small projects of 100 kilovolt-amperes or less, most utilities supply 120/208-volt three-phase four-wire, 120/240-volt three-phase four-wire, or 120/240-volt single-phase three-wire service. For larger projects, 277/480-volt three-phase four-wire service is usually available. Lighting loads are usually served with 120 volts or 277 volts from these systems. Low-voltage light sources which may be incorporated in the lighting system may be operated at their rated voltages by using small dry-type step-down transformers operating from any of the higher standard voltages.

Care should be exercised in the selection of voltage due to certain provisions of various codes which prohibit the use or restrict the use of higher voltage lighting devices in various occupancies.

Various techniques for switching, such as half switching four-lamp fluorescent luminaires, or row switching to provide various lighting levels should be employed for energy conservation.

The requirements of energy conservation suggest the use of short three-phase home run circuits in order to minimize energy losses. For wiring economics as well as energy conservation, high power-factor ballasts should be specified for discharge type lighting.

Outdoor lighting systems generally employ the same techniques for wiring system design. Longer distances involved require voltage drop calculations to be performed to assure adequate voltage at the luminaire. Systems will usually perform adequately when the voltage drop does not exceed 5 per cent at the farthest outlet. Higher voltage luminaires, such as those operated on 480 volts may generally be used outdoors when permitted by codes to reduce the size and quantities of circuit conductors.

In the United States 60-hertz power is almost universal for lighting systems. However, lighting designers and electrical consultants sometimes consider the use of high-frequency electric power for the operation of fluorescent lighting systems, especially when the advantages of high-frequency operation are important (see Section 8 of the 1981 Reference Volume). One such advantage might be the reduction of the excessive weight of normal ballasts used for 60-hertz power. Another might be to obtain the maximum light output of the fluorescent lamps being considered for the project. In earlier installations, frequencies of 360, 420, and 840 hertz were used. More recently, a 3000-hertz system has been used, which reportedly converts power from 60 to 3000 hertz at about 93 per cent efficiency, compared with over-all conversion efficiencies of only 80 to 85 per cent for the earlier model converters operating at the lower frequencies.

Total system efficiency, from system electrical watts input to system useful lumen output, must be considered to properly evaluate a lighting system's efficiency. The evaluation of partial systems often leads to erroneous conclusions. Lowest watts per luminaire does not guarantee lowest lighting energy use for the system.

The total electrical load required for the lighting system is first calculated in total watts or kilowatts. For purposes of selection of transformers, main switchboards, circuit breakers, and other similar electrical distribution considerations, kilowatts (kw) or kilovolt amperes (kva) are used. However, when converted to "watts per square meter (foot)," the term becomes more meaningful and useful not only to the lighting designer, from the standpoint of an economical lighting system, but also to the air-conditioning and heating engineers. The chart shown in Fig. 2-16 shows maintained average illuminance for each of a variety of luminaires based on their wiring capacity expressed in watts per square meter (foot). The data in this chart are only approximate, and are based on a room of average shape with a Room Cavity Ratio of 2.5 and for high reflectances of 80 per cent for the ceiling cavity, 50 per cent for the walls, and 20 per cent for the floor cavity. For more accurate results, data for specific luminaires, and for specific sized rooms and reflectances should be used. However, the chart is useful for quick appraisals when one type of luminaire is being considered versus other types, especially in early stages of the lighting design procedure.

Rooms with numerically higher Room Cavity Ratios will require higher loadings (watts per square meter (foot)); with lower Room Cavity Ratios, the loadings will be correspondingly reduced. With other parameters constant, the required loading for any luminaire in an area with another Room Cavity Ratio is inversely proportional to the Coefficients of Utilization associated with the Room Cavity Ratios of the two areas. Similarly, changes in reflectances will also affect the required loading in inverse proportion to the Coefficients of Utilization associated with the reflectance conditions.

The wiring system should be designed to provide maximum flexibility with adequate capacity for present and anticipated future needs. Overloading or excessive extensions of circuits, in addition to the hazards involved, results in a lowering of the light output of both incandescent filament and fluorescent lamps. As an example, the light output of an incandescent filament lamp is about three per cent less for each volt the lamp is operated under its rated voltage. Fluorescent lamp light output is also affected by undervoltage operation but not to so great an extent; starting difficulties and reduced lamp life may also result. For further information, see Section 8 of the 1981 Reference Volume.

With gaseous discharge sources, the wattage consumed and the current carrying capacity required will, in general, be greater than the rated wattage of the light sources. Provision for auxiliary wattage losses and for power factors lower than unity must be made in the wiring system. Mercury, metal halide, and high pressure sodium lamps require a warm-up period of several minutes. During this time, or with certain types of

water vapor pressure, mean radiant temperature and air velocity all affect the thermal environment. Both steady-state and non-steady state criteria for these stimuli must be satisfied.

Research performed by ASHRAE prior to 1974 resulted in the specification[22] of the comfort envelope as shown in Fig. 2-17. This specification covers a wide range of environmental applications such as offices, homes, schools, shops, theatres, etc. It applies for average clothing and activity.

ASHRAE recommends that the dry bulb temperature be adjusted for any change in mean radiant temperature by:

$$\text{ADBT} = \frac{\text{DBT} + \text{MRT}}{2}$$

where:

ADBT = adjusted dry bulb temperature
DBT = dry bulb temperature
MRT = mean radiant temperature*

The relative humidity should be maintained between approximately 20 per cent and 65 per cent while the air velocity, without regard to direction, in the occupied zone is controlled to less than 0.35 meters per second (70 feet per minute) at any point.

Additional research has been performed at The Institute of Environmental Research at Kansas State University under ASHRAE contract.[23, 24] These studies provide methods for varying the comfort envelope for lightly clothed persons and sendentary activity.[25]

Further research has been carried out by Fanger[26] beginning in 1966 at Kansas State University and continued at the Technical University of Denmark. From a comfort equation developed by Fanger it is possible to predict a combination of environmental factors that produce a "comfortable" environment for a clothed person performing any selected activity.

The details of all these methods of defining comfort levels are beyond the scope of this section. The reader should consult the references for more detailed information.

Lighting Load on Air Conditioning

Fig. 2-18 gives one example of interior load distribution in a particular office building. The amount of heat gain through the exterior masonry is small enough to be neglected. Fluorescent lighting is provided consuming 27 watts per square meter (2.5 watts per square foot). It is assumed that Venetian blinds will be lowered during corresponding sunny hours.

* The uniform surface temperature of an imaginary black enclosure with which man exchanges the same heat by radiation as in the actual environment.

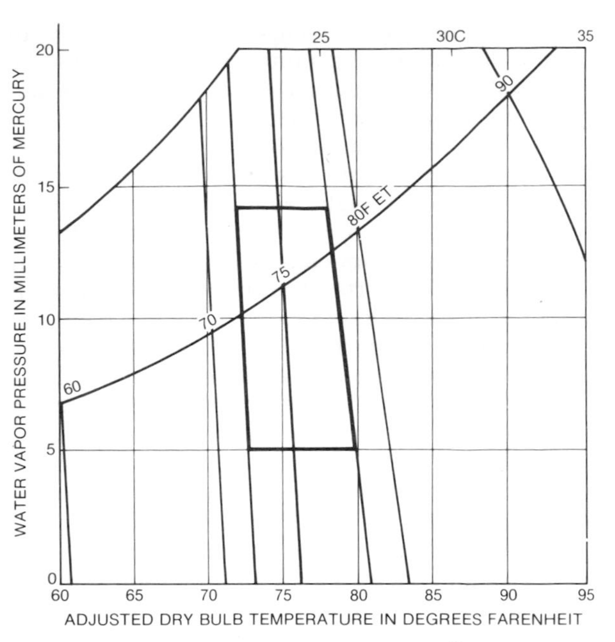

Fig. 2-17. Comfort envelope.[22]

In typical buildings, factors such as solar radiation through windows, heat transmitted through wall and roofs, and cooling of ventilation air, people, and machines comprise 50 to 80 per cent of the total cooling load. The refrigeration and/or air handling capacity required for lighting may be reduced if some of the principles described later in this section can be applied.

For the building of Fig. 2-18 the relative magnitude of all heat sources at the time of the total building peak cooling load is shown in Column A of Fig. 2-19. For comparison, analyses of the total cooling load for two other installations are shown in Columns B and C. The buildings are based on 27 and 32 watts per square meter (2.5 and 3 watts per square foot) respectively.

Electric Lamps as Heat Sources. Electric lamps are efficient converters of electric power to heat energy. Each watt of electric power consumed by a lamp generates 1 watt (3.4 British thermal units per hour) of heat, just as any electric heating device. The energy takes two principal forms: (1) conduction-convection energy, and (2) radiant energy (including infrared, light and ultraviolet). From this it is obvious that only a part of the energy generated by electric lamps is light. However, light itself produces heat. It does not heat air as convection sources do, but it raises the temperature of any surface which absorbs it.

A knowledge of the relative amount of each type of energy emanating from an electric lamp can be helpful in analyzing its performance and the effect it may have on thermal considerations, Figs. 2-20, 2-21, and 2-22 show approximate data for some representative fluorescent, incandescent and high intensity discharge lamps, respectively. These values are for lamps suspended in space under specific operating conditions.[27] Low energy ballasts will exhibit lower ballast losses; and although total lamp energy and bulb wall temperatures will differ, the lamp energy proportion may be considered typical for low energy lamps as well. Energy output for an individual luminaire in space, or for a system of luminaires installed in a room is likely to vary considerably from that for lamps alone.

Fig. 2-18. Example of Cooling Load Distribution in One Modern Office Building

Heat Source	Exterior Offices*				Interior Offices
	North	East	South	West	
Glass	46%	70%	65%	69%	—%
Lighting	30	18	20	17	39
Occupants	20	8	11	10	55
Miscellaneous	4	4	4	4	6
Totals	100%	100%	100%	100%	100%

* Per cent at time of maximum load in each office.

Fig. 2-19. Example of Cooling Load for Several Types of Installations

	Office Building A	Chain Store B	Clinic C
Glass	16%	6%	15%
Lighting	22	20	17
Roof and walls	1	12	33
Occupants	21	34	16
Ventilation	35	22	19
System power	5	4	—
	100%	100%	100%
Glass and lighting	38%	26%	32%

Fig. 2-20. Energy Output for Some Fluorescent Lamps of Cool White Color (Lamps Operated at Rated Watts on High Power Factor, 120-Volt, 2-Lamp Ballasts; Ambient Temperature 25 °C (77 °F) Still Air)

Type of Energy	40WT12	96 Inch T12 (800 mA)	PG17† (1500 mA)	T12 (1500 mA)
Light	19.0%	19.4%	17.5%	17.5%
Infrared (est.)*	30.7	30.2	41.9	29.5
Ultraviolet	0.4	0.5	0.5	0.5
Conduction-convection (est.)	36.1	36.1	27.9	40.3
Ballast	13.8	13.8	12.2	12.2
Approximate average bulb wall temperature	41 °C (106 °F)	45 °C (113 °F)	60 °C (140 °F)	

* Principally far infrared (wavelengths beyond 5000 nanometers).
† Grooves sideways.

Fig. 2-21. Energy Output for Some Incandescent Lamps

Type of Energy	100-Watt* (750-hour life)	300-Watt (1000-hour life)	500-Watt (1000-hour life)	400-Watt‡ (2000-hour life)
Light	10.0%	11.1%	12.0%	13.7%
Infrared†	72.0	68.7	70.3	67.2
Conduction-convection	18.0	20.2	17.7	19.1

* Coiled-coil filament.
† Principally near infrared (wavelengths from 700 to 5000 nanometers).
‡ Tungsten-halogen lamp.

Fig. 2-22. Energy Output for Some High Intensity Discharge Lamps

Type of Energy	400-Watt Mercury	400-Watt Metal Halide	400-Watt High Pressure Sodium	180-Watt Low Pressure Sodium
Light	14.6%	20.6%	25.5%	29.0%
Infrared	46.4	31.9	37.2	3.7
Ultraviolet	1.9	2.7	0.2	0
Conduction-convection	27.0	31.1	22.2	49.1
Ballast	10.1	13.7	14.9	18.2

Luminaires as Heat Sources. Performance characteristics of luminaires are well documented in terms of luminous efficiency, light control and candlepower distribution because equipment designers have been concerned primarily with the purposeful distribution of visible light. Now it is necessary to consider the total energy distribution of any luminaire destined to become a component of a building.

From Figs. 2-20, 2-21 and 2-22 it can be seen that the high percentage of energy converted by electric lamps is radiation lying predominantly in the near infrared or far infrared regions—the proportions depend on the light source. Because the properties of lighting materials are different in the range from visible to invisible radiation, it is important to consider the underlying physics.

Fig. 2-23 shows that some materials used in luminaires can be good reflectors of light and good absorbers of far infrared. Several materials used to transmit light show significant differences in the far infrared reflected.

Any quantitative analysis of luminaires as heat sources should assume conditions of temperature

Fig. 2-23. Properties of Lighting Materials
(Per Cent Reflectance (R) and Transmittance (T) at Selected Wavelengths)

Material	Visible Wavelengths						Near Infrared Wavelengths						Far Infrared Wavelengths							
	400 nm		500 nm		600 nm		1000 nm		2000 nm		4000 nm		7000 nm		10,000 nm		12,000 nm		15,000 nm	
	R	T	R	T	R	T	R	T	R	T	R	T	R	T	R	T	R	T	R	T
Specular aluminum	87	0	82	0	86	0	97	0	94	0	88	0	84	0	27	0	16	0	14	0
Diffuse aluminum	79	0	75	0	84	0	86	0	95	0	88	0	81	0	68	0	49	0	44	0
White synthetic enamel	48	0	85	0	84	0	90	0	45	0	8	0	4	0	4	0	2	0	9	0
White porcelain enamel	56	0	84	0	83	0	76	0	38	0	4	0	2	0	22	0	8	0	9	0
Clear glass-3.2 millimeters (.125 inch)	8	91	8	92	7	92	5	92	23	90	2	0	0	0	24	0	6	0	5	0
Opal glass-3.9 millimeters (.155 inch)	28	36	26	39	24	42	12	59	16	71	2	0	0	0	24	0	6	0	5	0
Clear acrylic-3.1 millimeters (.120 inch)	7	92	7	92	7	92	4	90	8	53	3	0	2	0	2	0	3	0	3	0
Clear polystyrene-3.1 millimeters (.120 inch)	9	87	9	89	8	90	6	90	11	61	4	0	4	0	4	0	4	0	5	0
White acrylic-3.2 millimeters (.125 inch)	18	15	34	32	30	34	13	59	6	40	2	0	3	0	3	0	3	0	3	0
White polystyrene-3.1 millimeters (.120 inch)	26	18	32	29	30	30	22	48	9	35	3	0	3	0	3	0	3	0	4	0
White vinyl-0.76 millimeters (.030 inch)	8	72	8	78	8	76	6	85	17	75	3	0	2	0	3	0	3	0	3	0

Note: (a) Measurements in visible range made with General Electric Recording Spectrophotometer. Reflectance with black velvet backing for samples (b) Measurements at 1000 nm and 2000 nm made with Beckman DK2-R Spectrophotometer. (c) Measurements at wavelengths greater than 2000 nm made with Perkin-Elmer Spectrophotometer. (d) Reflectances in infrared relative to evaporative aluminum on glass.

stabilization, constant voltage, and service position. In this state, total energy may not follow the distribution of light energy. However, it will be helpful to compare total energy distribution with the general classifications assigned to candlepower distribution curves.

Thermal distribution characteristics would narrow the CIE classifications to (1) semi-direct, (2) direct-indirect, and (3) semi-indirect, as illustrated in Fig. 2-24. Totally direct or indirect lighting classifications are unlikely in total energy distribution curves. Of the two general diffuse lighting classifications, direct-indirect would be more appropriate.

Several test methods have been employed to assess the *total energy* distribution from a particular luminaire. One involves an adaptation of photometric techniques. Two others involve calorimetry, including a continuous water flow calorimeter[28] and continuous air flow calorimeters.[29, 30] Though procedures and equipment varied widely, test results were of the same order of magnitude.[31]

Testing guides for determining the thermal performance of luminaires have been published by ADC,* IES and NEMA. IES approved a new test method in 1978 which considers the effect of plenum temperature and air return in the light output. The test also provides data on heat distribution and power input dependent upon return air flow through the luminaire.[32]

Lighting Systems as Heat Sources. Visual and thermal conditions are two of the most important considerations in a planned interior environment. Visual comfort is partly due to quantity and quality of illumination. Thermal comfort is the result of a proper balance in temperature,

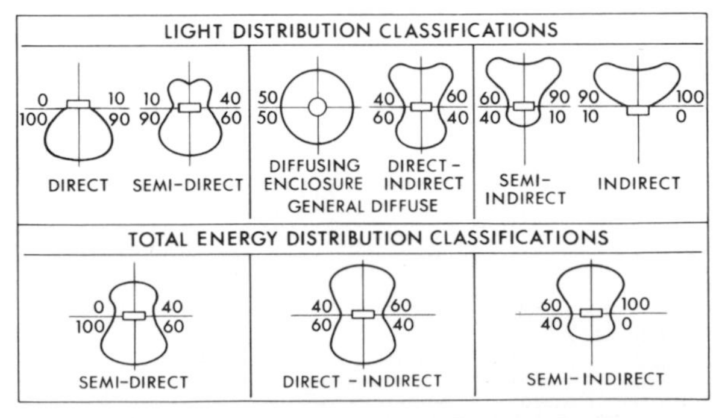

Fig. 2-24. Light distribution curves by CIE classifications of luminaires, compared with typical total energy distribution types. Upward and downward components are in percentage ranges.

* Air Diffusion Council.

relative humidity and air motion. If these factors are to be fully evaluated, consideration should be given to the total energy distribution of luminaires, their relationship to room surfaces, and to the type of conditioning system contemplated.

While total energy should ultimately be considered within the building envelope, comfort conditions will be primarily affected by that portion of the energy distributed into the occupied space. Thus, actual lighting and heating characteristics will be influenced by luminaire performance, ambient temperatures, surrounding materials and surface reflectances.

The *ASHRAE Fundamentals Handbook* covers the calculation of space load due to lighting for various luminaires and ventilation arrangements.[33] Generally, the instantaneous heat load from the lighting system is expressed using the relationship:

$$P_i = P_l \times \text{BLF} \times \text{CLF} \times \text{UF}$$

where

P_i = instantaneous heat load from the lighting system in watts (multiply by 3.41 to get Btu/h)

P_l = total lamp power in watts

BLF = ballast load factor—
incandescent, BLF = 1.00
rapid-start fluorescent, BLF = 1.08 to 1.30
high intensity discharge, BLF = 1.04 to 1.37

CLF = cooling load factor, a factor that allows for the type luminaire, furnishings, room envelope, length of time lights are on, etc. See reference 33.

UF = utilization factor, percentage of installed power in use expressed as a decimal. For commercial applications such as stores, UF is generally unity.

Luminaire mounting has an important role in the distribution of thermal energy. Fig. 2-25 illustrates typical heat flows for various types of ceiling-to-luminaire relationships. Total energy distribution involves all three mechanisms of heat transfer—radiation, conduction and convection. Although the illustration shows a fluorescent luminaire, HID luminaires exhibit similar patterns. The input of the suspended luminaire in Fig. 2-25 (A) would be convected and radiated in all directions to be reflected or absorbed and reradiated. Essentially, all of the input energy would remain within the occupied space.

Heat transfers from the surface mounted semi-direct luminaire in Fig. 2-25 (B) involve radiation, conduction and convection. Assuming good contact with the ceiling, upper surfaces of the luminaire will transfer energy to or from the ceiling by conduction. Since many acoustical ceiling materials are also good thermal insulators, it may be assumed that temperatures within the luminaire will be elevated. Thus, lower luminaire surfaces will tend to radiate and convect to the space below at a somewhat higher rate. Unless the ceiling material is a good heat conductor and can reradiate above, essentially all of the input energy will remain in the space.

A different situation exists when components of the system are separated from the space. The recessed luminaire in Fig. 2-25(C) distributes some portion of input wattage above the suspended ceiling. The actual ratio is a function of luminaire design and plenum and ambient conditions. For most recessed static luminaires, the

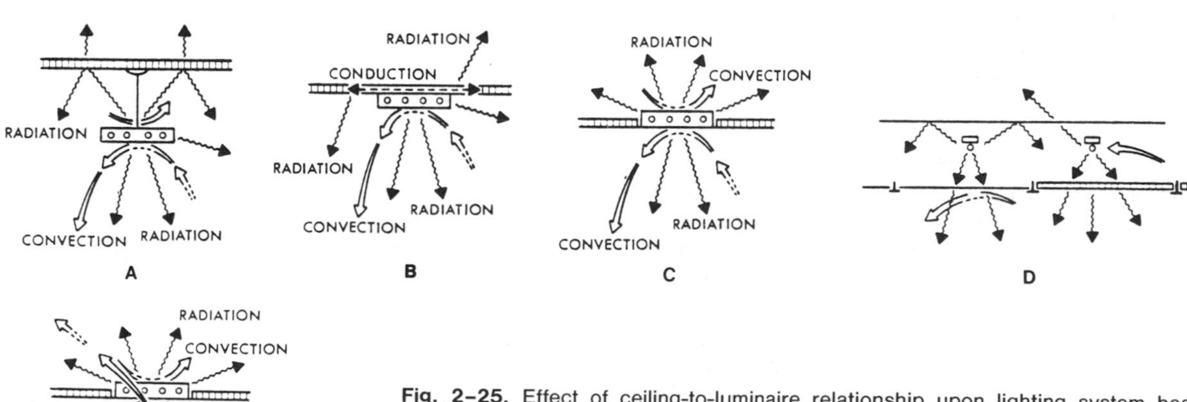

Fig. 2-25. Effect of ceiling-to-luminaire relationship upon lighting system heat transfer. Fluorescent luminaire has direct-indirect total energy distribution classification. (A) Suspension mounting. (B) Surface mounting. (C) Recessed. (D) Luminous and louvered ceiling. (E) Heat transfer luminaire.

ratio is very nearly 50 per cent above the ceiling and 50 per cent below.

Lighting systems of luminous and louvered ceiling types are illustrated in Fig. 2-25 (D). A similarity with the heat transfers of Fig. 2-25 (A) is noted. Although luminaires are separated from the occupied space, plastics and glass used in luminous ceilings are good absorbers of infrared. White synthetic enameled louvers are also good absorbers, whereas aluminum louvers reflect a high percentage of infrared. Unless some means of controlling the energy is employed, all of the electrical input energy also remains in the occupied space.

Heat transfer recessed luminaires are illustrated in Fig. 2-25 (E). Here, the convected and radiated component to the space have been reduced considerably while the upward energy has increased a proportional amount. Under certain conditions it is possible for the space load to consist almost entirely of light energy. The majority of the power input to the luminaire is directed upward where it can be captured by the system and be subject to some form of control. Laboratory tests conducted in accordance with IES procedures[32] will provide energy distribution data for evaluative purposes. However, the total system must be evaluated because heat removal to the plenum may raise plenum temperatures which cause conductive heat transfer back through the ceiling and floor to the space below and above adding thermal load back to the space.

Task/ambient systems have a different lighting energy distribution. Care must be exercised in the selection of the cooling load factor (CLF). Depending on the installation, it may be necessary to calculate task and ambient heat loads separately.

It is possible to have both systems completely within the space. This would be the case if suspended or surface-mounted luminaires were used for ambient lighting with task lighting being incorporated into the furniture or with suspended or surface-mounted luminaires being used for both. In this case, the entire input power is instantaneous space load.

With recessed luminaires utilized for ambient lighting and either suspended or furniture mounted for task lighting, the heat loads must be figured separately as only the task lighting load is entirely instantaneous space load. The recessed luminaire heat contribution may be considered less depending upon the CLF.

Systems can also utilize recessed luminaires for both task and ambient lighting. Here, both would impose a heat load which would be reduced by the CLF.

Benefits of Integrated Designs

The benefits of integrating building heat in lighting design are: (1) improved performance of the air conditioning system, (2) more efficient handling of lighting heat, and (3) more efficient lamp performance.

The control and removal of lighting heat can reduce heat in the occupied space, reduce air changes and fan horsepower, lower temperature differentials required in the space, enable a more economical cooling coil selection because of the higher temperature differential across the coil, and reduce luminaire and ceiling temperature thereby minimizing radiant effects.

The degree to which any of these benefits may be obtained depends on many variables such as the quantity of energy involved, the type of heat transfer mechanism, the temperature difference between source and sink, and the velocity and quantity of fluids and/or air available for heat transfer. However, in most applications luminaire temperature will be higher than room temperature so fluids at room temperature can be effective in heat transfer. Any unwanted heat that can be removed at room temperature or above can be removed much more economically than at lower temperatures.

The full benefits of integrated design can be achieved only through the combined efforts of a design team which should include architects, space planners, interior designers, electrical engineers, mechanical engineers, illuminating engineers and cost analysts.

LIGHTING FOR SAFETY

Importance. Safe conditions are essential to any inhabited space and the effect of light on safety must be considered. The environment should be designed to help compensate for the limitations of human capability. Any factor that aids visual effectiveness increases the probability that a person will detect the potential cause of an accident and act to correct it.

In many instances where illumination is associated with accidents, the cause is attributed to inadequate illuminance levels or poor quality of illumination. However, there are many less tangible factors associated with poor illumination which can contribute to many accidents. Some of these are: direct glare, reflected glare, and harsh shadows—all of which hamper seeing. Excessive visual fatigue itself may be an element leading toward accidents. Accidents may also be prompted by the delayed eye adaptation a person experiences when moving from bright surroundings into dark ones and vice versa. Some accidents which have been attributed to an individual's carelessness could have been partially due to difficulty in seeing from one or more of the above mentioned factors. The accidents might have been avoided through the use of good lighting principles.

Illuminance Levels. The lighting recommendations in Fig. 2-2 provide a guide for efficient visual performance rather than for safety alone; therefore, they are not to be interpreted as requirements for regulatory minimum illuminance levels.

Fig. 2-26 has been developed to list illuminance levels regarded as *absolute minimums for safety alone.* To assure these values are maintained, higher initial levels must be provided as required by the maintenance conditions. In those areas which do not have fixed lighting, localized illumination should be provided during occupancy by portable or material handling and vehicle mounted lighting equipment.

Other Factors. A visually safe installation must be free of excessive glare and of uncontrolled, large differences in luminances. Appropriate guides to limiting glare and adaptation effects are given earlier in this Section in discussions of luminance ratios and visual comfort. Maximum luminance ratios are important to avoid temporarily noticeable reductions in visibility because of changes in eye adaptation when alternately looking at areas of widely different luminances.

Fig. 2-26. Illuminance Levels for Safety*

Hazards Requiring Visual Detection	Slight		High	
Normal† Activity Level	Low	High	Low	High
Illuminance Levels				
Lux	5.4	11	22	54
Footcandles	0.5	1	2	5

* Minimum illuminance for safety of people, absolute minimum at any time and at any location on any plane where safety is related to seeing conditions.

† Special conditions may require different illuminance levels. In some cases higher levels may be required as for example where security is a factor. In some other cases greatly reduced levels, including total darkness, may be necessary, specifically in situations involving manufacturing, handling, use, or processing of light-sensitive materials (notably in connection with photographic products). In these situations alternate methods of insuring safe operations must be relied upon.

Note: See specific application reports of the IES for guidelines to minimum illuminances for safety by area.

Illumination Evaluation. Although the proper quality and quantity of illumination may be designed for safety in an area, it is necessary to know whether the design meets requirements. A standard procedure, titled "How to Make a Lighting Survey,"[34] has been developed in cooperation with the U.S. Public Health Service. This standard procedure is recommended for use in surveys of lighting for safety.

EMERGENCY LIGHTING

Consideration should be given to emergency lighting needs early in the planning stages of a building. Consultation between the owner and/or occupier of the premises, the architect, the lighting designer, the utility, and others concerned, should be arranged when, or perhaps before, the normal lighting planning is discussed. The installation contractor should be made aware of emergency lighting requirements at the earliest possible time.

Definitions. The following definitions are given for the terms used in this section:

Emergency lighting: Lighting provided for use when the power supply for the normal lighting fails, to insure that escape routes can be effectively identified and used.

Exit: A way out of the premises that is intended to be used at any time while the premises are occupied.

Emergency exit: A way out of the premises that is intended to be used only during an emergency.

Escape route: A route from a point inside the premises to an exit or emergency exit.

Normal lighting: All permanently installed electric lighting normally used when the premises are occupied.

Guides for the following are not provided here but are defined so that they are clearly excluded from this section.

Safety lighting: That part of emergency lighting that is provided to insure the safety of workers having to remain at work when the normal lighting fails.

Standby lighting: That part of emergency lighting that is sometimes provided to enable normal activities to continue.

Basic requirements for escape lighting are specified by federal codes and frequently are strengthened by local codes. The material that follows refers only to emergency lighting without regard for the type or location of the emergency power, which may be emergency motor driven generators, central battery systems, central inverter systems, unit inverters or unit equipment.

Emergency lighting is specified by the Life Safety Code (NFPA 101)[35] to be necessary in certain interiors where people work or meet, in order to enable them to leave the interior safely in the event that an emergency situation arises due to the failure of the normal power. References to that code, as well as any existing local codes, should be made at all times.

Design Requirements for Emergency Lighting.

When the normal lighting of an occupied building fails, irrespective of the cause, the emergency lighting is required to fulfill the following functions:

1. Indicate clearly and unambiguously the escape routes.
2. Provide illumination and a comforting visual environment along the escape routes sufficient to facilitate safe movement along them toward and through the exits and emergency exits provided.
3. Permit ready identification of all fire alarm call points and firefighting equipment provided along the escape routes under emergency lighting conditions.

Escape Route Indication. Signs are required to be illuminated in time of emergency to insure that from any point within the premises an escape route can be easily identified and followed in an emergency.

All normal exits should be illuminated at all times when the premises are occupied. This lighting should, practically speaking, be external to the exit signs themselves.

Where direct sight of an exit or emergency exit is not possible, a directional sign or series of signs should be provided. They should be so placed that a person following them will be progressed toward the nearest exit or emergency exit.

Exit signs cannot be counted on to be visible to many people at distances of more than 30 meters (100 feet), and should not be expected to be visible at longer intervals on long escape routes.

Illumination of Exit Signs. Either of the following methods of illumination may be used: (a) lamps external to the sign and (b) lamps contained within the sign. It is recommended that the method of illumination of exit signs described under (b) be used within any area where the normal lighting may be deliberately dimmed or extinguished, *e.g.*, places of entertainment.

In the event of failure of the supply to the normal lighting, escape route signs should receive the power needed for illumination from the emergency lighting supply. Power for exit signs should be unswitched or have the switch accessible only to authorized personnel.

Visual Impact and Legibility of Internally Illuminated Signs. Impact and legibility of exit signs are dependent upon luminance, size, viewing distance, contrast, positioning and uniformity.

Luminance: Where codes exist, an illuminance of 54 lux (5 footcandles) on the face of the sign is usually specified. Illuminance is an inappropriate parameter for internally illuminated signs. Currently research is being done along this line, but a luminance of 7 to 10 candelas per square meter (2 to 3 footlamberts) on the lighted area of the sign seems to be a reasonable level and parameter, because it appears to be adequate under emergency lighting conditions, is measurable, and provides better contrast under normal light.

Size: Letters must have at least a 19-millimeter (¾-inch) wide stroke and must be at least 150 millimeters (6 inches) high.

Viewing distance: In an emergency, an exit sign should not be expected to be useful at a distance greater than 30 meters (100 feet).

Contrast: Once other parameters have been met, this is a remaining important parameter. See luminance above. Transilluminated letters usually provide the best visibility. Color of letters is not an important point, so long as adequate light and contrast are provided. There seems to be little differentiation between dark letters on light background or illuminated (light) letters on dark or opaque background. Contrast is the important consideration.

Positioning: The location of the emergency exit sign will usually be determined by the desirable location under normal power conditions since, except for emergency exits, emergency exit signs mark the location of normal exit doors.

Uniformity: The exit sign face should be uniformly lighted, with a variation of not more than a factor of 2 above or below the average level over the lettered area.

All exit signs in a collective area should be of a similar color and design, as an aid to ready identification.

Externally Illuminated Exit Signs. Externally illuminated exit signs vary so greatly in design, material, color and printing that standards are difficult to establish. NFPA 101 requires 54 lux (5 footcandles) on the face of the sign. However, consideration must be given to contrast, glare, veiling reflectance, as well as reliability of the emergency power source for the external light, but the minimum letter size must adhere to that given above.

Egress Route Emergency Illumination

Illuminance. The horizontal illuminance of any escape route should be not less than 1 per cent of the average provided by the normal lighting, with a minimum average of 5 lux [0.5 footcandle] at floor level.

Illuminance Uniformity. Illuminance uniformity is more easily achieved by using a greater number of lamps with lower light output than by employing a lesser number of more widely spaced units with higher light output.

A uniformity ratio (E_{max}/E_{min}) of up to 20:1 along the center line of an escape route is desirable for safe movement. A value of 40:1 should not be exceeded.

Visibility of Hazards. By itself, illuminance is not a sufficient criterion of visibility, since it refers only to the light falling on a surface and not the amount reflected back to the eye. Luminance is really the only relevant measure.

It is recommended that all potential obstructions or hazards on an escape route be light in color with contrasting surroundings. Such hazards include the nosings of stair treads, barriers and walls at right angles to the direction of movement.

In restricted areas such as corridors, light-colored decoration throughout is an advantage and, under emergency conditions, prominent vertical surfaces can assist considerably in defining the escape route.

Location of Egress Luminaires. A luminaire should be provided for each exit door and emergency exit door and at points where it is necessary to emphasize the position of potential hazards, sufficient to light that area to a level of 30 lux [3 footcandles].

The floor area to be so lighted should be a square at the threshold of the point of egress that is double the width of the egress opening, or equal to the width of the corridor, whichever is less. Illuminance measurement should be on the horizontal.

Examples of such areas are:

1. Intersection of corridors.
2. Abrupt changes of direction of the egress path.
3. Staircases. Each flight of stairs should receive direct light.
4. Other changes of floor level that may constitute a hazard.
5. Outside each exit and emergency exit, and close to it.

Additional lamps, as required, should be located so as to ensure that the lighting throughout the escape routes complies with the recommendations for minimum illuminance and illuminance uniformity given above.

Windowless offices occupied by less than five people normally should not require emergency lighting, provided proper escape route light exists in the corridor.

Handicapped people and other special situations could be an exception.

A room nominally occupied by five or more people and not otherwise requiring emergency light, should have an illuminance at the door equal to the egress route, or a glass paneled door. Under these circumstances, solid doors should be avoided.

Illumination of Fire Alarm Call Points and Fire Fighting Equipment. Fire alarm call points and fire fighting equipment provided along escape routes should be illuminated either by emergency lighting or by normal electric lighting or daylight at all times while the premises are occupied.

Length of Time an Emergency Lighting System Should Operate Without Recharging

The time required to evacuate a premise will depend upon its size and complexity but it should normally be possible to complete an orderly evacuation, even of the largest premises, in less than one and one half hours. An owner, architect, or engineer, may agree that a lesser time is acceptable.

With battery operated emergency lighting, adequate light must be provided without the battery voltage dropping below 87½ per cent of rated voltage within the required time.

In an emergency, evacuation times may be considerably increased; for example, some of the escape routes may have been cut off, injured people may have to be found and possibly given on-the-spot medical treatment, etc. The time for which escape lighting is required to operate will, therefore, always be longer than the absolute minimum time required to evacuate the premises under ideal conditions.

Facilities designed especially for older and partially or wholly incapacitated people present special problems and special attention should be given to this parameter under these circumstances.

Power Supply Systems for Emergency Lighting

Emergency lighting is provided for use when the supply for the normal lighting fails and must, therefore, be powered by a source independent from that of the normal lighting. The recommendations here are confined to illumination by means of electric lamps. Furthermore, because this section is concerned primarily with permanently installed emergency lighting systems, the only power sources considered are motor driven electric generators and combinations of rechargeable secondary batteries together with suitable chargers.

The National Electrical Code section 700[36] specifically forbids the use of nonrechargeable (primary) batteries for emergency lighting.

Generator-Powered Systems. Emergency lighting systems should provide the required illuminance within a 10 second (NFPA101) period of the interruption of the normal lighting. If, therefore, such a system is to be powered by a generator, it is essential that the generator can be run up to its required output within the specified period, and that start-up be automatic on failure of the normal lighting. Care should be taken that batteries for generator starting are of a type specially designed for standby operation and are provided with suitable chargers.

Battery-Powered Systems. A battery-powered emergency lighting system utilizing suitable rechargeable secondary batteries may be designed for operation from a centrally located battery and charger combination (central system) or from batteries located at the lamps themselves (unit equipment). The battery/charger combination should, in each case, be so designed that, after the battery has been discharged for the specified duration of the category of the system, it should be capable of again supporting the emergency lighting system for one hour following a 24-hour recharge period.

All central battery powered units should be adequately ventilated.

Categories. Any emergency lighting system may be designed to supply the required load for any desired time. However, it must be designed to supply the minimum illuminances as established above.

Special areas: Almost every commercial building or apartment house normally occupied by 100 persons or more will have areas requiring special attention in the layout of the emergency lighting system. Examples are heavy machinery work areas, windowless stockrooms, restrooms and, particularly, stairwells. Emergency lighting for such areas only can be designed using good judgment and common sense.

The need for exterior emergency lighting to light escape routes away from the building, should not be overlooked.

Service and Maintenance

All emergency lighting systems should be tested and inspected at least every 30 days, no matter what type of emergency power is used.

For large installations where responsible semiskilled personnel are not available, a service contract with a responsible service organization should be provided. In some cases, this contract can be included in the specifications for the system, or it may be negotiated later between the owner and the service organization. Competent emergency lighting service organizations operate in most cities.

Criteria for Measurement of Emergency Lighting

Because of the very low illuminances provided by emergency lighting and because only escape routes need to be lighted, lux (footcandle) and watts per square meter (foot) are not suitable measuring criteria. Adequate visibility is really the only suitable criteria. However, at the present time there appears to be no better way to specify that than in illuminance or luminance.

Caution

Section 700 of the National Electrical Code[36] contains specifications for the installation of emergency lighting and these should be adhered to.

All emergency lighting equipment should carry

the label of a nationally recognized testing laboratory.

ECONOMICS

Economic analysis is an important tool to be used in making lighting design decisions. It is based on the economic impact of various alternatives, and assumes that the relevant variables are expressible in monetary units. In comparing a number of different lighting systems, for example, a life cycle cost study is not relevant unless all systems are functionally acceptable. There is no monetary expression that says one alternative doesn't provide enough light, or that another alternative is a system with a great deal of discomfort glare. Therefore, the analysis assumes that all alternatives are functionally equivalent.

Since money has a time value, a monetary unit (*e.g.*, dollar, peso, etc.) at one date is not directly comparable with the same monetary unit at another date. It is not sufficient, therefore, to determine the amounts of expenditures and receipts. It is necesary to determine the times of these cash flows. This is the essence of *life cycle costing*. All expenditures and receipts for the anticipated life of the system or length of the study period are expressed in terms of present worth.

For life cycle costing of a lighting system an evaluation is made of a series of payments which include initial cost, power costs, replacement lamp and ballast costs, cleaning and maintenance labor costs, etc. Life cycle costing is particularly useful in determining the most economical system when a series of costs are likely over a long period of time. With power costs, labor costs, and replacement parts increasing at a rapid rate, the system with the least initial cost may be the most expensive alternative over the life of the system.

For more specific information, see Section 3, Lighting Economics.

REFERENCES

1. Flynn, J. E., Spencer, T. J., Martyniuk, O., and Hendrick, C.: "Interim Study of Procedures for Investigating the Effect of Light on Impressions and Behavior," *J. Illum. Eng. Soc.*, Vol. 3, p. 87, October, 1973.
2. Flynn, J. E.: "A Study of Subjective Responses to Low Energy and Non-Uniform Lighting Systems," *Light. Des. Appl.*, Vol. 7, p. 6, February, 1977.
3. Flynn, J. E. and Spencer, T. J.: "The Effects of Light Source Color on User Impression and Satisfaction," *J. Illum. Eng. Soc.*, Vol. 6, p. 167, April, 1977.
4. Flynn, J. E., Hendrick, C., Spencer, T., and Martyniuk, O.: "A Guide to Methodology Procedures for Measuring Subjective Impressions in Lighting," *J. Illum. Eng. Soc.*, Vol. 8, p. 95, January, 1979.
5. Committee on Recommendations for Quality and Quantity of Illumination of the IES: "RQQ Report No. 1—Recommendations for Quality and Quantity of Illumination," *Illum. Eng.*, Vol. 58, p. 422, August, 1958.
6. Committee on Recommendations for Quality and Quantity of Illumination of the IES: "RQQ Report No. 6—Selection of Illuminance Values for Interior Lighting Design," *J. Illum. Eng. Soc.*, Vol. 9, p. 188, April, 1980.
7. *Guide on Interior Lighting,* CIE Publication No. 29 (TC-4.1) 1975.
8. *Uniform Framework of Methods for Evaluating Visual Performance Aspects of Lighting*, CIE Publication No. 19 (TC-3.1) 1972.
9. Committee on Recommendations for Quality and Quantity of Illumination of the IES: "RQQ Report No. 2 (1972)—Outline of a Standard Procedure for Computing Visual Comfort Ratings for Interior Lighting," *J. Illum. Eng. Soc.*, Vol. 2, p. 325, April, 1973.
10. Levin, R. E., "An Evaluation of VCP Calculations," *J. Illum. Eng. Soc.*, Vol. 2, p. 355, July, 1973.
11. Committee on Recommendations for Quality and Quantity of Illumination of the IES: "RQQ Report No. 3—An Alternate Simplified Method for Determining the Acceptability of a Luminaire from the VCP Standpoint for Use in Large Rooms," *J. Illum. Eng. Soc.*, Vol. 1, p. 256, April, 1972.
12. Petherbridge, P. and Hopkinson, R. G.: "A Preliminary Study of Reflected Glare," *Trans. Illum. Eng. Soc.* (London), Vol. XX, No. 8, 1955.
13. Finch, D. M.: "The Effect of Specular Reflection on Visibility, Part I—Physical Measurements for the Determination of Brightness and Contrast," *Illum. Eng.*, Vol. LIV, p. 474, August, 1959.
14. Chorlton, J. F. and Davidson, H. F.: "The Effect of Specular Reflection on Visibility, Part II—field Measurements of Loss of Contrast," *Illum. Eng.*, Vol. LIV, p. 482, August, 1959.
15. Allphin, W.: "Sight Lines to Desk Tasks in Schools and Offices," *Illum. Eng.*, Vol. LVIII, p. 244, April, 1963.
16. Crouch, C. L. and Kaufman, J. E.: "Practical Application of Polarization and Light Control for Reduction of Reflected Glare," *Illum. Eng.*, Vol. LVIII, p. 277, April, 1963.
17. Design Practice Committee of the IES: "Recommended Practice for the Specification of an ESI Rating in Interior Spaces When Specific Task Locations are Unknown," *J. Illum. Eng. Soc.*, Vol. 6, p. 111, January, 1977.
18. Kruithof, A. A.: "Tubular Luminescence Lamps for General Illumination," *Philips Tech. Rev.*, Vol. 6, p. 65, 1941.
19. O'Brien, P. F.: "Analytical Study of the Pencil Shadow-Caster," *Illum. Eng.*, Vol. 63, p. 183, April, 1968.
20. Cuttle, C.: "Lighting Patterns and the Flow of Light," *Light. Res. Technol.*, Vol. 3, p. 171, 1971.
21. Higbie, H. H.: *Lighting Calculations*, J. Wiley and Sons, Inc., New York, p. 429, 1934.
22. *Thermal Environmental Conditions for Human Occupancy*, ASHRAE Standard 55-74, The American Society of Heating, Refrigerating and Air Conditioning Engineers, Inc., New York, 1974.
23. Rohles, F. H. and Nevins, R. G.: "The Nature of Thermal Comfort for Sedentary Man," *ASHRAE Trans.*, Vol. 77, Part 1, 1971.
24. Rohles, F. H., Jr.: "The Revised Modal Comfort Envelope," *ASHRAE Trans.*, Vol. 79, Part II, p. 52, 1973.
25. *ASHRAE Handbook*, Fundamentals Volume, Chapter 8, American Society of Heating, Refrigerating and Air Conditioning Engineers, Inc., New York, 1977.
26. Fanger, P. O.: "Thermal Comfort," *Analysis and Applications in Environmental Engineering*, Danish Technical Press, Copenhagen, Denmark, 1970.
27. Committee on Light Sources of the IES: "The Effect of Temperature on Fluorescent Lamps," *Illum. Eng.*, Vol. LVIII, p. 101, February, 1963.
28. Bonvallet, G. G.: "Method of Determining Energy Distribution Characteristics of Fluorescent Luminaires," *Illum. Eng.*, Vol. LVIII, p. 69, February, 1963.
29. Mueller, T. and Benson, B. S.: "Testing and Performance of Heat

Removal Troffers," *Illum. Eng.*, Vol. LVII, p. 793, December, 1962.
30. Ballman, T. L., Bradley, R. D., and Hoelscher, E. C.: "Calorimetry of Fluorescent Luminaires," *Illum. Eng.*, Vol. LIX, p. 779, December, 1964.
31. Committee on Lighting and Air Conditioning of the IES: "Lighting and Air Conditioning," *Illum. Eng.*, Vol. LXI, p. 123, March, 1966.
32. Committee on Testing Procedures of the IES: "IES Approved Guide for the Photometric and Thermal Testing of Air Cooled Heat Transfer Luminaires," *J. Illum. Eng. Soc.*, Vol. 8, p. 57, October, 1978.
33. *ASHRAE Handbook*, Fundamentals Volume, Chapter 25, p. 14, American Society of Heating Refrigerating and Air Conditioning Engineers, Inc., New York, 1977.
34. Lighting Survey Committee of the IES: "How to Make a Lighting Survey," *Illum. Eng.*, Vol. LVII, p. 87, February, 1963.
35. *Life Safety Code*, National Fire Protection Association, Boston.
36. *National Electrical Code*, National Fire Protection Association, Boston.

Lighting Economics

SECTION
3

One way to put the various costs associated with the design, purchase, ownership and operation of a lighting system into perspective is to subject that system to a thorough economic analysis. Focusing on the initial cost of equipment, maintenance or energy costs may improperly bias a purchasing decision or subject the owner of a system to higher than necessary operating costs, and the potential value built into the design of the system may never be realized.

Lighting economic analyses have other rationales as well. For either new or existing systems they might be used for:

1. Comparing alternative systems as part of a decision making process.
2. The evaluation of maintenance techniques and procedures.
3. Determining the impact of lighting on other building systems.
4. Budgeting and cash flow planning.
5. Simplifying and reducing complex lighting system characteristics to a generally understandable common measure—cost.
6. Helping to determine the benefit of lighting (cost/benefit analysis).

Cost-of-Light Concept

From an economic standpoint, the process of providing the desired lighting involves the expenditure of money for a number of goods (lamps, luminaires, wire, etc.) and services (labor, electric energy, etc.). This was recognized early in the history of lighting and basic measures of lighting value have been developed based upon the idea of cost per unit of lighting delivered. This is the traditional "cost of light" and is commonly expressed as cost per lumen-hour. Because the cost per lumen-hour of typical general lighting systems is very small, the unit of dollars per million lumen-hours is usually used.

Cost-of-light data may be calculated for a luminaire by multiplying the average light output of the luminaire (lumens) by the life of the lamps (hours) and dividing the result into the total costs (owning and operating) of the luminaire for the same time period. One relationship[1] for this is:

$$U = 10/(Q \times D)[(P + h)/L + W \times R + (F + M)/H] \quad (1)$$

where

U = Unit cost-of-light in dollars per million lumen-hours.
Q = Mean lamp lumens.
D = Luminaire dirt depreciation (average between cleanings).
P = Lamp price in cents.
h = Cost to replace one lamp in cents.
L = Average rated lamp life in thousands of hours.
W = Mean luminaire input watts (lamps + ballast).
R = Energy cost in cent per kilowatt-hour.
F = Fixed or owning costs in cents per luminaire-year.
M = Cleaning cost in cents per luminaire-year.
H = Annual hours of operation in thousands of hours.

Lighting Cost Comparisons

The factors in the cost-of-light formula may be expanded and rewritten in a way more suitable for comparing one system to another.[2] A form which has been developed to help organize the data is shown in Fig. 3-1.

This is an example of an annual cost model in which both initial and recurring costs are put on a "per year" basis. For comparisons among systems to be valid, however, each lighting method must provide the same illuminance or the results must be normalized by dividing the various cost factors such as total capital expense per year (item 24), total operating and maintenance expense per year (item 46), and total lighting expense per year (item 49) by the maintained illuminance (item 14).

Either individual (spot) or group replacement

NOTE: References are listed at the end of each section.

Fig. 3-1. Lighting Cost Comparison (Annual Cost Model)

	Item	Lighting Method #1	Lighting Method #2
A. Installation Data			
Type of installation (office, industrial, etc.)	1	....	....
Luminaires per row	2	....	....
Number of rows	3	....	....
Total luminaires	4	....	....
Lamps per luminaire	5	....	....
Lamp type	6	....	....
Lumens per lamp	7	....	....
Watts per luminaire (including accessories)	8	....	....
Hours per start	9	....	....
Burning hours per year	10	....	....
Group relamping interval or rated life	11	....	....
Light loss factor	12	....	....
Coefficient of utilization	13	....	....
Illuminance, maintained	14	....	....
B. Capital Expenses			
Net cost per luminaire	15	....	....
Installation labor and wiring cost per luminaire	16	....	....
Cost per luminaire (luminaire plus labor and wiring)	17	....	....
Total cost of luminaires	18	....	....
Assumed years of luminaire life	19	....	....
Total cost per year of life	20	....	....
Interest on investment (per year)	21	....	....
Taxes (per year)	22	....	....
Insurance (per year)	23	....	....
Total capital expense per year	24	....	....
C. Annual Operating and Maintenance Expenses			
Energy expense			
Total watts	25	....	....
Average cost per kWh (including demand charges)	26	....	....
Total energy cost per year*	27	....	....
Lamp renewal expense			
Net cost per lamp	28	....	....
Labor cost each individual relamp	29	....	....
Labor cost each group relamp	30	....	....
Per cent lamps that fail before group relamp	31	....	....
Renewal cost per lamp socket per year**	32	....	....
Total number of lamps	33	....	....
Total lamp renewal expense per year	34	....	....
Cleaning expense			
Number of washings per year	35	....	....
Man-hours each (est.)	36	....	....
Man-hours for washing	37	....	....
Number of dustings per year	38	....	....
Man-hours per dusting each	39	....	....
Man-hours for dustings	40	....	....
Total man-hours	41	....	....
Expense per man-hour	42	....	....
Total cleaning expense per year	43	....	....
Repair expenses			
Repairs (based on experience, repairman's time, etc.)	44	....	....
Estimated total repair expense per year	45	....	....
Total operating and maintenance expense per year	46	....	....
D. Recapitulation			
Total capital expense per year	47	....	....
Total operating and maintenance expense per year	48	....	....
Total lighting expense per year	49	....	....

* Total energy cost per year $= \dfrac{\text{Total watts} \times \text{burning hours per year} \times \text{cost per kWh}}{1000}$

** See formulas (2), (3) and (4) in the text. They can be used to determine the most economical replacement method.
Note: Items 38, 39 and 40 may be eliminated and "washings" in items 35 and 37 changed to "cleanings".

of lamps can be handled by the model using the following equations:

Individual replacement

$$= \frac{B}{R}(c + i) \text{ dollars per socket annually.} \quad (2)$$

Group replacement (early burnouts

$$\text{replaced}) = \frac{B}{A}(c + g + KL + Ki) \quad (3)$$

dollars per socket annually.

Group replacement (no replacement

$$\text{of early burnouts}) = \frac{B}{A}(c + g) \quad (4)$$

dollars per socket annually.

where

B = burning hours per year.
R = rated average lamp life in hours.
A = burning time between replacements in hours.
c = net cost of lamps in dollars.
i = cost per lamp for replacing lamps individually in dollars.
g = cost per lamp for replacing lamps in a group in dollars.
K = proportion of lamps failing before group replacement (from mortality curve).
L = net cost of *replacement* lamps in dollars.

No general rule can be given for the use of group replacements; each installation should be considered separately. In general, group replacement should be given consideration when individual replacement cost i is greater than half the cost c and when group replacement cost g is small compared to i.

An iteration of equations (3) or (4) using various burning times between replacements (A) will indicate whether or not group relamping is economical and the best (lowest annual cost per socket) relamping interval.

The choice of a periodic relamping or cleaning interval will have a direct effect on the light loss factor (item 12). An economic analysis which includes a study of variable maintenance economics can, therefore, be used in the design phase of a project as a way to minimize system light output depreciation over time. As a result, for general lighting systems, fewer luminaires may be needed for a given maintained illuminance level and both initial as well as owning and operating costs can be reduced for the entire system over its life.

Life Cycle Costs

A second type of mathematical model known as the present worth or net present value analysis is becoming more widely used for lighting economic studies because of its power as a decision making tool. In contrast to the annual cost model in which all expenditures are put into a cost per year form, the present worth model is set up to take into account expenditures at the time they actually take place. It is, therefore, somewhat easier to consider all of the costs which are expected to occur over the life time or life cycle of the lighting system up to and including any salvage value.

All expenditures, whether they are periodic repetitive costs, such as for electric energy, cleaning, or lamp replacement, or nonperiodic costs such as damage repair or ballast replacement are factored in at the future time the expenditures are expected to occur. All costs are then transferred to the present time using relationships which take into account the time value of money via an interest factor or "opportunity rate." Tables[3, 4] are available to simplify the arithmetic, or the following seven basic equations may be used (terms used in these equations are: P = present or first cost; F = future worth; A = uniform annual cost; i = opportunity rate or interest; and y = number of years):

1. Uniform compound amount factor.

$$F = A\left[\frac{(1+i)^y - 1}{i}\right]$$

2. Uniform present worth factor.

$$P = A\left[\frac{(1+i)^y - 1}{i(1+i)^y}\right]$$

3. Uniform sinking fund factor.

$$A = F\left[\frac{i}{(1+i)^y - 1}\right]$$

4. Single present worth factor.

$$P = F\left[\frac{1}{(1+i)^y}\right]$$

5. Uniform capital recovery factor.

$$A = P\left[\frac{i(1+i)^y}{(1+i)^y - 1}\right]$$

6. Single compound amount factor.

$$F = P[(1+i)^y]$$

Fig. 3-2. Lighting System Cost Comparison (Life Cycle Costs—Present Worth Model)*

Life cycle cost analysis for ______

Area ______ m^2 (ft^2) Luminaire Layout ______ Luminaire Layout ______

I. Lighting and Air Conditioning Installed Costs (initial)		
1. Luminaire installed costs: luminaire, lamps, material and labor:	$ ______	$ ______
2. Total kW lighting:	____ kW	____ kW
3. Tons of air conditioning required for lighting:	____tons	____tons
4. First cost of air-conditioning machinery:	$ ______	$ ______
5. Variation in first cost of heating equipment:	$ ______	$ ______
6. Other differential costs:	$ ______	$ ______
7. Subtotal mechanical and electrical installed cost:	$ ______	$ ______
8. Initial taxes:	$ ______	$ ______
9. Total costs:	$ __(A1)	$ __(B1)
10. Installed cost per square meter (foot)	$ ______	$ ______
11. Watts per square meter (foot) of lighting:	__ watts	__ watts
12. Salvage (at *y* years):	$ __(As)	$ __(Bs)
II. Annual Power and Maintenance Costs		
1. Lamps: burning hours × kW × $/kWh	$ ______	$ ______
2. Air conditioning:	$ ______	$ ______
3. Reduction in heating costs fuel used: ____	$ ______	$ ______
4. Other differential costs (*i.e.*, air-conditioning, maintenance cost):	$ ______	$ ______
5. Cost of lamps: (No. of lamps ____ @ $____/lamp per *N*) (Group relamping every *N* years, typically every one, two or three years, depending on burning schedule.)	$ ______	$ ______
6. Cost of ballast replacement: (No. of ballast ____ @ $____/ballast per *n*) (*n* = number of years of ballast life.)	$ ______	$ ______
7. Luminaire cleaning cost: No. of luminaires ____ @ $____ each. (Cost to clean one luminaire includes cost to replace or clean lamps.)	$ ______	$ ______
8. Annual insurance cost:	$ ______	$ ______
9. Annual property tax cost:	$ ______	$ ______
10. Total annual power and maintenance cost:	$ __(Ap)	$ __(Bp)
11. Cost per square meter (foot):	$ ______	$ ______

Notes for Fig. 3-2.

II. 1. The number of burning hours and cost/kWh will depend upon occupancy and local power rates.

II. 2. Only that portion needed to remove lighting heat.

II. 3. Reduction in cost for fuel for heating equipment because of increased heat obtained from lighting system.

II. 8. Percentage of first cost (I.1.) (1 to 1.5 per cent) (compensating).

II. 9. Percentage of first cost (I.1.) (4 to 6 per cent) (compensating).

* Based on Reference 6.

7. The equations below give the present worth of an escalating annual cost (r = rate of escalation and A = annual amount before escalation).

$$\text{(a)}\quad P = \sum_{k=1}^{y_n} A \left[\frac{(1+r)^k}{(1+i)^k}\right]$$

$$\text{(b)}\quad P = A\left[\frac{(1+r)\,[(1+i)^{y_n} - (1+r)^{y_n}]}{(i-r)\,(1+i)^{y_n}}\right] \quad \text{if } i \neq r$$

$$\text{(c)}\quad P = A\,[y_n] \quad \text{if } i = r$$

System Comparison Based on Present Worth. In the process of comparing several alternative lighting systems, all costs might be converted into a single number representing the present worth of each system over its life. The system with the lowest over-all cost would then be readily apparent.

Public bodies have adopted this approach particularly for projects involving energy conservation and a number of calculational aids including a computer program in the public domain are available.[5]

Using the lighting system cost comparison outlined in Fig. 3-2, a comparison of lighting systems can be made based on present worth as follows:[6]

	System A	System B
First Costs	A1	B1
Power and maintenance costs	Ap	Bp
Salvage value	As	Bs

Terms = *y* years

Opportunity rate = *i* (*e.g.*, *i* = .09 means 9 per cent)

$$A_{T1} = A1 + Ap \cdot \left[\frac{(1+i)^y - 1}{i(1+i)^y}\right] - As \cdot \left[\left(\frac{1}{1+i}\right)^y\right]$$

$$B_{T1} = B1 + Bp \cdot \left[\frac{(1+i)^y - 1}{i(1+i)^y}\right] - Bs \cdot \left[\left(\frac{1}{1+i}\right)^y\right]$$

This present worth method is valid only if Systems A and B have equal lives.

In the event it is required to determine the number of years (y) at a given opportunity rate for a payout of one system over another the following formulas are given:

Let X_1 = difference between (present cost − salvage value) of the two systems ($A_1 - A_s$).

Let X_2 = difference between annual operating costs of the two systems (A_p).

Let i = opportunity rate.

Find $X_3 = \frac{X_2}{(i \cdot X_1)}$

Let $a = 1 + i$

$$b = \frac{X_3}{(X_3 - 1)}$$

Then $y = \frac{\ln b}{\ln a}$

Note: The above formulas are correct if $X_2/X_1 > i$; however, if $X_2/X_1 = i$, then both systems are equal; and if $X_2/X_1 < i$, then the increase in present cost (X_1) will never recover through the annual benefits (X_2).

REFERENCES

1. Merrill, G. S.: "The Economics of Light Production With Incandescent Lamps," *Trans. Illum. Eng. Soc.*, Vol. XXXII, p. 1077, December, 1937.
2. Barr, A. C. and Amick, C.: "Fundamentals of Lighting Analysis," *Illum. Eng.*, Vol. XLVII, p. 260, May, 1952.
3. Grant, E. L. and Ireson, W. G.: *Principles of Engineering Economy*, The Ronald Press Co., New York, N.Y. 1976.
4. *Life Cycle Cost-Benefit Analysis*, K. G. Associates, Box 7596, Inwood Station, Dallas, Texas 75209.
5. Ruegg, R. T. et al. *Life Cycle Costing.* A Guide for Selecting Energy Conservation Projects for Public Buildings. NBS BSS-113 National Bureau of Standards, U.S. Dept. of Commerce, Washington, D. C. 1978.
6. Design Practice Committee of the IES: "Life Cycle Cost Analysis of Electric Lighting Systems," *Light. Des. Appl.*, Vol. 10, p. 43, May, 1980.
7. DeLaney, W. B.: "How Much Does a Lighting System Really Cost?," *Light. Des. Appl.*, Vol. 3, p. 22, January, 1973.

Energy Management

SECTION 4

The concept of energy management has gained in importance since the early 1970's, when energy conservation became a concern, resulting in a close examination of the way buildings have been lighted and the criteria for future designs to use energy resources effectively and efficiently.

For *new buildings* there is a need to limit the amount of connected power available for lighting, while still enabling the designer to provide a lighting system suitable for the occupants' requirements. Power reduction, however, is only one aspect to be considered in a management program, since *energy = power × time.*

A savings in *energy* can be achieved by reducing either the amount of *power* consumed (watts or kilowatts), or by reducing the *time* the power is used (hours). Energy reduction options include modifying or replacing the lighting system with a more efficient one, using replacement components which use less power, or modifying the operating characteristics of the building to reduce kilowatt hours. If both *power* and *time* are reduced, *energy* savings potential is further increased.

Lighting in an *existing building* must first comply with the same power limit as in a new building and then a program must be developed to reduce energy. There are many opportunities in an existing facility to improve not only the operating efficiency of the building, and thereby save energy, but at the same time to improve the quality of the lighting.

LIGHTING POWER LIMITS (UPD PROCEDURE)

The following outlines the IES recommended procedure for establishing lighting power limits for new and existing buildings, using a Unit Power Density (UPD) procedure[1]. The procedure is consistent with that described in reference 2.

The procedure has been established for use with new and existing buildings to determine a power limit for a building's lighting system. *The power limit process should not be confused with the design process.* The procedure requires that each room or area in and around a building be identified and the task and general lighting requirements be determined. This limit process helps identify how buildings require different amounts of power to accommodate different types of tasks and densities of occupants.

The power limit is for the total facility; *i.e.*, the sum of power budgets for each individual space in the building and any exterior areas related to the facility. In the design process it may be desirable to exceed the power budget of a particular area. This can be accomplished by designing other areas below their budgeted power, which results in the total facility remaining within its power limit. The UPD procedure, therefore, does not limit the design of each area, but only the amount of power available for the total project, thereby providing design freedom within the calculated power limit.

The power limit process is identical for new and existing buildings. In some cases the limit initially established for a building will not change as the building usage undergoes minor variations. Major changes would, however, require that the budget be updated to incorporate the changing needs. The power limit for a facility as well as the operating instructions and architectural, mechanical and electrical plans are important documents for each building owner to have on file.

When insufficient information is known about the specific use of the building space (*e.g.*, number of occupants, space function, location of partitions), power limits are based on the apparent intended use of the building space.

The spaces and areas covered by the UPD procedure include:

NOTE: References are listed at the end of each section.

1. Building interior spaces that are heated and/or cooled.
2. Covered walkways, open roofed areas, porches, and similar spaces associated with buildings where lighting is required.
3. Building exteriors including entrances, exits, parking areas, driveways and outdoor storage and activity areas.

The procedure does not include the lighting power required for:
1. Theatrical productions, entertainment facilities, audio-visual presentations where lighting is an essential technical element for the function performed.
2. Outdoor athletic facilities.
3. Outdoor production and manufacturing activities not associated with the building facility.

Determination of a Base UPD

The UPD procedure involves the use of a "base UPD" (P_b), in watts per square meter or watts per square foot, which is applied in power budget calculations as shown later under Determination of a Power Limit, on page 4-6. Fig. 4-2 lists precalculated base UPD values for various tasks or areas, calculated using the general formula:

Base UPD (P_b)

$$= \frac{\text{Area} \times \text{Illuminance}}{\text{Luminaire CU} \times \text{Lamp Efficacy} \times 0.70}$$

Base UPD Criteria. The criteria used to precalculate the base UPD in Fig. 4-2 are the same as those used in reference 1. They are:

A. *Areas*
1. Task area—actual area of 4.65 square meters (50 square feet) per work station.
2. General area—area equal to the task area, but not to exceed 50 per cent of the total space area.
3. Non-critical area—area not designated as task or general areas, *i.e.*, room area less the sum of task area and general area.

B. *Illuminance*
1. For tasks—from Fig. 2-2.
2. For general lighting—⅓ of the task level, but not less than 200 lux (20 footcandles).
3. For non-critical lighting—⅓ of the average general lighting, but not less than 100 lux (10 footcandles).

C. *Light Source Efficacies*
Initial lumen output per watt input, including ballast losses:

Application	Efficacy (Lumens per watt)
Where moderate color rendition is appropriate and for outdoor spaces	55
Where good color rendition is appropriate	40
Where high color rendition is appropriate or total space is less than 4.65 square meters (50 square feet)	25
Where use of HID lamps under 250 watts or fluorescent lamps under 40 watts is appropriate	25

D. *Luminaire Coefficient of Utilization*
Values in Fig. 4-1 for representative luminaires and task criteria, based on:
1. Room dimensions.
2. Cavity height as distance from luminaire to work-plane.
3. Room cavity ratio as:

$$\text{RCR} = \frac{5 \times \text{Cavity Height (Length + Width)}}{\text{Length} \times \text{Width}}$$

4. Ceiling cavity reflectance − 80 per cent.
5. Wall reflectance − 50 per cent.
6. Floor cavity reflectance − 20 per cent.

E. *Light Loss Factor*
A value of 0.70 is used for all calculations except where it is impractical to control reflectances and a dirty atmosphere exists.

Fig. 4-1. Coefficients of Utilization of Representative Luminaires for Lighting Power Budgeting

Task Criteria	Room Cavity Ratio										
	0	1	2	3	4	5	6	7	8	9	10
For spaces with tasks subjected to veiling reflections and where visual comfort is important	55	55	50	45	40	36	32	29	26	26	26
For spaces without tasks, or with tasks not subjected to veiling reflections, but where visual comfort is important	63	63	57	51	46	41	37	33	30	30	30
For spaces without tasks and where visual comfort is not a criterion	70	70	63	57	51	46	41	37	33	33	33

Fig. 4-2. Base UPD Values for Lighting Power Limit Calculations.*

Task or Area	Base UPD Watts per Square Meter	Base UPD Watts per Square Foot	Note
Common Areas			
Boiler Room	7.53	0.7	d
Conference Room	13.99	1.3	a
Corridor	6.46	0.6	d
Electrical Equipment Room	6.46	0.6	d
Food Service Facilities			
Dining Areas	15.07	1.4	b
Kitchen	18.30	1.7	
Garage, Parking	2.15	0.2	d
General Assembly (Auditorium)	11.84	1.1	
Laboratories	34.44	3.2	
Library Room	23.68	2.2	
Lobby, Reception, Waiting	10.76	1.0	
Locker Room and Shower	6.46	0.6	d
Mail Room	30.14	2.8	
Material Handling (Bulk)	7.53	0.7	
Mechanical Equipment Room	6.46	0.6	d
Stairs	6.46	0.6	d
Storerooms or Warehouse			
Inactive	2.15	0.2	
Active Bulky	4.31	0.4	
Active Medium	6.46	0.6	
Switchboard and Control Room	18.30	1.7	
Toilet and Washroom	7.53	0.7	
Unlisted Spaces	2.15	0.2	d
Utility Room, General	4.31	0.4	
Office			
Accounting	34.44	3.2	f
Drafting	50.59	4.7	f
Filing (Active)	21.53	2.0	
Filing (Inactive)	8.61	0.8	
Graphic Arts	32.29	3.0	f
Office Machine Operation			
Computer Machinery	18.30	1.7	
Duplicating Machines	7.53	0.7	
EDP I/O Terminal (Internally Illuminated)	7.53	0.7	
EDP I/O Terminal (Room Illuminated)	18.30	1.7	
Typing and Reading	23.68	2.2	f
Residential			
Bath	46.28	4.3	d
Bedroom	15.07	1.4	a, d
Finished Living Spaces	23.68	2.2	d
Garage	5.38	0.5	d
Kitchen	43.06	4.0	d
Laundry	10.76	1.0	d
Unfinished Living Spaces	5.38	0.5	d
Commercial and Institutional			
Armories			
Drill	6.46	0.6	
Exhibitions	8.61	0.8	
Seating Area	4.31	0.4	
Art Galleries	17.22	1.6	a
Banks			
Lobby, General	24.76	2.3	
Posting and Keypunch	50.59	4.7	
Tellers' Stations	50.59	4.7	
Bar (Lounge)	11.84	1.1	b
Barber Shops and Beauty Parlors	40.90	3.8	
Church and Synagogues, Main Worship Area	24.76	2.3	a
Club and Lodge Rooms	11.84	1.1	
Courtrooms	9.69	0.9	
Depots, Air Terminals and Stations			
Baggage Checkroom	13.99	1.3	
Concourse (Main Thruway)	8.61	0.8	
Platforms	6.46	0.6	d
Ticket Counter	23.68	2.2	
Waiting and Lounge Area	8.61	0.8	
Hospitals			
Autopsy	34.44	3.2	
Central Sterile Supply	13.99	1.3	
Corridor, Special Areas			
Nursing Areas	6.46	0.6	d
Surgical & Lab Area	8.61	0.8	d
Critical Care Areas	40.90	3.8	
Cystoscopy Room	40.90	3.8	
Dental Suite	34.44	3.2	a
EKG and Specimen Room	8.61	0.8	
Emergency Outpatient	40.90	3.8	
Endoscopy Rooms (Non-Urologic)	30.14	2.8	
Examination and Treatment Rooms	21.53	2.0	
Fracture Room	21.53	2.0	
Inhalation Therapy Units	8.61	0.8	
Laboratories	34.44	3.2	
Linen Room	7.53	0.7	
Lobby (or Entrance Foyer)	21.53	2.0	
Medical Illustration Studio	59.20	5.5	
Medical Records Room	30.14	2.8	
Morgue	8.61	0.8	
Nurseries	40.90	3.8	
Nurses' Stations	15.07	1.4	a
Obstetric Delivery Suite			
Delivery Area	81.81	7.6	
Labor Rooms	9.69	0.9	
Postdelivery Recovery Area	11.84	1.1	
Occupational Therapy	13.99	1.3	
Patients' Rooms (includes bathrooms)	15.07	1.4	a, d
Pharmacy	32.29	3.0	
Physical Therapy	20.45	1.9	
Postanesthetic Recovery Room	40.90	3.8	
Pulmonary Function Lab	34.44	3.2	
Radiological Suite			
Preparation Area	26.91	2.5	
Special Procedures	53.82	5.0	
Treatment Room	5.38	0.5	
Solarium	6.46	0.6	
Surgical Induction and Hold Area	34.44	3.2	
Surgical Suite			
Operating Room, General	81.81	7.6	
Scrub and Cleanup Area	26.91	2.5	
Substerilizing and Instrument Area	7.53	0.7	
Utility Room, Work Area	13.99	1.3	
Waiting Area, General	6.46	0.6	
Hotels			
Bathrooms	13.99	1.3	a, d
Bedrooms	15.07	1.4	a, d
Entrance Foyer	11.84	1.1	
Lobby, General	11.84	1.1	a
Laundries			
Fine Hand Ironing	30.14	2.8	
Ironing, Weighing, Listing, Marking	13.99	1.3	
Machine and Press Finishing, Sorting	18.30	1.7	
Washing	7.53	0.7	
Library			
Audio Listening Areas, General	7.53	0.7	
Audiovisual Areas	18.30	1.7	
Book Stacks (Active)	9.69	0.9	
Book Stacks (Inactive)	4.31	0.4	
Book Repair and Binding	20.45	1.9	

See page 4-6 for footnotes.

Fig. 4–2. *Continued*

Task or Area	Base UPD		Note
	Watts per Square Meter	Watts per Square Foot	
Card Files	34.44	3.2	
Cataloging	23.68	2.2	
Microfilm Areas	23.68	2.2	
Reading Areas	23.68	2.2	
Municipal Building—Fire and Police			
Fire Engine Room	7.53	0.7	
Fireman's Dormitory	15.07	1.4	a
Identification Records	50.59	4.7	
Jail Cells	8.61	0.8	
Recreation Room	9.69	0.9	
Nursing Homes			
Administrative & Lobby Areas	15.07	1.4	
Chapel or Quiet Area, General	9.69	0.9	
Nurses' Station	15.07	1.4	a
Occupational Therapy	13.99	1.3	
Patient Care Unit (or Room), General	15.07	1.4	a, d
Pharmacy Area, General	17.22	1.6	a
Physical Therapy	20.45	1.9	
Recreation Area	15.07	1.4	
Post Offices			
Lobby	7.53	0.7	
Sorting, Mailing, etc.	30.14	2.8	
Restaurants-See Common Areas (Food Service)			
Schools			
Art	32.29	3.0	
Classrooms	23.68	2.2	
Dormitories	15.07	1.4	a
Drafting	34.44	3.2	
Home Economics	15.07	1.4	
Laboratories	30.14	2.8	
Lecture	23.68	2.2	
Music	18.30	1.7	
Sewing	45.21	4.1	
Shops	30.14	2.8	
Study Halls or Typing	23.68	2.2	
Service Stations, Auto	7.53	0.7	
Stores			
Alteration and Fitting	61.35	5.7	
Circulation	9.69	0.9	
Merchandise	40.90	3.8	
Sales Transaction	20.45	1.9	
Show Windows	93.65	8.7	
Stockrooms	7.53	0.7	
Wrapping and Packing	13.99	1.3	
Theaters and Motion Picture Houses	8.61	0.8	b
Industrial			
Aircraft Maintenance			
Docking and Maintenance	18.30	1.7	
Engine Overhaul	26.91	2.5	
Fabrication (Preparation for Assembly)	26.91	2.5	
First Manufacturing Operations (First Cut) Marking, Shearing, Sawing	13.99	1.3	
Aircraft Manufacturing			
Assembly—Sub and Final	26.91	2.5	
General Production	22.60	2.1	
Inspection Assembly	26.91	2.5	
Inspection Stock Parts	53.82	5.0	
Automobile Manufacturing			
Body Manufacturing			
Assembly	26.91	2.5	
Finishing and Inspecting	53.82	5.0	
Parts	18.30	1.7	
Chassis Assembly Line	26.91	2.5	
Final Assembly, Inspection Line	53.82	5.0	
Frame Assembly	13.99	1.3	
Bakeries			
General Production	10.76	1.0	
Hand Decorating	36.60	3.4	
Mechanical Decorating	19.38	1.8	
Mixing and Filling	13.99	1.3	
Book Binding			
General Production	18.30	1.7	
Embossing and Inspection	81.81	7.6	
Brewing			
General Production	7.53	0.7	
Filling (Bottles, Cans, Kegs)	13.99	1.3	
Candy Making			
Die Cutting and Wrapping	26.91	2.5	
General Production	13.99	1.3	
Sorting and Decorating	36.60	3.4	
Canning and Preserving			
Can Unscrambler	18.30	1.7	
Color Grading	81.81	7.6	
Initial Grading	13.99	1.3	
Inspection—Container	59.20	5.5	
Inspection—Food	81.81	7.6	
Labelling and Cartoning	7.53	0.7	
Preparation and Canning	26.91	2.5	
Chemical Works (See Petroleum)	7.53	0.7	
Clay Products and Cements			
Enameling and Glazing, Rough	36.60	3.4	
Fine Glazing	73.19	6.8	
General Production	7.53	0.7	
Cleaning and Pressing			
General Processing	13.99	1.3	
Pressing	39.83	3.7	
Repair and Alteration, Inspection & Spotting	59.20	5.5	
Cloth Products	53.82	5.0	
Clothing Manufacturing			
General Production	15.07	1.4	
Inspection, Pressing, Sewing, Cutting	59.20	5.5	
Piling up and Marking, and Shops	36.60	3.4	
Cotton Gin Industry	10.76	1.0	
Coal Tipple and Cleaning Plants	4.31	0.4	
Dairy Products			
Bottle Sorting	13.99	1.3	
Filling and Inspection	26.91	2.5	
General Processing	7.53	0.7	
Milk Equipment Washing	26.91	2.5	
Electric Generating Stations—Interiors			
Controlled Access Areas	22.60	2.1	
General Operation Area	6.46	0.6	
Laboratories	25.83	2.4	
Electrical Equipment Manufacturing			
General Production and Testing	26.91	2.5	
Explosives Manufacturing	7.53	0.7	
Flour Mills			
General Production	13.99	1.3	
Packing, Cleaning, Checking	7.53	0.7	
Product Control	26.91	2.5	
Forge Shops	13.99	1.3	
Foundries			
Core Making and Inspection	26.91	2.5	
Cupola Area	6.46	0.6	
Fine Inspection	53.82	5.0	
General Production	10.76	1.0	
Molding and Grinding	26.91	2.5	
Garage—Service			
Repair Area	26.91	2.5	
Traffic Area	6.46	0.6	
General Assembly and Production Area			
Medium	26.91	2.5	

See page 4–6 for footnotes.

Fig. 4-2. *Continued*

Task or Area	Base UPD: Watts per Square Meter	Base UPD: Watts per Square Foot	Note
Rough Easy Seeing	7.53	0.7	
Rough Difficult Seeing	13.99	1.3	
Glass Works			
Fine Grinding, Leveling, Polishing	34.44	3.2	
General Production	10.76	1.0	
Inspection, Etching and Decorating	67.81	6.3	
Glove Manufacturing			
General Production	26.91	2.5	
Pressing, Cutting, Sewing, and Inspection	59.20	5.5	
Hat Manufacturing			
Forming, Finishing Process, Sewing	59.20	5.5	
Preliminary Processing	26.91	2.5	
Inspection			
Difficult	30.14	2.8	
Fine	59.20	5.5	
Ordinary	15.07	1.4	
Iron and Steel Manufacturing			
Inspection			
Black Plate, Bloom and Billet Chipping	26.91	2.5	
Tin Plate and Other Bright Surfaces	67.81	6.3	
Motor Room	7.53	0.7	
Open Hearth Operations Area	6.46	0.6	
Rolling Mills Operations Area	10.76	1.0	
Tin Plate Mills	13.99	1.3	
Jewelry and Watch Manufacturing	67.81	6.3	
Leather Manufacturing and Working			
Cutting, Fleshing and Stuffing	13.99	1.3	
Finishing and Scarfing	30.14	2.8	
Grading, Matching, Cutting, Scarfing, Sewing	81.81	7.6	
Preparation Area	7.53	0.7	
Pressing, Winding, Glazing	44.13	4.1	
Machine Shops			
Medium Bench and Machine Work, Ordinary Automatic Machines, Rough Grinding, Medium Buffing and Polishing	26.91	2.5	
Rough Bench and Machine Work	13.99	1.3	
Meat Packing			
General Processing	26.91	2.5	
Slaughtering	7.53	0.7	
Metal Fabrication (Bulk)	6.46	0.6	
Paint Manufacturing	7.53	0.7	
Dipping, Simple Spraying, Firing	13.99	1.3	
Fine Hand Painting and Finishing	26.91	2.5	
Rubbing, Ordinary Hand Painting and Finishing Art, Stencil and Special Spraying	15.07	1.4	
Paper Box Manufacturing	13.99	1.3	
Paper Manufacturing			
General Production	10.76	1.0	
Inspection	26.91	2.5	
Rewinder	39.83	3.7	
Petroleum and Chemical Plants			
General Process Area	4.31	0.4	
Plating	7.53	0.7	
Polishing and Burnishing	30.14	2.8	
Poultry Industry			
Brooders and Hatcheries	7.53	0.7	
Egg Handling and Shipping	13.99	1.3	
Egg Processing	18.30	1.7	
Egg Production	6.46	0.6	
Fowl General Processing	22.60	2.1	
Fowl Unloading and Killing	6.46	0.6	
Printing			
Composing Room	30.14	2.8	
Electrotyping	20.45	1.9	
Photoengraving	20.45	1.9	
Printing Plants	54.90	5.1	
Type Foundries	20.45	1.9	
Rubber Goods Mechanical			
General Production	10.76	1.0	
Inspection	59.20	5.5	
Rubber Tire Manufacturing			
General Production	10.76	1.0	
Inspection, Cutting, Splicing	26.91	2.5	
Sawmills			
General Production	6.46	0.6	
Sorting Tables and Grading	26.91	2.5	
Sheet Metal Works			
General Production	13.99	1.3	
Inspection and Scribing	59.20	5.5	
Shoe Manufacturing, Leather	53.82	5.0	
Shoe Manufacturing, Rubber			
General Production	10.76	1.0	
Sole Rolling, lining, finishing	26.91	2.5	
Soap Manufacturing	10.76	1.0	
Stone Crushing and Screening	5.38	0.5	
Storage-Battery Manufacturing	13.99	1.3	
Structural Steel Fabrication	13.99	1.3	
Sugar Refining			
Color Inspection	81.81	7.6	
Grading	13.99	1.3	
Testing			
Extra-Fine Instruments, Scales, etc.	67.81	6.3	
General	18.30	1.7	
Textile Mills			
Drying and Finishing	40.90	3.8	
General Production	10.76	1.0	
Warping, Weaving, Grading	40.90	3.8	
Tobacco Products			
General Production	7.53	0.7	
Grading and Sorting	59.20	5.5	
Upholstering—Automobile, Coach, Furniture	26.91	2.5	
Welding			
General Illumination	13.99	1.3	
Woodworking—General	13.99	1.3	
Indoor Sports			d
Seating Area All Sports	4.31	0.4	
Badminton			
Club	6.46	0.6	c
Recreational	4.31	0.4	c
Tournament	8.61	0.8	c
Basketball			
College and Professional	15.07	1.4	c
College Intramural and High School	8.61	0.8	c
Bowling			
Approach Area	4.31	0.4	
Lanes	6.46	0.6	
Boxing or Wrestling (Ring)			
Amateur	26.91	2.5	c
Championship or Professional	53.82	5.0	c
Gymnasiums (Refer to Individual Sports Listed)			
Exhibitions, Matches	15.07	1.4	
General Exercising and Recreation	8.61	0.8	
Handball			
Club	8.61	0.8	
Recreational	6.46	0.6	
Tournament	15.07	1.4	
Hockey, Ice			
Amateur	15.07	1.4	c
College or Professional	30.14	2.8	c

Fig. 4-2. *Continued*

Task or Area	Base UPD: Watts per Square Meter	Base UPD: Watts per Square Foot	Note
Recreational	6.46	0.6	c
Skating Rinks	4.31	0.4	
Swimming			
Exhibitions	13.99	1.3	c
Recreational	7.53	0.7	c
Tennis			
Club	20.45	1.9	c
Professional	30.14	2.8	c
Recreational	15.07	1.4	c
Tennis, Table			
Club	8.61	0.8	c
Recreational	6.46	0.6	c
Tournament	15.07	1.4	c
Volleyball	6.46	0.6	c

Areas and Functions	Power Allowance: Watts per Linear Meter	Power Allowance: Watts per Linear Foot	Note
Exterior			e
Driveways			
Private (based on 2-lane width)	6.56	2.0	
Public (based on 2-lane width)	9.84	3.0	
Entrances With Canopy			
Decorative (Retail, Hotel, Theater, etc)	107.64	10.0	g
Utilitarian (Hospital, Office, Ind., etc)	43.06	4.0	g
Entrances Without Canopy	98.43	30.0	
Exits, With or Without Canopy	65.62	20.0	
Loading Area	3.23	0.3	g
Loading Doors	65.62	20.0	
Outdoor Production and Processing	4.31	0.4	g
Outdoor Storage	2.15	0.2	g
Parking Lots (Open)			
Private	20.0	20.0	h
Public	30.0	30.0	h

* Values are not for design purpose, since they do not take room size or shape into consideration.
[a] Includes 5.4 watts per square meter (0.5 watts per square foot) for special tasks.
[b] Allows additional lighting for clean-up.
[c] Gross floor area to include up to 3 meters (10 feet) surrounding the activity or playing area.
[d] Use RF = 1 for these spaces when determining power limit.
[e] These areas and activities are associated with the building under consideration.
[f] It is important to determine task area within the space by the size and number of work locations.
[g] Watts per square meter or watts per square foot.
[h] Watts per space.

Determination of a Power Limit

The lighting power "limit" of a facility is determined by totaling for lighting power "budgets" of individual rooms or spaces and that allowed for exterior areas and activities directly associated with the building facility under consideration. Figs. 4-3 and 4-4 are to be used to determine the power budgets and limit. There are three basic steps:

1. Determine the lighting power budget for each individual room or space.
2. Determine the lighting power limit of the building interior.
3. Determine the lighting power limit of the facility.

Calculation Procedure. The lighting power budget of an individual room or space is determined by the formula:

$$P_r = A_r \times P_b \times \mathrm{RF} \times \mathrm{SUF}$$

where

P_r = power budget of the room or space in watts
A_r = area of the room in square meters (square feet)
P_b = base unit power density (UPD) in watts per square meter (watts per square foot)
RF = room factor
SUF = space utilization factor

Base UPD. Base UPD (P_b) is the UPD which will provide sufficient power to satisfy the lighting requirements of the listed visual tasks for the space, assuming the power is utilized effectively in a large and unobstructed space. Base UPD's of various Task/Areas are given in Fig. 4-2.

For rooms with multiple Task/Areas, the base UPD is the weighted average of the individual task UPD's, using the following formula:

$$\text{Weighted Average UDP} = \frac{(\mathrm{UPD}_{t1}\, N_{t1}) + (\mathrm{UPD}_{t2}\, N_{t2}) + (\mathrm{UPD}_{t3}\, N_{t3}) + \ldots}{N_{t1} + N_{t2} + N_{t3} + \ldots}$$

where N_{t1} is the number of visual tasks type 1, etc.

For example, the weighted average UPD in a room with two locations at 2.5 watts per square foot and four locations at 1.5 watts per square foot is:

$$\frac{(2.5 \times 2) + (1.5 \times 4)}{2 + 4} = 1.82 \text{ watts per square foot}$$

Room Factor. Room factor (RF) is a multiplying factor (between 1.00 and 2.00) which adjusts the base UPD for spaces of various dimensions to account for the effect of room configuration on lighting efficiency. Room factors are given in Fig. 4–6. Values corresponding to the nearest room dimensions are used with interpolation as necessary.

Irregular shaped rooms, such as circles, triangles, hexagons, etc., are converted to the most appropriate rectangular shape with a similar area. The average ceiling height is used for uneven or sloped ceiling spaces.

When selecting RF, only the major room dimensions and ceiling height need be used, without concern for the type of lighting system installed or to be designed.

Space Utilization Factor. Space utilization factor (SUF) is a multiplying factor (between 0.4 and 1.0) which adjusts the allowable room power budget downward when the total area of all visual tasks within the room (A_t) is less than 50 per cent of the room area. Since the majority of properly-planned spaces generally utilize more than 50 per cent of the room area for work locations, this step may frequently be omitted (SUF = 1.0 is used). An SUF of less than 1.0 may be encountered in large general purpose spaces such as warehouses, lobbies, etc., where there are only isolated work stations. Fig. 4–3 lists the SUF to be used in task area ratios less than 1.0.

Determination of Task Areas (A_t). Task areas are determined by identifying the number of work locations (stations) at which a worker performs visual tasks. For lighting power determination purposes, each work location is considered as 4.65 square meters (50 square feet), unless the actual task area exceeds 4.65 square meters (50 square feet) in which case the actual task area is used.

Use of Fig. 4–3. Fig. 4–3 is used for listing the room identification, the visual tasks (one line per task), the number of work stations for each task, the total area of all the tasks in the room, room dimensions and the number of identical rooms or spaces. Room factors, base UPD's and calculated power budgets are also listed.

The lighting power limit for the building interior is determined by totalling the power budgets for the individual listed rooms or spaces and unlisted spaces. Listed spaces are the individual rooms listed in Fig. 4–3. Unlisted spaces are the difference between the building gross floor area* and the listed spaces. The lighting power budget for unlisted space is 2.15 watts per square meter (0.2 watts per square foot).

Use of Fig. 4–4. Fig. 4–4 is used to summarize the above determinations and the following:

The lighting power limit of the facility is determined by adding the lighting power allowances for supplementary and facade lighting, building entries, exterior parking, roadways and work areas to the lighting power limit for the building interior.

Supplementary and facade lighting—five per cent of the total building interior lighting power limit is allowed as a lighting power allowance for these purposes.

Building entries—an allowance as listed in Fig. 4–2 "Exteriors" is to be used.

Open parking and roadways—an allowance for open (uncovered) parking areas and roadways listed in Fig. 4–2 "Exteriors" is to be used.

Exterior work areas—areas directly associated with the operation of the facility, such as active storage, loading, staging and processing and production areas, with a power allowance as given in Fig. 4–2 "Exteriors" is to be used.

The UPD for the building interior is determined by dividing the lighting power limit for the building interior by the gross floor area* of the building.

The UPD for the building is determined by dividing the lighting power limit for the building interior and that associated with the building exterior, but excluding outdoor parking, roadway and other lighted areas, by the gross floor area of the building.

The UPD for the facility is determined by dividing the total lighting power limit of the facility by the gross floor area of the building.

Example of Use of the Procedure. An example of the use of the UPD procedure for determining the lighting power limit for a hypothetical composite building (see Fig. 4–5) is shown in Figs. 4–3 and 4–4.

* Gross floor area—the sum of the areas of the several floors of the building, including basements, mezzanine and intermediate-floored tiers and penthouses of headroom height, measured from the exterior faces of exterior walls or from the centerline of walls separating buildings.

Fig. 4-3. Form for Listing Data Used to Determine the Lighted Power Limit

Project: Composite Bldg. By: DLD Date: 3 Nov 80
Units of Length Used: Meters ______ or Feet ✓ Page 1 of 2

A	B	C	D	E	F	G	I	J	K	L	M	N	O
Room Name or Number	Description of Visual Tasks (One Type per Line) or General Use of the Room	Work Stations: Number	Work Stations: Area A_t = C × 4.65 (50) m² (ft²)	Room Dimensions m (ft): L × W / Ceiling Height	No. Ident. Rms.	Room Area, A_r m² (ft²) / H: Total Area of Identical Rms. A_T = FxG m² (ft²)	Room Factor RF	Space Utilization Factor: A_t/A_r (≥.5, <.5, <.4, <.3, <.2)	SUF (1.00, .85, .70, .55, .40)	Base UPD(P_b) W/m² (W/ft²): Individual Task	Base UPD(P_b) W/m² (W/ft²): Weighted Average	Power Budget of Room Watts = G · I · K · M	Power Budget for Total of Identical Rooms Watts = F · N
1A	Lobby	—	—	50 x 50 / 16	1	2500	1.20				1.0		3000
1B	Vestibule and Lobby	—	—	22 x 30 / 8	1	660	1.15				1.0		759
2A	Library 30% Stack, 70% Read	—	—	50 x 50 / 9	1	2500	1.05			0.9 / 2.2	1.8		4725
2B 2C	Office Reading	2	100	10 x 14 / 8	2	140 / 280	1.45	0.71	1.0		2.2	447	893
3A 4A	Retail Store 50% Merch. 50% Cir.	—	—	45 x 20 / 9	2	900 / 1800	1.15			3.8 / 0.9	2.4	2484	4968
3B 4B	Show Window	—	—	5 x 20 / 8	2	100 / 200	1.70				8.7	1479	2958
3C 4C	Stockroom	—	—	10 x 20 / 8	2	200 / 400	1.40				0.7	196	392
5A	Computer Equip. Room	—	—	70 x 45 / 9	1	3150	1.0				1.7		5355
6A	Mechanical Room	—	—	50 x 50 / 10	1	2500	1.0				0.6		1500
6C	Electrical Equip. Room	—	—	18 x 10 / 8	1	180	1.0				0.6		108
6D	Woodworking Shop	—	—	20 x 50 / 10	1	1000	1.20				1.3		1560
7A	Food Service Dining Area	—	—	20 x 30 / 9	1	600	1.20				1.4		1008
7B	Kitchen	—	—	20 x 10 / 9	1	200	1.50				1.7		510
8 A-D	Toilet	—	—	10 x 18 / 8	4	180 / 720	1.40				0.7	176	706
9A	General Office	25	1250	40 x 50 / 9	1	2000	1.05	0.63	1.0		2.2		4620
9B	Accounting Office	10	500	25 x 30 / 9	1	750	1.15	0.67	1.0		3.2		2760
9C	Drafting	4	200	20 x 25 / 9	1	500	1.25	0.40	0.70		4.7		2056
9D	Filing Inactive	—	—	10 x 15 / 8	1	150	1.45				0.8		174
9 E-I	Office Reading	2	100	10 x 10 / 8	5	100 / 500	1.60	1.0	1.0		2.2	352	1760
	Totals This Page					20590							39812

for a Building Interior (An Example of Its Use Is Shown)

Project: Composite Bldg. By: DLD Date: 3 Nov. 80
Units of Length Used: Meters ______ or Feet ✓ Page 2 of 2

A	B	C	D	E	F	G	I	J	K	L	M	N	O
Room Name or Number	Description of Visual Tasks (One Type per Line) or General Use of the Room	Work Stations: Number	Work Stations: Area A_t = C × 4.65 (50) m^2 (ft^2)	Room Dimensions m (ft): L × W / Ceiling Height	No. Ident. Rms.	Room Area, A_r m^2 (ft^2): H / Total Area of Identical Rms. A_T = FxG m^2 (ft^2)	Room Factor RF	Space Utilization Factor: A_t/A_r (≥.5, <.5, <.4, <.3, <.2)	Space Utilization Factor: SUF (1.00, .85, .70, .55, .40)	Base UPD(P_b) W/m^2 (W/ft^2): Individual Task	Base UPD(P_b) W/m^2 (W/ft^2): Weighted Average	Power Budget of Room Watts = G·I·K·M	Power Budget for Total of Identical Rooms Watts = F·N
10A	Classroom	—	—	40 x 40 9	1	1600	1.05				2.2		3696
11A	Basketball College + Hi Sch.	—	—	100 x 60 25	1	6000	1.0				0.8		4800
11B	Basket ball Seating	—	—	15 x 100 25	1	1500	1.0				0.4		600
11C	Storage Active Bulky	—	—	10 x 20 8	1	200	1.40				0.4		112
12A	Surgical Suite Operating Rm.	—	—	20 x 15 8	1	300	1.25				7.6		2850
12B	Scrub Up Area	—	—	15 x 10 8	1	150	1.45				2.5		544
12C	Nurses Station	—	—	10 x 13 8	1	130	1.50				1.4		273
12 D-F	Laboratory	—	—	10 x 14 8	3	140 420	1.45				3.2	650	1949
12 G-J	Patients' Rooms	—	—	10 x 18 8	4	180 720	1.0				1.4	252	1008
12 K-L	Office Reading	3	150	10 x 20 8	2	200 400	1.40	0.75	1.0		2.2	616	1232
12M	Hospital Lobby	—	—	22 x 20 8	1	440	1.20				2.0		1056
12N	Corridor	—	—	42 x 10 8	1	420	1.0				0.6		252
13A	Tennis Club	—	—	60 x 120 35	1	7200	1.0				1.9		13680
14A	Hotel Lobby	—	—	24 x 36 8	1	864	1.10				1.1		1045
14 B-C	Guest Room	—	—	12 x 20 8	2	240 480	1.0				1.4	336	672
14D	Corridor	—	—	50 x 8 8	1	400	1.0				0.6		240
14E	Corridor	—	—	80 x 8 8	1	640	1.0				0.6		384
15A	Garage-Service 35% Traffic, 65% Repair			32 x 60 20	1	1920	1.45			0.6 2.5	1.8		5011
Totals This Page						23784							39404

Fig. 4-4. Form for Listing Data Used to Determine the Lighting power Limit for a Facility (An Example of Its Use Is Shown)

Project: Composite Bldg. By: DLD Date: 3 NOV 80
Units of Length Used: Meters ______ or Feet ✓ Page 1 of 1

Part I—Summary of Building Interiors

Reference	Item	Power (watts)	Area m^2 (ft^2)	Guide	Notes
Q	Gross Floor Area of Building		47600	Exterior Building Dimensions (Less Covered Exteriors)	
R	Total Area Listed Interior Spaces		44374	Total of All Columns H (Fig. 4-3)	
S	Unlisted (Net) Building Area		3226	Q − R	
T	Power Limit—Unlisted Space	645		2.15 × S_{m2} (0.2 × S_{ft2})	
U	Power Limit—Listed Space	79216		Sum of Column O for All Pages	
V	Power Limit—Building Interior	79861		T + U	
W	Basic Building Interior UPD	1.68		V ÷ Q watts/m^2 (ft^2)	

Part 2A—Summary of Exteriors—Building

Reference	Item	Power (watts)	Area or Length m^2(ft^2) or m(ft)	Guide		Notes
X	Supplementary and Facade Lighting	3993		5% of Item V		
Y	Building Entries—with Canopies		540	Total of Exterior Areas Under Canopies		
Z	Power Limit—Entries with Canopies	2160		107.6 W/m^2 Decor. (10.0 W/ft^2)	43.0 W/m^2 Util. (4.0 W/ft^2)	
AA	Building Entries—without Canopies		22	Total Width m (ft)		
BB	Power Limit—Entries without Canopies	660		98.4 watts/linear m × AA (30.0 watts/linear ft × AA)		
CC	Building Exits & Loading Doors		24	Total Width m (ft.)		
DD	Power Limit—Exits & Loading	480		65.6 watts/linear m × CC (20.0 watts/linear ft × CC)		
EE	Total Building Exterior	7293		X + Z + BB + DD		

Part 2B—Summary of Exteriors—Roads—Grounds

Reference	Item	Power (watts)	Area or Length	Guide		Notes
FF	Number of Exterior Parking Stalls		100	Quantity, Not Area		
GG	Power Limit Exterior Parking	2000		Private: 20 W × FF	Public: 30 W × FF	
HH	Length of 2 Lane, Feeder Roadways		800	Total Length m (ft)		
II	Power Limit—Feeder Roadways	1600		Private: 6.56 W/m × HH (2.0 W/ft)	Public: 9.84 W/m × HH (3.0 W/ft)	
JJ	Total Area Exterior Processing		0	m^2 (ft^2)		
KK	Power Limit Exterior Processing	0		4.3 watts/m^2 × JJ (0.4 watts/ft^2 × JJ)		
LL	Total Area Exterior Storage Yds.		5000	m^2 (ft^2)		
MM	Power Limit Exterior Storage Yds.	1000		2.15 watts/m^2 × LL (0.2 watts/ft^2 × LL)		
NN	Total Roads & Grounds	4600		GG + II + KK + MM		

Part 3—Summary of Total Project

Reference	Item	Power (watts)	UPD	Guide	Notes
OO	Total Building Power Limit	87154		V + EE (watts)	
PP	Total Building UPD		1.83	OO ÷ Q watts/m^2 (watts/ft^2)	
QQ	Total Project Power Limit	91754		V + EE + NN (watts)	

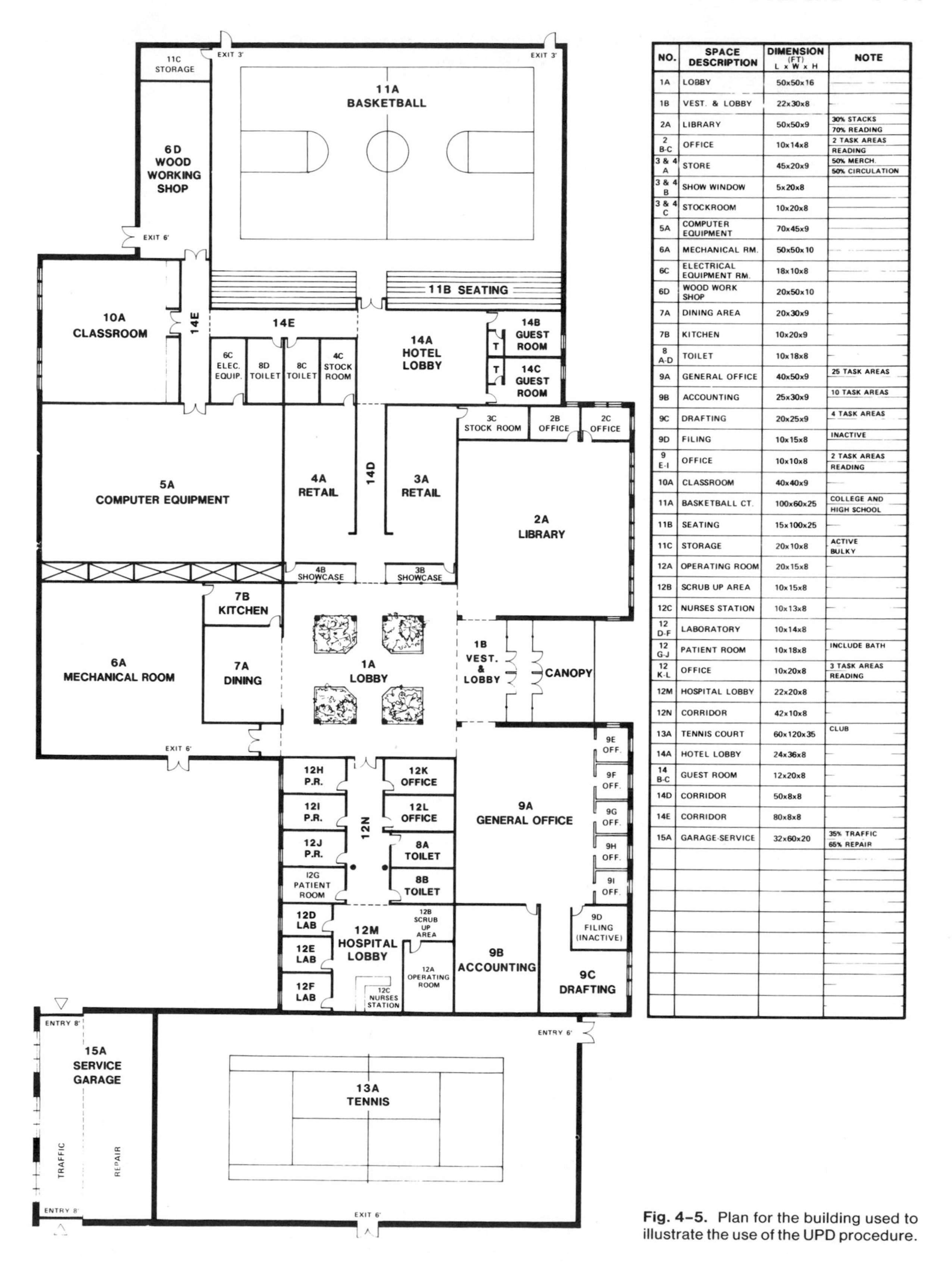

NO.	SPACE DESCRIPTION	DIMENSION (FT) L x W x H	NOTE
1A	LOBBY	50x50x16	
1B	VEST. & LOBBY	22x30x8	
2A	LIBRARY	50x50x9	30% STACKS 70% READING
2 B-C	OFFICE	10x14x8	2 TASK AREAS READING
3 & 4 A	STORE	45x20x9	50% MERCH. 50% CIRCULATION
3 & 4 B	SHOW WINDOW	5x20x8	
3 & 4 C	STOCKROOM	10x20x8	
5A	COMPUTER EQUIPMENT	70x45x9	
6A	MECHANICAL RM.	50x50x10	
6C	ELECTRICAL EQUIPMENT RM.	18x10x8	
6D	WOOD WORK SHOP	20x50x10	
7A	DINING AREA	20x30x9	
7B	KITCHEN	10x20x9	
8 A-D	TOILET	10x18x8	
9A	GENERAL OFFICE	40x50x9	25 TASK AREAS
9B	ACCOUNTING	25x30x9	10 TASK AREAS
9C	DRAFTING	20x25x9	4 TASK AREAS
9D	FILING	10x15x8	INACTIVE
9 E-I	OFFICE	10x10x8	2 TASK AREAS READING
10A	CLASSROOM	40x40x9	
11A	BASKETBALL CT.	100x60x25	COLLEGE AND HIGH SCHOOL
11B	SEATING	15x100x25	
11C	STORAGE	20x10x8	ACTIVE BULKY
12A	OPERATING ROOM	20x15x8	
12B	SCRUB UP AREA	10x15x8	
12C	NURSES STATION	10x13x8	
12 D-F	LABORATORY	10x14x8	
12 G-J	PATIENT ROOM	10x18x8	INCLUDE BATH
12 K-L	OFFICE	10x20x8	3 TASK AREAS READING
12M	HOSPITAL LOBBY	22x20x8	
12N	CORRIDOR	42x10x8	
13A	TENNIS COURT	60x120x35	CLUB
14A	HOTEL LOBBY	24x36x8	
14 B-C	GUEST ROOM	12x20x8	
14D	CORRIDOR	50x8x8	
14E	CORRIDOR	80x8x8	
15A	GARAGE-SERVICE	32x60x20	35% TRAFFIC 65% REPAIR

Fig. 4-5. Plan for the building used to illustrate the use of the UPD procedure.

Fig. 4-6. Room Factors (RF) for Lighting Power Limit Calculations†

Dimensions* (feet)		Ceiling Height (feet)										Dimensions (meters)	
W	L	8	8.5	9	10	11	12	14	16	18	20+	W	L
6	6											1.8	1.8
6	9											1.8	2.7
6	12	1.85										1.8	3.7
6	15	1.75	1.90									1.8	4.6
6	18	1.70	1.80	1.95								1.8	5.5
6	24	1.65	1.75	1.85								1.8	7.3
6	30	1.60	1.70	1.80								1.8	9.1
6	36	1.60	1.65	1.75	1.95							1.8	11.0
6	60	1.55	1.60	1.70	1.85							1.8	18.3
6	60+	1.45	1.50	1.60	1.75	1.90						1.8	18.3+
8	8	1.85										2.4	2.4
8	12	1.65	1.75	1.85								2.4	3.7
8	16	1.55	1.65	1.70	1.90							2.4	4.9
8	20	1.50	1.55	1.65	1.80	1.95						2.4	6.1
8	24	1.45	1.50	1.60	1.75	1.90						2.4	7.3
8	32	1.40	1.45	1.55	1.65	1.80	1.95					2.4	9.8
8	40	1.40	1.45	1.50	1.65	1.75	1.90					2.4	12.2
8	48	1.35	1.45	1.50	1.60	1.75	1.85					2.4	14.6
8	80	1.35	1.40	1.45	1.55	1.65	1.80					2.4	24.4
8	80+	1.30	1.35	1.40	1.50	1.60	1.70	1.90				2.4	24.4+
10	10	1.60	1.70	1.80								3.0	3.0
10	15	1.45	1.50	1.60	1.75	1.90						3.0	4.6
10	20	1.40	1.45	1.50	1.65	1.75	1.90					3.0	6.1
10	25	1.35	1.40	1.45	1.55	1.70	1.80					3.0	7.6
10	30	1.30	1.35	1.40	1.50	1.65	1.75					3.0	9.1
10	40	1.30	1.35	1.40	1.45	1.60	1.70	1.90				3.0	12.2
10	50	1.25	1.30	1.35	1.45	1.55	1.65	1.85				3.0	15.2
10	60	1.25	1.30	1.35	1.45	1.50	1.60	1.80				3.0	18.3
10	100	1.25	1.25	1.30	1.40	1.45	1.55	1.75	1.95			3.0	30.5
10	100+	1.20	1.25	1.25	1.35	1.40	1.50	1.65	1.85			3.0	30.5+
12	12	1.45	1.50	1.60	1.75	1.90						3.7	3.7
12	18	1.35	1.40	1.45	1.55	1.70	1.80					3.7	5.5
12	24	1.30	1.35	1.40	1.45	1.60	1.70	1.90				3.7	7.3
12	30	1.25	1.30	1.35	1.45	1.50	1.60	1.80				3.7	9.1
12	36	1.25	1.30	1.30	1.40	1.50	1.55	1.75				3.7	11.0
12	48	1.20	1.25	1.30	1.35	1.45	1.50	1.70	1.90			3.7	14.6
12	60	1.20	1.25	1.25	1.35	1.40	1.50	1.65	1.85			3.7	18.3
12	72	1.20	1.20	1.25	1.30	1.40	1.45	1.60	1.80			3.7	21.9
12	120	1.15	1.20	1.25	1.30	1.35	1.40	1.55	1.75	1.90		3.7	36.6
12	120+	1.15	1.15	1.20	1.25	1.30	1.35	1.50	1.65	1.80	1.95	3.7	36.6+
16	16	1.30	1.35	1.40	1.45	1.60	1.70	1.90				4.9	4.9
16	24	1.20	1.25	1.30	1.35	1.45	1.50	1.70	1.90			4.9	7.3
16	32	1.20	1.20	1.25	1.30	1.35	1.45	1.60	1.75	1.95		4.9	9.8
16	40	1.15	1.20	1.20	1.25	1.35	1.40	1.55	1.70	1.85		4.9	12.2
16	48	1.15	1.15	1.20	1.25	1.30	1.35	1.50	1.65	1.80	1.95	4.9	14.6
16	64	1.15	1.15	1.20	1.25	1.30	1.35	1.45	1.55	1.70	1.85	4.9	19.5
16	80	1.10	1.15	1.15	1.20	1.25	1.30	1.40	1.55	1.65	1.80	4.9	24.4
16	96	1.10	1.15	1.15	1.20	1.25	1.30	1.40	1.50	1.65	1.75	4.9	29.3
16	160	1.10	1.10	1.15	1.20	1.20	1.25	1.35	1.45	1.60	1.70	4.9	48.8
16	160+	1.10	1.10	1.10	1.15	1.20	1.25	1.30	1.40	1.50	1.60	4.9	48.8+
20	20	1.20	1.25	1.25	1.35	1.40	1.50	1.65	1.85	2.00	2.00	6.1	6.1
20	30	1.15	1.15	1.20	1.25	1.30	1.35	1.50	1.65	1.80	1.95	6.1	9.1
20	40	1.10	1.15	1.15	1.20	1.25	1.30	1.40	1.55	1.65	1.80	6.1	12.2
20	50	1.10	1.10	1.15	1.20	1.25	1.30	1.40	1.50	1.60	1.70	6.1	15.2
20	60	1.10	1.10	1.15	1.15	1.20	1.25	1.35	1.45	1.55	1.65	6.1	18.3
20	80	1.10	1.10	1.10	1.15	1.20	1.25	1.30	1.40	1.50	1.60	6.1	24.4
20	100	1.05	1.10	1.10	1.15	1.20	1.20	1.30	1.40	1.45	1.55	6.1	30.5
20	120	1.05	1.10	1.10	1.15	1.15	1.20	1.30	1.35	1.45	1.55	6.1	36.6
20	200	1.05	1.05	1.10	1.10	1.15	1.20	1.25	1.35	1.40	1.50	6.1	61.0
20	200+	1.05	1.05	1.05	1.10	1.15	1.15	1.20	1.30	1.35	1.45	6.1	61.0+
		2.4	2.6	2.7	3.0	3.4	3.7	4.3	4.9	5.5	6.1+	W	L
		Ceiling Height (meters)										Dimensions (meters)	

Fig. 4-6. *Continued*†

Dimensions* (feet)		Ceiling Height (feet)											
W	L	8	8.5	9	10	11	12	14	16	18	20+		
24	24	1.15	1.15	1.20	1.25	1.30	1.35	1.50	1.65	1.80	1.95	7.3	7.3
24	36	1.10	1.10	1.15	1.20	1.25	1.25	1.35	1.45	1.60	1.70	7.3	11.0
24	48	1.10	1.10	1.10	1.15	1.20	1.25	1.30	1.40	1.50	1.60	7.3	14.6
24	60	1.05	1.10	1.10	1.15	1.15	1.20	1.30	1.35	1.45	1.55	7.3	18.3
24	72	1.05	1.05	1.10	1.10	1.15	1.20	1.25	1.35	1.40	1.50	7.3	21.9
24	96	1.05	1.05	1.10	1.10	1.15	1.15	1.25	1.30	1.40	1.45	7.3	29.3
24	120	1.05	1.05	1.05	1.10	1.15	1.15	1.20	1.30	1.35	1.45	7.3	36.6
24	144	1.05	1.05	1.05	1.10	1.10	1.15	1.20	1.25	1.35	1.40	7.3	43.9
24	240	1.05	1.05	1.05	1.10	1.10	1.15	1.20	1.25	1.30	1.35	7.3	73.2
24	240+		1.05	1.05	1.05	1.10	1.10	1.15	1.20	1.25	1.30	7.3	73.2+
30	30	1.10	1.10	1.15	1.15	1.20	1.25	1.35	1.45	1.55	1.65	9.1	9.1
30	45	1.05	1.05	1.10	1.10	1.15	1.20	1.25	1.35	1.40	1.50	9.1	13.7
30	60	1.05	1.05	1.05	1.10	1.15	1.15	1.20	1.30	1.35	1.45	9.1	18.3
30	75	1.05	1.05	1.05	1.10	1.10	1.15	1.20	1.25	1.30	1.40	9.1	22.9
30	90		1.05	1.05	1.05	1.10	1.10	1.20	1.25	1.30	1.35	9.1	27.4
30	120		1.05	1.05	1.05	1.10	1.10	1.15	1.20	1.25	1.30	9.1	36.6
30	150			1.05	1.05	1.10	1.10	1.15	1.20	1.25	1.30	9.1	45.7
30	180			1.05	1.05	1.05	1.10	1.15	1.20	1.25	1.30	9.1	54.9
30	300				1.05	1.05	1.10	1.10	1.15	1.20	1.25	9.1	91.4
30	300+				1.05	1.05	1.05	1.10	1.15	1.20	1.20	9.1	91.4+
40	40	1.05	1.05	1.05	1.10	1.15	1.15	1.20	1.30	1.35	1.45	12.2	12.2
40	60		1.05	1.05	1.05	1.10	1.10	1.15	1.20	1.25	1.30	12.2	18.3
40	80				1.05	1.05	1.10	1.15	1.20	1.20	1.25	12.2	24.4
40	100				1.05	1.05	1.05	1.10	1.15	1.20	1.25	12.2	30.5
40	120				1.05	1.05	1.05	1.10	1.15	1.20	1.20	12.2	36.6
40	160					1.05	1.05	1.10	1.10	1.15	1.20	12.2	48.8
40	200					1.05	1.05	1.10	1.10	1.15	1.20	12.2	61.0
40	240					1.05	1.05	1.05	1.10	1.15	1.20	12.2	73.2
40	400						1.05	1.05	1.10	1.15	1.15	12.2	122.0
40	400+							1.05	1.10	1.10	1.15	12.2	122+
60	60				1.05	1.05	1.05	1.10	1.15	1.20	1.20	18.3	18.3
60	90						1.05	1.05	1.10	1.15	1.15	18.3	27.4
60	120							1.05	1.10	1.10	1.15	18.3	36.6
60	150							1.05	1.05	1.10	1.10	18.3	45.7
60	180							1.05	1.05	1.10	1.10	18.3	54.9
60	240								1.05	1.05	1.10	18.3	73.2
60	300								1.05	1.05	1.10	18.3	91.4
60	360								1.05	1.05	1.10	18.3	110.0
60	600								1.05	1.05	1.05	18.3	183.0
60+	600+									1.05	1.05	18.3	183+
		2.4	2.6	2.7	3.0	3.4	3.7	4.3	4.9	5.5	6.1+	W	L
		Ceiling Height (meters)										Dimensions (feet)	

* W = width, L = length.
† Where no value is listed on page 4-12, RF = 2.00. Where none appears on page 4-13, RF = 1.00.

ENERGY MANAGEMENT

The procedure described in the previous material establishes an upper limit of *power* for a building. This section deals with criteria for reducing *energy* consumption while maintaining the lighting quality and good practices described throughout this Handbook.

Energy Management Procedures

An existing building is evaluated from the standpoint of power and energy by the following methods:[3]

1. Building survey — an examination of the building to determine the connected power for lighting.
2. Power budget for rooms — determination of a power budget for each individual space in the building.

3. Power limit — determination of the total power used for the facility (interior and exterior).
4. Power limit analysis — comparison of the connected power in the building from 1 with the power computed in 2 and 3, and a modification of the existing lighting system to comply if the connected power exceeds the calculated limit.
5. Management program — the development of an energy management program to reduce energy when the connected power is equal to or less than the calculated limit, while maintaining good lighting practice.

Building Survey. An examination of the building to determine the connected power for lighting involves a survey of all spaces. The wattage of each connected luminaire including lamp watts, ballast watts and losses introduced by dimming devices is verified and totaled. The power for portable and supplementary lighting devices also should be included in the total. Similarly, any special use equipment, such as that needed for entertainment facilities and audio-visual presentations is included in the connected power total.

Power Budget and Limit Determination. By following the procedure previously explained (UPD procedure) the budgets for individual spaces and the limit for the building is calculated. Any building which averages 10.8 watts per square meter (1 watt per square foot) or less of connected power is in compliance with the power limit for lighting and is exempt from the procedure.

Power Limit Analysis. If the connected power for lighting in the building exceeds the calculated UPD limit, the lighting does not comply. Those spaces which are most inefficient in lighting power utilization are selected for modifications to reduce connected power. As many spaces as necessary are modified until the connected power is below the calculated limit, *i.e.*, until the building is in compliance.

Lighting Considerations and Ideas for Energy Management

Fig. 4-7 lists eight considerations for energy management, which are not only applicable to existing buildings, but should be reviewed for their impact on the design of a new building.[4]

In addition, a checklist of 52 energy management ideas is included in Fig. 4-8[3] as a guide for determining as many items as possible in developing a management program. Not all suggestions will necessarily apply to each building type and the user must review the entire list for an appropriate selection. Refer also to the energy discussions in other sections of this Handbook relating to specific applications.

Lighting Energy Management Form

The development of a management program for a building involves a comparison of the projected lighting use patterns (new buildings) or actual lighting use patterns (existing buildings) against the calculated power limit and the checklist of energy saving ideas.

This comparison can easily be facilitated by using the Lighting Energy Management Form shown in Fig. 4-9 and by referring to the checklist of energy saving lighting ideas in Fig. 4-8 and reference 5. The following are steps to follow in using Fig. 4-9:

1. List all spaces being considered, in the column titled *Space Description*. It is suggested that the same order of spaces, as used on the form in Fig. 4-3 be followed.
2. Record in the space provided under the *Budget* column, the lighting power budget calculated from UPD for each respective space.
3. This step involves the examination of either the new lighting design (new buildings) or the existing lighting system (existing buildings) and the recording of pertinent data for each space, in the space provided on the form, as follows:
   a. *Description — Existing Installation.* The line provided may be used to record brief descriptions of the number and type of luminaires, the number of lamps per luminaire, surface reflectance values, existing or predicted illuminance values, or other data pertaining to the designed (new building) or existing space under consideration.
   b. *Power-Load (W) — Connected.* The actual lighting power load designed or existing in each respective space, including ballast losses, should be recorded in this space.
   c. *Operating Hours (existing).* The number of hours per day that the designed lighting system in that space is expected to operate (new buildings), or the actual number of hours per day which the lighting system is being operated (existing building) are to be recorded in their respective spaces.
   d. *Hours per Week, Weeks per Year and Hours per Year (existing).* Again, the predicted (new buildings) or actual (existing

Fig 4-7. Lighting Considerations and Ideas for Energy Management

What to Consider	What Can Be Done in New Construction	What Can Be Done in Existing Spaces	Resource and Comments
1. The lighting needs—for productivity, safety and esthetics. a. Seeing tasks b. Seeing task locations c. Purposes of nonseeing task areas d. Illumination recommendations e. Uniform lighting f. Nonuniform lighting	Seeing tasks and their locations should be identified so that recommended illuminances can be provided for those tasks with less in surrounding non-critical seeing areas. Where there are no tasks, there is no visual need for task levels. Then safety and esthetics are of prime consideration. When tasks and locations can be identified, it may be possible to use a nonuniform lighting system, such as a nonuniform pattern of luminaires arranged to light work stations or built-in or supplementary lighting at work stations, coordinated with a general lighting system. When tasks and their locations cannot be identified, an illuminance can be selected for the expected tasks and a uniform pattern of luminaires installed with controls for lowering the level at specific points when no tasks are present.	Seeing tasks and their locations should be identified so that recommended illuminances can be provided for those tasks with less in surrounding non-critical seeing areas. Where there are no tasks, there is no visual need for task levels. Then safety and esthetics are of prime consideration. Since tasks and their locations can be identified, a lighting survey would show where illuminances are in excess of recommended maintained values. Lighting then can be adjusted to meet the recommendations. Careful consideration should be given at task locations that any change in the lighting system will not produce veiling reflections in the tasks.	Other sections of this handbook should be consulted for specific lighting needs.
2. Luminaires a. Effectiveness for task lighting b. Effectiveness for nontask lighting c. Efficiency d. Heat transfer capability e. Cleaning capabilities	In selecting a luminaire for task lighting, consideraton should be given to its effectiveness in providing high task contrast (minimum veiling reflections) and sufficiently high visual comfort (VCP). Luminaire light distribution and appearance are also important, particularly for esthetics, but consideration also should be given to efficiency. Luminaires with heat transfer capabilities should be considered so that the lighting heat can be utilized or removed and coordinated with the building thermal design and total building energy use. See Section 2. Luminaires that can be cleaned easily and those with low dirt accumulation will reduce maintenance needs and cost.	Review luminaire effectiveness for task lighting and efficiency, and if ineffective or inefficient consider luminaire or component replacement. Check to see if all components are in good working condition. Transmitting or diffusing media should be examined and badly discolored and depreciated media replaced to improve efficiency (without producing excessive brightness and unwanted visual discomfort).	Luminaire manufacturers' data are useful guides in determining luminaire effectiveness in terms of comfort (VCP), efficiency and ability to provide high contrast rendering. Energy use will be affected by luminaire light source and distribution characteristics as shown in one study (see Fig. 2-16). That study shows that for a medium size room, indirect, incandescent luminaires can consume over 5½ times the energy compared with a direct, fluorescent system, for the same hours of use and average illuminance (but not the same lighting quality).
3. Light sources (lamps and ballasts) a. Efficacy (lumens per watt) b. Color (chromaticity) c. Color rendering d. Lumen maintenance	Highest efficacy lamps and lamp ballast combinations should be used that are compatible with the desired light source color, color rendering capabilities, source size, life and light output depreciation. Compare sources being considered on basis of life cycle cost and energy use through life.	Where inefficient sources are used, consider relighting with more efficient sources, compatible with desired light source color and color rendering capabilities, based on life cycle costing. Consider reduced wattage fluorescent lamps in existing luminaires where 10 to 20 per cent reduction in illuminance can	Light source efficacy (lumens per watt) varies with lamp type and within types. Section 8 of the 1981 Reference Volume and Lamp manufacturers' catalogs will provide lumen output, lumens per watt, life and lumen maintenance data as well as cost for comparison purposes.

Fig. 4–7. *Continued*

	Consider multi-level ballasts for flexibility in achieving nonuniform lighting as needed in hour-to-hour operations throughout each day or with changing work layouts over the years. Consider reduced current ballasts where module size and/or luminaire spacing permit needed illumination to be achieved.	be tolerated. (With improved maintenance procedures—periodic cleanings, group relamping—as much light can be obtained as with standard lamps.) Consider multi-level ballasts for flexibility in varying lighting during occupied hours and cleaning periods, and reduced current ballasts where a reduction in illuminance can be tolerated.	
4. Daylighting a. Availability of daylight b. Fenestration (windows, sky-lights) c. Controls	Evaluate the daylighting potential—the levels and hours of availability—keeping in mind that glare from fenestration should be controlled to the same degree as from luminaires and that the heat gained or lost through fenestration needs to be coordinated with the building thermal system. Coordinate the electric lighting design with the daylighting so that glare, heat and illumination are controlled. Consider the use of "high performance" heat reflecting insulating glass in windows to minimize heat gain in summer and heat loss in winter, while permitting a view of the exterior.	If daylighting can be used to replace some of the electric lighting during substantial periods of the day, lighting in those areas should be dimmed or switched off. If control is not provided, consider adding controls based on life-cycle-costing. Evaluate the effectiveness of the existing fenestration shading controls (interior and exterior) for possible replacements or additions.	Sections 7 and 9 of the 1981 Reference Volume provide useful information on availability of daylight, daylight control systems, and design and evaluation methods. Recent data from manufacturers of fenestration materials and controls should be consulted.
5. Room surfaces	Work with the interior designer toward the specification of room surface and equipment reflectances at the higher end of recommended reflectances, not forgetting the importance of proper color schemes for esthetics. (Reflectances higher than those recommended may produce excessive luminance ratios and glare.)	Where the reflectances of room surfaces are lower than, or at the lower end of the recommended reflectance range, consider repainting using matte paints with reflectances toward the upper end of the range. When equipment is replaced, select light color finishes. (Reflectances higher than those recommended may produce excessive luminance ratios and glare.)	The use of higher reflectance finishes saves energy. Below is a chart showing the decreased energy needs for equal general illumination in a medium size room with improved reflectances. Reflectances are: Ceiling Walls Floor A 50 30 10—Not recommended B 80 40 20 C 90 60 40 The grayed area shows per cent energy savings. 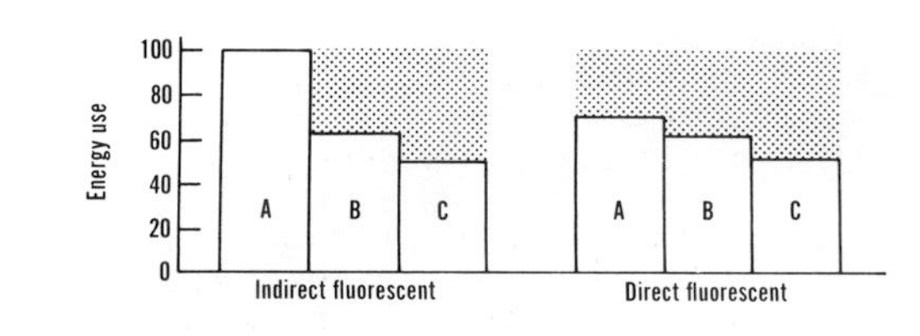

6. Maintenance procedures	Carefully consider a planned lighting maintenance program early in the design stage to allow for desired maintained levels using less equipment and less installed lighting wattage. Also consider the daylighting (fenestration and controls) maintenance program when planning on the daylight contribution to the desired illuminance levels. The owner must be committed to the maintenance program used in the lighting system design. If not, the lighting will be less than planned and will be an energy waster.	Reevaluate the present lighting maintenance program and revise it as necessary to provide desired maintained illuminances. This may allow some reductions in lighting energy.	The diagram below shows the effect of maintenance procedures on energy use. In system A, luminaires are cleaned and relamped every three years. In system B, luminaires are cleaned every year and one-third of the lamps are replaced every year. For equal maintained illuminance over a 12-year period, the more frequent cleanings and relampings in system B saves about 15 per cent energy use due to the need for less equipment initially. Maintain Illum. level; A; B; Energy saved; 0 3 6 9 0 3 6 9 12; Years
7. Operating procedures a. During working hours b. During building cleaning periods.	In spaces with tasks, consider switching arrangements so that only general surround lighting can be used when tasks are not performed; however, it is important that an adequate level of illumination be maintained for building cleaning. Allow for switching and switching ease to encourage turning lights off when not needed. Prepare a suggested lighting switching scheme, based on the design, to aid operations.	Analyze the lighting use during working and building cleaning periods. Institute an educational program to have workers turn off lights when they are not needed. Also have cleaners' schedules adjusted to minimize the lighting use, such as by cleaning fewer spaces at the same time, turning off lights in unoccupied areas. Where large floor or room areas are controlled by a single switch, consider adding more switching flexibility to turn off lights in areas when and where not needed.	
8. Space utilization	Where seeing task locations have not been specified, work with the office space planner to show locations where higher levels will occur to best take advantage of the lighting design and where daylighting can be used effectively. The use of open plan offices, when practical, increases room size and improves utilization of light. If new space is sparsely populated, consider locating employees with related work close together to efficiently provide the illumination needed for their tasks, and the remaining open area can be lighted to the lower level values for surrounding space and circulation areas.	Where it is found that workers are sparsely distributed consideration might be given to moving workers closer together and closing off unused space (with minimum heating, cooling and lighting). Also, an analysis of the existing lighting can show where tasks may be located to take advantage of the existing illumination provided and where ESI values are highest.	

Fig. 4-8. Checklist of Energy Saving Lighting Ideas

A. Lighting Needs. . .Tasks, Task and Luminaire Location, Illumination Requirements and Utilization of Space.

1. *Identify seeing tasks and locations* so recommended illuminances can be provided for tasks with less in surrounding areas.
2. *Identify seeing tasks where maintained illuminance is greater than recommended* and modify to meet the recommendations.
3. *Consider replacing seeing tasks with those of higher contrast* which call for lower illuminance requirements.
4. *Where there are no visual tasks, task illumination is not needed.* Review lighting requirements then, to satisfy safety and esthetics.
5. *Group tasks having the same illuminance requirements or widely separated* work stations, and close off unused space (with minimum heating, cooling and lighting).
6. *When practical, have persons working after-hours work in close proximity* to lessen all energy requirements.
7. *Coordinate layout of luminaires and tasks for high contrast rendition* rather than uniform space geometry. Analyze existing lighting to show where tasks may be relocated to provide better contrast rendition. Use caution when relocating tasks to minimize direct and reflected glare and veiling reflections in the tasks.
8. *Relocate lighting from over tops of stacked materials.*
9. *Consider lowering the mounting height of luminaires* if it will improve illumination, or reduce connected lighting power required to maintain adequate task lighting.
10. *Consider illuminating tasks with luminaires properly located in or on furniture* with less light in aisles.
11. *Consider wall lighting luminaires, and lighting for plants, paintings and murals,* to maintain proper luminance ratios in place of general overhead lighting.
12. *Consider high efficacy light sources* for required floodlighting and display lighting.
13. *Consider the use of open-plan spaces versus partitioned spaces.* Where partitions are tall or stacked equipment can be eliminated, the general illumination may increase, and the lighting system connected power may be reduced.
14. *Consider the use of light colors for walls, floors, ceilings and furniture* to increase utilization of light, and reduce connected lighting power required to achieve needed light. Avoid glossy finishes on room and work surfaces.

B. Lighting Equipment. . .Lamps and Luminaires.

15. *Establish washing cycles for lamps and luminaires.*
16. *Select a group lamp replacement time interval* for all light sources.
17. *Install lamps with higher efficacy (lumens per input watts)* compatible with desired light source color and color-rendering capabilities.
18. *In installations where low wattage incandescent lamps are used in luminaires, investigate the possibility of using fewer higher wattage (more efficient) lamps* to get the needed light. Lamp wattages must not exceed luminaire rating.
19. *Evaluate the use of R, PAR or ER lamps to get the needed light with lower watts* depending on luminaire types or application.
20. *Evaluate use of reduced wattage lamps when the illuminance is above task requirements,* and whenever luminaire location must be maintained.

B. Lighting Equipment *continued*

21. *Consider reduced wattage fluorescent lamps in existing luminaires* along with improved maintenance procedures. CAUTION: Not recommended where ambient space temperature may fall below 16 °C (60 °F).
22. *Check luminaire effectiveness for task lighting and efficiency,* and if ineffective or inefficient, consider luminaire and component replacement or relocation for greater effectiveness.
23. *Consider reduced-current ballasts* where a reduction in illuminance can be tolerated.
24. *Consider the use of ballasts which can accomodate high pressure sodium or metal halide lamps* interchangeably with other lamps.
25. *Consider multi-level ballasts* where a reduction in illuminance can be tolerated.
26. *Consider substituting interchangeable-type metal halide lamps on compatible ballasts* in existing mercury lighting systems. Two options: Upgrade sub-standard lighting in a mercury system with no increase in lighting power, or reduce lighting power by removing luminaires that may increase lighting levels above task lighting requirements.
27. *Consider substituting interchangeable high pressure sodium lamps on retrofit ballasts in existing mercury lighting systems.* Results: reduced connected lighting power with lamp substitution and more light.
28. *Consider using heat removal luminaires* whenever possible to improve lamp performance and reduce heat gain to space.
29. *Select luminaires which do not collect dirt rapidly* and which can be easily cleaned.

C. Daylighting.

30. *If daylighting can be used to replace some of the electric lighting* near the windows during substantial periods of the day, lighting in those areas should be dimmed or switched off.
31. *Maximize the effectiveness of existing fenestration-shading controls* (interior and exterior) or replace with proper devices or media.
32. *Use daylighting effectively by locating work stations requiring the most illumination nearest the windows.*
33. *Daylighting, whenever it can be effectively used, should be considered in areas when a net energy conservation gain is possible,* considering total energy for lighting, heating, and cooling.

D. Controls and Distribution Systems.

34. *Install switching for selective control of illumination.*
35. *Evaluate the use of low-voltage (24 volts or lower) switching systems* to obtain maximum switching capability.
36. *Install switching or dimmer controls to provide flexibility* when spaces are used for multiple purposes and require different amounts of illumination for various activities.
37. *Consider a sold state dimmer system* as a functional means for variable lighting requirements of high intensity discharge lamps.
38. *Consider photocells and/or timeclocks* for turning exterior lights on and off.
39. *Install selective switching on luminaires according to grouping of working tasks* at different working hours, and when not needed.
40. *Consider plug-in electrical wiring* to allow for flexibility in moving/removing/adding luminaires to suit changing furniture layouts.

Fig. 4-8. *Continued*

D. Controls and Distribution System *continued*

41. *Consider coding on light control panels and switches* according to a predetermined schedule of when lights should be turned off and on.

E. Lighting Maintenance Procedures.

42. *Evaluate the present lighting maintenance program,* and revise it as necessary to provide the most efficient use of the lighting system.
43. *Clean luminaires and replace lamps on a regular maintenance schedule.*
44. *Check to see if all components are in good working condition.* Transmitting or diffusing media should be examined, and badly discolored or deteriorated media replaced, to improve efficiency (without producing excessive brightness and unwanted visual discomfort).
45. *Replace outdated or damaged luminaires with modern luminaires* which have good cleaning capabilities, and which use lamps with higher efficacies and good lumen maintenance characteristics.
46. *Trim trees and bushes that may be obstructing luminaire distribution and creating unwanted shadows.*

F. Operating Schedules.

47. *Analyze lighting used during working and building cleaning periods* and institute an education program to have workers turn off lights when they are not needed. Inform and encourage personnel to turn off light sources such as: (a) Incandescent—promptly when space is not in use; (b) Fluorescent—if the space will not be used for five minutes or longer; (c) High Intensity Discharge Lamps—(Mercury, Metal Halide, High Pressure Sodium)—if space will not be used for 30 minutes or longer.
48. *Light building for occupied periods only,* and when required for security purposes.
49. *Restrict parking to specific lots* so lighting can be reduced to minimum security requirements in unused parking areas.
50. *Try to schedule routine building cleaning during occupied hours.*
51. *Reduce illuminance levels during building cleaning periods.*
52. *Adjust cleaning schedules to minimize the lighting use,* such as by concentrating cleaning activities in fewer spaces at the same time, and by turning off lights in unoccupied areas.

G. Post Instruction Covering Lighting Operation and Maintenance Procedures in All Management and General Work Areas.

buildings) hours and weeks of use of the lighting system, in the respective space, are to be recorded.

e. *Annual Energy Consumption (kWh) — Existing.* This value is the product of the *Connected* load and the *Hours per Year.*

4. This step involves:
   a. The direct comparison of the "designed" (new buildings) or "actual" (existing buildings) power load, as recorded under *Connected* to the budgeted power load as recorded under *Budget.*
   b. Areas or spaces within the building which demonstrate a designed or connected lighting load in excess of the budgeted lighting power are potential areas for lighting power savings through lighting system modification, retrofit or redesign.
   c. At this point in the process, references must be made to Fig. 4-8 and decisions made as to which changes to the lighting system will result in cost effective power reductions, without compromising good practice.
   d. The modifications decided upon for each space may be described in the space *Description—Proposed Modifications* opposite each respective space. The estimated capital cost of these modifications can be entered in the respective space under *Cost of Modification* and the modified power load recorded in *Power Load—Modified.* Modifications to the lighting circuitry which will allow automatic and/or selective switching "off" of lights not needed because of daylight or non-occupancy, should be included at this point.
   e. If there is insufficient room in the *Description* spaces, additional notes may be made on the reverse side and referred to by *Note* number.
5. This step involves:
   a. An appraisal of the *Operating Hours* and *Annual Energy Consumption* (existing) as recorded and calculated in steps 3.c. through 3.e. can be made, again referring to Fig. 4-8 and reference 5.
   b. If it is possible to effect a reduction in the number of hours of operation, by modifying maintenance or operation procedures, the form allows for the calculation of the revised energy consumption.
6. The form allows for:
   a. The totaling of several columns which in turn allows an appraisal of the energy use situation in the building as a whole.
   b. Conversion of the consumption (kWh) totals to dollars cost per year by applying the local power rates.
   c. Comparison of dollars savings per year in energy costs to the capital cost of effecting such savings and the pay-back period calculated.

Fig. 4–9. Lighting Energy Management Form

Space Description	Condition	Description	Existing Installation / Proposed Modifications	Power-Load (watts)			Operating Hours							Hours per Week	Weeks per Year	Hours per Year	Annual Energy Consumption (kWh)		Cost of Modification	Note
				Connected	Budget	Modified	M	T	W	T	F	S	S				Existing	Modified		
	Existing																			
	Modified																			
	Existing																			
	Modified																			
	Existing																			
	Modified																			
	Existing																			
	Modified																			
	Existing																			
	Modified																			
	Existing																			
	Modified																			
	Existing																			
	Modified																			
	Existing																			
	Modified																			
	Existing																			
	Modified																			
	Existing																			
	Modified																			
	Existing																			
	Modified																			
	Existing																			
	Modified																			
Totals →																				

MAINTENANCE OF LIGHTING SYSTEMS

Maintenance is the only way to continue the effectiveness of any lighting installation. Lighting maintenance includes all the means that can be used to keep the output of a lighting system as near to its initial level as is practical.

The use of *light loss factors* in planning installations is a necessary admission that no amount of recurring maintenance can keep the output of a system up to its initial level. The value of the light loss factor used indicates the amount of the expectedly uncontrollable depreciation and the amount of effort that is expected will be devoted to try to overcome this depreciation.

The objectives of recurring lighting maintenance are to maintain a desired illuminance level or illumination effect with minimum operation and maintenance costs and effective energy use.[6,7]

Trained maintenance people are needed to implement systematic maintenance plans. Some plants, commercial buildings, etc., can afford to equip, train, and supervise such personnel, although many find it advantageous to hire outside specialists. In medium and small installations, lighting maintenance contractors can supply planning services, equipment and manpower, and by more thorough work, may justify their existence with a resulting lower cost per unit of light delivered. In addition, contractors are often trained to analyze potential energy conservation opportunities available when implementing a lighting maintenance plan.

The following paragraphs enumerate and discuss causes of light loss; advantages of planned relamping and cleaning; and operating programs, methods, material and equipment. There also is a discussion of the remedies for mechanical and electrical difficulties that can develop in lighting systems.

Causes of Light Loss

It is important to recognize that the following factors contribute to the over-all loss of light (see Fig. 4-10 and Section 9, page 9-1 in the 1981 Reference Volume.[8]

1. Luminaire ambient temperature.
2. Voltage to luminaire.
3. Ballast factor.
4. Luminaire surface depreciation.
5. Room surface dirt depreciation.
6. Burnouts.
7. Lamp lumen depreciation.
8. Luminaire dirt depreciation.

The individual effect of each factor varies with the kind of work performed and the atmospheric location of the building. Air is dirtier in a foundry than in an air-conditioned office; the amount and type of dirt found in office air is different for an industrial area compared to suburbs; and black steel mill dirt is unlike some of the light-colored dust of a woodworking shop.

Lamp Lumen Depreciation. Light output of lamps decreases as the lamps progress through life. This decrease is called lumen depreciation and is an inherent characteristic of all lamps. (See Section 8 of the 1981 Reference Volume.) Losses due to this effect will be reduced by lamp replacement programs such as planned relamping.

Luminaire Dirt Depreciation. A significant amount of light loss can generally be attributed to dirt accumulation on luminaire surfaces. In addition to the kind and amount of dirt in the area, the amount of light loss depends on luminaire design, lamp type and shapes, and luminaire finish.

Ventilated lighting units tend to collect dirt less rapidly than those with closed tops.[9,10] This requires proper luminaire design and placement of the opening. The temperature difference between the lamp and surrounding air causes convection currents that help carry much of the dust and dirt through the luminaire rather than allowing it to accumulate on its reflector.

Dirt accumulation on a reflecting surface can be minimized if the reflector is sealed from the air, as in a dust-tight luminaire or a reflectorized lamp. A suitable protective shield should be used with reflectorized lamps to reduce glare and, if they are subject to moisture dropping on the bulb, to reduce lamp breakage.

Burnouts. Lamp burnouts contribute to loss of light. If lamps are not replaced promptly after burnout, average illuminance will be decreased proportionately. In some instances, more than just the faulty lamp may be lost, *i.e.*, when series sequence fluorescent ballasts are used and one lamp fails, both lamps go out. In some cases a burnout may cause a change in the electrical supply to other lamps.

Luminaire Surface Depreciation. Materials used in luminaire construction differ in their resistance to deterioration. Processed aluminum finishes tend to have a slow rate of depreciation.

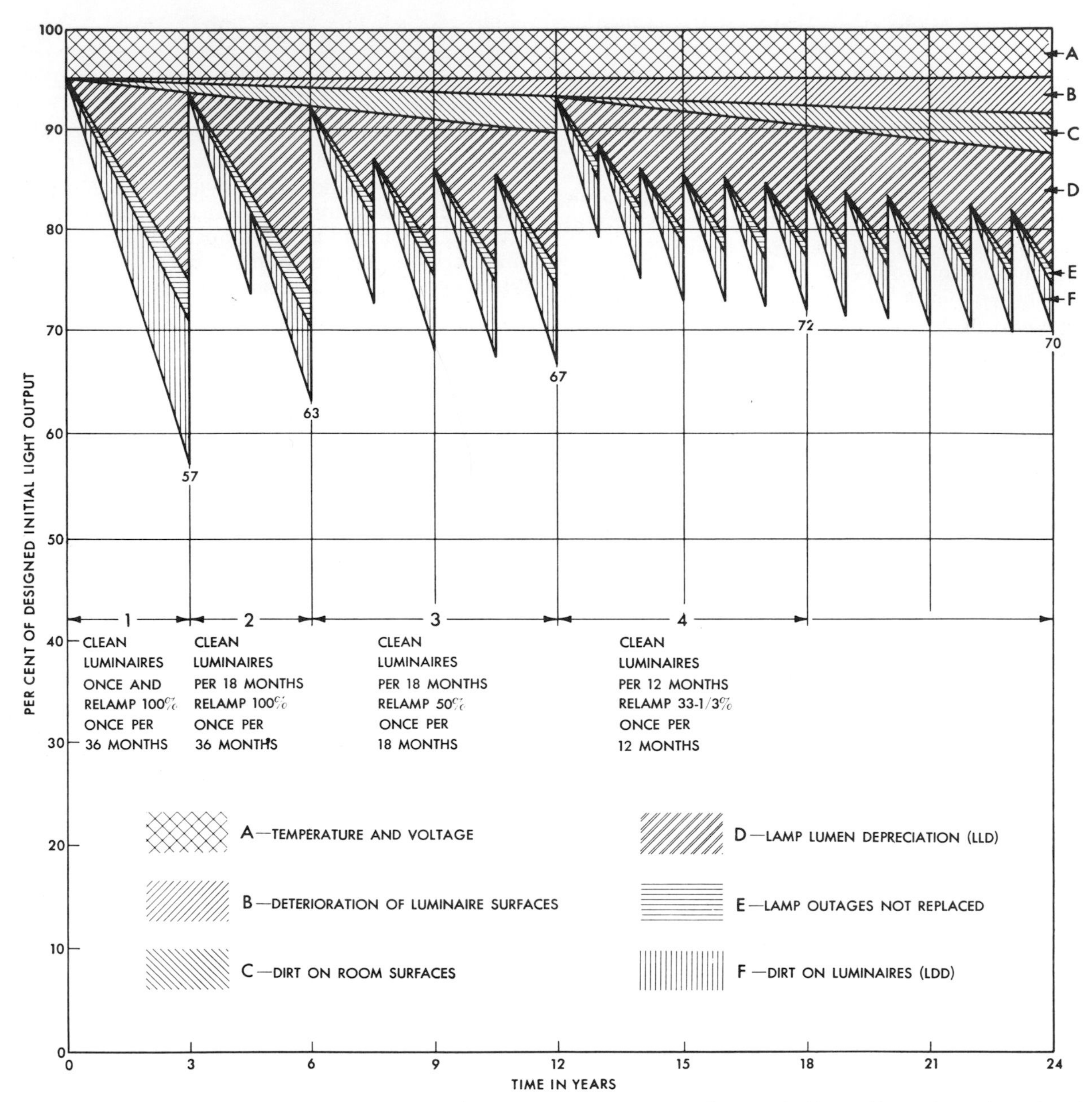

Fig. 4-10. Effect of light loss on illuminance level. Example above uses 40-watt T-12 cool white rapid start lamps in enclosed surface mounted units, operated 10 hours per day, 5 days per week, 2600 hours per year. All four maintenance systems are shown on the same graph for convenience. For a relative comparison of the four systems, each should begin at the same time and cover the same period of time.

Enamels, on the other hand, are usually easier to clean. In addition to the absorption of light, accumulation of dirt in certain luminaires may change the light distribution. For example, dirt accumulation on a specular aluminum reflector in a high bay luminaire can change the beam shape from narrow to wide. The loss in utilization here is much greater than the loss in efficiency.

The use of plastics in luminaire construction has increased markedly in the past decade. The types most widely used for light transmission and control are acrylics and polystyrenes followed by cellulosics, polycarbonates and vinyls. Over a period of time, the transmittance and color of all will change upon exposure to ultraviolet radiation and to heat. Acrylics are the most resistant to these changes. The other types of plastics are less resistant to varying degrees.

The rate of change in transmittance and color also depends upon the specific application: type of lamp used, distance of the plastic from the lamp, and ambient temperature of the plastic during the operating period of the luminaire. Use of improper cleaning materials and/or techniques can cause added changes in transmittance due to chemical action and/or scratching of the surface. Therefore, hard surfaced, chemically resistant materials are most resistant to changes from these causes.

Exactly how much change occurs in transmittance, reflectance, absorption and color of the plastics from application to application remains to be studied. For further information on plastics, see Section 6 of the 1981 Reference Volume.

Room Surface Dirt Depreciation. General practice in all lighting fields is to use finishes with high reflectance to balance brightnesses and to utilize light most efficiently. Room proportion and distribution of light from the lighting units determine the amount of light which strikes the walls and ceilings. Dirt collection on room surfaces tends to reduce the amount of reflected light. While periodic cleaning and painting of walls and ceilings is necessary in all installations, it should be done more frequently in those areas where a larger per cent of light is reflected by these surfaces.

It should be remembered that there may be instances of luminaire and room appreciation, *i.e.*, the bleaching out of wall paints, curtains, or anodized aluminum parts, or where light color dust may have higher reflectance than the room surface on which it collects.

Temperature, Voltage and Ballast Factor. There are other factors which may cause a variation in light output from day to day in addition to the causes discussed above. One factor is the variation in voltage. Another factor is ambient temperature. Fluorescent lamps are particularly affected by changes in temperature. See Section 8 of the 1981 Reference Volume for data on the effect of voltage and temperature on lamps. For specific information about a specific luminaire design and ballast factors, it is best to consult the manufacturer.

Relamping and Cleaning

Periodic Planned Relamping. A properly planned and implemented relamping program will arrest lamp lumen depreciation and prevent many burnouts, thereby improving and maintaining illuminance levels. This improvement can sometimes result in the installation of lower wattage lamps and thus reduced energy use. Reduction of burnouts gives an added advantage in saving the time and expense otherwise involved in spot burnout replacement.

Periodic Planned Cleaning. Clean luminaires and room surfaces produce these results:

1. *More light delivered per dollar.* Cleaning improves the light output of a system when the dirt on luminaires and room surfaces has a lower reflectance than do the surfaces themselves.

2. *Better energy management.* Clean luminaires allow for the use of lower wattage lamps or fewer luminaires.

3. *Pride in ownership.* Clean luminaires and room surfaces give noticeable evidence of pride of ownership and pride of good housekeeping. The appearance of the installation is improved. Only by cleaning may such improvement be gained.

4. *Improved morale.* Clean luminaires and room surfaces can improve appearance, thereby directly improving morale and indirectly improving production and sales.

5. *Reduced capital investment.* When a lighting system is periodically relamped and cleaned on a correctly planned schedule, it provides more light than when lamps are replaced only after burnout and when luminaires remain dirty. If lighting designers know that a planned maintenance program will be performed, they can design for a given average illuminance level using fewer luminaires. This will, of course, result in reduced capital investment as well as reduced operating cost and energy use.

Operations—Programs and Methods

Planning. The timing of relamping and cleaning should be in accordance with the plans of the lighting system designer. If intervals between operations are too long, excessive loss of light results. If intervals are too short, labor, equipment and lamps are wasted.

Lighting systems are becoming more complex. As a result, the requirements are increasing for labor and equipment to relamp and clean them. The size of a lighting installation is the greatest single factor in determining how efficiently it can be cleaned. For servicing a few luminaires, the purchase or use of specialized equipment is impractical, and the training of labor in special skills is hardly necessary. As the number of luminaires increases, the value of the savings which result from mechanization and employee training also increases.

Lighting systems are installed in so many different kinds of locations and atmospheres that luminaire accessibility and rate of dirt accumulation vary in almost every circumstance. Thus, programs of relamping and cleaning should be planned to fit each set of circumstances.

Performing. The sequence of cleaning steps will vary with types of luminaires and locations. For example, one person with a ladder, sponge and pail can handle, unaided, open strip units three meters (ten feet) from the floor. On the other hand, a sizeable crew with an elaborate scaffolding assembly may be required for a transilluminated ceiling 9 meters (30 feet) above the floor. In general, the operations of a two-person team represent typical methods and are as follows:

1. *Remove shielding material and lamps.* Louvers, plastic or glass panels, etc., and lamps are removed from luminaire and passed to the person at the floor.
2. *Make luminaire shock-free.* Care should be taken to prevent shock when working around electric sockets. Electrical circuit can be turned off or sockets can be covered with tape, dummy lamp bases, etc.
3. *Clean basic unit.* If required, heavy deposits of dirt can be removed first from top surfaces of channel, reflector, etc., by vacuuming, wiping or brushing. The entire unit then can be washed with a suitable solution, using brushes, sponges or cloths. The unit then should be rinsed to remove any residue of solution and dirt.
4. *Clean shielding material and lamps.* While the person on the ladder works on luminaire as in 3, the person at the floor takes shielding material and lamps to a cleaning station or cleans them at the ladder. Plastic materials should be allowed to drip dry after rinsing or be damp dried with toweling or some other material. Dry wiping can cause the formation of electrostatic charges. New lamps should be dry wiped before installation.
5. *Replace lamps and shielding material.* When the person on the ladder finishes cleaning, clean shielding and new or cleaned lamps are installed, passed by the person at the floor.

Incandescent and high intensity discharge luminaires usually do not require as many cleaning steps as fluorescent units but the same general method applies to cleaning all lighting equipment.

Cleaning Compounds

A knowledge of cleaning compounds is needed to determine the ones best suited to any particular cleaning application. Correct cleaning compounds, properly used, save time and money. The following information has been found to apply in most cases to the commonly used luminaire finishes:

Aluminum. Very mild soaps and cleaners can be used on aluminum and will not affect the finish, if the material is thoroughly rinsed with clean water immediately after cleaning. Strong alkaline cleaners should never be used.

Porcelain Enamel. This finish is not injured by non-abrasive cleaners. Detergents and most automobile and glass cleaners do a good job under average conditions.

Synthetic Enamel. Some strong cleaners may injure this finish; particularly in cases where the enamel is left to soak in the solution. Alcohol or abrasive cleaners should not be used. Detergents produce no harmful effects.

Glass. As with porcelain enamel, most non-abrasive cleaners can be used satisfactorily on glass. Dry cleaners are usually preferred on clear glass panels, but not on etched or sand blasted surfaces. Most detergents will work well under average conditions.

Plastics. Very often dust is attracted by a static charge developing on plastic. Most common detergents do not provide a high degree of permanence in their anti-static protection. In most areas, however, if the plastic is cleaned at least twice a year with a detergent, a satisfactory relief from static dirt collection is obtained. Destaticizers are available which have greater permanence than common detergents in this respect. Plastic should not be wiped dry after the application of a rinse solution.

Cleaning Equipment

Time, labor and expense of maintaining a lighting system can be greatly reduced by choosing maintenance equipment with features most suitable to the requirements of each system. Many different kinds of maintenance devices are available to facilitate the cleaning task. The choice of equipment will depend on several factors such as: mounting height, size of area, accessibility of lighting units, and obstacles in the area. Some available maintenance equipment are:

Ladders. Ladders are often used in lighting maintenance because their low weight, low cost and simplicity make them desirable for simple maintenance tasks. See Fig. 4-11. However, safety and mobility restrictions limit their use in many cases.

Scaffolding. Portable scaffolding generally has greater safety and mobility than ladders. See Fig. 4-11. More equipment can be carried and the maintenance man has a firm platform from which to work. In general, scaffolds should be light, sturdy, adjustable, mobile, and easy to assemble and dismantle. Special requirements often dictate the type of scaffolding which can be used, for example, for mounting on uneven surfaces or for clearance of obstacles such as tables or machines.

Telescoping Scaffolding. The telescoping scaffold provides a quick means for reaching lighting equipment at a variety of mounting heights. This equipment comes in various sizes that have platforms which can be raised or lowered either manually or electrically. See Fig. 4-12.

Lift Truck. Often the quickest and most efficient maintenance device is the lift truck. See Fig. 4-13. Although there are different types available, the method of operation is basically the same. The platform can be raised or lowered automatically and, in some types, the truck can be driven from the platform. While the initial investment for such equipment often is high, the maintenance savings often are large enough to make this device economical.

Fig. 4-11. Ladder and portable scaffold.

Fig. 4-12. Telescoping scaffold.

Fig. 4-13. Lift truck.

Disconnecting Hangers. Disconnecting hangers lower lighting units to a convenient work level, enabling the worker to maintain them with a minimum of equipment. See Fig. 4-14. When a lighting unit is raised into place, the hanger positions the unit and makes the proper electrical circuit connection automatically. An additional safety feature of this type of device is that the electrical circuit is disconnected when the luminaire is lowered.

Lamp Changers. Lamp replacement can often be simplified by the use of lamp changers. By gripping the lamps either mechanically or with air pressure, as in a vacuum type, the lamp changer can be used to remove and replace lamps. See Fig. 4-15.

Catwalks, Cranes, Cages, etc. Lighting maintenance can be incorporated as an integral part of the lighting system. This can be accomplished in many ways. Luminaires can be maintained from catwalks, cranes or maintenance cages (see Fig. 4-16). The catwalks and maintenance cages can be installed alongside each row of lighting units so that maintenance can be performed from them with safety, speed and efficiency.

Vacuum Cleaners and Blowers. A blower or vacuum cleaner is sometimes used to remove dust from lighting units. While some of the dirt can be removed in this way, the units still have to be washed for a thorough job. But, the periodic use of a vacuum cleaner or blower can prolong the cleaning interval.

Wash Tanks. It is desirable to have a wash tank specifically designed for lighting maintenance. Tanks should have both wash and rinse sections and be the proper size for the luminaire parts to be washed. Heating units, mounted in each section, are generally desirable to keep the cleaning solution hot. Louvers or reflectors can be set on a rack to drip dry after washing and rinsing while another unit is being cleaned. Special cleaning tanks have been designed for fluorescent luminaire parts and for flexible types of ceiling panels.

Ultrasonic Cleaning. This method removes foreign matter from metals, plastics, glass, etc. by the use of high frequency sound waves. Basic equipment consists of a generator, a transducer, and a suitable tank. The generator produces high frequency electrical energy which the tank mounted transducer converts to high frequency sound waves that travel through the cleaning solution. These waves cause the "cavitation" effect—formation of countless microscopic bubbles which grow in size, then violently collapse—thus creating a scrubbing action that forcibly and rapidly removes dirt from the material immersed in the solution.

Fig. 4-14. Disconnecting and lowering hangers (below left).

Fig. 4-15. Pole lamp changer (above right).

Trouble Shooting and Maintenance Hints

Lighting maintenance incorporates lamp replacement and cleaning, planned maintenance to prevent trouble, as well as repairs to the lighting components. While the operation of fluorescent and high intensity discharge lamps is more complex than incandescent filament lamps, trouble can generally be diagnosed and corrected quickly with simple test equipment.

In the discussions below, the references to Section 8 refer to the 1981 Reference Volume.

Preheat Starting Fluorescent Lamp Circuits.

Trouble Shooting. See Section 8 for diagram and explanation of the circuit.

1. Replace existing lamps with lamps known to be operative.
2. Replace existing starters with starters known to be operative. No blink type starters are recommended for replacement. Refer to Section 8

for description of various types and the features of each.
3. Check luminaire wiring for incorrect connections, loose connections or broken wires. Refer to the wiring diagram printed on the ballast.
4. Replace the ballast.

Maintenance Hints.
1. Deactivated lamps should be replaced as quickly as possible. Blinking lamps cause abnormal currents to flow in the ballast which will cause ballast heating and thereby reduce ballast life.
2. Blinking lamps will also reduce starter life.

Rapid Starting Fluorescent Lamp Circuits.

Trouble Shooting. Refer to Section 8 for the diagram and explanation of the circuit. Constant heater current is essential for proper starting of all rapid start type lamps. For 800 mA and 1500 mA types, the constant heater current is also essential for proper lamp operation.
1. If lamp requires 5 or 6 seconds to start, one cathode is probably not receiving the cathode heating current. This usually results in excessive end darkening of lamps after a short period of operation. With lamps out of the sockets, check heater voltages. This can be done with available testers which have a flashlight lamp mounted on a fluorescent lamp base. If a voltmeter is used, a 10-ohm, 10-watt resistor should be inserted in parallel with the meter. If proper voltage is available, check for poor contact between lampholder and base pins or contacts on the lamp. If no voltage is measured, check for open circuit (poor or improper connections, broken or grounded wires, open heater circuit on the ballast). Check for proper spacing of lampholders.
2. If one lamp is out and the other lamp is operating at low brightness or if both lamps are out, only one lamp may be deactivated. Refer to circuit diagram in Section 8 and note that 2-lamp circuit is a sequence starting, series operating design.
3. Replace the ballast.

Fig. 4-16. Maintenance cage.

Maintenance Hints.
1. Deactivated lamps should be replaced as quickly as possible. The 800 mA and 1500 mA lamps require both heater current and operating current for proper operation. If either is missing, poor starting or short lamp life will result. In the 2-lamp series circuit, one lamp can fail and the second lamp will operate at reduced current. This condition will reduce the life of the second lamp.
2. Lamps should be kept reasonably clean. The bulbs of all rapid start type lamps are coated with a silicone to provide reliable starting in conditions of high humidity. Dirt can collect on the lamp surface which could absorb moisture in high humidity atmospheres, thus nullifying the silicone coating, and prevent starting or cause erratic starting.

Instant Starting Fluorescent Lamp Circuits.

Trouble Shooting. Refer to Section 8 for the diagram and explanation of the circuit. Note that 2-lamp circuits can be either lead-lag or series-sequence design.
1. Replace existing lamps with lamps known to be operative.
2. Check lampholders for broken or burned contacts. Check circuit for improper or broken wires. Refer to wiring diagram on the ballast.
3. If ballast is suspected of being defective, replace with ballast known to be operative. Measurement of ballast voltages in the luminaire is difficult because the primary circuit of the ballast is open when a lamp is removed. Refer to circuit diagram in Section 8.

Maintenance Hints. Deactivated lamps should be replaced as soon as possible. In the 2-lamp series circuit, one lamp can fail and the second lamp will operate at low brightness. This condition will reduce the life of the second lamp and also will cause an abnormal current to flow in the ballast, giving rise to ballast heating and a reduction in ballast life.

Incandescent Lamps.

Maintenance Hints. Troubles with incandescent lamps are usually the result of misapplication, improper operating conditions or poor maintenance practice. Most problems can be avoided by applying the following maintenance hints:
1. *Over-voltage or over-current operation.* Rating of lamps should correspond with actual circuit operating conditions. Over-voltage or over-current operation may shorten lamp life drastically. For example, a 120-volt lamp operated on a 125-volt circuit suffers a 40 per cent loss in life. Refer to Section 8.
2. *Shock and vibration conditions.* Under such conditions, the use of vibration service or rough service lamps is recommended. The use of general service lamps under these conditions results in short life.
3. *Sockets.* Lamps of higher wattage should not be operated in sockets designed for a specified wattage or excessive lamp and socket temperatures may result. Excessive temperatures may affect lamp performance or may shorten the life of insulated wire, sockets, etc. Refer to Section 8.
4. *Luminaires.* Only the proper lamps for which the luminaire was designed should be used. Contact of any metal part of a luminaire with a hot lamp may result in violent failure of the lamp.
5. *Cleaning lamps.* A wet cloth should not be used to clean a hot lamp. A violent failure may occur with a wet cloth.
6. *Proper burning position.* Lamps should be operated in their proper burning position as specified by the lamp manufacturer. Operation of the lamps in the wrong position may cause a lamp to fail immediately or after short life.
7. *Replacing lamps.* Whenever possible, lamps should be replaced with power "off." Replacing lamps with power "on", particularly high voltage types, can result in the drawing of an arc between the lamp base and the socket.
8. *Tungsten-halogen lamps.* Lamps should be installed with the power "off." It is also recommended that the bulb be held with a clean cloth or tissue to avoid fingerprints on the bulb. Fingerprints may result in bulb discoloration and a subsequent reduction in light output. Follow lamp manufacturers' instructions on the carton of each lamp.
9. *Dichroic reflector lamps.* Certain lamps utilize a dichroic reflector designed to radiate heat back through the reflector portion of the bulb. Luminaires using these lamps should be ventilated, or otherwise designed to provide adequate cooling of the socket and wiring adjacent to the bulb.

Mercury Lamps.

Trouble Shooting.
1. Replace lamp with lamp known to be operative. Be sure operative lamp is cool as hot lamps will not restart immediately.
2. Check lampholder for proper seating of the lamp and for proper contact.
3. Check ballast name plate reading, particularly if low temperatures are involved.
4. Check ballast wiring. If a tapped ballast is used, be sure ballast tap matches supply voltage at the ballast.
5. Check circuit wiring for open circuit or incorrect connections.
6. Replace ballast.
7. If lamps fail prematurely, check for the following: *a. Lamp breakage.* Check lamps for cracks or scratches in the outer bulb. These can be caused by rough handling, by contact with metal surfaces in bulb changer or luminaire, or by moisture falling on overheated bulb. *b.* Bulb touching the luminaire, the lampholder or other hard surfaced object.
8. If arc tube is cracked, blackened or swollen early in life, or if connecting leads inside outer bulb are burned-up check for the following: *a. Over-wattage operation.* Check the ballast rating and the voltage at the ballast and if proper tap on ballast is used. *b. Excessive current.* Check if ballast is shorted. Check for possible voltage surges or transients on the supply line.

Caution: Do not replace with new bulb until circuit is checked and cause of trouble has been corrected.

Maintenance Hints.
1. If tapped ballasts are used, check should be made to be sure tap matches supply voltage at ballast. Low voltage will cause low light output, poor lumen maintenance and reduced lamp life. High voltage will cause short lamp life.
2. The circuit should be reasonably free from voltage fluctuations. Replacement ballasts should match the particular voltage, frequency and lamp type.
3. The proper lamp type should be used for the ballast in the installation. Certain lamps with the same wattage rating are available in two distinct lamp types, each designed for operation on specific ballasts having completely different electrical characteristics. Incorrect matching of lamp and ballast may result in short lamp life or lamps going on and off repeatedly.
4. Lamps should be handled carefully. Rough

handling can cause cracks and scratches in outer bulb which will result in short lamp life.

Metal Halide Lamps. Recommendations given for mercury lamps also apply to metal halide lamps; however, the following additional information is pertinent:

1. Until there is industry standardization of metal halide lamps, it is important to be sure that the ballast will operate a given variety and brand of metal halide lamp satisfactorily.
2. Time to restrike after an outage may be much longer than for a mercury lamp.
3. Metal halide lamps may have a vacuum in outer bulb area, as a result all metal halide lamps should be handled carefully. In some cases enclosed luminaires are recommended.
4. When first turned on, metal halide lamps may require up to two days to reach characteristic color and light output. Also, a change in burning position may cause a change in color and light output.

High Pressure Sodium Lamps. Recommendations for mercury lamp trouble shooting and maintenance apply. In addition, it is important to replace lamps with power "off" since ballast may have a high voltage spike. Handle vacuum-type bulbs carefully. Restriking time for hot lamps is generally very short—a minute or two.

Self-Ballasted Mercury Lamps. Items 1, 2, and 5 of mercury lamp trouble shooting section and item 4 of maintenance hints apply. In addition, supply voltage should match voltage rating for the lamp within manufacturer's approved tolerances. Self-ballasted mercury lamps take longer than mercury lamps to restrike following a current interruption.

REFERENCES

1. Energy Management Committee of the IES: *IES Procedure for Calculating Lighting Power Limits for New and Existing Buildings (Unit Power Density Procedure)—EMS-6*, Illuminating Engineering Society, New York, July, 1980.
2. *Energy Conservation in New Building Design, ANSI/ASHRAE/IES 90A-1980,* American National Standards Institute, New York. Energy Management Committee of the IES: *IES Recommended Lighting Power Budget Determination Procedure—EMS-1*, Illuminating Engineering Society, New York, September, 1978.
3. Energy Management Committee of the IES: "IES Recommended Procedure for Lighting Energy Management of Existing Buildings—EMS-4," *Light. Des. Appl.*, Vol. 9, p. 30, April, 1979.
4. "Energy Management and the Lighting of Office Buildings—EMS-5," *Light. Des. Appl.*, Vol. 7, p. 16, February, 1977.
5. Design Practice Committee of the IES: "An Interim Report Relating the Lighting Design Procedure to Effective Energy Utilization—EMS-2", *Light. Des. Appl.*, Vol. 5, p. 34, September, 1975.
6. Pierpoint, W.: "Energy Conservation from Lighting Maintenance," *J. Illum. Eng. Soc.*, Vol. 8, p. 195, December, 1969.
7. Clark, F.: "Accurate Light Loss Factors Contribute to Efficient Energy Use," *Light. Des. Appl.*, Vol. 3, p. 31, October, 1973.
8. Clark, F.: "Light Loss Factor in the Design Process," *Illum. Eng.*, Vol. 63, p. 575, November, 1968. Clark, F.: "Accurate Maintenance Factors—Part Two (Luminaire Dirt Depreciation)," *Illum. Eng.*, Vol. LXI, p. 37, January, 1966. Clark, F.: "Accurate Maintenance Factors," *Illum. Eng.*, Vol. LVIII, p. 124, March, 1963.
9. Steiner, J. W.: "Practical Reduction of Dirt Accumulation in High Wattage Luminaires," *Illum. Eng.*, Vol. XLVIII, p. 184, April, 1953.
10. Sell, F. W.: "Ventilation—The Key to Self Maintenance," *Illum. Eng.*, Vol. XLVIII, p. 500, September, 1953.
11. Christensen, M.: "A Method for Determining Minimum Cost of Relamping," *Illum. Eng.*, Vol. LVII, p. 712, November, 1962.
12. Mangold, S. A.: "Lighting Economics Based on Proper Maintenance," *Light. Des. Appl.*, Vol. 4, p. 6, August, 1974.

Office Lighting

SECTION 5

The purpose of office lighting is to provide for effective visual performance with optimum use of energy. It is important, therefore, to analyze the controllable factors which contribute to visibility: the *task*, the *lighting* and the *environment*.[1] While in practice these factors are closely interrelated, they are individually treated in this section for practical considerations, with appropriate references to other applicable sections.

OFFICE TASKS

General Considerations. In modern office operations, the eyes are used at close range for severe visual tasks such as reading duplicated material, handwriting, typed carbon copies and fine print. The visibility of such work is often poor and more attention should be devoted to its improvement (see Fig. 5-1). The visibility of the details of a task or object is determined by its *contrast* with the background, *luminance*, *size* and *time* of viewing. Each factor is sufficiently dependent upon the magnitude of the others that a deficiency in one may, within limits, be compensated by augmenting one or more of the others.

Contrast. Each critical detail of a seeing task must differ in luminance or in color from the surrounding background to be visible. Visibility is at maximum when the luminance contrast (and color contrast, if present) of details with the background is greatest. See Section 3 of 1981 Reference Volume.

To illustrate conditions encountered in most office seeing tasks, the printing in this paragraph has been shaded from a light gray at the left to a solid black at the right. The left is representative of the poor quality of characters resulting from penciled longhand and shorthand notes, the use of worn machine and typewriter ribbons, poor carbon paper, and many duplicating methods. Care should be taken to use darker inks, ribbons, carbons and pencils which will produce results as represented by the quality printing at the right.

Visibility may also be reduced with lower reflectance of the background. To illustrate this factor, the background of this paragraph has been shaded. Most office tasks, generally considered as black print on white paper, actually consist of shades of dark gray against shades of light gray. Care should be taken to avoid the use of thin papers which are likely to lack opacity and have a low reflectance. Light tints of color are sufficient for the necessary identification of forms and need not appreciably reduce the reflectance of paper as do dark and highly saturated colors. Identification can also be obtained through the use of a colored band across the top of white paper.

Luminance. To be visible the visual task must be illuminated. This can be demonstrated by attempting to read in a dark room. The visual contrast discussed in the previous paragraph must have some luminance in order for it to become visible. The eyes' sensitivity to contrast increases significantly with increased luminance over the range of 0 to 1000 candelas per square meter (0 to 300 footlamberts).[2] Older eyes require greater luminance than younger eyes for an equal level of visibility.[3]

Size. Both printing and writing vary considerably in size. Type sizes in general use range from 6-point to 12-point type. Within limits, as size increases, visual performance improves.

The use of this 6-point type is not desirable for continuous reading even by a person with normal vision.

This 8-point type may be regarded as the minimum size tolerable for good readability.

NOTE: References are listed at the end of each section.

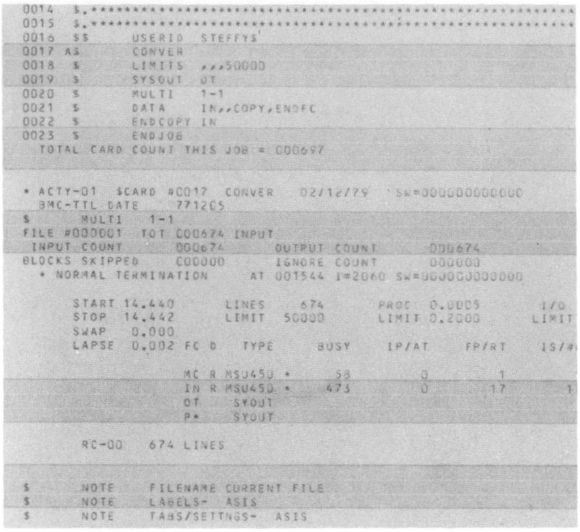

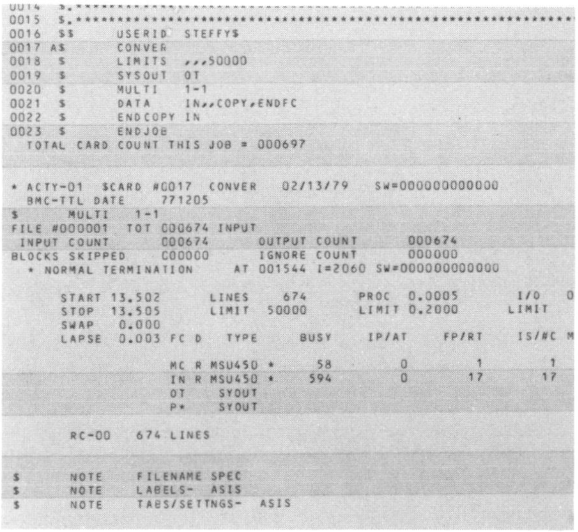

Fig. 5-1. The computer printout tasks shown here are esentially identical in content. The printing ribbon, however, was changed just after the top printout was made to provide a much clearer printout shown in the bottom photograph. This simple machine maintenance procedure can help produce material having better contrast, and thus easier to read.

A more reasonable type size and, fortunately, one finding ever-increasing use for prolonged or continuous reading is this 10-point size.

Studies indicate the desirability of this 12-point type size for continuous reading over long periods of time.

Time. The eye assimilates details one at a time; that is, the eye focuses on a detail, assimilates it and then moves to the next detail. If visibility is poor due to small size of detail, low contrast or low luminance, the rate of assimilation decreases and work takes longer. If the rate of assimilation is to be maintained under conditions of poor visibility, accuracy of assimilation will suffer. Accuracy, however, is more important for some tasks than for others. For instance, in reading it is not necessary to assimilate every letter in order to understand the meaning, whereas when working with figures a "3" mistaken for an "8" can be crucial.

LIGHTING CRITERIA

Office lighting should be evaluated in terms of its effect on people and their performance. A great deal is known about the human response to light and more knowledge is being gained through both vision research and experience with lighting installations. Specifically, lighting will affect:

1. Ability to see visual tasks with speed and accuracy.
2. Visual comfort.
3. Visual environment or the pleasantness of a space in which one lives and works.

All of these are needed to achieve the best performance.

It is convenient to organize lighting criteria into two basic considerations: (1) quality and (2) quantity. However, they are not independent considerations, and both must be taken into account in any lighting design.

Quality of Lighting

There should be no compromise with the quality of lighting in offices since quality can provide equal productivity with the use of less electrical energy. See Section 4 for energy management procedures. There are three main elements to be considered in providing quality in office lighting, namely: visual comfort or the level and extent of the luminances in the normal field of view, luminance ratios in the normal field of view, and veiling reflections present on the seeing task (or conversely the degree of contrast rendition provided).

Visual Comfort. Excessive luminances in the normal field of view can cause discomfort, be distracting and should be avoided. Research indicates that discomfort is associated with the stress produced in the muscle that reduces the size of the pupil when relatively high luminance is encountered.[4]

Tests of a large group of people have determined the relative degree of discomfort caused by varying degrees of glare from room luminances. The data have been reduced to a rating system termed Visual Comfort Probability (VCP), which can be computed for various lighting systems. See Section 2 of this Volume and Section 9 of the 1981 Reference Volume. The VCP value is a prediction of the percentage of people who will be expected to find a lighting system acceptable in terms of discomfort glare when located at a specific point in a specific installation, and can be used to compare lighting systems. Higher VCP values indicate a higher probability of comfort, but the following values should be observed: (1) a VCP of 70 or more; (2) a ratio of maximum-to-average luminance of no more than five to one (preferably three to one) at 45, 55, 65, 75 and 85 degrees from nadir, crosswise and lengthwise; and (3) maximum luminaire luminances, crosswise and lengthwise, of no more than the following values:

Angle Above Nadir Degrees	**Maximum Luminance** Candelas per square meter	Footlamberts
45	7710	2250
55	5500	1605
65	3860	1125
75	2570	750
85	1700	495

Luminance Ratios. Two separate effects are influenced by the luminance ratios within the field of view. (a) transient adaptation and (b) disability glare.

Transient Adaptation. The eyes adapt themselves for optimum vision when moving from one luminance level to another by a photochemical reaction within the eye and by a change in pupil size. This compound effect is termed transient adaptation, and takes a finite time for completion. See Section 3 of the 1981 Reference Volume. Thus, if there are significant differences in appreciable areas of the visual environment the visual assimilation may be slower as the eyes move from one luminance level to another.

Disability Glare. Glare sources close to or in the line of sight may cause stray light within the eye which in turn is superimposed upon the retinal image (see Section 3 of 1981 Reference Volume). This alters the luminances of the image and its background and reduces the contrast so that visibility may be reduced.[5, 6]

To limit the effects of transient adaptation and disability glare the luminance ratios should not exceed the following:

Task to immediate surround, 1 to 1/3
Task to remote dark surfaces, 1 to 1/10
Task to remote light surfaces, 1 to 10.

Veiling Reflections and Contrast Rendition. As discussed in the above paragraph *contrast*, the degree of contrast between the details of the visual task and the task background affects visibility. The degree of contrast depends in part upon the relative locations of the light sources, the task and the eyes. If when viewing a visual task an image of a luminaire happens to be reflected in the details of the task, the details will assume some of the brightness of the luminaire surface. Contrast will be reduced and visibility will be impaired. This effect is often called "veiling reflections"because the reflections of the luminaire in the task seem to interpose a veil over the task. See Fig. 5-2.

Usually one luminaire among many which may be lighting a specific task causes a reflection in the task and is, therefore, responsible for causing veiling reflections and contrast loss. The area on the ceiling where a luminaire would cause most of these reflections is termed the "offending zone"(see Fig. 2-8, page 2-28). In a private office, luminaires may be placed to avoid the "offending zone." In open office areas, however, the luminaire in the "offending zone" for one worker may be contributing quality light to another. Research has shown that office tasks are viewed most often at about 25 degrees from the vertical.[7] Since the angle of reflection is equal to the angle of incidence most veiling reflections will be caused by illumination incident at 25 degrees from the vertical. Thus, luminaires which emit less light at about 25 degrees from the vertical and more light at greater angles from vertical may be helpful. Care should be taken, however, to prevent discomfort glare by limiting the light output at angles greater than 50 or 60 degrees from vertical. Luminaires which polarize a substantial proportion of the light at 25 degrees from the vertical may also be helpful.

For a discussion of the factors causing contrast loss from veiling reflections, guidelines for reducing the effect of veiling reflections and evaluation methods, see Section 2.

Reflected Glare. The adverse visual condition of reflected glare is almost as annoying as direct glare. Usually it is caused by a mirror image of the light sources reflected from highly polished wood or glass-covered desk tops (see

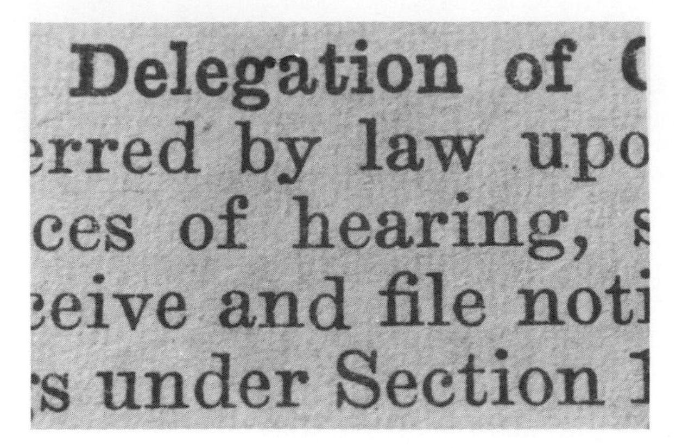
Delegation of
rred by law upo
ces of hearing,
eive and file noti
s under Section

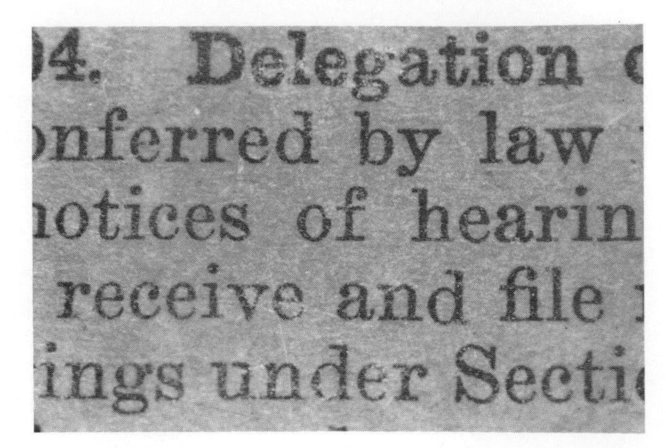
4. Delegation
nferred by law
otices of hearin
receive and file
ings under Secti

Fig. 5-2. (Left) An enlarged example of printing on a soft paper with matte black ink. Illumination is provided by a spotlight behind the camera. Note that there are very few highlights in the black letter strokes and that the texture of the paper can be observed. (Right) The same soft paper with matte black ink. Lighting in this case is with a spotlight in front of and above the task where it is in a position to cause the maximum veiling reflection effect. Note the many highlights in the black stroke and on the paper as well. These highlights in both areas are specular reflections which tend to veil the task and reduce contrast.

Fig. 5-3). It can be reduced by the use of matte surfaces (see Fig. 5-4) and by carrying out the procedures for reducing veiling reflections on the task. Large-area low-luminance luminaires are used when specular surfaces cannot be avoided. Glare shields should be added to office machines and computer equipment to reduce offending specular light source images.

Shadows. Shadows cast on the visual task reduce the luminance of the task, and may impair effective seeing. In addition, when shadows are sharply defined at or near the task, they may be annoying.

Shadows will be softened if the light comes from many directions. High-reflectance matte finishes on room surfaces become effective secondary light sources and materially reduce shadows by reflecting a significant amount of diffused light into shadowed areas.

Pronounced shadows will be caused by concentrating, small area sources, such as incandescent filament downlights. Lighter shadows will result from using many wide distribution downlights. Softer edge shadows are produced by large area light sources, such as fluorescent luminaires or indirect lighting systems.

Quantity of Illumination

Fig. 2-2 suggests ranges of illuminances for efficient visual performance of given functions. Knowledge of the visual tasks expected and their importance in the operation of the office is critical to the correct application of lighting levels. Similarly, consideration must be given to the occupants, their expected performance, and their desired reaction to the office environment. Discussions on illuminance selection are found in Section 2.

Illuminance will vary with the life of the lighting system (see Section 9 of the 1981 Reference Volume for determination of light loss factors) and may also vary with the location of the task within the office environment. These light loss factors must be considered carefully, so that desired lighting levels are maintained throughout time and space (see *task lighting* in this section).

Methods are available for predicting illuminance and luminance on planes as well as at a point. Section 9 of the 1981 Reference Volume has a complete discussion of such methods.

LUMINOUS ENVIRONMENT FACTORS

The eyes function most comfortably and efficiently when luminance ratios within the entire field of view are not excessive. A comfortable balance of luminances in the room may be achieved by selecting luminaires in accordance with the recommendations discussed in the paragraphs under *Quality of Lighting* and by employing matte finishes for room and furniture surfaces with reflectances as recommended in Fig. 5-5. By observing these recommendations, the luminance ratios generally will be within the limit established as being desirable and practical. To avoid blandness and to produce visual emphasis in the environment, however, luminances may be deliberately unbalanced within the ratio limits recommended on page 5-3.

Finishes

Room Finishes. The room surfaces (ceilings, walls and floors) are important factors in determining the luminance ratios between the lighting equipment and its surroundings and between the task and its more remote surroundings. The use of matte finishes having the recommended reflectances helps prevent excessive luminance ratios and specular reflections.

The reflectances of room surfaces have a considerable effect upon the utilization of light. By reflection, the ceiling, walls and floor act as secondary large area light sources and if finished with the recommended reflectances, will increase the utilization of light and reduce shadows.

The reflectances recommended for room surfaces shown in Fig. 5-6 permit a wide range of color choices. Many attractive colors can be found that will meet the recommended reflectances.[8]

Wall reflectances should generally fall within the range of recommended reflectances; however, under certain conditions, higher or lower reflectances are satisfactory or even desirable. The ceiling finish may, for example, be carried down the walls to the level of pendant luminaires having a high upward component; utilization of this upper wall surface could result in an increase of as much as 10 per cent in work-plane illuminance.

Small areas of the room may have reflectances higher or lower than permitted by the ranges given in Fig. 5-5. If these areas are thought of as color accents and restricted to about 10 per cent of any person's visual field they will not affect the efficiency of the lighting system or the significant environmental luminance ratios. The environment can then be made to look considerably more pleasant and interesting.

Window shielding media should have approximately the same reflectance as that recommended for the walls.

Office Equipment Finishes. For a person working at a desk, the top surface may occupy most of the visual field. The task is usually of high reflectance, and desk tops having the reflectance recommended in Fig. 5-5 are necessary to prevent the luminance ratio between the task and its surround from exceeding the limits recommended on page 5-3. Matte finishes are essential to minimize reflected glare.

It is important, too, that the vertical surfaces of desks and file cabinets have finishes in the recommended range of reflectances. They will generally occupy a significant portion of the visual field. Since these surfaces receive less illumination than adjacent horizontal surfaces, light finishes are necessary to make them appear moderately bright.

Business machines often occupy a central part of the visual field, even when operated by the touch system. Finishes having the recommended reflectances are essential for seeing comfort; shiny surfaces should be eliminated, for even a

Fig. 5-3. Streaks of light are reflected images of two continuous rows of luminaires. Very severe glare condition is produced by top which is both dark and polished.

Fig. 5-4. The reflected glare from luminaires disappears when a piece of light diffuse linoleum is placed over the dark, polished desk top.

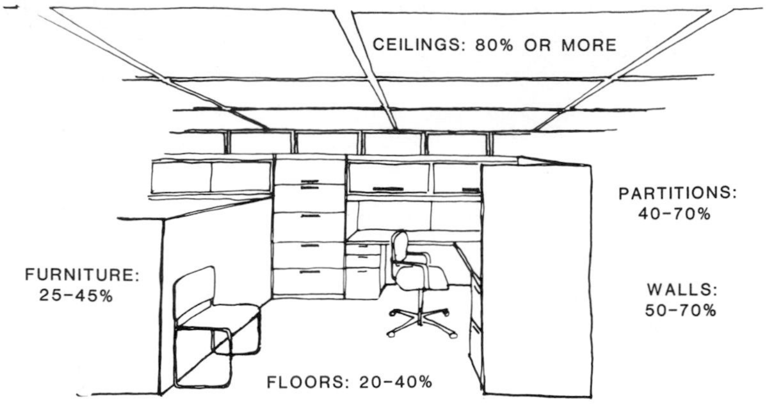

CEILINGS: 80% OR MORE
PARTITIONS: 40-70%
FURNITURE: 25-45%
WALLS: 50-70%
FLOORS: 20-40%

Fig. 5-5. Reflectances recommended for room and furniture surfaces in offices.

small amount of specular trim can be distracting and annoying even though it may not be viewed directly. See Figs. 5-6 and 5-7.

Color

Color and Office Surfaces. People respond to the colors they see in their surroundings, and in the office where workers are exposed to an environment for long periods, the color in that environment can have an effect on their performance, positively or negatively, consciously or unconsciously (see Section 5 of the 1981 Reference Volume).

Small offices can be made to appear larger and less crowded if walls, woodwork and furniture placed against walls, are in the same hue or have similar reflectance. A contrasting color or a lighter or darker value of the same color may be used at some point or points in the room. This may be in the draperies, the upholstery for chairs or sofas, or with any added wall decoration such as pictures. Small touches of accent color will give vitality and dramatic interest to any large or small office area. The color selected for large areas should have reflectances as shown in Fig. 5-5.

Color and Light Sources. There are two separate and distinct application considerations with respect to color and light sources, namely: chromaticity, or source-color, and color rendering. See page 2-31.

Chromaticity refers to the color of the source when viewed directly. For clear, untinted incandescent lamps this depends largely on the temperature of the filament. For fluorescent and high intensity discharge lamps there is a large variation in chromaticity depending on the components used for the discharge and on the phosphor coatings. The chromaticity depends on the combination of spectral compositions from the arc components and phosphors used in these lamps.

It is possible, however, for two different sources with different spectral power distributions to have the same chromaticity—that is they will look alike.

Color rendering refers to the appearance of colored objects when illuminated by a particular light source. More specifically "good" color rendering often indicates that colors viewed by the light of a specific light source closely match their appearance under some reference source such as daylight, or in some cases incandescent light. The color rendering index may be used to rank the color rendering capabilities of a group of specific lamps (see Section 5 of the 1981 Reference Volume).

The color rendering capabilities of fluorescent lamps depend on the mixture of phosphors used for the lamp coating. Those used for standard warm white and cool white provide fair color rendering for offices with reasonably high lumen output. Those used for the deluxe warm white and cool white provide very good color rendering but provide less lumen output. Other mixtures using rare-earth-activated phosphors provide very good color rendering with almost the same lumen output as the standard lamps. Mercury and metal halide lamps are available with color improving phosphor coatings to provide better color rendering than standard lamps. There are other lamps which do not render colors as well as the lamps mentioned above but provide more lumens per watt.

The color rendering provided by various lamps can be compared by installing them in identical boxes or rooms, each having an identical display of colored objects so that comparisons can be made. The choice of light source depends on the importance given to color rendering balanced [9] against the initial cost, the lamping and maintenance cost, and the energy cost of the lighting system. The selected light source should be used when choosing colors and finishes for interior designs.

LIGHTING SYSTEMS

All interior lighting systems are included in one of the following six classifications: indirect, semi-direct, general diffuse, direct-indirect, semi-direct and direct. A complete description of these six classifications, and their lighting distributions, is given in Section 1.

Lighting Equipment Selection. The selection of particular types of luminaires from the above classification depends upon a number of physical and esthetic considerations, among which are:
1. Ceiling height and type.
2. Ceiling module dimensions (if modular).
3. Ceiling appearance desired.
4. Energy constraints.
5. Structure fire protection requirements.
6. Budgetary limitations.
7. Equipment maintenance program.

Illumination Uniformity. It may be neither necessary nor desirable for the lighting level to be uniform throughout an entire office space.[10] If the recommended luminance ratios are observed, task lighting levels need only be maintained at the work locations with lower levels occuring between work stations. However, if the density of work locations is high, uniform lighting levels may be a practical lighting technique.

Non-uniform lighting layouts may necessitate the use of flexible wiring systems to facilitate the relocation of ceiling luminaires when work stations are moved.

Fig. 5-6. Dark surfaces may have reflectances of only five to seven per cent, resulting in an uncomfortable working environment if the task involves a white or light-colored task medium. Also these dark surfaces can cause decreased worker performances due to the large luminance ratios.

Fig. 5-7. Light-colored furniture materials provide an attractive, comfortable visual environment for many office situations .When using high vertical surfaces, their reflectances should be in the range of 40 to 70 per cent. These surfaces should also be matte to avoid glare discomfort.

Perimeter Lighting

Whichever type of lighting equipment is used, consideration should be given to luminaire spacing and placement at the perimeter wall areas of the room. See Fig. 5-8.

The use of recommended reflectances for walls is important to the final illumination design goal—especially for providing sufficient illuminance at perimeter desk locations. Since the illuminance may diminish at the walls, a possible solution is to increase the wattage of perimeter lamps, increase the number of the lamps in the perimeter luminaires, or the total number of luminaires that parallel walls. Another alternative would be to provide a separate perimeter lighting system, such as a cove, soffit, valance, cornice or equipment which has been designed to specifically illuminate walls. High wall-luminances should be avoided to minimize glare.

Fig. 5-8. The fluorescent wall wash helps to enlarge this narrow reception space. Visual interest is also added to this casual area because of the relatively high wall luminance. Note how the high luminances at the far end of the corridor (left of center) attract the eye, and can be used to move people from one area to another.

Accent Lighting Systems

Washing the walls with light helps to create an interesting luminous environment as shown in Fig. 5-8. Spot-lighting pictures or other focal points, or lighting large wall areas as well, will do much to add visual interest to general and private offices.

Task and Ambient Lighting

The concept of task and ambient lighting is to provide desired levels for tasks only where required, while maintaining a lower illuminance in the surrounding areas.

Task Lighting. The term "task lighting" became popular in connection with furniture integrated lighting. This type of lighting system may be more accurately termed "task oriented lighting" and may be defined as lighting designed to relate to a specific task location and orientation.

Quality task oriented lighting should have minimum direct glare, provide uniform and shadowless illumination, and incorporate provisions for reducing veiling reflections. In reducing veiling reflections, it is generally helpful to diminish the light from in front of the viewer, and increase the light from the side. This can be accomplished by luminaire positioning or luminaire optics. Luminaires which vertically polarize a significant proportion of light coming from the offending zone may also be helpful.

Task oriented lighting can be accomplished by using ceiling luminaires or by using "local" luminaires close to the task. When using ceiling luminaires they should be located so that most of their light falls on specific visual task areas. Luminaires should be located away from the "offending zone" to avoid veiling reflections so that good contrast rendition may be obtained. For closely spaced work stations in open offices it is difficult to avoid the offending zone, and luminaire light control becomes particularly important. With high dividing screens close to the work surface or overhanging shelves or cabinets some of the ceiling light will be blocked and shadows may fall on the work surface.

Task oriented lighting, using local luminaires for direct illumination of each task area from close range is well adapted for "systems furniture" and eliminates shadows on the work surface. In many cases this may save energy and increase task visibility. The luminaire may be mounted in the furniture (furniture integrated), on a desk stand, on a floor stand, or suspended from the ceiling. It is generally not practical to position local lighting behind and to the sides of the task, in order to minimize veiling reflections. A luminaire location in front of the task may be in the "offending zone" and is likely to cause veiling reflections when viewing the task directly ahead or to the side. For this reason the light from local task oriented luminaires must be carefully controlled to reduce veiling reflections.[11]

Ambient Lighting. In addition to the local task oriented light, ambient (or general) light is required. Quality ambient lighting should provide safe levels of illumination for traffic circulation throughout the space. The luminance ratios between the task area and the surrounding area should be within the guidelines described on page 5-3.

Ambient light may be provided by ceiling mounted direct luminaires, by indirect luminaires suspended below the ceiling, or by furniture integrated or free standing indirect luminaires which illuminate the ceiling. The ambient lighting is likely to be used in large open spaces where a large area of the ceiling will be in the field of view. Direct luminaires should have low luminance in the normal field of view. Indirect luminaires should not produce excessive ceiling luminances. In addition, indirect luminaires should minimize direct glare.

Systems generally employed in task and ambient lighting fit one of the following classifications:

1. Direct task and direct ambient.

2. Indirect task and indirect ambient.
3. Direct task and indirect ambient.

The luminaires can be recessed, surface mounted, pendant mounted or portable. Portable equipment may be integrated with open office furniture systems. Luminaires should relocate easily as the furniture is moved about. Certain local codes may restrict the use of the portable equipment and should be investigated prior to planning installations in those cities.

Daylighting

A psychological well-being is felt by many people when an exterior view is available. The daylight contribution (see Section 7 of the 1981 Reference Volume) should be carefully evaluated and should always be coordinated with a planned electric lighting system.

Fenestration has at least three useful purposes: (1) for the admission, control and distribution of daylight; (2) for a distant focus for the eyes which relaxes the eye muscles; and (3) to eliminate the dissatisfaction many people experience in completely closed-in areas. An adequate electric lighting system should always be provided because of the wide variation in daylight. The basic requirements of quantity, luminance ratios, and the reflectances of the principal architectural and work surfaces are the same whether the lighting is daylight, electric or a combination of the two.

Thermal Environment Considerations

Systems are available for control of the heat produced by lighting. Such systems may use the plenum above recessed luminaires to collect heat or may involve use of special air handling luminaires and associated ducts to remove lighting heat. For pendant mounted, furniture mounted and free standing lighting systems it is difficult to effectively extract the heat of the light.

For many sizeable buildings, the lighting system is an important component of the mechanical system. Frequently, in northern regions, the heat from lighting contributes significantly to the heating of a building during the winter. This is a clear indication that the lighting designer must interface with the mechanical designer to assure a balanced, energy conserving environmental system. See Section 2 and the *ASHRAE Systems Handbook* for more detailed information concerning light and the thermal environment.

LIGHTING FOR SPECIFIC AREAS

Offices and Office Spaces

Open Offices. Open or open-plan offices are areas which usually accommodate workers in a common space with no floor-to-ceiling partitions. In these areas, luminance limitations (see page 5-3) are critical to task performance and human comfort. Because of the variety of functions, and the size of the visual field, seeing is usually continuous and difficult. For these reasons, lighting quality and seeing comfort should never be compromised. Details of visual tasks should be of reasonable size and exhibit good contrast. Finishes, colors, and other environmental factors should be selected to assure full benefit from the lighting and enhance the tasks themselves.

Open offices can be extremely flexible environments, and to permit such flexibility, lighting systems should provide comfortable and efficient task performance for all viewing directions (see *task lighting*). For successful open office environments, the lighting system should not only make the space visually interesting and functional, but should be considered for its energy and acoustical attributes. Consideration should be given to future office functions, and the possibility of subdivision to smaller, private offices. Also, because of the large areas associated with open offices, care should be taken to correctly illuminate perimeter walls (see perimeter lighting).

Private Offices. In private offices where the desk is in a fixed location, or the practical limit of flexibility is a small area, uniformity of horizontal illuminance is not required. Lighting should be designed to adequately cover the desk area and other work space zones. The remaining room luminances, especially wall luminances, can be provided by supplementary lighting. This approach can result in proper task illumination, balanced room luminances, and an interesting visual environment.

Lighting system flexibility for private offices can be obtained by switching alternate rows of lamps in luminaires or by dimming, thereby providing several illuminance levels. Floor or table lamps strategically placed can provide a relaxing conversation area, adding warmth and pleasantness to the space.

Conference Rooms. Visual tasks in conference rooms vary from casual to difficult viewing.

Two or more systems should be planned:
1. A general lighting system, in which control is provided by switching of lamps, or by dimmers to vary the illumination.
2. A supplementary lighting system, consisting of downlighting, for slide projection and other low-level illumination requirements.
3. A perimeter wall wash lighting system.

Business Machines and Computer Rooms

When lighting for business machines and computer rooms, care should be taken to evaluate the great variety of tasks that are performed. Lighting should be provided to suit the common critical task at each location. Accuracy is an important consideration because mistakes can become costly.

Business Machines. Many types of equipment fall into the general category of business machines. In general, the guidelines in Fig. 2-2 should provide for a sufficient quantity of illumination. However, great care must be taken to avoid veiling reflections on transparent covers over readout areas and on surfaces of keys on equipment, such as desk top calculators. Large area, low luminance luminaires will help to minimize this problem, or locating luminaires properly with respect to the machine can eliminate the problem.

The reading task on machines with self-luminous presentations, such as microfilm projectors and equipment with cathode ray tube projection screens, may present some difficulty if these units are installed in areas having higher levels of illumination. Provision should be made to enable the illumination to be reduced by switching or dimming. Reflections can be reduced by reorienting or shielding the projector so that the reflected images of luminaires or windows are not seen on the screens.

Computer Rooms. Computer operations can range from a time-sharing terminal up to a complex of several rooms including the machine room, key punch area and programming office. The time-sharing terminal and key-punch operators may have to read fine print or pencil writing, often on colored paper, while typing in data and reading data coming out, printed on white or colored paper and carbon copies. The programmer is usually working with printed forms and pencil writing.

In a large system, a lower illuminance may be practical in the machine room than in the keypunch room. In this case, the readout data is usually analyzed in the office of an accountant or engineer. The machine room operator has a variety of short-term seeing tasks such as making logbook entries, reading labels on magnetic tapes, reading of input cards, typing, or correcting malfunctions caused by jammed cards or tapes.

Some manufacturers of equipment require that where an oscilloscope is used for analysis a lower illuminance level be available when operators are servicing and making tests and adjustments. The lighting should be arranged so that by switching out circuits or dimming the room illumination, good screen visibility will be accomplished. Servicing within large computer cabinets will require supplementary lighting.

Care should be taken in the placing of luminaires in relation to equipment with high vertical-information-faces so that images of luminaires do not produce veiling reflections. Indicator lights and lighted push buttons will need to be sufficiently bright so that they can be easily seen under general illumination.

Drafting Rooms

Visual requirements for drafting demand high quality illumination since discrimination of fine detail is frequently required for extended periods of time. Significant gradation of shadows along T-squares and triangles reduces visibility. Harsh directional shadows from drawing instruments and hands may reduce efficiency. Lighting systems which avoid reflections are most important in providing maximum contrast.

Quality of Lighting. The selection of large luminous area systems, indirect, semi-direct, or other forms of over-all ceiling lighting will minimize shadows. When ceiling heights or energy constraints do not permit the use of these systems, direct lighting systems can be effectively applied when the drafting board is illuminated from the sides. In such a system, the absence of any luminaire in the offending zone also minimizes veiling reflections and reflected glare. Referring to Fig. 2-8, note that the offending zone moves with board angle changes, and with eye movements relative to the task surface area.

Supplementary Lighting. Supplementary lighting equipment with movable support arms, may be attached to the drafting table, allowing the worker to position the light for critical task requirements or to overcome shadows or reflections.

Files

Files present the particular problem of work surfaces that are vertical, inclined and horizontal. In active filing areas, the work is likely to be of protracted nature and the visual task is of more than average severity. Where the room is primarily devoted to files, consideration should be given to luminaires designed and located to provide illumination on vertical surfaces. Where files are located in a general office environment, consideration may be given to local illumination at the files, with individual switching located nearby.

Reception Rooms

The receptionist should be provided with lighting in keeping with the tasks assigned. Frequently, general office work is part of the assignment and the visual requirements are more exacting than those for the casual occupant. Luminance ratios in the receptionist's visual field should be limited to low values and either a uniform amount of light should be provided throughout the room or more light provided at the work area, commensurate with the severity of the work assignments. The latter requires particular attention to luminance patterns on walls, floors and desk tops.

If a telephone switchboard with signal light indicators is present, it is important that the general lighting not "wash out" the signal lights (particularly if they are located on a horizontal plane). Care should also be taken to prevent reflections of bright light sources in the signal light caps.

Rest Rooms

Uniform illumination is not required in rest rooms. Luminaires should be located to provide enough light in the vicinity of the mirrors for adequate illumination of the face. See page 10-5. Other luminaires should be located so that their maximum light output will be concentrated in areas of the urinals and toilet stalls. Concentration of light in these areas has a tendency to encourage more effective cleanliness. Because rest rooms are at times subject to abuse, consideration should be given to vandal-resistant luminaires as well as distribution and esthetic capabilities.

Public Areas

Public areas in a building generally include entrance and elevator or escalator lobbies, corridors and stairways. Since many people move through these areas, lighting considerations should include safety requirements (see page 5-13) and luminance differences with adjacent areas, in addition to the appearance of the space. Lighting systems in public areas are required to remain illuminated for long time periods, if not continuously. Therefore, serious consideration should be given to energy conserving systems.

Since many public areas are egress areas, a complete auxiliary lighting system is required to cope with power outages and system failures. See Emergency Lighting in Section 2. These same auxiliary systems can serve as security lighting.

Entrance Lobbies. First impressions of office buildings are often made in entrance lobbies. The lighting environment should be esthetically pleasing, complement the architecture, and fulfill primary visual requirements. The illumination should provide safe and attractive transition from the exterior to the interior. Consideration must be given to adaptation conditions (see Section 3 of the 1981 Reference Volume) thereby avoiding hazardous areas. Adaptation requirements may necessitate a "day" lighting system and a "night" lighting system. This will depend on fenestration and exterior lighting conditions.

To add interest, to aid circulation and draw attention to functional areas, nonuniform lighting is often desirable in entrance lobbies. The use of various light sources in combinations will add character to the entrance space. Where large expanses occur, consideration may be given to supplementary adjustable lighting equipment for displays or art.

The lighting of vertical surfaces in lobbies will help to define the space. For this reason, high reflectance surfaces are appropriate. To provide sparkle, or to make an arcitectural statement, the building designer may select specular materials for the lobby. These include polished marble, chrome, stainless steel, aluminum, mirrors, glass and ceramics. Reflections in these materials must be considered when establishing a lighting system. A well designed lighting system will enhance the interior architecture, as the lighting system is an architectural element.

Corridors. Corridor illumination should provide at least one fifth the illuminance level of adjacent areas. (Illuminances are calculated for the floor plane.) This illumination does not cause an uncomfortable degree of eye adaptation upon entering the corridor from adjacent areas. Also, eye adaptation greatly depends on surface luminances.

Wall finish reflectance values should equal or exceed those of adjacent areas. Linear luminaires oriented crosswise to the corridor generally make it appear wider. Continuous linear luminaires located adjacent to the side walls provide high wall luminance, generally giving a feeling of spaciousness. Corridors, being a path of exit, should be provided with emergency lighting.

Elevator Lobbies. Classified as a casual seeing area, high luminance differences are acceptable. Contrasting luminous areas and geometric patterns of light may be designed to add interest. Higher levels of safety lighting should be provided at the elevator threshold to call attention to possible differences in elevation between the elevator cab and floor level.

Elevators. Lighting levels equal to those provided in the building corridors should be provided in elevators. Bright ceilings and walls will give the feelings of increased size to the confined space of an elevator cab. One method of accomplishing this is through the use of a luminous ceiling. The lighting in an elevator should always be connected to the building's emergency power supply to help avoid possible panic in the event of elevator power failure or malfunction. Elevator car interior finishes should be light in color.

Escalators. Good lighting should be provided on the escalator treads. The illuminance should be increased at the location where a person steps on and off the moving treads. The luminaires in this critical area should be arranged in such a manner that a shadow will not be cast on the treads by the person entering or leaving the escalator. It may, in some instances, be necessary to provide shielded supplementary luminaires in the balustrade. A colored luminous strip at the edge of the stair tread is commonly used and aids in seeing the moving tread quickly. Escalator finishes should be of a light, nonspecular type.

Stairways. The stair treads should be well illuminated and the luminaires so located and shielded that persons neither cast shadows on the stairs nor encounter glare at eye level. The positioning of luminaires should provide ease of maintenance as ladders are difficult to use in stairways. Emergency lighting should be provided in all public stairways. Although the lighting requirements are the same for all stairways, the lighting design solutions may be different.

DESIGN CONSIDERATIONS

The primary consideration in design of a lighting system is the creation of a desirable visual environment. However, the system must also be compatible with acoustical, thermal, spatial and esthetic requirements of each design area. The optimum total environment can be attained only through cooperative efforts by the owner, architect, engineer and specialized consultants to integrate all design components into a final solution. See Section 1.

The following items (not in order of importance) should be considered when selecting and comparing lighting systems:

1. Economics. Establish initial and life cycle costs of the system. See Section 3.
2. Energy. Determine total lighting system energy consumption. Review energy code compliance and effects on mechanical system. See Section 4.
3. Task visibility. Review task characteristics, task location with respect to luminaire, and luminaire photometrics to be certain lighting system provides adequate visibility and illuminance. Visibility can be compared from system-to-system by using equivalent sphere illumination (ESI) analyses.
4. Visual Comfort. Determine visual comfort probability, if applicable, and review luminance ratios.
5. Architecture. Review the architectural intent of the project and lighting system compatability. Recognize that some luminaire layouts appear acceptable in plan, yet may produce visual clutter when viewed in three dimensions.
6. Color. Select colors based on lamp color rendering or if colors have previously been determined, select lamp to enhance room colors, task and skin tones.
7. Acoustics. Select ballasts with an "A" sound rating. For open office environments, review systems speech privacy compliance with acoustician.[12]

The annual cost of lighting systems may be compared by considering:

1. Annual depreciation of the initial installed cost over a selected period of time (ten to twenty years).
2. Annual energy cost (including the impact on heating and cooling costs).
3. Annual maintenance costs.
4. Annual rearrangement costs.

Users Responsibility for Servicing the Lighting System. If the lighting system is to perform as designed, information is needed for the user to assume responsibility. The design maintenance plan should be communicated in writing to assure wise and efficient use of investment and energy. As a bare minimum this plan should include cleaning and relamping schedules.

Safety. Safe working conditions are essential in any office area, and the effect of lighting on safety must be considered. See Section 2 for a discussion of lighting for safety.

The illuminance recommendations in Fig. 2-2 provide a guide for efficient visual performance rather than for safety alone; therefore, they are not to be interpreted as requirements for regulatory minimum illuminances. On the other hand, the illuminances in Fig. 2-26 represent levels for safety alone.

Emergency Lighting. Emergency lighting insures the safety of a building's occupants when the normal lighting system fails. The illumination provided by the emergency lighting system should permit an orderly accident-free exit from the building during the emergency condition. If exiting is not required, the system will operate to provide security and comfort to the remaining occupants until the general lighting can be restored. See Section 2 for a discussion of emergency lighting.

Energy Management

Section 4 deals with criteria for reducing energy consumption while maintaining the lighting quality and good practices described in this section. The following are some specific considerations for offices.

Lighting Quality. Various lighting systems having the same energy input may produce different levels of visibility. It is therefore, important to consider the previously discussed factors which affect visibility so that the desired level may be provided with minimum energy expenditure.

Lighting Flexibility. Lighting for task visibility should be located only where tasks occur with lower illuminance elsewhere (one-third to one-fifth of task illuminance). Since task locations may be unknown at the time the lighting design is considered, or may change later, flexibility is important for minimum energy use. Flexibility may be achieved by relocating luminaires or by switching luminaires.

Relocating Luminaires. Some ceiling systems allow relocation of recessed luminaires. To take advantage of this flexibility the luminaires may be interconnected with cable connectors which plug into the luminaire and may be arranged without changing the building wiring. Furniture integrated lighting should be capable of rearrangement within the furniture so that changes can be accommodated.

Switching Individual Luminaires. Rather than moving luminaires, they may be switched off or on for an indefinite period to accommodate changes in task locations. For this purpose a switch may be incorporated in each luminaire.

Switching Groups of Luminaires. Energy can be saved by using switches so that small groups of luminaires can be controlled. This is important where cleaning operations occur after hours or where a few workers may work overtime. Group switching may also be used to switch off luminaires near windows during daylight hours.

Variable Level Lighting. Luminaires which use several lamps may be wired with two switch legs so that different combinations of lamps can be switched on to vary the lighting level. Thus, energy may be saved during daylight hours or after work for office cleaning. The lighting level may be varied for different areas depending on the severity of the visual tasks performed.

When fluorescent lamps are used, a two level ballast may be employed to reduce the energy input and the light output in areas where the visual tasks are less demanding. Dimming may also be employed to save energy as described above. Some fluorescent dimming systems do not reduce energy input throughout their dimming range.

Time Switching. Time switching can be used to ensure that lighting is switched off when not needed. In its most sophisticated form, a computer controlled system with individual relay switches in luminaires may be used for time control of each luminaire. A manual override at local switches can be provided to allow for an overtime worker. Such a system also provides flexibility for switching individual luminaires from a central location, for office changes throughout the system.

Luminous Efficacy. The luminous efficacy of light sources measured in terms of lumens per watt should be constantly reviewed and balanced against the desirability of their other characteristics. A high efficacy may be linked to poor color rendering or a low efficacy to excellent beam control.

Ballast Losses. Some energy is converted into heat within the ballast transformer. The "loss" can be reduced by using high quality materials in the transformer resulting in lower watts input for the same lumen output.

Ballasting. Two one-lamp ballasts cause a greater loss than one two-lamp ballast. Therefore, when one-lamp fluorescent luminaires are used it may be an advantage to tandem wire them so that one two-lamp ballast accommodates two luminaires.

Reducing Lumen Output. Ballasts and lamps for fluorescent luminaires may be used which reduce the lumen output with a proportional reduction in watts input.

Room Surfaces. Room surface finishes should be in the upper portion of the suggested reflectances (see Fig. 5-5).

Life Cycle Costing. This accounts for the cost of owning and operating a lighting system over a finite period, usually 10 to 20 years. By analyzing in this way, the energy savings possible by using the above ideas can be methodically compared with the added cost of owning an energy efficient lighting system.

REFERENCES

1. Committee on Office Lighting of the IES: "American National Standard Practice for Office Lighting,"*J. Illum. Eng. Soc.*, Vol. 3, p. 3, October, 1973.
2. CIE Committee E-1.4.2: "Recommended Method for Evaluating Visual Performance Aspects of Lighting," CIE Report No. 19, January, 1971.
3. Blackwell, O. M. and Blackwell, H. R.: "Individual Responses to Lighting Parameters for a Population of 235 Observers of Varying Ages," *J. Illum. Eng. Soc.*, Vol. 9, p. 205, July, 1980.
4. Fry, G. A. and King, V. M.: "The Pupillary Response and Discomfort Glare," *J. Illum. Eng. Soc.*, Vol. 4, p. 307, July, 1975.
5. Holladay, L. L.: "The Fundamentals of Glare and Visibility," J. Opt. Soc. Amer., Vol. 12, p. 271, April, 1926.
6. Stiles, W. S.: "Recent Measurements of the Effect of Glare on the Brightness Difference Theshold," *Compte Rendu*, Commission Internationale de l'Eclairage, Saranac Inn, New York, 1928.
7. Crouch, C. L. and Kaufman, J. E.: "Practical Application of Polarization and Light Control for Reduction of Reflected Glare," *Illum. Eng.*, Vol. LVIII, p. 277, April, 1963. Crouch, C. L. and Buttolph, L. J.: "Visual Relationships in Office Tasks," *Light. Des. Appl.* Vol. 3, p. 23, May, 1973.
8. Color Committee of the IES: "Color and the Use of Color by the Illuminating Engineer," *Illum. Eng.*, Vol. LVII, p. 764, December, 1962.
9. Flynn, J. E. and Spencer, T. J.: "The Effects of Light Source Color on User Impression and Satisfaction," *J. Illum. Eng. Soc.*, Vol. 6, p. 167, April, 1977.
10. Flynn, J. E.: "A Study of Subjective Responses to Low Energy and Nonuniform Lighting Systems," *Light. Des. Appl.*, Vol. 7, p. 6, February, 1977.
11. Florence, N.: "The Energy Effectiveness of Task-oriented Office Lighting Systems," *Light. Des. Appl.*, Vol. 9, p. 28, January, 1979.
12. *Test Method for the Direct Measurement of Speech Privacy Potential (SPP)*, PBS-C.1, General Services Administration, Public Building Service, August, 1972.

Educational Facilities Lighting

SECTION 6

There are continuing challenges and opportunities for the lighting designer in the field of educational facility lighting. New educational buildings will result from new needs, while others will be replacements for obsolete structures. There are many existing educational buildings which are in need of retrofitting to improve the visual environment and to reduce energy usage; however, it is not sufficient simply to replace old lighting equipment—more efficient luminaires and innovative lighting designs, developed to meet new needs and to reduce energy consumption, also are required.

This section deals specifically with the lighting of educational facilities, but for other related information there are appropriate references to other sections.

The Goal of Educational Facility Lighting

The goal of educational facility lighting is to provide an optimum visual environment for both student and instructor in each school situation that is supportive of the learning processes. This can only be secured in a visual environment where the occupants of the space can see their visual tasks accurately, quickly and comfortably. Uniformity in lighting systems throughout an educational facility does not assure optimum visual performance in every area because of the great variety of visual tasks found in any school situation. These visual tasks are increasingly varied as the pupil moves upward from the elementary school through high school to the college level.

A visual environment supportive of the learning processes must also be visually comfortable and satisfy the psychological and emotional needs of learners. Proper lighting can enhance pleasant and attractive surroundings, provide a feeling of spaciousness and delineate areas when suitably utilized as a part of the architectural design. Effectively used, lighting can attract and hold attention, stimulate learning and influence behavior in a positive way.

NOTE: References are listed at the end of each section.

Energy Conservation

Energy can be better utilized in providing an optimum visual environment by a careful selection of highly efficient light sources and luminaires, along with the use of daylight, to provide the proper quantity of high-quality, comfortable lighting (see Section 4, Energy Management). Daylight must, however, be carefully controlled if it is to be effectively used (see Section 7 of the 1981 Reference Volume).

The lighting designer's responsibility is to select luminaires and related equipment, such as ballasts and controls, which suitably affect the over-all efficiency of both the lighting system and the heating, ventilating and air-conditioning systems. Proper switching arrangements can make it possible for the occupants of each space to conserve energy by making effective use of both electric light and daylight. The designer must also provide written information for each building to enable administrators, instructors, students and custodians to operate the lighting system in the most visually effective and energy conserving manner consistent with visual effectiveness. Finally, the designer must prepare written instructions for the maintenance of the lighting system, covering such major items as: (1) the regular cleaning of lamps and luminaires, (2) the specific type of lamps to be used for relamping, (3) group relamping, and (4) general surface maintenance, including repainting. If such information is not made available, neglect will result in decreased illumination while using the same amount of energy.

VISUAL TASKS

Visual tasks in educational facilities differ in size, viewing direction and distance, and contrast.[1] Major critical tasks are reading and writing, commonly requiring prolonged and close attention. There are both near and far visual tasks, of small and large size, on matte and glossy surfaces. Students are often required to adapt from reading at a desk to reading from a chalkboard; from looking almost straight down to looking along or above the horizontal.

Research indicates that reading and writing tasks, with high contrast (such as is provided by 6- to 12-point black type on suitable white book stock) have adequate visibility at relatively low illuminance levels, while tasks with poor contrast, such as pencil writing (see Fig. 6-1), sewing with black thread on dark cloth, or reading badly reproduced material, need higher levels. Poorly reproduced materials may necessitate excessively high levels for good visual response; thus, it is worthwhile to secure quality duplicating equipment and to provide skilled operation. The alternatives are either a compensating increase in illuminance or an unsatisfactory visual situation.

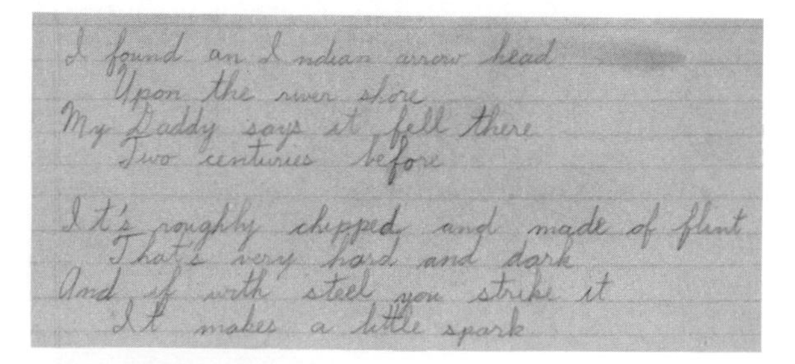

Fig. 6-1. Example of ink writing (upper) and pencil writing (lower) made by the same sixth-grade student. Note the difference in contrast between the ink and pencil writing.

Educators have a basic responsibility to improve the visibility of tasks to be performed by students. This may be accomplished by providing tasks having maximum contrast and minimum specularity and by properly designed illumination. With regard to the major school tasks of reading and writing, the steps which educators can take to improve visibility include:

1. The use of felt-tipped pens instead of ball-point pens which often have glossy (specular) ink. Pencils should be soft lead, No. 2 or softer, carefully selected so that the marks on the paper are not "shiny" and provide good contrast with the paper.
2. The minimum size of type recommended for textbooks is 10-point Bodoni or the equivalent. Younger pupils will require a larger type size. Adequate spacing between the lines of type and the use of low-gloss ink also contribute greatly to readability.
3. Matte paper should be used in textbooks and workbooks, for tablets and notebooks, and for tests and other reproduced materials. Paper should have a good degree of opaqueness, high diffuse reflectance and low specular reflectance.
4. A proper combination of high-grade chalk, chalkboard, and illumination should be provided. White chalk is more visible than other light-colored chalks on most colors of chalkboard. Chalkboards should be kept clean and periodically restored or resurfaced. Teachers should be encouraged to use large letters and figures when writing on chalkboards. Supplementary lighting is advised for chalkboards to provide adequate vertical illumination. Lighting for horizontal tasks provides less than one-half the same illuminance on vertical surfaces.

QUALITY AND QUANTITY OF ILLUMINATION

Quality and quantity of illumination are so interdependent that lighting, apparently adequate in quantity, is of little value if illumination of proper quality is not provided. The quality aspect of educational facility lighting embraces those factors which allow a visual task to be performed quickly and accurately and which render the visual environment comfortable and pleasing. Those factors involve considerations of luminance, luminance ratios, light distribution, task specularity and diffusion, location of lighting equipment, color and shadows. Reflected glare, veiling reflections and excessive luminance in the field of view all detract from lighting quality. See Section 2, Lighting System Design Considerations.

Quality of Illumination

Quality of illumination implies that all luminances contribute favorably to visual performance, visual comfort, ease of seeing, safety and esthetics in relation of the visual tasks involved. There should be no compromise with quality in educational facility lighting if good visual performance is expected. The distribution, size and range of brightnesses within an environment are extremely important, since they affect comfort, eye adaptation and contrast rendition.

Luminance Ratios. The luminance relationships of the various surfaces in the normal field of view in a space must be kept within limits. When the eye fixates on a task an adaptation level is established. As the eye shifts from one luminance, such as a book, to another luminance, such as the chalkboard, it must readapt to the new luminance level. If there is much difference between the two levels a period of time is required for the eye to accommodate itself to the new situation. Further, if the difference is too great, the reaction will be discomfort, attended by a transient change in pupillary opening and adaptation to the new level. In order to avoid this, where large surfaces are involved, the difference in luminance between adjacent surfaces should be kept within acceptable limits.

In general, for good visual performance the luminance of any significant surface normally viewed directly should not be greater than five times the luminance of the task (see Fig. 6-2). No large area, regardless of its position in the room, should be *less* than one-third the luminance of the task. The luminance of surfaces immediately adjacent to the visual task is more critical in terms of visual comfort and performance than that of more remote surfaces in the visual surround. Surfaces, such as desk tops, immediately adjacent to the visual task, should not exceed the luminance of the task, but should be at least one-third the luminance of the task. The difference in luminance between adjacent surfaces in the visual surround should be kept as low as possible.

The general approach in providing low luminance ratios over the entire visual field is to limit the luminance of luminaires and of fenestration and to increase the luminance of all interior surfaces by providing suitable reflectances and distribution of light.

Reflectances. Walls, including tackboards and large cabinets or cupboards mounted on the wall, should have nonspecular surfaces with from 40 to 60 per cent reflectance (see Fig. 6-3). Blinds or drapes, like the walls, should be light colored, with similar reflectances. Walls beside windows should also have very high diffuse reflectances, to avoid excessive luminance ratios between the windows and the wall surface. That portion of the wall above the level of the luminaires should have a reflectance of 80 per cent. The ceiling should be as nearly white as practicable and nonspecular, for this surface is most important in reflecting light downward toward tasks on desk tops. It is also necessary to avoid obvious brightness differences between the ceiling and the lu-

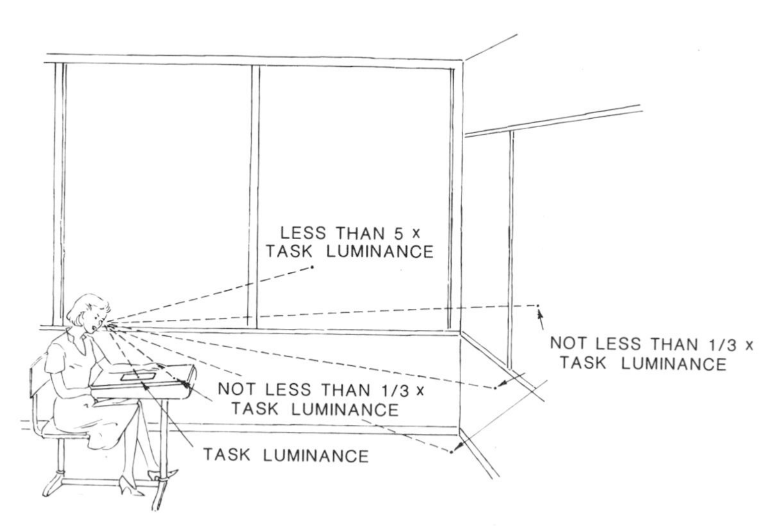

Fig. 6-2. In a classroom the luminance of significant surfaces should not differ greatly from the visual task. The luminance of the surface immediately surrounding the task should be less than the task but not less than one-third the task luminance. The lowest acceptable luminance of any significant surface should not be less than one-third the task luminance. The highest acceptable luminance should not be greater than five times the task luminance.

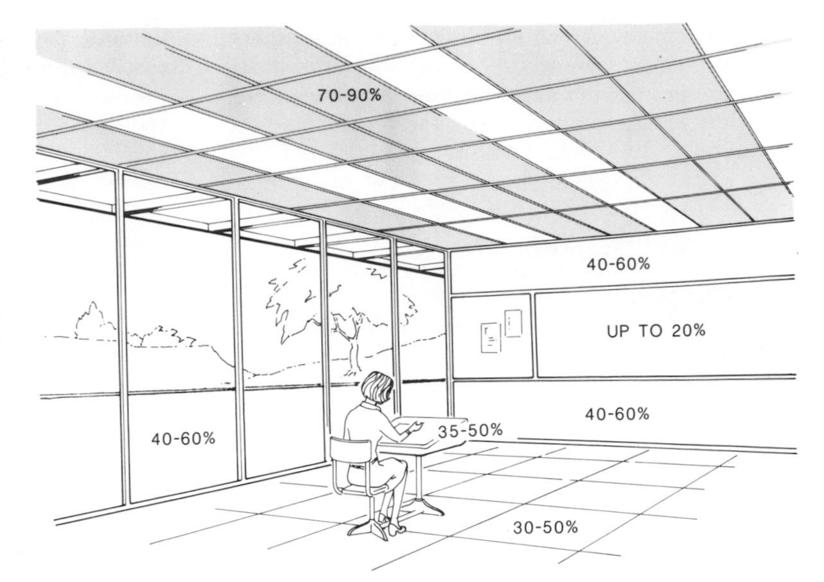

Fig. 6-3. Recommended reflectances for surfaces and furnishings in the classroom. (Note control media is used at windows to reduce exterior luminances so that they are in balance with interior luminances.)

minaires, whether located within or below the ceiling. Ideally, the ceiling should have a luminance greater than, or at least equal to, that of the side walls. It is desirable to have the luminance of the side walls at least one-half that of the immediate surround.

Floors provide the secondary background for desk-top tasks, thus floors should have a non-specular reflectance as high as possible. Floor coverings, whether carpet, resilient title or some other material, should be light in color for maximum non-specular reflectance. Floors or floor coverings should be lower in luminance than the walls. While there are floor coverings with soil-resistant finishes, there are no soil-resistant colors and dark colors become soiled as readily as do the light, higher reflectance colors.

Direct Glare. Any educational space should be free from direct glare. When light sources, electric or daylight, are too bright they produce direct glare. As the brightness is increased, glare sources become progressively distracting, uncomfortable and, finally, produce disability glare such that seeing is actually impaired. In order to prevent direct glare windows must be either located outside the normal field of view or provided with means of control. Shades, blinds, louvers, baffle systems or roof overhangs can be used (see Section 7 of the 1981 Reference Volume). Unless sunlight is desired, it usually should be prevented from entering a space, since it can produce areas of excessively high luminance within the space itself. Luminaires having too high a luminance for their environment will also produce discomfort glare when viewed directly or when reflected from specular surfaces. Appropriate control devices, such as louvers, baffles, etc. must be incorporated into the luminaires.

Visual Comfort Probability. The Visual Comfort Probability (VCP) rating system, which predicts the relative freedom from discomfort glare of a lighting system, is used in evaluating lighting systems in many educational facility spaces. The system takes into account the size and shape of the space; surface reflectances; illuminances; luminaire type, size and light distribution; luminance of the entire field of view; observer location and line of sight; and individual differences in glare sensitivity. The visual comfort probability (VCP) rating of a lighting system expresses the per cent of people who, if seated in the most undesirable location will be expected to find it acceptable. A VCP of 70 or more is considered acceptable in learning spaces. That is, when located at the worst position in the room, approximately 70 per cent of the occupants would sense no discomfort producing glare. This does not mean that the lighting installation is thus judged to be good; it simply means that discomfort glare is not a problem for a major portion of the observers. See Section 2, page 2-24, for a further discussion.

Reflected Glare. Reflections of light sources from specular surfaces in the field of view are referred to as reflected glare. The luminance reflected may originate from luminaires, windows or other sources. These reflections are, in effect, secondary high luminance sources. They are often uncomfortable and distracting and, when they appear near the visual task, reduce task visibility. Reflected glare can be reduced by using non-glossy finishes on furniture, equipment and room surfaces and by using low luminance sources. This does not mean that "sparkle" must be eliminated. A small patch of sunlight, a brilliant view, or a well-decorated Christmas tree may be relevant at a particular time and should not be considered as unacceptable glare. Some sparkle may be a permanent feature, but much will be transitory, providing a welcome variation in the visual environment.

Veiling Reflections. Substantial losses in task contrast, and in visibility and visual performance, can result when light is reflected directly from a specular part of the task itself. These reflections act as if a luminous *veil* were superimposed over the task and are thus differentiated from other reflections in the visual field (see Section 2, page 2-26). This apparent veil between the task and its immediate background varies greatly in degree, but always results in some loss of contrast. The veiling reflection may be quite apparent, as it is when a page in a magazine printed on glossy paper becomes practically obscured when viewed under light from a concentrated source, or it may be quite subtle, as when typed material on matte paper is just slightly obliterated or veiled by the reflection of a large light source from specular ink. Often veiling reflections cannot be detected simply by the eye, but their effect is eventual diminution of visual performance.

Contrast losses caused by veiling reflections may be evaluated in terms of Equivalent Sphere Illumination (ESI). See Section 2, page 2-30. For information concerning ESI calculations and measurements, see Sections 9 and 4, respectively, in the 1981 Reference Volume.

Color. Color in classrooms and other educational spaces is an important consideration for the lighting designer, the interior designer and the architect. Light colors should predominate in educational facilities, both for quality and quantity of illumination and for the positive psychological effects upon students and teachers. Ceiling, wall, floor and furniture colors which conform to the reflectances and to the luminance ratios recommended on page 6-3 will necessarily be the lighter colors. Light colors carried to an extreme, however, may lead to bland and uninteresting results. Touches of accent color or limited areas of relatively bright color, not conforming to the general recommendations given above, can give vitality and dramatic interest to a space, if used with discretion. The design team must be consulted as to suitable colors, finishes and reflectances of furniture and equipment for each educational space.

Where color rendition is important, in such spaces as elementary classrooms, art rooms and clothing classrooms, it is advisable that light sources be selected which will make materials appear as nearly as possible as they do under daylight. Where accurate color rendition is less important, light sources having a more limited spectrum may be employed. Where the light sources has a limited spectrum it must be utilized with caution, with proper consideration given to the activities to be carried on in the space. This is particularly true of spaces from which daylight is largely excluded, or which are used as much at night as during the day.

Shadows. Shadows cast on visual tasks reduce the luminance of the task and may impair effective seeing. Often, when sharply defined or too near the task, such shadows may be particularly annoying. This does not mean that shadows must be completely eliminated, but that shading and shadows must not be excessively dense or confusing. On the other hand, there are many areas in which three-dimensional surfaces are viewed and shadows produce desirable modeling, varying from relatively indistinct shadows to stark shadows producing intentional dramatic effects. Shops, sewing rooms, food rooms, art rooms and many other spaces will require lighting which makes provision for modeling shadows in varying degrees.

Quantity of Illumination

In an era of energy management an attempt to increase the visibility of a task simply by increasing illuminance, rather than by providing illumination quality, is wasteful and may involve the presence of such undesirable effects as discomfort glare or veiling reflections. For these reasons it is necessary to determine required illuminances, in lux or footcandles, after a careful consideration of the tasks involved, the need for both speed and accuracy, and the ages of the observers, as well as other factors. For situations in which veiling reflections can be expected, ESI may be used as a tool in determining the effectiveness of controlling veiling reflections and as part of the evaluation of lighting systems.

Because it is uncommon for learning spaces to contain a single visual task, the determination of appropriate illuminances must begin by evaluating each visual task in terms of such variables as size, contrast and time. Then illuminance can be selected in relation to such factors as the most demanding visual task which occupies an important part of the time spent in the space and the effects of the lighting system on other aspects of the total environment.

The lighting designer must decide which tasks can be adequately lighted within the framework of the budget and other design factors. The usual approach is to select a "commonly occurring, most difficult task" and to provide an adequate level for that task. Reading pencil writing is most often the task selected. Where reading printed materials is the commonly occurring, most difficult task, lower levels would be considered adequate. As other more difficult tasks are attempted, assuming a constant illuminance level, a smaller percentage of the students will be provided with adequate light to reach a given performance level.

In some cases, selection of the commonly occurring, most difficult task in a classroom or other teaching station can result in uneconomically high illuminances. In such cases it is preferable to provide a level which is adequate for the less demanding tasks and to provide increased illuminance at each specific task location where a higher level is needed. Thus the ambient lighting in the space provides adequate general illumination, with task lighting only at specific points. Drafting tables and chalkboards, for example, need the higher levels, but the remainder of the space need not be lighted to the same level, but should not exceed the 3 to 1 ratio discussed on page 6-3. School shops, sewing rooms, art classrooms, areas for the partially sighted, and many other educational spaces can and should be adequately lighted by means of a proper combination of task and ambient lighting.

Illuminance Levels. Illuminance recommendations for only a few tasks found in educational facilities are listed in Fig. 2-2 on page 2-6. In order to select suitable illuminance values for the many other tasks and areas it is necessary to consult other portions, including Part I of Fig. 2-2. From this table a suitable range of illuminance values for design purposes can be determined. It is necessary to select an appropriate design value within the given range and then to modify it by using a suitable weighting factor (see page 2-21). When equivalent contrast ($\tilde{C}$) values are available, it is possible to use them (see Fig. 2-3) and to apply the appropriate weighting factors. In either case the result is the selection of an appropriate illuminance value within the range given.

LIGHTING SYSTEMS FOR EDUCATIONAL FACILITIES

Daylighting

There are numerous aspects of daylight that make its use in educational facilities desirable as a light source and valuable both psychologically and esthetically. With rising energy costs, daylight must be considered as an important source when planning new buildings or retrofitting old ones. The recent trend to utilize educational facilities for evening classes, community functions, and other events has resulted in long hours of lighting use and, therefore, electric lighting is a significant energy consumption factor. Daylight utilization can reduce the total electric usage and net annual energy costs. It can also reduce peak power demand in many areas, assisting energy conservation and further reducing costs.

Utilizing Daylight. Those concerned with providing educational facility lighting should (1) understand and appreciate the opportunities and constraints inherent in daylighting to insure careful planning for its proper utilization, (2) relate the use of solar energy for daylighting to the optimization of the building design in the closely related areas of heating and cooling, and (3) analyze the resulting life-cycle cost benefits. For discussions of daylighting and daylighting design, see Sections 7 and 9 of the 1981 Reference Volume.

The desirability of windows to give contact with the outside world is underscored by surveys which indicate that people actively desire daylight within their buildings. Direct or indirect sunshine evokes positive feelings related to its warmth and brightness. The desire for daylight seems to be related to latitude, for those in the higher latitudes express a marked preference for sunlight inside their buildings, while those nearer the tropics feel the need to exclude direct sunlight and excessive glare and heat.

A means of directing additional light into classrooms is the technique of "beam sunlighting" where direct rays from the sun are reflected by appropriate equipment, such as venetian blinds, to provide daylight meeting suitable brightness limitations.

Well-designed fenestration serves a number of purposes in educational facilities:

1. It provides for the admission, control and distribution of daylight.
2. It provides a distant focus for the eye, relaxing the eye muscles.
3. It eliminated the dissatisfaction felt by some people when an exterior view is not available.
4. If admitted horizontally into classrooms or other spaces, daylight is quite effective for the performance of visual tasks which might be subject to veiling reflections.

Since daylight is a variable source, it must be supplemented by an adequate electric system. Various control mechanisms for electric light have been proposed. The simplest control consists of separate switching for each row of luminaires parallel to the window wall. These, under the control of an alert teacher, can provide considerable improvement over using electric light continuously during all class hours. More sophisticated electronic controls have been proposed. These include those which automatically turn off each row of luminaires as the need for electric light decreases or which adjust the light levels in proportion to the daylight levels. At the present such controls may not be economically feasible for use in the typical school situation. Dependence upon manual control is the alternative. Automatic dimming systems may result in consuming even less energy than an on-off system. Supplementary chalkboard lighting is mandatory with either system.

Control of Daylight. Means of controlling potential glare from the sun include exterior architectural appendages, such as screens, overhangs, or awnings, and interior devices such as shades, blinds, drapes and solar-reflecting or reducing glazing materials. Some degree of light control glazing is desirable, but a transmittance lower than 50 per cent may be uneconomical. This glazing will provide reasonably satisfactory light and heat control, particularly in conjunction with other control devices. Some sort of drapes or blinds should be used with such glazing. Blinds or drapes can be managed by the occupants so as to compensate for changes in daily orientation and for seasonal differences in daylight. A window management procedure should be furnished for manual control situations.

Daylighting Design Evaluation. Daylight utilization must always be coordinated with the total design elements: heating, cooling, ventilation, electric lighting, etc. It must be evaluated using life-cycle cost benefit analyses to assure proper trade-offs between the various elements of the total design.

Electric Lighting

Light Source Selection. At the heart of any electric lighting system is the lamp itself. Of the three main categories, fluorescent, high intensity discharge and incandescent, the most frequently used in classrooms is the fluorescent lamp. Fluorescent lamps have many advantages including high efficacy—about four times that of an incandescent lamp—low brightness and long life—about 20 times that of the incandescent (see Section 8 of the 1981 Reference Volume). For linear applications the 1200-millimeter (4-foot) and 2400-millimeter (8-foot) lamps are the most popular. In 600-millimeter (2-foot) square modules the, U-shaped lamp is the most popular.

The standard cool-white lamp and reduced wattage lamps of similar color give reasonable color rendition and are generally satisfactory for most school tasks. Deluxe and other specialized high color rendering lamps can be employed where superior color rendering is required. These lamps are available in both "cool" and "warm" white color temperatures and are suitable for art, fashion, science and other classroom areas where critical color viewing is important.

High intensity discharge (HID) lamps have long been used in school gymnasiums and outdoor applications. Their high efficacy, compact size and long life make them the most economical and practical lamps for use in high ceiling interior spaces and for almost all exterior applications. Care should be taken in the selection and mounting of HID ballasts to minimize "hum" or "buzzing" in quiet areas such as libraries and study halls.

Incandescent lamps are excellent supplemental light sources for the classroom, but should not be used for general lighting. The optical compactness of incandesent lamps allows for a high degree of light control making them excellent choices for highlighting, display lighting, and for featuring instructional aids in the classroom. They are also easily and instantly switchable and inexpensively dimable.

For more information on the characteristics of each type of light source see Section 8 of the 1981 Reference Volume.

Luminaire Selection. In selecting appropriate luminaires for any educational space the lighting designer should be aware of the purpose for the lighting, *i.e.*, for visibility, or to attract attention, establish a mood, create interest, unify or separate spaces, etc. The designer should consider the appropriate mode of lighting, such as the following:

1. Uniform lighting—permits flexibility of user location; visually bland.
2. Non-uniform lighting—restricts user locations; visually interesting.
3. Directional (small source) lighting—emphasizes occupied zone; imparts modeling and sparkle; lighting level uniformity varies depending on spacing.
4. Indirect lighting—emphasizes overhead zone; produces uniform lighting distribution in occupied zone.
5. Supplementary lighting—can modify visual effect of any of above general lighting systems and focus attention on teaching aids.

Additional factors to be considered in selecting light sources and luminaires can be found in reference 2 and in the discussions above on Quality of Illumination.

SPECIFIC APPLICATIONS

The modern educational facility, whether an elementary school, a comprehensive high school or a university building, contains so great a variety of lighting needs that it is not possible to discuss each in detail within this section. A few specific applications are reviewed here to emphasize the unique requirements of educational facility lighting. See reference 1 for further information.

Lighting the Classroom

Classrooms or "teaching stations" vary from conventional classrooms to any space in an educational building in which a class group and a teacher meet.

Regular Classrooms. Lighting the regular or "academic" classroom is the most common lighting design problem. These classrooms are those in the elementary school in which pupils spend a major portion of their school day, those in the middle and senior high schools in which pupils study the academic subjects, and similar classrooms found throughout the college and university campus. Such classrooms have a commonality of lighting requirements and require that the entire space be provided with an adequate lighting level, with variations resulting from the interplay of daylight and electric light which are subject to reasonable control.

Regular classroom floor areas range from 61 to 83 square meters (660 to 900 square feet) or more. Windows range from a "vision strip" the length of one wall, but only a meter (a few feet) high, to a full floor-to-ceiling window wall. Some classrooms are windowless or have a relatively small window in one corner. Between these extremes is the typical classroom with windows reaching almost to the ceiling above a solid spandrel usually 760 to 910 millimeters (30 to 36 inches) high.

Classrooms of the future may have somewhat different window areas to minimize heat loss and gain and electric lighting loads and thus conserve energy. Double glazing will be needed to conserve energy and yet permit windows adequate to provide the desired daylighting. Thus the lighting designer must work closely with the architectural and mechanical designers to integrate the effects of the heat loss and gain produced by the lighting (both electric and daylight) and the need for heating, ventilating and cooling the classroom.

Open plan arrangements found in many schools require types of furniture and equipment which are also suited to more conventional classrooms. Many schools utilize moveable partitions and portable chalkboards, desks, demonstration tables and book shelving. These can be relocated and regrouped as needed. The lighting system should be equally flexible and responsive, and be adequate for the most frequently encountered, most difficult seeing task. It should be so arranged that the locations of pupils' desks will not be dictated by the lighting, for regular rows of desks are rarely found today, at least in the lower grades. The system should have controls to take advantage of daylight by switching off unneeded luminaires.

Classroom Illuminance Levels. The required classroom illuminance is dependent upon the tasks or task to be performed. The task in most regular classrooms, which occupies at least one-third of each pupils' time, involves writing with a No. 2 or harder pencil on matte tablet paper. Lighting for this task should be provided wherever students are required to write for any prolonged period, such as completing a spelling test or taking notes during a college lecture. If the most-common, most-difficult task to be encountered is reading good quality printed material the appropriate level for that task is adequate. See Fig. 2-2.

Luminaire Selection. Once the desired lighting level has been determined the resulting power budget (see Section 4) becomes one of the limiting parameters. In addition, the factors discussed above in Quality of Illumination provide other parameters, such as visual comfort, veiling reflections and color. Then the type of luminaire selected becomes a function of the ceiling height. In high-ceiling spaces suspended luminaires provide for some reflection from the ceiling and provide downlighting. Luminous ceilings and well-designed indirect lighting systems provide good light effectiveness factors wherever they can be utilized. A great many school classrooms, however, have low ceilings which necessitate ceiling-mounted luminaires. Then recessed luminaires are less obtrusive and tend to have lower brightnesses than surface-mounted luminaires of the same size.

Placement of Luminaires. Classroom luminaires can be arranged in a variety of patterns, the simplest being three evenly spaced rows parallel to the window wall. Depending upon the type of luminaire selected, however, other patterns may be much more effective. Perimeter arrangements offer a number of advantages. Perimeter luminaires should be positioned no more than 0.6 to 0.9 meters (2 to 3 feet) on center from the wall. Peripherally deployed luminaires, bringing light at wide angles from several directions, including that reflected from the walls, can help reduce veiling reflections. In addition, there is more uniform illumination in the room and a better brightness balance. Perimeter lighting may also result in somewhat improved brightness and visibility of chalkboards, tackboards or teaching aids displayed on the wall. The improved wall brightness also makes the room more pleasant.

Special Area Lighting. In addition to the general lighting of the classroom there are a number of areas which often should be provided with supplemental lighting. Among these are chalkboards, carrels, and locations to which attention is to be directed.

Perimeter lighting, even though it may wash the walls with light, often fails to illuminate a chalkboard properly. Supplementary lighting, separately switched, may be the best means of adequately lighting the chalkboard to the levels required without veiling reflections.

Modern classrooms, particularly at the elementary and middle school levels, and libraries may contain carrels. Carrels reduce the effectiveness of general room lighting because the carrel walls create light barriers.[3] By introducing light at the proper height at the sides of the carrel a desired illuminance can be secured without veiling reflections. If a luminaire is improperly mounted directly in front of the student, as it usually is, the lighting is largely ineffective, for it contributes to veiling reflections and so reduces contrast. A small, properly shielded fluorescent lamp on one side of the carrel results in improved visibility and an adequate quality and quantity of light. With a lamp on each side of the carrel the improvement is even more marked. Care should be taken in the selection of the carrel, for the interior surfaces, surrounding the task area as well as the task area itself, should be light-colored matte surfaces with a reflectance of 35 to 50 per cent.

Special area lighting can also be employed in the classroom to direct attention. In a laboratory it can be used to direct attention to a demonstration; in other classrooms it can be used to light a small area such as an easel or materials placed on a table. In order to utilize such special lighting effectively it is necessary to lower the level of the general illumination in the room.

Lighting Controls. Control or both daylight and electric light to achieve proper illuminance levels and brightness balances is an important part of classroom lighting design. Daylight can be controlled by various types of blinds, either venetian blinds or roller shades, by architectural features such as overhangs, and by exterior elements, including landscaping. Glass having the desired low transmission characteristics also assists in daylight control. Drapes which are not opaque are also of value, even when blinds are also employed.

The simplest means of controlling electric light is the separate switching of each group of luminaires. A further refinement can be provided by utilizing luminaires with three lamps, so that either one, two, or three can be used by proper switching. Dimmers can be used, but may not be practical in the typical classroom at this time.

It is necessary to adjust the controls for both daylight and electric light throughout the day. Lighting designers should provide teachers and administrators with instructions for the proper use of the available controls. It may be advisable to provide a brief printed set of directions, sealed in a protective cover, and mounted adjacent to the light switches in each classroom.

Lighting for Audiovisual Presentations. Television, motion pictures, and slides and film strips are used extensively in classrooms.[4] For effective viewing it is necessary to reduce or to turn off the general overhead lighting. The room should be darkened, but not blacked out. (Black-out drapes are not recommended for use in the classroom.) Careful adjustment of drapes or blinds can usually provide adequate daylight control. Any needed light can be provided by the regular luminaires if those over the screen can be turned off, while others are operated at the lowest level available. An alternative is supplementary lighting, such as controlled downlighting or adjustable spotlights. This is particularly important if students are expected to take notes during the audiovisual presentation.

Task/Ambient Lighting

Numerous areas in schools have lighting design criteria different from those of regular classrooms. These criteria may put a greater emphasis on the discrimination of color, three-dimensional modeling, or variations in brightness. While it may be necessary to provide relatively high uniformly-distributed levels in regular classrooms, it is practical, in the interest of wise energy use, to light many teaching stations for the illuminance suitable for the generalized activities being carried on and then provide supplementary lighting where it is needed. The ambient lighting should be suitable for reading printed material. The supplementary lighting, often in the form of "task lighting," is provided at points where a higher light level is needed. This task lighting can be supplied by small sources concentrated on these areas. At times this supplementary lighting includes a directional component to provide modeling or to serve some other need. There are many opportunities in schools to provide task/ambient lighting and thus reduce the overall energy requirements.

Sewing Rooms. One of the most difficult visual tasks is that of sewing. The process of seeing fine stitching on cloth where the thread matches the cloth, is inherently difficult and fatiguing. The stitching is seen in part by the reflected glint from the thread. Task lighting is needed which can provide the higher levels required. A directional component to the light is often required. Such task lighting cannot usually be provided by ceiling mounted luminaires, so some sort of portable or machine-mounted lamps may be required.

Shops. The lighting of each school shop should follow the best industrial lighting practice for the types of activities provided in the shop, with special emphasis on assisting the effectiveness of manual operations and on making all elements of danger visually obvious (see Section 9). Special paint colors should be adopted for machine controls and for machine parts which represent a hazard for the student.[5] Important work points on each machine should be painted to make them stand out.

The ambient lighting in each shop should be supplemented by task lighting, either fixed or portable. Supplementary task lighting can be mounted on many types of machines. In other cases portable or hand-held lamps, such as the familiar trouble-lamp used in auto repair, may be required. Task lighting should be provided wherever moving machinery is in use.

Art Rooms. Because the appearance of colors in an art room is paramount, the light sources used should render colors accurately. Lamps with high color rendering capability will be required, even though they may be of lower efficacy (see Section 5 of the 1981 Reference Volume). This is true of both the ambient and the task lighting. Supplementary lighting from concentrating directional sources is useful with displays and models, for it provides improved visibility and creates desired highlights and shadows needed for modeling purposes. Adjustable luminaires can be used to provide this type of lighting. In short, lighting for the art room should be as versatile in many respects as that for a small theater stage.

Classrooms for the Handicapped

Certain classrooms may be specifically designed for handicapped pupils.[6] Those handicapped by impaired vision or hearing should be provided with high quality illumination, adequate in quantity. In most cases partially seeing individuals benefit from high illuminance levels. (There are some exceptions, however, and special provisions should be made for those who need lower levels. Information on such special cases should be provided by the school authorities and the lighting designer may find it necessary to arrange specialized lighting.) Supplementary lighting on chalkboards, charts and displays will be needed. Special task/ambient lighting may be of value in some situations.

Pupils having impaired hearing often depend upon speech reading (lip reading) for much of their understanding. It is necessary that the speaker's face be well lighted. The illumination should provide sufficient modeling for the movements of the lips and other facial features to be readily perceived. This is not unlike the need for modeling in an art room where three-dimensional objects, such as sculptures, are to be viewed.

Learning Resources Centers and Libraries.

Libraries in schools range from very simple rooms in the elementary school (with a reading area surrounded by book stacks) to the very complex learning resources centers in high schools and the general, special and technical libraries in colleges and universities. Many libraries include a reading area requiring uniform illumination, adequate for reading printed materials, plus stack areas with special lighting to make possible the reading of information on the spines of shelved books. Libraries may include such spaces as the circulation desk and card catalogs, conference and seminar rooms, display and exhibition areas, microform viewing areas, audiovisual rooms, technical processing areas and offices. Some may have typing rooms for the students' use. Each of these areas present different and, often, unique lighting problems.

Reading, by far the most common visual task in the reading and reference area, requires that the lighting be suited to the wide range of materials to be read. Type smaller than that on a newspaper page will be encountered. Considerable handwriting is performed, so lighting adequate for reading pencil writing should be provided. Room surface reflectances recommended for school classrooms will usually provide satisfactory luminance ratios. Some special area of task lighting will be required, for example, adjacent to the machines in the microform viewing area, so that the machines themselves can be located in an area of reduced lighting. If proper precautions are not taken in the reading areas, veiling reflections may greatly reduce visibility. Additional information on the special problems of large libraries will be found in Section 7.

Physical Education Spaces

Physical education spaces in schools are usually the largest spaces available in the building. For this reason they may be planned and used as multifunction spaces to accommodate large groups for many activities not related to physical education and athletics.

Multipurpose Rooms. The elementary multipurpose room is often planned to provide for physical education classes, after school recreation, dining and auditorium activities, including musical and dramatic rehearsals and presentations. Community activities are also accommodated. Usually there should be no windows in the multipurpose room (or they should be controllable), for daylight invariably produces unwanted reflections and, often, blinding glare. The lighting should be appropriate for the games to be played and should be adaptable to levels suited to assemblies and other activities. If there is a stage, either built-in or portable, suitable stage lighting of a simple type should be specified.[7]

Gymnasium. The high school gymnasium or college field house is a multifunction space. Besides physical education and athletics it is used for graduations, assemblies, dances, concerts and community meetings. The resulting diversity of seeing tasks and lighting needs dictates a choice of lighting levels. A good design technique is the provision of the highest lighting level needed with flexible circuitry to assure a range of lower levels. Supplementary low-level lighting may also be appropriate. Portable or temporary lighting equipment should be included so that special effects can be obtained for the non-athletic events.

High intensity discharge sources or other high efficacy light sources may be suitable for gymnasium lighting. These should be supplemented by sources which restart more rapidly and are suited to the other uses of the gymnasium. Gymnasium luminaires should be covered with protective grids which may reduce light output to such an extent that it must be compensated for in the design.

See Section 13 for further information on sports and recreational area lighting.

College and University Applications

The recommendations given above for lighting typical classrooms and other school spaces are applicable to community college, college, and university classrooms. Though more typical of the college or university, the following may be found in some high schools.

Lecture Rooms. Lecture room capacities range from 50 or less to several hundreds. Floors may be flat, just slightly sloping or steeply ramped. The problems of lighting lecture rooms become more complex with an increase in size and the demand for good observation of demonstrations. A lecture room should have a general lighting system which is flexible enough to provide at least two illuminance levels, the higher for note-taking and a subdued level for demonstrations. An intermediate level is also desirable. When downlighting is used it should be very carefully designed to avoid loss of visibility due to veiling reflections.

If a demonstration table is to be used, directional downlights should be located within a 40- to 60-degree angle above the horizontal in relation to the location of the lecturer. This minimizes glare and provides good lighting for the speaker's face. Such lighting can be arranged to allow use of an audiovisual screen at the same time that the lecturer is speaking.

Dormitory Rooms. Dormitory rooms are commonly provided with two systems of illumination, one of relatively low level for general illumination and one of a higher level of task lighting for study purposes. Direct-indirect table or floor lamps are suited to reading and study. If study carrels are used the luminaires should be mounted at the sides as discussed above. Illuminances on desk tops should be adequate for pencil tasks. When the room is shared by two or more students it should be possible for one to retire, after turning off the ambient lighting, and for the other to continue to study by the task illumination alone. This situation may partially violate the desirable luminance relationships between the task and the visual surround, but it is a necessary condition of dormitory life.

Corridors and Stairs

There is no one illuminance level applicable to all corridors or all stairways. Each must be considered in relation to the occupants of the building and their needs.

Corridors. Corridors are the transition areas from the higher luminances of the out-of-doors to the lower luminances of the interior spaces of a school during daylight hours. Corridors adjacent to entrances can be lighted to a slightly higher level with daylight and electric lighting to aid in the transition. This will also serve to indicate the location of exits at night. Corridors should be adequately lighted to promote safety and discipline. Corridors lined with lockers require higher levels than those used simply for passage. Supplementary lighting may be required at the positions where monitors or security personnel are stationed. When corridors are used also as work or study areas, as is true in some elementary schools, they become extensions of the classrooms and so should be as well lighted as the classrooms themselves.

Corridor lighting also provides an opportunity to add visual interest to school environments and importance to displays, bulletin boards and posters. Special attention to the lighting of architectural elements will add to the pleasantness and visual vitality of corridors.

Stairs. The lighting of stairwells must be very carefully planned so that the edge of each step is properly illuminated, there are no high luminances in the line of vision, and landings are adequately lighted. Corridor lighting seldom illuminates stairs and stairwells adequately, so it is important that corridor and stair lighting be coordinated to eliminate dark areas, especially where there are smoke-screen doors.

Emergency Lighting. The illumination provided by the emergency lighting system should be adequate to permit an orderly, accident-free exit from anywhere in the building. Sufficient light must be provided in the exit route to make it readily identifiable. Emergency general lighting should be provided for windowless classrooms and offices, and for auditoriums, cafeterias, multipurpose rooms, and other large-group spaces. The lighting must be so planned that panic will be avoided and safe evacuation assured. This can only be accomplished if all routes to the outside are clearly evident. National, state and local safety codes must be complied with.

The emergency lighting equipment should consist of the regular exit signs with emergency light sources and luminaires on special circuits which will be activated immediately in case the normal electric power serving the space fails. The exit illumination should provide adequate light and focus attention on egress passageways and exits. Angles and intersections of corridors, stair landings and exit doors should have emergency illuminated directional signs.

Care in specifying emergency lighting units suited to the locations is very important. Emergency units should be protected in school situations where mischief-prone youngsters may find the unit or its test switch an attractive target. The units selected should be capable of resisting such actions and remain in working condition.

See Section 2 for further information on emergency lighting.

Outdoor Lighting and Security Lighting

Many educational buildings are used after dark, thus it is important to give careful consideration to the various aspects of outdoor lighting.

Building facades, approaches, and outdoor activity areas should be illuminated both for the activity itself and for general safety, as well as protection against vandalism and theft. Lighting of outdoor areas such as parking lots, roadways, and athletic fields are discussed in Sections 13 and 14. Outdoor lighting and security lighting are so closely related that they must be considered together. Often the same installation will serve both purposes.

Outdoor Lighting. Outdoor lighting for schools should facilitate legitimate nighttime approach and entry, whether on foot or by vehicle, provide security for the building and its contents, and make the decorative aspects of the building visible. In many cases the latter will be largely achieved by proper provision for the other two. Walks, driveways, internal streets and parking lots should be illuminated at night. They may be lighted using conventional walkway and roadway lighting patterns with special lighting for parking lots. Entrance and exit areas of the building should be high-lighted with efficient light sources. This lighting, if properly located, will tend to make the area about the building attractive.

Security Lighting. In some cases total darkness may prove to be the best security measure, so the need for "security" lighting must first be established. When its use is indicated, the most effective security lighting for a school building is facade lighting. While this type of lighting can result in decorative effects when properly applied, this is not its primary purpose. Such lighting, supplied by sources located away from the building, facilitates security by (a) providing direct viewing of persons and of the structure, (b) allowing indirect viewing of the enlarged shadows cast by intruders as they approach the lighted surface of the building, (c) permitting observation of intruders in silhouette against the lighted background before they are in range of the illumination from the luminaires, and (d) reducing or eliminating the glare distraction which often accompanies floodlighting mounted anywhere on the building itself.

Facade lighting for security purposes can be placed on poles, trees, adjacent walls or special pedestals. Often low-level luminaires on pedestals can be hidden behind plantings so that the sources are not readily apparent. Available high efficacy sources, such as high intensity discharge and low pressure sodium, are well suited to the type of dusk-to-dawn security lighting required. Combined with properly located luminaires, the result can be an attractive, efficient and functional nighttime security lighting system. Lamps need not provide high color rendition to be effective, for their color may actually improve the appearance of a building. See Section 9, page 9-67 for additional information on security lighting.

REFERENCES

1. *American National Standard Guide for School Lighting*, ANSI/IES RP-3, 1977, American National Standards Institute, New York. (Committee on School and College Lighting of the IES: "American National Standard Guide for School Lighting," *Light. Des. Appl.*, Vol. 8, p. 12, February, 1978.
2. Committee on Design Practice of the IES: "Factors to be Considered in Lighting Design," *Light. Des. Appl.*, Vol. 4, p. 38, April, 1974.
3. LaGiusa, F. F. and McNelis, J. F.: "Guide for Evaluating the Effectiveness of Supplementary Lighting for Study Carrels," *Light. Des. Appl.*, Vol. 1, p. 6, November, 1971.
4. Committee on School and College Lighting of the IES: "Guide for Lighting Audiovisual Areas in Schools," *Illum. Engl.*, Vol. LXI, p. 477, July, 1966.
5. *American National Standard Safety Color Code for Marking Physical Hazards*, ANSI Z53.1-1978, American National Standards Institute, New York.
6. Herron, P. L. and LaGiusa, F. F.: "Brightness Variations Affect Deaf Learners' Attention," *Light. Des. Appl.*, Vol. 5, p. 30, February, 1975.
7. Committee on Theatre, Television and Film of the IES: "Lighting for Theatrical Presentation on Educational and Community Proscenium-Type Stages," *Illum. Eng.*, Vol. 63, p. 327, June, 1968.

Institution and Public Building Lighting

SECTION 7

Banks, churches and synagogues, hotels and motels, food service facilities, libraries, and museum and art galleries are usually considered to be institutions or public buildings. The lighting of the spaces peculiar to these buildings is included in this section.

Appropriate design illuminances for areas and tasks in institutions and public buildings are given in Fig. 2-2, page 2-5, in terms of illuminance categories, and within this section in illuminance where specific values have been found to be effective based on criteria other than visual task performance.

For the lighting of office areas, merchandising areas, and exteriors of institutions and public buildings, see Sections 5, 8, and 12, respectively. For general information on interior lighting design and energy management, see Sections 1, 2 and 4.

BANKS

The various functions and tasks that occur in a bank are mainly the same as those that occur in offices, *i.e.,* conference areas, accounting, general and private offices, bookkeeping, etc. (see Section 5, Office Lighting). There are, however, several specific areas with special banking functions where the lighting needs may be different.

Specific Areas

The following are several specific banking areas to be considered (illuminance recommendations for areas and associated seeing tasks are given in Fig. 2-2):

NOTE: References are listed at the end of each section.

Lobbies. Historically, bank lobbies have had very high ceilings, but today, because of high building costs, they are no more than 3.5 to 4.5 meters (12 to 15 feet) high. Where very high ceilings exist, the use of high intensity discharge lamps should be considered, not only for energy utilization but also for the economy of relamping and other maintenance procedures.

Special attention should be given to writing areas so that there is adequate illumination for the activities performed there. When promotional incentives, such as merchandise, are located in the lobby, there should be provisions for highlighting to create a point of interest (see Section 8, Lighting Merchandising Areas).

Tellers' Stations. The most active areas in a bank are the tellers' stations. Here the lighting should provide for fast, accurate transactions. Because of highly polished material for the deal plate, there is a tendency for reflections from lobby lights. One way to reduce reflections is to utilize a low brightness luminous ceiling. Recessed downlight luminaires directly over the deal plates should be avoided as they tend to cause shadows and discomfort glare for the teller.

The interior lighting at a *drive-up window* should be similar but due to sloping of the window glass, luminaires behind the drive-up teller should be of low brightness to minimize any reflections on the glass itself.

Probably the most neglected area of bank lighting is the *outdoor drive-up area.* The outside lighting there should be about the same magnitude as the interior to avoid a mirror effect on the glass, looking out from the drive-up teller's position. The lighting should be designed to light the person in the vehicle and the drive-up unit, not the top of the car. In addition to the "visual" drive-up teller facilities, where the teller visually sees the client, there are television or remote units requiring lighting for camera needs.

Security Lighting. Security lighting should be incorporated in accordance with the Bank Protection Act of 1968, whereby adequate night lighting and exit lighting and lighting on the vault are on at all times. Also, there must be adequate interior lighting for an alarm camera system. Lighting requirements for cameras and films used should be evaluated.

CHURCHES AND SYNAGOGUES[1]

Skillfully used lighting can make worship services more meaningful and enhance the architectural design of the space. The lighting can mold and give depth, and can subdue or accentuate, or perhaps change its accent, as the service proceeds. In certain interiors it can add a fourth dimension: a suggestion of the infinite.

For good energy management, consideration should be given to the use of the least wattage to create a desired mood or to perform visual tasks, and the use of adequate controls and maintenance procedures. See Section 4.

Entrances

In the entrance vestibule or narthex, the lighting should enable the quick recognition of faces, facilitate the taking of notes of names and requests, and provide a transition between the exterior and main worship area brightness. Diffuse illumination should be used so that faces appear pleasantly lighted and are not made to appear lined or strained by highly directional harsh sources.

Fig. 7-1. Illuminances Currently Recommended for Churches and Synagogues*

Area	Illuminance	
	Lux	Footcandles
Altar, ark, reredos	1000[b]	100[b]
Choir[a] and chancel	300[b]	30[b]
Classrooms	300	30
Pulpit, rostrum (supplementary illumination)	500[b]	50[b]
Main worship area[a]		
Light and medium interior finishes	150[b]	15[b]
For churches with special zeal	300[a]	30[a]
Art glass windows (test recommended)		
Light color	500	50
Medium color	1000	100
Dark color	5000	500
Especially dense windows	10000	1000

* Maintained target values on tasks.
[a] Reduced or dimmed during sermon, prelude or meditation.
[b] Two-thirds this value if interior finishes are dark (less than 10 per cent reflectance) to avoid high luminance ratios, such as between hymnbook pages and the surround. Careful planning is essential for good design.

Main Worship Areas

There are vast differences in the service and liturgy of the many faiths and denominations and the lighting designer should be familiar with their customs in order to assure proper lighting emphasis at the proper time. Fig. 7-1 lists suggested illuminance values for design based on needs during various types of religious services.

General Lighting. There should be appropriate general lighting for reading, moving about, visual social contact, and to help the worshipper relate to the structure and its features. In many churches there is a trend away from the traditional service of listening, watching and meditating, to a service that includes more participation. Higher general lighting in the space can encourage the feeling of being part of a body of people. Such participation also means more reading—requiring particular attention to light at the pew.

Often there are two components to the general lighting: (1) direct lighting for the pews, and (2) indirect lighting to relieve shadows and to create desired brightnesses on the structure. Sometimes, indirect lighting provides all the general illumination—particularly if lighting equipment cannot be mounted on or in the ceiling.

Lighting from one direct point source creates dark shadows and specular reflections. This may be desirable to highlight an object, but it is undesirable where people are attempting to read or follow printed material. Also it can make the leader of a service appear unpleasant through deep eye and other facial shadows. An overlap of light from direct sources or the use of indrect lighting with direct light will soften shadows and reduce specular reflections.

Accent Lighting. Certain parts of the worship area become central at different times in the service, and when so, they should be highlighted. Those areas may be where the worship leader, the choir, the Torah, the communion table, the stations of the cross, the Bible and the Ark are located. Controlled beams of light should be used that will properly render the features in these areas and not create glare for those participating. This will mean careful choice of beam spread, intensity and location of spotlighting.

The location and orientation of the congregation should be kept in mind so that any directional lighting will not create glare. Particular attention to the shielding of directional lighting is needed for the church-in-the-round.

Controls. Lighting can help shift attention and emphasis during a service. By switching or dimming, the appropriate changes can be made in the brightness of different parts of the worship area. When dimming is used it produces these changes much more subtly than switching. This is particularly true of general lighting or lighting of large features.

Fig. 7-2. Tent-type church. Exposed rafters conceal equipment which is aimed down and forward for lighting pews and chancel. Light from coves diffuses, balances and supplements the downlighting to create reverent mood desired during a service.

Church Architecture and Lighting

In the nave or main auditorium the quantity of light, and use of patterns of light and shade, vary widely with different architectural styles. See Figs. 7-2 through 7-5. The lighting designer should consult closely with the architect to understand the purpose behind the architectural style being used and to develop the lighting approach for it. They should cooperate through the following stages of translation: (1) the architect's concept of the space, (2) the brightness patterns desired and (3) the lighting equipment needed. For a further discussion of the relationship of light to architecture, see Section 1.

If lighting equipment is to be concealed in or behind a structural element, space is often limited. If care is not taken, the results may be uneven illumination and excessive brightness from spill light on exposed surfaces adjacent to the equipment. For example, incandescent lighting in a small cove or cornice should use a large number of small devices rather than a few high output units unless very compact, sophisticated optics are used; otherwise the adjacent ceiling or wall could be unevenly and excessively bright. When their color, dimming and starting capabilities are acceptable high intensity discharge and fluorescent sources should be considered. See Fig. 7-3 for one fluorescent application.

Fig. 7-3. As in Fig. 7-2, the ceiling design hides luminaires from the seated congregation. Incandescent and fluorescent sources (inset) provide varied lighting effects during a service. As the wiring is semi-exposed, lighting changes can be easily made in future years.

Fig. 7-4. Unsymmetrical design. Floodlights across court at left illuminate the ceiling of nave and part of chancel. Other exterior floodlights, small in size, aim light down at a pool in the court so that reflected rays play faintly over the nave ceiling. Chancel and pew lighting come from behind the ceiling beams.

Fig. 7-5. Colonial church. The lighting designer worked with the interior designer to arrive at the feeling and the image required. Tradition and a worshipful atmosphere are preserved. After dark three systems of nave lighting provide effectively for evening worship. On rainy days, most or all of the evening lighting is used. On sunny mornings, interior lighting may be used to retain attention indoors.

Lanterns or other suspended decorative luminaires may be effectively used with many architectural styles. If they are to produce direct general illumination, however, care should be taked that there is sufficiently wide distribution of light for good coverage but without discomfort glare. This may be impossible if appearance dictates a very low suspension; in this case, other sources should be used to provide the illumination and the suspended equipment used as a luminous decorative element.

The reflectance of some large surfaces in the worship area—usually wood—may be very low. Such surfaces should be lighted to make them visible and to relieve an otherwise "too-dark" atmosphere, but not so much as to make them brighter than they would be expected to appear normally.

Art Windows

As in the architectural considerations for lighting, the lighting designer should work closely with the art window designer to determine the desired appearance of the lighted window. In all stained and art windows, the density, diffusion and refractive qualities of the glass or plastic will determine the light source luminance and size to be used.

It is not necessary to achieve a perfectly flat or uniform lighting effect. In fact, it is often not desirable. It is almost always necessary to set up a trial lighting system to see how the glass responds to different lighting. In such trials, a great deal of equipment may be necessary both to get sufficient light on the glass and to have enough different lighting approaches to examine.

Generally the lighting of art windows serves two main purposes: (1) for viewing from inside during nighttime services, and (2) for viewing from outside for passing traffic.

Viewing From Inside. The window can be lighted with outside floodlighting units if the glass has sufficient diffusion and refracting qualities (from irregularities on the surface of the glass and within the glass). If the glass is not extremely diffuse, the units should be located so that they are not seen through the glass and do not produce visual "hot spots." Clear stained glass needs a luminous background such as a closed light box around the outside of the window.

Viewing From Outside. The floodlighting approach above also can be used, but with equipment located inside. Spots of brightness may be more difficult to avoid, however, since lighting equipment is usually most conveniently located on the ceiling and the viewer is usually below the window. A larger number of lower intensity floodlights can make the spots of brightness less apparent. For clear stained glass a movable screen or drape can be used on the inside, lighted (either transilluminated or lighted from the window side) to form a luminous background for clear glass; it can be moved away for times of viewing from the inside.

HEALTH CARE FACILITIES[2]

The lighting of health care facilities presents many problems involving a wide range of seeing conditions. Optimum seeing conditions should be provided for doctors, nurses, technicians, maintenance workers and patients. For a better appreciation of the principles involved, a review should be made of Section 3, Light and Vision and Section 5, Color, of the 1981 Reference Volume; and Section 11, Interior Lighting Design, and Section 2, Lighting System Design Considerations, in this volume.

Many activities in health care facilities are not

related directly to patient care but are necessary as supportive institutional functions. Areas such as business offices and laundries are not discussed in this section (See Sections 5 and 9). Some of the activities are identical or similar to ones in other institutions. These include libraries and kitchens. There will be some locations in which there is overlap in recommendations. For example, the patient room may be similar in its lighting requirements to the hotel room, *when it is used for minimal care patients,* yet the lighting must be considered differently in the patient room for the sick, the aged or the infirm.

Illuminance recommendations for health care facilities are given in Fig. 2-2, page 2-6. Where higher illuminances from localized lighting are required as in surgery, obstetrics, dentistry, emergency treatment and autopsies, it is desirable to insure comfortable lighting conditions by limiting luminance ratios between the task and other areas in the normal field of view; *i.e.,* the luminances between the task and adjacent surrounding should be limited to 1 to ⅓, between task and remote darker surfaces to 1 to ⅕ and to remote lighter sufaces to 1 to 5. To help achieve these reflectances, room surfaces should be within the following percentage ranges: ceilings, 80 to 90; walls, 40 to 60, furniture and equipment, 25 to 45, and floors, 20 to 40.

Types of Facilities

Health care facilities usually include acute general hospitals, chronic general and chronic specilized institutions for the care of the physically and mentally ill, and the extension of these services into facilities which offer more than the patient's own residence in professional care.

In describing good practice in lighting such institutions, the designer should take into account not only the immediate objectives, but also the services which might be required in the future. For example, a facility designed as an extended care unit in conjunction with an acute care hospital may find its beds recertified as acute care beds.

The Acute Care Hospital. While an acute care hospital might be faced with all of the diverse lighting design considerations in a complete multidisciplinary one, there may be some which will not. Obstetric and pediatric hospital sections are being allocated to certain hospitals and abandoned in others. This trend is increasing and results in greater specialization in each hospital. This will result in greater demand upon the support facilities.

Another trend is the expansion of outpatient services, particularly for those functions previously considered in-hospital ones. This does not mean complete abandonment of such functions for in-hospital patients, but it means a major reduction of these functions with a reduction of space allocation. Conversely, the planning of facilities for these and other activities such as laboratories will be partially moved to freestanding clinic buildings or office buildings designed for physicians and dentists. All of these require special illumination.

There are also constant transitions in the instrumentation of medical, surgical and dental practice. The computer and its application to radiology may entirely alter that specialty's requirements.

Where once there were large multibedded open wards now there are either single or double occupancy rooms. Where, in intensive care areas, there were multiple bed spaces individualized by curtains, there are now either semi-enclosures with glass observation windows or cubicles acting as open bed bays from a central hall or work space. By federal guideline these must have access to windows to afford the patient access to daylighted surroundings for orientation. The night and task illumination, however, presents the same problems for the designer.

The Chronic Hospital. The chronic long care facility is fortunately largely disappearing. The psychiatric institution is being replaced by mental health units often situated in general acute hospitals. The contagious disease hospitals, including the sanitaria for tuberculosis are also on the wane. This means provision for psychiatry and contagion in the acute care hospital.

The Extended Care Facility. The institutions which are proliferating are the extended care facilities: the nursing homes. Most of these are largely inhabited by an aging population. Many of these persons have vision difficulties: cataracts, yellowed lenses, aphakia or presbyopia. They, therefore, pose special problems for the lighting specialist.

Other Facilities. Free-standing office buildings, clinic buildings and medical teaching facilities also deserve consideration as they form an appreciable and growing part of health care. Every physician's office suite should contain lighting equipment that will provide that physician with the quantity, quality and directionality to permit performance of all functions with ease.

Lighting Objectives

In recent years there have been many changes in lighting concepts and in solutions to lighting problems. Basic research has increased our knowledge of visual requirements, industry has provided new equipment for producing light and modifying its quality, and there is a greater concern for energy conservation and management (see Section 4).

These years have also seen great growth in the medical techniques which have created new challenges for the lighting designer to provide the best lighting for the new visual tasks. For example, there is the problem of constant patient observation in intensive care units containing monitoring equipment, which must be constantly under meticulous visual and auditory surveillance, yet the illumination must be unobtrusive while being fully adequate so that the observer will not be visually fatigued.

A new appreciation of the sensibilities of the patient is another dimension to be considered by the designer. Although the lighting should serve the demands of the medical and nursing attendants, it also should be suited to the comfort needs of the patient. The patient must feel at ease in this environment. The illumination in multi-bed rooms should be designed to be unobtrusive to one roommate while remaining adequate for the other.

Lighting Design Considerations

A hospital is a very complex institution and has almost an infinite number of functions being carried out by persons who are normal or sick.

In designing the lighting system for a new or renovated space, consideration should be given to the needs of the occupant of that space—the visual tasks to be performed, the desired appearance of that space and energy and economic restraints. The recommendations that follow are for visual effectiveness and have been established based upon the state-of-the-art. It is recognized that there is still a need for further research in many areas.

The personnel working in the diagnostic and therapeutic facilities encompass a wide variety of ages, as do the patients, and, consequently the lighting should be planned to be adequate for all. The exact need for good color rendering appears to be obvious in most task-related areas of the hospital.

Task Lighting. In areas where visual tasks are performed, from surgical procedures to patient reading, the tasks are the focal point with less in the surround. Lighting for task performance depends upon (1) the importance and delicacy of the particular task from both the standpoint of time allowed and accuracy required without undue fatigue, (2) upon the person performing the task and (3) upon the task itself. See page 2-3.

Seeing is a dynamic activity and eyes do not remain fixed upon a single point, but move to all parts of the task and beyond it. For this reason, it has been suggested that where task levels are high, as in surgery, consideration be given to three zones of lighting—the highest in the operative field, a second of lower level surrounding the table, and a third peripheral one grazing the wall.

Hospital areas are designated for specific activities and, for these, some definitions of lighting requirements both quantitative and qualitative can be suggested.

Patients' Rooms—Adult. The patients' room lighting problems are to reconcile the need for lighting at various times by various people and usually to provide such lighting as simply and economically as possible. The patient, nurse, doctor and housekeeping personnel require different illuminance levels, in the same room, to accommodate their individual needs. This range of lighting is needed for a variety of nursing services; it should be provided in a way that is not objectionable to other patients in the same room and that caters to the lighting needs and whims of patients whose only field of view may be the ceiling.

Nursing Services. Since the primary purpose of the hospital is to restore the patient to health, lighting for nursing services and critical examinations is common to nearly all hospitals. The variation is in the provision for patient comfort during convalescence. This may vary greatly, depending upon the health and mobility of the patients, the quality of services supplied by the hospital, whether the hospital is public or private, and perhaps most important, on whether a room is for single or multiple occupancy.

Routine Nursing. For control of the general lighting level in patients' rooms to create a soft light for the patients' comfort, the use of variable-control dimmers located at the door of the patients' room is suggested. The nurse should never have to search for light to read charts and thermometers. Should more lighting be needed, the patient's reading light may be used. The luminance of luminaires and nearby surroundings should be less than 310 candelas per square meter (90 footlamberts) as usually seen from any normal reading or the patient's bed position.

Luminaires to meet these conditions should have low luminance. One or more such luminaires in a single or multiple occupancy room may be needed to provide general lighting 760 millimeters (30 inches) above the floor for normal use. To prevent excessive spottiness of general lighting, the installation should provide a lighting level ratio of not more than 1 to 5 on a horizontal plane 760 millimeters (30 inches) above the floor within a radial distance of 2.4 meters (8 feet) from the maximum level on that plane.

Observation of Patients. There should be provision for local low-level illumination of a color quality that will provide for proper diagnosis of the patients appearance. There should be lighting at each bed and its floor area so that the nurse may frequently observe the patient and equipment, such as drainage tubes and containers, during the night with minimum disturbance to patients. This light should be switched at the door, and may also be controlled by a dimmer. When the observation lighting must be left on all night, or when higher levels are needed, temporary screening from other patients may be necessary.

Night Lighting. Wall-bracket combination lighting units for patients' use frequently incorporate a night light with switch at the bed. A night light of this arrangement is desirable for the occasional use by patient or nurse. However, when it is left on continuously, the luminance produced in the surrounding field of darkness is sometimes a source of annoyance to patients wishing to sleep.

For continuous use, the night light recommended incorporates a low-brightness luminaire with louvered or refractive cover, flush wall type, installed so that its center is approximately 360 millimeters (14 inches) above the floor to direct a low illuminance along the floor where it is needed for walking or moving about in the room.

The important criterion for night lighting is limiting the source luminance. This luminance should not exceed 70 candelas per square meter (20 footlamberts) for continuous use, or 205 candelas per square meter (60 footlamberts) for a short time.

Critical Examination. The lighting for critical examination of the patient should be of a color quality that will not distort the color of skin or tissue and of a directional quality to permit careful inspection of surfaces and cavities. The examination lighting should, however, be confined to the bed area and should provide the recommended lighting in the center of a circular area 0.6 meter (2 feet) in diameter and at least half as much at the outer edge, when measured at a distance of not less than 0.6 meter (2 feet) from the lamp enclosure.

Patient Use. Patient use implies control by the patient for reading, visiting, self-care or viewing television. This control must be limited so as to prevent annoyance to other patients (see Fig. 7-6).

The reading light should provide light at the normal reading position, assumed to be 1.14 meters (3 feet 9 inches) above the floor. To allow the patient freedom to turn in bed without moving out of the reading light zone, the area of the

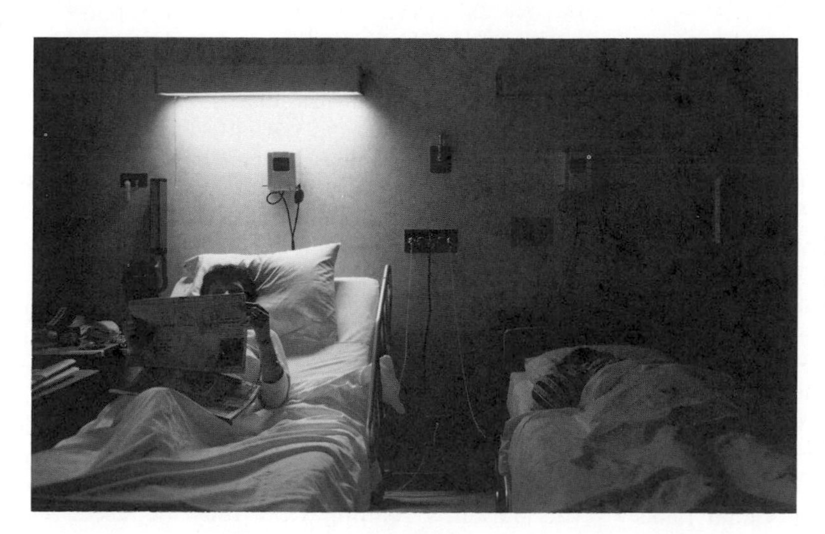

Fig. 7-6. Patient room lighting in a multiple occupancy accommodation. Note one patient reading while another sleeps under reduced illumination.

Fig. 7-7. Both under counter task lighting and ceiling general lighting are used to illuminate this nurses' station. The low luminance general lighting system was used so that patients in surrounding rooms would not see areas of high luminance at night.

reading plane lighted by an adjustable type of unit should be approximately 0.3 square meter (3 square feet), and for a nonadjustable unit the area should be approximately 0.7 square meter (6 square feet). To provide a reasonable degree of uniformity of light over these recommended areas, the lighting level at the outer edge of each area should not be less than two thirds of the lighting level at the center of the area. To provide comfortable lighting conditions for reading, the luminance in candela per square meter (footlamberts) on the ceiling, provided by some means of general lighting, should be at least equal to the illuminance in $1/\pi$ lux (footcandles) on the reading matter.

The luminance of the reading lamp and of any surface illuminated by it, as normally seen from any usual reading or the patient's bed position, should be less than 310 candelas per square meter (90 footlamberts). This condition is admittedly hard to secure and entails careful choice of luminaires and built-in limitations to its movements.

Housekeeping. A very important consideration is the lighting for housekeeping functions. Housekeepers need to see dust and dirt and to remove it. The housekeeper must be able to see beneath the furniture and to have oblique lighting over horizontal surfaces to observe dust.

Nursing Stations. In most hospitals each nursing unit is coordinated around a nursing station (see Fig. 7-7). At this point, charts are stored, read and written. Thus a desk or shelf is invariably provided, usually against some type of counter or below a hung cabinet. Lighting mounted beneath this counter should provide for the task lighting. It should be so arranged that it supplements the over-all illumination of the station.

Some of this lighting will be in continuous use, night and day. It is well to consider this in the lighting plan for the station. Usually, although by no means universally, when the nursing station is not visible from any of the patient accommodations, general ceiling sources remain lighted during the night hours.

The luminaires beneath counters, which are placed so that a person sitting at the desk is shielded from glare, should not be within the patient's direct view.

As the nurse must make frequent trips from the station to the patient facilities as well as to service locations, the corridors between should have transition lighting; a higher level during the day and switched or dimmed to a lower level at night. For safety, the illumination at the nursing stations is usually on an emergency auxiliary lighting system.

Critical Care Areas. The term critical care is replacing many of the former names such as intensive care. Critical care areas are especially designed for the very ill, and may be highly specialized or be quite flexible in their acceptance of patients. These accommodations have been designed for postsurgical patients, coronary disease, respiratory disease, burns, acute childhood and neonatal problems, isolation units, neurosurgical units, etc. Basically all of these require physical and instrument monitoring, and the capability for mounting emergency methods for resuscitation, hemorrhage and other situations which can be anticipated.

The illumination should enable the observer to note the prominence of veins on the neck and, if possible, the presence of yellow tints in the patients' eyes. Lighting should be avoided with a predominance of any color which may give the patients' complexion a false appearance. Thus, only improved color fluorescent lamps should be used. See Fig. 7-8.

While the demands for visual tasks in these units may be great, the psyche of the patient must also be carefully considered in planning. For example, the minimum requirements of construction of the Health Resources Administration (79-1450) require the provision of windows to enable *each* patient to be cognizant of the outdoor environment. Yet the provision of illumination by this means is not important.

The general lighting should be capable of dimming. It should be located so that neither the prone patient nor the one sitting with an elevated backrest will be subjected to glare. In addition to general lighting there should be lighting for examinations by the physician. Also, some type of surgical task light should be readily available for emergency procedures.

Most of these facilities contain a handwashing area.

Monitoring devices (see Fig 7-9) should be studied so that there will be adequate illumination for reading them. This also includes a review of their placement and whether or not they are internally illuminated.

Children's (Pediatric) Section. The child admitted to the hospital for the first time may feel dwarfed by its huge size and depressed by the concentration of suffering. Strange equipment may be frightening and alarm ill patients or intensify anxiety. For this reason the childrens' department should be provided with ample space for things for the children to do which will be diverting and educational. The lighting should be planned with this in mind, but should be similar to adult areas.

The use of daylight is essential. There should be a light and sunny atmosphere. Corridors should be pleasant with warm colors and surfaces, and diffused lighting used. However, spots of lighting patterns with interesting views shorten times of waiting and distances for travel down hospital corridors. Arrangements for vary-

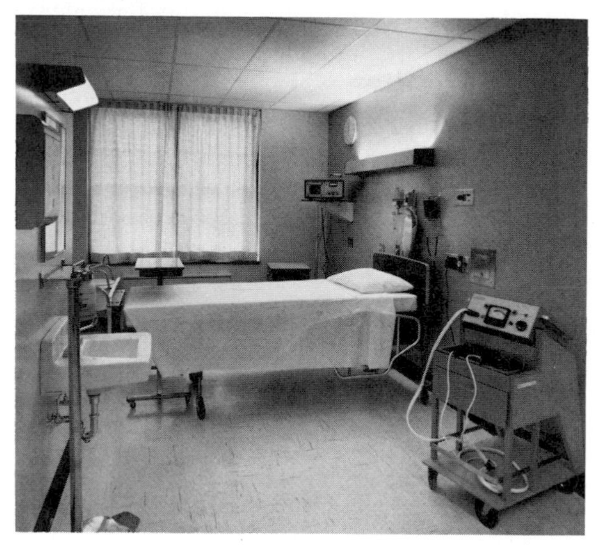

Fig. 7-8. Critical care room. Wall brackets contain two fluorescent lamps for indirect general lighting, one fluorescent lamp as a downlight for reading, and an incandescent night-light for surveillance from the nurses' station. Two 325-watt tungsten-halogen lamps in ellipsoidal reflectors are also provided for indirect examination light.

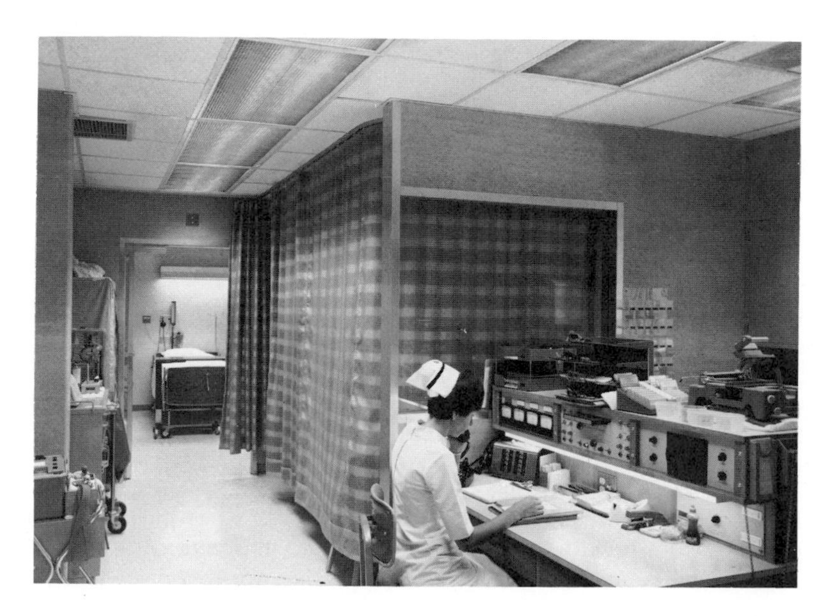

Fig. 7-9. Nursing station in critical care unit. Note the lighting beneath the counter and out of the patient's view. Also, monitoring devices are easily visible.

ing the lighting by multiple switching or dimming is often worthwhile.

Children play and sit on the floor, and often use it as a table. For this reason the lighting should be planned for reading, looking at pictures, drawing and other visual activity on the floor.

Nurseries. Nursery lighting should be designed so that infants in cribs and in incubators can be observed easily (see Fig. 7-10). General room illumination should be reducible for casual observation of the infants. Luminaires for general lighting should be of such a type or so installed that the luminance of any luminaires, ceiling or wall surface, as seen from working or normal bassinet position, would be less than 310 candelas per square meter (90 footlamberts).

There is often a need for higher level than that for the general lighting for careful observation, but it is not kept at this level too long in order to avoid retinal overexposure, for the infant does not have the ability to roll over or employ adult protective mechanisms. This must be taken into account when planning the illumination.

In order to recognize minor changes in the color of the skin and sclera, light sources should be specially chosen. It is preferable that these sources be in the higher kelvin ranges and have a relatively flat spectral power distribution.

There are special publications which should be referred to for information of the treatment of infantile jaundice with fluorescent light, and particularly to the precautions which are recommended for therapy,[3] and to the use of ultraviolet bactericidal barriers in pediatric sections.[4]

Mental Health Facilities. The facilities for mentally or emotionally disturbed patients are generally either of the open type in which the patients are unrestricted or of the closed variety where access is controlled. Either facility may house patients who are considered to be under maximum security. For this type of patient the lighting should be designed to be inaccessible to the patients to protect them from injuring themselves or others, and yet designed to avoid a prison-like environment. Lighting should be provided by non-adjustable recessed, ceiling luminaires, not only out of reach of the patient, but protected from access to thrown or other objects. These should be controlled by key switches preferably mounted in hallways outside of the detention area.

Most mental health facilities today handle other than just the severely disturbed patient. Regardless of the type of patient proper lighting depends on knowledgeable selection of patterns and areas of illumination most helpful and least disturbing. A generalized, basic guideline is that the lighting of these facilities should provide interest, warmth, definition of spaces, and illumination for tasks and safety.

Surgical Holding Areas. These areas (see Fig 7-11) are designed primarily for the retention of patients, nearly always supine on a wheeled stretcher (gurney) after they have had a sedative premedication. They are retained in this area out of the traffic stream for periods of from a few minutes to as much as 30 or more minutes.

The patient's eyes should not be exposed to a luminance of more than 100 candela per square meter (30 footlamberts). Most of the time a subdued slumber type of illumination is advisable, and designed to be out of the line of sight of the recumbent patient, but a higher level is needed for supervision and observation.

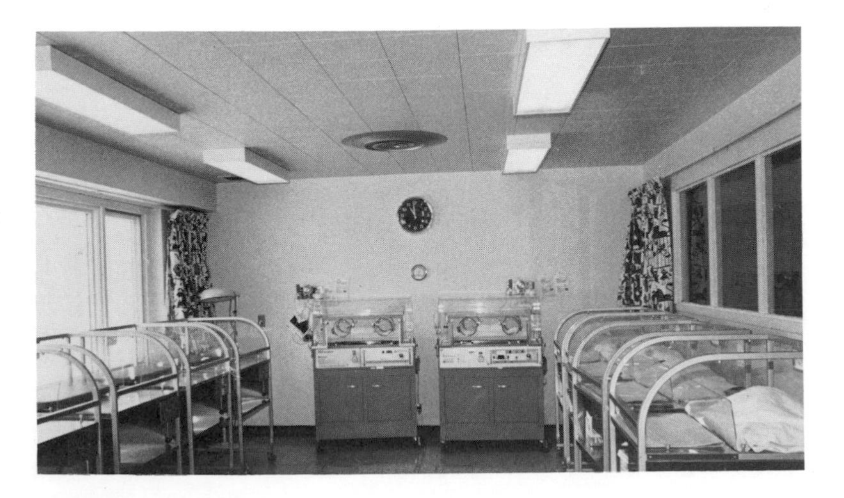

Fig. 7-10. Infant nursery. Windows at right permit relatives to view the babies.

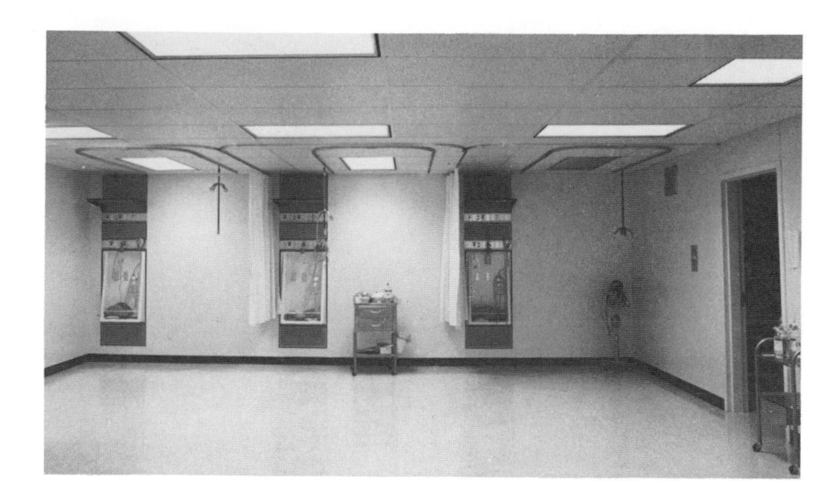

Fig. 7-11. Surgical holding area where patients are kept before being taken into the operating room. Note each unit here is separately dimmable (surgical recovery similar).

The holding area is not usually designed for surgical induction; however, some hospitals will use it as such, and the planner must give such an arrangement additional consideration. Some type of lighting is useful which will facilitate the starting of intravenous lines, and other pre-anesthetic activities such as shaving, etc. This purpose might well be served by flexible wall hung bracket luminaires. One patient's preparation will not then disturb another waiting patient.

Surgical Induction Room. The patient is transferred here from a stretcher to an operating table, the anesthesia started, needles placed into the patient's veins, and the patient maneuvered into a variety of positions by manipulations of the operating table. Connections to a variety of monitoring devices are attached to the patient. The positioning of the patient might take a considerable time, but for this the patient is usually already partially anesthetized.

Ideally the patient is brought into this room under subdued light. The anesthesia can be induced after placement of a needle in the vein. For this, a task light of some type must be available. Once the patient is unconscious, the illumination can be increased to serve the needed staff tasks. The capability of again reducing the light in the room should be available while the anesthesiologist inserts a tube into the trachea (windpipe). This is accomplished utilizing a lighted laryngoscope. This device provides only a little light and thus a low ambient level is preferred.

Surgical Suite.

Operating Room. The lighting of the operating room is perhaps the most important in the hospital, not in the number of people to be satisfied but in the importance of the work done there. There should be no dense shadows to prevent the surgeon from seeing past his own hands and instruments, nor to prevent him from adequately seeing the patient's tissue, organs and blood exactly as they are. Sometimes he must see into deep body cavities, natural or artificial. To enhance physical comfort for the surgical team, heat reaching the back of the surgeon's head and neck from the overhead surgical light must be minimized. The surgeons must be able to work for hours, if necessary, without any discomfort and must be able to glance to and from their work without having to take time for their eyes to adjust to large differences in luminance. Even more important than the comfort of the surgeon and the surgical team is the safety of the patient. Body tissues exposed during an operation must not be excessively heated or dried.

Colors and reflectances of operating and delivery room interior surfaces, draping and gown fabrics should be somewhat as follows: ceilings, a nearwhite color with 90 per cent or more reflectance; walls, non-glossy surfaces of *any* light color with 60 per cent reflectance; floor reflectance preferably in the range of 20 to 30 per cent but may be as low as eight per cent depending on the limited selection of flooring materials available; and fabrics for gowns and surgical drapes should be colored, usually a dull shade of blue-green, turquoise or pearl gray with 30 per cent or less reflectance. Surgical instruments should be of a nonreflecting matte finish to minimize reflected glare in the area of the operative cavity. Any plastic materials used in draping should also be of matte finish.

Equipment such as that for x-ray, anesthesia and ventilation competes with the lighting system for the limited ceiling space available. Therefore, to achieve desired general levels, it is necessary to carefully plan the location and arrangement of the lighting system. Due to the variety of surgical procedures, it is highly desirable to allow for control of the general lighting system to suit visual requirements of the surgeon and staff. The general illumination in the operating room should provide a uniformly distributed level with provisions for reducing the level. Luminaires should be equipped with elements giving diffusion to the light and to prevent glare.

As levels of general lighting have become higher, luminance balance has assumed greater importance. To achieve this, luminance ratios (not illuminance ratios) between areas of appreciable size within view of the surgeon and his team, should be no greater than 1 to 3 between the wound and the surgical field and 5 to 1 between the surgical field and the instrument table. The surgical field to the room's lighter surfaces also should be no greater than 1 to 5. Visual comfort is probably greatest when there are no excessively bright reflections in the field.

When fluorescent luminaires are utilized in the surgical suite they should be designed to reduce electromagnetic interference to a level that will not interfere with operation of delicate electronic equipment in a life support system. This generally requires welded construction to minimize radio frequency leakage through openings, lenses with an electrically grounded conductive coating, and radio frequency filters to reduce the radio energy getting into the electric wiring.

The appearance of the patient should not change significantly when viewed under either the surgical light or the general room illumination. This is best achieved by matching the spectral power distributions to the two types of lights; however, usually it is only practical to match the color temperatures. For example, if the main surgical light has a color temperature of 4000 K, general room illumination should be provided by fluorescent lamps with similar color temperature—in this case, deluxe cool white fluorescent lamps. In all cases fluorescent lamps should be of the improved color type.

The surgical task lighting system (see Fig. 7-12) should be capable of providing a minimum of 27 kilolux (2500 footcandles) directed to the center of a 500 square centimeters (78 square inches) (or larger) pattern on a surgical table with the top 990 millimeters (39 inches) from the floor. This pattern is defined as an area within which the illuminance tapers from center to edge so that at the edge it is no less than 20 per cent of that at the center. For ceiling-suspended surgical lighting systems, the illuminance and patterns are measured 1070 millimeters (42 inches) from the face of the lamp cover glass, if a cover glass is used, or the lower edge of the outer reflectors in a multiple reflector unit with individual covers over each lighting source.

The above is intended as minimum for general surgical procedures. In many specialized instances higher illuminances, various pattern sizes and shapes, and control of level are desirable. Variable pattern sizes are provided by moving the light closer to or farther from the patient. Some lights provide, in addition, a focusing con-

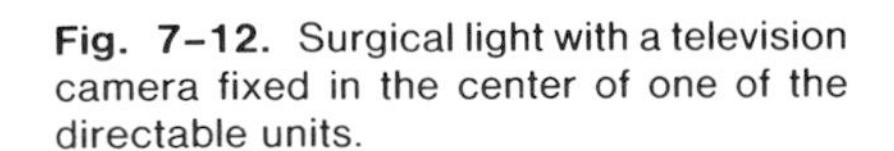
Fig. 7-12. Surgical light with a television camera fixed in the center of one of the directable units.

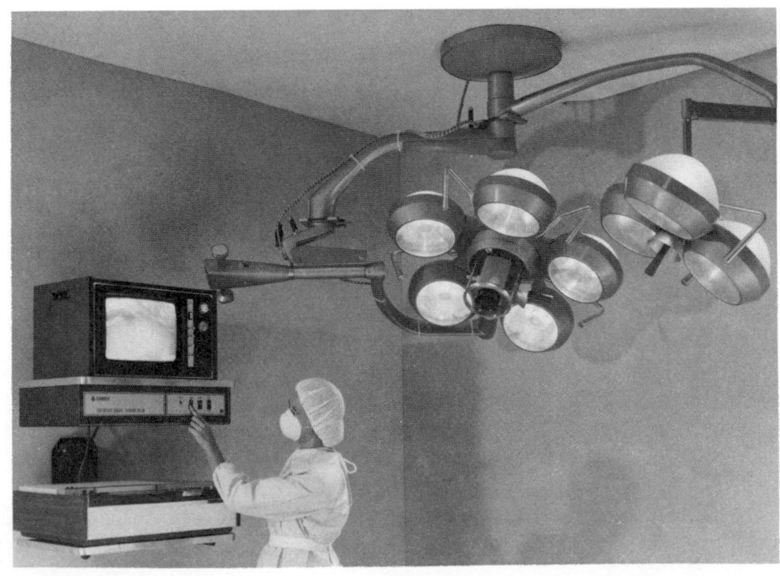

trol which varies pattern size. Users should determine the depth of field required for their work and evaluate the luminaires available that will give a useable pattern over the depth of field required. All illuminance measurements should be made with a color and cosine-corrected light sensing element that will indicate the average level over a 38-millimeter (1½-inch) diameter. To prevent obscuring shadows from the surgeon's hands, head, and instruments, the light should reach the operating area from wide angles. For test purposes, the light should provide a level of 10 per cent of the unshadowed level inside and at the bottom of a tube 50 millimeters (2 inches) in diameter and 76 millimeters (3 inches) long, finished flat black inside, from a distance of 1070 millimeters (42 inches) when the beam is obstructed by a disk 254 millimeters (10 inches) in diameter, 580 millimeters (23 inches) above the operating table, and normal to the axis of the tube. This means that, in the testing meter, a 38-millimeter (1½-inch) sensor is used. It is a very sensitive test which means that the black finish must have no specularity, and the disk and tube must be correctly positioned. See Fig. 7-13.

Protection should be given against a total lamp failure, for example, by multiple lamps in a single lighthead, or by multiple lightheads, etc.

The radiant heat produced by surgical lights must be minimized for protection of surgically exposed tissues and the comfort and efficiency of the surgeon and assistants. For most operations the radiant energy in the spectral region of 800 to 1000 nanometers should be kept at a minimum. This is the energy of infrared absorption by flesh and water and hence results in a noticeable heat to the surgeon, or more important may cause drying of exposed tissues. Current research suggests that in certain neurosurgical or intestinal procedures on delicate, thin, dry or abnormal tissue, the user of surgical lights should take care not to exceed approximately 25,000 microwatts per square centimeter at maximum intensity in the light pattern. The manufacturer of surgical lighting should provide information on conditions under which his equipment can exceed these energy levels. An irradiance factor (μW cm^{-2}) per lux or footcandle would be helpful to determine the total irradiance of the lighting system.

For general surgery, the light from the luminaire should have a color within an area described by a five-sided polygon on the CIE chromaticity diagram (see Section 5 of the 1981 Reference Volume). The range of CIE coefficients is appropriately defined by the following x and y values:

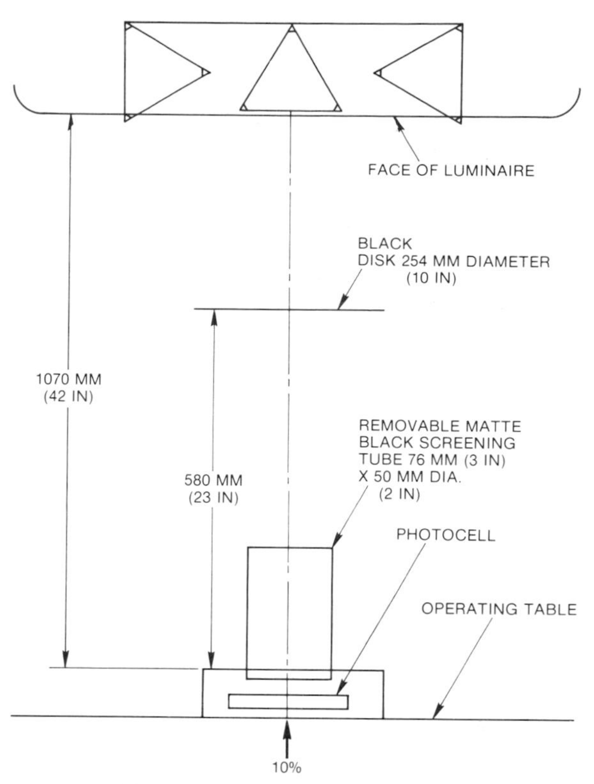

Fig. 7-13. The test for shadow reduction. Distance and sizes of objects are as shown above. Ten per cent of the incident light should be seen at the bottom of the tube.

x	y
.310	.310
.400	.375
.400	.415
.375	.415
.310	.365

When plotted, the above points will result in correlated color temperatures between 3500 K and 6700 K. Spectral power distribution should be so designed as to provide color rendering satisfaction to the surgeon.[5]

Secondary only to its unusual optical quality is the flexibility designed into the surgical lighting unit. This may be mechanical on units suspended from the ceiling or by electrical switching arrangements in stationary units in the ceiling. Directionality is sometimes achieved by permitting the scrubbed surgeon to adjust a sterile handle, but the asepsis of this technique has been questioned. If handles are used, they should be demountable for sterilization, smooth to avoid glove puncture, and have a guard to prevent contact with a nonsterile area.

The requirements for directional flexibility in

the main task lighting system will vary with the surgeons and the type of surgical procedures to be performed and limited by the "five-foot" rule[6] imposed for the use of flammable anesthetic agents. (NFPA-56A-1973 permits breaching the "five-foot" rule when operating rooms are clearly defined to avoid all use and presence of flammable agents.) The requirements for orthopedic operations differ greatly from those for cardiovascular and neurosurgical which, in turn, differ from those of the gynecologist. Thus the selection of the lighting system cannot be simply defined and the prospective purchaser must be aware of the limitations of all of the equipment.

Two-team surgery is now frequently practiced for many procedures. For example, one team may remove a vein from the thigh while another implants it into the heart. Additional lightheads or satellite units may extend from the primary luminaire mounting. On the other hand the use of two or more luminaires on one surgical field is to be employed only with care. See above.

Supplemental surgical task illumination is of two main types: those with a beam encompassing the entire field and those which operate by directing light through a glass or plastic fiberoptic bundle. The former, if they are to be used in an operating room where flammable anesthetic gases are employed, must be explosion proof or limited in movement to 1.5 meters (5 feet) above the floor. The fiberoptic light source and all of its electrical system must be in accordance with NFPA 56A.[6] In addition supplemental headlamps, self-lamped, often project an image of a tungsten filament, without filtering, and should be tested for illumination safety in a manner similar to fiberoptic headlamps (see below). These units could produce six times the irradiance of surgical lighting for equal illuminance.

Freestanding lights must conform with safety from tipping, as prescribed in NFPA 56A, and must have a reasonable 'memory' for retaining their position. No part of the portable wide-beam lamp housing should project below 1.5 meters (5 feet) from the floor. The entire unit must be grounded through a third wire in the flexible cable.

A fiberoptic unit, for use in a sterile field must be capable of sterilization or be encased in a waterproof and sterile static-free barrier. At the exit face of the fiberoptic device the irradiance should be no more than 25,000 microwatts per square centimeter.

Low voltage lighting equipment (less than 8 volts) may be used in accordance with NFPA 56A, if supplied by an individual isolating transformer "connected by an anesthetizing location cord and plug" or from dry cell batteries or transformer above the "five-foot" level. Isolating transformers should have a grounded case and core.

The anesthesiologist usually sits behind a tent of surgical drapes which prevent an accurate view of the patient's face color. Most of the monitoring instruments and dials which must constantly be observed are hard to see. Their cover glasses usually are very reflective and may well produce veiling reflections or reflected glare. Furthermore, the anesthesiologist should be shielded from the operating task light and shielding of the general illumination system and other special lighting considerations also may be needed.

Scrub Room. Scrub areas and corridors adjacent to the operating room are areas where personnel can accommodate their eyes to the illuminances of the operating room. While scrubbing before an operation, the surgical team should be exposed to light of the same level that they will encounter in the operating room. This will not only insure a good job of scrubbing, but will allow them to enter the operating room fully light adapted. These illuminance levels will also promote a cleaner scrub area and consequently more aseptic conditions. This same reasoning holds for corridors leading to the operating room. These are the areas where the surgical team adapts to the operating room environment.

Not all operations are performed under higher illuminance levels, and in fact, many are carried out at very low ambient levels. For those the light should be reduced, by switching or dimming, to a low level, just sufficient to observe the clock and the technique of scrubbing. Adaptation from low to higher illuminances is very rapid, but dark adaptation is very slow so that the level should be kept low for endoscopic surgery.

Special Lighting. The lighting requirements for photography and television in the operating room have been given relatively little consideration.

Microsurgery may be televised and special cameras are built to be adapted to operating microscopes. When these are used, special attention should be directed toward heat production on the tissues being observed as associated beam splitting devices require increased light.

Surgical lights are available with color television cameras built into the lights (see Fig. 7-12). This automatically insures that the camera is aimed at the surgical site. These cameras are usually equipped with motorized zoom, focus and iris adjustments for the lens. Thus, all camera control can be done remotely. The color balance of the camera must be adjusted optically or electronically to match the spectral distribution of the light.

Specialized Operating Rooms.

Eye Surgery. Rooms for eye surgery contain some type of fixed pedestal or columns connected with an operating microscope. This equipment may contain its own luminaires and frequently beam splitting devices to permit viewing by more than one person. There may be camera or television equipment also attached. A separate ruby laser may also be present as well as an electromagnet for removing ferrous foreign material from the eye.

The general room illumination is planned to give the same level as in the general operating room. The surgeon will, at times however, require less general illumination and may prefer almost complete darkness. Therefore, a method for reducing the illumination becomes mandatory in the eye room. Separate lighting may be necessary for the anesthesiologist so that he may observe equipment.

Ear, Nose and Throat Surgery. The requirements for this specialty are identical with that of the eye surgery. Microscopic surgery is used for operations on the inner ear.

Neurosurgery. In general this operating room is no different in its visual requirements than the general surgery operating room. Some neurosurgeons prefer to use headlamps, frequently fiberoptic ones. Recently surgical microscopes have been employed in a darkened room. These operating microscopes contain their own illumination and may be ceiling or wall mounted. Neurosurgeons often require a horizontal rather than a vertical beam of light. Thus, luminaires that can be brought as low as the codes allow are needed.

Orthopedic Surgery. In general the orthopedic operating room visual needs are no different than those of the general surgery but better facilities for x-ray equipment may be necessary. The type of x-ray equipment and mounting needs to be coordinated with the lighting systems. Particular attention should be paid to flexibility of the luminaires, for orthopedic surgery frequently requires unique positioning on the side of the operating table for low level lighting of the patient's hip. Fluoroscopy with image intensification and television screening will permit the use of a room which is not darkened. Extra negatoscopes are required.

The orthopedic surgeon also uses the surgical microscope.

For implantable joint replacement, the orthopedist sometimes employs laminar airflow chambers, and surgical luminaires pose a problem both from their disturbance of the laminarity of the airflow and from the convection currents they cause. These situations are very difficult to avoid as the necessity for illumination of the surgical task is paramount.

Postanesthetic Recovery Room. This is an area of meticulous monitoring and equipment observation, plus the capability for carrying out certain emergency procedures. Color recognition for changes in the patient's skin must be facilitated. The lighting should be variable so that presentations on the face of oscilloscopes (electroencephelographic and electrocardiographic) can be recognized.

Cystoscopy Room. Cystoscopy is normally carried out in a dark room. The cystoscope is however introduced in a lighted room. For female procedures a gynecologic examining light should be provided. If flammable anesthetics are *not* used in the area (which is usual) the light should be available at the level just above the sitting urologist's shoulder.

The darkening of the room should be possible by switching or dimming. The low level should be adequate for the anesthesiologist to see his equipment and to recognize the patient's color. A surgical lighting capability should be available for the performance of some operative procedures. It should be centered in such a way as to illuminate the lower end of the cystoscopic table.

Obstetric Delivery Suite.

Labor Rooms. Monitoring apparatus is often applied to the patient to observe uterine contractions and the heart tones and responses of the unborn child. This is usually recorded on paper and must be observed by attendants. Examinations performed in this room are usually manual and will not require visual control. However, observation of the patient includes blood pressure measurement and observation of the patient's general status; therefore, good color rendering lighting is preferred so that any cyanosis (blueness) will be obvious.

Delivery Area. The area for delivery scrub should be identical in its illumination with the surgical scrub area. The general illumination of the delivery room should be achieved by recessed luminaires providing light throughout, the same as in the operating room.

The task light should be capable of focusing and produce a level of 27 kilolux (2500 footcandles) at its center. Ideally it should be capable of being centered over the shoulder of a sitting obstetrician. Mounting should be in accordance with NFPA 56A. Explosion-proof portable units are also available. In some institutions the anesthesiologist will ban the use of flammable anesthetic agents in the delivery suite and remove the explosion hazard.

A special lighting plan should exist for the area in which the newborn infant is resuscitated. The lighting should have good color rendering capability, particularly for cyanosis and jaundice.

General Radiographic/Fluoroscopic Room. In the modern unit, most fluoroscopy is now performed with image intensification with visualization on a television screen; therefore, complete darkness is no longer imperative. Overall illumination is necessary for cleanup. The general lighting should have a dimming capability as different radiologists prefer working under different levels of ambient lighting. Often the overhead lighting is extinguished by the foot switch on the radiographic unit so that the ambient light will go out when the fluoroscopy is in progress. Light localizing devices are sometimes used and are assisted by low general lighting.

The competition for ceiling space is great in these rooms and the placement of ceiling luminaires is very important.

Laboratories.

Specimen Collecting (Venipuncture) and Donor Areas for the Blood Bank. Lighting should be provided for the site of the venipuncture, at the height of the arm of an armchair. Veins are often best seen in other than flat light; therefore, ceiling luminaires or task lights should be placed to provide oblique illumination. The walls in this area should be of pastel shades of low reflectance for donor comfort and reassurance.

Tissue Laboratory. Lighting in a tissue laboratory should have an excellent color rendering quality. Of particular interest is that there are usually two counter heights [760 and 910 millimeters (30 and 36 inches)] involved, one to be used sitting, and another at standing bench level. The same lighting arrangements are valuable in the room devoted to the preparation of cytology specimens. Backgrounds for microscope viewing are best dark in color and of very low reflectance to avoid glare.

Microscopic Reading Room. The pathologists spend a considerable portion of their day in reading microscopic material. For this purpose the tables upon which the microscopes are placed are usually at a 810-millimeter (32-inch) level from the floor and the tabletop is of low reflectance often in a mahogany or walnut finish. Room lighting should be adjustable for long-time viewing.

Central Sterile Supply. The inspection area of the central processing department should have general lighting and in special areas where delicate instruments and other equipment are inspected, illumination should be increased.

Dental Suites. In the dental operatory the luminance differences between the patient's

Fig. 7-14. Current Recommended Illuminances in Lux and Footcandles, on Tasks, for Emergency or Continuity Service (for Use When Normal Service is Interrupted)*

	Lux	Footcandles
Exit Ways		
Corridors leading to exits, at floor	30	3
Stairways leading to exits, at floor	30	3
Exit direction signs, on face of luminaire	50	5
Exit doorway, at floor	30	3
Operating Room, surgical table	27000	2500
Operating Room, emergency table	22000	2000
Delivery Room, obstetrical table	27000	2500
Recovery Rooms for operating rooms and obstetrical suites	100	10
Nurseries, infant, 760 millimeters (30 inches) above floor	100	10
Nurseries, premature, 760 millimeters (30 inches) above floor	100	10
Nurseries, pediatric, 760 millimeters (30 inches) above floor	20	2
Medication Preparation Area, local	300	30
Nurses' Station	50	5
Pharmacy	50	5
Blood Bank Area	50	5
Central Suction Pump Area	50	5
Telephone Switchboard, face of board	50	5
Central Sterile Supply, issuing area	50	5
Psychiatric Patient Bed Area	20	2
Main Electrical Control Center	50	5
Hospital Elevator—Exit Lighting	50	5
Stairwells	50	5
Life Safety Areas (Life Support Areas)	50	5
Cardiac Catheter Laboratories	100	10
Coronary Care Units	300	30
Dialysis Units	200	20
Emergency Room Treatment Areas	500	50
Intensive Care Units	300	30

* These are minimum lighting levels. It is particularly desirable that they be increased to as near the levels normally provided in these areas as the available capacity of the emergency electrical supply will permit.

mouth and face, patient's bib, the instrument tray and the surrounding areas should be no greater than 3 to 1.

Lighting should be provided at the level of the patient's face and the instrument tray. Lighting inside the mouth, or oral cavity, should be supplied from a luminaire easily adjustable to exclude high luminance in the patient's eyes and at the same time provide such lighting as is needed by the dentist to see fine details over long periods of time. This light should have color characteristics and level suitable for the dentist to judge the matching of colors of teeth and fillings, and occlusions of dentures in any place within the mouth. The dentist must be able to judge accurately the depth of drillings and the preparation for retention of fillings.

A luminaire for producing such a penetrating light, relatively free of shadows at the oral cavity, must produce a convergent beam, and at a distance of about 1 meter (3 or 4 feet) should be capable of lighting a semicircular area with a cut-off to exclude the bright light from the patient's eyes.

Prosthetic work in the laboratory requires speed, accuracy and close inspection. Therefore, a general level should be provided, with a supplementary lighting at the workbench, and at one or more points depending on the number of people using the laboratory at any one time.

Examination and Treatment Rooms. For examination and nonsurgical treatment there should be general lighting with supplementary lighting on the table. Also, there should be a special lamp for vaginal inspection.

Emergency Outpatient. The emergency outpatient suite should be generally self-sufficient to handle most cases without resorting to the rest of the hospital. Fixed ceiling-mounted directional luminaires or portable lights that provide lighting at the center of the operating area, with a lower level of general illumination are usually adequate for examination and emergency surgery.

Autopsy Room and Morgue. Good lighting is imperative in the dissecting room. A surgical type of dissection must be performed, yet it is done in the open rather than in the restricted cavity as in the surgical exposure. Therefore, the highest levels of surgical illumination are not needed to overcome the losses in deep cavity lighting. While some of the dissection may be meticulous and tissue planes must be visualized, the meticulous placement of sutures and the meticulous placement of instruments to control bleeding from fine blood vessels is not necessary. Therefore, some of the contouring so advantageous in living surgery can be sacrificed.

The task light of the autopsy room can therefore be a nonadjustable large unit with good color rendering lamps augmented by spotlights providing illumination at a level of the autopsy table 760 millimeters (30 inches) above the floor. Surgical lights are not necessary in this room. A single spot with filters to greatly reduce infrared radiation is valuable for the skull portion of the autopsy. Additional lighting for a scale placed over a counter is also valuable.

Pharmacy. The pharmacy should be well illuminated so that labels and fine print of precautionary literature supplied with the medications can be read. Illumination should be provided at the workbench level 910 millimeters (36 inches) from the floor.

Emergency Lighting. Emergency lighting is needed to perform two categories of essential tasks: the task of evacuation under adverse conditions and the task of providing life support services to the patient who cannot be evacuated. These two categories may be thought of as requiring two lighting systems. The first is emergency light of relatively low level to provide adequately for ambulatory mobility of patients and staff, and the second is of higher level and in most applications equal to that provided by the regular lighting system.

With the increased usage of electrical power in the operating room and critical care areas there is a need to increase the reliability of the electrical service to these areas. The regular room lighting becomes the emergency lighting whenever the power supply to the critical care areas switches from the normal source to the emergency source. See NFPA 76A for information relating to essential electrical systems for hospitals.

The remaining areas of the hospital should have low level emergency lighting to give the levels recommended in Fig. 7-14.

HOTEL/MOTEL AND FOOD SERVICE FACILITIES

In designing lighting for hotels, motels and food service facilities the first task is to identify those things which the staff and users want or need to see. Both groups must be able to see and comprehend their environment, to move about and work within it. In addition, they should find it *enjoyable* to do so. In such facilities as hotels and restaurants, the psychological effects of lighting are particularly important. The lighting design becomes a marketing tool by creating a successful, attractive, comfortable and functional environment, but only if it is integrated with the over-all architectural design concept. Lighting which is inappropriate in terms of quality or quantity can ruin an otherwise successful installation. Using an appropriate combination of daylight and electric lighting, the designer can develop and reinforce almost any visual mood and satisfy the visual needs in any space by day and by night. The lighting system must be compatible with acoustic, thermal, spatial and esthetic requirements and objectives for each area. A successful total environment requires a cooperative effort by the owner, facility manager, architect, engineers, interior designer and specialized consultants who work to integrate all concepts into a harmonious final solution. In hotels and restaurants, where architectural treatment is critical, the lighting designer must seek to strike an appropriate balance between efficiency and esthetics while considering energy management. See Section 4.

The following general objectives should be addressed:

1. Harmony with the architectural and decorative character of the facility.
2. Provision of high quality illumination for visual tasks.
3. Control of glare and luminance ratios.
4. Provision of adequate quantity of illumination.
5. Cost optimization to maximize net revenues, including first costs, operating costs and maintenance costs. See Section 3.

Consideration must be given to the desired appearance of each space and to the seeing tasks to be performed. The factors which affect task visibility and performance are discussed in Section 3 of the 1981 Reference Volume. All potential hazards such as changes in floor level should be well lighted for protection of guests and staff. If thought out in advance, such highlighting of potential hazards can be made part of any decorative scheme.

Hotel and restaurant spaces often have different visual tasks at different times. Function rooms, for example, are used for dining, meetings, lectures, conferences, classroom applications, exhibitions and entertainment. To accommodate all these different uses with a variety of illuminance levels and distribution patterns, several lighting systems may be required with multiple switching and dimming.

Design Considerations for Specific Locations

Specific visual tasks and design considerations for different areas are discussed below. Illuminance recommendations are given in Fig. 2-2 and Fig. 2-26. The levels selected from Fig. 2-2 should be based on an evaluation of the needs of the occupant and on management experience. The levels in Fig. 2-26 represent minimums for safety alone.

Exterior and Site. The total exterior lighting system should identify the facility and create a favorable visual impression for welcoming patrons (see Fig. 7-15). Building facade lighting and marquee, walkway and parking lighting should be coordinated with signage to produce an effective coherent over-all impression. Grounds of buildings should be lighted:

1. To merchandise the property (when warranted).

Fig. 7-15. Halos of clear decorative lamps operated at reduced voltage define the room towers of this hotel and, combined with the uplighted allee of trees framing the entrance drive and the line of glittering sources over the entry, creates a distinctive image for the facility.

Fig. 7-16. By day, shafts of sunlight make this clear-glazed atrium dramatic. The high level daylighting is beneficial to the large trees. At night, the elevators become moving illuminated sculptures against a neutral sculptural background created by the gentle glow from fluorescent coves concealed in the handrails of the balconies and bridges. Uplighted trees and low mushroom lights reinforce the exterior imagery created by careful selection and detailing of materials. Pendant high intensity discharge downlights, required for tree maintenance, are only turned on late at night after the public has left the space.

2. To provide for the safety of guests and property, especially in parking areas and along pedestrian paths. It is also essential to make areas around steps, walkways and entries feel safe—which means the elimination of threatening shadows as much as it does the provision of an adequate level of light.
3. To eliminate areas which would otherwise be inviting to vandalism, or pose a problem in terms of security.
4. To make accessible to the handicapped all areas of barrier-free design.

All *entryways* should be well lighted to make them "landmarks" which may be used safely by guests. Lighting should be used to provide orientation and to reinforce intended traffic patterns. Marquees, portes-cochère, drive-ups, registration areas and unloading areas should be lighted in such a way as to distinguish them clearly from surrounding areas.

Lighting of *parking areas* should provide for visual security and physical safety. See Section 14, Roadway Lighting. Luminaire placement should be coordinated with buildings and plantings in such a way as to minimize shadows. Light sources should provide adequate color rendition for easy vehicle identification.

Public Spaces. The *lobby* typically establishes the main design themes for the facility, and houses a variety of functions which can be differentiated and enhanced with appropriate lighting techniques: elevator lobby, reception desk, lounge areas, bell captain's desk, etc. (see Fig. 7-16). The *entrance foyer* is a transition space between the outdoors and interior space, so foyer lighting should promote a sense of security and welcome while allowing adaptation between high and lower illuminances. In the *lounge area* both casual and prolonged reading tasks must be anticipated, though these can usually be accommodated with relatively low illuminances. A more residential treatment may be appropriate to create an inviting ambiance.

Many visual tasks are performed at the registration desk. To make this area easy to locate and to use (see Fig. 7-17), the designer may choose a high general lighting level to accom-

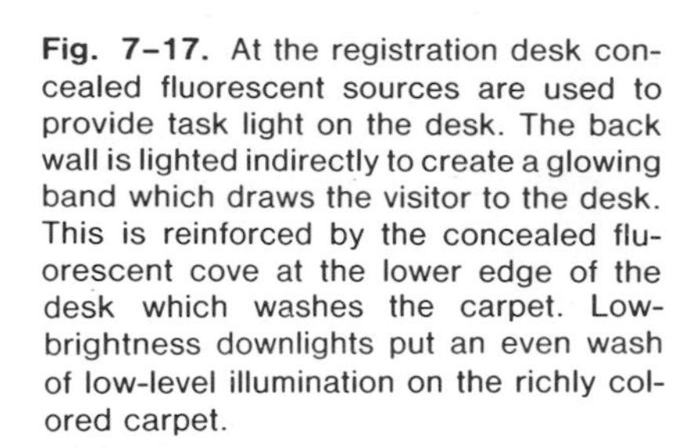

Fig. 7-17. At the registration desk concealed fluorescent sources are used to provide task light on the desk. The back wall is lighted indirectly to create a glowing band which draws the visitor to the desk. This is reinforced by the concealed fluorescent cove at the lower edge of the desk which washes the carpet. Low-brightness downlights put an even wash of low-level illumination on the richly colored carpet.

modate all tasks; however, a lower over-all level with a system of local task lighting should be considered. Care should be taken that the lighting be compatible with surrounding areas. (See Section 5 for the lighting of office areas).

Areas for storage of luggage, etc., should be lighted so that labels and other means of identification can be quickly and easily seen. In enclosed storage areas the color rendering qualities of the light should be adequate to permit correct identification of stored items. If not separated from the main lobby, storage areas should have somewhat higher illuminance levels than the general ambient.

The lighting for *elevator lobby areas* should be designed to orient people to the elevators and should enable them to read directional signage and instructions and select the proper signal controls for elevator call. Internally illuminated signage and controls should be considered.

Corridor lighting should illuminate room numbers, room name identification signs, and the locks in doors. Lighting should be designed to make the passage through hallways, on stairs and to elevators a pleasant and safe experience. Lighting should make guests feel secure. It should call attention to circulation modes such as elevators and vending areas. The tunnel effect associated with long corridors should be minimized.

Suitable lighting for *shops, newsstands and other specialized services* may require quite sophisticated equipment and controls. Such lighting should be considered as part of the over-all interior lighting scheme (see Fig. 7-18), and should not be so bright as to dominate adjacent public areas unless that is the intent of the design team. It may be possible to use display lighting which remains on 24 hours a day to light adjacent corridors, eliminating corridor luminaires.

In *public lavatories*, visual tasks include grooming, which requires shadowless illumination on both sides of the face. Color rendition is important. Lounge areas in restrooms require only low levels of light.

In most cases lighting in *ballrooms, function and meeting rooms and conference areas* should be related to the over-all design themes for the hotel. See Fig. 7-19. These rooms are used for meetings, exhibits, dancing, dining and other functions, which makes it important to provide a variety of lighting levels and effects. If decorative luminaires such as chandeliers are used, at least one or two supplementary lighting systems will usually be required. Dimming and multiple switching should be provided, organized and clearly labeled for easy operation by banquet and function personnel. Lighting must be adequate for critical tasks such as reading and note-taking. Adjustable accent lighting should be provided at speaker's areas, head-table locations and likely locations for displays. Outlets for local lighting should be provided in exhibit areas. It should be remembered that highest lighting levels may be required for set-up and cleaning purposes.

Guest Rooms. The guest room is one of the major commodities of a motel or hotel. Since it is frequently used for small business conferences, flexibility needs to be a part of the lighting plan. General illumination from ceiling or wall-mounted luminaires provides a background for task lighting, aids in housekeeping and gives a feeling of cheer, as well as providing the needed flexibility for nonresidential uses. To establish an inviting, home-like atmosphere a variety of lighting equipment, some decorative in appearance, is usually needed. Visual tasks which need consideration in the guest room include: reading in chair or bed, desk work, television viewing, and grooming at the mirror in both the bathroom and at the dresser. See Section 10. The small entry foyer which is typically part of the guest room should have its own source of general illumination which reflects light from the ceiling or walls. Recessed incandescent luminaires are not usually suitable because the distribution of light is too narrow. Often the foyer lighting can be designed to illuminate closets, luggage storage and/or grooming areas as well (see Fig. 7-20). Switches with lighted handles are a convenience for guests in unfamiliar surroundings. Low wattage switch-controlled night lights should be installed in each guest room, usually in the bathroom so that guests do not leave other lights on all night long. Mirror lighting generally provides adequate illumination in bathrooms. If, however, there are separate compartments for toilet, tub or dressing, each space should have a separately switched source of general illumination adequate for safety when the door or curtain is closed.

Entertainment and Food Service Spaces. Entertainment and food service spaces within the hotel/motel and restaurant industry are complex and energy-intensive areas in which lighting plays a key role in establishing the mood or atmosphere. The success of lighting effects depends on the appropriateness of the illuminance level, color of light, luminaire candlepower distribution, type of luminaire and its locations in relation to the architecture, and source size. Well-shielded downlights, for example, can create a pleasing sparkle in reflective objects such as table settings, as well as an intimate feeling.

Fig. 7-18. In this commercial corridor, all "corridor lighting" is provided with spill light from displays.

Fig. 7-19. (Left) Sections of tubing were flocked and hung from the structure over this ballroom to create a richly colored decorative ceiling in scale with the size of the room. Downlights, accent lights and air registers are concealed in and between the tubes. The large contemporary chandelier is made of planes of woven wire mesh, highlighted from lines of R-type lamps above. All circuits are dimmer controlled. (Right) Concealed in the decorative ceiling of this function room are several independently controlled lighting systems. Effects include wall washing, accent lighting for art and speakers, and neutral downlighting for projection and note-taking. The ceiling is fabricated of bronze acrylic cubes, the walls of which multiply the sparkle of incandescent sources.

On the other hand, indirect lighting or large-area diffuse sources such as fluorescent luminaires typically create a brighter looking space and call more attention to the whole room. The lighting of any feature in the dining area will need special attention. These can range from the highlighting of a picture or sculpture to a full luminous wall, with effects ranging from dramatic to open and friendly. The luminaires themselves may, if used decoratively, become distinctive features in their own right. Many suspended decorative luminaires, regardless of shape, size or style, have a general diffuse distribution that can produce dull, uniformly lighted spaces when used as the sole source of illumination. Unless low-wattage lamps are used, permitting these luminaires to act only as luminous ornaments, and supplementary lighting is provided, the luminance of such suspended luminaires can be uncomfortably high.

Fig. 7-21. In this bar, an indirect cove provides a gentle glow from the fabric ceiling. Clear decorative sources provide sparkle and light the walls. A central chandelier over the bar, consisting of downlighted plastic balls, makes the theme design statement for the room.

Fig. 7-20. A fluorescent luminaire concealed in the header of the closet door in this guest room foyer provides light for both closet and foyer (there is no wall above the luminaire). In the room beyond, a ceramic shade covers an inexpensive porcelain socket creating a decorative pattern of projected light and providing indirect ambient room illumination as well as reading light for the chair below.

In a multi-foodservice facility, switching, supplemental systems and/or the use of dimmers may be required to make the same space feel suitable for breakfast, lunch and dinner. Variation in illumination and level may be needed, both to change the mood for different times of the day, and to permit a higher level for clean-up than would be desirable for dining. Dimming control is preferable to switching since a smooth transition between levels is desirable.

The success of display lighting is best measured by how well it helps to sell merchandise. Food displays should be so lighted that attention is attracted to them and the details are clearly seen. Color rendition is even more important over fresh foods than it is over packaged foods. Heat from luminaires can be a major consideration over fresh, cooled or frozen foods. (See Section 8 for the lighting of merchandising areas.) A good rule is that the level used in food displays should be at least twice that in surrounding areas.

The mood established by lighting can vary from subdued and relaxing to bright and lively, depending on the type of facility and the intended clientele. Dining spaces are usually grouped into three categories: intimate, leisure and quick service.

The "intimate" type (see Fig. 7-21) consists of those areas where people congregate as much to visit, be entertained, and show off as to eat and drink. These include cocktail lounges, night clubs, and some dining rooms and restaurants. These spaces characteristically have a subdued atmosphere with low luminances throughout, accented with subtly lighted feature elements. Lighting must be well-controlled in terms of level and distribution.

The "leisure" type refers to most restaurants and many dining rooms—places where eating is the most important activity and time is often a factor. A restful, but interesting atmosphere is called for. Lighting should generally be unobtrusive except where decorative luminaires or highlighted features are used as part of the theme decor. Moderate illuminance levels are typical. Good control of glare is required.

The "quick-service" class includes lunchrooms, cafeterias, snack bars, coffee shops and franchise menu types, where the diner and management are both intent on fast service and quick customer turnover. Higher lighting levels and uniform distribution can be used to suggest a feeling of economy and efficiency.

Kitchen and Food Preparation Areas. Well-designed lighting helps to create a bright, hygienic atmosphere in a kitchen[7] and, by revealing dirt and the presence of debris, it can stimulate good housekeeping. Food preparation involves peeling, slicing, dicing and cutting operations, both by machine and by hand. These are obviously hazardous operations and lighting for safety must be a strong consideration. See Section 2. Good quality lighting can reduce accidents, reveal spills which make floors slippery, and emphasize hazardous areas. In kitchen and associated support areas there is a need to eliminate shadows and to provide illumination on both vertical and horizontal surfaces. While kitchens contain difficult and demanding tasks which may require relatively high light levels, it is important that luminaires be arranged and shielded so as not to create glare or "blasts of light" into adjacent intimate dining areas when kitchen doors are opened. Color rendering is important in food preparation and inspection areas.

Visibility is reduced by great variations in luminance in the task surround, and direct and reflected glare can be significant obstacles to employee comfort, productivity and safety. Therefore, exposed lamps in direct luminaires should not be used. Although glare can be controlled in direct luminaires by effective shielding of the lamp, indirect or semi-indirect lighting is preferable because it turns the entire ceiling into a large, low-brightness area source (see Fig. 7-22). Light colored walls will further diffuse the general lighting, reducing shadows. Because vertical surfaces of equipment and furnishings typically occupy a significant portion of the visual field, expecially in kitchens, light finishes are recommended for these surfaces. Horizontal surfaces such as table tops in restaurants and equipment tops in kitchens are very important because they serve as backgrounds for critical tasks. Whenever possible, matte finishes are preferable because they minimize reflected glare. Reflected glare can be exceedingly objectionable, and can produce discomfort and fatigue. Stainless steel kitchen equipment is a common offender in this respect. Matte or brushed finishes combined with careful placement of luminaires and good glare control can minimize these problems. The same principles apply elsewhere. Lighting near specular surfaces such as mirrored ceilings or glazed walls must be very carefully worked out if one is to avoid unintended reflections of sources.

Fig. 7-22. Indirect fluorescent valances are arranged around the venting hoods to define the work islands in this commercial kitchen. High reflectance surfaces are used throughout. Sources are shielded from adjacent dining spaces.

All luminaires in kitchens should be so located and designed so that those entering or leaving the kitchen area will not see bare lamps or high brightness. This is particularly important when the adjacent dining area has lower light levels. Even the relatively low luminance of unshielded fluorescent lamps can produce a period of reduced visibility during adaptation. All luminaires should be designed for ease of cleaning and relamping, and should have high reflectance permanent surfaces. In areas such as bakeries and dishwashing areas which have inherent dust or moisture conditions, the use of enclosed dust-proof or vapor-tight luminaires is recommended. Where open type fluorescent luminaires are located directly over food storage, preparation, service or display areas, plastic sleeve protectors should be used to prevent glass and phosphor from falling into the food in case of breakage.

In *receiving and storage areas* lights should be installed in the aisles rather than near the walls, so that stacked shelves do not block the illumination.

Support Areas. Support areas include such spaces as key shops, plan rooms, paint shops, etc. The visual tasks in such areas are often demanding and sometimes dangerous. Where flammable materials are stored or used, explosion-proof luminaires should be used. Task lighting using localized luminaires should be considered as an alternative to general higher level lighting.

Refuse Areas. The maintenance of safe and sanitary conditions in refuse areas is extremely important. Illumination should permit all hazardous or unsanitary conditions such as slippery spots, dirty waste, and evidence of insects, rodents, or mold to be seen. Corners and other out-of-the-way places should be well lighted.

Merchandising Areas, Offices, Laundry and Valet Areas, Indoor and Outdoor Recreation Areas. For lighting recommendations for these spaces see Sections 8, 5, 9 and 13 respectively.

Selection of Architectural Finishes. The factors which influence how well people perceive a task or space are discussed in detail in Section 2. They include luminance and its distribution, contrast, size, color, form, texture, familiarity, and length of time for viewing. Some specific recommendations can be made here with regard to hotel, motel and food service facilities.

Because areas in these facilities typically include adjacent areas with very different lighting levels, adaptation can be a problem for both staff and patrons. The designer must, therefore, provide appropriately lighted transition spaces. For example, a patron who has been registering at the front desk usually adapts to a relatively high luminance. Turning back into the lobby, the person may be momentarily unable to distinguish the location of steps or to recognize faces if the difference in luminance is too great. As the potential hazard or task difficulty increases it becomes more and more important to keep luminance ratios within recommended limits.

Room surfaces exert an important influence on luminance ratios between luminaires and their surroundings, and between tasks and their backgrounds. On large surfaces the use of matte finishes with recommended reflectance, helps to prevent excessive luminance ratios and undesirable specular reflections. Light-colored matte surfaces serve as effective secondary light sources which can materially reduce shadows. Soft shadows generally accentuate the form and depth of objects, supporting rather than hindering the process of perception.

Selection of Light Sources. Both daylight and electric lighting systems are used in hotels, motels and restaurants. Each generic source has its own characteristics. The systems should be designed to complement each other.

Most hotel and food service spaces should have windows. The opportunity to look through windows can be psychologically satisfying and permits relaxation when the eyes can shift focus occasionally from nearby to distant objects. However, windows can bring large areas of high luminance into the field of view, causing discomfort, thus good control of glare is important. The designer should not locate very brightly lighted areas next to dimly lighted interior spaces without allowing adequate transition spaces between. Proper daylight control (see Section 7 of the 1981 Reference Volume) is preferable to increasing the level of electric lighting.

Enhancement of public spaces and provision of proper task lighting are the two principal considerations in the design of lighting systems. There are three basic electric light sources in use today: incandescent, fluorescent and high intensity discharge (HID). Source selection depends on the particular requirements of each space, the economics, and the personal preference of the system designer and the facility operator. Generally, for public spaces, the use of incandescent filament or one of the improved-color white fluorescent lamps is recommended.

In these facilities, the chief advantages of incandescent lighting are low initial cost, good color-rendering properties, and the excellent optical control inherent in such a small source. Because they are so easily controlled by dimmers, it is recommended that all incandescent luminaires in public areas be dimmer-controlled. Drawbacks of incandescent lamps are their relatively short life and low efficacy.

In support areas, where optical control and

good color rendition may be less critical, the designer should consider the more efficient and long-lived fluorescent and HID sources. When using mercury, metal halide, high or low pressure sodium, or fluorescent sources, the designer should avoid three common pitfalls: ballast hum, low ambient temperature effects and inadequacy of color rendition. Ballasts can generally be mounted remotely from critical areas if ballast noise will be objectionable. When using fluorescent sources outdoors and in unheated spaces such as garages, only lamps and ballasts rated for low ambient temperature should be used. The designer should personally verify the appropriateness of color temperature and color rendition of each source selected.

Color of Surfaces and Light. In any hotel or restaurant space, the color of the environment will affect both patrons and workers—positively or negatively, consciously or unconsciously, according to the harmony of the scheme and the expectations of the viewers. While no hard and fast rules exist, it is generally accepted that strong colors are relatively stimulating while less intense colors are more restful and tend to expand the perceived size of a space. Whatever the colors selected, it is imperative that they be evaluated under the light source or mix of sources which will be used in the finished space, since light sources vary significantly in their color-rendering qualities. See Section 5 of the 1981 Reference Volume. See also page 2-31. The use of colored light is often overlooked as a design tool. Strong colors of light can create interesting effects when surfaces are illuminated for decorative purposes, but should not be used to light food or people because of the inherent and undesirable color distortion which will result.

Emergency Lighting. In public facilities such as hotels, motels and food service establishments the designer must provide lighting for public safety during emergency conditions, without either disorienting or panicking the users. Emergency systems for public facilities should be designed to provide short-duration lighting for evacuation and safety of guests and staff. See Section 2. Longer duration emergency lighting may be required at hazardous locations, for security purposes, and to assure continuity of critical operations. A common mistake to avoid is the installation of permanently-on emergency luminaires in restaurants. Since these luminaires cannot be switched off or dimmed at night they are sure to disrupt an intimate dining atmosphere. While it may be possible in a small facility to meet emergency lighting needs with independent battery-powered units, a central emergency generator or battery installation may be required in a major hotel or motel. Options, of which the designer should be aware, include double-circuiting of luminaires and the use of transfer relays to provide power to emergency-only sources during power failures.

Safety. Safe working and living conditions in hotels, motels and food service facilities are dependent on good lighting. See Section 2.

LIBRARIES[8]

Libraries have a variety of seeing tasks. Among them are: (1) reading matter, (2) browsing or searching through book stacks or storage areas, (3) studying at a carrel or other work surface, (4) viewing microform or computer retrieval systems, (5) meeting or conferring, (6) general office and clerical work and (7) repair and inspection work. These tasks along with general illumination for circulation spaces or audio booths, special lighting for audio-visual areas and accent lights for exhibits and displays provide a variety of lighting problems.

Seeing Tasks in Libraries

Reading is by far the visual task performed most often in a library. Reading tasks vary from children's books printed in 10 to 14 point type on matte paper, to newspapers printed in 7 point type on low contrast off-white pulp paper, to law books with long paragraphs in condensed type, to rare books with unusual type faces printed on old paper. There are also handwriting tasks involving pencils and pens. Details about the general principles which must be considered to provide the quantity and quality of illumination needed for these tasks may be found in Sections 2 and 5. In addition, illuminance recommendations are found in Fig. 2-2, page 2-7.

A task that is fairly unique to the library is that of browsing and/or searching in a stack or other form of storage space. In public spaces material may be on low shelves, on tables, on racks, in bins, etc., which are very accessible and have limited quantities of items to view. However, the vast majority of books, magazines and reference materials are stored in shelving that is tightly spaced and up to 2.5 meters (8 feet) high or in compact shelving with limited aisles. The task involves reading a title or author's name assisted by perhaps a numbering system applied to the material. The books or other material are often well used or old causing the title or other means of identification to be of very poor contrast.

When a library is associated with an educational institution, areas used for studying involve both reading and writing tasks. Such areas may have several work stations or individual work stations such as a study carrel. Task lighting is often provided at these locations, and veiling reflections should be minimized at these study locations.

Lighting Systems

A variety of lighting systems are used in libraries. Many libraries make use of daylight through windows or skylights. In all cases the luminance comfort recommendations should be the same as for offices and educational facilities. See Sections 5 and 6.

In areas where architectural features are dominant, design concepts may require sacrifice of efficiency for esthetics when translating the architect's concepts into practical lighting designs. In areas that do not have dominant architectural features, the lighting systems should be selected to provide comfortable seeing conditions with more emphasis placed on economics and luminaire design features. See Figs. 7-23 and 7-24.

Fig. 7-23. Two lighting techniques are combined in this high-ceiling college library. General lighting is provided by a high intensity discharge lamp downlighting system. The chandeliers, with low-wattage lamps, are used as decorative elements in keeping with the original architect's design concept.

For library lighting applications, there are three basic types of light sources in use today: incandescent, fluorescent and high intensity discharge. (See Section 8 of the 1981 Reference Volume for light source data.) No one lighting system can be recommended exclusively. Each system has qualities that match the requirements for a given situation. The first consideration in choosing a lighting system should be to allow the library user to see efficiently and without distraction. The second should be the appearance of the installation within the architectural and decorative design concepts of the library. The third consideration is for the energy efficiency of the system.

In general it is desirable to provide sufficient illumination for the most common seeing task performed in an area. However, if a more difficult seeing task is being performed in a small portion of that area, additional illumination should be provided by providing additional overhead luminaires or by supplementary lighting equipment located in relation to the specific seeing task. Higher illuminances should also be provided in areas that will be used by persons with impaired vision. When relighting existing traditional-type library reading rooms, the use of supplementary lighting equipment consistent with the decorative treatment of the room is sometimes required. It is especially important to avoid direct and reflected glare and to avoid veiling reflections when using supplementary lighting equipment.

Specific Areas

Reading Areas. Reading areas in libraries, including main reading rooms and reference rooms, occur throughout almost the entire library. Reading is usually performed on either side of long tables, in lounge chairs, in study carrels or at the circulation desk. Care should be taken to locate the luminaires to avoid veiling reflections on the seeing tasks and to use luminaires that reduce the luminance in the direct glare zones.

Individual Study Areas (Carrels). Individual study areas or carrels may be found in almost any public area of the library building, such as main reading rooms, enclosed individual rooms

Fig. 7-24. Low ceiling library, where open access stacks are lighted with rows of fluorescent luminaires at right angles to the stacks.

and stack areas. One of the most serious lighting problems for carrels are the shadows produced by dividing walls. To avoid shadows it is desirable to provide lighting from as many directions as practicable. Special care should be taken to avoid veiling reflections especially from localized luminaires.

Shelving and Stack Areas. This area applies to shelving and storage units for all types of materials in addition to books. The visual tasks in book stacks are very difficult; for example, it is necessary to identify the book by number and author on the lowest shelf. As a result of studies made of typical books at actual viewing angles, it is recommended that, when practical, non-glossy plastic book jackets should be used rather than glossy; large and legible non-glossy lettering should be used for authors' names, book titles and index number. Dark book, shelf and floor surfaces reflect very little light; therefore, the use of light colored surfaces should be encouraged.

Open Access Stacks. Open access stacks are open to the public for finding their own books or for browsing. Book stacks are usually arranged in rows with continuous rows of fluorescent luminaires located along the center of each aisle. An alternative is to locate luminaires at right angles to the stacks (see Fig. 7-24). Obtaining maximum illumination on the lower shelves is the greatest concern.

Limited Access or Closed Stacks. These stacks are used primarily by library personnel. The aisles are usually narrower which increases the problem of obtaining illumination on the lower shelves. Compact shelves may also be used for limited access or closed stacks. Luminaires controlled by delayed time switches may be considered for these stack areas.

Card Catalogs. Individual files of card indexes are usually located in the main reading rooms. Location of overhead general lighting luminaires at right angles to the file cards rather than parallel to them will provide slightly better illumination on the vertical surfaces of the cards.

Circulation Desks. Circulation desks are usually located near the entrance to the main reading room. Often the general overhead lighting system will provide sufficient illumination for the desk; however, if not, sufficient supplementary illumination should be provided, which may be from an architectural element that will identify the circulation desk.

Conference and Seminar Rooms. Conferences are frequently scheduled in libraries, and groups hold seminars on occasion. In addition to general overhead lighting, provision should be made to illuminate the speakers and their materials at the lectern and at the seminar table. Several illuminance levels should be provided for the multiple type use of this space.

Display and Exhibition Areas. Many libraries have display and exhibition areas. These may be in glass covered horizontal cases or may be mounted on vertical walls or dividers. See Museums and Art Galleries, page 7-33 for lighting such displays.

Audio Visual Rooms. There is an increasing use of listening areas for lectures, music and other recorded material. These areas are either small rooms with individual reproducing equipment, or large rooms where head receivers may be plugged into circuits or carrels. Small rooms have poor utilization of light because more light is absorbed by the wall surfaces and, therefore, require a closer spacing between luminaires. Lighting similar to that required for carrels in a large room is also needed for the audio carrel system.

Lighting for CRT and Mircroform Viewing Areas. Computers and microform materials permit much larger holdings of newspaper files, rare books, special collections and technical publications. Microform materials include rolls and cartridges of microfilms on strips, aperture cards containing single frames of microfilm and microfiche cards or sheets containing a series of highly reduced micro-images. One of the most difficult seeing tasks is reading a screen filled with a printed page located under a general lighting system needed for other tasks in the area. (Reflections, diffuse and specular, tend to wash out the already poor image on the screen.) When notes must be taken over long periods of time, it is desirable to provide illumination on the note pad, but controlled to reduce reflections on the screen.

Higher illuminances are needed for files of microforms than are needed for viewing. Where viewers must be placed in reading areas or work areas with higher level general illumination and no controlled lighting or dimming is available, machines should be selected that are hooded and have screens which are treated to reduce reflections. A small luminaire should be provided between viewers to illuminate a fixed or sliding shelf in front of each machine for note taking. Such a luminaire should be moveable so that it may be individually located to accommodate right or left-handed operators.

Offices. Office areas in libraries should be illuminated in accordance with the recommendation for Office Lighting in Section 5.

Rare Book Rooms. Higher illuminances are recommended for rare book rooms because of the poor quality of printing often found in many rare books; however, lighting techniques such as those used in Museums and Art Galleries should be used for books displayed in glass cases. These would include means of reducing the amount of deleterious radiation.

Archives. Archives are for the storage and examination of public documents of all kinds. This would include legal documents, minutes of meeting, legislative actions and other historical papers. Pencil writing, small letters and condensed type are used in many of these documents.

Map Rooms. Map rooms have both storage and reading areas. Storage of maps involves the use of deep cabinets which, in turn, requires aisles sufficiently wide enough to open drawers for access to the maps. Maps mounted on vertical surfaces require vertical surface lighting.

Fine Arts, Picture and Print Rooms. See Museums and Art Galleries for the proper lighting of displays, paintings and art objects.

Group Study Rooms. Sometimes a group of 4 to 6 students is assigned to a project to be solved by consulting among themselves, and isolated rooms may be provided for this purpose. Techniques used for classroom lighting are recommended for these rooms. See Section 6, Educational Facilities Lighting.

Overnight Study Halls. Sometimes students prefer to work all night when preparing for examinations, and libraries may provide a portion of the building that can be isolated for this purpose. Lighting for these areas is similar to that required for Reading Areas or Individual Study Areas.

Entrance Vestibules and Lobbies. Lighting in entrance vestibules and lobbies should create an atmosphere suitable for the particular type of library. The lighting may emphasize the architectural features and provide a smooth transition to the functional areas.

MUSEUMS AND ART GALLERIES

A museum is defined by a leading organization (the American Association of Museums) as "an organized and permanent non-profit institution, essentially educational or esthetic in purpose, with professional staff, which owns and utilizes tangible objects, cares for them, and exhibits them to the public on a regular schedule." For illuminances recommendations, see Fig. 2-2, page 2-8.

A museum's highest responsibility is to the study and care of its collection and to the collection's effective public display. Thus, lighting has been considered to be the third, or perhaps the second, most important responsibility of the curator or designer, for lacking effective lighting, the most interesting collection and tasteful displays are ineffective. Lighting, however, can cause or accelerate degradation of certain kinds of museum and art gallery objects and this should be kept in mind.

Damage to Museum and Gallery Objects

The principle risks associated with museum and art galley objects are due to:

1. Vandalism.
2. Excessive heat and humidity, especially rapid changes.
3. Chemical attack, such as from acidic matte boards.
4. Airborne pollutants, including vapors and dirt.
5. Improper handling or marking.
6. Radiant energy: infrared, light and ultraviolet.
7. Biological attack: insects, fungi, etc.

Each of these sources of damage must be dealt with by the curator and designer, often with the assistance of a consulting specialist; however, only the problems caused by radiant energy are covered here, along with guidance on minimizing the damages it can cause. All damage cannot be eliminated, but it can be restricted to an acceptable degree. For example, a suit of a famous 18th century American was displayed for 8 years under good controlled conditions with a maximum illuminance of 60 lux [6 footcandles] and no ultraviolet radiation—the dye was of poor quality and suffered noticeable fading.

Materials Subject to Light Damage. Essentially all materials of organic origin will change due to the absorption of light or its related energy. Among pigments, vegetable dyes should be treated with great care. Writing inks are also subject to extreme fading. The lower the reflectance of a colorant, the greater the energy absorbed and the greater the risk of deterioration. The material on which the pigment is placed and the thickness with which it is applied also affect its susceptibility. Prints and watercolor painting are more vulnerable than oil paintings.

The most fragile of the common fabric materials is silk. All other natural fibers have greater resistance, while the synthetic fibers, except nylon, are most resistant to destruction. Color, however, affects the degree of risk.

Other materials which can suffer degradation are paper, leather, fur, feathers, plastics and wood.

Minimizing Damage. To minimize damage:
1. The source of light should contribute as little heat as possible to the displayed object.
2. Humidity in the building and in display cases should be stabilized, whether lights are on or off.
3. All ultraviolet radiation should be removed by filters, such as UJ-3 acrylic sheet.
4. Total exposure, in terms of lux (footcandles) × hours, should be held to the absolute minimum as light damage is the product of illuminance and time (see Fig. 7-25). For highly susceptible materials, light should be fully excluded except during actual viewing.
5. For continuous exposure, the most precious and delicate objects should have no more than 50 lux [5 footcandles]. Surrounding areas of the museum will have to contribute to the viewer's accommodation to this low illuminance.
6. Daylight is potentially far more damaging than most electric light sources. Direct sunlight or sky light should not be allowed to reach susceptible materials. Daylight may be used if redirected by louvers, blinds or similar treatment onto upper wall or ceiling surfaces to indirectly illuminate a gallery. A paint containing zinc oxide or titanium dioxide will absorb ultraviolet radiation.
7. Period houses should have a ultraviolet absorbing plastic or film placed over the glass in all windows. Light can then be controlled with shutters, blinds or drapes, as may be appropriate. Louvered window screening is another possible control medium for daylight.
8. Display time for susceptible specimens can be shortened by rotating objects seasonally, putting copies on display part of the time, covering cases with opaque lids which the viewer can lift at will, or providing local switching for viewer operated lights (with or without timers, as appropriate). See page 19-31 for addition information on fading and bleaching.

Principles of Museum and Gallery Lighting

High contrast of light and color produce tension and drama; over-all soft lighting and pastel colors create a mood of relaxation. Either treatment carried to extremes can produce fatigue or

Fig. 7-25. Recommended Total Exposure Limits in Terms of Illuminance-Hours Per Year to Limit Damage to Light-Susceptible Museum and Art Gallery Objects

Objects	Lux-Hours Per Annum	Footcandle-Hours Per Annum
Highly susceptible displayed materials—silk, art on paper, antique documents, lace, fugitive dyes	120,000*	12,000*
Moderately susceptible displayed materials—cotton, wool, other textiles where the dye is stable, certain wood finishes, leather	180,000**	18,000**

* Approximately 50 lux (5 footcandles) x 8 hours per day x 300 days per year.

** Approximately 75 lux (7.5 footcandles) x 8 hours per day x 300 days per year.

NOTE: These illuminances, if carefully applied, will not result in worse than just perceptible fading in the stated materials in ten years exposure. *All wavelengths shorter than 400 nanometers should be rigidly excluded.*

boredom. Museum curators and designers will usually be aware of these effects and will use their galleries appropriately. A museum may have a sculpture court or a hall devoted to light stable artifacts which could be daylighted or lighted with large general lighting luminaires providing excellent visibility. Drama may even be added with appropriate strong directional lights, but not always without a problem. One problem could be the need for an extreme change in illumination required in an adjacent space where light must be restricted because of perishable artifacts.

Superior gallery planning or visitor routing is the ordinary solution to this problem, but frequently these cannot be done. In this case, the brightly lighted space may need to be dimmed. The ratios of adjacent spaces in terms of luminance should not exceed 10 to 1 and preferably be less. This ratio of luminances is not merely the ratio of the two illuminance readings taken with an illuminance meter, but the apparent visual impact on normal viewers as they make the transition from one space to the next. This would allow the mathematical ratios to be violated by clever placement of a brightly lighted object or surface in an otherwise dark room. For example, if a daylighted court were devoted to showing frontier artifacts such as plows, wagons and locomotives, and adjacent space were devoted to native American objects such as fur, robes and feathers, an introductory display of replica costume pieces on a manikin could be very strongly illuminated to attract visitors to the dimly lighted room and to assist in the visual transition.

Museum display lighting tends to be principally oriented at objects or artwork. This would lead one to expect high drama in all exhibits, but this is rarely true. Background surfaces pick up spill and reflected light and modify the appearance and impact of the space. True focussed lighting only occurs when all surfaces are relatively dark. Such exhibits must be used sparingly since they tend to fatigue the visitor. As in all things, variety increases interest.

In some instances the light and the lighting equipment are part of the exhibit statement. These occasions are exceptions. Normally, the purpose of the lighting is to provide exhibit visibility and enhancement in the most unobtrusive manner possible.

Light Sources

Modern electric light sources for museum and art gallery lighting fall into three broad categories: incandescent, fluorescent, and high intensity discharge. The latter group is very efficient in generating light at low cost, but their spectra will usually make them unsuitable for most interior display lighting. Experienced lighting designers will use these sources in limited areas—for exterior floodlighting and protective lighting these lamps can be effective and economical.

Fluorescent Lighting. Fluorescent sources offer the following lighting advantages to museums:

1. High efficacy, resulting in lower energy use and costs and lower heat output.

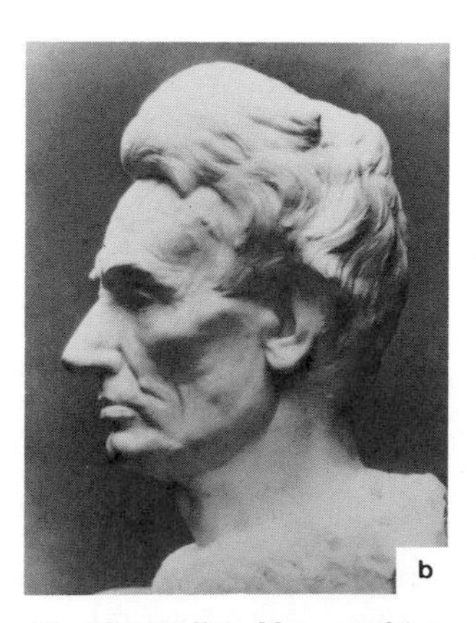

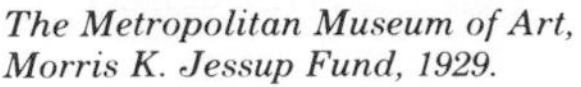

The Metropolitan Museum of Art, Morris K. Jessup Fund, 1929.

The Metropolitan Museum of Art, Rogers Fund, 1912.

The Metropolitan Museum of Art, Rogers Fund, 1917.

Fig. 7-26. Lighting of sculpture. a. Concentrating sources, alone, from front left. b. Total overhead diffuse lighting conforms to expression of features. c. Low diffuse lighting and strong concentrated accent adds to stern expression. d. Light concentration from upper right. Strong overhead diffuse lighting aids in viewing the details of this complex sculpture.

2. Excellent range of colors (including over 20 different white lamps).
3. Excellent color appearance, when the correct lamps are chosen.
4. Long lamp life, reducing maintenance effort and cost.
5. Good size variety.

Fluorescent lamps are inherently large, soft light sources and are incapable of generating true beams of light. Creation of highlights and shadows for revealing three dimensional form and texture is not practical with fluorescent lamps. They *are* ideally suited to case lighting, dioramas, transparencies, luminous elements, creating false windows and virtually all practical working areas of a museum. They should also be used for stairway and other nightlights because of their economies. Wherever general illumination is required, fluorescent will usually be found to be effective.

Incandescent Lamps. Incandescent sources can be subdivided into four broad categories for museum and gallery use:
1. General service lamps in the familiar "A" and "PS" bulbs. They require directing of the light by reflectors and/or lenses. They are principally used for illumination of areas rather than specific objects. Most museums can fulfill all their needs with four of five sizes of these lamps.
2. Reflectorized lamps of the R, ER and PAR types, flood and spot distributions in a large range of sizes. See Section 8 of the 1981 Reference Volume. There are some varieties of low-voltage lamps available, but most are intended to operate directly at line voltage. Blown soft glass R and ER lamps are relatively inexpensive with reasonable life ratings, and are generally highly useful. The "field" of the light beam is usually quite smooth and the beam edge is not pronounced.

Molded hard glass PAR lamps, in flood and spot, have a beam narrower than the corresponding R lamp beam. The field of the beam may often contain irregularities and the beam edge is considerably more noticeable than with R lamps. Life rating is generally the same as R lamps and cost is slightly higher. For longer beam throws and higher drama, the PAR spot is superior to the R spot. One particular caution; these lamps are relatively heavy and should not be mounted above glass cases or fragile objects.
3. Low voltage lamps are available in general service and reflectorized types. Small low-voltage lamps are useful in simulating candle or oil lamp light in period rooms. The most useful low voltage lamps are the 25- and 50-watt spot and flood types. These produce excellent shaped beams with very little spill light. The beams can be projected considerable distances and the lamps are physically compact, allowing easy concealment. They have two disadvantages: their maintenance and that they require transformers. They are, however, energy efficient.
4. Tungsten-halogen lamps. They are: more efficient, whiter in color, more compact, longer lived than standard lamps of the same type. They are also hotter, more costly, produce *double* the ultraviolet radiation and are difficult to filter. These qualities suggest their use in intense light on metal or stone objects, but preclude their use in many other museum areas.

Specific Museum Lighting Problems

Museum exhibit design work can be divided into the categories shown below and further divided into "permanent" and "temporary." Many museums have spaces intended for changing exhibits and these require great flexibility. After discussing the most basic design problems, techniques of flexible lighting will be described for solving these problems in short-term exhibits.

Large Three-Dimensional Objects. Whether sculpture or machinery, or costume manikins or cannon, an important aspect of such objects is their mass. Lighting must reveal their plasticity and as much surface detail as is appropriate to the object. Total uniformity of lighting is rarely called for. There should be some distinction of strong and weak illumination, such as a spot light from one side and a flood light from the other. A degree of shadow is desirable (see Fig. 7-26). The shadow should not be so dark as to conceal significant details, or to produce distortions in appearance, but the shadow is a significant visual clue to the solidity of the object.

If not objectionable from the standpoint of honesty, subtle color differences may be used in the lighting from different directions. For example, a portrait bust might be placed near a window with a spot light directed at the other side. Even more subtle is the difference between a standard incandescent lamp and a tungsten-halogen lamp, but the difference is sufficient to improve surface texture in an object that must have almost equal illuminance from all directions.

When large objects are placed in open space, light will usually need to be directed sharply downward to avoid glare to the observer on the opposite side of the object. If some degree of upward lighting is required to avoid overly harsh shadows, this can frequently be achieved by providing a high reflectance base, such as white linoleum, to bounce the light upward. Small mirrors, strategically placed also may serve, if they are not distractions themselves. Direct lighting from the bottom is possible, but care must be observed that nearer surfaces are not overlighted, producing the unpleasant effect of shadows running upward.

Where the object is placed in a niche and the viewing will be only from one direction, the lighting can be as theatrical as desired. Lighting that strikes the objects directly from the front, at the same angle that the object is viewed, will tend to flatten surface features. This could be desirable for revealing the colors of a tapestry, while concealing the weave, but it would be undesirable for most sculpture.

Flat Displays on Vertical Surfaces. Paintings, prints and documents fit into this very important category. There are essentially three approaches to this display problem: light the entire space uniformly, light the vertical surfaces, or light only the objects. The first approach assumes that the object will be prominent in its environment without lighting assistance. This may well be the case where the objects are in high contrast to the wall surface. Many of our classical museums were designed in this manner, but the trend is to provide some additional focused light to enhance paintings or text materials. The National Gallery in Washington employs skylights, but has recently installed a system of spotlights to add warm color to the artworks and to provide more focus on them.

A gallery space may be the ideal setting for the objects exhibited and supplemental lighting might seem artificial. This is true of period rooms and houses, so that specific objects are rarely highlighted in these spaces.

Lighting only the vertical surface is often a very attractive method of displaying flat objects. Several commercial "wall-washer" luminaire systems are available in all lamp types. Fluorescent types offer very smooth continuous illumination and can provide good uniformity vertically, when used according to the manufacturer's directions. The appeal of the displayed object is again related to the contrast with the background. Incandescent wall-washers have optical devices to provide lighting higher on the wall than a common downlight. This minimizes scalloping and provides a more even wash on the wall for a predetermined area. Such wall washers are normally mounted quite close to the wall and the manufacturer's installation directions should be followed closely. Since such units are providing light at a steep angle, shadows of frames, shelves, etc., can be troublesome.

Lighting vertical objects individually, the third technique, permits more dramatic separation of objects, including lighting from varying angles and at varying illuminances. Where correctly used, this method is highly effective. Where incorrectly used, glare or shadows can seriously detract from the finished appearance (see Fig. 7-27). Fig. 7-28 shows preferred angles for lighting flat vertical displays. If the display has considerable relief or heavy frames, the angle can be

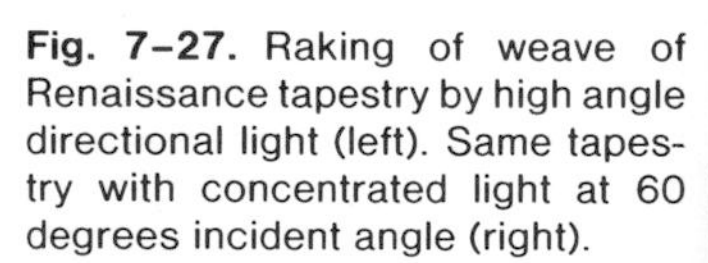

Fig. 7-27. Raking of weave of Renaissance tapestry by high angle directional light (left). Same tapestry with concentrated light at 60 degrees incident angle (right).

The Metropolitan Museum of Art, Bequest of George Blumenthal, 1941.

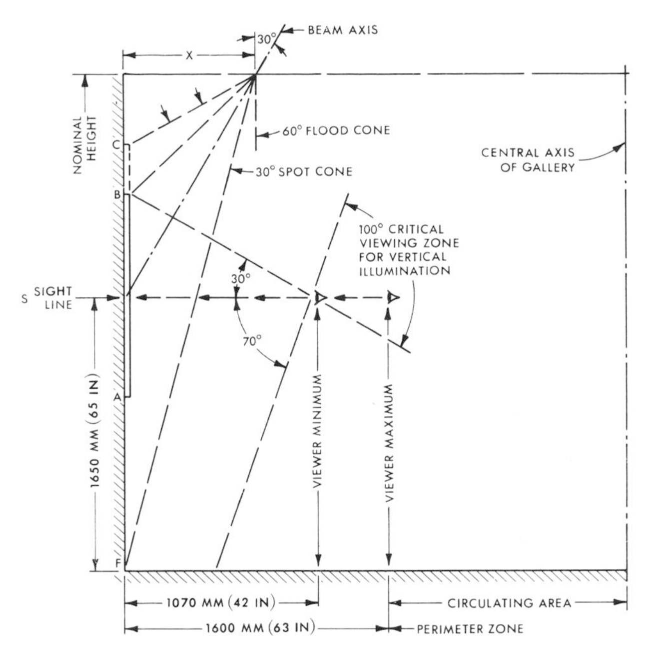

Fig. 7-28. Model perimeter (viewing) zones at nominal ceiling height. Model based on: (1) primary diffuse component (of vertical illuminance) at approximately 40 per cent of horizontal illuminance at S, (2) height of wall-hung display, (3) ideal utilization of beam cones, and (4) minimum effective viewing distance relative to a nominal height of object (A to B = 1320 millimeters (52 inches) for a 30-degree cone, A to C = 1650 millimeters (65 inches) for a 60-degree cone). To calculate viewing zones for higher objects, increase horizontal dimensions 38 millimeters (1.5 inches) for each 25-millimeter (1.0-inch) increase in height of object.

$$X = (\text{ceiling height} - \text{eye height})\ (.577)$$

for an aiming angle of 30 degrees from the vertical.

greater and distance X increased. If the display is mounted high and has substantial glossy areas, the reverse is appropriate. Unless there is a contrary reason, normal eye height is assumed to be 1650 millimeters (65 inches) above the floor.

Display Cases. Built-in display cases are generally provided with customized lighting which is integral to the case. While any surface could be the location for the lights, certain objects dictate specific lighting approaches. Glass, for example, looks very handsome when back or bottom lighted. Objects with little surface detail, but interesting shapes may be silhouetted. Most objects, however, will be top, or top-front, lighted, since this is most effective and natural. In these categories, the lighting units should be fully concealed. Fluorescent lighting is particularly appropriate to this purpose since the source delivers more light with less damaging heat and can be readily filtered against ultraviolet radiation. The entire case is then lighted with considerable uniformity and objects are made to assume importance by their position and the contrast with the case background. In most light finished cases there will not be substantial variation in the light vertically, due to the multitude of interior reflections, including the front glazing. Dark finished cases will have much more top-to-bottom variation. In these, it may be necessary to overcome this by arranging the smaller items high in the case and the larger, crude items toward the bottom. Occasionally this will require side lighting, particularly where the case is taller than wide.

Display case luminaires (light boxes) do not need to extend the entire length of the case and, in fact, will produce a very noticeable end scallop if they do. Normally, fluorescent lamps should end 100 to 200 millimeters (4 to 8 inches) from the end walls. In a recessed case, the light box can be completely out of sight in the top front of the case and no lens or louver is required for concealment (though a sheet of clear UF-3 plastic should be installed).

Most often, the light box will need to be in the viewer's line of sight and concealment of the light source is important. A low brightness lens or louver is then mandatory. The parabolic-wedge louver is highly favored for this, but it will produce a partial dark area at the top walls of the case because of the cut-off of the louver. In tall cases, or in those where the display begins somewhat below the top, this is a desirable feature, but in others, it may be preferred to use an angled louver to provide the lamp shielding while still providing light to the top of the rear panel.

Incandescent spotlights can be provided in cases to add warm color or to highlight specific items. Jewelry has more sparkle and can be placed on dark backgrounds when the items are individually spotlighted. The light box then requires greater depth, both for the bulkier lamps and to allow for greater heat dissipation. The use of plastics for glazing is limited because of the heat in the light beam.

In all instances of built-in cases, provision must be made for easy access to the lighting compartment. This access should not be through the case interior, thus risking damage to the displayed objects. Recent designs of fluorescent ballasts have greatly reduced their heat output and have made the risk of ballast leakage almost disappear; it is therefore no longer necessary to mount ballasts remote from the case. Remote ballasting requires more complex wiring and creates greater problems when replacements are needed.

Freestanding cases also can be lighted externally by the room illumination or by spotlights. If the general room illumination serves the cases, they will be undifferentiated from the remainder of the space and will have lesser impact. Spotlights on cases can give the color and emphasis that the display deserves; however, there will be a shadow of the top edge of the case somewhere within the case, unless the lights are directly above, and there will be a reflection of light from the top of the case onto the ceiling. By careful positioning of the case or the lights, these problems can be overcome.

The most severe problem with all cases is reflections in the viewing glass. This is most noticeable when a relatively dark case faces an area of high brightness. The only complete solutions are to use either specially curved glass (so-called invisible glass) at great cost in money and space, or to place all cases opposite dark walls. Other partial solutions are: to slope the front glass of the case outward to that only the floor is reflected to the viewer, and then the floor must have a relatively dark, matte surface; to place cases at right angles to each other so that there is at least one favorable viewing position; or to keep all case interiors very bright in appearance. In period rooms, natural settings, etc., it may be possible to omit the front glazing if other suitable protection of the displayed objects can be achieved; *e.g.*, railings and alarms, or a taut wire screen. A truly non-reflective glass is available, but at very high cost and only in limited dimensions. Areas of cases below eye level will normally reflect the floor to the viewer.

Table cases always reflect the ceiling area opposite, and this portion of the ceiling should be free of lights and relatively dark to avoid reflections. Table cases against a wall will reflect the wall directly above the case, thus restricting the use of this for vertical displays or text. Sloping cases will present troublesome reflection problems unless the designer makes a preliminary sketch in profile showing the normal viewing positions and the areas that will be reflected to the viewer. Such a sketch, done accurately, will anticipate problems and show the designer where the lights must be mounted to minimize both shadows and reflections.

The above three categories (large objects, wall displays and cases) cover virtually all museum displays; however, another category could be called "environments." This includes houses or period rooms which are not treated as cases—any exhibit where the public walks into the actual display. For such spaces, naturalism will enhance the ambience, but the problem of object degradation complicates this. Genuine antique or natural items must be protected against the very light that the original owner lived by. Windows will need to have ultraviolet filtering, and it may also be necessary to provide drapes or blinds to reduce total lighting levels. In this case, the guide or docent will be instructed to open the blinds for better lighting and close them as the visitors leave. Concealed lights can be controlled by switches or an automatic program device to point out objects in the space with short duration spots of light.

Halls or Galleries for Changing Exhibits. If a flexible exhibit space is consistently used for the same type of display, such as a changing exhibit of prints, then it can be lighted in a semipermanent manner, positioning lights on regular spacings, permitting the changes with variable lamp wattages, beam patterns and luminaire aiming. Alternatively, track can be installed parallel to the areas of interest at appropriate distances from the surface or area to be lighted and a selection of track units kept available to provide the necessary variety.

Many changeable spaces are not used in a predictable manner from one show to the next, and this would call for considerably higher degrees of variability than can be accomplished with simple straight runs of lighting track. One solution uses a "track-on-track" approach, allowing lighting units to be placed virtually at any point in a space. This calls for a heavy investment in built-in and auxiliary lighting hardware and will always result in a certain degree of ceiling clutter. The aesthetics of track lighting systems are important to the interior museum environment. Luminaire types should have compatible housings and be finished in neutral colors to blend with ceiling architecture. The other extreme is to merely provide sufficient electrical service above a dropped ceiling and reconstruct the ceiling for each exhibit, locating all lights in the new ceiling design. A variation in this last, is to provide a modular ceiling grid with lighting units mounted in standard panel sizes. These would be removed and replaced at will. Grid patterns of lighting track can be installed, with flexibility dependent on the closeness of the grid spacing.

When preparing service for such a changeable space, it is desirable to allow at least 50 watts per square meter (5 watts per square foot) in the panel and cable sizes, *though this high loading will probably never be fully utilized at any given time.*

Standard receptacles should be provided on relatively close spacings, about 2.5 meters (8 feet) on center in all permanent partitions, and additional provision should be made to provide electrical service in the middle of the area, either through the floor or the ceiling.

Conservation and Restoration Shop. Non-display portions of museums should be designed in accordance with the best modern techniques. One unique museum space which may require special attention, is the conservation and restoration shop. Such a space should have moderately high illuminances with special localized task lighting units available. All the fixed lighting should be color balanced according to the preference of the working conservator and should also be fully filtered against ultraviolet. Special ultraviolet portable examination lights will usually be required for specific tests.

Entrance Foyer or Lobby. An entrance foyer or lobby should be regarded as a vision conditioning space in which the visitor's eyes are permitted to adjust from daylight levels to the lower lighting in most galleries and museums. It is useful to have some control over the general illumination level in such foyers to adjust to the changing need.

Color

In general, the color of exhibit lighting is intended to achieve the highest degree of verisimilitude for the material displayed. This would suggest that daylight tonalities should be used for all items that were found in nature, or that were created under daylight. A major problem arises here. The high illuminance levels that nature provides are not permitted and the cool tones of daylight become gloomy and cold at low levels. See Fig. 2-11. The color of the noonday sky is about 5000 K. The correlated color temperature of a deluxe cool white fluorescent lamp is only about 4100 K, and yet the light could appear excessively cold in many exhibit circumstances. In fact, where illuminances are at or below 200 lux [20 footcandles], light source colors should rarely exceed 3400 K. Exhibits should be prepared, or checked under the light source that will be used for display.

Actual tinting of the illuminant color should be used with great discretion, and only with the concurrence of the curator concerned. Where this is done, it is preferred to use glass filters over incandescent lamps rather than colored lamps, first because this is more economical and second because the range of colors in glass filters permits more freedom than the limited number of lamp colors available.

Providing color with fluorescent lamps is relatively easy in the more popular sizes (such as the 40-watt lamp). See Fig. 8-115 in the 1981 Reference Volume. Where more than one lamp is used, a combination of types can provide almost unlimited color choice. Interposing colored acrylic filters can further extend color possibilities, and for short term exhibits, theatrical color gels are appropriate.

Maintenance

One of the larger problems of the museum administrator is providing good maintenance on restricted budget. Lighting, requiring constant attention and being highly visible, is a very significant part of this problem.

The designer of museum lighting can ease the burden by following the listed precepts:

1. Use as few types and styles of lamps as practical.
2. Use as much inherently long-lived types of lighting as practical.
3. Avoid lamps which require disassembly of the luminaire for changing.
4. Don't use special purpose lamps where general service lamps can serve.
5. Provide ready accessibility to all lighting equipment.
6. Where possible, use fewer large lamps instead of many small lamps (this will materially increase efficiency as well).
7. For incandescent lamps, use those with medium screw base, unless there is a demonstrable advantage to a lamp with a different base type.
8. Provide adequate storage space for lamps and maintenance supplies.
9. Provide ladders, towers, hand-tools and cleaning materials; any material device which will expedite lighting maintenance.
10. Treat the lighting maintenance person as the skilled, important person which the position justifies.

Energy Management

Museums will require more incandescent lighting than most other categories of public building. This will not justify careless or wasteful lighting practices. To achieve higher lighting energy efficiency, designers must think as follows: "Lamp and luminaire *X* is the most efficient available. Will this work effectively for the exhibit lighting problem? No. Then lamp and luminaire *Y* is the next most efficient system. Will that work?" and so forth until they arrive at the most satisfactory lighting solution.

In general, museum and gallery illuminances will not be high and lighting energy consumption should be quite moderate. Over and above conserving energy, maintaining a low lighting energy use contributes to better specimen conservation, since there will be less drastic temperature changes as lights are turned on and off. See Section 4.

REFERENCES

1. Church Lighting Subcommittee of the Institutions Committee of the IES; "Church Lighting," *Illum. Eng.*, Vol. LVII, p. 67, February, 1962.
2. Health Care Facilities Subcommittee of the Institutions Committee of the IES: "Lighting for Health Care Facilities," *Light. Des. Appl.*, Vol. 8, p. 19, June, 1978 and p. 36, July, 1978.
3. Sisson, T. R. C.: "Visible Light Therapy of Neonatal Hyperbilirubinemia," *Photochem. Photobiol. Rev.*, Vol. 1, p. 241, Plenum Publishing Corp., New York. Kethley, T. W. and Branch, K.: "Ultraviolet Lamps for Room Air Disinfection," *Arch. Environ. Health*, Vol. 25, p. 205, September, 1972.
4. *Occupational Exposure to Ultraviolet Radiation: Criteria for a Recommended Standard*, (HSM 73-11009, NIOSH) U.S. Government Printing Office, 1976.
5. Beck, W. C., Schreckendgust, J. and Geffert, J.: "The Color of the Surgeon's Task Light," *Light. Des. Appl.*, Vol. 9, p. 54, July, 1979.
6. *Use of Anesthetics in Hospitals,* NFPA-56-A-1978 National Fire Protection Association, Boston, MA, 1978.
7. Subcommittee on Kitchen Lighting of the Institutions Committee of the IES: "Lighting for Commercial Kitchens," *Illum. Eng.*, Vol. LI, p. 553, July, 1956.
8. Library Lighting Subcommittee of the Institutions Committee of the IES: "Recommended Practice of Library Lighting," *J. Illum. Eng. Soc.*, Vol. 3, p. 253, April, 1974.

Lighting Merchandising Areas[1]

SECTION
8

New lighting techniques, equipment and more efficient light sources present the lighting designer with the tools to meet the challenge of the ever changing requirements of the merchandising world and the spiraling cost of energy.

Sophisticated consumers and the general lack of trained sales personnel make it essential to present merchandise under lighting that will aid in sales and a reduction in merchandise returns.

Consideration should be given to the quality, quantity and effectiveness of light on the task or displayed merchandise that will contribute to a pleasant and secure environment in which to do business.

OBJECTIVES OF MERCHANDISE LIGHTING

There are three primary objectives of lighting merchandising areas: to attract the customer, to initiate purchases and to facilitate the completion of the sale (and minimize returns).

Lighting to Attract the Customer. The first step in the merchandising process is to attract the customer to the merchandise and merchandising space. Light attracts. The quantity, quality, and effect of the light reaching the merchandise and the appearance of the area—show window or store interior—are determining factors in the effectiveness of the sale of the merchandise.

Lighting to Initiate the Purchase. Buying decisions start when the customer is visually intrigued. The actual purchase is not accomplished until the customer can visually evaluate the merchandise and read labeling through adequate illumination.

Lighting to Complete the Sale. Adequate lighting at the point of sale is necessary to complete the transaction. It should enable sales personnel to quickly and accurately perform their sales duties, such as: registering sales, preparing sales slips, reading prices and packaging.

There are numerous factors which must be considered in the lighting design for merchandising spaces to achieve the above objectives. In general, these fall within four design considerations: appearance of the space and occupants, appearance of the merchandise and graphics, merchandising operation and methods, and physical and environmental aspects of the space and merchandise. See Section 1 for a general discussion of lighting design.

LIGHTING DESIGN CRITERIA

The fundamental factors affecting the appearance of merchandising spaces, sale of merchandise and performance of tasks are: brightness and brightness distribution produced by luminance and luminance ratios; merchandise or task size, contrast, color, form and texture; and time for viewing.

Factors in Seeing Merchandise

There are four fundamental factors that affect visibility—the size of the details to be seen, contrast against their background, the amount of time available for viewing, and the luminance of the details and their background.

Size. The size of things to be seen varies considerably. As size increases, visibility in-

NOTE: References are listed at the end of each section.

creases and seeing becomes easier up to a certain point. Also, when the size is small, visibility can be improved by increasing illuminance.

Contrast. To be visible, each critical detail of merchandise and seeing task must differ in luminance or color from its background. Visibility becomes maximum when the luminance contrast of details with their background is greatest—*e.g.*, the luminance contrast between stitching in black on light fabric is high. Conversely, light stitching on light fabric has a very low luminance contrast. Where poor contrast conditions exist, visibility can be improved by increasing the illuminance or by using a more effective lighting system.

Time. It takes time to see. The time taken to decipher a given message and to take action on it is an important measure of efficiency. If it takes longer to see, the task becomes less pleasant and assimilation of knowledge more difficult. Increasing the illuminance will help to increase task performance speed by reducing the time it takes to see and comprehend the information.

Fig. 8-1. Linen household items here are of a higher luminance than the surroundings and attract attention.

Luminance. As the luminance of merchandise and tasks is increased, seeing is made easier. For example, under equal lighting conditions, the pattern on a light-colored suit is easier to see than that on a dark-colored suit, because the light-colored suit is brighter. The dark-colored suit can be made equally visible by increasing the illuminance (luminance) until it is equally bright.

Interrelation. All four factors of size, contrast, luminance and time are interrelated. But since merchandise or task size and usually its contrast are fixed, improvements in visibility and visual performance can be made through improvements in the characteristics of the lighting.

Luminance and Luminance Ratios

Luminance Ratios. When customers scan the merchandising space, their eyes become adapted to the average luminance of the entire space. The average luminance to which the customers adapt depends upon the luminance and area of the surfaces seen (wall, ceiling, floor, merchandise and store fixtures). As the customers look at merchandise of a different luminance, there is a sudden loss of ability to see the details of the merchandise, until their eyes readapt. Seeing quickly anywhere in the merchandising space (the merchandise or seeing task) means that the ratio of luminance should be no more than five-to-one. On the other hand, for the customer to be attracted to the merchandise, the merchandise and displays should be at least three times the luminance of the surround. Small areas of higher or lower luminance within the merchandising space will not affect the average adaptation appreciably but may create interest (see Fig. 8-1).

Surface Reflectances and Textures. Surface textures may vary from mirrored (specular) to totally diffuse (matte). The brightness of matte surfaces, such as carpeting, appears uniformly bright in all directions viewed and is independent of direction and area of the impinging light source. For surfaces not totally diffuse, such as flatware, the brightness varies with the direction of viewing, the location and size of the light source and its luminance, and the degree of specularity.

Where reflectances of the surfaces within the merchandising areas can be selected, they should be chosen so that the luminances produced are within the limits outlined. When these cannot be selected, the luminances and luminance ratios can be controlled by varying the illuminances on the surfaces—*i.e.*, darker surfaces should be provided with higher levels or lighter surfaces with lower levels.

Merchandise Surfaces. There are no limitations on the reflectance of merchandise displayed as a result of the variety of products displayed. To keep within the luminance ratios recommendations, the luminance can be controlled by varying the illuminance on the merchandise—*i.e.*, less illuminance may be required on light-colored merchandise.

Characteristics of Light and Lighting

Distribution and Direction of Light. Lighting systems and luminaires are described in Section 1 according to the percentage of total luminaire output emitted above and below the horizontal. Luminaires such as spotlights, floodlights, and wall washers for specific merchandise highlighting are not classified in the same manner, but are described according to performance characteristics—beam spread, focusing ability, sharpness of cut-off, etc. Performance characteristics are dependent on the light source type, its housing and auxiliary devices (louvers, lenses, reflectors, diffusers).

Lighting systems provide either diffuse or directional light. Form in objects (modeling) and texture in surfaces can be revealed by directional lighting such as from small concentrated light sources—for example, incandescent and high intensity discharge lamp fixed and adjustable downlights. Diffuse light from wide distribution downlights or large area light sources, such as fluorescent luminaires or indirect lighting systems, tends to reduce the variations that relate to form and texture. Three-dimensional merchandise, for example, should generally be displayed with some directional lighting, but with significant diffusion to relieve the harshness (see Fig. 8-2). The appearance of customers and sales personnel (facial forms and expressions) are best

(a)

(b)

(c)

(d)

Fig. 8-2. (a) Diffuse lighting provides uniform illumination without special emphasis. (b) Backlighting gives form definition through silhouetting, but without revealing details such as surface finish, color or other inherent qualities. (c) Highlighting alone produces surface and color detail but fails to separate the object from its surroundings. Harsh shadows may also be a resulting problem. (d) A balanced lighting system of the three techniques—general diffuse, backlighting and highlighting—reveals the important details of the object and places it in context with its lighted surroundings.

treated with a similar combination of directional and diffuse lighting.

Difficult seeing conditions may occur under directional lighting when harsh shadows are produced on seeing tasks. Shadows cast on the visual task reduce the luminance of the task, and may be annoying when sharply defined at or near the task. High reflectance matte finishes on room surfaces become effective secondary light sources to materially reduce shadows by reflecting a significant amount of diffused light into shadowed areas. Softer shadows will generally result when light-colored room finishes are used.

Color. A knowledge of the principles of light and color can allow for the individual taste of different people, different environmental conditions, or changing fashions. See Section 5 in 1981 Reference Volume. Some responses to different colors and color combinations are almost universal, even though personal taste may vary with climate, nationality, age or sex, and psychological attitudes. Color, as one of the most powerful of all merchandising tools, can attract attention, create an ambiance, stimulate sales or guide customers.

A positive psychological attitude can be introduced in special merchandising areas by creating a warm or cool color ambiance. It is important, however, that the customer have a good appearance; hence, the light falling on the customer should render skin tones acceptably as well as enhance the merchandise.

Color of Area Surfaces. In merchandising areas where customers are exposed to an environment for long periods, the color in the environment can have an effect on their buying attitudes—positively or negatively, consciously or subconsciously.

The color of the merchandise should be considered in the selection of colors of large area surfaces. The use of large areas of strong colors could clash with the color of the merchandise displayed and could adversely affect the color of reflected light reaching the merchandise. In general, where the merchandise is colorful and varied, background color should be of light reflectance and neutral. Should stronger surface color be desired, the light source can provide this. Small areas of accent color will, however, give vitality and dramatic interest.

Color of Light Sources. The color of area surfaces and merchandise is a function of the color of the light source and the pigment of the surfaces and objects viewed. For example, red surfaces are seen as red when the light source contains red. A "white" light such as daylight or sunlight contains energy throughout the visible spectrum, ranging from violet through blue, green, yellow, orange and red.

Fluorescent, incandescent and high intensity discharge lamps, as well as daylight, are all considered as sources of "white" light; however, these "white" light sources vary significantly in the amount of energy in each portion of the spectrum. See Section 8 of the 1981 Reference Volume.

It is desirable for a lighting system to show merchandise under the same apparent illuminance level and with the same color appearance as it will have where it is ultimately to be used. This will stimulate sales and minimize customer dissatisfaction and returns.

Fig. 8-3. Veiling reflections on specular transparent display cases can obscure the merchandise. Angled glass can reflect the adjacent aisle scene, obscuring the meat products within the case (left). Similarly, a horizontal counter can display images of the ceiling luminaire to compete with the jewelry (right).

Reflections

Reflections of light sources in merchandise or sales tasks can reduce the contrast between the details to be seen and the background. They are often unrecognized and can reduce the ability to see by obscuring the details of the merchandise and seeing tasks. These "veiling" reflections can also be caused by reflections of light sources in specular transparent surfaces, such as countertops, show windows, wall cases and packaging materials, where the visibility of the merchandise is obscured. See Fig. 8-3.

In some instances, specular reflections in merchandise can enhance its appearance.

Veiling Reflections. Reflections that reduce the contrast between the details of the merchandise or task and the background produce losses in visibility and visual performance. The reflection may be readily discernable as in the case of coated or glossy packaging material (*e.g.*, blister packs), but frequently it is so indistinct as to be undetected even though serious losses in visibility occur (*e.g.*, pencil written sales receipts).

Veiling reflections should be avoided and can be controlled by the lighting system design (see Section 2) or the location of merchandise.

Enhancing Reflections. Points of brightness from light source images reflected in shiny surfaces appear as highlights or streaks of brightness and are often used as elements of sparkle to enhance the appearance of merchandise. The designer may deliberately locate a luminaire to produce reflections from a surface; *e.g.*, a light source reflected in a silver tureen emphasizes the form of the object as well as its patina.

Glare

When overly bright light sources, window or lighted surfaces are within the visual field of the shopper or clerk and can be seen directly or by reflection, they may produce glare that will attract attention away from the merchandise displayed. When they are near the line of sight of the merchandise or sales tasks, they may reduce the ability to see.

Direct Glare. Where large area luminaires (more than 0.37 square meters (4 square feet)) are utilized and visual comfort is a requirement, the VCP concept of glare evaluation applies. See Section 2. For small luminaires, the average luminance of the total luminous area should not exceed 1710 candelas per square meter (500 footlamberts). Luminaires or luminous elements near the line of sight should not be so bright as to distract from or compete with the merchandise (see Fig. 8-4) or impede circulation flow.

Fig. 8-4. Direct glare from unshielded or improperly directed luminaires and reflected glare from specular surfaces can distract from the merchandise, emphasizing the ceiling plane and creating visual confusion.

Reflected Glare. Glare from reflections of light sources or large areas can be reduced by the use of high-reflectance matte surfaces and by carrying out the procedures for reducing veiling reflections on the task. Large-area, low-luminance luminaires are used when shiny surfaces cannot be avoided. Glare shields or louvers should be attached to luminaires that may be reflected in glass lenses of viewing machines and in computer equipment at registers to reduce offending reflections.

Illuminances

Fig. 8-5 lists current recommended illuminances for merchandising areas. It is anticipated that the values in this table will be modified periodically as further study indicates need for a change.

These values do not refer to average levels throughout the store, since in many types of stores it is desirable and practical to provide nonuniform lighting between aisles and merchandise locations, as well as for accent lighting, creation of contrast, etc.

To assure these levels at any time, the probable standard of maintenance should be reflected in the lighting calculations.

Fig. 8-5. Currently Recommended Illuminances for Merchandising Areas

Areas or Tasks	Description	Type of Activity Area*	Illuminance**	
			Lux	Foot-candles
Circulation	Area not used for display or appraisal of merchandise or for sales transactions	High activity	300	30
		Medium activity	200	20
		Low activity	100	10
Merchandise† (including showcases & wall displays)	That plane area, horizontal to vertical, where merchandise is displayed and readily accessible for customer examination	High activity	1000	100
		Medium activity	750	75
		Low activity	300	30
Feature displays†	Single item or items requiring special highlighting to visually attract and set apart from the surround	High activity	5000	500
		Medium activity	3000	300
		Low activity	1000	150
Show windows				
Daytime lighting				
General			2000	200
Feature			10000	1000
Nighttime lighting				
Main business districts-highly competitive				
General			2000	200
Feature			10000	1000
Secondary business districts or small towns				
General			1000	100
Feature			5000	500

* One store may encompass all three types within the building

High activity area — Where merchandise displayed has readily recognizable usage. Evaluation and viewing time is rapid, and merchandise is shown to attract and stimulate the impulse buying decision.

Medium activity — Where merchandise is familiar in type or usage, but the customer may require time and/or help in evaluation of quality, usage or for the decision to buy.

Low activity — Where merchandise is displayed that is purchased less frequently by the customer, who may be unfamiliar with the inherent quality, design, value or usage. Where assistance and time is necessary to reach a buying decision.

** Maintained on the task or in the area at any time

† Lighting levels to be measured in the plane of the merchandise

The recommended levels in Fig. 8-5 are intended as a guide. Current practice indicates a strong trend to variations in illumination for different kinds of stores and for various departments within a store. These considerations, plus those of competition, need for distinctive effects, store design, etc., may make it desirable to vary the levels at specific locations over the values shown.

Where merchandise is displayed in one location and appraised in another (such as taking items out of a showcase to show a customer) it is desirable not to exceed a 3-to-1 ratio in illuminances between the two locations.

During the day, reflections in the glass of show windows determine the levels needed to enable customers to see through them effectively. Higher or lower values than recommended may be desirable, depending on such considerations as severity of reflections, reflectance of merchandise and backgrounds, and illuminances in competitive store windows.

For illuminance recommendations relating to sales transactions and tasks in service or support areas, see Fig. 2-2.

Energy

The basis for lighting energy conservation in the design and operation of merchandising facilities is described in IES (ASHRAE) standards for new and existing buildings.[2] These standards identify a process to determine the power budget for each total store building, which is the upper limit of power to be used by the lighting systems to be energy efficient. This process does not restrict the lighting system design or operation, since it only establishes a limit which can be utilized as desired. Beyond establishing a power budget, the standards also identify methods of achieving energy conservation through an energy management program. See Section 4 for the budget procedure and energy management programs. These apply to the entire building and exterior spaces. The following energy management considerations apply specifically to the merchandising spaces.

In looking for ways to more efficiently utilize energy, the objectives of merchandise lighting as stated earlier should be firmly kept in mind to avoid inappropriate decisions which while producing the desired energy efficiency might lead to ineffective merchandising results. On the other hand energy evaluations may produce ways to improve the quality of the illumination, lower operating costs, as well as lower energy use.

Lighting Requirements. Merchandise lighting levels in Fig. 8-5 are measured in the plane in which the merchandise is displayed. A frequent check should be made because of the flexibility of space use, changing displays and display planes and if levels are higher than recommended a reduction is encouraged.

The type of activity area initially selected from Fig. 8-5 should be reevaluated to determine if the category description is still accurate. In some cases rearrangement of store space may have resulted in different usage (*e.g.*, from medium to low activity) and illuminance may then be accordingly reduced.

Ratios between illuminance on displays and illuminance in customer appraisal location should be determined. If ratios exceed the recommended 3-to-1, levels should be reduced to fall within the recommendation.

Light Sources (Lamps and Ballasts). A periodic review of sources being used and a familiarity with new types is advised to determine if more efficient types may be substituted. In evaluating the efficacy of lamps consideration should be given to not only lumens per watt but also to candlepower because of application situations where small displays requiring special emphasis are lighted from distant luminaire locations. Color rendition of merchandise is also of importance in selecting light source substitutions and may be a deciding factor in changing from one light source family to another, or within one category.

Luminaire Layout and Control. Each time displays are changed the lighting system should be checked for appropriate aiming angles to obtain full benefit on the merchandise. Specific highlighting units not required for the changed display should be removed if mounted on a track or other flexible installation, or shut off if part of a permanent installation.

Consideration should also be given to controlling the lighting system so that a low level of lighting may be provided for nighttime security surveillance purposes, *e.g.*, perimeter lighting or a portion of the general lighting. Frequently a low level on a wall at the rear of a store will be sufficient to provide a view into the store and silhouette any intruder.

Area Surfaces. Lighter finishes should be used for greater utilization of reflected light, but excessive luminance ratios should be avoided, keeping in mind the suitability of background surfaces for the merchandise.

Maintenance Program. Coordination between maintenance and display personnel is important so that equipment is maintained (lamped, cleaned, positioned) to produce efficient display lighting solutions.

Operating Procedures. Accent and display lighting is for customers—to attract and aid in appraisal of the items presented. It should therefore be turned off during hours of non-use by customers, including cleaning periods.

Entrances and windows require a higher illuminance in daytime because of daylight competition than at night when surrounded by a lower ambience. It is recommended that levels in both areas be reduced during evening hours of operation.

Space Utilization. Consideration should be given to the use of the space so that the most energy efficient lighting solutions may be achieved. For example, spill light from feature displays may be sufficient for delineating circulation areas without the necessity of providing separate aisle lighting systems.

LIGHTING SYSTEMS CONSIDERATION

Both daylight and electric lighting systems are used in merchandising areas, but each has its own specific characteristics and considerations.

Daylighting

In open-front stores, or where there may be windows, or skylights, it is necessary to avoid large differences in luminance between daylighted areas and interior areas illuminated to recommended levels. This may be accomplished by controlling the daylight, rather than by increasing the level of electric illumination.

The amount and distribution of daylight received in store interiors depends on the orientation and total area of windows, their light transmission properties, and the relationship of the window height to the room width. See Section 7 of the 1981 Reference Volume.

Comfortable seeing conditions in merchandising areas result from careful consideration of the types of glass used in windows, the method and degree of shading the windows, and the reflectance values of the area surfaces.

Draperies, shades, baffles or louvers should be used for windows in areas where sky luminance or sunlight becomes uncomfortable or glaring to persons within. Horizontal or vertical overhangs outside the windows can eliminate glare from direct sunlight. Sales personnel should be oriented so that bright windows are not within the normal field of view, and shadows are not cast on reading material.

Electric Lighting

Light Sources. For merchandising applications, there are three basic types of light sources in use today: incandescent filament, fluorescent and high intensity discharge. Each light source type has certain advantages, and the proper selection will depend upon the particular requirements of the installation, the economics, and perhaps some personal preference of the system designer or owner. See Section 8 of the 1981 Reference Volume for a discussion and data on light sources.

Incandescent Filament Lighting. The chief advantages of incandescent filament lighting are its low initial cost, good color rendering properties, and good optical control capabilities. Disadvantages are shorter lamp life and lower lamp efficacy (lumens per watt) as compared to the other light sources. Included in the family of incandescents are the tungsten-halogen lamps, having a much better light output maintenance characteristic and longer lamp life than standard incandescent filament lamps, and low voltage lamps having good beam control. In addition, both the tungsten-halogen and low voltage lamps can be compact in physical size and of a shape that results in small luminaires.

Fluorescent Lighting. Many merchandising areas are illuminated with fluorescent light sources. A fluorescent lighting system provides higher lumens per watt, long lamp life, and good color rendition depending on lamp color selection. For indoor applications, louvers and prismatic or diffusing covers are desirable for use with fluorescent luminaires to provide lamp protection as well as maximum shielding. Essentially a tubular light source, fluorescent lighting may be controlled to some extent; however, it is difficult to control the distribution of light emitted lengthwise from the lamp.

High Intensity Discharge Lighting. The family of high intensity discharge lamps includes mercury, metal halide and high pressure sodium. Although each of these lamp types has its own specific characteristics, they have the following characteristics in common: long lamp life and high luminous efficacy when compared with incandescent lamps; compact source size, which allows for good optical control; and a time delay and slow build-up of light output when the lighting system is first energized or when there is a power interruption. Because of this delay characteristic, it is essential to include incandescent or fluorescent lighting.

In areas where color rendition is important, improved-color phosphor-coated mercury lamps are recommended rather than clear mercury lamps. It should be noted, however, that phosphor-coated lamps provide medium to wide beam spreads. In comparison to mercury lamps, the metal halide lamp provides higher luminous efficacy, but has a shorter life. These lamps also have good color rendition.

The high pressure sodium lamp has a higher luminous efficacy than the metal halide lamps and an excellent light output maintenance characteristic. Color acceptability is fair, but all colors are recognizable with these lamps. High pressure sodium lamps are used primarily for outdoor lighting.

Luminaires. No one lighting system can be recommended exclusively. Each system has qualities that may match the requirements for a given situation. The first consideration, however, should allow the customer to see efficiently and without distraction to produce sales; the second should be the appearance of the installation within the architectural and decorative design of the store.

Among the factors that affect the selection of a luminaire are:

1. The type of light source to be used.
2. The illumination performance that it will provide such as light distribution.
3. The proper luminance ratios for appearance and efficiency.
4. The structural factors and materials used.
5. The effectiveness of heat dissipation.
6. The modular size.
7. The appearance.
8. The quality of product.
9. The economics.

Two luminaires may have the same general appearance, but differ in performance. Comparisons using distribution curves and data from photometric tests obtained by qualified testing laboratories are the effective way to determine if the luminaires will provide equivalent lighting results.

Acoustical and Thermal Factors

Today's store frequently requires integration of lighting with acoustical and thermal aspects. Acoustical treatment of ceiling surfaces can be incorporated in the majority of stores. The reflectance of the acoustical material is important to the lighting scheme.

Heat from light sources and luminaires can have a great effect on the air conditioning and heating systems in merchandising areas. Air handling luminaires—supply, return, heat removal—may be effectively used in merchandising areas as part of the comfort control system and may also improve the efficiency of the lighting system. See Section 2.

Economics

The total cost of a lighting system is the sum of owning and operating charges. While initial investment may in some cases be a dominant factor in selecting specific luminaires or lamp types, there are capital expenses (amortization, interest, taxes, and insurance) that also should be considered. See Section 3.

Maintenance

All lighting systems depreciate in light output with the passage of time. See Section 4. Lighting equipment and room surfaces should be well maintained if reasonable efficiency and appearance are to be obtained, and consideration should be given to the accessibility of luminaires for cleaning and relamping in high mounting areas.

Fading, Bleaching and Spoilage

When the merchandiser displays a product, the color stability of merchandise should be considered. Not all products have the same color stability and products fade or change chemical composition because of varying environmental reasons.

Fading of merchandise may be caused by exposure to high illuminances for extended periods of time. Other factors that could contribute to fading are duration of environmental exposure, spectral distribution of radiation, moisture, temperature, chemical composition of merchandise, saturation of dye in merchandise, and composition of weave of fabrics. See Section 19.

LIGHTING INTERIOR SPACES

Merchandising Spaces

Merchandising spaces can be conducive to initiating and completing sales transactions. Each of the following factors should be considered in the design and lighting for merchandising spaces:

1. Type and characteristics of merchandise.
2. Location of merchandising area within the store.
3. Ambient illuminance in adjacent areas.
4. Size and shape of space.
5. Surface reflectances, colors and textures.
6. Flexibility requirements.
7. Size and location of graphics.
8. Method of display—racks, gondolas, counters, etc.
9. Method and location of sales transactions.
10. Location of merchandise displays, including feature displays.
11. Traffic patterns.

Lighting Methods for Merchandising Spaces

Once the type of store, class of merchandise to be handled, and clientele desired are determined, the lighting should be designed in keeping with their character. The lighting design should consider all surfaces in the customers' fields of view. Merchandise should dominate the scene.

There are three basic approaches to the lighting of merchandise areas in stores—the general pattern system, the specific system and the flexible system. Each system should have supplemental lighting to attract attention to featured displays, to influence traffic circulation and to create added interest.

General Pattern System. The general pattern system employs a pattern of luminaires to provide general lighting with or without display lighting throughout the sales area without regard

Fig. 8-6. Fluorescent and incandescent luminaires used in a general pattern system.

to the location of the merchandise (see Fig. 8-6). The system should include switching or dimming controls for flexibility of space use and for efficient energy utilization. If neither display lighting nor switching or dimming controls are used, there will be a lack of area emphasis on focal points.

Specific System. The specific system employs a layout of luminaires determined by the location of the merchandise displays (store fixtures, showcases, gondolas, etc). It is tailored to emphasize the merchandise and delineate sales areas (see Fig. 8-7).

Flexible System. The flexible system employs a pattern of electric outlets of continuous or individual type for nonpermanent installation of luminaires. These may be wired for multiple circuit application and/or control.

This system may be used for general pattern lighting or for specific lighting and offers the added advantage of interchangeability of luminaire types to create lighting tailored to the merchandise display.

Fig. 8-7. Specific lighting systems relate to the merchandise displays. They may be concealed as shown here to avoid competing with the merchandise. Flexibility may be an additional desirable feature of the system.

Feature and Supplementary Lighting

Lighting on vertical displays and wall cases, because of the favorable viewing angle, is important because these are prime profit centers. The proper balance of general or specific and feature and supplementary lighting is dependent on the type of merchandise, methods of presentation and type of store. Certain portions of the merchandising area should be given special consideration as to the most effective supplementary lighting methods to attain the lighting results desired. Each must receive individual consideration in lighting design, selection of lighting equipment and illuminances. Specific consideration should be given to placement and aiming the light sources at angles to prevent direct and reflected glare from reaching the eyes of customer and sales personnel. The following is a discussion of those merchandising areas that will generally require supplementary lighting.

Counter Lighting. Counter lighting is a form of accent lighting in which merchandise on the tops of counters, or point of sale at counter tops, receives three to five times the circulation area illuminance. This is usually accomplished with high intensity directional downlight equipment.

Lighting at Mirror. Lighting at the mirror is important because shoppers finally appraise hats, dresses, shoes, cosmetics and hairdos in terms of color, fit and how well the personality is complemented. When a buyer appraises wearing apparel, the face is generally observed first. The following factors should be considered when lighting mirrors:

1. The face should be softly lighted with light sources that flatter skin tones, and from a direction that minimizes harsh lines. Overhead lighting directed toward hair can add sheen and color.
2. The sales item should be adequately lighted over its entirety.
3. The lighting should be of a quality consistent with the illumination under which the merchandise will be worn—*i.e.*, outdoors for beach and sportswear, indoors for evening wear.

Light sources can be used to direct light into the mirror and then back onto the subject if the source brightness cannot be seen by the customer. It is possible with fixed viewpoint to control the lighting, especially in fitting rooms or special selling alcoves. Side lighting can be used and confined directly to the garment, especially in departments where there are coats and furs. In the case of triplicate mirrors, such lighting can be reflected from the wing mirrors. Controlled downlighting can also be used effectively in confined selling spaces such as fitting rooms. The reflectance, color and illumination of the background are important. See Fitting Rooms, page 8-12.

Showcase Lighting. Another element for emphasis in merchandise lighting is to call attention to merchandise displayed in showcases. Generally, showcase lighting is three times the illuminance required for circulation area lighting. Fluorescent lamps may be employed for a continuous line of light, and to minimize the heat created in enclosed spaces. The major objective of showcase lighting is to attain maximum light on merchandise without obstruction from lighting equipment; therefore, small-diameter lamps are generally preferred. Despite the general use of fluorescent lamps, incandescent filament "showcase" lamps are frequently used for more acceptable color rendition. They show merchandise as it will be worn or seen in warm light and may create sparkle for the display of jewelry, glassware and other similar merchandise. For a curved or irregular case, cold-cathode tubing can be bent to conform to the shape of the case. See Fig. 8-8.

Modeling Lighting. The form and texture of merchandise may be more apparent through the use of directional lighting supplementing the general diffuse lighting needed for the over-all effect. However, light should not be directed too obliquely, since objectionable shadows may be cast.

Wall Case Lighting. Wall case lighting can be considered in three categories: (1) the freestanding vertical display mounted against a wall; (2) the encased, open-front, wall-mounted display; and (3) the glass-door, wall-mounted display case.

Accent lighting of the freestanding vertical display offers the greatest freedom in expression to the lighting designer. It may be accomplished by flush, surface-mounted or suspended adjustable luminaires, strategically located to produce highlights and shadows to create a three-dimensional display. Colored lamps in lieu of clear lamps may further dramatize or call attention to the displayed merchandise.

The open-front, wall-case display follows the lighting methods of the freestanding vertical displays. The system should be planned to project light within the encased area. In this type of display, added flexibility of design can be accomplished by using adjustable units installed at strategic points at angles that avoid veiling reflections around the outer edge of the case or within the encased area.

Display cases with glass doors present a different problem—namely, the merchandise displayed behind the glass panel is obscured by surface reflections from the glass. Since this becomes virtually a show window problem, the best way to overcome annoying reflections is to increase the lighting level within the case. Spotlighting can accomplish this; however, extended periods could cause fading.

Rack Lighting (Clothing). Rack lighting should be designed to attract customers and for easy evaluation of the merchandise. Racks located in large, cased wall areas may have concealed baffled light sources above racks. Where linear light sources are used, the color should

Fig. 8-8. Fluorescent sources beneath diffusing material produce a transilluminated display for a cosmetic counter showcase. Perfume bottles appear to glow and float.

Fig. 8-9. A uniform wash of light over the length of the hanging garments facilitates customer evaluation of the item and eases the task of reading pertinent data on tags. Fluorescent sources are housed here in the top of the wall case.

render color in the same way as the ambient lighting in the fitting rooms. The lighting system chosen should be one that fully illuminates the articles of clothing from the standpoint of color and texture. A lighting level should be provided that will permit quick and discerning customer selection (see Fig. 8-9).

In the open rack areas, flush or surface-mounted adjustable ceiling downlights should be directed obliquely onto the displayed merchandise. The lighting level on the clothing should be greater than that of the general or ambient lighting of the aisles between racks. In the aiming of the downlights, caution must be exercised to avoid directing the light beam into the eyes of customers viewing clothing on the opposite side of the rack or at adjacent racks. The use of louvers, baffles or lenses helps to alleviate this situation.

Perimeter Lighting. Perimeter lighting is an asset to a store environment, contributing to a sense of pleasantness and adding to the visibility and visual impact of displays at the walls (see Fig. 8-10).

Other Store Spaces

In addition to the lighting consideration for merchandising areas, attention should be given to those for other spaces utilized by customers and/or store personnel. The following is a list of those spaces:

1. **Fitting Rooms.** The selling potency of fitting rooms is of tremendous importance. This is where the final, critical decision to buy is made. Every effort should be made to hold and motivate the customer to complete the sale. The lighting systems in these small spaces should create a feeling of relaxed security and pleasant anticipation.

Background finishes should be matte, simple and light in color to avoid color distortion or distract from merchandise.

Light sources should be compatible in color rendering with those in the selling space to insure that initial customer attraction to the merchandise is continued when a close personal evaluation is made.

Careful choice and placement of overhead luminaires will add to the vibrancy of color, en-

Fig. 8-10. A store, seen from the outside, utilizing perimeter lights to produce vertical surface luminances for attraction and to overcome potential veiling reflections in the windows.

Fig. 8-11. Exterior lighting of a large department store to attract and lead the customer to the entrance.

hancement of texture, sheen or glitter of hair and materials, and create modeling effects. Lighting at the mirror should be used to compliment and soften facial shadows. Vertical illumination should extend far enough down to enable customer to easily evaluate full-length garments. See page 10-5.
2. **Alteration Rooms.** Sewing and Pressing tasks are involved.
3. **Stock Rooms.** Methods of storage shelves, bins, racks, etc.
4. **Wrapping and Packaging.** Sample displays, wrapping and boxing are involved.
5. **Toilets, Washrooms, and Locker Rooms.** Lighting at mirror, sanitary maintenance, lounge facilities, and locker lighting are important.
6. **Offices.** See Section 5.
7. **Food Service Facilities.** See Section 7.
8. **Escalators, Elevators, and Stairways.** Safety, traffic pattern and use, graphics, and emergency provisions are important.

EXTERIOR SPACES

The role of outdoor lighting at stores and shopping centers is numerous and varied; it should attract customers to the center and then to specific stores (see Fig. 8-11); identify key areas such as entrances, exits, parking and the various stores; facilitate safe passage of motorists and pedestrians on the grounds; contribute to effective security and surveillance of people and property, and visually unify the shopping area, providing a positive contribution to the visual environment. See Section 12.

When the potential customer arrives in the vicinity of the shopping center (or freestanding store) there is a pattern of progression to arrive at the point of purchase. Lighting plays a major role in leading the shopper from one zone to another and eases the identification process through each step—from locating the shopping center site, entrance to center, parking area and store (all vehicular circulation) to locating the store entrance, departments and finally the merchandise (all pedestrian circulation).

SHOW WINDOWS

The show window can be a powerful attraction, providing the link between the potential customer passing by and the merchandise within the store. Each of the following factors should be considered in the design of show-window lighting:
1. Type and characteristics of merchandise.
2. Location of show window—outdoor or enclosed mall area, urban or suburban, solo or shopping center.
3. Night and/or day use and associated ambient illuminances (see Fig. 8-5), including the nature of the competition.
4. Open back or enclosed.
5. Size and shape.
6. Contour and slant of show-window glazing—brightness from daytime and nighttime reflections.
7. Interior surface reflectances and colors.
8. Flexibility requirements.
9. Size and location of display graphics.

See Section 9 of the 1981 Reference Volume for calculation techniques.

REFERENCES

1. Committee on Merchandising Lighting of the IES: "Recommended Practice for Lighting Merchandising Areas," *Light. Des. Appl.*, Vol 6, p. 6, June, 1976.
2. *American National Standard, Energy Conservation in New Building Design*, ANSI/ASHRAE/IES 90-80, American National Standards Institute, New York. Proposed American National Standard, *Energy Conservation Existing Buildings—Commercial*, ASHRAE 100.3 (IES/EMS-4.3), American National Standards Institute, New York.

Industrial Lighting*

SECTION 9

The purpose of industrial lighting is to provide energy efficient illumination in quality and quantity sufficient for safety and to enhance visibility and productivity within a pleasant environment.[1]

Industry encompasses seeing tasks, operating conditions and economic considerations of a wide range. Visual tasks may be extremely small or very large; dark or light; opaque, transparent or translucent; on specular or diffuse surfaces; and may involve flat or contoured shapes. With each of the various task conditions, lighting must be suitable for adequate visibility in developing raw materials into finished products. Physical hazards exist in manufacturing processes and, therefore, lighting must contribute to the utmost as a safety factor in preventing accidents (see page 2–44). The speed of operations may be such as to allow only minimum time for visual perception and, therefore, lighting must be a compensating factor to increase the speed of vision (see Section 3 of 1981 Reference Volume).

Lighting must serve not only as a production tool and as a safety factor but should also contribute to the over-all environmental conditions of the work space. The lighting system should be a part of an over-all planned environment.

The design of a lighting system and selection of equipment may be influenced by many economic and energy related factors (see Section 3, Lighting Economics, and Section 4, Energy Management). Economic decisions in regard to the lighting system should not only be based on the initial and operating costs of the lighting, but also on the relationship of lighting costs to other plant producing facilities and costs of labor.

For emergency lighting guidelines, see page 2–45.

FACTORS OF GOOD INDUSTRIAL LIGHTING

Quantity of Illumination

The desirable quantity of light (illuminance) for an installation depends primarily upon the seeing task, the worker and the importance of speed and accuracy in performing the task (see page 2–3).

* The information in this Section serves as a guide in designing lighting for various industries. Additional information can be obtained from the *American National Standard Practice for Industrial Lighting*, ANSI/IES RP-7-1979,[1] and IES study reports on specific industries. See references at the end of this section.

Illuminance recommendations for industrial tasks and areas are given in Fig. 2-2, page 2-9. In addition, in several instances industry representatives have established tables of single illuminance values which, in their opinion, can be used in preference to employing Fig. 2-2. However, illuminance values for specific operations can also be determined using illuminance categories of similar tasks and activities found in Fig. 2-2 and the application of the appropriate weighting factors in Fig. 2-4. In either case, the values given are considered to be target maintained illuminances. If it is desired to determine the illuminance produced by an existing installation, the measurement procedure outlined in Section 4 of the 1981 Reference Volume should be followed.

To insure that a given illuminance will be maintained, it is necessary to design a system to give initially more light than the target value. In locations where dirt will collect very rapidly on luminaire surfaces and where adequate maintenance is not provided, the initial value should be even higher. For typical light loss data and a further discussion see Sections 8 and 9 of the 1981 Reference Volume.

Where workers wear eye-protective devices with occupationally-required tinted lenses that materially reduce the light reaching the eye, the illuminance for individual tasks should be increased accordingly.

Fig. 9-1. Work is performed on delicate meter mechanisms in an assembly room. Recommended room-surface reflectances help achieve good luminance relationships.

Quality of Illumination

Quality of illumination pertains to the distribution of luminances in the visual environment. The term is used in a positive sense and implies that all luminances contribute favorably to visual performance, visual comfort, ease of seeing, safety and esthetics for the specific visual task involved. Glare, diffusion, direction, uniformity, color, luminance and luminance ratios all have a significant effect on visibility and the ability to see easily, accurately and quickly. See Section 3 of the 1981 Reference Volume. Certain seeing tasks, such as discernment of fine details, require much more careful analysis and higher quality illumination than others. Areas where the seeing tasks are severe and performed over long periods of time require much higher quality than where seeing tasks are casual or of relatively short duration. See Fig. 9-1.

Industrial installations of very poor quality are easily recognized as uncomfortable and are possibly hazardous. Unfortunately, moderate deficiencies are not readily detected, although the cumulative effect of even slightly glaring conditions can result in material loss of seeing efficiency and undue fatigue.

Direct Glare. When glare is caused by the source of lighting within the field of view, whether daylight or electric, it is described as direct glare.

To reduce direct glare in industrial areas, the following steps may be taken: (1) decrease the

Fig. 9-2. Recommended Maximum Luminance Ratios

	Environmental Classification*		
	A	B	C
1. Between tasks and adjacent darker surroundings	3 to 1	3 to 1	5 to 1
2. Between tasks and adjacent lighter surroundings	1 to 3	1 to 3	1 to 5
3. Between tasks and more remote darker surfaces	10 to 1	20 to 1	†
4. Between tasks and more remote lighter surfaces	1 to 10	1 to 20	†
5. Between luminaries (or windows, skylights, etc.) and surfaces adjacent to them	20 to 1	†	†
6. Anywhere within normal field of view	40 to 1	†	†

* Classifications are:
A—Interior areas where reflectances of entire space can be controlled in line with recommendations for optimum seeing conditions.
B—Areas where reflectances of immediate work area can be controlled, but control of remote surround is limited.
C—Areas (indoor and outdoor) where it is completely impractical to control reflectances and difficult to alter environmental conditions.
† Luminance ratio control not practical.

luminance of light sources or lighting equipment, or both; (2) reduce the area of high luminance causing the glare condition; (3) increase the angle between the glare source and the line of vision; (4) increase the luminance of the area surrounding the glare source and against which it is seen.

Unshaded factory windows are frequent causes of direct glare. They may permit direct view of the sun, bright portions of the sky or bright adjacent buildings. These often constitute large areas of very high luminance in the normal field of view.

Luminaires that are too bright for their environment will produce glare: *discomfort* glare or *disability* glare, or both. The former produces visual discomfort without necessarily interfering with visual performance or visibility. Disability glare reduces both visibility and visual performance and is often accompanied by visual discomfort. To reduce direct glare, luminaires should be mounted as far as possible above the normal line of sight. They should be designed to limit both the luminance and the quantity of light emitted in the 45- to 85-degree zone because such light, likely to be well within the field of view, may interfere with vision. This precaution includes the use of supplementary lighting equipment.

There is such a wide divergence of industrial tasks and environmental conditions that it may not be economically feasible to recommend a degree of quality satisfactory to all needs. The required luminance control depends on the task, the length of time to perform it, and those factors which contribute to direct glare. In production areas, luminaires within the normal field of view should be shielded to at least 25 degrees from the horizontal—preferably to 45 degrees.

Luminance and Luminance Ratios. The ability to see detail depends upon the contrast between the detail and its background. The greater the contrast, difference in luminance, the more readily the seeing task is performed. However, the eyes function most comfortably and more efficiently when the luminances within the remainder of the environment are relatively uniform. Therefore, all luminances in the field of view should be carefully controlled. In manufacturing there are many areas where it is not practical to achieve the same luminance relationships as are easily achieved in areas such as offices. But, between the extremes of heavy manufacturing and office spaces lie the bulk of industrial areas. Therefore, Fig. 9-2 has been developed as a practical guide of recommended maximum luminance ratios for industrial areas.

To achieve the recommended luminance relationships, it is necessary to select the reflectances of all the finishes of the room surfaces and equipment as well as control the luminance distribution of the lighting equipment. Fig. 9-3 lists the recommended reflectance values for industrial interiors and equipment. High reflectance surfaces are generally desirable to provide the recommended luminance relationships and high utilization of light. They also improve the appearance of the work space. See Fig. 9-4.

In many industries machines are painted such that they present a completely harmonious environment from the standpoint of color. It is desirable that the background be slightly darker than the seeing task. It appears desirable to paint stationary and moving parts of machines with contrasting colors to reduce accident hazard by aiding identification.

Veiling Reflections. Where seeing task details are specular, care should be taken to mini-

Fig. 9-3. Recommended Reflectance Values (Applying to Environmental Classifications A and B in Fig. 9-2)

Surfaces	Reflectance* (per cent)
Ceiling	80 to 90
Walls	40 to 60
Desk and bench tops, machines and equipment	25 to 45
Floors	not less than 20

* Reflectance should be maintained as near as practical to recommended values.

Fig. 9-4. A pleasant environment is provided for die-makers in a machine shop. Supplementary lighting is used to maintain close tolerances of die production. Carpet, green plants, and a modern wall treatment contribute to making a stimulating area.

mize veiling reflections which will decrease task visibility. See Section 2.

Reflected Glare. Reflected glare is caused by the reflection of high luminance light sources from shiny surfaces. In manufacturing processes this may be a particularly serious problem where critical seeing is involved with highly polished surfaces such as polished sheet metal, vernier scales and critically machined metal surfaces.

Reflected glare can be minimized or eliminated by using light sources of low luminance or by orienting the work so reflections are not directed in the normal line of vision. Often it is desirable to use reflections from a large-area, low-luminance luminaire located over the work. The section that follows on supplementary lighting covers in detail the solutions to such problems.

In special cases it may be practical to reduce the specular reflection (and the resultant reflected glare) by changing the specular character of the offending surface.

Distribution, Diffusion and Shadows. Uniform horizontal illuminance (where the maximum level is not more than one-sixth above the average level; or the minimum, one-sixth below) is frequently appropriate for specific industrial interiors where tasks are closely spaced and where there are similar tasks requiring the same amount of light. In such instances, uniformity permits flexibility of functions and equipment and assures more uniform luminances. Alternate areas of extreme luminance differences are undesirable because it tires the eyes to adjust to them.

Maintaining uniformity between contiguous areas which have significantly different visibility (and illumination) requirements might be wasteful of energy; for example, a storage area adjacent to a machine shop. In such instances, it is prudent to design and apply nonuniform lighting between those areas. It may be accomplished by using luminaires of different wattage and/or by adjusting the number of luminaires per unit area. Local lighting restricted to a small work area is unsatisfactory unless there is sufficient general illumination.

Harsh shadows should be avoided, but some shadow effect may be desirable to accentuate the depth and form of objects. There are a few specific visual tasks where clearly defined shadows improve visibility and such effects should be provided by supplementary lighting equipment arranged for the particular task.

Color Quality of Light. For general seeing tasks in industrial areas there appears to be no effect upon visual acuity by variations in color of light. However, where color discrimination or color matching are a part of the work process, the color of light should be very carefully selected. One example is in the printing industry, covered in detail on page 9-41. Color, of course, has an effect upon the appearance of the work space and upon the complexions of personnel. Therefore, the selection of the lighting system and the decorative scheme should be carefully coordinated.

Fig. 9-5. Trimmer room, finishing department in a paper mill. The use of high-reflectance room finishes helped to create an environment with relatively uniform luminances. Luminaires with high intensity discharge lamps are mounted 6.4 meters (21 feet) above the floor.

General Considerations of Design for Lighting Industrial Areas

The designer of an industrial lighting system should consider the following factors as the first and all-important requirements of good planning.

1. Determine the quantity and quality of illumination desirable for the manufacturing processes involved.
2. Select lighting equipment that will provide the quantity and quality requirements by examining photometric characteristics, and mechanical performance that will meet installation, operating and actual maintenance conditions.
3. Select and arrange equipment so that it will be easy and practical to maintain.
4. Balance all of the energy management considerations discussed in Section 4 and economic factors including initial, operating and mainte-

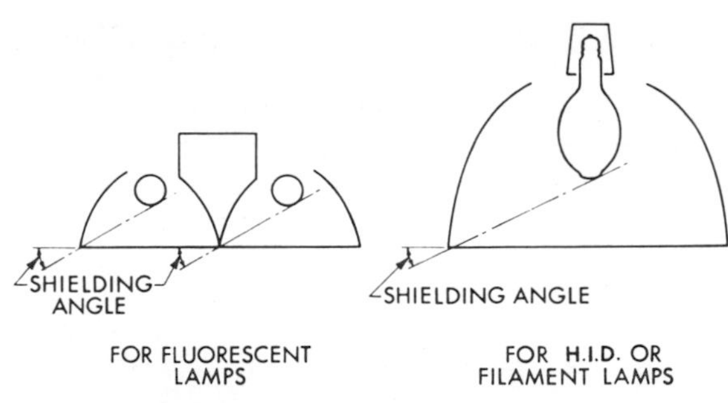

Fig. 9-6. Luminaires require adequate shielding for visual comfort. This is particularly important for higher luminance sources. An upward component also contributes to visual comfort by balance of luminances between luminaires and their backgrounds. Top openings help minimize dirt accumulation.

nance costs, versus the quantity and quality requirements for optimum visual performance. The choice of the electric distribution system may affect over-all economics.

Types of Lighting Equipment. The manner in which the light from the lamps is controlled by the lighting equipment governs to a large extent the important effects of glare, shadows, distribution and diffusion. Luminaires are classified in accordance with the way in which they control the light. Fig. 1-10, page 1-13 gives the standard CIE classifications for interior lighting equipment.

Most industrial applications call for either the direct or semi-direct types. Luminaires with upward components of light are preferred for most areas because an illuminated ceiling or upper structure reduces luminance ratios between luminaires and the background. The upward light reduces the "dungeon" effect of totally direct lighting and creates a more comfortable and more cheerful environment as shown in Fig. 9-5. Industrial luminaires for fluorescent, high intensity discharge and incandescent filament lamps are available with upward components. See Fig. 9-6. Good environmental luminance relationships can also often be achieved with totally direct lighting if the illuminances, and room surface reflectances are high.

In selecting industrial lighting equipment, it will be noted that other factors leading to more comfortable installations include:

1. Light-colored finishes on the outside of luminaires to reduce luminance ratios between the outside of the luminaire and the inner reflecting surface and light source.
2. Higher mounting heights to raise luminaires out of the normal field of view.
3. Better shielding of the light source by deeper reflectors, cross baffles or louvers. This is particularly important with high-wattage incandescent filament or high intensity discharge sources and the higher output fluorescent lamps.
4. Selecting light control material, such as specular or nonspecular aluminum or prismatic configurated glass or plastic that can limit the luminaire luminance in the shielded zone.

Top openings in luminaires generally minimize dirt collection on the reflector and lamp by allowing an air draft path to move dirt particles upward and through the luminaire to the outer air. Therefore, ventilated types of luminaires have proven their ability to minimize maintaince of fluorescent, high intensity discharge, and incandescent filament types of luminaires. Gasketed dust-tight luminaires are also effective in preventing dirt collection on reflector surfaces.

Direct Lighting Equipment. Direct industrial lighting equipment distributions vary from wide to narrow (see Fig. 9-12 in the 1981 Reference Volume).

The wide distribution types are comprised of porcelain-enameled reflectors and various other types of diffuse white reflecting surfaces. Aluminum, mirrored-glass, prismatic glass, and other similar materials may also be used to provide a wide distribution when the reflector is designed with the proper contour. This type of light distribution is advantageous in industrial applications where a large proportion of the seeing tasks are vertical or nearly vertical.

Narrow distributions are obtained with prismatic-glass, mirrored-glass and aluminum reflectors. This type of light distribution is useful where the mounting height is approximately equal to or greater than the width of the room or where high machinery and processing equipment necessitate directional control for efficient illumination between the equipment.

In making a choice between wide and narrow distribution equipment on the basis of horizontal illuminances, a comparison of coefficients of utilization for the actual room conditions involved will serve as a guide in selecting the most effective distribution. Care should be taken to use such coefficients based as close as practical to actual ceiling, wall and floor reflectance as well as actual room proportions.

If, however, it is desired to determine illuminances at specific points, then a point calculation method should be used to obtain accurate results (see Section 9 of the 1981 Reference Volume). This is particularly true for high mounting heights.

Other Types of Direct Lighting Equipment. Where low reflected luminance is a neces-

Fig. 9-7. A low-bay area used for meter-assembly work. Fluorescent luminaires with 30 per cent upward light are mounted 2.2 meters (7 feet, 6 inches) above the work-plane.

sity, large-area types of low-luminance luminaires should be used. Such a luminaire may consist of a diffusing panel on a standard type of fluorescent reflector, an indirect light hood, or a completely luminous ceiling.

Semi-Direct Lighting Equipment. This classification of distribution is useful in industrial areas because the upward component (10 to 40 per cent) is particularly effective in creating more comfortable seeing conditions. A variety of fluorescent and high intensity discharge luminaires of this distribution are available and designed specifically for industrial application. See Figs. 9-7 and 9-8.

While the semi-direct type of distribution has a sufficient upward component to illuminate the ceiling, the downward component of 90 to 60 per cent of the output contributes to good illumination efficiency, particularly where ceiling obstructions may minimize the effectiveness of the indirect component.

Industrial Applications of Other Distribution Classifications. The general diffuse, semi-indirect and indirect systems are suitable for industrial applications where a superior quality of diffused, low-luminance illumination is required and where environmental conditions make such systems practical. An example of such applications includes the precision industries where a completely controlled environment is important, including lighting, air conditioning, and carefully planned decoration.

Factors of Special Consideration

Lighting and Space Conditioning. With the use of higher illuminances, it is often practical to combine the lighting, heating, cooling and atmospheric control requirements in an integrated system. The lighting system can often provide most of the energy during the heating period. When cooling is required, much of the lighting heat can be removed by the air exhaust system. See Section 2 for further details.

High Humidity or Corrosive Atmosphere and Hazardous Location Lighting. Enclosed gasketed luminaires are used in non-hazardous areas where atmospheres contain non-inflammable dusts and vapors, or excessive dust. Enclosures protect the interior of the luminaire from conditions prevailing in the area. Steam processing, plating areas, wash and shower rooms, and other areas of unusually high humidity are typical areas that require enclosed luminaires. Severe corrosive conditions necessitate knowledge of the atmospheric content to permit selection of proper material for the luminaire.

Hazardous locations are areas where atmospheres contain inflammable dusts, vapors or gases in explosive concentrations. They are grouped by the *National Electrical Code* on the basis of their hazardous characteristics, and all electrical equipment must be approved for use in specific classes and groups. Luminaires are available specifically designed to operate in these areas, which are noted in Article 500 of the *National Electrical Code* as Class I, Class II and Class III locations.

For definitions of luminaires used in these areas, such as, *explosion-proof, dust-tight, dust-proof, and enclosed and gasketed,* see Section 1 of the 1981 Reference Volume.

Abnormal Temperature Conditions. Low ambient temperatures must be recognized as existing in such areas as unheated heavy industrial plants, frozen food plants and cold storage warehouses. Equipment should be selected to operate under such conditions and particular attention should be given to lamp starting and light output characteristics if fluorescent equipment is considered. With high intensity discharge equipment, temperature variation has practically no effect on light output, but the proper starting characteristics must be provided. With incandescent filament lamp equipment, neither the starting nor the operation is a problem at low temperature.

Abnormally high temperatures may be common at truss height in foundries, steel mills, forge shops, etc. Caution should be observed in selecting lighting equipment for mounting in such locations. It is particularly important to consider the temperature limitations of fluorescent and high intensity discharge ballasts under such conditions. Often ballasts should be remotely located at a lower and cooler level or special high temperature equipment should be used. The reduction in fluorescent lamp output at high operating temperatures should be recognized. See Section 8 of the 1981 Reference Volume.

Maintenance. Regular cleaning and prompt replacement of lamp outages is essential in any well-operated industrial lighting system. It is important for the lighting designer to analyze luminaire construction and reflector finish and also to make provisions for maintenance access so the system can be properly serviced. Another point that should be considered is that it may often be necessary to do the servicing during the plant operating hours. Further details on maintenance, access methods and servicing suggestions are found in Section 4.

SUPPLEMENTARY LIGHTING IN INDUSTRY[2]

Difficult seeing tasks often require a specific amount or quality of lighting which cannot readily be obtained by general lighting methods. To solve such problems supplementary luminaires often are used to provide higher illuminances for small or restricted areas. Also, they are used to furnish a certain luminance, or color, or to permit special aiming or positioning of light sources to produce or avoid highlights or shadows to best portray the details of the task.

Before supplementary lighting can be specified it is necessary to recognize the exact nature of the visual task and to understand its light reflecting or transmitting characteristics. An improvement in the visibility of the task will depend upon one or more of the four fundamental visibility factors—luminance, contrast, size and time. Thus, in analyzing the problem, the engineer may find that seeing difficulty is caused by insufficient luminance, poor contrast (veiling reflections), small size, or that task motion is too fast for existing seeing conditions.

The planning of supplementary lighting also entails consideration of the visual comfort of both those workers who benefit directly and those who are in the immediate area. Supplementary equipment must be carefully shielded to prevent glare for the user and his associates. Luminance ratios should be carefully controlled. Ratios between task and immediate surroundings should be limited as recommended in Fig. 9-2. To attain these limits it is necessary to coordinate the design of supplementary and general lighting.

Luminaires for Supplementary Lighting

Supplementary lighting units can be divided into five major types according to candlepower distribution and luminance. These are:

Fig. 9-8. This installation utilizes high intensity discharge lamps in high-bay luminaires which also allow for uplight. A few fluorescent units are also provided for safety lighting while the HID lamps are restarting following power interruptions.

Type S-I-directional: includes all concentrating units. Example are a reflector spot lamp or units employed concentrating reflectors or lenses. Also included in the group are concentrating longitudinal units such as a well-shielded fluorescent lamp in a concentrating reflector.

Type S-II-spread, high-luminance: includes small-area sources, such as incandescent or high intensity discharge. An open-bottom, deep-bowl diffusing reflector with a high intensity discharge lamp is an example of this type.

Type S-III-spread, moderate-luminance: includes all fluorescent units having a variation in luminance greater than two-to-one.

Type S-IV-uniform-luminance: includes all units having less than two-to-one variation of luminance. Usually this luminance is less than 6800 candelas per square meter (2000 footlamberts). An example of this type is an arrangement of lamps behind a diffusing panel.

Type S-V-uniform-luminance with pattern: a luminaire similar to Type S-IV except that a pattern of stripes or lines is superimposed.

Portable Luminaires

Wherever possible, supplementary luminaires should be permanently mounted in the location to produce the best lighting effect. Adjustable arms and swivels will often adapt the luminaires to required flexibility. Portable equipment (see Fig. 9-9), however, can be used to good advantage where it must be moved in and around movable machines or objects such as in airplane assembly, garages, or where internal surfaces must be viewed. The luminaires must be mechanically and electrically rugged to withstand possible rough handling. Lamps should be guarded and of the rough-service type. Guards or other means should protect the user from excessive heat. Precautions should be taken to prevent electrical shock.

Classification of Visual Tasks and Lighting Techniques

Visual tasks are unlimited in number, but can be classified according to certain common characteristics. The detail to be seen in each group can be emphasized by an application of certain lighting fundamentals. Fig. 9-10 classifies tasks according to their physical and light controlling characteristics and suggests lighting techniques for good visual perception.

It should be noted when using Fig. 9-10 that the classification of visual tasks is based on the prime and fundamental visual task characteristics and not on the general application. For example, on a drill press the visual task would be the discernment of a punch mark on metal. This could be specular detail with a diffuse, dark background, classification A-3(b) in Fig. 9-10. Luminaire Types S-II or S-III are recommended. S-II on an adjustable arm bracket is a practical recommendation due to space limitations. Several or all of the luminaire types are applicable for many visual task classifications and the best luminaire for a particular job will depend upon physical limitations, possible placements of luminaires, and the size of the task to be illuminated.

Special Effects and Techniques

Color as a part of the seeing task can be very effectively used to improve contrast. While black and white are the most desirable combinations for continual tasks such as the reading of a book, it has been found that certain color combinations have a greater attention value. Black on yellow

Fig. 9-9. A portable supplementary unit provides localized lighting on the task.

Fig. 9-10. Classification of Visual Tasks and Lighting Techniques

Part I—Flat Surfaces

Classification of Visual Task	Example		Lighting Technique	
General Characteristics	Description	Lighting Requirements	Luminaire Type	Locate Luminaire
A. Opaque Materials				
1. Diffuse detail & background				
a. Unbroken surface	Newspaper proofreading	High visibility with comfort	S-III or S-II	To prevent direct glare & shadows (Fig. 9-11a)
b. Broken surface	Scratch on unglazed tile	To emphasize surface break	S-I	To direct light obliquely to surface (Fig. 9-11c)
2. Specular detail & background				
a. Unbroken surface	Dent, warps, uneven surface	Emphasize unevenness	S-V	So that image of source & pattern is reflected to eye (Fig. 9-11d)
b. Broken surface	Scratch, scribe, engraving, punch marks	Create contrast of cut against specular surface	S-III or	So detail appears bright against a dark background
			S-IV or S-V when not practical to orient task	So that image of source is reflected to eye & break appears dark (Fig. 9-11d)
c. Specular coating over specular background	Inspection of finish plating over underplating	To show up uncovered spots	S-IV with color of source selected to create maximum color contrast between two coatings	For reflection of source image toward the eye (Fig. 9-11d)
3. Combined specular & diffuse surfaces				
a. Specular detail on diffuse, light background	Shiny ink or pencil marks on dull paper	To produce maximum contrast without veiling reflections	S-III or S-IV	So direction of reflected light does not coincide with angle of view (Fig. 9-11a)
b. Specular detail on diffuse, dark background	Punch or scribe marks on dull metal	To create bright reflection from detail	S-II or S-III	So direction of reflected light from detail coincides with angle of view (Fig. 9-11b)
c. Diffuse detail on specular, light background	Graduations on a steel scale	To create a uniform, low-brightness reflection from specular background	S-IV or S-III	So reflected image of source coincides with angle of view (Fig. 9-11b or d)
d. Diffuse detail on specular, dark background	Wax marks on auto body	To produce high brightness of detail against dark background	S-III or S-II	So direction of reflected light does not coincide with angle of view (Fig. 9-11a)
B. Translucent Materials				
1. With diffuse surface	Frosted or etched glass or plastic, lightweight fabrics, hosiery	Maximum visibility of surface detail	Treat as opaque, diffuse surface—See A-1	
		Maximum visibility of detail within material	Transilluminate behind material with S-II, S-III, or S-IV (Fig. 9-11e)	
2. With specular surface	Scratch on opal glass or plastic	Maximum visibility of surface detail	Treat as opaque, specular surface—See A-2	
		Maximum visibility of detail within material	Transilluminate behind material with S-II, S-III or S-IV (Fig. 9-11e)	
C. Transparent Materials				
Clear material with specular surface	Plate glass	To produce visibility of details within material such as bubbles & details on surface such as scratches	S-V and S-I	Transparent material should move in front of Type S-V, then in front of black background with Type S-I directed obliquely. Type S-I should be directed to prevent reflected glare
D. Transparent over Opaque Materials				
1. Transparent material over diffuse background	Instrument panel	Maximum visibility of scale & pointer without veiling reflections	S-I	So reflection of source does not coincide with angle of view (Fig. 9-11a)
	Varnished desk top	Maximum visibility of detail on or in transparent coating or on diffuse background		
		Emphasis of uneven surface	S-V	So that image of source & pattern is reflected to the eye (Fig. 9-11d)

Fig. 9–10. *Continued*

Classification of Visual Task	Example		Lighting Technique	
General Characteristics	Description	Lighting Requirements	Luminaire Type	Locate Luminaire
Part I—*continued*				
D. *Continued*				
2. Transparent material over a specular background	Glass mirror	Maximum visibility of detail on or in transparent material	S-I	So reflection of source does not coincide with angle of view. Mirror should reflect a black background (Fig. 9–11a)
		Maximum visibility of detail on specular background	S-V	So that image of source & pattern is reflected to the eye (Fig. 9–11d)
Part II—Three-Dimensional Objects				
A. Opaque Materials				
1. Diffuse detail & background	Dirt on a casting or blow holes in a casting	To emphasize detail with a poor contrast	S-III or S-II or	To prevent direct glare & shadows (Fig. 9–11a)
			S-I or	In relation to task to emphasize detail by means of highlight & shadow (Fig. 9–11 b or c)
			S-III or S-II as a black light source when object has a fluorescent coating	To direct ultraviolet radiation to all points to be checked
2. Specular detail & background a. Detail on the surface	Dent on silverware	To emphasize surface unevenness	S-V	To reflect image of source to eye (Fig. 9–11d)
	Inspection of finish plating over underplating	To show up areas not properly plated	S-IV plus proper color	To reflect image of source to eye (Fig. 9–11d)
b. Detail in the surface	Scratch on a watch case	To emphasize surface break	S-IV	To reflect image of source to eye (Fig. 9–11d)
3. Combination specular & diffuse a. Specular detail on diffuse background	Scribe mark on casting	To make line glitter against dull background	S-III or S-II	In relation to task for best visibility. Adjustable equipment often helpful
				Overhead to reflect image of source to eye (Fig. 9–11b or d)
b. Diffuse detail on specular background	Micrometer scale	To create luminous background against which scale markings can be seen in high contrast	S-IV or S-III	With axis normal to axis of micrometer
	Coal picking	To make coal glitter in contrast to dull impurities	S-I, S-II	To prevent direct glare (Fig. 9–11b)
B. Translucent Materials				
1. Diffuse surface	Lamp shade	To show imperfections in material	S-II	Behind or within for transillumination (Fig. 9–11e)
2. Specular surface	Glass enclosing globe	To emphasize surface irregularities	S-V	Overhead to reflect image of source to eye (Fig. 9–11d)
		To check homogeneity	S-II	Behind or within for transillumination
C. Transparent Materials				
Clear material with specular surface	Bottles, glassware—empty or filled with clear liquid	To emphasize surface irregularities	S-I	To be directed obliquely to objects
		To emphasize cracks, chips, and foreign particles	S-IV or S-V	Behind for transillumination. Motion of objects is helpful (Fig. 9–11e)

is most legible and the next combinations in order of preference are green on white, red on white, blue on white, white on blue, and finally black on white.

The color of light can be used to increase contrast by either intensifying or subduing certain colors inherent in the seeing task. To intensify a color, the light source should be strong in this color; to subdue a color the source should have relatively low output in the particular color. For example, it has been found that chromium plating imperfections over nickel plating can be emphasized by using a bluish color of light such as the daylight fluorescent lamp.

Three-dimensional objects are seen in their apparent shapes because of the shadows and highlights resulting from certain directional components of light. This directional effect is particularly useful in emphasizing texture and defects on uneven surfaces.

Silhouette is an effective means of checking contour with a standard template. Illumination behind the template will show brightness where there is a difference between the contour of the standard and the object to be checked.

Fluorescence under ultraviolet radiation is often useful in creating contrast. Surface flaws in metal and nonporous plastic and ceramic parts can be detected by the use of fluorescent materials.

The detection of internal strains in glass, mounted lenses, lamp bulbs, radio tubes, transparent plastics, etc., may be facilitated by transmitted polarized light. The nonuniform spectral transmittance of strained areas causes the formation of color fringes that are visible to an inspector. With transparent models of structures and machine parts, it is possible to analyze strains under operating conditions.

Inspection of very small objects may be greatly simplified by viewing them through lenses. For production work the magnified image may be projected on a screen. Because the projected silhouette is many times the actual size of the object, any irregular shapes or improper spacings can be detected readily. Similar devices are employed for the inspection of machine parts where accurate dimensions and contours are essential. One typical device now in common use projects an enlarged silhouette of gear teeth on a profile chart. The meshing of these production gears with a perfectly cut standard is examined on the chart.

It is sometimes necessary to inspect and study moving parts while they are operating. This can be done with stroboscopic illumination which can be adjusted to "stop" or "slow up" the motion of constant speed rotating and reciprocating machinery. Stroboscopic lamps give flashes of light at controllable intervals (frequencies). Their flashing can be so timed that when the flask occurs an object with rotating or reciprocating motion is always in exactly the same position and appears to stand still.

AIRCRAFT/AIRLINE INDUSTRIES[3]

The aircraft and airline operations covered below consist of aircraft manufacturing and aircraft maintenance. The illuminance recommendations listed in Fig. 9-12 represent those established by aircraft/airline industry representatives and in their opinion can be used in preference to employing Fig. 2-2; however, values can still be determined using Fig. 2-2 for similar tasks and activities.

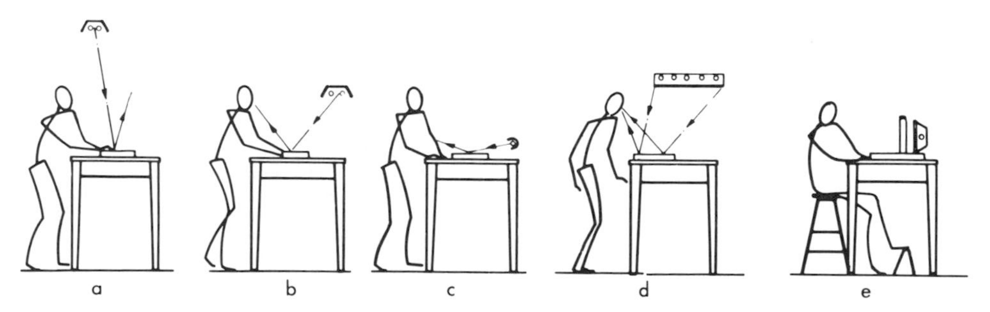

Fig. 9-11. Examples of placement of supplementary luminaires: a. Luminaire located to prevent veiling reflections and reflected glare; reflected light does not coincide with angle of view. b. Reflected light coincides with angle of view. c. Low-angle lighting to emphasize surface irregularities. d. Large-area surface source and pattern are reflected toward the eye. e. Transillumination from diffuse source.

Fig. 9-12 Illuminance Values Currently Recommended by Industry Representatives for Aircraft Maintenance and Manufacturing[1] (Maintained on Tasks)

Area and Task	Illuminance on Task: Lux	Illuminance on Task: Footcandles
Aircraft Maintenance		
Close up		
Install plates, panels, fairings, cowls, etc.	750	75
Seal plates	750	75
Paint (exterior or interior of aircraft) where plates, panels, fairings, cowls, etc., must be in place before accomplishing	750	75
Stencils, decals, seals, etc., where final paint coat needed before applying	750	75
Final "fly-away" outfitting (trays, loose gear, certs, includes final cleaning)	750	75
Docking		
Position doors and control surfaces for docking	300	30
Move aircraft into position in dock	500	50
Attach grounding wires and other safety equipment	300	30
Jack and level aircraft	750	75
Shore aircraft	750	75
Position ramps, walk-overs and other work facilities and equipment	750	75
Systems deactivation and safety locks installed	750	75
Removals—prior to power shutdown	750	75
Reposition doors, flaps, etc. after docking	750	75
Maintenance, modification and repairs to airframe structures		
Jacking and shoring not accomplished during docking phase	750	75
Remove any carrier or energy transmission portions or systems	750	75
Remove any linings, insulation, blankets, etc., to expose structure	750	75
Remove any minor structures (brackets, clips, angles, boxes, shelves, etc.) that attach to, obstruct or cover up major structure to be replaced, modified or repaired	750	75
Remove any sealant necessary to expose structures	750	75
Remove any major structural members (spars, stringers, longerons, circumferentials, etc.) that will be replaced with new ones	750	75
Install new structural members	750	75
Sealant installation after structural member adjustment, modification or repair	1000	100
Install any linings, insulation, blankets, etc.	750	75
Prime paint exterior	750	75
Top coat paint exterior	1000	100
Modifications or repairs to systems		
Install those carrier or energy transmission portions of systems previously removed which do not require modifications. (Elec. wires, hyd. lines, ducts, fuel lines, cables, etc.)	750	75
Modify any energy transmission or carrier portion of a system or add new ones previously nonexistent (Electrical, mechanical, liquid, pneumatic)	750	75
Repair any carrier or energy transmitting portions of any systems	750	75
Post overhaul—ramp	50	5
Predocking		
Convert hanger to fit incoming plane	300	30
Check, operate, pre-inspect, record	750	75
Install safety devices (lockpins-sleeves, etc.)	750	75
Drain tanks, relieve struts	500	50
Remove any plates, doors, cowls, fairings, etc. required for precleaning	750	75
Install protective covers and masking	750	75
Strip paint	750	75
Clean	750	75
Install personnel protective devices (sharp edge covers, people barriers, etc.)	300	30
Preparation for dedock		
Remove shoring	750	75
Remove workstands, ladders, etc.	750	75
Close aircraft doors and position control surfaces	300	30
Let aircraft down off jacks	750	75
Dedock	750	75
Preparation for maintenance and modification		
Check, operations, recordings—required in dock prior to power shutdown	750	75
Draining	750	75
Shutdown aircraft power systems	750	75
Remove plates, panels, cowls, fairings, linings, etc., for accessibility	750	75
Vent, purge, flush, etc., any tanks, lines, systems, etc., drain systems, drain and cap off not previously accomplished	750	75
Precleaning prior to removals	750	75
Disconnect lines, cables, ducts, linkages, etc., required for accessibility	750	75
Remove components	750	75
Install protective covers, masking, or devices	750	75
Dock cleaning and/or stripping required for inspection, later modification, maintenance and/or painting	750	75
Sand painted areas	750	75
Area inspection		
Ordinary	500	50
Difficult	1000	100
Highly difficult	2000	200
Specialty shops		
Instruments, radio	1500	150
Electrical	1500	150
Hydraulic and pneumatic	1000	100
Components	1000	100
Upholstery, chairs, rugs	1000	100
Sheet metal fabrication, repairs, welding	1000	100
Paint	1000	100
Parts inspection	1000	100
Plastics	1500	150
System operations and functional checks requiring aircraft power systems activation to perform		
Activate any aircraft power system	300	30
Block areas for operationals	750	75
Functional check of any system that prohibits other operations or actions within that system	750	75
Operational or functional check of any system not a part of or requiring sequential accomplishment	750	75
Test sequentially required operation of systems	750	75

Fig. 9-12 *Continued*

Area and Task	Illuminance on Task		Area and Task	Illuminance on Task	
	Lux	Footcandles		Lux	Footcandles
Aircraft Maintenance					
Release areas after operationals	300	30	Re-install components of systems that do not require a system check or ring-out before component installation	750	75
Nonpressure lube after operations	300	30	Install components requiring preliminary checks and ring-outs	750	75
Cleaning after operations	750	75	Hook up systems (wires, lines, pipes, ducts, cables, etc.) other than rigging	750	75
System repairs after operations and close up preparation			Physically block areas for dangerous operations	750	75
Repairs after system operationals	750	75	Rig cable systems that do not require sequential rigging	750	75
Corrosion treatment	750	75	Rig cable systems in step sequences	750	75
Apply masking	750	75	Operate any system for checking that can be operated from power source other than aircraft power	750	75
Painting and/or chromating	750	75	Clear blocked area	300	30
Removing masking	500	50	Reconnect lines to aircraft systems	750	75
Final inspections prior to close			Install cavity or tank covers or plates necessary to filling	750	75
Ordinary	500	50	Precheck before filling	500	50
Difficult	1000	100	Fill tanks. Service or lube tanks, struts, accumulators, etc.	500	50
Highly difficult	2000	200	Static leak checks	500	50
System restoration or new system component installation			Pressure check systems from pressure sources external to the aircraft	750	75
Hook up any lines, cables, ducts, panels and insulation to be covered by later component installation	750	75			
Install any components previously removed that must be in place for others to attach to, or subsequent components which when installed would obstruct or cover	750	75			
Paint	750	75			
Paint preparation and clean up	500	50			
Aircraft Manufacturing					
Fabrication (preparation for assembly)			General		
Rough bench work and sheet metal operations such as shears, presses, punches, countersinking, spinning	500	50	Rough easy seeing	300	30
Drilling, riveting, screw fastening	750	75	Rough difficult seeing	500	50
Medium bench work and machining such as ordinary automatic machines, rough grinding, medium buffing and polishing	1000	100	Medium	1000	100
Fine bench work and machining such as ordinary automatic machines, rough grinding, medium buffing and polishing	5000[a]	500[a]	Fine	5000[a]	500[a]
Extra fine bench and machine work	10000[a]	1000[a]	Extra fine	10000[a]	1000[a]
Layout and template work, shaping and smoothing of small parts for fuselage, wing sections, cowling etc.	1000[a]	100[a]	**First manufacturing operations (first cut)**		
Scribing	2000[a]	200[a]	Marking, shearing, sawing	500	50
Plating	300	30	**Flight test and delivery area**		
Final assembly such as placing of motors, propellers, wing sections, landing gear	1000	100	On the horizontal plane	50	5
			On the vertical plane	20	2
			General warehousing		
			High activity		
			Rough bulky	100	10
			Medium	200	20
			Fine	500	50
			Low activity	50	5
			Outdoor receiving and storage areas		
			Unloading	200	20
			Storage		
			High activity	200	20
			Low activity	10	1

[a] Obtained with a combination of general lighting plus specialized supplementary lighting. Care should be taken to keep within the recommended luminance ratios (see page 9-3). These seeing tasks generally involve the discrimination of fine detail for long periods of time and under conditions of poor contrast. The design and installation of the combination system must not only provide a sufficient amount of light, but also the proper direction of light, diffusion, color and eye protection. As far as possible it should eliminate direct and reflected glare as well as objectionable shadows.

Aircraft Manufacturing

Aircraft manufacturing is comprised basically of three main functions: fabrication, assembly and flight test. The lighting problems encountered in the areas where these functions are carried out are in many cases similar to those found in other manufacturing plants. Fabrication of parts and processing of materials is accomplished in open bay, medium height buildings, similar to structures used by other related industries. Assembly and flight test functions are not necessarily similar to other manufacturing processes; therefore, special lighting applications may be required in many instances.

Fabrication. Lighting techniques for many of the general fabrication and processing operations are covered under other headings in this Section, such as, machining metal parts and sheet metal parts. In aircraft manufacturing, these same machines and operations are used for more precise work than in many other industries. The massive problem of weight reduction in aircraft design has included many special operations in the manufacturing and processing of many parts from minute size to large metal skins.

An over-all general lighting level may be required for work areas with supplementary lighting for the most critical seeing tasks (see Fig. 9-13).

Sub and Final Assembly. This phase of aircraft manufacturing has special requirements not usually found in other types of manufacturing. The age of the so-called "air bus" has necessitated the construction of hangar-type buildings with clear bay areas exceeding 23,000 square meters (250,000 square feet) and truss heights of more than 24 meters (80 feet) from floor level. See Fig. 9-14. The lighting problems in buildings of this size are not confined to the engineering and design concepts but include the task of maintenance and lamp replacement. The use of either a system of catwalks or travelling bridge cranes should be relied upon to allow access to the lighting units. In some cases, mobile telescoping cranes can be used to reach luminaires from the floor below, but the heights involved and obstructions on the floor make this method of maintenance generally impractical.

Fig. 9-13. Typical area for component fabrication.

Fig. 9-14. Aircraft undergoing subassembly operation.

One of the special problems in lighting certain tasks in small or large assembly areas is that although the lighting is designed to specific task levels as if the areas are completely open, the areas are in reality seldom completely open. The lighting from overhead systems is often reduced by large assembly equipment which obstructs the overhead general lighting. Typical examples of these specific assembly tasks are riveting, bolting, and hydraulic and electric work. They are all done in confined areas and special or supplementary lighting usually is needed.

Aircraft Painting. Aircraft must be painted indoors to minimize the adherence of dust particles, moisture and other undesirable materials. Large open bay buildings are needed for priming and painting airplanes of the jet age. Conventional paint booths are used for processing small subassembly parts.

True color rendition is critical. The finished product is normally stationed outdoors, and the paint colors are based on their appearance on airfields or aloft; therefore, the light sources in the paint booth should be chosen carefully for color rendering qualities.

Flight Test and Delivery. Flight testing is the final phase of aircraft manufacturing and is conducted outdoors on concrete ramps and taxi strips. The mounting height is critical not only from the standpoint of adequate dispersion of light, but to provide clearance for maneuvering the aircraft and to meet *National Electrical Code* requirements for hazardous areas. It should be noted that in most cases aircraft have been fueled when reaching this phase of production.

Maximum height of floodlight poles may be limited by the proximity of adjoining airfields, traffic patterns and maintenance facilities. Federal Aviation Administration airport manuals and directives should be checked prior to determining location and height of poles.

Explosion-proof portable equipment may be required to provide supplementary lighting under wings and fuselage areas to illuminate access hatches and landing gear fittings.

Illumination of the interior of the aircraft is usually accomplishd by energizing the ship's electrical system with external 400-hertz ground power and utilizing the cockpit and cabin lighting systems.

Aircraft Engine Maintenance

The maintenance of an aircraft jet engine consists essentially of the disassembly, cleaning, inspection, repair, assembly and testing of the engine.

Disassembly. Turbojet and turbofan engines are designed to break into sections and subsections that may be separated or joined by three different basis techniques or variations thereof. These techniques are known as vertical assembly, horizontal assembly and combined assembly.

The design of the engine lends itself to dismantling into subassemblies, simultaneous operations, and a proper sequence of operations to quickly reach those engine sections that will be needed first at assembly. Normally this would be the compressor sections.

Overhaul of Accessory Components. The overhaul and testing of accessories and components is a large enough operation to warrant a completely separate shop. The testing of the components involves several expensive test stands. In addition, the fire and explosion hazards present special problems in the planning of an accessory component overhaul and test shop.

Inspection. For engine maintenance, inspection falls into two broad categories: line inspection and parts inspection. Line inspection covers the quality of assembly and test. Parts inspection covers the serviceability of parts. Parts inspection is subdivided into two categories generally called crack detection and table inspection. Crack detection employs the aid of magnetic flux and penetrant dyes or fluorescent particles to detect cracks and defects not easily visible to the naked eye. In general, table inspection covers the visual and dimensional checks required to determine the serviceability of parts.

Cleaning. The function of cleaning of engine parts for maintenance is to permit proper inspection to determine serviceability. At the present time chemical cleaning is the most widely accepted choice of the industry. A pitfall in the selection of a good cleaning process is the tendency to assume that the process that makes the parts look the cleanest, brightest or smoothest to the eye is the best process. However, some processes in achieving this look may make the visual task more difficult by smearing cracks or leaving a crack or surface in such a condition that a fluorescent penetrant inspection is not possible.

Repair. The majority of the parts rejected by the parts inspectors may be returned to service after a suitable repair. The range of repairs and reconditioning of parts is such that a fully equipped overhaul shop will have a plate shop, a machine shop and a weld shop. Owing to the size of jet engine parts, large tanks, ovens and machine tools are required. However, except in the larger maintenance facilities, only a minimum of such equipment is needed.

Spare Parts. Beyond the simple storage of parts, spare parts supply has many related tasks of procuring, recording, shipping, receiving, preserving of parts, etc.

Marshalling. Marshalling is the term applied to the making up of the engine parts and/or subassemblies into complete sets prior to assembly. Usually the first marshalling area is located adjacent to the parts inspection and minor rework area where stock chasers and expediters hunt down missing parts, expedite repair work and procure new parts as necessary to make up an engine into assembly groups ready for subassembly.

Assembly. The difference between assembly and disassembly is that assembly work requires higher quality workmanship, and miscellaneous inspections and checks. In general, it can be expected to take about three times as many man-hours as disassembly to accomplish. Probably the most critical assembly operation is the assembly and balance of the compressor and turbine rotors. In order to attain the desired freedom from vibration, it is necessary to weight blades to obtain an even distribution of forces in the disks; statically balance disk and blade assemblies; and, finally, dynamically balance the rotor assemblies.

Airframe Maintenance Operations

Aircraft are designed so that maintenance work can be performed in distinct, different levels of complexity. The work is either inside or outside the hangar, and the aircraft is either jacked up or on its tires and may require auxiliary electrical power. The overhaul consists of cleaning, disassembly for inspection, disassembly for replacement of parts, repair, replacement of parts, inspection, assembly and test of aircraft. Also, while the aircraft is in the dock for inspection, other work is done that falls within this time restraint. Other work includes component replacements, modifications and repairs of defects.

Disassembly. Certain work performed to gain access to functional parts of the aircraft is done so frequently as to be routine.

1. *Cabin.* Remove and overhaul (at "component shop") chairs, rugs, class dividers, curtains, galleys and toilets. On those items where a match fit is required, the items are tagged to facilitate reinstallation in the exact location.
2. *Fuel tanks.* Remove access plates from all tanks and install purging equipment to purge fuel vapor from tanks.
3. *Structural inspection.* Remove plates and components as required to perform the necessary inspections.
4. *Engines.* The maintenance work required on engines is so repetitive that removals of plates and components for access is nearly a fixed routine.
5. *Paint stripping.* Although paint stripping is done primarily in conjunction with decorative painting of the airplane, some paint is protective only. In these cases, a paint-stripping operation may be performed to bare the metal for an inspection.

Maintenance, Modification and Repairs to Airframe Structures. Normally, the "maintenance of structures" refers to protection from corrosion, removal of corrosion or replacement of decorative parts such as cabin wall panels. "Modifications and repairs" are a much more elaborate process. They frequently involve specialized shoring to support the structural members being worked on. Invariably, components and outer skin must be removed to get at primary structures. The modification and repair process usually involves either the installation of doublers to strengthen members or replacement of these members.

Modifications or Repairs to Systems; System Restoration; New System Component Installation. The inspection of "systems" is somewhat different than that of "structures." "System" refers to the landing gear system, the flap system, etc., or to a larger entity such as the entire hydraulic system. Some systems-inspection tasks are performed before the system is disassembled; but usually the inspection is performed after all replacements have been made to the system and it is intact once again.

System Operational and Functional Checks Requiring Aircraft Power Systems Activation. After each system has been repaired and restored, it must be tested to see that it performs according to specifications. These are called operational checks and almost always come at the end of the overhaul. While the operational check is conducted by a mechanic with an inspector controlling the test from the cockpit, several mechanics, stationed at strategic locations throughout the aircraft, observe conditions during the test. As the test proceeds, the observers report to the cockpit what is happening. During the test, adjustments may be made and lock fasteners installed to secure the system in proper operation condition.

System Repairs After Operationals and Close Up Preparation. The successful accomplishment of system operational checks signals the end of the overhaul. At this point everything of major consequence has been done. Only cleaning up and the closing of access doors and plates remain. One significant operation performed at this time is the application of protective paint. This requires cleaning, masking, painting and removing masking—primarily performed on high-strength steel parts of the landing gear. Before closing of doors or plates, an inspection determines that no debris has been left inadvertently in an area and that all wires are hooked up.

AUTOMOTIVE INDUSTRY

The automotive industry consists of many facilities devoted to parts manufacturing or assembly. Each facility is dynamic: processes and systems are frequently modified, added, removed or renovated because of technological changes and marketplace demands. The visual tasks at these facilities are dynamic, too: the product is frequently in motion. Thus, the visual tasks include the product, the manufacturing process, the process equipment and the transitions between work stations.

Several support activities have visual tasks. These activities include, but are not limited to, maintenance, machine repair, tool and die work, equipment calibration, utility generation and distribution and construction work.

General, localized general and supplementary lighting techniques are used at automotive facilities. The selection, design and operation of lighting systems are involved, in part because of the many light sources and variety of lighting systems (particularly luminaires) that can be applied. Each system is unique in its design, application and operation. No single system is ideal for the full range of this industry's lighting needs.

Illuminance Levels. Recommended maintained illuminances for automotive parts manufacturing and assembly facilities are shown in Fig. 9-15. The values listed represent a consensus of industry representatives on the IES Automotive Subcommittee. Should designers/owners elect, they may use the procedures for selecting illuminances given in Section 2.

Fig. 9-15 Illuminance Values Currently Recommended by Industry Representatives for Automotive Industry Facilities (Maintained on Tasks)

Activity	Illuminance	
	Lux	Footcandles
Coal yards, oil storage	5	0.5
Exterior inactive storage, railroad switching points, outdoor substations, parking areas	15	1.5
Inactive interior storage areas, exterior pedestrian entrances, truck maneuvering areas	50	5
Elevators, steel furnace areas, locker rooms, exterior active storage areas	200	20
Waste treatment facilities (interior), clay mold and kiln rooms, casting furnace area, glass furnace rooms, HVAC and substation rooms, sheet steel rolling, loading docks, general paint manufacturing, plating, toilets and washrooms	300	30
Frame assembly, powerhouse, forgings, quick service dining, casting pouring and sorting, service garages, active storage areas, press rooms, battery manufacturing, welding area	500	50
Control and dispatch rooms, kitchens, large casting core and molding areas (engines), machining operations (engine and parts)	750	75
Chassis, body and component assembly, clay enamel and glazing, medium casting core and molding areas (crankshaft), grinding and chipping, glass cutting and inspection, hospital examination and treatment rooms, ordinary inspection, maintenance and machine repair areas, polishing and burnishing, upholstering	1000	100
Parts inspection stations	1500	150
Final assembly, body finishing and assembly, difficult inspection, paint color comparison	2000	200
Fine difficult inspection (casting cracks)	5000	500

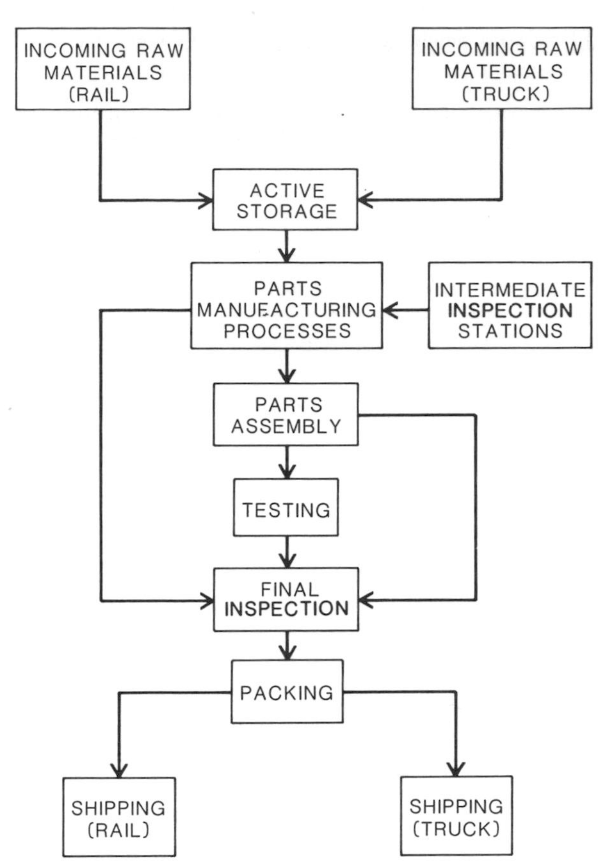

Fig. 9-16. Flow chart showing major activities in an automotive parts manufacturing facility.

Automotive Parts Manufacturing

The flow chart in Fig. 9-16 identifies the major steps in a parts manufacturing process. The major operations-related seeing tasks, and typical lighting systems are as follows:

Incoming Raw Materials: Raw materials are delivered to manufacturing facilities by truck or rail shipment. Both open-top and closed-top vehicles are used. *Seeing Task:* To identify the materials and correlate the material and shipping documents. *Lighting:* General lighting with supplementary lighting for trailer or rail car interiors.

Active Storage Areas: Raw materials are unloaded in the receiving areas by lift trucks and/or overhead cranes. They are transported to the active storage areas and/or directly to the production process by the same means. *Seeing Task:* To identify the materials (labels, markings, etc.) from the cab of an overhead crane or lift truck and to move the materials and deposit them at a designated location. *Lighting:* General illumination.

Parts Manufacturing Processes: Parts facilities manufacture several different products using many unique processes. The designer should refer to other portions of this section for major activities which occur in an automotive manufacturing plant: *i.e.*, machining, sheet metal shops, castings. *Lighting:* General lighting with supplementary lighting in areas or on equipment requiring increased illuminance levels.

Parts Assembly: In many manufacturing plants, individual components are assembled into subassemblies (wiring harness, engine, transmission, leaf spring, carburetor, etc.). The assembly processes combine manual, semi-automatic and automatic activities. *Seeing Tasks:* To select, orient, install and fasten a component to the subassembly. *Lighting:* General lighting with supplementary lighting added to specific work stations.

Testing: Highly diversified and complicated procedures and test equipment determine compliance with design specifications for the many mechanical, electrical and electro-mechanical subassemblies. Testing activities are manual, semi-automatic and automatic. *Seeing Task:* To secure subassembly to test fixture; to make electrical and/or mechanical connections; to run test and read gauges, meters, etc.; to perform mechanical and/or electrical adjustments as required; to complete test report; to disconnect and remove the subassembly from the test fixture. *Lighting:* General and supplementary lighting.

Final Inspection: Determines if the manufactured part or subassembly is in total compliance with the design specification. *Seeing Task:* To visually inspect the part or subassembly for specification compliance and to insure that all intermediate inspections and tests are satisfactory. *Lighting:* General lighting with supplementary lighting as required to inspect the part or subassembly.

Packing: Parts are manually or semi-automatically placed in boxes, metal shipping containers, or shipping racks for shipment. *Seeing Task:* To identify the part and place it in a destination-designated shipping container or rack. *Lighting:* General area lighting.

Shipping: Parts are usually shipped to assembly plants and/or warehouses by enclosed rail cars and trucks. Lift trucks are generally used to load these vehicles. *Seeing Tasks:* To identify a shipping container and/or rack by part and destination and load it into the designated rail car or truck. *Lighting:* General lighting with portable supplementary lighting provided for the rail car or truck trailer interior.

Automobile Assembly

The automotive assembly process is continuous and requires both general and supplementary lighting to illuminate the many difficult seeing tasks along the major portion of the assembly line. The maze of ducts, piping, supporting steel and conveyors around the assembly line reduce the effectiveness of the general lighting system. The flow chart shown in Fig. 9-17 identifies the major steps in the assembly operation. The major operations, related seeing tasks and typical lighting systems are as follows:

Body Framing Area: Metal parts are placed in large jigs or fixtures and automatically or manually welded. *Seeing Task:* Alignment of mating parts in jigs, welding and inspection of welding tips. *Lighting:* Localized general lighting with luminaires positioned slightly behind the operator.

Body Soldering Area: Joints (between welded parts), dents and scratches are filled with solder to give a smooth appearance to welded body parts. *Seeing Tasks:* To see all welds, joints, dents and scratches and to cover with solder. *Lighting:* Luminaires are positioned on both sides of line and oriented above and behind worker. Usually, luminaires are built into exhaust canopy over this area.

Metal Finishing Area: Raw metal and solder are ground and polished to the desired surface contour. *Seeing Tasks:* To inspect all welds and soldered areas and to grind and polish to desired contour and smoothness. *Lighting:* Luminaires are mounted parallel or perpendicular along both sides of the assembly line.

Body Inspection Area: The body is inspected to locate all metal defects. Defects are marked for correction so that when paint is applied, the surface will be uniform. *Seeing Task:* To locate and mark all metal body defects (dents, scratches, high spots, etc.) so that they can be repaired before painting. *Lighting:* Illumination is provided along both sides of the assembly line. Luminaires are mounted at an angle with the horizontal to provide illumination for the hood, roof and upper vertical surfaces. Luminaires are also positioned to illuminate the lower vertical surfaces and quarter panels.

Body Repair Area: Defects that have been marked by inspection personnel are repaired. *Seeing Task:* To see and distinguish the various marks that identify the location and nature of the defect. Also, to see that surface of body is smooth and uniform after grinding or polishing. *Lighting:* Same quantity and quality of illumination as that for Body Inspection Area.

Painting Area: Paint is applied to all body surfaces. *Seeing Tasks:* To see identification marks so that the proper color paint is applied; to see that paint is applied completely and evenly over the entire surface; to eliminate running and insufficient coverage. *Lighting:* Luminaires are installed in spray booth walls and ceiling so that illumination is provided over all horizontal and vertical body surfaces.

After-Paint Inspection Area: Inspect for proper paint coverage and for defects not identified and corrected prior to painting. *Seeing Task:* To detect any irregularities in body surface, color and insufficient coverage. *Lighting:* Luminaires usually mounted at an angle, along both sides of body and perpendicular to assembly line. Luminaires should be positioned to enable inspectors to see defects by observing distortion of the reflected light-source image in the specular body surface.

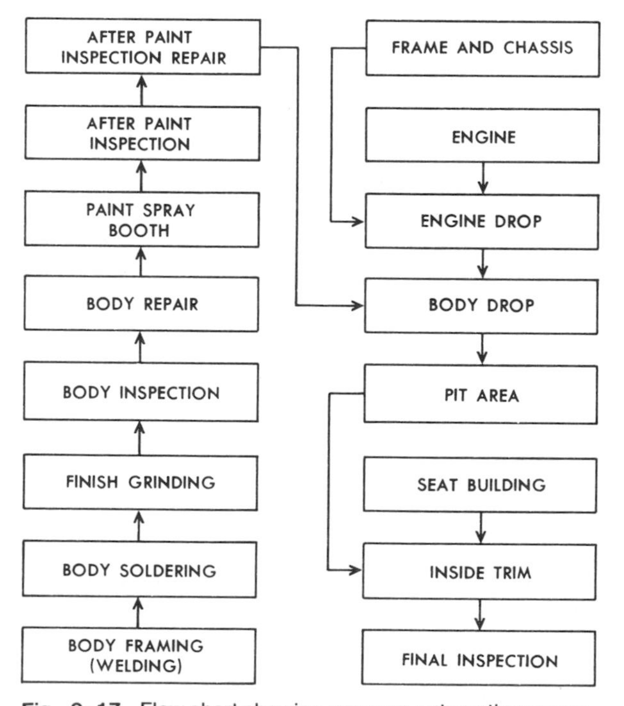

Fig. 9-17. Flow chart showing areas on automotive assembly lines.

Frame and Chassis Area: The frame of the automobile is started down the assembly line. The gas tank, exhaust system and suspension are all assembled on the frame during this phase of the assembly operation. *Seeing Task:* There are no critical seeing tasks in this area. Parts are generally large and alignment is usually all that is required. *Lighting:* General or local lighting.

Engine Drop Area: The engine is placed on the frame, fuel lines connected, and miscellaneous other parts bolted into their proper position. *Seeing Task:* Relatively large parts are bolted in place with hand tools. *Lighting:* Luminaires on both sides of line angled to provide horizontal and vertical illumination.

Body Drop Area: The body is lowered onto the chassis and fastened into position. *Seeing Task:* The alignment of the body is a very severe seeing task. *Lighting:* General and vertical illumination is required. The illumination must be sufficient to ensure proper positioning of the body and to bolt it in place with hand tools.

Pit Area: The body is bolted to frame. Other operations to undercarriage are accomplished by workers with hand tools. *Seeing Task:* To see that bolts are properly installed and tightened. *Lighting:* Luminaires are usually recessed in the walls on both sides of pit to provide light over worker's shoulder.

Inside Trim Areas: Seats, head liner, instrument panel and other interior accessories are installed. *Seeing Task:* To identify proper positions and fasten seats, instrument panel, accessories and interior head liner. *Lighting:* Luminaires are located along both sides of assembly line. General illumination in this area will satisfy seeing requirements inside the automobile.

Final Inspection Area: Automobile is inspected for all defects (body, trim, etc.), so that repairs can be effected. *Seeing Task:* The most critical task is identification of body defects and surface damage which may have occurred during the assembly process. *Lighting:* Luminaires are installed above and to the side of the automobile so that all surfaces can be inspected. Inspection personnel spot body defects by viewing distorted reflected lamp images in specular surface.

BAKERIES

Visual tasks in bakeries are not severe. Most hand operations are almost automatic with little attention to detail. General lighting systems provide adequate illumination for most functions; a few will need supplementary lighting. For selection of illuminance levels, see Fig. 2-2.

Lighting is also an aid to sanitation, safety and morale.[4] Cleanliness, of great importance in all food-producing establishments, is much easier to maintain in well-lighted interiors. In bakeries a large part of production areas is occupied by trucks, racks, mixing bowls, etc. Adequate illumination can help prevent injuries because of congestion in these areas and enhance the safety of employees operating moving machinery, working adjacent to hot surfaces and handling hot pans.

Mixing Room. Flour, normally stored in bins directly above the mixers, can be weighed and sifted directly into the mixing bowl. Sometimes it is conveyed to scale hoppers located above the mixers. Standard dial or beam scales are used to weigh the flour, with meters to measure the liquid ingredients. Some mixing rooms have a side bench where additional ingredients are mixed and weighed. General illumination with vertical illumination on the face of the shelves is required. For vertical mixers, supplementary lighting should be provided to light the inside of the mixing bowl.

Fermentation Room. From the mixing room, bread dough or sponges are taken in large troughs (holding 550 to 680 kilograms [1200 to 1500 pounds]) to the fermentation room which is maintained at about 27 °C (80 °F) and 80 per cent humidity. Since little attention is paid to the dough during fermentation, only sufficient light is required to insure safe handling of the equipment. High humidity requires the use of enclosed and gasketed luminaires. Experiments have shown that ultraviolet radiation can control wild molds in the fermentation room.

Make-Up Room. The dough is divided, shaped and panned preparatory to baking. Modern bakeries use production-line procedures. Operations are largely mechanical and require very little handling.

Proofing Room. During proofing the panned dough rises to its final shape. Local general illumination is required directly in front of the tray racks to permit a quick inspection of the panned dough. Because of the low ceilings in most proofing rooms, symmetrical, angular-directional luminaires or luminaires with asymmetric distribution can be used; because of high humidity, enclosed and gasketed luminaires are recommended.

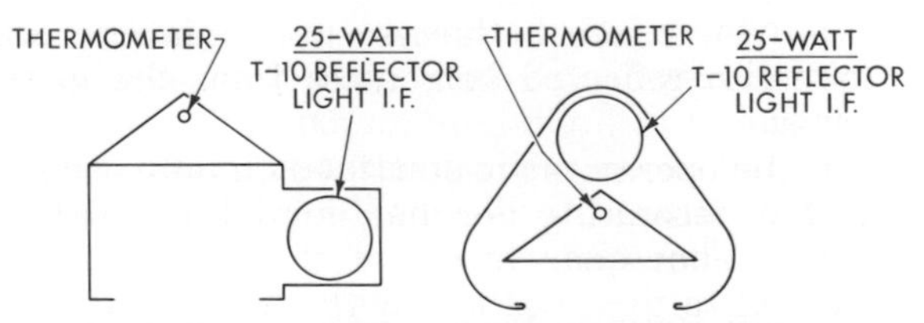

Fig. 9-18. Two suggested methods for illuminating thermometers.

Oven Room. Ovens present a special lighting problem because (1) many of them are finished black-and-white and (2) they are lighted on the interior by lamps with high-temperature bases in glass enclosures. The front of the oven should be illuminated to balance the luminances of the lighted oven opening to keep luminance ratios within a 10 to 1 ratio. Luminaires must be mounted so that tray trucks placed near the oven mouth do not cause shadows when the oven is being filled or emptied.

Fillings and Other Ingredients. Preparation of fillings involves accurate weighing and mixing various ingredients. Many ingredients are perishable and must be carefully inspected. Luminaires should be placed to provide illumination for cleaning the inside of the mixing bowl.

For cooking fruit fillings, an exhaust fan and canopy are usually provided above the kettles. Auxiliary luminaires can be mounted inside the canopy to provide good shielding and to properly direct the light. Enclosed and gasketed luminaires protect the lamps.

Tables for fruit washing, cutting and mixing are normally placed near the windows. Venetian blinds, shades, etc. control outside glare for any workers who must face such windows.

Decorating and Icing. In many large bakeries, icing is mechanically applied to the products. The lighting problem is similar to that in the make-up room.

Hand decorating and icing, however, require greater skill and care. Because of the detail involved in hand decorating and the necessity for working rapidly, quick and accurate seeing is required. Lighting on the decorating benches can be best provided by industrial luminaires.

Properly shielded fluorescent luminaires will minimize reflected brightness from the glossy icings.

If the operators are stationed on only one side of the decorating benches, luminaires may be placed above the heads of the operators and parallel to the benches. With this location illumination of the vertical surfaces of the product will be improved and the slight shadows produced will bring out the detail of the decoration. Also, reflections from the highly-reflective surfaces of the icing will be directed away from the operator. The luminaires may be tilted slightly toward the decorating benches if the benches are so located that a single lighting unit is not expected to illuminate more than one bench.

If operators are located on both sides of the benches, the luminaires should be placed across the benches and between the operators.

The color of the light sources in the decorating and icing department should match that under which the product will be purchased and/or consumed.

Illumination of Scales and Thermometers. Illuminated instruments permit quick, accurate reading of the graduations; however, one should be careful to avoid direct or reflected glare from luminaires or windows. See Fig. 9–18. Where local luminaires are impractical, suitable shielded spotlights are recommended.

Wrapping Room. Wrapping is done by automatic machines. Interruptions may be caused by torn wrappers which clog the mechanism, etc. General lighting expedites locating and removing obstructions, and cleaning, oiling and adjusting the machines.

Germicidal (bactericidal) lamps can be used to control mold in packaged bakery goods.

Storage. Correct illuminance level promotes good housekeeping and tends to decrease storage losses.

Because storage rooms frequently have low ceilings, attention should be paid to proper shielding of lamps. Correct placing of the units with regard to the bins, platforms, etc., is essential to prevent blocking light from the working areas.

Shipping Room. The finished product is delivered to the shipping room by tray trucks or conveyors. Shipping room tasks include keeping records, loading trucks and, in some places, maintaining and cleaning the trucks.

Truck Platform. Truck bodies are enclosed and are loaded from the rear door. It is necessary to use appropriately located angle or projector equipment permanently attached to walls, columns or other facilities to provide adequate seeing within the truck body. Sufficient illumination is necessary in the garage for cleaning the trucks.

CANDY MANUFACTURING

To comply with stringently enforced pure food laws and to foster good will, progressive candy manufacturers conscientiously promote cleanliness and efficient plant operation. For selection of illuminance levels, see Fig. 2–2.

Chocolate Making. In the manufacture of chocolate, the cacao beans are first toasted and then passed through shell-removing machines. The bean is conveyed by gravity to the crushers which press out liquid cacao butter. After milling and mixing with powdered milk and confectionary sugar, the pulverized beans are pressed through a series of rollers and then mixed with the cacao butter in a conche.

In a large plant, many of these gravity-fed operations utilize portions of two or three floors with conveyors or chutes passing through the floors. There is very little handwork because practically all processes are confined to the inside of hoppers, refiners, conches and other machines. Consequently, few difficult seeing tasks are encountered; however, the periodic setting of the five rollers of the roller mill is a difficult task. Supplementary lighting (having a predominant vertical component) should be included. See Fig. 9–19.

Chocolate Dipping. Dipping is performed in various sections of large plants because it facilitates the manufacture and minimizes the conveyance of the different fillings. Dipping tables are usually located symmetrically. The operator sits beside a depressed section of the table. Drippings from the operator's fingers set a design on top of the decorated candy. The dipper must see the relative position of the drippings over the confection in order to insure a neat and orderly design. General illumination should be provided in each dipping room.

Cream Making. Glucose, which is the base for most creams and fillings, is cooked, beaten by paddles, remelted and recooked to increase its viscosity. It is flavored, beaten again, and finally pressure-formed in plaster-of-Paris molds. The seeing task in cream making is of moderate severity. General illumination is recommended.

Kiss Wrapping. A kiss-wrapping production line consists of many individual kiss-wrapping machines arranged on both sides of a belt conveyor. General illumination should be provided over the entire area, with supplementary lighting at the critical seeing points. These vary in location with the type of wrapping machine.

Gumdrop and Jellied Form Making. In this process, plaster-of-Paris patterns are used to make smooth molds of fine-milled cornstarch. The molds are symmetrically arranged in shallow wooden trays. These are moved by a belt conveyor into such a position that one row of molds is placed directly under a series of injectors which automatically place the proper quantity of syrup in each. This operation is repeated until all the molds in the tray are filled. In mold-fillers for gumdrops and similar candies, the automatic injectors which press the fluid candy into the molds are kept clean by an attendant.

Luminaires with concentrating distributions should be hung above the equipment and directed toward the molds.

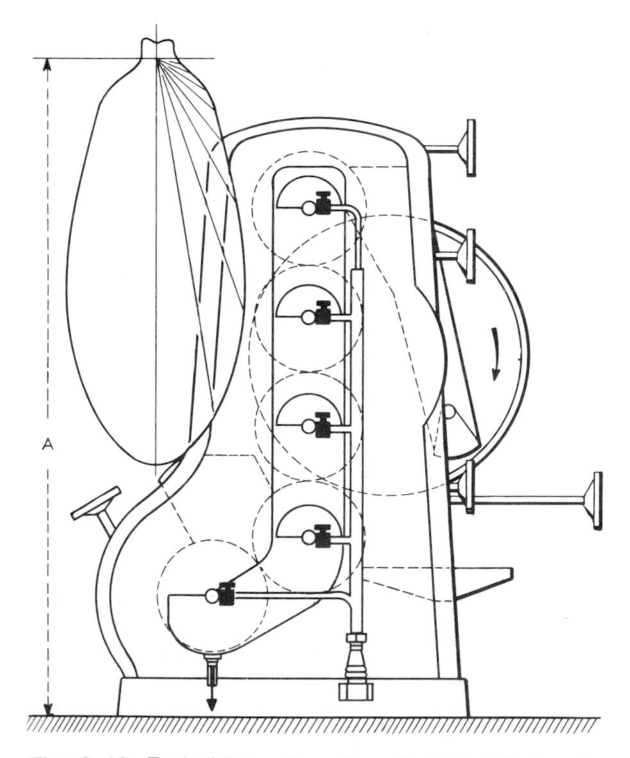

Fig. 9-19. Typical five-roller refiner. Periodic adjustments are made at the five rollers. Light should be distributed so as to illuminate the entire refining area. A = 2.4 meters (7.75 feet).

Hard Candy Making. In the manufacture of hard candy, sugar is cooked, flavored and placed in a semi-solid state on water-cooled tables where a batch is kneaded into an oblong shape. Fillings are added. The batch then is worked into a cylinder about 250 millimeters (10 inches) in diameter and 1.8 meters (6 feet) long. After tapering in a heated canvas hammock; the point is fed through a die-casting machine (see Fig. 9-20) which automatically shapes and cuts the candy.

General illumination should be provided for ingredient mixing and cooking. Supplementary lighting increases the illuminance at the die-casting machine. These luminaires should be located between the operator and the die-cutting machine. To minimize specular reflections in hard candy, luminaires with a large, low-luminance, luminous surface should be centered 1.2 meters (4 feet) above each hand-mixing table. Continuous fluorescent-lamp luminaires also may be used.

Assorted Candy Packing. There are three methods of packing candy:

1. *Progressive method.* Candy is placed in simple containers in front of the operators who sit on each side of a long table with a centrally-located conveyor belt.
2. *Stationary method.* Long flat tables, 900 millimeters (36 inches) high and 900 millimeters (36 inches) wide, are used. Directly over the center of the table a stock rack, 450 millimeters (18 inches) wide, is suspended from the ceiling or fastened to the table so that its bottom is 450 millimeters (18 inches) above the top of the table. The operator removes eight or ten different types of candy from the rack and packs them in a box.
3. *Circular method.* Not as common as the other two, a ring table, 900 millimeters (36 inches) high and 450 millimeters (18 inches) wide, is used. The outside diameter is about 1.8 meter (6 feet); the inside, about 0.9 meters (3 feet). The operator sits on a swivel stool in the center. Candy for packing is placed on the circular table—one kind to a container. By rotating the stool 360 degrees an operator is able to pick a complete assortment.

Special Holiday Mold Candy Making. Holiday candy usually is made on the north side of the building where daylight is available to aid in the necessary hand artistry. At window tables, operators with small artist's brushes decorate molded candy with a thin, colored mixture of cream filling. Because of the intricate positions and fine detail with which decorations must be placed on the confection, the seeing task is severe.

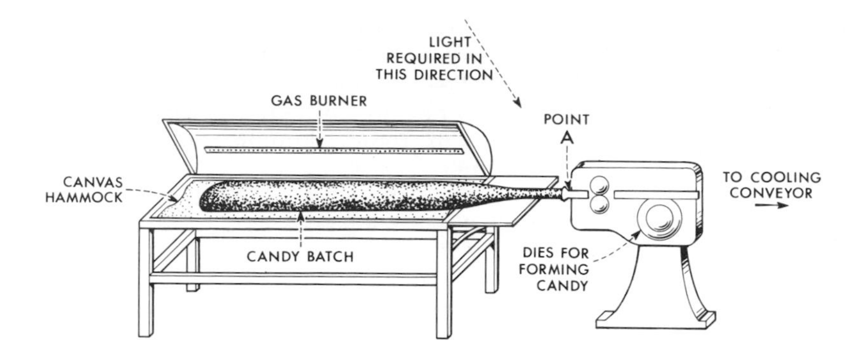

Fig. 9-20. Hard-candy-forming machine. The batch is revolved slowly in the canvas hammock. Heat is applied for surface glazing. The operator tapers one end to enter the dicing machine at point A, which cuts and forms in one operation and delivers the pieces to a cooling conveyor.

Box Making and Scoring. In many candy factories, containers and boxes are made on the premises in a department divided into two main sections: one devoted to making standard boxes; the other, to special boxes.

The first operation in making boxes is mechanical scoring. Care should be taken that the frame which holds the knives in position does not cast a shadow on the flat cardboard surface. All light sources should be located between the operator and the frame of the scorer to avoid shadows under the frame holding the scorers.

Flat cardboard is fed over rollers at the front of the machine. The first set of scorings is made by circular knives. In manufacturing these boxes, scorings must also be made at right angles to the original scorings.

After the cardboard has been scored, it is conveyed to a box-forming machine which bends the cardboard at the scorings, applies the gummed corner supports, and automatically shapes the container. The machine is pedal-controlled, and all work is accomplished on a horizontal plane. The tool and forming-die complete this work.

Most box-cover papers have high reflectances compared to those of the machine backgrounds. The luminance ratio may be as high as 5 to 1.

General illumination is recommended for box manufacturing.

CARGO HANDLING AND PORT SHIPPING

The pertinent areas and seeing tasks peculiar to port shipping facilities are described below along with suggested specific general lighting recommendations.

Recommendations are based upon a study of existing installations and practices and represents a consensus of the members of the Cargo Handling and Port Shipping Subcommittee. Areas that would be considered common to the shipping industry and other manufacturing operations are not included.

Port Cargo Handling Facilities

Port cargo handling facilities can be classified by type and divided into general areas which have different seeing tasks. By separately considering each type of facility, and by further breaking down each type into areas involving specialized seeing tasks, specific illuminances that cover most variations can be recommended.

Many port areas are used for one type of cargo handling only—such as containers, bulk or automobiles—while other port areas may be used for a combination of operations.

Types of Facilities.

Bulk Cargo—These commodities are received by truck or rail for loading aboard ship or are discharged from the ship, using cargo handling devices that include dump pits, rail car shakers and rotating dumpers, suction devices, conveyor systems, gantry cranes fed by conveyors through pouring spouts, or clamshell or magnet operations.

Container/Automobile—This cargo is received by truck or rail for loading aboard ship or is discharged from the ship, using shipboard or dockside cranes or vehicles on bridge-type ramps.

General Cargo—Includes all types of cargo such as break bulk (small packages, boxes, bagged), packaged lumber and steel products, palletized loads, refrigerated cargoes not in containers, machinery, and similar cargoes not specifically defined.

Oil Transfer Terminals—(Illumination requirements for these facilities are provided by the US Department of Transportation, Coast Guard enforced regulations under 46 CFR 126.)

Areas of Operation.

Facility Entrance—Controlled or designated entrances and exits such as locations where pedestrians and/or vehicle traffic have access to the facility.

Open Dock/Storage Yard—Areas where cargo is placed prior to loading or after discharge from the ship. Includes traffic lanes and means of access to the dock area adjacent to the ship.

Front (Dock Adjacent to Ship, High Line)—The location where cargo is landed from the ship or placed for hoisting/driving aboard the ship.

Transit Shed/Stuffing Station—An indoor cargo storage facility used prior to loading or after discharge for the receiving and delivering of cargo.

Employee Parking—Designated areas which may be located within the facility or immediately outside the facility entrance.

Low Line—Area usually on landside of transit shed with a truck loading dock.

Dumping Pit—Where bulk commodities are received from rail hopper cars or trucks, either by bottom drop, rotary dumping, scoop, clamshell or magnet.

Conveyor Transfer System—Moves and piles bulk commodities using endless belts, pockets or scoops.

Safety

Fig. 2-26, page 2-45 lists minimum illuminances for safety, alone. To assure these values are maintained, higher initial levels must be provided as required by the maintenance conditions. In those areas which do not have fixed lighting,

Fig. 9-21. Categorization of Port Cargo Handling and Shipping Facilities with Recommended Illuminances for Safety

Area	Activity	IES Class	Illuminance	
			Lux	Foot-candles
General Cargo				
Employee Parking	Pedestrian traffic, security	Slight Hazard/Low Activity	5	0.5
Facility Entrance	Pedestrians access, traffic control, security	Slight Hazard/Low Activity	5	0.5
Open Dock Area	Equipment operator moving cargo with machine. Dockman piling cargo, setting blocks, etc.	Slight Hazard/Low Activity	5	0.5
Transit Shed	Placing cargo, piling cargo, building loads, hand handling	Slight Hazard/High Activity	10	1.0
Front	Landing/hoisting loads, equipment operators, frontman	Slight Hazard/High Activity	10	1.0
Transit Shed	Inactive, security only	Slight Hazard/Low Activity	5	0.5
Low Line	Receiving/delivering of cargo from trucks, rail cars	Slight Hazard/Low Activity	5	0.5
Container/Automobile				
Employee Parking	Pedestrian access, security	Slight Hazard/Low Activity	5	0.5
Facility Entrance	Truck traffic, pedestrian walkways, weighing scales, security	Slight Hazard/High Activity	10	1.0
Storage Yard, Open Dock	Equipment operator moving cargo	Slight Hazard/Low Activity	5	0.5
Transit Shed/Stuffing Station	Loading/discharging containers, piling cargo, equipment operations	Slight Hazard/High Activity	10	1.0
Front/Container	Landing/hoisting cargo, securing/releasing chassis devices, pedestrian vehicle traffic	High Hazard/Low Activity	20	2.0
Front/Automobile	(Same as Front/Container)	Slight Hazard/High Activity	10	1.0
Walkways Through Traffic Lanes	Pedestrian traffic, vehicle operations	Slight Hazard/Low Activity	5	0.5
Perimeter Walkways	Pedestrian foot traffic, security	Slight Hazard/Low Activity	5	0.5
Transit Shed/Stuffing Station	Inactive, security only	Slight Hazard/Low Activity	5	0.5
Bulk Cargo				
Employee Parking	Pedestrian traffic, security	Slight Hazard/Low Activity	5	0.5
Facility Entrance	Pedestrian access, traffic control security	Slight Hazard/Low Activity	5	0.5
Open Dock Area	Moving rail cars, truck dump traffic	Slight Hazard/Low Activity	5	0.5
Dumping Pit	Opening hoppers, rotary and shaking operations	Slight Hazard/High Activity	10	1.0
Conveyor System Point of Operation/Transfer	Observing flow of cargo, control belt system	Slight Hazard/Low Activity	5	0.5

local illumination needs to be provided during occupancy by portable luminaires or by luminaires mounted on material-handling or other vehicles. A visually safe installation must also be free of excessive glare and of uncontrolled large differences in luminances within the area.

Based on the recommended illuminances for safety, the port cargo handling and shipping facilities can be categorized generally as in Fig. 9-21.

Tasks

The following is a description of the tasks performed by workers in General Cargo and Container Operation facilities. These descriptions may be helpful in determining illuminance values for design purposes when using Fig. 2-2.

General Cargo.

Frontman. The frontman's function includes securing and removing stevedore gear to vessels' hoisting equipment or cranes using shackles and hooks; landing cargo or slinging cargo; placing hooks, bridles, clamps and slings to hoist cargo or removing them from loads. Work is performed in vicinity of fork lifts, tow tractors, pipe trucks and other vehicles which service the front. Items vary in *size*—from hand-held to large bundles and loads weighing many tons. *Contrast* of work area varies, depending upon the vessel and types of cargo. *Time* to perform tasks depends on the cycle of operation, usually every 2 to 5 minutes.

Dockman. The dockman's duties include placing blocks, cones and stickers for landing loads from fork lifts and shed or open dock storage areas; hand-handles cargo in palletizing or sorting; works directly with fork lift operators; loads and unloads rail cars. *Size* of objects varies. *Contrast* depends on types of commodities; steel products would provide very little contrast. *Time* to perform task of placing blocks: approximately once every 5 minutes. Hand-handling and palletizing may be continuous.

Combination Lift Driver. Equipment operator for fork lifts, Ross carriers, tow tractors and other miscellaneous haulage equipment ranging from small to very large. Picks up and transports cargo in terminal and shed area, driving in forward and reverse. Often carries wide and bulky objects which impare visibility. Must maneuver in close quarters and coordinate vehicle handling with other equipment operators and personnel working on foot. Equipment is operated at slow rate of speed. Equipment is not usually provided with driving lights. *Size* of objects, generally, will be those items which are at least as large as a pallet-board load. *Contrast* between piles of cargo and the load and other structures will usually be good. *Time* of performing this task will be relatively constant.

Clerk. This employee performs seeing tasks related to the paper work necessary to cargo documentation. It involves the reading of lot numbers, seals and other identifying marks on cargo and cargo packaging and directing equipment operators to the proper locations for the pickup and delivery of such cargo. The task requires being able to properly read all types of documentation, including computer printout and carbon copies. This employee works in the vicinity of equipment operators and is required to make written notations and pencil- and chalk-marks in performing duties. *Size* of the objects varied according to the type of cargo being handled, as well as for conventional paper documents. *Contrast* will generally be good, depending on the types of cargo and papers. *Time* for performing the duty is relatively constant, including the switching from paper work to terminal operation.

Foreman. This employee is responsible for the supervision of other employees, handles paper work pertaining to payrolls and hatch logs, is responsible for utilization of testing equipment such as carbon monoxers, and must be able to observe all operations. *Contrast* of his seeing task depends on the types of cargo. *Time* of these tasks is relatively constant.

Watchman. The watchman is responsible for security and traffic control in longshore operations. Duties may include vehicle, bicycle and foot patrols; handling cargo or objects other than a clipboard or keys; working in the vicinity of other employees and equipment operators. *Size* of tasks related to type of cargo. *Contrast* may be poor in patrolled areas of perimeter. *Time* is relatively constant.

Container Operations.

Dockman. The duties of this employee are to place and remove lashing materials on vans being hoisted to and from the vessel and to release and secure locking devices on the chassis in a truck-tractor operation. The work is usually located in the congested area on the front. Object *size* includes large hand tools and lashing wires. *Contrast* is not very good. *Time* to perform task depends on the cycle of operation—usually every 1 to 2 minutes.

Semitractor. The truck-tractor operator handles long chassis with loaded and empty containers by driving a yard hustler tractor. The required skills include the maneuvering of these large rigs in close quarters and within the traffic pattern established in container yards. *Size* of the objects is large, except for the vehicle controls which generally are lighted within the cab. *Contrast* should be generally good. *Time* of performing these duties is relatively constant.

Straddle Truck. The operator drives a piece of equipment located 3 to 9 meters (10 to 30 feet) above the ground; handles large objects with this machine, such as long containers; must locate and identify spots within the yard, and vans to be handled by reading numbers; and must operate in the vicinity of other employees and equipment. *Size* of the objects is large, except for cab controls. *Contrast* generally will be good. *Time* for the operation is relatively constant.

Crane Operator. Crane operators handle a variety of container cranes; however, they usually are similar to a bridge-type gantry crane, either rubber-tire or rail-track mounted. They must operate from a height up to 27 meters (90 feet) above ground, and are responsible for the safe movement of long containers around other employees and equipment and for the loading and unloading of chassis. *Size* of the objects is large. *Contrast* will usually be good. *Time* of the cycle is dependent upon the skill of the operator, generally every one-half to two minutes.

Clerk. The clerk generally operates a small vehicle and is responsible for the paperwork of receiving, delivering and loading of containers aboard ship. This involves the use of various documents and notations. The clerk is also required to maneuver throughout the yard—directing other operators to the location of their respective cargoes. *Size* of the objects will be generally that of paperwork items. *Contrast* will be generally good; however, the clerk's ability to maneuver in the yard will be somewhat limited by contrast situations. *Times* during which a clerk performs his duties vary greatly.

Watchman. The watchman may be located at the gate entrance or be on roving patrol throughout the yard. Duties include security and traffic control within the yard, and both operates equipment and works on foot around other equipment operators. *Size* of the objects will be limited to that of a clipboard. *Contrast* is dependent upon situations in the yard. *Time* of a watchman's job cycle is relatively constant.

Foreman. The foreman is responsible for the foregoing duties being properly performed as well as for the paperwork and for maintaining testing procedures when necessary. The foreman is often provided with a small vehicle. The foreman also works in the congested area around the crane, as well as throughout the yard. *Size* of the objects handled is usually limited to small paperwork items. *Contrast* in the container yard is generally good. *Time* of the foreman performing his duties is irregular.

CLEANING AND PRESSING

For selection of appropriate illuminances for the tasks involved in the operations tabulated below, see Fig. 2-2.

Operations. The operations in dry-cleaning plants are functionally divided as follows:

1. Receiving.
2. Checking and sorting.
3. Dry-cleaning.
   a. Naphtha-solvent process
   b. Synthetic-solvent process
4. Steaming.
5. Examining and spotting.
6. Laundry or wet cleaning.
7. Repair and alteration.
8. Machine finishing.
9. Hand finishing.
10. Final inspection.
11. Shipping.

Receiving. Soiled items are received from pickup trucks at a receiving platform. Garments are transferred from motor truck to hand truck and wheeled to the checking and sorting tables.

Checking and Sorting. A checker reads special instructions on a driver's ticket (written in pencil) attached to incoming garments, and pins identification tags to each garment. The penciled notations are difficult to read, and contrast, as well as the handwriting, often is poor. Pockets are searched for matches or articles of value. The sorter divides the garments into synthetics, cottons, silks and woolens; dark or light colors; and other necessary classifications.

Dry Cleaning by the Naphtha-Solvent Process. Because naphtha is flammable, cleaning is effected in a separate building or in a section of the plant separated by a firewall. Lighting equipment suitable for Class I, Group D is mandatory.

There is no attempt in the washing and drying room to determine whether the cleaning has removed all spots; thus, there are no other difficult seeing problems. The explosion-proof luminaires must be located so the washer, extractor and drying-tumbler interiors are well illuminated when the covers are thrown back. Distribution must be such as to properly light pressure and flow gauges on the filters and in the piping. The time of washing is largely determined by the clarity of the naphtha coming from the washer. This dirt can best be seen in silhouette against a white background while it is passing through the filter gauge.

Dry Cleaning by the Synthetic-Solvent Process. A nonflammable solvent is used in a closed system. The seeing tasks are related to loading and unloading the cylinder and reading the temperature, pressure and flow gauges. Light should be so directed into the cylinder and toward the gauges to preclude an image of the source.

Examining and Spotting. The dry-cleaning process removes practically all oil and grease out of stains unless they are firmly ground into the fabric. Nearly always, however, some remaining spots must be removed by chemicals or steam.

Many stains have characteristic colors by which they may be identified by a skilled "spotter." This experienced worker is trained to detect, classify and remove all types of spots with the proper chemicals. The critical visual task is detecting and identifying the spot.

After washing, there is little contrast between the spots and the material. The reflectance of many materials has a strong specular component. Low-contrast stains are particularly difficult to see on such materials as taffeta, gabardine, crepe and many of the synthetic varieties.

The spectral characteristics of the light source affect the appearance of the garment. (A blue garment appears blue because it reflects more blue light than any other color; similarly, a red garment reflects more red light than any other color.) In addition it will affect the eyes' state of chromatic adaptation (see Section 5 in the 1981 Reference Volume).

It is practical to examine individual colors of garments for low-contrast stains by increasing the contrast of the color such that the stains become more visible. This can be accomplished effectively by employing a complementary-color light source. Deluxe warm white fluorescent lamps are effective for inspection of low-contrast stains on blue and green garments; daylight fluorescent lamps, for yellow and red garments. The combination of deluxe warm white and daylight lamps can be used effectively on black and brown garments.

A low-luminance, large-area luminaire is recommended over spotting tables. This could be a luminaire equipped with an acrylic or glass diffuser (opal or polarized) housing two deluxe warm white fluorescent lamps controlled by one switch, and two daylight fluorescent lamps controlled by a second switch. Switches should be located so as to be convenient to the spotter. Separate or combined switching can be applied as needed for the color of the garment being examined.

Fig. 9-22 illustrates recommended mounting locations for these luminaires over hand spotting boards and steam spotting tables.

Laundry or Wet Cleaning. On some garments, there may be so many large and widely distributed spots that it is uneconomical to use the spotting method to remove them. Each cleaning plant maintains a small laundry that employs a conventional cylindrical washer, centrifugal extractor and drying tumbler. The lighting problem is very similar to that of the dry-

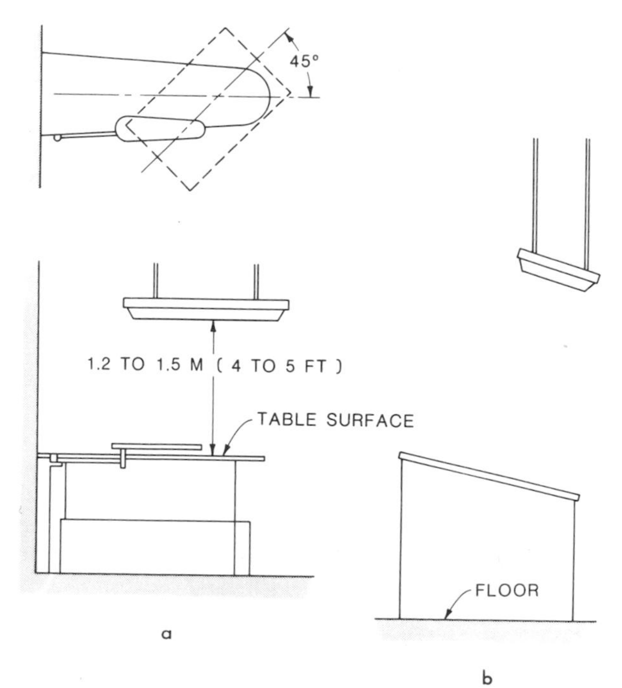

Fig. 9-22. Plan and elevation of installation for inspection lighting for (a) spotting boards and (b) steam tables.

cleaning operation. Vapor-proof luminaires are recommended. They should be located so as to light the interior of the machines.

Repair and Alteration. Here, there is both hand and machine sewing, often with dark thread on dark material. For handwork a well-diffused light is preferred. Supplementary illumination is recommended at the needle point of the machines.

Machine Finishing. Machine presses usually are lined up in a row for convenience and for minimum cost in the steam-piping installation. The operator combines speed with good workmanship. Each garment is moved several times as a small section is finished and another moved onto the buck (working surface). The worker sees that all wrinkles are eliminated. The buck of the press should be uniformly illuminated without shadows from the head of the press or from the worker's body. A continuous row of fluorescent luminaires extending over the ends of the presses is recommended. This technique illuminates the working area on the buck and the clothes racks, aisles and machine space.

A most difficult task is preventing double creases in trouser legs. A concentrating reflector at the rear of the buck causes a crease to cast a shadow, making it more easily discernible.

Hand Finishing. Hand finishing (ironing) boards usually are installed in rows spaced 1 to 1.5 meters (3½ to 5 feet) apart. Improvements in machines are gradually decreasing the volume of handwork; however, the hand iron still achieves the best results on lightweight materials. The hand finisher observes that wrinkles are eliminated and that garments are completely pressed and not scorched, and that minor defects are corrected.

The seeing task is moderately critical: careful handling of the iron is required for pleats, shirring, ruffles and trimming.

Final Inspection. Garments on individual hangers are delivered to the final inspector, either on portable racks or by a power-chain conveyor. Each garment is hung on an overhead support that will rotate easily. The inspector examines the garment carefully, watching for inferior finishing, for spots, for damage done to the material during cleaning and for the completion of any customer-ordered repairs or alterations.

Most of the critical visual work is done at short range with the garment at approximately a 45-degree angle with the vertical. The lighting requirements are about the same as for spotting, and similar luminaires with two colors of lamps and separate switching are recommended. To increase the vertical-plane illumination, the luminaire should be tilted parallel to the plane of the garment.

Shipping. The shipping section covers the garments with protective bags, attaches the original ticket and loads the delivery trucks. Identifying tags often are difficult to read because the ink is partially washed out during the cleaning process.

COTTON GINS

Areas included in the operation of a cotton gin contain overhead equipment, gin stands, lint cleaners and bale press. These areas should be illuminated with a well planned general lighting system to provide the recommended illuminances in Fig. 2-2.

The function of a cotton gin is to separate seed and trash from cotton lint, gather the seed, dispose of the trash, and place the lint in a form that may be easily handled and transported.

Overhead Equipment. The overhead equipment in a cotton gin includes separators, driers, cleaners, stick machines and feeders. These machines are automatic and do not require an operator. Catwalks are usually provided around the overhead equipment for maintenance purposes. General lighting should be provided for safety and maintenance.

Gin Stand and Lint Cleaners. The gin stand separates cotton seed from the cotton lint. The gin stand and the adjacent control console are operated by ginners, who start the gin and adjust the suction conveyor to regulate the flow of cotton into the gin stands. They examine ginned cotton for foreign material and adjust spiked drums and saw blades within the gin stands. The seeing tasks for the lint cleaner are similar.

The seeing tasks at the gin stand, lint cleaner and control panel are on a vertical plane and, therefore, the general lighting system should provide a high vertical component. It is suggested that supplementary lighting be mounted in the seed channel inside the gin stand, to help the ginner to observe the rate of flow of the cotton seed.

Bale Press. The bale press is a hydraulic press that compresses the cotton into a bale that is about 1500 by 1100 by 610 millimeters (60 by 42 by 24 inches) and weighs about 230 kilograms (500 pounds). Most bale presses are operated by two pressmen, who spread a piece of burlap or bale covering on the bed of the chamber of the press. Then they start the equipment that dumps and tamps bulk cotton into the chamber. They observe the level of cotton in the chamber and stop the loading equipment when the chamber contains the required amount of cotton. The chamber is then positioned under the press ram. The pressmen then start the press to compress the cotton. One pressman inserts metal bands through the channel in the upper ram of the press to the pressman on the opposite side of the compressed bale. The second pressman returns the band through the channels in the lower ram to encircle the bale. The pressman then buckles the ends of the bands together. General lighting should be provided in this area. Seeing tasks are on both the horizontal and vertical planes.

DAIRY AND POULTRY INDUSTRIES

The drastic changes that have taken place in agriculture in the last decade or so have greatly affected the tasks to be performed in and around farm buildings. The tasks have not only changed in nature, but the time available to perform them has been reduced, in most cases. New lighting concepts to keep pace with these developments have necessitated research that will provide the basic data on lighting requirements. Because of their universal application, Dairy Farm Lighting and Poultry Lighting have been selected by the IES-ASAE (Joint Farm Lighting) Committee to be the first to receive special attention. The recommendations under Dairy Farm Operations and Lighting for the Poultry Industry are based on the results of these studies. Lighting for other types of farming will be the same where the tasks are similar.

Quantity of Illumination. The illuminance required to perform visual tasks on the dairy farm vary through wide extremes. For example, relatively high levels may be required in the farm shop, or where inspections are made for cleanliness and disease control, such as in milking operations. But relatively low levels are adequate on walkways between buildings, or in storage areas where seeing tasks may be much less critical. Illuminance recommendations for the various dairy and poultry farm operations are listed in Fig. 2-2.

Quality of Illumination. Specifying the quantity of light in no way establishes the quality of a lighting installation. Some important factors of the lighting installation are uniformity, glare, color and environment (see page 9-2).

Illuminance uniformity may be expressed as a ratio of the maximum value to the minimum value over an area. Based on this, satisfactory uniformity ratios for farms vary from 1.5 to 1 for critical seeing task areas, to 5 to 1 for less critical areas. In general, with greater mounting heights and closer spacing of luminaires, better uniformity can be achieved.

The quality of the lighting installation can be greatly influenced by certain environmental factors. Room surfaces should have matte finishes of high reflectance to help prevent excessive luminance ratios. The ceilings, walls and floors can increase the utilization of the light within the room by acting as a secondary light source of large area. Luminaires which have some light directed upward toward a ceiling having a relatively high reflectance will help create a more comfortable visual environment. Recommended reflectance values for farm interiors are consistent with those of other industrial areas. See Fig. 9-3.

Codes. The use of lighting equipment on the farm is governed by many federal, state and local codes. Some public health codes specify minimum illuminances required in the milking and milk handling areas for maintenance of health standards. These public health lighting requirements are normally below those levels shown in Fig. 2-2, since they are concerned only with sanitation. The recommendations in Fig. 2-2 were selected for efficient performance of the visual task in addition to the concern for sanitation.

The modern farm includes many types of occupancies which may be wet, damp, corrosive, dirty, surrounded by combustible materials, or saturated with gasoline fumes. Therefore, it is important to follow the *National Electrical Code* and any local regulation which may be in effect when installing lighting equipment.

Dairy Farm Operations[25]

Dairy farms vary in size, location, amount of investment and degree of mechnization; however, there are some areas common to every dairy facility. These are the housing areas, including community, maternity, and calf pens; the feeding areas; the milking operation areas; the milk handling equipment and storage areas; and the feed storage areas.

Livestock Housing. The two basic housing systems are stall barn and free stall. The stall barn is a structure for sheltering dairy cattle and/or young stock, where the adult animals are confined to stalls by means of stanchions, straps, halters or chains during part of the year, and usually for milking purposes. There may be one or more rows of stalls and pens. Roughages and concentrates may be fed in mangers at the individual stalls. All, part or none of the feeds and bedding may be stored in the structure. The stall barn is sometimes referred to as a stanchion barn or warm-housing.

Free stall barns may be insulated and mechanically ventilated or may be a cold housing system with natural ventilation. The cattle are free to move from the stalls to the feeding area as there are no feed mangers in front of the stalls. At milking time, the lactating animals are passed through a milking parlor. Other dairy animals may be in separate pens, lots or buildings.

Adequate lighting is required to observe the condition of the animals and to detect hazards to the livestock and operator. Portable supplementary lighting units can be used to examine or treat individual animals when required.

Milking Operation. Both systems, stall barn and free stall, are quite different in layout, but each contains similar basic areas of operation which require lighting. In the stall barn, the milking operation is performed in the stall. In the free stall, the animals are milked in a milking parlor. There are various milking parlor layouts, but all contain milking stalls, an alley for cow entrance and exit, and an operator's area for use by the personnel performing the milking operation.

Adequate lighting is required to determine cleanliness of cow, to detect undesirable milk, to handle milking equipment readily, and to detect dirt and foreign objects on the floor. Recommended illuminances should be provided at cow-edge of gutter and on the floor. Supplementary lighting should be available to determine cleanliness of the udder and to clean and examine all parts of the udder.

Milk Room. The milk room is a room with one or more sections for handling raw milk, wholly or partly enclosed by the structure in which the cows are milked. The milk house is the same as a milk room except that it is not a part of, but may or may not be connected with, any other structure. The milk room or milk house would be the same for stall barn and free stall systems.

Illumination is required for the operator to move about readily and safely and to determine floor cleanliness. In the washing area it is necessary to detect dirt and other impurities on the milk handling equipment. Supplementary portable ultraviolet luminaires should be available in this area to aid in the detection of milkstone on the equipment.

Lighting is also necessary to adequately inspect the bulk tank interior for cleanliness. Additional lighting may be required to illuminate dipstick or scale.

Feeding Operation. The feeding area is a part of a barn, shed or open lot where cows are fed roughages and water, and sometimes concentrates. This area may or may not include feed storage. Feed storage and processing areas include haymow, silo, silo room, grain bins, concentrate storage areas, and feed mill and mixing areas.

Adequate lighting is required in the feeding area for the detection of foreign objects in the grain, hay or silage.

In forage storage areas, lighting is required for safety of the operator in moving about, but adequate lighting is required for the detection of foreign objects in the grain, hay or silage.

Luminaires should be mounted at the top of the silo near the ladder chute for ease in luminaire cleaning and lamp replacement.

In grain and concentrate storage areas adequate lighting is required to inspect amount and condition of grain. When grain is suspected of being moldy, containing foreign objects, etc., samples should be inspected under higher illuminance levels. Adequate lighting is required to read concentrate labels and again higher illuminance levels are required for critical inspection for impurities and spoilage.

Illumination is required in the processing area to see to move about readily, and to read labels, scales and equipment dials. If machine repairs are necessary, additional light should be supplied by portable luminaires or daylighting.

Miscellaneous Areas. Other buildings and/or areas, common not only to dairy farms but to all types of farms, include: the machine storage areas, such as the garage and machine sheds; the farm shop; the farm office; restrooms; and the pump house. (See earlier portions of this Section for industrial type areas and Section 5 for Offices.) Many exterior areas on the farm require lighting such as paths, steps, outdoor storage

Fig. 9-23. Typical laying houses of the floor type (left) and cage type (right).;

areas, feed lots, shop aprons and building entrances. Security lighting is also required in many exterior areas (see page 9-67).

Poultry Industry Operations[6]

Poultry farms and processing plants vary in size, layout and degree of mechanization; however, there are some areas common to most facilities. The following are brief discussions describing the operations in the major areas.

Laying Houses. There are basically two types of laying houses used on most poultry farms, the floor type house and the cage type house. See Fig. 9-23. In the floor type houses, the hens may move about freely over the floor area. In the cage type houses, hens are confined in cages with three to four hens per cage. A common size cage for housing four birds is 300 millimeters (12 inches) wide, 460 millimeters (18 inches) deep, and 400 millimeters (16 inches) high. Both the floor and cage system may range in size from a few thousand to several hundred thousand birds. Depending on the degree of mechanization, feed is mechanically conveyed to the birds in both type systems. Water is also provided continuously to both systems. Eggs are normally collected twice per day using a hand push cart, mechanical self-propelled cart, or automatic egg collectors.

General lighting provided by a lighting circuit separate from the circuit used to stimulate production and growth is needed for feeding, inspection and cleaning. Localized lighting is also needed for reading charts and records and to read thermometers, thermostats and time clocks.

Egg Handling, Packing and Shipping. The egg handling area may be located in a separate or adjoined building. When the eggs arrive at the egg processing area, they are either stored in a refrigerator area or loaded directly onto a washer. After the eggs have been washed, they are sorted and graded. Rough shells, cracked shells, and dirty or stained eggs are removed at the grading table. Candling equipment is used inside an enclosed booth to sort eggs that have internal defects such as blood spots or meat spots. If the more modern equipment is used, the eggs are sorted according to size by a machine and placed in an egg carton for shipment. See Fig. 9-24. The cartons are held in refrigerated storage until they are ready for shipment to retail outlets.

General lighting is needed to keep the area clean, detect any unsanitary conditions, and to

Fig. 9-24. Eggs being sorted according to size by machine and placed in cartons for shipment.

examine and grade eggs. In the loading platform and egg storage areas, general lighting is needed for safe operation of mechanical and loading equipment.

Raw Egg Processing. Eggs which are to be marketed as liquid, frozen or powdered are processed in this area. The area must meet the sanitary requirements of a public food preparation area as set up by the health department. Cracked eggs, eggs with shell defects and shell stains, plus grade A eggs, are utilized in the processing of liquid eggs. The eggs are broken out of their shells and pumped into a holding tank. The liquid eggs for interstate shipment are pasteurized and packaged. Some states also require that all broken-out eggs be pasteurized for intrastate shipment. Also, liquid eggs may be frozen at the processing area and shipped in a frozen state.

General lighting is needed to provide for cleanliness for food preparation.

Hatchery. Fertile eggs are brought to the hatchery and loaded onto trays to be placed in incubators. The incubators maintain a temperature required to develop the embryo and the chick is hatched in twenty-one days. When hatched, baby chicks that are to be grown for layers are sorted according to sex. The male chicks are disposed of and only the females are marketed. Most of the female chicks have their combs dubbed, a process of clipping off the top of the comb which results in the mature bird having a smooth comb. This helps to prevent the adult bird from catching her comb in the wires of the laying cages. Broiler chicks are handled in much the same manner except they are not sex-sorted or dubbed. Broilers are not normally reared in cages due to the increased tendency of bruising from the cages.

General lighting is needed for operators to move about readily and safety and for cleaning. Supplementary lighting is required for inspecting and cleaning inside incubators, at dubbing stations to prevent cuts and injuries to chicks, and for sex sorting (should be provided in a closed area to prevent excessive luminance ratios between task area and the immediate surrounding area).

Poultry Processing Plant. The birds are brought to the plant in crates of about 20 per crate. Crates are unloaded and the birds hung by the feet on a continuously revolving overhead carriage. They pass by an area where they receive a slight electrical shock which stuns them just before killing. They move through a bleeding area and then into the scalding tanks. Feathers are then removed by machine and the birds move on to the processing area. All birds are government inspected for wholesomeness after eviscerating, are thoroughly washed, inspected again, and sorted according to grade. Generally, the processed birds are then packed in ice and shipped to retail outlets.

General lighting is needed for cleanliness, inspection and sanitation. Supplementary lighting is required to detect diseases and blemishes (vertical illumination if the birds are hanging).

ELECTRIC GENERATING STATIONS

Electric power generating stations may be classified by their types of primary energy: fossil fuel, hydroelectric and nuclear. This classification is convenient because while such spaces as turbine buildings, transformer yards and switchyards have similar lighting problems, there are few similarities in lighting the various types of energy-handling areas. For illuminance recommendations see Fig. 2-2.

Switchyards, Substations and Transformer Yards[7]

These areas are occupied by station transformers; oil, gas or air circuit breakers; disconnect switches; high-voltage buses; and their auxiliary equipment. In addition, there often is a small building housing transmission line and substation relays, batteries and communication equipment. Lighting outdoor substations pre-

Fig. 9-25. Control room of a large generating station illuminated with a louverall ceiling.

Fig. 9-26. Control board illuminated by fluorescent lamp directional units tilted at 12 degrees and mounted in an arc around the board. General room illumination is provided by recessed fluorescent luminaires mounted in parallel rows.

sents a peculiar problem because the seeing tasks are frequently a considerable distance above eye level and contrast varies greatly.

Two kinds of lighting are required: general illumination for the movement of occupants and directed illumination for seeing fans, bushings, oil gauges, panels, disconnect switch jaws, station bus and other live parts.

Especially designed stanchion or bracket mounted substation luminaires are available to provide the required upward lighting component, although floodlights and spotlights are often used. Because of their long life, high intensity discharge lamps help to reduce the maintenance problem caused by the difficult access to lighting equipment.

Control Rooms and Load Dispatch Rooms[8]

The control room is the nerve center of the power plant and must be continuously monitored. Its lighting must be designed with special attention to the comfort of the operator: direct and reflected glare and veiling reflections must be minimized, and luminance ratios must be low.

Along with ordinary office-type seeing tasks, it is necessary to read meters—often 3 to 4.5 meters (10 to 15 feet) away. Reflected glare and veiling reflections must be eliminated from the meter—especially those with curved glass faces.

While the practice is not standardized, most control room lighting involves one of two general categories: diffuse lighting or directional lighting. Diffuse lighting may be from low-luminance, luminous indirect lighting equipment, solid luminous plastic ceilings or louvered ceilings (see Fig. 9-25). Directional lighting (see Fig. 9-26) may be from recessed troffers which follow the general contour of the control board. (These luminaires must be accurately located to keep reflected light away from the glare zone.) Illumination for the balance of the room may utilize any type of low-luminance general lighting equipment.

Giving operators full control of illuminance levels through the use of multiple switching or dimming systems increases lighting flexibility. *Load dispatch rooms* resemble control rooms and may be illuminated similarly.

Turbine Buildings[7, 9]

Turbine buildings usually have medium-to-high ceilings. Seeing tasks include general inspection, meter and gauge reading and pedestrian movement. Where there is detailed maintenance or inspection, higher-level portable lighting equipment is recommended.

In low- and medium-bay areas, high intensity discharge or fluorescent industrial luminaires suspended below major obstructions are appropriate for general illumination (see Fig. 9-27). If there are structural limitations it may be necessary to use floodlights mounted on walls or platforms. Supplementary lighting is recommended for vertical illumination on such equipment as control panels, switchgear and motor control centers. Luminaires with an upward component contribute to improved visual comfort.

For high-bay areas (generally 7.6 meters (25 feet) or higher) it may be appropriate to use either high intensity discharge or fluorescent luminaires. Lighting continuity must be maintained during any power interruptions. (See Emergency Lighting, page 2-45.)

Fossil Fuel Areas[7]

Fossil fuel plants (those which use gas, coal, oil or lignite) require facilities to receive, store and transport the fuel. Coal handling usually involves extensive outdoor storage areas. High intensity discharge floodlights will provide the required illumination for stacker, reclaimer and bulldozer operations. Post-top luminaires often provide the light required for both pedestrian traffic and conveyor inspections. Crusher houses, transfer towers, thaw sheds, unloading hoppers and other enclosures usually require indoor industrial luminaires. When these areas are classified as "hazardous", they should be lighted with equipment that meets the applicable provisions of the *National Electrical Code.*

Oil-fired stations have large oil storage tanks and many pumps which must be lighted for inspection, and may have barge or tanker unloading facilities, whereas gas-fired stations usually have gas metering areas. Illumination for these areas should be provided by floodlights. It is mandatory to comply with the requirements of the *National Electrical Code.*

Boilers on fossil-fuel stations have a series of indoor and outdoor platforms. These platforms (and their associated stairs and landings) should be illuminated for safe pedestrian passage and for inspection of burners, pumps, valves, gauges, soot blowers, etc. Where no overhead structure is available for the support of general-purpose industrial luminaires, stanchion supported luminaires may be used. The extreme heat associated with certain indoor boiler areas necessitates the use of luminaires that are designed to operate in high ambient temperatures.

Fig. 9-27. View of condenser well lighted from general lighting system in turbine room ceiling. Lighting layout keyed to special acoustical treatment of ceiling and light tile walls. Note visitors' gallery.

Hydroelectric Stations[7]

The turbine building of a hydroelectric station, commonly called the "power house", has lighting requirements generally the same as those previously outlined under Turbine Buildings.

Lighting of the intake and discharge areas, where applicable, is best accomplished with floodlighting.

Nuclear Stations[10]

The selection of lighting equipment for nuclear stations is often limited by factors other than economy or efficiency. There are extensive station areas—especially the containment building around the reactor and fuel storage facilities—where there may be a restriction on the kinds of lamps and metals used in the luminaires; thus, luminaire and lamp selection should be coordinated with the appropriate authorities.

Emergency Lighting[7, 10]

In certain areas, if the normal system fails, standby lighting is needed for the continued performance of critical functions and for safe building egress. Emergency lighting (see Section 2) should be considered for control rooms, first aid rooms, turbine rooms, exit stairs and passages, battery rooms and emergency generator rooms.

In nuclear stations it is recommended that emergency lighting be provided in all areas subject to contamination, especially the decontamination room and laboratory area.

FLOUR MILLS

Inspection of the material in process is by a combination of visual examination and feel—usually examined in the hands of the operator.[11] Sufficient illumination (see Fig. 2-2) of daylight

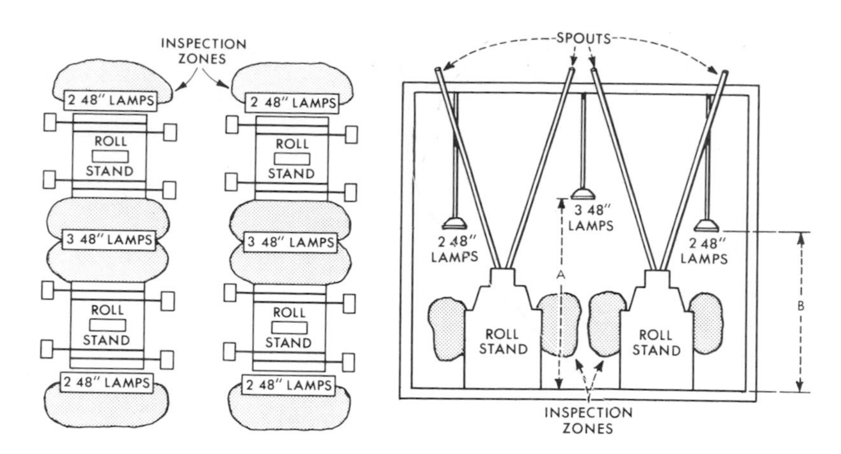

Fig. 9-28. Recommended arrangement of luminaires for roll floor: (left) plan view and (right) elevation. A = 2.7 meters (9 feet). B = 2.3 meters (7.5 feet). Lamps are fluorescent 1200 millimeters (48 inches) long.

quality enhances accuracy of judgment. Numerous obstructions, such as spouts, shafting, etc., encountered above the equipment (as on the roll floor) need not interfere with the location of luminaires. See Fig. 9-28.

All lighting equipment in Class II, Group G locations must meet Underwriters Laboratories approval for those locations. Such equipment will also reduce lighting maintenance caused by accumulation of dust, dirt and grease. If possible, luminaires should deliver some upward component for illumination of overhead line-shafting, belting, etc.

Roll, Sifter and Purifier Floors. Illumination from "daylight" fluorescent lamps provides satisfactory lighting for inspection. The contrast between the flour and the impurities is higher under daylight quality of illumination. Fluorescent luminaires can produce uniform illuminance without dark shadows and high luminance differences. Fig. 9-28 shows a suggested arrangement of dust-tight fluorescent lighting equipment for roll floors.

Packing Areas. Color quality is not important. Supplementary illumination may be necessary for the scale dials and sewing heads.

Product Control. Illumination requirements here are similar to those for inspection on the roll, sifter and purifier floors.

Miscellaneous Areas. Lighting is required for use in the *bins* to inspect material level. Lightweight, portable lighting equipment should comply with Class II, Group G.

Floodlighting from permanently located units is recommended around the *cleaning screens* and *man lift* areas.

FOUNDRIES[12, 13]

The design, construction and operation of modern foundries have provided for improvements in materials handling and in dust control. It is not uncommon to find semi-automatic production conducted in a comparatively clear atmosphere.

Although materials handling and atmospheric conditions have been improved, it is necessary to carefully select: (1) proper light sources, (2) luminaires with acceptable performance characteristics and (3) locations for luminaires that assure ease of installation and maintenance.

Metal castings are made in a variety of sizes and shapes, from a few ounces to many tons. Some are made to very close tolerances, others require less accuracy. The lighting requirements for foundry operations vary with the required accuracy and the severity of the seeing task. See Fig. 2-2.

Melting, molding and core-making usually involve equipment with nonspecular surfaces. Where such work is done in high bay areas, high intensity discharge luminaires may be installed without introducing reflected glare.

Maintenance may be minimized by the use of reflector lamps, ventilated luminaires or gasketed luminaires. Some of the latter have filters which permit "breathing" but prevent the ingress of dust. It is prudent to install the least practicable quantity of luminaires which will provide the recommended illuminance levels.

Core Making. Three general methods are used to form sand cores—hand ramming, machine ramming and blowing. Rammed cores are formed by packing sand in the core boxes by hand or by machine. Core-blowing machines use compressed air to inject the sand into the core box and pack it tightly.

The most critical seeing tasks in core making are: (1) inspecting empty core boxes for foreign material or sand and (2) inspecting the cores for such defects as missing sand or heavy parting lines.

The severity of the seeing task varies with the size of the cores and/or with the degree of tolerance. Core making is a rapid and continuous operation which required an almost instantaneous inspection at frequent intervals.

Contrast is fairly good between light sand and the metal boxes; it is extremely poor between brown sand and orange-shellacked boxes or between black sand and black boxes. Contrast and seeing conditions may be improved by finishing the inner surfaces of wooden boxes with white paint.

Bench tops having a light, natural wood finish are both practical and desirable. They can be kept clean and will provide a comfortable visual environment with low luminance ratios. Benches lighted with ventilated, "up-light" industrial fluorescent luminaires provide good visibility. Centering the luminaires on a line above and parallel to the worker's edge of the bench will minimize reflected glare and shadows.

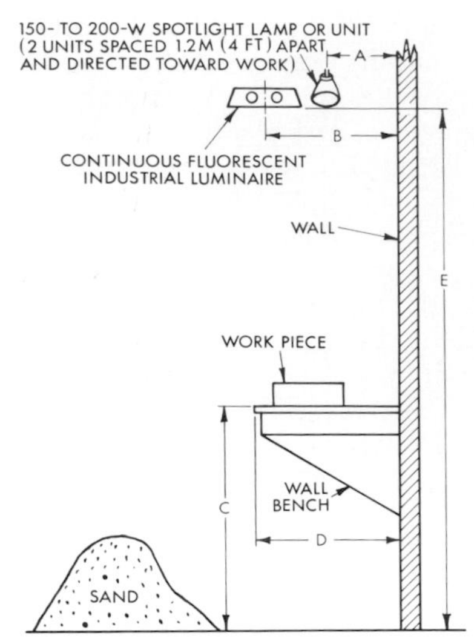

Fig. 9-30. A lighting layout for core or bench molding (wall area). A = 300 (12), B = 560 (22), C = 910 (36), D = 610 (24) and E = 2130 millimeters (84 inches).

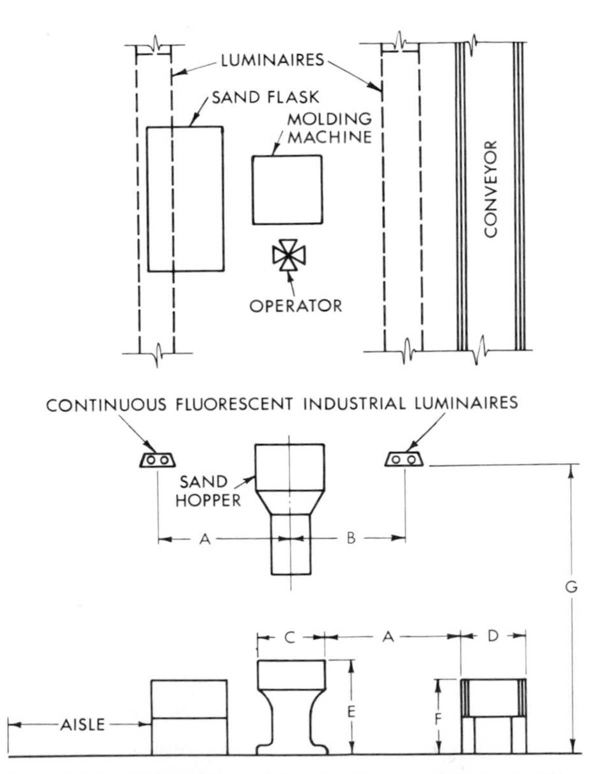

Fig. 9-29. Lighting layout for molding machines used for small castings. A = 1070 (42), B = 910 (36), C = 560 (22), D = 510 (20), E = 760 (30), F = 610 (24) and G = 2440 millimeters (96 inches).

Molding. The visual tasks in forming molds from treated sand are: (1) inspecting the pattern for foreign material, (2) setting the pattern in the flask and packing sand around it, (3) removing the pattern and inspecting the mold for loose sand and for accuracy of mold contour, (4) inserting core supports and cores (operator must be able to see the core supports) and (5) smoothing mold surfaces, checking core position and checking clearance between parts. The critical seeing tasks are: (1) inspecting the mold and (2) placing the cores (and chaplet supports, if employed).

The size and detail of the seeing tasks may vary. The smallest task has a visual angle of about 10 minutes—corresponding to the size of separate grains of sand. A defect involving the misplacement of only 5 or 6 grains of sand will cause imperfections in small castings. The more exacting seeing tasks are repetitive and of interrupted short-time duration.

Lighting should be designed for the intermittent, critical seeing of materials which have low reflectances and unfavorable contrasts. The varying depths of mold cavities demand adequate vertical illumination which does not produce harsh shadows. Fig. 9-29 illustrates an example of good lighting practice for molding-machine areas where small castings of 460-millimeter (18-inch) maximum dimension are made.

Deep pit molds require additional consideration in planning proper lighting. The walls of the pit may block some of the light from the general lighting system and result in shadows and lower luminance, especially on the vertical surfaces of the molds. The pit areas will benefit by the installation of additional general lighting luminaires, located to avoid conflict with materials-handling equipment.

Supplementary lighting is sometimes recommended for locations where sand is supplied from overhead ducts and conveyors (Fig. 9-30); however, it is usually preferable to install a general lighting system which satisfies these requirements.

Charging Floor. The weighing and handling of metal for charging furnaces is a simple, non-exacting task; thus, safety of the worker is more important than accuracy of seeing.

Pouring. The most crucial seeing task is that of directing the molten metal into the pouring basin. If the flow is not directed accurately, splashing may occur and cause injury to the workers. It can also displace sand in the mold and become a reason for a rejected casting.

Proper general illumination contributes to safety. The eyes of the workers often become adapted to the bright, molten metal contrasted with dark surroundings. This adaptation may cause difficulty in seeing any obstructions on a poorly illuminated, dark-color floor. Adequate lighting reveals such obstructions.

To improve visibility within the mold, contrast is sometimes increased by placing white "parting sand" around the opening. When weights are used, the opening in the weight indicates the general location of the pouring basin.

Shake-Out. Castings are removed from the molds and freed of sand in the shake-out area. The most critical seeing task occurs during removal of the gates and risers. Where a ventilation hood is used over the shake-out grate, the latter should be illuminated with a supplementary lighting system.

Heat-Treating. In malleable iron foundries, small castings are prepared in annealing ovens. The primary lighting requirement is for safety and for the packing of castings (to prevent warping during annealing).

Sandblasting or Cleaning. Three methods are used for cleaning castings: (1) sandblasting in a blast room, (2) sandblasting in a cabinet or on a rotary table and (3) eroding by friction in a tumbling barrel. The principal tasks are: (1) handling castings, (2) directing (manually) the sandblast stream and (3) inspecting the castings to see that they are clean.

Where workers wear goggles or helmets, additional lighting should be provided to compensate for the reduction in visibility.

The use of blast cleaning requires that the room general lighting be supplemented by luminaires which are mounted outside the cabinet but which project their light inside through tempered-glass windows (see Fig. 9-31).

Grinding and Chipping. Excess metal is usually removed from castings by: (1) breaking off the greater part of fins, sprues, risers and gates with a hammer; (2) chipping remaining projections with a hand or power chisel; and (3) grinding to a finish. Cleaning by a tumbling operation may have removed all or most of the excess metal from small castings.

Grinding operators remove excess metal and fins from castings by grinding to a contour, to a mark or to a gauge. Protective glasses worn by the operators often become fogged and the seeing task becomes fairly severe. For stationary grinders, both general and supplementary lighting systems should be used. It is good practice to locate the center line of the supplementary luminaires approximately 150 millimeters (six

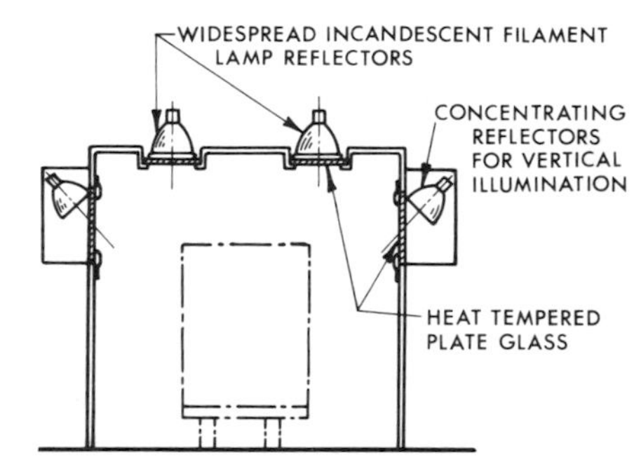

Fig. 9-31. Supplementary lighting for a sand blast house.

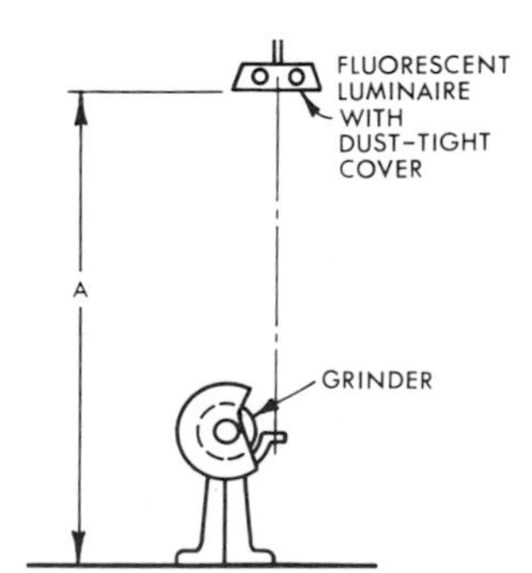

Fig. 9-32. Lighting for a stand grinder. Luminaire mounting height A is 2.1 to 2.4 meters (7 to 8 feet).

inches) from the edge of the wheel on the side toward the operator. See Fig. 9-32.

Inspection. Quality control is largely a function of visibility. A casting meets the specified tolerances when: (1) patterns are carefully checked against the drawings; (2) flasks are inspected for fit; (3) cores and molds are inspected for size, accuracy, and alignment; (4) core clearances are gauged prior to mold closing; (5) castings are checked against templates and gauges; and (6) surfaces are inspected and defective castings are culled.

Inspections are generally conducted at intermediate stages during the manufacture of the product. The inspections at some stages are either combined with the functional operation or are performed in the same area. The type of inspection will dictate the proper quality and quantity of illumination.

A typical inspection is that of cores by the core-maker prior to baking. Later, the castings may be inspected and scrapped by the shake-out handlers or by the grinder operators—avoiding subsequent wasted labor on defective parts. Small castings are frequently inspected and sorted simultaneously.

In sorting areas, a simple, general lighting system of ventilated fluorescent industrial luminaires may be mounted 1.2 meters (4 feet) or more above the sorting table or conveyor. Atmospheric and maintenance conditions will determine the type of luminaires (open, enclosed or filtered) to be used.

For "medium" inspections, fluorescent luminaires may reduce reflected glare and improve diffusion of light.

"Medium fine" and "fine" inspection sometimes require special lighting equipment. The use of a criss-cross grid of fluorescent luminaires is one method which will provide optimum diffusion, eliminate shadows and augment the illuminance level. This arrangement is also useful for putting light into cavities. The mounting height should be 1.5 meters (5 feet) or more above the inspection level.

Seeing into deep cavities and tubular areas may require the use of small, shielded, portable luminaires.

FRUIT AND VEGETABLE PACKAGING

Lighting design in this industry should be based on full knowledge and consideration of the tasks to be performed in the specific packaging plant, for each food has special characteristics of origin, composition, size and color.

There is a substantial difference between inspecting bunches of grapes for fancy-pack shipping and inspecting peaches that are to be peeled by an automatic peeling machine. Therefore, the general lighting principles outlined below need to be adapted to meet varied conditions for each specific application. Many of the lighting situations encountered in the food processing industry are common to other industrial problems discussed in this section, but lighting here can contribute to sanitation as well as safety, production, plant appearance and morale. The general flow diagram of a typical food processing operation is shown in Fig. 9-33.

Although most seeing tasks may be illuminated by a general lighting system, sorting and canning belts and other critical tasks should be illuminated by supplementary or local lighting designed specifically for each individual task.

Fig. 2-2 gives illuminance recommendations for canning and preserving. The recommendations for supplementary lighting are based on lighting studies for specific kinds of canning. The supplementary lighting recommendations given for apricots and peaches may be used as a guide for other light-colored products. Those for tomatoes are suggestive for medium-to-dark products; those for olives represent values for most dark-colored products.

Receiving

Raw produce is received primarily by truck in field boxes on large bins. These are handled by fork lifts or gondolas which dump the produce or unload it by floating it out with water. Techniques vary according to product. Most receiving activities are outdoors and normally are performed during daylight hours; however, when the produce ripens there is a peak period that requires lighting for night operation.

The combination of heavy truck traffic, storage facilities and conveyor machinery requires three types of lighting systems:

1. *General floodlighting* systems should be provided for area lighting. Mounting height should be as high as practicable (preferably 9 meters (30 feet) minimum) to reduce distracting glare that may prevent a driver from seeing pedestrian traffic.

2. *Supplementary lighting* should be provided

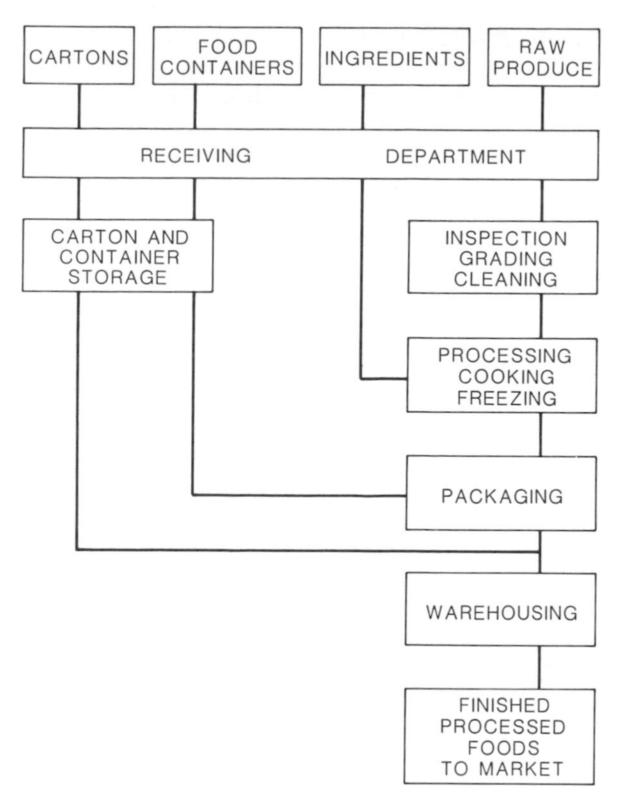

Fig. 9-33. General flow diagram of a typical food-processing operation.

at bin dumping or box unloading areas. Unloading boxes and bins (often with overhead cranes) presents an additional hazard to operating personnel. This lighting should be adequately shielded. Each luminaire should be limited to a maximum luminous intensity of 5000 candelas to abate distracting glare.

3. *Local lighting* may be used under platforms and conveyors that cause shadows. This will vary with each installation according to product requirements and plant configuration.

Grading and Sorting

Generally the first processes encountered are preliminary inspection or grading, depending on whether it is a shipping or processing operation. The procedure is usually the same, with the produce passing on a continuously-moving belt in front of the viewer. This method is almost universally used from initial inspection through the more critical grading procedures. The number of stations will vary, as will the degree of inspection—simple for crushed juice operations and more involved for fancy pack, whole or sliced applications.

Fresh-pack shipping is usually a single operation in that there is no cutting, peeling or similar operations. The inspectors look for foreign material, sunburn, rot, insects, color and general appearance. In the case of some product, such as dark-colored bunches of grapes, because the dark colors lack contrast and because of the unique shape of the bunches, light is required to penetrate into the cavities between the grapes. This lighting can result in reflected glare from either the belt or the fruit. Therefore, luminaires should be selected with care.

An important factor in shipping fresh produce is the ripeness of the product. To insure proper color and firmness at the point of delivery, produce is often picked at various stages of ripeness. Usually, cool-white fluorescent lamps will produce satisfactory color rendering. However, different colors of fluorescent lamps in three- or four-lamp luminaires, plus supplementary incandescent luminaires, add further color rendering capabilities.

General sorting and grading for processing require various stages of inspection, often during the transferring of produce from one belt to another for supplemental sorting and grading. This transfer can cause special problems such as additional reflections. Sorting belts and trays vary greatly from plant to plant and from product to product.

The speed of the conveyer belt, the size and color of produce, and the number of graders have a significant effect on the required lighting level. Considering the difference between small, dark olives and large, light colored peaches, for the grader to observe the same number of items on a 600-millimeter (24-inch) wide belt, either the belt must move slower or the illuminance must be increased to a level that permits maximum visibility of detail. Any increase above the recommended values should be determined by experimentation with the speed of the belt and the wet or dry condition of the product.

The lamps in all luminaires mounted above food products should be enclosed in a non-breakable material to prevent glass from falling into the product.

Preparation for Canning or Packaging

The preparation departments contain a variety of tasks: peeling, cutting, chopping, pitting, etc. These operations are manual, automatic or a combination of the two. The extremes may run from cauliflower coring (operators place each head of cauliflower on whirling cutter blades) to pitting machines (which automatically remove the pits from olives). This range of tasks requires an equal range of lighting applications. For cutting operations, the lighting level must be high enough to permit good visibility during rapid movement together with low reflected glare. See Fig. 2-2. Where cutting or pitting is done by machine, a lower illuminance with good shielding will be adequate. But when inspection is required at these machines (or immediately following either operation), the illuminance should be increased to the recommended level for inspection and grading. This point in the processing line—just prior to the filling of containers—probably has the widest range of seeing tasks in the food processing plant.

Container Filling

No matter what the container may be, basic lighting recommendations are similar. Good general lighting plus supplementary lighting (1) provide lighting to protect personnel from rapidly moving machinery, and (2) allow inspection of filled containers with occasional hand topping-off or additional filling. Here, higher levels may be desirable to allow adequate seeing among the complex filling mechanisms.

While this operation requires very little visual observation during normal operation, during periods of adjustment or maintenance this same higher level is desirable to obtain maximum efficiency. It would be very awkward and time consuming to use portable lighting for this purpose.

Very few plants devote any time or personnel for empty container inspection because of the automation and volume. Containers are seldom re-used. It is assumed that cans coming directly from the manufacturer are sterile and uncontaminated either by human hands or foreign material. When empty container inspection is required, a system of overhead luminaires and mirrors should be employed to reveal all sides of the container and to detect such defects as chips, cracks and air bubbles in glass containers or to detect open seams, cracks, solder pellets or damage to cans. Inspection booths may be necessary to eliminate specular reflections from surrounding light sources.

Carton Lighting

In general, no special lighting is required to inspect empty cartons. They are received flat, in bulk, and are shaped by machines prior to filling. Plastic bag containers come in rolls and are expanded prior to filling. Broken or torn bags can be detected by normal inspection lighting as described under Supplementary Lighting (see page 9-7).

Frozen Food Processing, Packing and Storage

This branch of the industry has all the problems of inspection and sorting plus the additional problems of temperature, air movement and moisture.

Lamp temperature and lumen output should be considered. In addition, luminaires should be selected to afford maximum protection from moisture and corrosion. Fluorescent lamps operated at cold temperatures suffer from low lumen output as discussed in Section 8 of the 1981 Reference Volume. Enclosed luminaires or jacketed lamps (and low-temperature rated ballasts) can be utilized to solve both the problems of temperature and sanitation. High intensity discharge lamps, generally not affected by temperature, should be considered for cold-room lighting. Factors such as luminance, distribution and effects of voltage interruptions are important.

GRAPHIC ARTS—PRINTING

The graphic arts industry is one of the oldest industries. Its history includes that period when daylighting was a major source of interior illumination.

Experience in the industry has shown, however, that a modern, well designed lighting installation, excluding daylighting, has a beneficial effect upon quantity and consistent quality of work. See Fig. 2-2 for illuminance recommendations.

Receiving Area

The most difficult tasks are those of reading markings on shipments, labels and bills of lading.

General illumination will provide sufficient light for these tasks and for the operation of hand operated and automatic fork lift trucks, as well as for general traffic in the area.

Supplementary illumination may be necessary for the interior of transport carriers bringing material to the plant. Angle or projector type luminaires may be utilized, but care must be taken to avoid glare from these sources. If the conveyances are deep, reel-type or other suitable portable equipment may be necessary.

Yard or loading dock lighting should be installed for night operation.

Stock Room Area

Identification marks on the sides of large bulky materials, rolls of paper, and crates or boxes require vertical illumination.

Additional illumination should be provided over the aisles where high piles of stock interfere with general lighting.

Local building code requirements should be checked as to permissible luminaires for lighting areas where volatile materials are stored.

Copy Preparation Area

Functions preliminary to printing take place here (paste up, layout, art, designing, etc.) as do decisions relative to point size and style of type, size and placements of cuts (pictures, drawings and charts) and size of page. Colors are specified. (Color work requires special lighting techniques. See Color Appraisal Area below.)

Well-diffused, glare-free illumination is essential.

Composing Room Area

In printing plants where type composition is by machines such as Ludlow, Monotype and Linotype, the operator uses a keyboard similar to that of a typewriter. General illumination from well shielded luminaires with good upward component is recommended.

Compositors performing hand tasks incur special lighting problems. In inspecting type, the visibility varies with the number of times the type is run (impressions made), its reflectance, and the contrast between its face and shoulder, and its spacing and size. Lighting should be provided by large-area, high-luminance sources. See Fig. 9-34. The high luminance, reflected from the shoulder, gives the necessary contrast between the mirror-like shoulder and the type face. Research has indicated that the downward luminance of the large-area source should be in excess of 10,000 candelas per square meter [3000 footlamberts]. Care must be taken to select and locate the luminaires to provide a well diffused glare-free system throughout the composing room.

Proofreading is also done in this area. Well shielded diffuse illumination is recommended.

Color Appraisal Area

In 1959 a joint report was prepared by the Illuminating Engineering Society and the Research and Engineering Council of the Graphic Arts Industry[15] to define a standard light source for the color appraisal of reflection type material. In 1972 a more comprehensive report, incorporating the CIE color rendering system, was issued.[16] It suggests a light source (luminaire) with a correlated color temperature of 5000 K for appraisal of color quality and 7500 K for appraisal of color uniformity. Both sources should have a color rendering index of 90 or higher.

Fig. 9-34. Composing room lighted for high visibility of type. It was determined by visibility tests that optimum conditions would result from well-shielded, large-area fluorescent luminaires with fairly high downward luminance and good upward component (40 per cent in this case).

Plate Preparation Area

The visual task is severe and prolonged.

The galley (camera) area needs only enough general illumination for traffic. Light source glare must be shielded from the camera lens. Packaged lighting units for photography are usually furnished by the camera supplier.

Stripping and opaquing are done on a luminous-top table. The table design should provide good, low-luminance diffusion to assure visual comfort. Any part of the luminous area not covered by the negative should be masked.

General overhead, low-level lighting for traffic in this area should be so located to eliminate table-top reflections.

Plate making requires low levels of illumination. Higher levels are injurious to plates processed over a considerable length of time. Colored lamps are frequently utilized.

Pressroom Area

The pressroom is usually a large, high ceiling area (necessitated by the dimensions of the equipment which reduces the utilization of light).

The tasks can be divided into three groups:

1. Tasks with type for make-ready, register and correction of errors in both the proof press and the production press. Included are the movement of semi-finished products from one press to another, the movement of finished sheets from presses to other departments, and the movement of raw materials to the presses.
2. Mechanical functions such as adjustment of the presses, installing frames, and cylinders on the presses, adjustment of ink fountains for the inking rollers, and the feeding of the paper.
3. The inspection of semi-finished and finished products. (If this involves color appraisal, see above Color Appraisal Area.)

General illumination is recommended, using a luminaire with good shielding and with a minimum of 10 per cent upward component.

Ink and drying compounds in the atmosphere create a difficult maintenance condition; therefore, ease of maintenance is an important item in luminaire selection. Supplementary lighting should be used wherever required and this can be determined only by careful examination of the equipment. Low-mounted luminaires, tilted on an angle to penetrate recesses in the presses, may be necessary. Pressmen need a large-area high-luminance light source (as recommended above for inspecting type).

Bindery Area

After printing is completed, if binding or any other bindery function is required, the finished product of the printing plant becomes the raw product of the bindery.

Practically all production involves hand labor: collating, folding, stapling, stitching, gluing, backing and trimming pages. In many operations, critical seeing is not important to many functions, but speed is. Diffuse general lighting should be provided by well-shielded luminaires with an upward component.

In book binding, there are frequent additional operations which are more tedious and exacting, such as: corner rounding, indexing, cover imprinting and applying gold leaf or gold ink. Additional luminaires (or closer spacing) are necessary to provide a higher level.

Shipping Area

The tasks are similar to those in the receiving area but include wrapping, packaging, labeling, typing and weighing.

It is a more critical area than receiving because the finished product is being sent to the customer. It is important that the product is not damaged, that it is sent to the proper address, and that it is shipped promptly.

LOGGING AND SAWMILL INDUSTRIES[17,18]

The conversion of living trees into lumber involves two basic industries: the manufacturing of logs (logging) and the manufacture of lumber (sawmilling).

Logging, traditionally an activity during daylight hours, presents the worker with basically simple seeing tasks—the handling of trees and logs with adequate time to see necessary detail. Some functions of the logging industry are accomplished under electric lighting.

Sawmilling has become a high-speed production process almost entirely under electric lighting. There is a wide range of visual tasks, from assessing the lumber content of logs on the mill log deck to the discrimination of fine detail and poor contrast as the log is broken down into finished lumber.

Since the seeing tasks in both logging and sawmilling involve viewing logs and lumber from the side as well as from above, a relatively high vertical illumination component is required. In general, a 2 to 1 ratio of horizontal-to-vertical illumination is recommended.

For illuminance recommendations see Fig. 2-2.

Logging

Yarding. This is the moving of logs from the logging area to the logging road. Logs yarded under electric lighting will have been felled, limbed and bucked during daylight hours. The general illumination must be adequate to enable the logger to move about safely over rough terrain covered with a confusion of logs and limbs. This general illumination will normally be provided by floodlights mounted either on poles or masts attached to movable platforms or trunks, or on the boom of the yarding machine.

Mounting heights in excess of 15 meters (50 feet) are difficult to obtain in the usual logging operations; thus, the effective area of yarding is limited to about 180 meters (600 feet) from the yarding machine or spar.

The general illumination must be supplemented by light from portable spotlights attached to the loggers' hard hats and by headlights and spotlights mounted on trucks, tractors and specialized yarding and loading equipment.

Loading. Loading requires adequate general illumination to enable the operator of the loading machine to hoist the logs deposited by the yarding operation from the "landing" to the waiting trucks or railroad flat cars. The illumination should also be adequate for the safe movement of men throughout the area and to enable the "head loader" to direct the operation efficiently. General lighting in the loading area is usually provided by floodlights mounted on the boom of the loading machine and/or on special poles or masts mounted on movable platforms or trucks.

Supplementary illumination, provided by portable spotlights attached to the loggers' hard hats, is required for strapping or chaining the logs securely to the trucks and stamping or painting logs and for general utility operations in the loading area.

Unloading. Lighting requirements for unloading are similar to those for loading. As in loading, the floodlight mounting height governs the lighting effectiveness. Both sides of the loaded truck or flatcar should be illuminated to facilitate the release of cable or chain clamps.

Supplementary lighting mounted on the workers' hard hats will not normally be required.

Dry-Land Sorting. The sorting of logs as to species, size and grade on land rather than in water is increasing in popularity in the industry. The logs are usually sorted on a conveyor system which allows the operators to cut the logs to suitable lengths and to route them to stock piles or "bins," from where they are taken (usually by loaders) to either a log dump on the water or to a stock pile on land.

Floodlighting must provide adequate visibility for safe movement of workmen using heavy loading equipment, and for releasing tie-down cable clamps. Seeing is most critical at the log handling conveyor system where the sorting is carried out at a relatively high speed. The sorting machine operator must be able to quickly recognize the species of log and any major defects.

Water Sorting and Booming. The water sorting and booming operation starts at the log dump area where logs are unloaded or dumped from trunks or flat cars. Often a large crane is used to off-load logs, sometimes as a large "bundle." Adequate illumination assists the crane operator and enables workmen to safely release tie-down cables. Once the logs are in the water, they are moved according to species and size by boom operators or boom men to specified storage areas and assembled into booms or rafts.

Lighting in the log dump, sorting and booming areas is usually provided by floodlights mounted on poles or towers at the shore line or on dolphins in the water.

Following transport to the sawmill boom storage area, booms are dismantled and the logs moved to a sorting and grading area prior to moving up the log-haul or jackladder. The seeing tasks in this operation are generally more critical than during boom assembly. Boom men and the boom boat operators must be able to quickly and safely recognize log species and spot cables and chains. Floodlights mounted on poles or dolphins should be located to minimize direct glare and shadows, and should provide relatively high lighting levels in key areas where logs are graded and scaled or where cables are released from log bundles.

Log Haul, Side Lift, Log Deck. Seeing tasks involve recognition of large objects such as cables, pieces of steel or rocks (which could damage saws or equipment) as the logs move to the barkers and the head-saw. The movement of the logs is accelerated on the log deck, requiring a higher illuminance for safety and productivity.

Debarkers. Bark is removed by rotating knives in mechanical debarkers and by high-pressure water jets in hydraulic debarkers. The debarker operator must clearly see all surfaces of the logs to determine when the bark is being removed effectively. Lighting will normally be provided by luminaires mounted on the mill frame or on poles if the barkers are exposed to the weather.

Sawmilling

Head Saw. Converting logs into lumber commences at this point. The head sawyer must see the complete log clearly and quickly while making decisions on the cuts to make in order to obtain the best use of the log. Illumination in both the horizontal and vertical planes must be adequate for quick recognition of major defects in the log. Seeing requirements for using either band or circular saw head rigs were found to be more critical than for mills processing smaller logs using a gang saw head rig.

General Sawmill Lumber Processing Areas. Once the log has passed through the head saw, the large slabs or cants are progressively passed through the edger, re-saws, trim saws, rough grading station and the green chain. The seeing tasks are similar and involve the machine operators' high-speed recognition of lumber characteristics and defects. A relatively high luminance in both horizontal and vertical planes is required—particularly on the in-feed side of the machines.

Wastewood Collecting and Conveyors (Basement). Seeing tasks in these areas are generally less demanding and involve clearing blockages in conveyors and routine maintenance. Conveyor in-feeds to chippers require a higher illuminance to ensure safe working conditions and to enable the operators to feed material into the chipper.

Planer Mill. The planing or sizing and surfacing operation is usually located in a separate building, particularly if the rough-sawn lumber is dried before planing. The planing machine is often enclosed in a soundproof room because of the high-frequency noise produced by the planer heads. The lighting system should enable the operator to move safely around the planer, to spot imperfections in the planed lumber caused by nicked planer blades, and to do routine maintenance work during shutdown. A portable trouble light is often required for setting up the planer heads or repairing the machine.

Lumber Grading. Lumber is graded or sorted in the sawmill and the planer mill to meet quality standards established by the industry and grading agencies or to meet customer specifications. Usually, lumber moves past the grading station on a conveyor system at a rate of up to 30 pieces per minute. Graders must recognize a great number of defects in the lumber which, in turn, enable them to determine the correct class or grade of lumber for each board. The grader must be able to spot defects at distances of up to 7.3 meters (24 feet) and must examine all four surfaces of each board.

Grading of rough sawn-lumber requires a relatively high illuminance, primarily on the horizontal plane. The higher illuminance will gener-

Fig. 9-35. Grading planed lumber under a combination fluorescent and incandescent filament lighting system. The 150-watt R-40 floodlamps are mounted in a shielded trough, directing light on the lumber at an angle of about 20 degrees to accentuate surface defects.

ally be provided by supplementary lighting and should illuminate a section of the conveyor system of 3 meters (10 feet) minimum width for the full depth of the conveyor. Improved color mercury, fluorescent or metal halide lamps in well-shielded industrial luminaires should be used for this system.

The grading of planed lumber requires a relatively high illuminance with a strong unidirectional component aimed at an angle of from 20 to 45 degrees below horizontal (usually provided by a row of incandescent reflector or projector lamps) to enable the grader to quickly recognize surface defects in the lumber. See Fig. 9-35.

Filing Room. The primary functions in the filing room are the grinding or filing of the teeth of both circular and band saws, the grinding of planer or sticker knives, the swedging of saw teeth, and the levelling or trueing of saws. The seeing task in sharpening saws or the planer knives is to determine that the sawtooth or the planer knife has been ground to a sharp and accurately shaped cutting edge. The filer recognizes proper grinding by observing the reflection of the light source on the ground surface of the saw teeth. Low brightness luminaires (such as fluorescent industrial units) are recommended to minimize reflected glare.

MACHINING METAL PARTS[19]

Machining of metal parts consists of: the setting up and operation of machines such as lathes, grinders (internal, external and surface), millers (universal and vertical), shapers, drill presses; bench work; and inspection of metal surfaces. The precision of such machine operation usually depend upon the accuracy of the setup and the careful use of the graduated feed indicating dials rather than the observation of the cutting tool or its path. The work is usually checked by portable measuring instruments, and only in rare cases is a precision cut made to a scribed line. The fundamental seeing problem is the discrimination of detail on plane or curved metallic surfaces.

Visibility of Specific Seeing Tasks*

Convex Surfaces. The discrimination of detail on a convex surface, such as reading a convex scale on a micrometer caliper, is a typical seeing task. The reflected image of a large-area low-luminance source on the scale provides excellent contrast between the dark figures and divisions and the bright background without producing reflected glare. The use of a nearly-point source for such applications results in a narrow, brilliant (glaring) band that obscures the remainder of the scale because of the harsh specular reflection and loss of contrast between the figures or divisions and the background. See Fig. 9-36.

* See also Fig. 9-10.

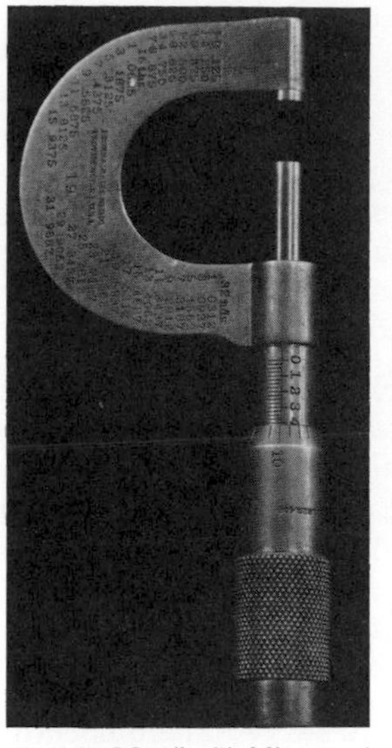

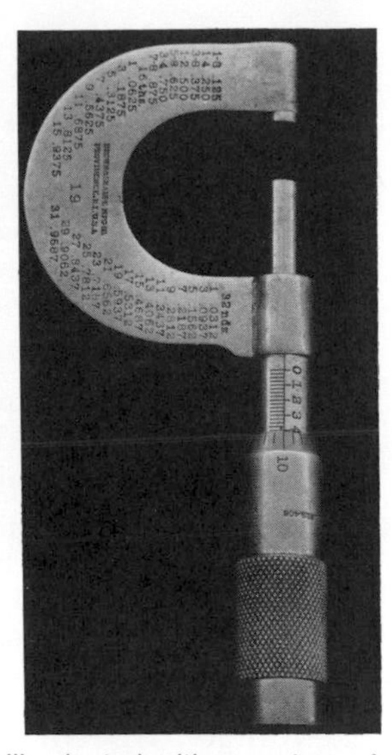

Fig. 9-36. (Left) Micrometer illuminated with a system of small, bright sources is seen with bright streak reflections against a dark background. (Right) When illuminated with a large area, low luminance source, the micrometer graduations are seen in excellent contrast against a luminous background.

Flat Surfaces. In viewing a flat surface, such as a flat scale, the seeing task is similar to that in reading a convex scale. With a flat scale, however, it is possible, depending on the size, location and shape of the source, to reflect the image of the source either on the entire scale, or only on a small part of it. If the reflected image of the source is restricted to too small a part of the scale, the reflection is likely to be glaring.

Scribed Marks. The visibility of scribed marks depends upon the characteristics of the surface, the orientation of the scribed mark and the nature of the light source. Directional light produces good visibility of scribed marks on untreated cold rolled steel if the marks are oriented for maximum visibility, such that the brightness of the source is reflected from the side of the scribed mark to the observer's eye. Unfortunately, this technique reduces the visibility of other scribed marks. Better average results are obtained with a large-area low-luminance source. If the surface to be scribed is treated with a low reflectance dye, the process of scribing will remove the dye and expose the surface of the metal. Such scribing appears bright against a dark background. The same technique is appropriate for lighting specular or diffuse aluminum. In this case, the scribed marks will appear dark against a bright background.

Center-Punch Marks. A visual task quite similar to scribing is that of seeing center-punch marks. Maximum visibility is obtained when the side of the punch opposite the observer reflects the brightness of a light source. A directional source located between the observer and the task provides excellent results when the light is at an angle of about 45 degrees with the horizontal.

Concave Specular Surfaces. The inspection of concave specular surfaces is difficult because of reflections from surrounding light sources. Large-area, low-luminance sources provide the best visibility.

Lighting for Specific Visual Tasks

In the machining of small metal parts, a low-luminance source of 1700 candelas per square meter (500 footlamberts) has been found to be desirable. The size of the source required depends on the shape of the machined surface and the area from which it is desired to reflect the brightness. The techniques applicable to specular reflections can also be applied to semi-specular surfaces.

Flat Specular Surfaces. The geometry for determining luminous source size is illustrated in

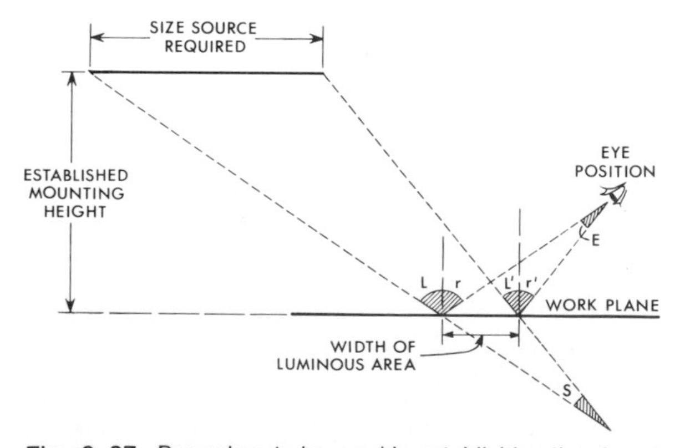

Fig. 9-37. Procedure to be used in establishing the size of source necessary to obtain the desired area of reflected brightness when applied to a flat specular surface.

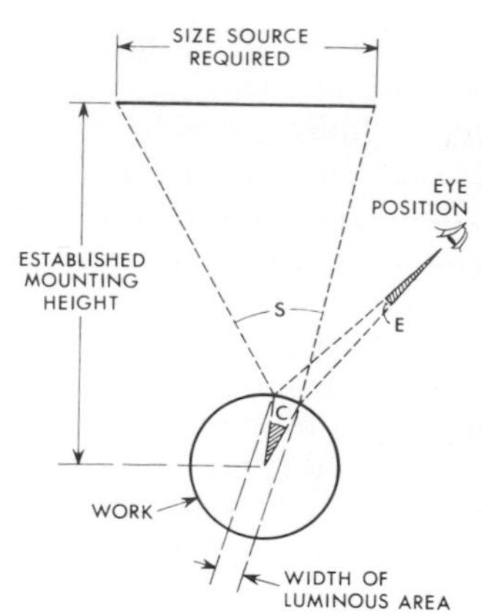

Fig. 9-38. Procedure to be used in establishing the size of source necessary to obtain the desired area of reflected brightness when applied to a convex specular surface:

$$S = 2C + E$$

where S = angle subtended by source on work, C = angle subtended by luminous area on work, and E = angle subtended by luminous area at eye.

Fig. 9-37. First, draw lines from the extremities of the surface that is to reflect the source, to the location of the observer's eye, forming angle E. At the intersections of these lines with the plane of the surface, erect normals to that plane, forming angles r and r'.

At these same intersections and on the other side of the normals, construct lines to form angles L and L' equal to angles r and r'.

Project these lines to the established luminaire location to define luminaire width; extend them in the opposite direction until they intersect, forming angle S equal to angle E.

Convex Specular Surfaces. Determine the appropriate width of the luminous area of the convex surface. See Fig. 9-38. Draw lines from the location of the observer's eye to the edges of surface's luminous area, forming angle E. Erect normals at intersections of lines with the surface.

At these intersections and on the other side of the normals, construct lines to form angles equal to those to the eye (procedure same as that for flat surfaces, above). Project lines (as for flat surfaces) to define luminaire width. The same general procedure can be applied to concave surfaces.

General Lighting

There is an evident advantage from the use of the large-area low-luminance sources for most of the visual tasks in the machining of metal parts. The ideal general lighting system is one having a large indirect component. While both fluorescent and high intensity discharge sources are used for general lighting, fluorescent luminaires, particularly in a grid pattern, are preferred if servicing is not difficult. High-reflectance room surfaces improve utilization of illumination and visual performance.

MEN'S CLOTHING[20]

Recommended illuminance ranges for the general tasks in manufacturing men's clothing are listed in Fig. 2-2. Although this section describes the tasks that occur in the manufacture of men's clothing, many are similar in the manufacture of women's and children's clothing. The same illuminance ranges are suggested for these similar tasks.

Definitions and Flow Chart

The descriptions that follow explain the process through which the cloth must pass to become a finished garment. These are diagrammed in the flow chart shown in Fig. 9-39. The key letters at the left of the chart are used parenthetically after each process name in the description.

Preparation of Cloth (A to H).

Receiving (A). Receiving shipment of cloth from mills.

Opening (B). Opening bolts of cloth.

Examining (C). Checking and marking of all material for imperfections in texture or color.

Sponging (D). Shrinking of cloth by applying moisture and drying. There are two types: steam sponging and cold water (London) sponging.

Decating (E). Setting the cloth by means of steaming. (This is not done to all materials.)

Winding (F). Rerolling the cloth.

Storing (G). Holding cloth until needed in cutting room.

Measuring (H). Measuring cloth. (An accurate record is sent to cutting room.)

Patterns (I). Making of patterns for different styles.

Cutting (J to Q).

Piling up (J). Placing the desired number of layers of cloth, one on top of the other.

Marking (K). Marking the patterns on the cloth with chalk.

Cutting (L). Operating the cutting machine.

Fitting (M). Bundling of similar pieces of cloth.

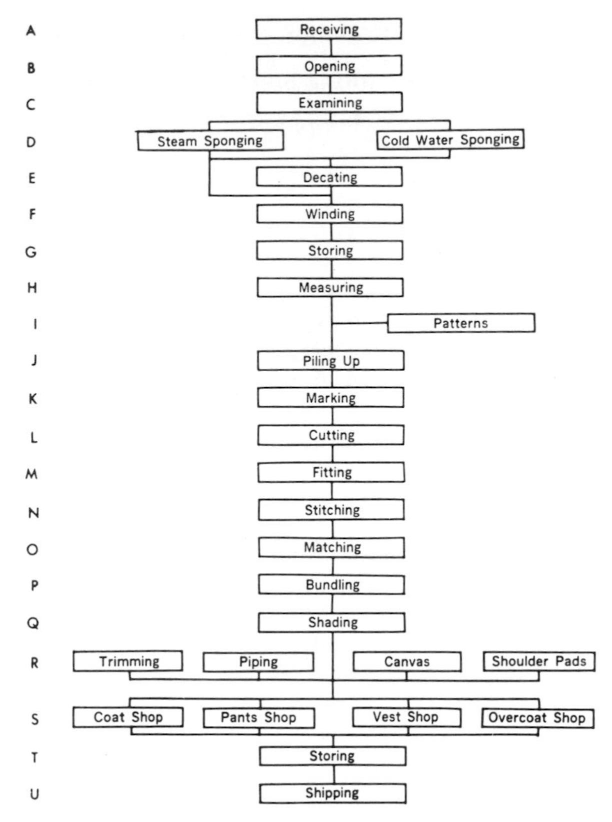

Fig. 9-39. Flow chart for the manufacturing of men's clothing.

Stitching (N). Marking cloth for pockets, etc.

Matching (O). Selecting pockets, etc., of same color and patterns.

Bundling (P). Tying up bundles and moving to trimming room.

Shading (Q). Marking of lot number on cloth.

Preparation (R). Trimming, piping, canvas, shoulder pads, etc., for different garments.

Sewing (S). Manufacturing of different garments. The operations can be broken down into the following:

1. Machine sewing.
2. Hand finishing and sewing (standard or sitting).
3. Hand pressing.
4. Jumper pressing.
5. Machine pressing.
6. Inspection.

There are over 100 operations in making a coat, all of which can be classified in the above categories.

Handling of Finished Garments (T and U).

Storing (T). Storing finished garments.

Shipping (U). Shipping finished garments.

Lighting Design for Specific Tasks

Preparation of Cloth (A to H), Examining. Of the eight steps in this processing division, only examining requires special lighting.

Historically, examining was done by daylight, preferably north sky light. However, with need for increased and continuous production, it has been necessary to use electric lighting for this important task. Daylight is not always dependable, and it is difficult to obtain the necessary illumination in many buildings.

The cloth is examined and is inspected for shading and defects. Lighting equipment should provide uniform illuminance.

Patterns (I). There is no special lighting required.

Cutting (J to Q). General lighting or localized general lighting should provide the base lighting for the different operations in this section. If more than one operation is performed in the same area, the higher illuminance required will govern the design. Supplementary lighting on the cutting machine will provide the necessary increase in illuminance levels over the general lighting. If the cutting is done by hand, the localized general lighting will have to be increased.

Preparation (R). There is no special lighting required.

Sewing (S). General lighting should provide the base lighting for the general operation in this section. Supplementary or localized general lighting will provide the required illuminance for the different tasks.

Handling of Finished Garments (T and U). There is no special lighting required.

MINE INDUSTRY

The mining industry offers many diversified environmental conditions for lighting. Underground mining, surface mining and processing facilities are unique areas that contain their own special lighting problems.*

Underground mines contain miles of infrequently used haulage and travel areas as well as high-activity restricted work areas. The area in which material is being extracted from its natural deposit is a highly mobile area, and it usually advances as the material is extracted and transported to the surface. Extraction is accomplished by drilling and blasting or by ripping the material with rotating steel drums laced with carbide bits. Extracted material is then loaded onto the haulage equipment, which is either rail, conveyor belt, water slurry, pneumatic or elevator, or by any combination, depending upon the mining conditions. The haulage equipment that bridges the distance between those permanent conveying installations and the advancing extraction work areas is usually diesel- or electric-powered vehicles or mobile conveyor systems.

Surface facilities include storage areas, loading points, maintenance shops, railroad yards, offices, bath houses and both elevated and surface walkways. The mined material is stored in silos, storage barns or yard areas. Transportation is usually via conveyor belts. Large buildings contain facilities to wash, size, refine, separate and process the mined material.

Surface mining is conducted in quarries, open pits and by other types of stripping operations. Some of the equipment, larger than a house, consumes more energy than a small town. The operation may cover many square kilometers (miles) of active workings; or it may be confined to the small area of a mountain ledge. Haulage is usually by large trucks with limited visibility or by overland conveyors.

Underground Mining

Illumination of underground mines involves many diverse areas that are covered in other portions of this handbook. Some of the work areas in underground mining are: maintenance shop, electrical substations, track switch points, track loading points, pumping stations, storage rooms, loading platforms and battery-charging stations.

Providing illumination for underground work areas requires particular thought to worker mobility. In maintenance shops or medical facilities (where personnel are present for extended periods of time), standard lighting practices are

* US federal government regulations covering illumination requirements in the mining industry are contained in the *Code of Federal Regulations, Title 30, Mineral Resources.*

appropriate. Areas such as track loading points (where personnel visit only a few times during a working shift) *must* be illuminated to a very low luminance level. Personnel are usually adapted to the luminance produced from their cap lamps when they approach a lighted work area (for safety it is critical that they experience minimum adaptation); hence, these areas should be illuminated to approximately the same level as provided by the individual cap lamps. In addition to maintaining an over-all low illuminance level, care must be used in the selection of the luminaire. The mounting height may be restricted in many areas underground; therefore, a diffused, low-level light source is desirable.

Coal Mine Face Illumination (Coal Extraction Areas). Coal mine face illumination is applied to the most difficult lighting environment in the world. There are five distinct contributing conditions:

1. Surface reflectance, usually less than five per cent, almost eliminates secondary reflections and indirect lighting.
2. Mounting-height restrictions and job tasks place the luminaires in the workers' direct line of sight.
3. Suspended dust and water vapor cause backscattering which obscures task surfaces.
4. Mounting positions restrict the size, location and light distribution of the luminaires.
5. Luminaires must meet the safety requirements of the US Department of Labor's Mine Safety and Health Administration for use in hazardous locations (methane, air-and-coal-dust atmospheres).

These problems require unique lighting systems for each size of entry and for each type of machine being used in that entry. With the luminaires mounted at or below eye level, discomfort and disability glare present real problems and are dominant design considerations in any lighting system.

Typical entry dimensions are 1.5 meter (5 feet) high, 6 meters (20 feet) long and 6 meters (20 feet) wide. Luminaire mounting positions are usually available only in the center 2.4 meters (8 feet) of the length (on board the mining equipment). Work tasks and safety hazards can be located anywhere and on any surface within the entry. The most difficult problems involve the low mounting heights and the varied positions of mining personnel in the entry.

One advantage in underground lighting is that most tasks (usually involving large tools or equipment) do not require the higher illuminances that are needed to see fine detail. Also, many hazards in the face area involve equipment motion that can easily be detected at low illuminance.

When a lighting system is designed, installed and has worker acceptance, the job may not be completed. The face is an area that can move 30 meters (100 feet) a day and change in height from 600 millimeters (24 inches) to 3.7 meters (12 feet). Roof conditions can force an entry to decrease the mining width from 9 meters (30 feet) to 4.3 meters (14 feet). Lighting design must be flexible enough to adapt to changing conditions and still provide minimum illuminances for safety.

Other Extraction Areas. The natural deposit of materials usually dictates the most economic method of mining. Salt, for example, is often found in large nodes or domes which may result in mining areas hundreds of meters (feet) high and hundreds of meters (feet) wide. High intensity discharge sources can be mounted in elevated positions to provide good general illumination. The high reflectance of salt helps to provide uniform lighting.

Other materials lie in horizontal, vertical or sloping veins. Each requires a unique approach to mining and illumination. In many instances, only hydraulic or pneumatic power is available to generate electricity to operate the lighting system in the extraction areas. These power sources limit the design choices of the lighting engineer. However, the same difficult problems of maintaining very low illuminances, of restricted mounting heights and locations, and of utilizing luminaires with low brightness are important in lighting design.

Surface Mining

Stripping Operations. Surface stripping involves equipment that operates 24 hours a day, six days a week. Illuminances must be sufficient so that tasks can be performed safely around the equipment and in the stripping area. If the work area is limited, high-wattage, mast-mounted luminaires can provide good general illumination. Examples of tasks in such an area are scraping top soil prior to stripping or spreading top soil during reclamation. The area being worked in either of these instances is limited, and the general illumination would help in equipment control and contribute to general safety—especially when used with large equipment such as scrapers or trucks, and/or with machines operated in close proximity to each other.

Equipment. Large excavating equipment requires a variety of luminaires for safe lighting. Interior work areas should use heavy-duty, shock-mounted luminaires for general walkway lighting. Tasks should be accented according to the safety conditions surrounding the accomplishment of those tasks. Exterior work areas enable the lighting designer to use high intensity discharge sources mounted on the equipment superstructure. (These luminaires should also include heavy shock mounting.) Large excavating equipment—either draglines, shovels or wheel excavators—is subject to high vibration during normal operation. Shock mounting will increase the life of the luminaires and lower the maintenance costs. This is particularly important on the booms of draglines and shovels where high impact loading on luminaires shortens lamp life.

Other equipment used in surface mining, such as front end loaders, bulldozers, scrapers, haulage trucks and drill rigs, is usually diesel powered. To supply high intensity discharge lighting for these types of equipment, special inverter ballasts or rotary converters develop the required voltages. This equipment, too, requires shock-mounted luminaires for adequate lamp life.

When lighting interior tasks in control rooms, general lighting may not be appropriate for gauges, meters and controls. In many instances, the operator has a dual task. One is to monitor the gauges and meters and to operate the controls. The other task is to monitor the equipment operation exterior to the control room. An example is the dual functions of a dragline operator who sits in a control room and controls the operation of the bucket 45 meters (150 feet) away from the mainframe of the machine. Controls and indicators are usually self-illuminated and all interior control room lighting is very subdued.

Surface Facilities

Interior. Surface facilities wash, crush, size, separate, mill and perform other processes on the mined material. These facilities are usually large, damp, dusty, noisy and subject to vibrations. Within the buildings are many support structures and large pieces of equipment that restrict light distribution. Luminaires should be adequate in number and in distribution to eliminate shadows that could cause safety problems. For ease of cleaning, moisture-proof and dust-proof luminaires will allow the use of water sprays during maintenance shifts.

Exterior. Exterior areas around surface facilities are used for equipment storage, mined material storage, train and truck loading and unloading, travelways and conveyor haulage systems. Many of these areas require lighting for safety hazards. Unloading areas, for example, contain open dumping pits, surge bins, crushers, belt feeders, conveyor belts and walkways. General lighting or specific task lighting can be used for illuminating these tasks and their safety hazards.

PETROLEUM, CHEMICAL AND PETROCHEMICAL PLANTS[21]

A petroleum, chemical or petrochemical plant converts raw material (gas, liquid or solid) into a usable product. At each plant there is some form of receiving and shipping, storage, removal of impurities and processing. It is common to find several different process units in a single plant. Frequently, several processes are combined in an integrated unit.

The modern plant is a highly-automated, continuous-process operation. Each unit is controlled from a local control house by one or two operators. A central control house may be used instead of unit control houses to operate several process units. It is apparent that there are very few people in a modern plant.

The seeing tasks in the processs units are reduced to very basic operations such as turning a valve, starting a pump, taking a sample, or just walking through a unit to sense some disorder. More critical seeing tasks require supplementary local illumination.

Most modern continuous process plants have preventive maintenance programs scheduled during daytime shifts. When unusual maintenance is required at night, portable illumination may be necessary.

Petroleum, chemical and petrochemical plants are restricted to the employees. Many areas are further restricted to personnel specially trained and assigned to the area. Most visual tasks are greatly reduced by sophisticated automatic control systems. Many areas require illumination only for the safe movement of personnel; many have only daytime occupancy and do not require lighting.

Most processes involve elevated temperatures and pressures and are designed for the continuous flow of vapor, liquid or solid from one vessel to another. Many of these materials are highly toxic and highly flammable. For these reasons,

Fig. 9-40. Illuminances Currently Recommended by the Petroleum, Chemical and Petrochemical Industry Representatives.[a]

Area or Activity	Illuminance Lux (Foot-candles)	Elevation Millimeter (Inches)
I. **Process areas**		
A. General process units		
Pump rows, valves, manifolds	50 (5)	Ground
Heat exchangers	30 (3)	Ground
Maintenance platforms	10 (1)	Floor
Operating platforms	50 (5)	Floor
Cooling towers (equipment areas)	50 (5)	Ground
Furnaces	30 (3)	Ground
Ladders and stairs (inactive)	10 (1)	Floor
Ladders and stairs (active)	50 (5)	Floor
Gage glasses	50 (5)[b]	Eye level
Instruments (on process units)	50 (5)[b]	Eye level
Compressor houses	200 (20)	Floor
Separators	50 (5)	Top of bay
General area	10 (1)	Ground
B. Control rooms and houses		
Ordinary control house	300 (30)	Floor
Instrument panel	300 (30)[b]	1700 (66)
Console	300 (30)[b]	760 (30)
Back of panel	100 (10)[b]	760 (30)
Central control house	500 (50)	Floor
Instrument panel	500 (50)[b]	1700 (66)
Console	500 (50)[b]	760 (30)
Back of panel	100 (10)[b]	900 (36)
C. Specialty process units		
Electrolytic cell room	50 (5)	Floor
Electric furnace	50 (5)	Floor
Conveyors	20 (2)	Surface
Conveyor transfer points	50 (5)	Surface
Kilns (operating area)	50 (5)	Floor
Extruders and mixers	200 (20)	Floor
II. **Nonprocess areas**		
A. Loading, unloading, and cooling water pump houses		
Pump area	50 (5)	Ground
General control area	150 (15)	Floor
Control panel	200 (20)[b]	1100 (45)
B. Boiler and air compressor plants		
Indoor equipment	200 (20)	Floor
Outdoor equipment	50 (5)	Ground
C. Tank fields (where lighting is required)		
Ladders and stairs	5 (0.5)	Floor
Gaging area	10 (1)	Ground
Manifold area	5 (0.5)	Floor
D. Loading racks		
General area	50 (5)	Floor
Tank car	100 (10)	Point
Tank trucks, loading point	100 (10)	Point
E. Tanker dock facilities[c]		
F. Electrical substations and switch yards[d]		
Outdoor switch yards	20 (2)	Ground
General substation (outdoor)	20 (2)	Ground
Substation operating aisles	150 (15)	Floor
General substation (indoor)	50 (5)	Floor
Switch racks	50 (5)[b]	1200 (48)
G. Plant road lighting (where lighting is required[d]		
Frequent use (trucking)	4 (0.4)	Ground
Infrequent use	2 (0.2)	Ground
H. Plant parking lots[d]	1 (0.1)	Ground
I. Aircraft obstruction lighting[e]		
III. **Buildings**[d]		
A. Offices (See Section 5)		
B. Laboratories		
Qualitative, quantitative and physical test	500 (50)	900 (36)
Research, experimental	500 (50)	900 (36)
Pilot plant, process and specialty	300 (30)	Floor
ASTM equipment knock test	300 (30)	Floor
Glassware, washrooms	300 (30)	900 (36)
Fume hoods	300 (30)	900 (36)
Stock rooms	150 (15)	Floor
C. Warehouses and stock rooms[d]		
Indoor bulk storage	50 (5)	Floor
Outdoor bulk storage	5 (0.5)	Ground
Large bin storage	50 (5)	760 (30)
Small bin storage	100 (10)[a]	760 (30)
Small parts storage	200 (20)[a]	760 (30)
Counter tops	300 (30)	1200 (48)
D. Repair shop[d]		
Large fabrication	200 (20)	Floor
Bench and machine work	500 (50)	760 (30)
Craneway, aisles	150 (15)	Floor
Small machine	300 (30)	760 (30)
Sheet metal	200 (20)	760 (30)
Electrical	200 (20)	760 (30)
Instrument	300 (30)	760 (30)
E. Change house[d]		
Locker room, shower	100 (10)	Floor
Lavatory	100 (10)	Floor
F. Clock house and entrance gatehouse[d]		
Card rack and clock area	100 (10)	Floor
Entrance gate, inspection	150 (15)	Floor
General	50 (5)	Floor
G. Cafeteria		
Eating	300 (30)	760 (30)
Serving area	300 (30)	900 (36)
Food preparation	300 (30)	900 (36)
General, halls, etc.	100 (10)	Floor
H. Garage and firehouse		
Storage and minor repairs	100 (10)	Floor
I. First aid room[d]	700 (70)	760 (30)

[a] These illumination values are not intended to be mandatory by enactment into law. They are a recommended practice to be considered in the design of new facilities. For minimum levels for safety, see Fig. 2-26. All illumination values are average maintained levels.
[b] Indicates vertical illumination.
[c] Refer to local Coast Guard, Port Authority, or governing body for required navigational lights.
[d] The use of many areas in petroleum and chemical plants is often different from what the designation may infer. Generally, the areas are small, occupancy low (restricted to plant personnel), occupancy infrequent and only by personnel trained to conduct themselves safely under unusual conditions. For these reasons, illuminances may be different from those recommended for other industries, commercial areas, educational areas or public areas.
[e] Refer to local FAA regulations for required navigational and obstruction lighting and marking.

most process streams are contained entirely within closed piping systems and vessels. Outdoor luminaires are appropriate to such equipment and construction.

Illuminance recommendations are listed in Fig. 9-40. They represent those established by petroleum, chemical and petrochemical industry representatives and in their opinion can be used in preference to employing Fig. 2-2; however, values can still be determined using Fig. 2-2 for similar tasks and activities.

Corrosive Areas. A variety of corrosive chemicals is generally present in each plant. The usual methods to protect against these are to use metals that resist attack, special surface preparation, epoxy finishes, polyvinyl chloride coatings or nonmetallic parts. In addition to these protections against the corrosive conditions, it is quite common to hose down an area. Further, outdoor plants are exposed to the elements of rain, snow, fog, high humidity and salt-laden sea air. Luminaires should be selected that are protected against the pertinent corrosive element.

Classified Areas. Some areas may be exposed to the release of flammable gases, vapors or dusts. The *National Electrical Code*[22] requires that these areas be classified and sets forth rules for the type of liminaire that may be installed. These liminaires must be approved for the Class, Group and Division in which they are to be used. Improper application of a lighting unit can result in fire and/or explosion.

Classification of these areas within a plant must be made prior to selection of equipment. A general classification is shown in Fig. 9-41.

Lighting designers should investigate the feasibility of floodlighting outdoor Hazardous Locations by locating non-explosion-proof floodlights beyond the hazard boundaries.

General Practice. Once the environmental conditions of hazardous location, corrosive vapors, and other ambient atmospheric conditions of moisture, temperature, etc., have been considered, lighting the task follows accepted industrial practice.

The outdoor process unit, storage areas, loading and unloading and other areas can be effectively illuminated by combinations of high-wattage floodlights and low-wattage local luminaires (the latter for shadowed areas). The use of exterior floodlighting includes preventing light pollution or spill light that will cause annoyance outside the facility.

The industry presently uses high intensity discharge lamps for process and other industrial areas. Fluorescent lamps are utilized in control rooms, switch rooms, shops and administration areas.

Luminaires within reach of personnel, or where exposed to breakage, should always be equipped with strong metal guards.

Outdoor Tower Platforms, Stairways, Ladders, Etc. Luminaires should provide uniform illumination and be shielded from direct view of persons using these facilities. Enclosed-and-gasketed or weather-proof luminaires equipped with refractors or clear, gasketed covers may be used for reading gauges. Luminaires above top platforms or ladder tops should be equipped with refractors or reflectors. Reflectors may be omitted on intermediate platforms around towers so that the sides of the towers will receive some illumination and the reflected light therefrom will mitigate sharp shadows.

Special Equipment. Special lighting equipment may be needed for such functions as illuminating the insides of filters or other equipment whose operation must be inspected through observation ports. If the equipment does not include built-in luminaires, concentrating-type reflector luminaires should be mounted at ports in the equipment housing.

Portable luminaires are utilized where manholes are provided for inside cleaning and maintenance of tanks and towers. Explosion-proof types (where hazardous conditions may exist) with 15-meter (50-foot) portable cables are connected at industrial receptacles (either explosion-proof or standard) located near tower manholes or at other locations.

Fig. 9-41. Area Classifications (Based on the 1978 *National Electrical Code*)

Flammable	Flammable Mixture	Classification	Basic Type of Fixed Luminaire†
Gas	Normally hazardous	Class I, Division 1*	Explosion-proof
	Occasionally hazardous	Class I, Division 2	Enclosed and gasketed
Dust	Normal	Class II, Division 1	Dust-ignition proof
	Occasionally hazardous	Class II, Group G Division 2 only	Enclosed and gasketed
Fibers or flyings		Class III	Enclosed and gasketed

* Group and temperature markings shown on the luminaire establish its classification.

† The terms *explosion proof, dust-ignition proof* and *enclosed and gasketed* are types of construction only. The Class, Group, Division and operating temperature must be known to select the appropriate luminaire.

For portable lighting units, there is only Class I or II; Division 1 listed construction is permitted.

RAILROAD YARDS[23]

The lighting of railroad yards, storage areas and platforms is essential to personnel safety, to expedite operations, and to reduce pilferage and damage to equipment. Illuminance recommendations for these functions are listed in Fig. 2-2.

Because light is absorbed by moisture, smoke and dust particles, the amount of absorption must be considered even in an apparently clean atmosphere, especially when luminaires are located at a distance from task and/or the task is viewed from a distance. For example, if the atmospheric transmittance were 80 per cent per 3-meter (10-foot) distance and the task were viewed from 30 meters, (100 feet), the illuminance on the task would have to be increased by a factor of 10 to obtain the same visibility as at 100 per cent atmospheric transmittance.

Seeing Tasks

Railroad yards can be divided into general areas which have different seeing tasks.

Retarded Classification Yard. The large and often highly-automated retarder classification yard, with its supporting yards and servicing facilities, presents different seeing tasks.

Receiving Yard. Seeing tasks throughout the area consist of walking between cars, bleeding air systems, opening journal box covers, observing air hoses, safety appliances, etc.

Hump area. Seeing tasks are diversified. The scale operator and hump conductor are usually required to check each car number. Illumination on the underneath surfaces of the cars and on the running gear is necessary for inspectors' ready and precise inspection of a car that is in motion. There also should be enough light at the tops of cars to permit judgment of height. Car uncouplers should be able to see the coupling mechanism. The hump conductor, car inspector and the car uncoupler should have specifically directed lighting of a higher level than that provided by the general lighting in other parts of the hump area.

Control Tower and Retarder Area. Modern retarder classification yards are computer controlled and equipped with various methods for determining car speed, "rollability," track occupancy, etc. These devices automatically set retarders to permit a car to roll from the hump to its proper position in the yard without action by the control tower operator. In other, less automated yards, it may be necessary that the operator check the extent of track occupancy, gauge the speed of the car coming from the hump and manually set the amount of retardation to be applied to the car. Even in the automated yard, the operator may also be required to do this manually in the event one or more of the automatic features fail. In many yards, the control tower operator is expected to check the car number against a switching list and to see that the car goes to the correct track. Accordingly, it is essential that the operator be able to quickly and accurately identify the moving car.

Under clear atmospheric conditions, it is important that there be no *direct* light projected toward the operator (this may cover a considerable angle). However, under adverse atmospheric conditions (dense fog, for example) it is general practice to utilize auxilary lighting equipment on the side of the tracks opposite the control tower to reveal the outlines of cars in silhouette. In this situation, the tower operator cannot check car numbers, but can observe and regulate the movement of the cars.

Head End of Classification Yards. The operator should be able to see that cars entering the classification yard actually clear switch points and clearance points so that following cars will not be impeded or perhaps wrecked.

Body of Classification Yard. Frequently, the operator must be able to see the body of the yard sufficiently to determine the extent of track occupancy. On some railroads, men are required to move along cars in the body of the classification yard to couple air hoses, close journal box covers, etc.

Pull-Out End of Classification Yard. In this area, switchmen are required to walk along tracks to determine switch positions and, if necessary, to operate them. Illumination should provide safe walking conditions along the switch tracks.

Departure or Forwarding Yard. Some railroads make up departing trains by having cars pulled from the classification yard into a departure yard. Here, minor repairs made to the cars avoid the delay of switching cars to a repair track. Air hoses may be coupled, journals lubricated and journal boxes closed, and any necessary tests or inspections effected. Lighting should be sufficient to permit work to progress with a minimum amount of auxiliary or portable lighting.

Hump and Car Rider Classification Yards. The seeing tasks here and around the hump, are considerably different from those in the retarder yard. Around the hump area, a yard clerk should be able to read car numbers. Cars must be uncoupled, and car riders must be able to see grab irons, ladders, etc., sufficiently to climb safely onto the cars. Switchmen along the lead track must have visibility adequate for walking safely along the lead track and to operating switches. Car riders on any cars rolling into the yard should be able to see any cars on the track ahead so that they can brake adequately to reduce impact and prevent damage to lading. The rider must then be able to see to get off the car and walk along yard tracks to the hump.

Flat Switching Yards. The only seeing requirements in most of these yards are for safe walking by switchmen around the switches at the head end and pull-out end, and for pulling pins and/or throwing switches. A yard supervisor may also be required to read car numbers at the head end of the yard in order to assign cars to their proper tracks. A locomotive pushes cars into the body of the yard. In most instances, the locomotive headlight furnishes sufficient light for the locomotive engineer. Illumination is needed at clearance points to insure no interference with cars going to adjoining tracks. General lighting is recommended in the area of the switches at both the head end and pull-out end of the yard. If a supervisor must read car numbers, local lighting should be added.

Intermodal Facilities. There has been rapid growth in the hauling of highway trailers and standard containers loaded on special railroad flatcars. There are both several types of rail equipment and methods of loading and unloading the trailers and containers. Also, many railroads operate large automobile transloading facilities. Security lighting is very important at all intermodal facilities.

Trailer-on Flatcar Yards. These provide a ramp leading from the ground level up to the body level of the flatcars. To load the cars, each trailer is backed up the ramp by a standard highway tractor, then backed or pushed from one flatcar to the next until it is on its prescribed car, working from the back car forward (or vice versa when unloading). Mechanized loading methods are used in most large facilities to lift and pivot the trailer onto or off the sides of the flatcars. When the trailers are loaded, most railroads use special tie-down equipment and methods to secure them for shipment. At most trailer-on-flatcar loading and unloading facilities, there is a parking area for trailers that are either waiting to be loaded or that have been unloaded.

The tractor operator must be able to see to back-up or drive along the tops of the flatcars, uncouple the tractor when loading, couple the tractor when unloading, and to pull-off. The trailers must be tied down to the flatcars when loading or unfastened when unloading. To do this, one must be able to see beneath the trailers at the tie-down points. Tractor operators must be able to see to back-up and to couple a trailer parked in the parking area when preparing to load a trailer on a flatcar; conversely, they must be able to see to park and uncouple the trailer in the parking area when unloading. Clerks checking the yard walk between the trailers and must read the numbers on them.

Container-or Trailer-on-Flatcar Yards. Cranes load or unload demountable containers and trailers from flatcars. Usually, trailers are lined up parallel to a row of flatcars. A crane straddling both trailers and flatcars lifts the demountable containers or trailers and places them on the cars. There is usually a parking area for trailers at intermodal facilities.

The crane operators must see to pick up containers (1) from any part of the trailer parking yard and place them in precise locations on the flatcars; or (2) from the flatcars and place them in precise locations on the trailers. In addition to general area lighting, local lighting from luminaires located near the top of the four corners of the crane should provide light on all sides of the vehicles.

Vehicle Loading and Unloading Facilities. Transporting automobiles on railroads is accomplished by using special multilevel railroad cars or vertically-packed cars for small, compact automobiles. On multilevel cars, automobiles are secured by a special tie down. They are loaded or unloaded from ramps that move from track to track and that can be adjusted to different levels.

The vertically-packed car has bottom-hinged sides that open outward to become a platform. The automobiles are driven or pushed onto the platform and secured by brackets on the automobile frame that engage hooks built into the platforms.

Both types of loading and unloading facilities usually have areas for parking those automobiles.

Security lighting is particularly important in these parking areas.

Lighting should be provided at the tie-down spots at each level. The drivers should be able to drive the automobiles on or off the multilevel cars, up and down the ramps, and to or from the parking areas. To reduce over-the-road vandalism, many automobile rack railcars are now equipped with sides. Special lighting may be needed for visual inspection to determine if there has been any kind of damage during shipment.

Lighting Systems

Two different systems of lighting are commonly used to illuminate railroad yards: Projected (Long Throw) Lighting and Distributed Lighting. Each has its advantages under specific yard situations.

In general, the principles in lighting railroad yards are the same as those for other outdoor locations; however, it is necessary to observe railroad regulations with respect to the location of any lighting equipment above or adjacent to the tracks.

Projected Lighting System. The function of this system is to provide illumination from a minimum of locations throughout the various work areas of the yard. See Fig. 9-42.

Advantages are:
1. Use of high poles on towers reduces number of mounting sites.

Fig. 9-42. A projected high mast lighting system using 1000-watt, metal halide luminaires to illuminate retarder area and head end of classification yard.

Fig. 9-43. A distributed lighting system using 400-watt high pressure sodium luminaires mounted on 17-meter (55-foot) wooden poles to illuminate the receiving yard.

2. Light distribution is flexible. Both general and local lighting are readily achieved. (Aiming of projectors, however, may be more critical.)
3. The projectors are effective over long ranges.
4. Maintenance problems are restricted to a few concentrated areas.
5. Physical and visual obstructions are minimized.
6. Electrical distribution system serves a small number of concentrated loads.

Distributed Lighting System. Distributed lighting differs from the projected technique in that luminaires are at many locations rather than at a relatively few. See Fig. 9-43.

Advantages are:
1. Good uniformity of illuminance on the horizontal.
2. Good utilization of light.
3. Reduction of undesirable shadows.
4. Less critical aiming.
5. Lower mounting heights. (Floodlight maintenance is facilitated.)
6. Reduced losses because of atmospheric absorption and scattering.
7. Electrical distribution system serves a large number of small, distributed loads.

Security Lighting. Security lighting is an auxiliary to task lighting and is important in railroad facilities to reduce theft and vandalism. Two basic systems (or a combination of both) may be used to provide practical and effective protective lighting: lighting the boundaries and approaches, or lighting the area and structures within the general boundaries.

Boundary lighting involves directing light toward approaching trespassers so that they can be seen by security personnel. This light also serves as a glare source that handicaps the trespassers' visibility; the guard is not so affected because the light comes from behind him.

RUBBER TIRE MANUFACTURING[24]

Illuminance recommendations for the manufacturing of rubber tires are listed in Fig. 9-44 along with those for mechanical rubber goods. These recommendations represent those established by rubber tire industry representatives and in their opinion can be used in preference to employing Fig. 2-2; however, values can still be determined using Fig. 2-2 for similar tasks and activities.

Fig. 9-44. Illuminance Values Currently Recommended by Industry Representatives for the Manufacture of Rubber Tires and Mechanical Rubber Goods (Maintained on Tasks)

Area/Activity	Illuminance	
	Lux	Foot-candles
Rubber tire manufacturing		
Banbury	300	30
Tread stock		
General	500	50
Booking and inspection, extruder, check weighing, width measuring	1000[a]	100[a]
Calendering		
General	300	30
Letoff and windup	500	50
Stock cutting		
General	300	30
Cutters and splicers	1000[a]	100[a]
Bead Building	500	50
Tire Building		
General	500	50
At machines	1500[b]	150[b]
In-process stock	300	30
Curing		
General	300	30
At molds	750[b]	75[b]
Inspection		
General	1000	100
At tires	3000[a]	300[a]
Storage	200	20
Rubber goods—mechanical		
Stock preparation		
Plasticating, milling, Banbury	300	30
Calendering	500	50
Fabric preparation, stock cutting, hose looms	500	50
Extruded products	500	50
Molded products and curing	500	50
Inspection	2000[b]	200[b]

[a] Localized general lighting.
[b] Obtained with a combination of general lighting plus specialized supplementary lighting. Care should be taken to keep within the recommended luminance ratios.

Processes and Flow Chart

The following descriptions pertain to the processes that convert raw materials to finished tires. Fig. 9-45 is a flow chart showing the major areas.

Banbury Area (1). Bales of synthetic and natural rubber are removed from storage and split into small pieces. Various grades of rubber are blended to meet specific compound requirements and are softened together in a plasticator. The resulting blend of rubber and precise amounts of carbon black and pigments are emptied from a conveyor into a Banbury machine. After mixing, this "Master Batch" drops from the Banbury into a Pelletizer where it is cut into small marble-like pellets of uniform size and shape to facilitate cooling, handling and processing. Sulphur and accelerators are added to the pellets in a final mixing process; and, after a trip through the Banbury mixer, the "Final Batch" drops onto an automatic mill where it is rolled into a thick continuous sheet and conveyed to a "wig-wag" loader which places it on a skid, ready for further processing.

When needed for the preparation of treads, sidewalls, plies or beads, a batch of compounded rubber stock is conveyed to a warming mill where it is kneaded and heated to make it more workable. It is then removed in continuous strips on conveyor belts to the machines which turn out the components.

Tread Stock (2). Rubber stocks are conveyed from warming mills and are fed into two opposing feed hoppers at a tread extrusion machine. The tread and sidewall are usually extruded through one die in a continuous strip which is then stamped with an identifying code and check weighed. The entire underside of the tread and sidewall unit is coated with cement.

The continuous unit is conveyed and cooled over a series of belts and on to a tread cutter where it is cut to specific lengths for use in different sizes and types of tires. The individual tread and sidewall units are then conveyed to booking stations where they are inspected, placed on racks and trucked to storage for aging before going to the tire assembly operation.

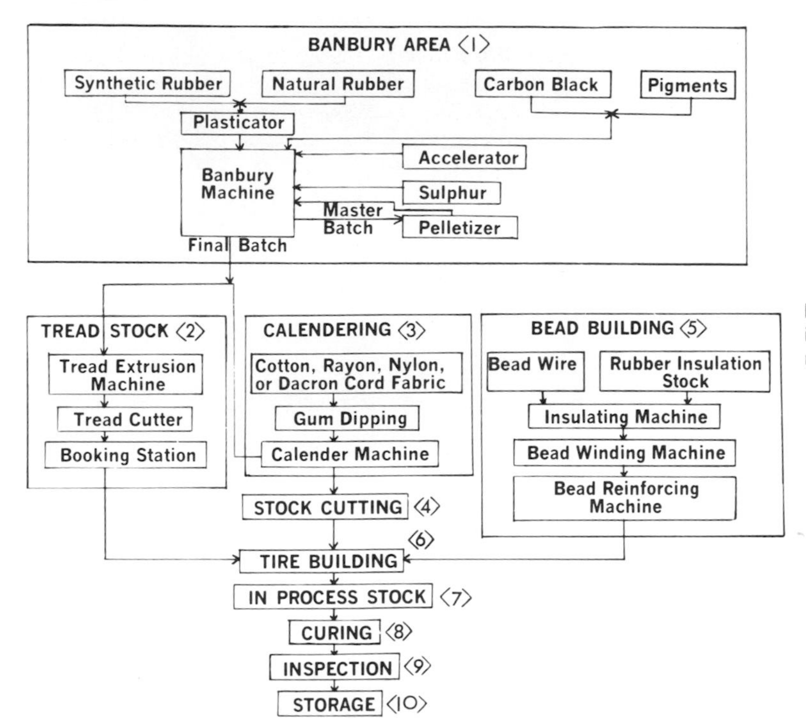

Fig. 9-45. Flow chart for the manufacturing of rubber tires. Numbers in parentheses refer to sections of the text.

Calendering (3). Hugh rolls containing cotton, rayon, nylon, dacron or other cord fabric are taken to a gum-dipping operation where the cord fabric is dipped in a special compound, heat treated and dried under tension.

The fabric passes on to calendering machines where rubber stock from warm-up mills, fed automatically into the calender rolls, is pressed onto both sides and between the cords of the fabric as it passes between the rolls. The fabric goes over cooling drums and is rolled into cloth liners to prevent sticking.

Square-yard weights and widths are recorded at the control board. Many quality checks, including use of the Beta Gauge and statistical quality control charts, aid the calender operations. Accuracy of these measurements is very important to insure a quality product.

Stock Cutting (4). From calendering, the rolls of fabric are trucked to the stock cutting operation where the fabric is cut on the bias (diagonally across the cords). Cut to specified widths and bias angles, these pieces, known as plies, pass from the cutting table, are turned end to end and precision-spliced into continuous strips ready for tire assembly.

Bead Building (5). Copper-plated high tensile strength steel wire is brought in on large reels. A specified number of strands are brought together from the reels and guided through a die in the head of a small tube machine. Here, a special rubber insulation stock is squeezed around and between the separate strands of wire.

The rubber-covered wire strip is then run into a bead winding machine, where it is wound on a chuck or collapsible ring. The machine automatically winds the proper number of turns, cuts the wire and ejects the wound bead on to a storage arm.

In the next operation, bead reinforcing, the wound bead wire is covered with a strip of calendered fabric called a "reinforce." The reinforce not only ties the bead into the tire but also plays a vital part in giving the tire added strength and stability.

Tire Building (6). All the parts of a tire are brought together in the tire assembly operation and are combined on a semi-automatic precision machine with a collapsible drum.

In Process Stock (7). Green tires are accumulated along with other parts still in process, such as fabrics and calendered stock.

Curing (8). Green tires are sprayed automatically with lubricants, inside and out, to aid in molding. They are then conveyed to long batteries of automatic forming and curing presses where they are transformed into tires. See Fig. 9-46.

Inspection (9). Tires are subjected to a series of quality checks. Passenger tires arriving from the curing room pass along a conveyor to automatic trimmers and on to spray painting machines for identification symbols. The tires are then conveyed to inspectors who check each tire inside and out.

The entire tread is examined to see that its angles are sharply molded so that they will grip the road. Sidewalls are inspected for gouges, pinvents and whitewall adhesion. The bead is examined for wrinkles or any defects that might prevent the tire from seating tightly and securely when mounted on a wheel or in any other way affect the service of the tire. Tire inner liners are inspected for any separations.

Finished Tire Storage (10). Tires are stored for later shipment.

Fig. 9-47. High bay large press shop lighted with 1000-watt improved color mercury lamp units.

Lighting Design for Specific Tasks

Banbury Area (1). When the Banbury machine is manually charged, the ingredients for the batch are made up, weighed and inserted into the charging hopper. The operator should be able to read temperature recorders, time clocks and scale dials during each machine cycle.

General lighting should be provided with particular attention to locations where the operators perform their task. Also, it should be noted that there will be high dirt accumulation on luminaires and room surfaces because there are large quantities of carbon black in the air.

Fig. 9-46. Forming and curing presses where "green" tires are fed in to be formed and vulcanized.

Tread Stock (2). General lighting is required in this area, with a higher level at the extruder; over the check weighing and width measuring station; and over the booking and inspection station where the tread is inspected for physical defects, weight, length and width.

Calendering (3). General lighting should be provided in the calendering area with a higher level at the letoff and windup, for the crew to handle the fabric; and at the instrumentation panel and over the measurement and inspection area, where the operator looks for such imperfections as bare cords.

Stock Cutting (4). This area should have both general and local lighting, with a higher level from the local units over the splicing table. The operator should be able to inspect fabric for bare cords, to make manual splices, or to set up an automatic splicer machine.

Bead Building (5). General lighting is needed in the bead room to permit proper machine setup, to see coatings on the wire, and to effect reinforcing.

Tire Building (6). General lighting should be provided throughout the entire area with supplementary lighting at the tire building machines to enable the operator to properly feed the materials onto the drum and to inspect laps.

In Process Stock (7). General lighting is needed for reading tags or color codes on items in this active storage area.

Curing (8). In addition to general lighting, supplementary lighting is required at the molds. Supplementary luminaires should be so positioned that molds may be readily inspected during cleaning and coating.

Inspection (9). This area should be provided with general and supplementary illumination or localized general at each inspection stand. Luminaires should be positioned on both sides of the tires to illuminate both the inside and outside of the casings. Luminaires should be shielded and installed in locations to eliminate as much glare as possible.

Finished Tire Storage (10). See In Process Stock.

SHEET METAL SHOPS

Visual tasks in the sheet metal shop are often difficult because: sheet metal (after pickling and oiling) has a reflectance similar to the working surface of the machine, resulting in poor contrasts between the machine and work; low reflectance of the metal results in a low task luminance; high-speed operation of small presses reduces the available time for seeing; bulky machinery obstructs the distribution of light from general lighting luminaires; and noise contributes to fatigue.

Fig. 9-48. Large press with built-in supplementary lighting to facilitate setup and operation.

Fig. 9-49. Low bay press shop with fluorescent lighting. Note: upward component and center shield for better comfort.

Punch Press

The seeing task is essentially the same for a large press as it is for a small press except that with a small press less time is available for seeing. The shadow problem, however, is much greater with the large press. With either, the operator must have adequate illumination to move the stock into the press, inspect the die for scrap after the operating cycle is completed, and inspect the product. Where an automatic feed is employed, the speed of operation is so great that the operator has time only to inspect the die for scrap clearance.

The general lighting system in press areas should provide the illuminance selected from Fig. 2-2. This is necessary for the safe and rapid handling of stock in the form of unprocessed metal, scrap, or the finished products. In the large press areas, such as that shown in Fig. 9-47, illumination should be furnished by high-bay lighting equipment or by a combination of high-bay and supplementary lighting as shown in Fig. 9-48. For low-bay areas (Fig. 9-49), the illumination should be supplied by luminaires having a widespread distribution to provide uniform illuminance for the bay and the die surface area. Where the mounting height exceeds 6 meters (20 feet), careful consideration should be given to maintenance costs.

The operator's ability to inspect the die is more directly related to the reflected brightness of the die surface than to the amount of light incident upon it. For example, a concentrated light placed on the operator's side of the press

Fig. 9-50. Highlights and shadows on the die of a small press are produced by supplementary or general lighting at the rear of the press. The operator can quickly inspect the open die for loose pieces of scrap.

and directed toward the die may produce results much less satisfactory than a large-area source of low luminance placed at the back or side of the press. See Fig. 9-50. The luminance required for optimum visibility of the die has not been established; consensus suggests that 1700 candelas per square meter [500 footlamberts] is satisfactory.

Paint applied to both the exterior and the throat surfaces of a press contributes to the operator's ability to see. The reflectance of the paint selected for the exterior of the press should be not less than 40 per cent. This treatment of vertical surfaces on the exterior provides for maximum utilization of light from the general lighting system. Similarly, the paint selected for throat surfaces should have a reflectance of 60 per cent or higher.

Shear

The operator must be able to see a measuring scale in order to set the stops for gauging the size of "cut." When the sheet has to be trimmed, either to square the sides or to cut off scrap from the edges, the operator must be able to see the location of the cut in order to minimize scrap.

The general lighting system should provide the illuminance selected from Fig. 2-2 in the area around the shear for the safe feeding of the sheets at the front, collecting the scrap at the back, and stacking the finished pieces in preparation for removal.

Local lighting, as shown in Fig. 9-51, produces a line of light to indicate where the cut will be made and the amount of scrap that will be trimmed. It also provides lighting for the oil-soaked gloves on the hands of the operators to enable the operator who is responsible for pressing the foot-release bar to see quickly that all hands are clear of the guard.

SHOE MANUFACTURING[25]

Recommended illuminances for typical shoe manufacturing operations are given in Fig. 2-2.

Seeing Task Groups. Leather and rubber shoe manufacturing processes may be separated into three groups according to difficulty of seeing tasks.

1. Simple seeing tasks include:

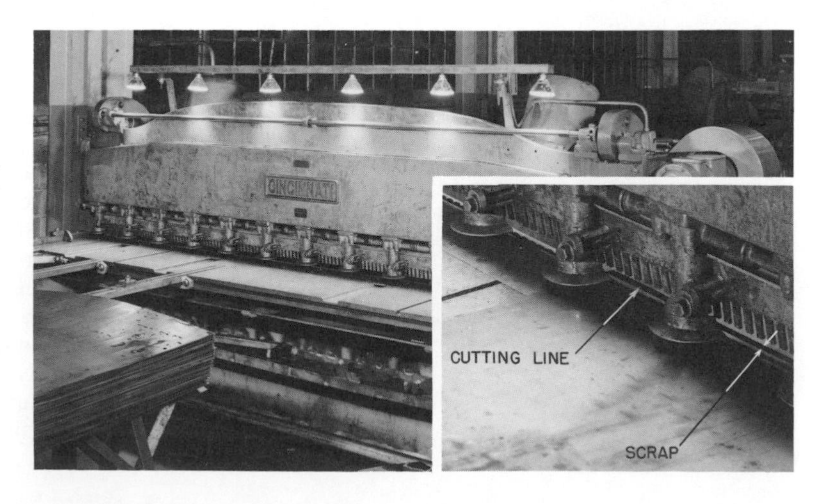

Fig. 9-51. Shear. Six 150-watt floodlamps provide a narrow line of light on the metal sheet to indicate where the cut will be made. Sockets are mounted in channel supported by springs. A 2-lamp fluorescent ventilated industrial luminaire will also provide good illumination and is less subject to short lamp life due to vibration.

Leather: storage, staying, sole laying, beveling, nailing, heel scouring, burnishing, spraying, box making, dinking, last racks, lasting, pulling over, trimming, channeling, heel breasting, edge setting.

Rubber: washing, compounding, calendering.

2. Seeing tasks of average difficulty include the mechanized operations:

Leather: skiving and splitting, treeing, welting, rough rounding, perforating, buttonholing, eyeletting on both light and dark materials, certain types of bench work.

Rubber: sole rolling milling, completed stages of compounding.

3. Seeing tasks of considerable difficulty include:

Leather: cutting, bench work, stitching, inspection, rounding, sole stitching, fine edge trimming on both light and dark materials.

Rubber: cutting, making, calendering.

Leather-Shoe Manufacturing

Sole Department. Leather, sorted for grain and thickness, is stored in 1.5- to 2-meter (5- to 6-foot) piles on low platforms arranged with passageways between them.

For grading according to color, there may be some advantage from illumination with spectral characteristics similar to daylight.

Beam Dinkers. A beam dinker uses dies to stamp soles and insoles from hides. It consists of a heavy cast-iron frame and a large beam that exerts pressure through a vertical motion on a cutting die. There is some hazard of finger injury in operating the machine; therefore, special attention must be given to this location with an adequate, well-diffused general lighting system and supplementary lighting to prevent shadows.

To avoid casting shadows of the beam on the platform, all luminaires in the area occupied by the beam dinkers should be placed at the operator's side of the machine.

Last Storage. Last storage bins usually are located in a segregated section of the sole department. Generally there is a 1-meter (3-foot) aisle between the bins. Luminaires should be mounted over the aisles in a manner to illuminate the inside of the bins so that lasts may be selected.

Upper Department. An upper department generally is divided into the following sections: (1) sorting; (2) trimming, cutting, and staying; (3) lining; (4) upper cutting; (5) marking and skiving; and (6) assembling.

When an order is received for a certain grade of shoes, the sorting department grades the leather as to color and quality. A well diffused, general lighting system is best; a daylight color quality is desirable.

Skilled workers then split each piece of leather into as many sheets as possible and cut out individual parts for uppers. This work generally is done on tables 760 to 910 millimeters (30 to 36 inches) above the floor.

The various pieces are counted and marked with job numbers. Skiving, which consists of the mechanical thinning of edges of the uppers (so that they can be turned over to present a finished appearance), is the next step. Assembling consists of bringing together the various parts which make up the uppers (lining, stay, vamp, counter, toe, tip, etc.). Uniform illuminance is recommended.

Stitching Department. The following operations are typical: (1) lining; (2) tip; (3) closing and staying; (4) boxing; (5) top stitching; (6) buttonholing and stamping; and (7) toe closing.

These are difficult seeing tasks. A high level of uniform general illumination should be provided plus supplementary lighting on the machines.

Making Department. The making department in the average plant is subdivided according to operations: (1) vamping; (2) welt-bottoming; (3) bottoming; (4) heeling; (5) turning; and (6) standard, screw, nail or pegged shoe making.

In some plants this department is called the *gang room* and occupies an entire floor. A general lighting system plus local lighting should be installed to mitigate shadows of overhanging machine parts.

Lasters, sole layers, levelers and nailing machines may be illuminated by diffused light. The source location is not critical. Other machines in the making department should be illuminated from the rear and to the right of the operator. Vertical as well as horizontal illumination is most important.

Finishing Room. Shoes are inspected and faults are corrected. Treeing machines are used to iron out wrinkles. From here the shoe goes to final inspection and then to the packers and shippers.

Rubber-Shoe Manufacturing

Typical operations include: (1) washing; (2) compounding and milling; (3) cutting and calendering; (4) drying; (5) sole rolling and cutting; (6)

making; (7) varnishing and vulcanizing; and (8) packing and shipping.

In the washing department crude rubber is cut by band saws. General illumination is recommended for washing and cutting; also for the compound and mill area. Where hoods are placed over the compounding machines, local lighting should be provided by luminaires installed under the hoods with the reflector directed at the point of work.

As materials pass over the cutting and calendering machines, the coating must be applied correctly. Calenders, especially the three- or four-roller type, should be lighted by luminaires on both sides of the machines. Well diffused light will avoid sharp shadows and glare.

After cutting or gumming, rolls go to the drying room where they are dried by steam heat. Where this department is confined to the center of the building and has no direct general ventilation, there is an explosion hazard. Explosion-proof lighting units are recommended. Supplementary lighting equipment provides light at both front and rear ends of the sole-rolling machine.

In the sole and upper cutting department, operators work rapidly with sharp knives. Uniform illumination throughout the area (from luminaires mounted as high as possible) is recommended. Some plants use beam dinkers. These should be lighted in the manner described above.

The making department is the most important. All parts are supplied, cut to shape, to bench workers who use cement to attach and complete a shoe. There may be a shelf or rack over the center of the bench, extending its entire length. The lasts are placed on this shelf. If low-mounted luminaires are placed over this shelf, the shelf causes a sharp shadow on the working areas of the bench. A general lighting installation is recommended. In the varnishing and vulcanizing areas, uniform illuminance is recommended.

STEEL MILLS

The manufacture of steel is an integrated process involving three major steps: the blast furnaces refine the iron ore into pig iron; the pig iron is transferred to either an open hearth furnace, basic oxygen furnace, or electric furnace where it is converted into steel; and the steel ingots are sent to various rolling mills where they are formed into sheets, bars, rails, structural forms, etc. See Fig. 9-52. Each one of these processes can be considered as a separate operation. They may be located anywhere from a few hundred meters (yards) to many kilometers (miles) apart.

Regardless of any other consideration, enough lighting equipment should be provided to satisfy minimum safety requirements. Also, there are certain areas of critical seeing where higher levels and proper quality are important. See Fig. 9-53, for recommended illuminances representing those established by iron and steel industry representatives, which in their opinion can be used in preference to employing Fig. 2-2; however, values can still be determined using Fig. 2-2 for similar tasks and activities. Beyond these values there are some examples of far-reaching management policies which also recognize good lighting as a useful tool to aid the production of increased outputs of higher quality material. Since considerable capital expenditures are authorized for lighting systems, the selection of the most practical and trouble-free lighting system pays off when preliminary planning is complete in every detail.

First costs of lighting systems are not a true measure of the systems' effectiveness, especially when steel mills operate at full capacity. Prime study should be given to the annual operating costs of these systems. In virtually all cases, lighting equipment which performs efficiently

Fig. 9-52. Steel mill area producing cold rolled sheet in coils. Deep reflectors shield the lamps in normal field of view. Left outer row of reflectors are located close to the wall to maintain better illumination along the edge of the space.

Fig. 9-53. Illuminance Values Currently Recommended by Representatives from the Iron and Steel Industry (Maintained on Tasks)

Area/Activity	Illuminance	
	Lux	Foot-candles
Open hearth		
Stock yard	100	10
Charging floor	200	20
Pouring slide		
Slag pits	200	20
Control platforms	300	30
Mold yard	50	5
Hot top	300	30
Hot top storage	100	10
Checker cellar	100	10
Buggy and door repair	300	30
Stripping yard	200	20
Scrap stockyard	100	10
Mixer building	300	30
Calcining building	100	10
Skull cracker	100	10
Rolling mills		
Blooming, slabbing, hot strip, hot sheet	300	30
Cold strip, plate	300	30
Pipe, rod, tube, wire drawing	500	50
Merchant and sheared plate	300	30
Tin plate mills		
Tinning and galvanizing	500	50
Cold strip rolling	500	50
Motor room, machine room	300	30
Inspection		
Black plate, bloom and billet chipping	1000	100
Tin plate and other bright surfaces	2000*	200*

* The specular surface of the material may necessitate special consideration in selection and placement of lighting equipment, or orientation of work.

with dependable and trouble-free operation will prove economically sound for steel mill application.

The following characteristics are common to steel mill building construction as well as the operation of the industry and each has its influence on the design of practical lighting systems:

1. All production areas have overhead cranes which must be used to move raw materials and finished products.

2. The raw materials and finished products are of such composition and size as to require considerable head room, with the result that lighting units must be installed 9 meters (30 feet) or higher above the floor.

3. Arrangement of equipment on floor plans influences the sizes and spans of overhead cranes, and the flow of products from raw material to finished form dictates the locations of lighting outlets to a degree seldom, if ever, found in any other industry.

4. The range of ambient temperatures where lighting equipment must operate runs from −18°C (0°F) in unheated rolling and beam mills during winter months, to soaking pit and hot top areas where readings of 60°C (140°F) are common above crane cabs in parts of June, July and August.

5. Atmospheric conditions are poor at best even in the most modern mills where well-planned ventilating systems are installed. The basic production methods which employ a succession of heating and cooling cycles make for conditions which induce fast rising air currents which carry metal dust, products of combustion, and oil vapors upward to locations where lighting equipment must be installed.

6. Instead of finished ceilings with high reflectances of 60 to 80 per cent, there are open truss frameworks below metal roof decks; instead of smooth, light colored walls with 30 to 50 per cent reflectances, there are corrugated metal sidings and a succession of columns marking the boundaries between bays; and finally, instead of floors with smooth finishes rated 10 to 30 per cent reflectance, there are pieces or stacks of dark colored material and equipment in most locations, or black dirt and unswept concrete where parts of floor space are visible.

In many areas, the high mounting heights favor the use of 400-watt and 1000-watt high intensity discharge lamps. Special consideration should be given to maintaining lighting continuity in case of momentary or long-term power outages. See Emergency Lighting in Section 2. The choice of light source will depend upon the power supply available. High intensity discharge lamps installed on cranes served by direct current should use appropriate ballasts.

Open Hearth[26]

Stock Yard. Scrap metal, ore and limestone are loaded by overhead cranes. Generally, overhead lighting should be provided. In outdoor areas in which no overhead framing is available, directional lighting units having wide distribution can be installed on the underside of crane girders.

Charging Floor. Charging boxes are emptied into the furnaces by machine and alloy materials added by shovel or machine. In addition to general lighting, localized lighting should be provided on the instrument panel which controls the furnace operation. Additional lighting should be provided over the small material storage bins. This usually can be best accomplished by mounting directional lighting equipment on the steel columns or on the underside of the crane girders.

Pouring Side. The molten metal is an area of very concentrated high luminance. When the furnaces are tapped and the ladles emptied into molds there are clouds of fumes and vapor. General overhead lighting is recommended. Auxiliary lighting may be provided for the slag pits in either of two ways: directional lighting equipment can be mounted permanently on the steel columns above the platform level which is part of the charging floor area, or, if this is not feasible, portable units mounted on a standard having either a tripod or flat platform with castors can be set on the platform above the slag pit area. At the ingot pouring platform supplementary lighting is necessary and can be obtained by mounting directional lighting equipment on the columns back of the platform. Portable lighting is not satisfactory due to low mounting height and glare from pour and rear directional light sources.

Mold Yard. The ingot molds are cleaned, coated, sprayed and stored. This is a very dirty area with a high concentration of fumes from the spraying operation. General overhead lighting with supplementary lighting on the underside of the crane is recommended. Consideration should be given to suitable maintenance features such as dirt-resistant design, dust-tight covers, etc.

Hot Top Relining and Repairing. General overhead lighting should be provided. Supplemental lighting at the point of hot top repairs and mold relining is also necessary. Directional lighting equipment can be mounted on the overhead beams; for close inspection of the lining it is necessary to use extension cords.

Hot Top Storage. The operator must be able to see in order to place a thin steel gasket accurately between the hot top and mold. General overhead lighting plus lighting units under the crane are required.

Checker Cellar. Brick repair work is sometimes done in this low-ceiling area. Overhead lights should be located wherever possible and portable work lights provided.

Buggy and Door Repair. This is a repair shop for charging boxes, buggies and furnace doors. Operations include rough machine and bench work, cutting and welding. General lighting is required.

Stripping Yard. The molds are removed from the ingots by a stripping machine. For indoor locations high-bay general lighting is required. Outdoors, directional lighting equipment should be mounted under crane girders; overhead units mounted on messenger cable are also satisfactory.

Scrap Stock Yard. In this outdoor area with overhead cranes, scrap is cut by torch and shears and then stored. Standard industrial lighting equipment can be mounted on messenger cable strung between the crane-supporting steel structure with floodlights on towers or poles erected beside the crane runway.

Mixer Building. Molten metal from blast furnaces is poured into a tilting vessel which acts as a reservoir. The vessel, called a mixer, has a capacity of 300 to 1500 tons. Owing to the very large mixers and the bins containing raw material mounted above them, it is almost impossible to provide general illumination over the entire area. Lighting equipment should be located where possible and convenient. Supplementary lighting should be provided for important areas such as at the top of mixers, pouring level, walkways and stairwells.

Calcining Building. This is a very dirty high-bay area in which limestone is stored and crushed. General overhead lighting is not advisable due to the height of crushing machinery and bins for feeding them. Industrial lighting units should be provided at convenient locations on all working platforms or levels.

Skull Cracker. In this outdoor area old ingot molds and solid material, which freezes on the bottom or sides of ladles, are broken up. Directional lighting equipment should be mounted on the underside of crane girders. Lighting is also necessary under the cranes.

Cinder Dump. Furnace cinders are stored outside. General lighting from overhead or directional lighting equipment and additional units under the cranes are necessary.

Basic Oxygen Furnace Plant

The lighting techniques are much the same as with the open hearth. Mounting heights are very high with servicing accomplished from cranes or disconnect and lowering hangers. High concentration of graphite particles, dust, etc., dictates use of ventilated luminaires, gasketed enclosed luminaires or reflector lamps for this facility.

Continuous Casting

The continuous casting process takes molten steel and produces the required cross sections of slabs or blooms and, therefore, bypasses the soaking pits and blooming or slabbing mills. The tundish area can be lighted similar to other hot metal handling facilities. Vertical type machines will usually require directional lighting from the sides. These luminaires can be mounted on building columns or structural supporting members of the casting units.

TEXTILE INDUSTRY

The textile industry is generally composed of firms that accept input materials of staple and continuous-filament fibers, convert these to yarn intermediates, yarn and fabric, and then finish that fabric for the apparel industry or others who convert the fabric to an end-use article. The term "textile" originally referred only to woven fabrics. Today, the term refers to fabrics produced by weaving, knitting, tufting and non-woven processes. The end products, as well as the type of input fiber, determine the exact nature of the intermediate processing steps.

Staple fiber yarn production involves alignment, attenuation, evening and twisting of relatively short, fine natural and synthetic fibers. *Filament* yarn production involves converting synthetic fiber producer raw stock to a yarn with bulk, stretch and texture—the operations occurring mainly in a segment of the industry known as "throwsters" (dating back to those who twisted and folded continuous-filament silk).

The description of processes given below identifies major process areas, and relates yarn production terminology to the most common yarn-processing systems; *e.g.*, the cotton system. Terms* used in American-modified woolen yarn and worsted yarn production are included. The descriptive organization follows the flow of input fiber through to the finished fabric.

A fiber, *e.g.* cotton, is of very light weight and of very small diameter. The yarns produced are also of small diameter—in the order of hundredths of a millimeter (thousandths of an inch)—and contain from one hundred to several thousand fibers in a cross section depending on yarn size. The processing equipment, however, is large and heavy; for economic reasons, it operates at high speeds. Lighting requirements are dictated by the need to have operators interface with the flow of fiber into yarn, yarn into fabric, and through fabric inspection as part of quality and production control. Typical operator duties include identifying and repairing broken ends, recognizing off-quality production and defects.

For illuminance recommendations see Fig. 2-2.

* Textile terminology may be obtained from the *Man-Made Fiber and Textile Dictionary*, published by the Celanese Corporation; and *Textile Terms and Definitions*, published by the Textile Institute in Manchester, England.

Staple Fiber Preparation

Stock Dyeing, Tinting. Coloration may be added to synthetic fibers prior to yarn and fabric manufacture to produce either a single-finished shade or special effects in color blends (such as some of the heathers used in carpets and woolens). When post-dyeing of fabric is scheduled, fugitive tints may be added to the fiber to identify and separate the various process streams through the plant. General lighting is adequate except in a localized special area for color matching. Most wool preprocessing (*e.g.*, degreasing, scouring) requires similar lighting except that wool sorting (and cotton grading) requires a higher level including specialized lighting.

Yarn Manufacturing

Opening and Picking (Chute Feed). Staple fiber is shipped to a mill in bales weighing approximately 230 kilograms (500 pounds) and of a density of 320 to 480 kilograms per cubic meter (20-30 pounds per cubic foot). The opening process reduces fiber density by producing small tufts, blends various lots of fiber, and removes large particles of trash. Some mills convert continuous-filament bundles of fiber (tow) to a staple or short-length form (top). Typically, opened fiber stock may be sent directly to the next process by pneumatic chute; or it may be further processed through a picker to produce a card lap. The picker removes more trash, blends and evens out variations in linear density of the stock. Lighting needs are less than that for general production areas (the degree of automation dictates how much less). Employee-controlled additional lighting for maintenance may be feasible.

Carding (Non-Woven Web Formation). The purpose of cards, whether roller top (long-staple fiber), granular top (synthetics) or flat top (cotton), is to clean and attenuate the stock to produce a yarn intermediate termed "sliver." "Slivers" are a loose collection of fibers of long length and of a diameter of 25 to 76 millimeters (one to three inches) (cotton vs. woolen). This process normally requires a moderate lighting level as well as additional supplementary lighting during machine maintenance, which includes setting the critical space between heavy components to less than 0.03-millimeter (0.001-inch) tolerance.

Drawing (Gilling, Pin Drafting). The drawing process aligns fibers, blends, reduces variation in linear density and produces a sliver generally of the same linear density as the card sliver. Normally, there are two passes of drawing to remove fiber hooks and improve fiber parallelization. Woolen processes which are designed to produce optimum yarn bulk, by-pass this and subsequent processes prior to spinning. Lighting needs are similar to those of carding.

Combing. This process is used only for fine, long-staple cottons and worsteds. The slivers are formed by a "sliver lapper" into a lap (blanket) form for input to the comber. Combing achieves an optimum level of parallelization of the fibers, cleaning and removal of short fiber from the stock. The comber is subject to very sensitive adjustments. Comber output is a sliver similar to the card sliver or drawing sliver. While moderate lighting is adequate during regular operation, additional switched or other supplementary lighting is needed for maintenance.

Roving (Slubbing, Fly Frame). The roving frame attenuates sliver input by a factor of ten and produces an output intermediate also termed "roving." Because the roving intermediate is unable to support its own weight as a parallel fiber bundle, a minimum amount of twist is added to facilitate material handling. Both the roving bobbin and a large device termed a "flyer" rotate at high speed. Lighting needs will be at or somewhat above that needed in carding or drawing.

Spinning (Cap Spinning, Twisting, Texturing). Yarn is produced from "roving" on the spinning frame. See Fig. 9-54. Final attenuation (as high as a factor of 100) occurs along with the twist. Twist provides the final yarn strength. In texturing processes, twist is inserted in a continuous-filament thread line at rates up to one million turns per minute. Spun or textured yarn diameter may be as small as 0.07 millimeter (0.003 inch). Because the operator must detect thread-line faults, lighting should be accentuated along the alleys between machines and concentrated on the thread lines. Coarse yarns are less demanding.

Yarn Preparation

Yarn preparation covers several operations including winding, quilling, sizing, warping, warp tie-in or drawing-in and twisting (a process similar to spinning). In general, the object is to change the medium of materials handling from a relatively low-mass spinning bobbin to a higher-mass package. Sizing involves treating the yarns superficially with a protective coating. Because of the small diameter of the yarn and the need to distinguish between closely-spaced threading-up paths, these processes generally re-

Fig. 9-54. Spinning room with continuous rows of luminaries running perpendicular to machines to prevent machine shadows.

Fig. 9-55. Weave room of a textile mill. Note rows of luminaires are perpendicular to looms to prevent harsh line shadows from upper parts of loom.

quire somewhat greater lighting levels than those of carding or drawing. Typically, concentration along thread lines is needed.

Fabric Production

The areas involved here are weaving (see Fig. 9-55), knitting, tufting (carpets) and inspection. Lighting needs are above that of the general production area. Each of these processes deals with a product of significant added value; thus, machine inspection is critical. Additionally, each machine contains fine or closely-spaced parts which are serviced regularly by the operator.

Finishing

Fabric Preparation. The various preprocessing steps of fabric finishing include (among others) desizing, scouring, bleaching, singeing and mercerization. These processes remove impurities and surface imperfections from fabric in preparation for dyeing. Moderate lighting is adequate for these areas.

Fabric Dyeing (Printing). Color and patterns are added at the dyeing and printing processes. Whether by continuous dye range or batch processes (*i.e.*, jet, beck or jig), the process area generally needs moderate lighting levels. At color-matching and inspection stations, local lighting at several times the normal levels are needed. Color temperature of the light source is important in metameric color matching (see Section 5 in 1981 Reference Volume). After dyeing, fabric is typically dried on a tenter frame. Some minor inspection follows drying.

Fabric Finishing. The final surface characteristics of a fabric are generated in these post-processing steps which may include calendering, sanforizing, sueding; or the application of chemical treatments to provide flame-proofing, soil resistance and softening. During such processing, normal lighting is sufficient. In final inspection, both color and illuminance are critical.

SECURITY LIGHTING

For the purposes of this section security lighting pertains to the lighting of building exteriors and surrounding areas—out to and including the boundaries of the property.

Security lighting contributes to a sense of personal security and to the protection of property. This may be accomplished through:

1. *Surveillance lighting*—lighting to detect and observe intruders.
2. *Protective lighting*—lighting to discourage or deter attempts at entrance, vandalism, etc. It may lead a potential intruder to believe detection highly possible and so not attempt entry.
3. *Lighting for safety*—lighting to permit safe movement of guards and other authorized persons.

Criteria of Effectiveness. An effective security lighting system should:

a. Discourage intruders.
b. Make detection highly probable should entry be effected.

c. Avoid glare that handicaps the guards and annoys passing traffic and the workers and legitimate occupants of adjacent properties.
d. Provide adequate illuminances. The illuminance depends upon the accessibility and vulnerability of the property and whether the surveillance is by eye or by electronics. See Fig. 9-56.
e. Provide low illuminance levels on guard posts, television cameras and other electronic or sensing locations to render their positions harder for the intruder to pinpoint.
f. Provide special treatment for sensitive locations (entrances and exits, railroad sidings, alleys, roofs of abutting buildings, wooded areas and water approaches).
g. Provide complete reliability. A single lamp outage should not result in a dark spot vulnerable to entry.
h. Provide convenient control and maintenance.

Surveillance Lighting

Surveillance of Large Open Areas—Standard System. The longer the *time* available to see intruders, the more likely they will be detected. A wide strip of flat land around a plant of 30 meters (100 feet) is desirable; or a large yard, free of places for intruders to hide, provides sufficient time for seeing a running intruder—assuming recommended illuminance (Fig. 9-56).

Floodlights mounted on poles or on buildings 10 meters (30 feet) or higher will produce sufficient illuminance on the ground to visually detect movement at ground level.* A floodlight directed at an angle of 45 degrees from nadir will provide approximately equal illuminances on horizontal and vertical surfaces at the same point.

Surveillance of Large Open Areas—Glare System. A variant to the system discussed directly above is to mount the floodlighting equipment at eye level and directed outward and far enough away from buildings to permit an area of darkness for guards. This provides very little illumination on the ground; however, it forces the intruder to face glaring lights which obscure the location of guards. The illuminances recommended for the Standard System apply—but the values are those falling on the faces of approaching trespassers.

Surveillance of Confined Areas. Higher illuminance is needed if only a short time is available to detect an intruder, as in areas between buildings, areas with obstructions, roadways, walkways, etc. Average minimum illuminance (see Fig. 9-57) should be on any plane that may require observation. The illumination should be as uniform as practicable.

* The illuminance for closed-circuit television surveillance is contingent upon camera sensitivity.

Fig. 9-56. Recommended Illuminances* to Detect and Observe Intruders (Surveillance Lighting)

	Illuminance	
	Lux	Footcandles
Large Open Areas—Standard System		
Average illuminance throughout the space, minimum at any time	2	0.2
Absolute minimum illuminance at any point or time	0.5	0.05
Large Open Areas—Glare System	Same as above	
Surveillance of Confined Areas		
Average illuminance throughout the space, minimum at any time	5	0.5
Absolute minimum at any point or at any time	1	0.1
Surveillance of Pedestrian and Vehicular Entrances		
Average illuminance throughout the space, minimum at any time	10	1.0
Absolute minimum at any point or at any time	2.5	0.25
Television Surveillance	Contact manufacturer of TV camera for required illuminance	

* Recommended illuminances on vertical surfaces in direction of guards and 1 meter (3 feet) above ground and on any other plane on which surveillance is related to seeing.

Surveillance of Pedestrian and Vehicular Entrances. The recommended average illuminance for entrances is shown in Fig. 9-56. Shadows should be minimized. Uniformity of lighting should be good and the minimum illuminance at any time and at any critical plane should not be less than one-fourth of that value.

Television Surveillance. The manufacturer of the television camera to be used should be contacted for the required illuminance. Some sensitive TV cameras used for surveillance give acceptable pictures with about 10 lux [1 footcandle] or less; however, some conventional equipment may require higher levels.

Protective Lighting

To discourage the potential acts of criminals,

the lighting system should provide an illuminance of at least 5 lux (0.5 footcandles). This requirement can be realized by choice of luminaire distribution and/or by aiming the lighting equipment. To help achieve the desired objective, surfaces should be as light in color as practicable and the area should be free of hiding places.

Lighting for Safety

Minimum lighting is needed in a designated area to allow reasonable movement of vehicles and people and/or operation of machinery without fear of accident, collision or other unexpected hazard to the health and welfare of the operators and other personnel. See Section 2 and Fig. 2-26. In the event the illuminances for safety are higher than those required for surveillance, the level required for *safety* should take precedence.

Lighting Systems

Light Sources. The characteristics of light sources are covered in Section 8 of the 1981 Reference Volume. Consider using the most efficient (highest in "lumens per watt") light source that meets other requirements.

Luminaires. Selecting the most appropriate luminaire for a specific security application includes considering purpose, area, size and characteristics, pole placement (or other mounting locations), hours of use, surrounding-area luminance, luminaire brightness control, visual environment and esthetics.

Luminaires may be evaluated in terms of their efficiency and/or their light distribution. Inefficient equipment is wasteful; improper distribution contributes to poor illuminance uniformity.

Four types of luminaires are most often used:

1. Floodlights.
2. Roadway luminaires.
3. Enclosed-and-gasketed luminaires with control lenses and reflectors.
4. Searchlights.

Floodlights. A floodlight is specified by wattage and beam spread. The latter, expressed in degrees, defines the angle at which the luminous intensity is equal to a stated per cent (usually 10 per cent) of the maximum. A common method of roughly classifying beam widths is by the terms "narrow," "medium" or "wide." More specifically, they are specified in terms of distributions (see footnote on page 13-13).

Floodlights may be either "open" or "enclosed." The latter is equipped with a cover glass to exclude rain, dust or other airborne contaminants. They may be further modified by glare-controlling shieldings such as louvers or visors.

The appropriate beam width will depend upon the area to be covered, the angle at which the beam strikes the area, the distance of the floodlight from the area and the desired beam overlap. Beam widths are contingent upon the contour and finish of the reflector, the refraction of the lens and the type of lamp. See Floodlight Calculations, Section 9 of the 1981 Reference Volume.

Roadway Luminaires. Light distribution may be symmetrical or asymmetrical and may have many degrees of vertical light control. A symmetrical distribution is one in which the distribution of light is approximately the same at any vertical plane throughout 360 degrees around the luminaire. See Section 14 for distribution classifications.

Enclosed-and-Gasketed Refractor Luminaires. These have the lamp enclosed with a refractor cover. They are generally utilized for perimeter lighting as well as for certain areas about water approaches, air approaches or where there is pedestrian traffic.

Searchlights. These are generally employed in a system of lighting for surveillance rather than for deterrence. They use small sources (usually incandescent filament lamps) which provide full light output immediately and permit very concentrated beam distribution. Searchlights are commonly rated by diameter of reflector, specific lamp type and lamp wattage. Searchlights are frequently mobile and supplement fixed lighting. See Section 20.

Specific Applications

Boundaries and Approaches. The glare system may be considered for isolated fence boundaries that are sufficiently remote from adjacent working areas, roads, building areas, etc. A patrol road or path 9 meters (30 feet) or more back from the pole line will remain in relative darkness which, in combination with the glare projected outside the property, makes it difficult for would-be intruders to see into the property. Intruders up to 90 to 120 meters (300 to 400 feet) from the fence line may be visible to the roving patrol. Mounting the units approximately 0.3 meter (1 foot) from the poles with short, adjustable brackets and at low mounting heights is permissible in these cases where glare is not projected into adjoining properties. Care should be taken in properly aiming and leveling these units to avoid excessive lighting of the inner side or the top of the fence line.

Where closed-circuit-television surveillance is employed, it is important to provide enough illuminance to meet the requirements of the electronic equipment and yet *not* allow an intruder or trespasser sufficient visibility to determine the camera locations.

Non-isolated fence boundaries differ from those described above in that the width of the lighted area inside the plant property line (or area to be lighted) is increased and the width of the lighted area outside the property line is materially decreased. This is particularly important in avoiding glare for adjoining properties. Another very critical area would be that where the lighted property is adjacent to a highway or railroad right-of-way. Projecting brightness and glare into these areas can cause serious safety problems. Residential property that adjoins the lighted property must also be considered: glare in these areas may reduce property value. Methods of obtaining lighting levels in these areas (while controlling the brightness of the luminaire) include the use of visors on louvers—or that of special luminaires with positive cut-off and which use lamps of small source size.

Fenced areas require uniform lighting over approaches and up to the building. A suggested method would be the parapet mounting of luminaires that illuminate the building facade and the walkway along the building. Should there be a parking lot nearby, a different distribution scheme or luminaire could be utilized to light the face of the building, walkway and the parking area—provided the parking area is sufficiently narrow and close enough to the building.

Water-front boundaries present very special problems for lighting and observation. Consideration must be given to approval by the US Coast Guard (or other pertinent government agency) involved in navigable channels or waters. Non-navigable water fronts of less than 4.6 meters (15 feet) wide are similar to (and may be handled much the same as) fenced or unfenced land boundaries. Significant attention must be given to the fact that water becomes an exceptionally good reflecting surface if proper projection angles are not chosen for the applied lighting equipment.

Thoroughfares that traverse open areas and are not close to buildings within industrial properties are commonly lighted with roadway luminaires. See Section 14.

A low level of general illumination serves as security lighting for storage and marshalling yards (large open areas). Lighting may be by groups of floodlights mounted on single poles dispersed through the area; or, if the area is sufficiently small, several floodlights mounted on a single pole.

Should the area involve storage racks sufficiently elevated above the traffic area, lighting equipment can be mounted adjacent to or on these structures to provide guidance and delineation lighting.

Piers and Docks. Lighting for piers and docks may be accomplished by either floodlights or streetlighting luminaires. The area beneath the pier flooring may be lighted with small-wattage floodlights arranged to best advantage with respect to piling. In addition, supplementary lighting with searchlights is recommended for guard towers, so positioned as to permit search of the water area.

REFERENCES

1. *American National Standard Practice for Industrial Lighting*, ANSI/IES RP-7-1979, American National Standards Institute, New York (Sponsored by the Illuminating Engineering Society of North America).
2. Subcommittee on Supplementary Lighting of the Committee on Lighting Study Projects in Industry of the IES: "Recommended Practice for Supplementary Lighting," *Illum. Eng.*, Vol. XLVII, p. 623, November, 1952.
3. Subcommittee on Aircraft Industry of the Industrial Lighting Committee of the IES: "Lighting for the Aircraft/Airline Industries—Manufacturing and Maintenance," *J. Illum. Eng. Soc.*, Vol. 4, p. 207, April, 1975. Subcommittee on Aircraft Industry of the Industrial Lighting Committee of the IES: "Lighting for the Aircraft/Airline Industries—Airframe Maintenance," *Light. Des. Appl.*, Vol. 8, p. 41, June, 1978.
4. Subcommittee on Lighting in Bakeries of the Committee on Lighting Study Projects in Industry of the IES: "Lighting in Bakeries," *Illum. Eng.*, Vol. XLV, p. 387, June, 1950.
5. IES-ASAE (Joint Farm Lighting) Committee: "Lighting for Dairy Farms," *Illum. Eng.*, Vol. LXII, p. 441, July, 1967.
6. Joint Farm Lighting Committee of the IES and ASAE: "Lighting for the Poultry Industry," *Illum. Eng.*, Vol. 65, p. 440, July, 1970.
7. Subcommittee on Electric Generating Stations of the Industrial Lighting Committee of the IES: "Lighting Outdoor Locations of Electric Generating Stations," *J. Illum. Eng. Soc.*, Vol. 4, p. 220, April, 1975.
8. Committee on Lighting of Central Station Properties of the IES: *Lighting of Central Station Properties; Part I: Lighting of Control Rooms; Part II: Lighting of Load Dispatch Rooms*, Illuminating Engineering Society, New York, 1951. Subcommittee on Lighting of Indoor Locations of the Committee on Lighting of Central Station Properties of the IES: "Lighting Indoor Locations of Central Station Properties," *Illum. Eng.*, Vol. LII, p. 423, August, 1957.

9. Subcommittee on High Bay Lighting of the Committee on Lighting of Central Station Properties of the IES: "Lighting of Central Station High Bay Areas," *Illum. Eng.*, Vol. L, p. 395, August, 1955.
10. Subcommittee on Electric Generating Stations of the Industrial Lighting Committee of the IES: "Nuclear Power Plant Lighting," *J. Illum. Eng. Soc.*, Vol. 5, p. 107, January, 1976.
11. Subcommittee on Lighting in Flour Mills of the IES: "Lighting for Flour Mills," *Illum. Eng.*, Vol. XLIV, p. 691, November, 1949.
12. Subcommittee on Lighting for Foundries of the Committee on Lighting Study Projects in Industry of the IES: "Lighting for Foundries," *Illum. Eng.*, Vol. XLVIII, p. 279, May, 1953.
13. Ruth, W., Carlsson, L., Wibom, R. and Knave, B.: "Work Place Lighting in Foundries," *Light. Des. Appl.*, Vol. 9, p. 22, November, 1979.
14. Subcommittee on Lighting in the Canning Industry of the Committee on Lighting Study Projects in Industry of the IES: "Lighting for Canneries," *Illum. Eng.*, Vol. XLV, p. 45, January, 1950.
15. Color Committee of the Graphic Arts Subcommittee of the Industrial Committee of the IES: "Lighting for the Color Appraisal of Reflection-Type Materials in Graphic Arts," *Illum. Eng.*, Vol. LII, p. 493, September, 1957.
16. *American National Standard Viewing Conditions for the Appraisal of Color Quality and Color Uniformity in the Graphic Arts*, ANSI PH 2.32-1972, American National Standards Institute, New York.
17. Subcommittee for Lighting in the Logging and Sawmill Industries of the Industrial Lighting Committee of the IES: "Lighting in the Logging and Sawmill Industries," *J. Illum. Eng. Soc.*, Vol. 6, p. 67, January, 1977.
18. Subcommittee on Sawmill Lighting of the Industrial Lighting Committee of the IES: "Lighting for Sawmills: Redwood Green Chain," *Illum. Eng.*, Vol. LII, p. 381, July, 1957.
19. Committee on Lighting for the Machining of Small Metal Parts of the IES: "Lighting for the Machining of Small Metal Parts," *Illum. Eng.*, Vol. XLIV, p. 615, October, 1949.
20. Subcommittee on Lighting in the Clothing Industry of the Industrial Lighting Committee of the IES: "Lighting for the Manufacturing of Men's Clothing," *Illum. Eng.*, Vol. LVII, p. 379, May, 1962.
21. Petroleum, Chemical, and Petrochemical Subcommittee of the Industrial Lighting Committee of the IES: "Lighting for Petroleum and Chemical Plants," *J. Illum. Eng. Soc.*, Vol. 6, p. 184, April, 1977.
22. *National Electrical Code*, National Fire Protection Association, 60 Batterymarch Street, Boston. Latest edition.
23. Subcommittee on Outdoor Productive Areas of the Industrial Lighting Committee of the IES: "Railroad Yard Lighting," *Illum. Eng.*, Vol. LVII, p. 239, March, 1962.
24. Subcommittee on Lighting for the Rubber Industry of the Industrial Committee of the IES: "Lighting for Manufacturing Rubber Tires," *Illum. Eng.*, Vol. 64, p. 112, February, 1969.
25. Committee on Industrial and School Lighting of the IES: "Report on Lighting in the Shoe Manufacturing Industry," *Trans. Illum. Eng. Soc.*, Vol. XXXII, p. 289, March, 1937.
26. Subcommittee on Lighting in Steel Mills of the Committee on Lighting Study Projects in Industry of the IES: "Lighting for Steel Mills—Part 1: Open Hearth," *Illum. Eng.*, Vol. XLVII, p. 165, March, 1952.
27. Subcommittee on Cotton Gin Lighting of the Industrial Committee of the IES: "Lighting for Cotton Gins," *J. Illum. Eng. Soc.*, Vol. 1, October, 1971.

Residential Lighting[1]

SECTION 10

The purpose of this Section is to serve as a guide to those designing lighting for interior living spaces—not residential alone. Many areas in commercial and industrial buildings may be treated as living spaces; waiting rooms, reception areas and private offices are among the many that fall into this category.

To avoid restricting the imagination this Section defines the lighting problems in detail, but no specific solutions are offered. It presents first the design objectives, then the quality and quantity criteria to be used in the design. This is followed by lighting methods and equipment available for the design. There is also a section on energy considerations.

DESIGNING THE LUMINOUS ENVIRONMENT

Light is an element of design which has certain characteristics that affect the mood and atmosphere of the space—influencing the emotional responses of the people who occupy the space. The definition and character of space is greatly dependent on the distribution and pattern of illumination. Luminaires themselves have dimensional qualities that may be used to strengthen or minimize architectural line, form, color, pattern and texture.

The design objectives of lighting residential type interiors are to provide illumination for the activity in the area, be it a difficult seeing task or casual entertainment; and blend these solutions in such a manner that they are in harmony with those basic esthetic and emotional sentiments associated with this type of interior. There should be brightness patterns to avoid monotonous "even" light and permit interesting interplay of light and shadow. For areas used for activities which do not involve close vision, illuminance levels are less critical and substantially stronger brightness patterns (higher luminance ratios) may be employed. Scintillating, exciting effects, or more subdued, intimate ones, may be geared to the desired mood.

Certain basic information is needed if the designer is to clearly identify the design problem and develop solutions that meet the occupants' needs.

Design Factors.

1. Number and age of people using the area.
2. Type of visual activities performed in the area.
3. Location and size of the task area if there is a specific one.
4. Frequency and duration of use.
5. Architectural style or period of the space and furnishings.
6. Size and scale of the space.
7. Desired mood to be created.
8. Structural constraints.
9. Building and electrical code requirements.
10. Power budget restraints.

Lighting Criteria.

1. Quality of ilumination—visual comfort considerations, absence of glare, color of light and its rendition of colors.
2. Quantity of illumination (illuminance) to be provided.
3. Daylight availability.
4. Type of equipment that will do the job.

NOTE: References are listed at the end of each section.

LIGHTING CRITERIA

Quality of Light

Lighting can be described as being "soft" or "hard." Soft or diffused light minimizes shadows and provides a more relaxing and less visually compelling atmosphere. When used alone the effect can be lacking in interest, like an outdoor scene on an overcast day. The artful use of hard light can provide highlights and shadows that emphasize texture and add beauty to form, as a shaft of sunlight may. An effect of brilliance or sparkle is obtained from small unshielded sources such as a bare lamp or a candle flame. Such sources are seldom used as the prime origin of illumination; they are generally decorative, and must be supplemented by other means of illumination. The glitter of crystal, jewels, and polished brass, and the luster and sheen of table settings and some types of surface materials create a sense of life and gaiety. Occasionally it is desirable to have more than one lighting system in a given space, so that distinct changes in atmosphere and mood can be created.

Zone	Luminance Ratios
2—Area adjacent to the visual task	
a) Desirable ratio	⅓ to equal to task†
b) Minimum acceptable ratio	⅕ to equal to task†
3—General surrounding	
a) Desirable ratio	⅕ to 5 times task†
b) Minimum acceptable ratio	⅒ to 10 times task†

† Typical task luminance range is 40 to 120 candelas per square meter [12 to 35 footlamberts] (seldom exceeds 200 candelas per square meter [60 footlamberts]).

Fig. 10-1. Seeing zones and luminance ratios for visual tasks.

Luminance Ratios. In general, an individual's visual field is considered to consist of three major zones (see Fig. 10-1). First is the task itself (zone 1 of Fig. 10-1), second is the area immediately surrounding the task (zone 2), and third is the general surroundings (zone 3). The brightnesses visible in all three zones are important; unsuitable relationships among them can cause distraction, fatigue, and even difficulty in seeing. For visual comfort in the performance of such tasks as studying, reading, sewing or any activity demanding close vision (especially if continued for some time) the luminance ratios in Fig. 10-1 should not be exceeded.

Achieving Luminance Ratios. There are a number of ways in which luminances in the visual field may be brought into balance. In general, they involve the following:

1. Limitation of the luminance of luminaires.
2. For daytime activities, control of high luminance of window areas by blinds, shades or draperies; use of high-reflectance colors on walls adjacent to fenestration.
3. For nighttime activities, use of light-colored covering for windows in the field of view.
4. Use of materials and finishes of favorable reflectance for large areas of room surfaces and furnishings.
5. Provision of additional lighting in the visual surround from other sources not specifically directed to the task.

Room Surface Reflectance. Every colored room surface reflects some portion of the light it receives, and this portion, expressed in percentage of the total light, is its reflectance.

To attain recommended luminance ratios of the visual task to the immediate surround and the more remote surround as listed in Fig. 10-1, use of matte finish surfaces with reflectances shown in Fig. 10-2, are recommended.

Surface Colors and Color Schemes. Colors of objects often appear to change with surface finish. Specular or mirror reflection from glossy surfaces may, in extreme cases, increase the chroma and darkness of the sample at one angle and wipe out all color at other angles, as well as

Fig. 10-2. Reflectances for Interior Surfaces of Residences

Surface	Reflectance (per cent)
Ceiling	60 to 90
Curtain and drapery treatment on large wall areas	45 to 85
Walls	35 to 60*
Floors	15 to 35*

* In areas where lighting for specific visual tasks takes precedence over lighting for environment, minimum reflectances should be 40 per cent for walls, 25 per cent for floors.

cause distracting glare. A matte finish reflects light diffusely, and thus gives an object the appearance of its "natural" color. Deeply textured finishes, such as velvet and deep-pile carpeting, cause shadows that make materials appear darker than smooth-surfaced materials, such as satin, silk, or plastic laminates, of the same inherent color.

The quality of light reaching a surface is the average of the light that reaches it from all points in the environment. If daylight from a bright blue sky is reflected from the green leaves of a tree into a room with a red rug and pink walls, the color of the light falling on an object in the room is not that of the daylight, but is the result of the daylight as modified by these several reflections. As light is reflected or bounced around a room, inter-reflections from large areas of a single color can cause the "amplification" of chroma or saturation of the wall color and cause it to appear more vivid than the original small sample from which it was selected.

There are no "rules" for color harmony. Certain generalities have been arrived at, however, through experience. These include the principle of order, in which there is a recognizable pattern in the color relationship; the principle of similarity, in which the same hue family or same value group is used; and the principle of familiarity, such as national identification (red, white and blue), seasonal identification (red and green), or personal identification (pink and blue). These principles merely indicate the general direction in which esthetic judgement may be applied with some safety. More sophisticated color schemes tend toward subtle relationships of color and cannot be analyzed to conform to a given axiomatic pattern. See Section 5, page 5-18, in the 1981 Reference Volume for a discussion of the use of color.

A well designed lighting system cannot be planned independently of the color and character of the surfaces it is to illuminate. The effect attained depends upon the way surfaces within an area reflect and absorb light. Thus ceilings, walls, floors and major furnishings become an integral part of the lighting design.

Fading and Bleaching. When selecting fabrics and furnishings, their color stability should be considered. Not all fabrics have the same color stability and they may fade or change chemical composition because of varying environmental reasons. See page 19-31 for additional specific information.

Task Surfaces. Visual comfort and ease of seeing are also influenced by a task which has glossy or specular surfaces. Some areas of a glossy task may contain veiling reflections from light sources or luminaires. Other areas may contain reflected glare. Veiling reflections and reflected glare cannot always be completely eliminated. Reduction of the light source or luminaire luminances or more careful placement of the source (or the task) tends to alleviate the condition. Reflected images are further "softened" if part of the light on the task comes "indirectly" from walls, ceilings, or other luminaires. See Section 2.

Quantity of Light

Visual activities in living spaces range from simple orientation to performance of difficult seeing tasks. Activities requiring the seeing of small details having poor contrast with their background, particularly if the colors are dark, is much more difficult visually and needs a higher illuminance level than work involving relatively large details of good contrast with the background. Simple orientation requires a much lower illuminance level.

Recommended illuminance values are listed in Fig. 2-2 and in Figs. 10-3 through 10-20 for general and seeing task lighting. Values are given as three-step ranges of maintained illuminance in both lux and footcandles, and are intended to be target values for design and evaluation purposes. See page 2-21 for guidance in selecting an appropriate value within each range.

Maintained values represent the illuminance just before lighting equipment is cleaned and relamped, that is, when the equipment is providing the least light due to dirt accumulation and lamp light output depreciation.

LIGHTING METHODS

General Lighting

Areas with Visual Activities. Residence lighting is planned on the basis of activities, not on the basis of rooms. Many seeing tasks can be, and are, performed in almost any part of the home. The designer must determine what activities are to be carried on in each room, and provide light for them first. The individual task areas are then tied together by means of a general or environmental lighting system. This is essential in order to prevent a spotty effect, to maintain recommended luminance ratios in the field of view, to provide light throughout the interior for safety in moving about and for general housekeeping activities, and to avoid the necessity for rapid eye adaptation required by excessive differences in illuminances between adjoining rooms.

In some cases, notably in utilitarian areas such as kitchens and laundries, a general lighting system can be designed to supply all, or nearly all, of the illumination needed for visual activities. However, the relatively high level and uniform distribution of general lighting necessary in a kitchen would be unacceptable, practically and esthetically, in a living area. The equipment most commonly used to light room surfaces and create a satisfactory background for visual work includes general diffusing ceiling luminaires, wall lighting elements, indirect portable lamps, and coves. In small rooms general illumination may even be supplied by the luminaires employed for specific task lighting—the mirror lighting in a bathroom, for example, or a number of well-designed open-top floor and table lamps used to supply reading light in a living room.

Areas Devoted to Relaxation. For periods when no close visual work is being done in a living area, or for the occasional room designed solely for relaxation, a low level of general lighting creates a pleasant atmosphere for conversation, watching television, or listening to music. Uniformity of illumination need not be an objective here. On the contrary, moderate variations create a more attractive pattern of light. Any of the methods mentioned above, plus a wide range of possibilities limited only by the imagination and ingenuity of the designer, are applicable. The primary considerations are comfort and esthetic satisfaction.

Halls and Stairways. Changes in eye adaptation determine the proper illuminance levels here. If the hall or stairway adjoins an interior area with a higher level, the illuminance should be one-fifth that of adjacent areas. Wall luminances are most important in creating a sensation of lightness, and in affecting adaptation. Therefore luminaires should direct light to the walls, and wall and floor finishes should have high reflectance values. Lighting in the entry hall should be flexible, so that adjustment can be made for eye adaptation in the daytime and at night. On stairs, lighting of the top and bottom steps is especially critical for safety. Differences in color and directional quality of light can help to emphasize treads. Under no circumstances should a luminaire or open-top portable lamp be located where a person descending the stairway can look directly at the light source.

Garages. The major areas where light is needed in a garage are on each side of the car or between cars, especially over the front and rear. Luminaires are usually located slightly to the rear of the trunk area, and approximately in line with the front wheels. A portable trouble light should be provided for repair work.

Closets. Light sources in closets should be located out of the normal view, generally to the front of the closet and above the door. In walk-in closets a ceiling luminaire should be mounted in the center of the traffic area so that shelves do not block the lighting of the garments.

Lighting for Specific Visual Tasks

In providing the recommended illuminance on the task plane, which corresponds to Zone 1 in Fig. 10-1, the essentials of good lighting quality must not be overlooked: proper luminance ratios in Zones 1, 2 and 3 wherever serious visual work is done, precautions against veiling reflection, use of high-reflectance surfaces and the limitations on various types of luminaires noted in the paragraphs that follow under Luminaires on page 10-19. Not many of the illuminance levels listed in Fig. 2-2 and Figs. 10-3 through 10-20 can be practically achieved by general lighting alone and local lighting alone is seldom satisfactory from the standpoint of comfort. In most circumstances, a combination of local and general lighting is needed. Daylighting may also contribute.

The recommendations which follow in Figs. 10-3 through 10-20 are organized for each specific task on the basis of (1) statement of the task, (2) description and location of the task plane, (3) recommended illuminance levels, (4) special considerations and requirements of the task and (5) typical equipment locations.

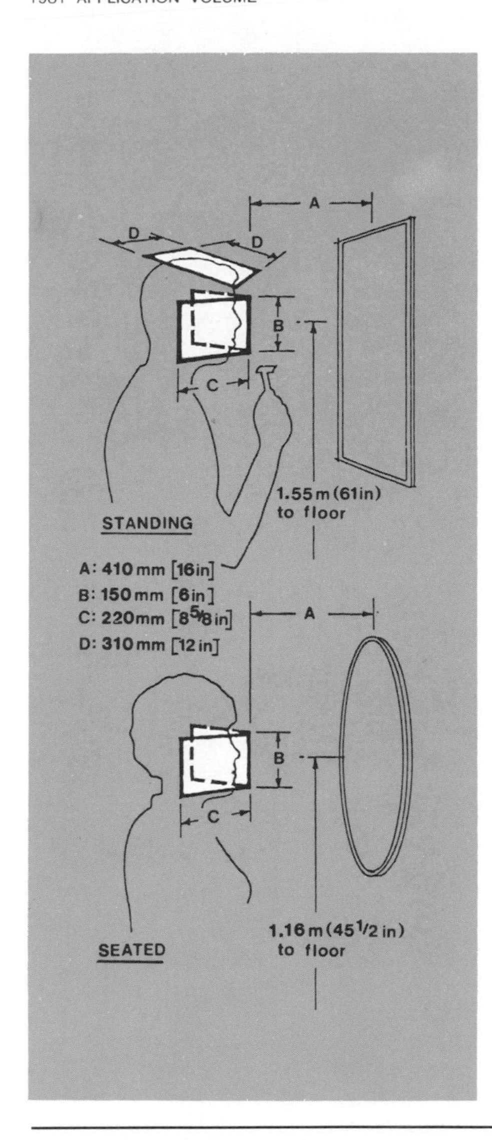

Fig.10-3. Grooming.

1. The task: The chief tasks are shaving and makeup. Because the apparent distance of the face or figure as viewed in the mirror is twice its actual distance from the mirror, and because the details to be seen in shaving or makeup are usually small and of low contrast with their background, the visual task may be critical. Skin and hair reflectance can be quite low, below 30 per cent, and the speed and accuracy can be critical for a fastidious individual rushed for time.

2. Description of the task planes: Standing position—the task area consists of two 150 by 220 millimeter (6 by 8⅝ inch) planes at right angles with each other, converging at a point 410 millimeters (16 inches) out from the mirror, and centered vertically 1550 millimeters (61 inches) above the floor. They represent the front and sides of the face. A third plane 310 millimeters (12 inches) square, its front edge also 410 millimeters (16 inches) out from the mirror, is tilted up 25 degrees above the horizontal and represents the top of the head. Seated position—the two facial planes are identical in size and position to those mentioned above, except that the center of the planes is 1160 millimeters (45½ inches) above the floor. No top-of-the-head plane is considered for seated grooming.

3. Illuminance recommendations (maintained on the task plane at any time): 200, 300 or 500 lux [20, 30 or 50 footcandles].

4. Special design considerations: Lighting equipment at a mirror should direct light *toward the person* and not onto the mirror. The luminances of surfaces reflected in the mirror and seen adjacent to the face reflection should not be in distracting contrast with it.
a) Adjacent walls should have a 50 per cent or higher reflectance.
b) Luminaires should be mounted outside the 60-degree visual cone, the center line of which coincides with the line of sight.
c) No luminaire should exceed 2100 candelas per square meter (600 footlamberts) in luminance, *i.e.*, an illumination meter held against it should not read more than 6500 lux (600 footcandles).

5. Typical equipment locations: Wall-mounted linear or nonlinear luminaires over the mirror. Wall-mounted linear or nonlinear luminaires over and at the sides of the mirror.
Combination of wall- and ceiling-mounted luminaires flanking the mirror and over the head of the user.
Structural devices (such as soffits) extending the length of the mirror.
Portable luminaires with luminous shades flanking the mirror.
Ceiling-pendant luminaires with luminous sides flanking the mirror.
NOTE: If grooming is performed in a seated position, the relationship of the luminaires to the face should remain as specified above for standing.

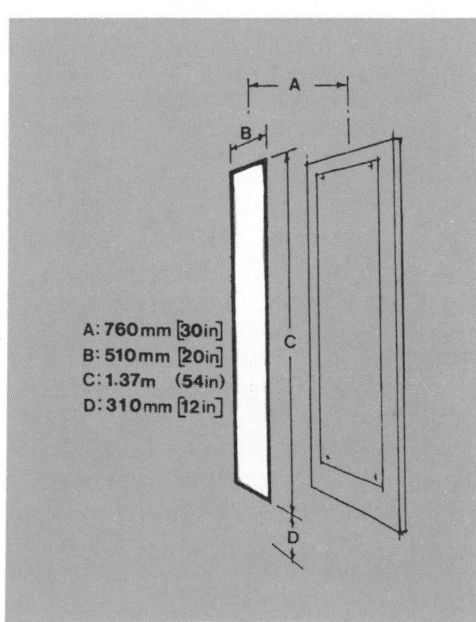

Fig. 10-4. Full-length mirror.

1. The task: The task is the alignment of clothing, commonly with reflectance between 30 and 70 per cent, and casual over-all appraisal. Speed and accuracy may or may not be important.

2. Description of the task plane: The task area is a plane 510 millimeters (20 inches) wide by 1370 millimeters (54 inches) high with the lower edge 310 millimeters (12 inches) above the floor; it is centered on and parallel with the mirror, 760 millimeters (30 inches) from the mirror surface.

3. The illuminance recommendations (maintained on the task plane at any time): 200, 300 or 500 lux [20, 30 or 50 footcandles].

4. Special design considerations: Lighting equipment at the mirror should direct light *toward the person* and not onto the mirror. The luminances of surfaces reflected in the mirror and seen adjacent to the face reflection should not be distracting.
a) Luminaires should be mounted outside the 60-degree visual cone, the center line of which coincides with the line of sight.
b) No luminaire should exceed 2100 candelas per square meter (600 footlamberts) in luminance, *i.e.*, an illumination meter in contact with the surface should not read more than 6500 (600).

5. Typical equipment locations: Vertical linear luminaires wall-mounted beside the mirror. The same, supplemented by wall-mounted or ceiling-mounted over-mirror luminaires.

Fig. 10-5. Desk.

1. The task: The tasks range from casual reading or writing to prolonged difficult reading, handwriting, typing and drawing. When studying, the task may involve fine print and close detail. Task reflectances are usually between 30 and 70 per cent. Speed and accuracy may not be important for the casual tasks, but may be important or critical for study.

2. Description of the task planes: The primary task is a plane 360 by 310 millimeters (14 to 12 inches) parallel with the desk top. The bottom edge of the task plane is 76 millimeters (3 inches) from the front edge of the desk. A secondary task plane for reference books, large drawings, etc., measures 610 millimeters (24 inches) deep by 910 millimeters (36 inches) wide with the front edge at the front of the desk top.

3. Illuminance recommendations (maintained on the task plane at any time): a) Primary task plane—casual reading or writing—200, 300 or 500 lux [20, 30 or 50 footcandles]; b) Primary task plane—prolonged study using difficult tasks—500, 750 or 1000 lux [50, 75 or 100 footcandles]; c) Maximum illuminance on the primary plane should not exceed the minimum by more than 3 to 1; d) Secondary task plane—½ the primary task plane but not less than 200 lux [20 footcandles].

4. Special design considerations: Equipment should be located so that shadows are not cast on the task area by the user's hand. The surface of the desk top should be nonglossy and light in color (30 to 50 per cent reflectance). The luminance of any luminaire visible from a normal seated position should be no more than 510 candelas per square meter (150 footlamberts), no less than 170 (50).

5. Typical equipment locations: Desk-mounted (1 or more luminaires). Wall-mounted (1 or more luminaires). Ceiling-mounted (1 or more luminaires). Floor mounted.

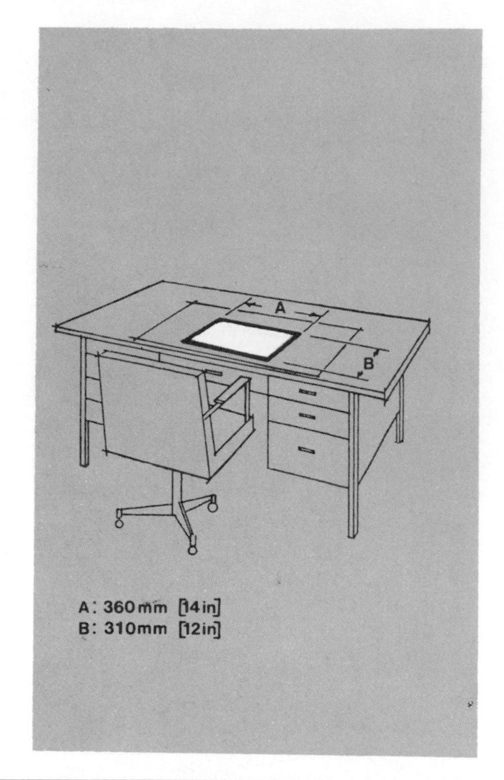

Fig. 10-6. Reading in bed.

1. The task: The majority of people who read in bed are only casual readers, perhaps reading for a few minutes before going to sleep. They are often interested in closely confining the light distribution so as not to disturb another occupant of the room. Such lighting arrangements are not satisfactory for comfortable reading over a long period. The following recommendations are for the individual who reads for a more extended period, or for the person who performs very critical eye tasks while confined to bed. The normal materials vary from books and magazines to pocket editions and to newspaper print with reflectances of 30 to 70 per cent. Speed and accuracy will vary from not important for leisure reading to important to critical for critical tasks.

2. Description of the task plane: The task plane is 310 by 360 millimeters (12 by 14 inches) tilted at an angle of 45 degrees from the vertical. The center of the task plane is 610 millimeters (24 inches) out from the headboard or wall and 310 millimeters (12 inches) above the mattress top. There are no customary reading positions or habits. These recommendations are based on the assumption that the reader is in an upright or semi-reclined position.

3. Illuminance recommendations (maintained on the task plane at any time): Normal reading—200, 300 or 500 lux [20, 30 or 50 footcandles]. Serious prolonged reading or critical eye work—500, 750 or 1000 lux [50, 75 or 100 footcandles].

4. Special design considerations: Equipment should be located so that no shadows are cast on the reading plane by the head or body, and so that the luminaire does not interfere with a comfortable position.

5. Typical equipment locations: Wall-mounted directly in back of or to one side of the user (both linear and non-linear designs).
Luminaire on bedside table or storage headboard.
Pole-type luminaires (floor-to-ceiling or table-to-ceiling).
Ceiling-mounted: a) Suspended: adjustable or stationary. b) Surface-mounted: directional or nondirectional. c) Luminous-area panel: surface mounted or recessed. d) Recessed: directional or nondirectional.
Luminaire incorporated into furniture design.

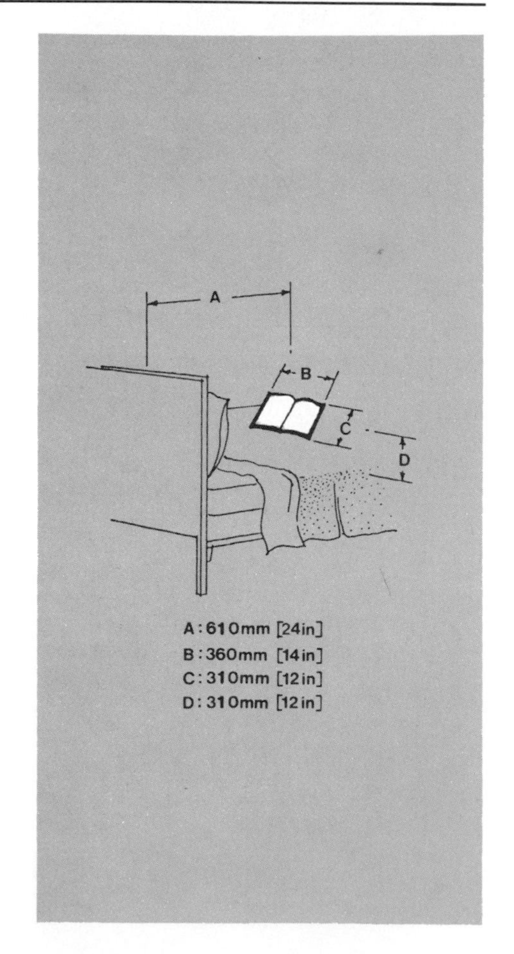

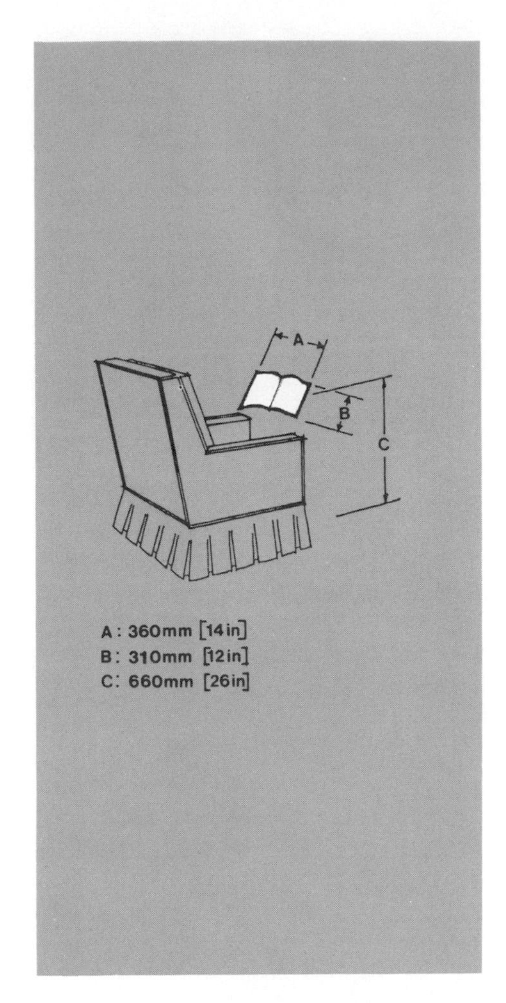

Fig. 10-7. Reading in a chair.

1. The task: Typical reading tasks in a home encompass a wide range of seeing difficulty, from short-time casual reading of material with good visibility (large print on white paper) to prolonged reading of poor material (small type on low-contrast paper). The majority of tasks have reflectances between 30 and 70 per cent or higher. Speed and accuracy may not be important for casual reading, but may or may not be important for prolonged reading.

2. Description of the task plane: The task plane measures 360 millimeters (14 inches) wide by 310 millimeters (12 inches) high with the center of the plane approximately 660 millimeters (26 inches) above the floor. The plane is tilted at 45 degrees from the vertical. The reader's eyes are approximately 1 meter (40 inches) above the floor.

3. Illuminance recommendations (maintained on the task plane at any time): a) Normal reading of books, magazines, papers—200, 300 or 500 lux [20, 30 or 50 footcandles]; (b) Prolonged difficult reading or handwriting, reproductions, poor copies—500, 750 or 1000 lux [50, 75 or 100 footcandles].

4. Special design considerations: Normal seated eye level is 970 to 1070 millimeters (38 to 42 inches) above the floor and is a critical consideration when the light source is to be positioned beside the user. The lower edge of the shielding device should not be materially above or below eye height. This will prevent discomfort from bright sources in the periphery of the visual field and yet permit adequate distribution of light over the task area. Variations in chair and table heights necessitate selection and placement of equipment to achieve this relationship for each individual case.

5. Typical equipment locations:
Table-mounted, floor-mounted and wall-mounted alongside or behind the user.
Ceiling-mounted
a) Suspended beside or behind the user
b) Surface—large-area
c) Recessed—large-area
d) Large luminous element
Directional small-area luminaires may be used (wall, ceiling, pole-mounted) in combination with other larger-area diffusing luminaires.

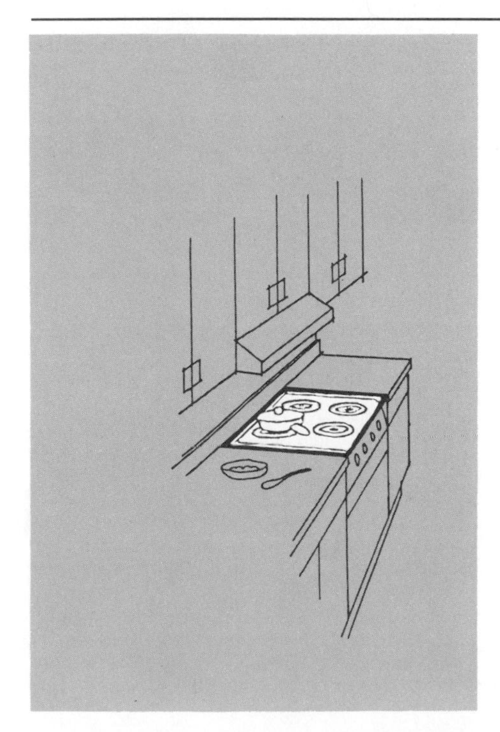

Fig. 10-8. Kitchen range.

1. The task: Typical seeing tasks at the kitchen range are the determination of the condition of foods in all stages of the cooking process (color and texture evaluation) and reading range controls, instructions and recipes. Food often has a reflectance of less than 30 per cent. Speed and accuracy is not important for noncritical tasks, but can be critical for some difficult preparation tasks.

2. Description of the task plane: The task area is a range top. Generally this is located 910 millimeters (36 inches) above the floor.

3. Illuminance recommendations (maintained on the task plane at any time): Food preparation and cleaning involving difficult seeing tasks—500, 750 or 1000 lux [50, 75 or 100 footcandles]. Other, noncritical tasks—200, 300 or 500 lux [20, 30 or 50 footcandles]. Variable intensity controls, such as multiple position switching and dimming equipment, can be utilized to lower the illuminance when there are no difficult seeing tasks.

4. Special design considerations: Reflected glare is inherent in the shiny finish of utensils and range tops. Some reduction in the luminance of reflected images may be obtained by the use of diffuse luminaires and/or sources. Color-rendering qualities of light source are especially important in the kitchen. Lighting equipment installed on the range by the manufacturer seldom is located high enough to direct light into cooking utensils.

5. Typical equipment locations:
Range hood.
Ceiling—recessed, surface-mounted or pendant.
Structural—lighted soffits, wall brackets, canopies.
Underside of wall cabinets.

Fig. 10-9. Kitchen counter.

1. The task: Typical seeing tasks at the kitchen counter include reading fine print on packages and cookbooks, handwritten recipes in pencil and ink, and numbers for speeds and temperatures on small appliances. Other tasks include measuring and mixing, color and texture evaluation of foods, safe operation of small appliances, and cleanup. Task reflectances are usually less than 70 per cent and often below 30 per cent. Speed and accuracy may not be important for noncritical tasks but critical for the difficult preparation and cleaning tasks.

2. Description of the task plane: The task area is a plane 510 millimeters (20 inches) deep (starting at the front edge of the counter) and the length of the counter used.

3. Illuminance recommendations (maintained on the task plane at any time): Food preparation and cleaning involving difficult seeing tasks—500, 750 or 1000 lux [50, 75 or 100 footcandles]. Other, noncritical tasks—200, 300 or 500 lux [20, 30 or 50 footcandles]. Variable intensity controls, such as multiple position switching and dimming equipment, can be utilized to lower the illuminance when there are no difficult seeing tasks.

4. Special design considerations: Although there are a great many ways in which counter surfaces may be lighted, luminaires are commonly mounted under the wall cabinets, above the counter. These may be well shielded because of the cabinet structure itself, but if not, there is need for added shielding. Also, care should be exercised to see that the luminances of the luminaires are comfortable to other users of the room particularly in seated positions.

5. Typical equipment locations:
Under-side of wall cabinets.
Ceiling—recessed, surface-mounted or pendant.
Structural—lighted soffits, wall brackets, canopies.

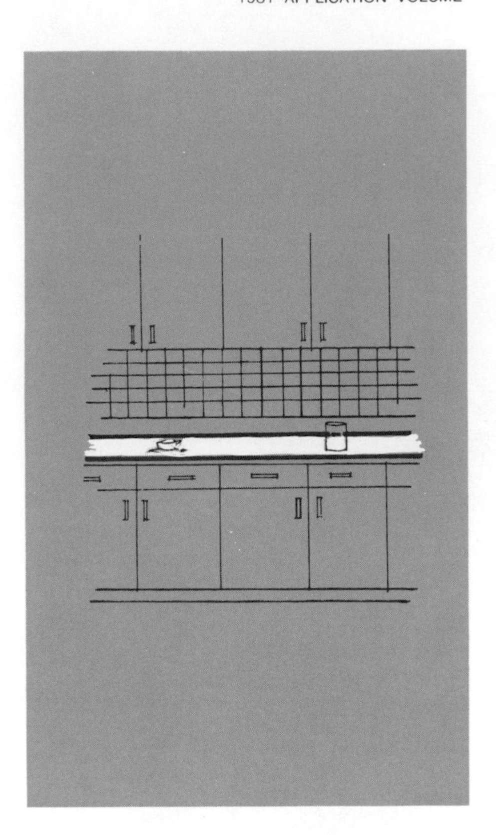

Fig. 10-10. Ironing.

1. The task: The basic visual task in ironing is the detection and removal of wrinkles from garments and the detection of possible scorch. The majority of fabrics have a reflectance of between 30 and 70 per cent, but dark clothes are less than 30 per cent. Speed and accuracy may be important.

2. Description of the task plane: The task plane is 310 by 660 millimeters (12 by 26 inches) and varies in its height, depending upon the ironing board. In general such boards are adjustable from 560 to 910 millimeters (22 to 36 inches) in height, to be used either standing or sitting. The seat of the average stool is located 610 millimeters (24 inches) above the floor, which places the average seated individual's eye level at 1350 millimeters (53 inches). In a standing position the eye level is 1550 millimeters (61 inches).

3. Illuminance recommendations (maintained on the task plane at any time): 200, 300 or 500 lux [20, 30 or 50 footcandles].

4. Special design considerations: A light source with directional quality may frequently reveal shadows cast by small wrinkles, creases, etc., to the advantage of the user. Ironing and television viewing are often done at the same time. Under these circumstances it is important to ensure a good balance among the luminances of task, television screen and other room surfaces in the line of sight.

5. Typical equipment locations:
Ceiling-mounted:
a) Suspended, adjustable
b) Fixed, directional
c) Fixed, nondirectional
d) Combination of luminaires (general diffusing plus directional component).

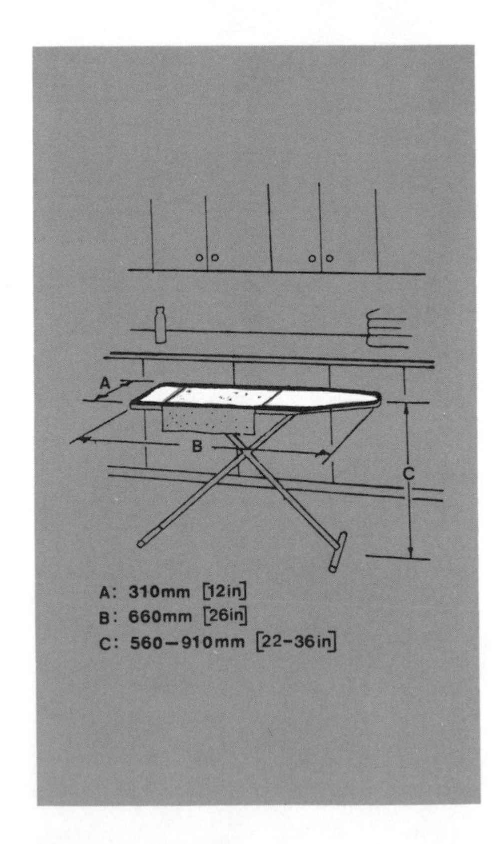

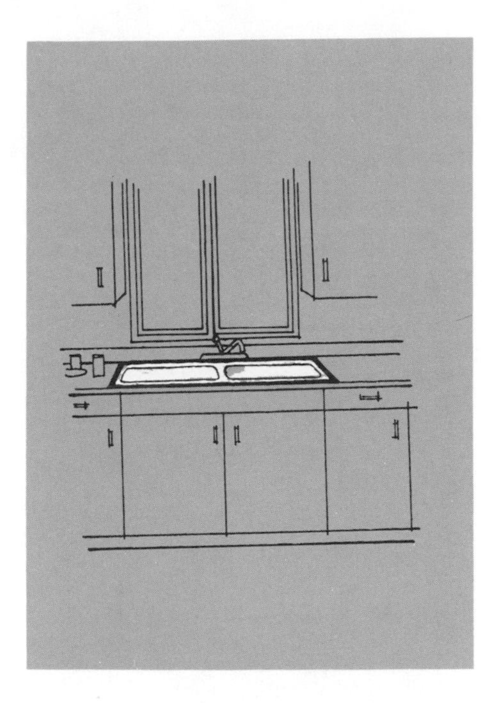

Fig. 10-11. Kitchen sink.

1. The task: Typical seeing tasks at the kitchen sink involve cleaning and inspection of dishes and utensils, evaluation of color and texture of foods in preparation, reading and measuring. Task reflectances and speed and accuracy needs are similar to those at the counters.

2. Description of the task plane: The task area measures 510 by 760 millimeters (20 by 30 inches). It is usually at a height of 910 millimeters (36 inches) above the floor.

3. Illuminance recommendations (maintained on the task plane at any time): Food preparation and cleaning involving difficult seeing tasks—500, 750 or 1000 lux [50, 75 or 100 footcandles]. Other, noncritical tasks—200, 300 or 500 lux [20, 30 or 50 footcandles]. Variable intensity controls, such as multiple position switching and dimming equipment, can be utilized to lower the illuminance when there are no difficult seeing tasks.

4. Special design considerations: Color-rendering qualities of the light source are particularly important in kitchen illumination. The limited space available for luminaire mounting at the sink location increases the possibility of shadows being cast on the work plane by the head or body of the user.

5. Typical equipment locations:
Ceiling—recessed, surface-mounted or pendant.
Structural—lighted soffits, wall brackets, canopies.
Underside of wall cabinets.

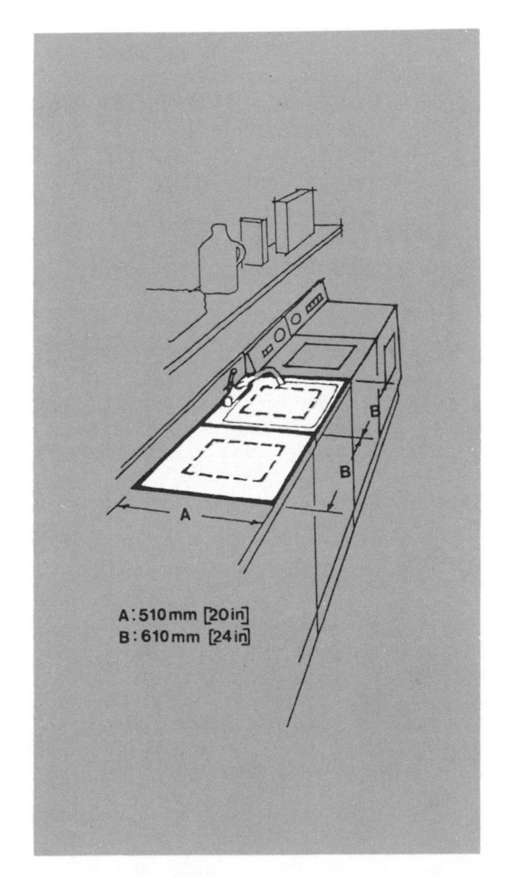

Fig. 10-12. Laundry.

1. The task: Preparation area—the tasks are sorting of fabrics by color and type, determination of location and type of soil, prewash treatment, tinting, bleaching and starching.

Tub area—the tasks are soaking, handwashing, tinting, rinsing, bleaching and starching.

Washing machine and dryer area—the tasks are loading, setting of dials and controls, removal of clothes.

Speed and accuracy may be important at the preparation and tub areas but not important at the washer and dryer. Reflectance of fabrics and packaging vary widely but most are within 30 to 70 per cent.

2. Description of the task planes: Preparation area—the general task area is 510 by 610 millimeters (20 by 24 inches) with a critical seeing area 310 by 310 millimeters (12 by 12 inches). Tub area—the task area is 510 by 610 millimeters (20 by 24 inches) on a single laundry tub with a critical seeing area 310 by 310 millimeters (12 by 12 inches) in the center. Washing machine and dryer area—this task area has no definable boundaries and can be illuminated by the general room lighting.

3. Illuminance recommendations (maintained on the task plane at any time): Task planes at both preparation and tub areas and at the washer and dryer—200, 300 or 500 lux [20, 30 or 50 footcandles].

4. Special design considerations: Totally direct and not highly diffused light sources can contribute to the visibility of certain laundry tasks. In most laundry locations task lighting equipment also must provide the general room illumination. Luminaires should in this case be designed and positioned to illuminate the ceiling and side walls for comfortable luminance relationships.

5. Typical equipment locations: Ceiling-mounted (pendant, surface or recessed) linear or non-linear luminaires centered over the front edge of the laundry equipment. As above, supplemented by wall-mounted or cabinet-mounted linear luminaires.
Large-area luminous panels.

Fig. 10-13. Machine sewing.

1. The task: The small detail and low contrast between thread and material usually involved in machine sewing make it a visually difficult task. The degree of difficulty varies with thread and stitch size, reflectance of materials and contrast between thread and fabric. Speed and accuracy may be critical.

2. Description of the task plane: The primary task area is a plane 150 millimeters (6 inches) square located so that the needle point is 50 millimeters (2 inches) forward from the center of the back edge.
The secondary task area of less critical seeing measures 310 by 460 millimeters (12 by 18 inches) with the needle point centered on the shorter dimension and 150 millimeters (6 inches) in from the right hand edge.

3. Illuminance recommendations (maintained on the primary task plane at any time):
a) Dark fabrics, less than 30 per cent reflectance (fine detail, low contrast)—1000, 1500 or 2000 lux [100, 150 or 200 footcandles].
b) Prolonged periods (light to medium fabrics, 30 to 70 per cent reflectance)—500, 750 or 1000 lux [50, 75 or 100 footcandles].
c) Occasional periods (coarse thread, large stitches, high contrast thread to fabric, higher than 30 per cent reflectance)—200, 300 or 500 lux [20, 30 or 50 footcandles].
d) Maximum level on the primary task plane should not exceed the minimum by more than 3 to 1.
e) The minimum level on the secondary task plane should not be less than ⅓ of the minimum on the primary task plane, but not less than 200 lux [20 footcandles].

4. Special design considerations: Equipment should be located so that shadows are not cast on the task area by the user's hand. Most sewing tasks involve low-sheen materials which minimize veiling reflections and allow the use of light with a moderate directional component to increase the visibility of threads by casting slight shadows.

5. Typical equipment locations (does not take into account the light built into the machine):
Wall-mounted directly in front of the user. (Both linear and nonlinear sources.)
Ceiling-mounted (location of luminaire and/or machine should avoid the possibility of the user's head blocking out light or casting a shadow on the task).
1) Suspended-adjustable
2) Fixed, directional (surface or recessed) or track mounted
3) Fixed, nondirectional
4) Luminous area
Floor-mounted or pole-mounted.

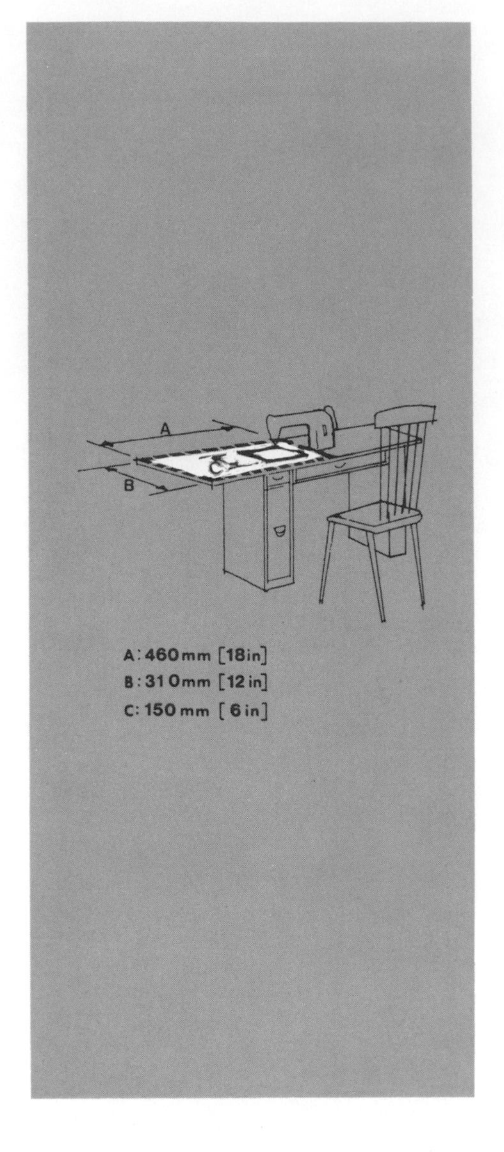

Fig. 10-14. Hand sewing.

1. The task: The seeing task encompasses a wide range of difficulty from coarse threads to fine; from light materials to very dark; from high contrast to virtually no contrast at all. Speed and accuracy may be important.

2. Description of the task plane: The task area is a plane 250 millimeters (10 inches) square tilted at 45 degrees toward the eye. The plane is centered at 760 millimeters (30 inches) from the floor. Eye position is approximately 1000 millimeters (42 inches) from floor.

3. Illuminance recommendations (maintained on the task plane at any time):
a) Dark fabrics, less than 30 per cent reflectance (fine detail, low contrasts)—1000, 1500 or 2000 lux [100, 150 or 200 footcandles].
b) Prolonged periods (light to medium fabrics, 30 to 70 per cent reflectance)—500, 750 or 1000 lux [50, 75 or 100 footcandles].
c) Occasional periods (coarse thread, large stitches, high contrast thread to fabric, greater than 30 per cent reflectance)—200, 300 or 500 lux [20, 30 or 50 footcandles].
d) Maximum illuminance on the task plane should not exceed the minimum by more than 3 to 1.

4. Special design considerations: Equipment should be located opposite the hand being used, so that shadows are not cast on the task area.

5. Typical equipment locations:
Floor-mounted or pole-mounted.
Ceiling-mounted:
a) Suspended, adjustable
b) Fixed, directional (surface or recessed)
c) Fixed, luminous-area
d) Combination luminaire (general diffusing plus directional component)
Wall-mounted (both linear and non-linear) sources located beside or behind the user

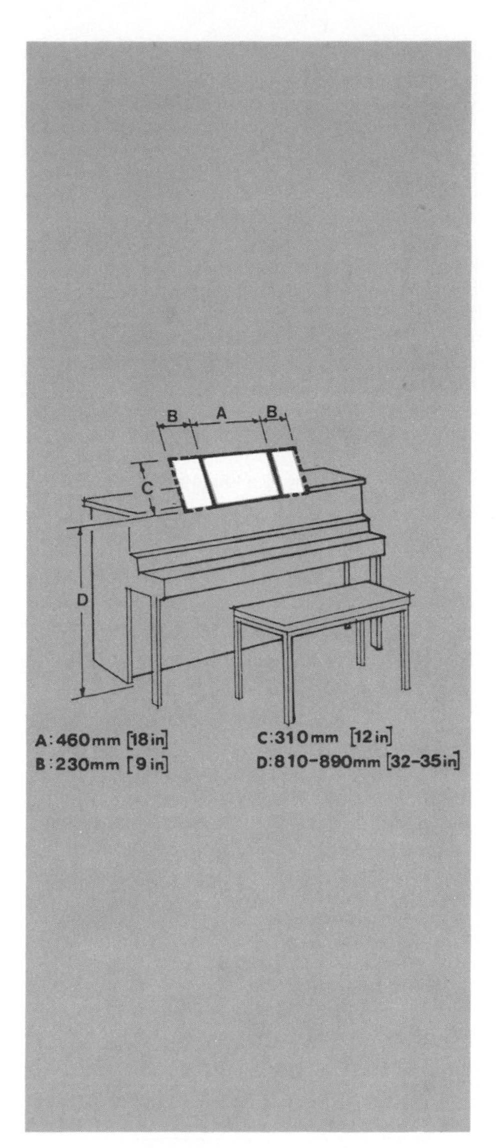

Fig. 10-15. Music study at piano or organ.

1. The task: The task is the reading of musical scores usually 30 to 70 per cent reflectance or higher ranging from very simple ones with large notes and staff lines to very difficult substandard-size scores with notations printed on the lines. Speed and accuracy may not be important for very simple scores, but important for advanced and critical for professional.

2. Description of the task plane: The primary task plane is on the music rack in an area 310 by 460 millimeters (12 to 18 inches); it is tilted back from the viewer about 17 degrees. The lower edge is 810 to 890 millimeters (32 to 35 inches) from the floor. The secondary plane includes an additional 230 by 310 millimeters (9 by 12 inches) on each side of the primary plane. The average piano keyboard, 1220 millimeters (48 inches) long and 710 millimeters (28 inches) above floor, is also a secondary plane. NOTE: These dimensions will vary greatly with electric organs and miniature pianos. The eye is approximately 1200 millimeters (47 inches) above the floor.

3. Illuminance recommendations (maintained on the task plane at any time):

a) Simple scores—200, 300 or 500 lux [20, 30 or 50 footcandles].
b) Advanced scores—500, 750 or 1000 lux [50, 75 or 100 footcandles].
c) When score is substandard size, and notations are printed on the line—1000, 1500 or 2000 lux [100, 150 or 200 footcandles].

The maximum illuminance should not exceed the minimum by more than 3 to 1. The minimum on the secondary task plane should not be less than one-third of the minimum on the primary task plane.

4. Special design considerations: The instrument is in the best position for control of luminance values if the player faces a wall.

5. Typical equipment locations: Ceiling-mounted or recessed above ceiling:

1) Directional source
a) Should be adjustable to strike the plane of the task at about 90 degrees.
b) Should be located above the user's head to avoid a shadow of his body.
c) Should be located and shielded to prevent glare to other persons occupying or passing through the area.
d) Downlights are not desirable; distribution is not good, and reflected glare and veiling reflections may be a problem.
2) Large-area nondirectional source
a) The luminance should be within the comfort range and esthetic considerations of the room.

Mounted on the instrument:
1) Uniformity of distribution over the task plane may be difficult to achieve.
2) There should be no luminous part within the user's field of view having a luminance of more than 170 candelas per square meter (50 footlamberts).

Pole-type luminaires: Because of the directional quality of this type of light source, it is possible to get acceptable illuminances on the task as well as general surround light. Care should be exercised to avoid glare to other occupants in the room as well as veiling reflections on the task area.

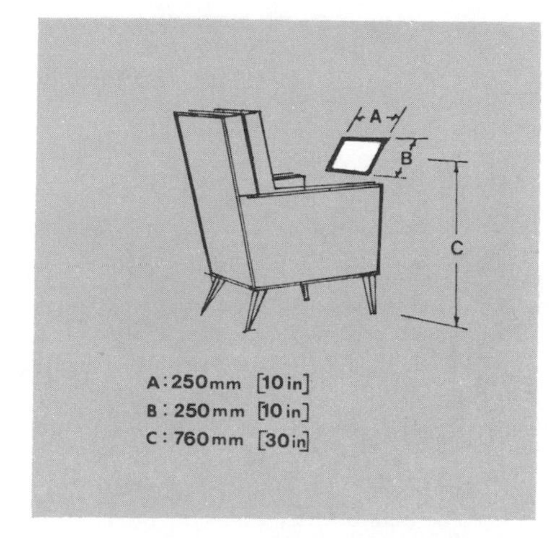

Fig. 10-16. Dining area.

1. The task: The task is essentially one of creating the desired mood or atmosphere for dining. Therefore, the illuminance, the luminances in the room, and the choice of luminaire are largely matters of personal taste.

2. Description of the task plane: The entire tabletop should be considered as the task plane. If carving, serving, etc. is done at a separate location, this becomes another task area and can generally be equated with the task plane for Kitchen Counter.

3. Illuminance recommendations (maintained on the task plane): 100, 150 or 200 lux [10, 15 or 20 footcandles]. Variable-intensity controls, such as multiposition switching and dimming equipment, can often add to the enjoyment of dining-area lighting by adapting the level of illuminance to the particular occasion. Room surface reflectances will influence the selection of the level.

4. Special design considerations: Lighting with a strong downward component will accent the table setting, creating attractive focal highlights. However, this type of distribution, if used alone, will render faces poorly, causing harsh shadows. Strong downward lights should be kept away from people's faces (*i.e.*, confined within the perimeter of the table itself), or should be well balanced by indirect light from the table top, walls or ceiling. The nature of the table top may also influence the choice of lighting distribution: downlighting may cause annoying specular reflections from glossy table tops such as glass or marble, and if all the light is directed on the table a colored tablecloth may appreciably tint the light by reflection.
Exposed sources such as unshielded low-wattage bulbs can often be tolerated, especially if some general lighting is provided and the background is not too dark. The darker the walls, the more general lighting is required to keep luminance relationships in the room within a comfortable range.
In situations where the dining table is moved from time to time, a flexible means of mounting the luminaire is desirable.

5. Typical equipment locations:
Task Area: Over-center-of-table luminaires:

Fig. 10-17. Multipurpose table.

1. The task: The task includes both the creation of the desired mood or atmosphere for dining (see Fig. 10-16 Dining) and provision for other general eye tasks such as sewing activities, reading, hobbies and playing table games. Task reflectances are generally 30 to 70 per cent and speed and accuracy is usually not important.

2. Description of the task plane: The entire table top must be considered as the task plane.

3. Illuminance recommendations (maintained on the task plane): For casual or relatively easy work or play activities such as card or table games, simple hobbies or crafts, sewing-pattern layout and cutting, casual reading—200, 300 or 500 lux [20, 30 or 50 footcandles]. For prolonged or difficult seeing tasks performed at the table, see recommendations listed for the particular task.

4. Special design considerations:
a) A broad distribution pattern of light is required to illuminate the entire table top rather uniformly. To minimize veiling reflections, the light sources should have a high degree of diffusion and/or substantial indirect components.
b) Flexibility of distribution and level: It is difficult for a single static light source to provide drama and atmosphere for dining and widespread, diffused lighting (at a higher level) for other table activities. Therefore, a multipurpose table usually requires more than one lighting system or a means of switching from one effect to another.

5. Typical equipment locations:
Task Area: Over-center-of-table luminaires:
a) Recessed (a group of recessed units is generally required)
b) Surface-mounted

Fig. 10-18. Workbench hobbies.

1. The task: Activities carried on at the workbench include woodworking (sawing, hammering, vise operation, planning, assembling parts, drilling, etc.) and craft hobbies. The majority of task reflectances may be below 30 per cent and speed and accuracy can be critical for power-tool operation.

2. Description of the task plane: The task plane area is 510 millimeters (20 inches) wide and 1220 millimeters (48 inches) long, 910 millimeters (36 inches) above the floor. (Home workbenches vary in length. The task plane extends the full length of the bench.)

3. Illuminance recommendations (maintained on the task plane at any time):
Ordinary Seeing Tasks—200, 300 or 500 lux [20, 30 or 50 footcandles]. Hobbies vary greatly in visual difficulty, and often require additional illumination and consideration of directional quality. The information below will provide some guidance for the more difficult tasks.
Difficult Seeing Tasks—500, 750 or 1000 lux [50, 75 or 100 footcandles]. * Leather work, ceramic enamelling, pottery, mosaic, wood carving, block cutting (linoleum-wood).
* Model assembly, electrical and electronic assembly, fly tying.
Critical Seeing Tasks—1000, 1500 or 2000 lux [100, 150 or 200 footcandles]. * Metal engraving, embossing, lapidary—gem polishing, jewelry making.

4. Special design considerations: Luminance balance within the visual field:
a) The wall immediately behind the workbench should have a reflectance above 40 per cent. This reflected light is necessary to provide good illumination on the task plane as well as eye comfort from the standpoint of luminance differences.
b) Additional room illumination should be provided to contribute to the luminance balance in the visual surroundings when the worker faces into the room.
c) If the individual is facing a window while working: 1) daylight glare should be controlled by blinds or shades, and

a) Recessed (a group of recessed units is generally required)
b) Surface-mounted
c) Pendant, generally mounted so that the bottom of the luminaire is 760 to 910 millimeters (30 to 36 inches) above the table top.
Area Surrounding the Task:
a) Luminous ceiling or large luminous area
b) Luminous wall
c) Cornice
d) Valance
e) Cove
f) Brackets, *e.g.*, linear fluorescent or decorative incandescent
g) Recessed luminaires, *e.g.*, incandescent downlight or wall washers
h) Ceiling-mounted luminaires, *e.g.*, shallow large-area types
i) Pendant luminaires, *e.g.*, small pendants
j) Table lamp
k) Floor lamp or torchere

---

c) Pendant, generally mounted so that the bottom of the luminaire is 910 millimeters (36 inches) above the table top.
d) Track mounted luminaires
Area Surrounding the Task:
a) Large luminous area
b) Luminous wall
c) Cornice
d) Valance
e) Cove
f) Brackets, *e.g.*, fluorescent or decorative incandescent
g) Recessed luminaires, *e.g.*, incandescent downlights or wall washers
h) Ceiling luminaires, *e.g.*, shallow large-area types
i) Track mounted luminaires
j) Pendant luminaires, *e.g.*, small pendants
k) Table lamp and floor lamp or torchere
l) Chandelier and chandelier with downlight

---

2) light-colored window coverings should be used at night. The light source should be so positioned that its image reflected in glossy materials is not visible to the user in normal position, unless desired for some special application.

5. Typical equipment locations:
Ceiling-mounted track or suspended linear luminaire running parallel with the task plane.
Ceiling-mounted or suspended non-linear luminaires in symmetrical arrangement over work area to provide uniform distribution of light.
Wall- or shelf-mounted luminaire or luminaires directly in front of user.
Luminous ceiling area.
Portable equipment to provide added directional lighting for special conditions as indicated in the table above.

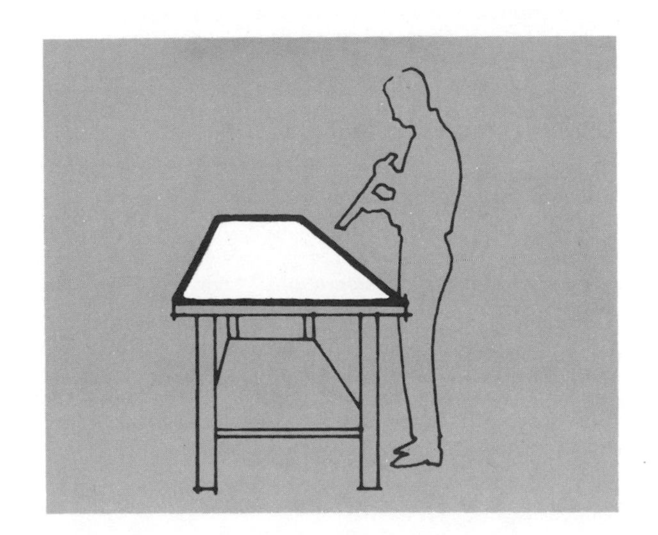

* These tasks require large-area low-luminance reflections in order to see fine detail which shows up as an interruption in a surface sheen.

Fig. 10-19. Table tennis.

1. The task: The basic game of table tennis, although it always has the same rules and table size, varies greatly in its visual difficulty according to the skill of the players. Lighting recommendations here are for "recreational play" where the speed of play is not high and the players' skill is minimal. At this level the game is often called "ping-pong."

2. Description of the task plane: In "recreational play" the task-plane area is to be considered as the 1.5- by 2.7 meter (5- by 9-foot) table only.

3. Illuminance recommendations (maintained on the task plane at any time): 200 lux [20 footcandles].

4. Special design considerations: Although the general guides for luminance balance within the visual field still hold true in table tennis, the background surfaces seen by the player should not be too light or they will not provide sufficient contrast with the white ball for easy visibility. Wall and ceiling surfaces must not have strong distracting patterns. Spottiness or uneven distribution of light can cause seeing difficulty.

In table tennis the celling plane is the major part of the visual field, the luminance ratios at the ceiling become more important than usual.

5. Typical equipment locations: Ceiling-mounted linear sources with the center lines of the luminaires crosswise of the table, located approximately 0.3 meter (one foot) in from each end of the table plus one or more in each runback area.

Ceiling-mounted linear sources lengthwise of the table, centered over the outer edges of the table and extending into runback areas.

Ceiling-mounted non-linear sources arranged in a symmetrical pattern over the entire task area.

Large-area luminaires of *low luminance* symmetrically located to cover the task area, *i.e.*, equipped with louvers or other material providing a minimum of 45-degree shielding.

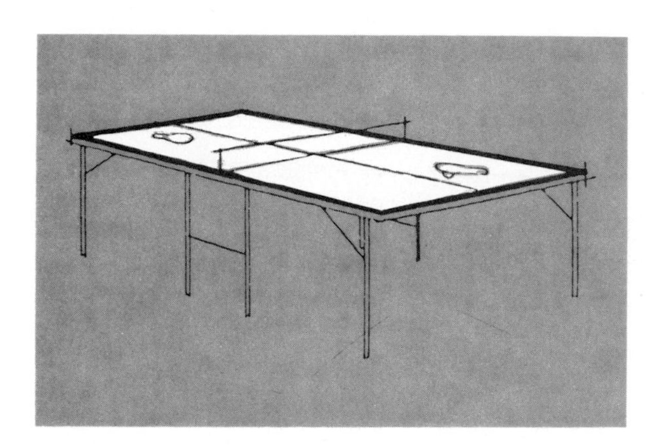

Fig. 10-20. Easel hobbies.

1. The task: Easel hobbies include painting, sketching, collage.*

2. Description of the task plane: The task area is a surface up to 0.9 meter (3 feet) square. The plane of the task is inclined from the vertical to suit the user and the task. Average eye height—1220 millimeters (48 inches) sitting, 1580 millimeters (62 inches) standing. There is also a palette and often an object being copied. The locations of these are not fixed.

3. Illuminance recommendations (maintained on the task plane at any time): 500, 750 or 1000 lux [50, 75 or 100 footcandles].

4. Special design considerations: Luminance balance within the visual field

a) Reflectance of wall surfaces should be above 35 per cent.

b) Additional room illumination should be provided to contribute to the luminance balance in the visual surroundings when the painter looks away from his work.

c) If the person faces a window while working: 1) Glare should be controlled by blinds or shades and 2) Light-colored window coverings should be used.

The light source should be so positioned that its image reflected in glossy materials is not visible to the artist in normal working position, unless specifically desired. For instance, large low-luminance reflection may be required in order to see fine detail in a glossy paint surface; the detail shows up as an interruption on a surface sheen.

A general recommendation would be to paint under the light source by which the painting will ultimately be seen.

5. Typical equipment locations:

Ceiling-mounted track, or suspended linear luminaires running parallel with the task plane.

Ceiling-mounted or suspended non-linear luminaires in a line to provide uniform distribution of light.

Portable equipment to provide added lighting for fine detail.

* Unlike most tasks which can be described in relatively exact terms, easel hobbies include widely divergent activities. In many cases task reflectance can be less than 30 per cent and speed and accuracy can be important.

The definition of a "standard painting" would be absurd. If it is assumed that the artist wishes to see small applications of nearly identical colors, a standard task may be described as the application of a spot of color 6.4 millimeters (¼ inch) in diameter on a background of the next nearest color in a scale comprising 1800 different colors (Munsell).

Lighting for Decorative Effect

Variation and areas of interest created by interplay of light and shadow are essential components of esthetically pleasing residence lighting design. Lighting can create decorative accents to enhance the appearance of art objects and/or create a focus of interest in a particular area. The following paragraphs discuss design considerations and suggests techniques for a variety of decorative possibilities.

Paintings, Tapestries and Murals.

Design Considerations.

1. Lighting equipment should be placed so that the light rays reach the center of the painting at an angle of 30 degrees with the vertical. This usually prevents specular reflections in the direction of the viewer's eye from frame, glass, or surface of the picture, and also avoids disturbing shadows of frame, heavy paint texture, etc.
2. A study of the sight-lines of people seated and standing anywhere in the room should be made, to ensure that no unshielded sources are in view.
3. Excessive luminance difference between the lighted object and surrounding areas is undesirable. Higher illuminances applied for extended periods may cause deterioration of the paint surface.
4. Above all, the primary consideration is the intent of the artist.

Typical Techniques.

1. Entire picture wall lighted by cornice or wall-wash equipment.
2. Individual frame-mounted luminaires.
3. Individual framing spotlights.
4. Individual spot or floodlamps not confined to picture lighting.
5. Lighting from below by luminaires concealed in decorative urns, planters, mantels, etc.

Sculpture.

Design Considerations.

1. A sculpture is a three-dimensional object in space. A certain amount of specular reflection is often pleasurable; experimentation with diffused and/or a directional source or sources will help to determine the most acceptable solution of a specific situation.
2. A luminance ratio between 2 and 6 usually results in a good three-dimensional effect with transparent shadows. If the modeling ratio is reduced below 2, the lighting becomes too "flat" and solid objects appear two-dimensional. If the ratio is above approximately 6, the contrast tends to become unpleasant, with loss of detail in the shadows or in the highlights.

Typical Techniques.

Adjustable spots, floods, individual framing spots, back lighting.

House Plants, Display and Growth.

Design Considerations for Decorative Purposes Only.

1. Lighting that provides for enhancement of appearance may not necessarily be suitable for plant growth. Plants may have to be rotated to a growth area from time to time to keep their beauty.
2. Plants tend to grow toward the light. This should be kept in mind when locating equipment.
3. Light sources, particularly incandescent, should not be located too close to plant material because the heat may be detrimental.
4. Backlighting of translucent leaves often will reveal leaf structure, color and texture.
5. Front lighting of opaque leaves often will reveal leaf structure, color and texture.
6. Silhouetting foliage can add a dimension.
7. The necessary fire safety precautions should be taken if artificial plants are used.
8. Live plants are not static; they change in height and bulk.

Typical Techniques.

1. Incandescent downlights recessed, surface-mounted or suspended above planting area.
2. Luminous panel or soffit using fluorescent and/or incandescent sources.
3. Silhouette lighting with luminous wall panel, lighted walls, concealed up-lights.
4. Low-level incandescent stake units.
5. Planting racks containing fluorescent and/or incandescent lamps especially selected for plant growth.
6. Luminaires recessed in earth to provide uplight.

Niches.

Design Considerations.

1. The lighting method is determined by what is to be displayed in the niche.
2. The luminances of shielding media and interior niche surfaces should be carefully controlled to avoid excessive difference from surrounding areas.

Typical Techniques.

1. Incandescent spotlights from one or more directions.
2. Incandescent or fluorescent sources concealed vertically or horizontally at the edges.
3. Luminous sides, bottom, top, back or combinations.
4. Lamps concealed behind object to create silhouette effect.

Bookcases.

Design Considerations.

1. The distribution of light should cover the faces of books and other objects on all shelves.
2. Luminance ratios between the bookcase area and the surrounding wall surfaces should be kept within comfortable limits.

Typical Techniques

1. Adjustable spot or flood lamps or wall washers aimed into the shelves from the ceiling in front of the bookcase.
2. Lighted bracket, cornice or soffit extending in front of the shelves.
3. Tubular sources concealed vertically at the sides of the bookcase or horizontally at the front edge of individual shelves.
4. Lighting concealed at the back of the bookcase for silhouette effects.

5. Some bookcases may require modification such as cutting the shelves back or using glass or transparent plastic shelves, to permit an ample spread of light over the entire bookcase.

Furniture.

There is a growing number of furniture manufacturers who include some lighting in the design of their products, such as: display cabinets and bookcases, shelving systems and planters.

These are excellent to use for they provide not only ways to emphasize objects of decorative value, but can contribute to the lighting ambience. Their portability is an added advantage.

Caution should be exercised that the light source be shielded and is not visible from any angle nor reflected from the display and that the luminance of any part of the display is not excessive.

Fireplaces.

Design Considerations.

1. Luminance relationships between the surrounding area and the fireplace facing of brick, stone, wood, etc. should be carefully studied.
2. It may be desirable to emphasize the mantel as a decorative wall treatment and as the focal point of interest in the room. Texture, color and orientation in the room may be an important consideration.

Typical Techniques.

1. Cornice and wall brackets.
2. Recessed or surface-mounted downlights close to the fireplace surface for grazing light.
3. Adjustable spots or floods (ceiling-mounted) aimed at decorative objects or mantel, or grouped together to light the entire fireplace wall.
4. Recessed or surface-mounted wall-washers for even distribution of light over the entire fireplace surface.
5. Lighting equipment concealed in the mantel to light pictures or objects above and/or surfaces below.

Textured Walls and Draperies.

Design Considerations.

1. Light directed at a grazing angle emphasizes textured surfaces.
2. Luminaires close to walls usually create patterns of light distribution on the walls. Some of these may be desirable, others unacceptable, depending on circumstances and individual preferences.

Typical Techniques.

1. Recessed or surface-mounted downlights located close to the vertical surface to direct grazing light.
2. Cornice, valance, wall brackets or wall-washers.
3. Individual adjustable incandescent luminaires.
4. Concealed up-lighting equipment.

Luminous Panels and Walls.

Design Considerations.

1. Luminous elements larger than about 1.5 square meters (15 square feet) should not exceed 170 candelas per square meter (50 footlamberts) average luminance when they will be viewed by people seated in a room.
2. Luminous elements of smaller size (or larger elements seen only in passing) may approach 690 candelas per square meter (200 footlamberts).
3. Variable brightness controls are valuable in adjusting the luminance of a panel to the desired level for specific activities.
4. Uniformity of a luminous surface may or may not be desirable, and is dependent on the spacing of the light sources.
5. Over-all pattern, shielding devices, structural members, or other design techniques can add interest and relieve the dominance and monotony of large evenly luminous panels.

Typical Techniques.

1. Fluorescent lamps uniformly spaced vertically or horizontally in a white cavity.
2. Fluorescent lamps placed at the top and bottom or on each side of a white cavity.
3. Random placement of incandescent lamps (perhaps with color) in a white cavity.
4. Uniform pattern of incandescent lamps in a white cavity.
5. Incandescent reflector lamps concealed at the top or bottom of a white cavity reflecting light from the back surface of the cavity.
6. For outdoor luminous walls, floodlamps directing light to a translucent diffusing screen from behind.

Other Architectural Elements.

Many residences have especially distinctive architectural features that require special lighting. Because of the variety of problems it is difficult to list design considerations that apply to all. However, the following general principles should be observed in all cases:

(a) All light sources should be concealed from normal view.

(b) "Unnatural" lighting effects should be avoided.

(c) Lighting patterns that conflict with or distract from the architectural appearance of the element being lighted should be avoided.

Luminaires as Decorative Accents.

Design Considerations.

1. Luminance of luminous parts can be critical.
2. Luminaires should not introduce distracting or annoying light patterns.
3. Where exposed lamps are used for a desired decorative effect, dimmer controls to provide high or low luminances should be considered.

Typical Techniques

1. Decorative chandeliers, wall sconces, lanterns and brackets.
2. Decorative portable lamps.
3. Pendant luminaires.

Lighting Outdoor Living Spaces

Lighting planned to fit the range of activities in outdoor living spaces such as food preparation, dining, table games, etc., requires the same illumination considerations as when planning for these activities indoors. Because of the all-purpose use of such spaces, luminance values are apt to be much greater than surrounding areas.

Care should be taken to provide comfortable brightness patterns, keeping in mind that luminance differences can be extreme against the black of night.

Switches and dimmers can provide needed control of brightness patterns as well as adjustment to mood or use of the area. The structure, walls, ceiling and the degree of enclosure, determines to a great extent the type of equipment used and its placement.

Grounds

In garden scenes, the choice of what to light goes to the best in intrinsic beauty, colorfulness, form and composition. Garden sculpture, pools, or special landscape features influence choice of area as does its location with relation to points of viewing.

Guiding Principles. Non-uniform floodlighting should be provided to form pleasing patterns in light, shade, shadow and color. The over-all may be subdued, stimulating or dramatic, depending on the effect desired and the manner of light application. Maximum luminance should not in general exceed 35 candelas per square meter (10 footlamberts) for the focal center of interest, unless the scene is viewed from a fully lighted terrace or interior. The highest luminance should usually be 2 to 5 times more than other features—avoiding black "holes" in the scene. When recommended illuminances exist, it is wise to provide low (10 to 20 lux [1 to 2 footcandles]) over-all floodlighting to unify the composition. The eyes of viewers, neighbors, and passers-by must be protected from the glare of lamps; even low wattage ones, if bare or poorly shielded, are offensive. Heavy ground cover, shrubs, hedges and foliage may provide "natural" shielding and daytime concealment.

Techniques. Subject matter should be analyzed for texture, form, line, reflectance and background to determine what and how to give emphasis. Head-on floodlighting tends to make objects flat; modeling is obtained by lighting from both sides with more light from one side than the other. Light, striking a surface at a very narrow angle, emphasizes texture. Translucency, depth, form, and pattern are best accentuated by silhouetting, with very little light on the front side.

Equipment. Weatherability of equipment is vital for continued outdoor use. Aluminum, brass, copper or stainless steel are generally applicable. When equipment is not concealed, it should be in keeping with the surround in color, form or concept.

Color. In addition to tinted, colored bulbs, color is obtained by tinted cover glass. Tinted colored light of the same hue as the object lighted heightens it. Yellow or low-wattage bulbs tend to deaden grass and foliage which is enhanced by blue-green or blue-white sources. Cool colors add depth, and pale blues simulate moonlight.

Walkways and Entrances

Walkways. In general, lighting equipment for walkways should be located high up in trees, on poles, or on buildings, so that the luminous parts are above the normal line of sight; or luminaires should be located quite low above the walkway so that luminous parts are below the line of sight. Low-mounted luminaires usually should be rather closely spaced for adequate coverage. Intermediate-height luminaires (in the line of sight) are the poorest choice for walkway lighting, but often are required for appearance or other reasons. Extreme care should be exercised in the selection and placement of such luminaires.

Steps. All considerations on the height of luminaires for lighting walkways apply to the lighting of steps as well. It is important to call attention to both treads and risers. This can be done by:

1. Using light reflecting surfaces.
2. Building lighting into risers or side walls.
3. Using low or high-level equipment placed to light both vertical and horizontal surfaces.

Doorways. Light is needed at doorways for safety of passage and also for identifying callers. Therefore, it is important in entranceways to get light on a caller's face, by either direct or reflecting techniques. Lighted house numbers, visible from the street, are thoughtful courtesies. Numerals should be 76 millimeters (three inches) high with 13 millimeters (one-half inch) stroke to be visible at 23 meters (75 feet).

LIGHT SOURCES

Electric Light Sources

The most common electric light source in use in the home is the incandescent filament lamp of the general service, reflectorized, tungsten-halo-

Fig 10-21. The use of a variety of lighting systems in one room provides flexibility for the desired mood or for the visual tasks to be performed. Recessed luminaires with a symmetrical distribution dramatically accent the chairs and coffee table in the view at the upper left. In the upper right, light for reading is provided by portable lamps at the couch and one chair, and by a pendant direct-indirect luminaire at the other chair. Asymmetric recessed luminaires accentuate the texture of the brick fireplace wall while a lighted cornice highlights the vertical louvered window treatment in the view at the lower left. Together, as in the view at the lower right, all lighting systems provide an interesting, livable environment.

gen, showcase and other special shapes, and low voltage types. The next most common is the fluorescent lamp of the straight, circline and U-shaped types. High intensity discharge lamps are being used primarily for outdoor application.

See Section 8 in the 1981 Reference Volume for data on each type of lamp.

Daylight

Daylighting in the residential design is desirable as it may provide the quantity and quality of light desired for many purposes as well as contribute to the economics and esthetics of the space.

The use of daylight allows for some reduction of electricity for lighting purposes and thus reduces the energy use. The degree to which less energy may be used for the lighting will depend upon the availability of daylight, orientation of the building, size and location of windows and doors, and the desired illuminance levels.

For new construction, electric light should be designed to blend with daylight in order to light interior areas and to provide suitable lighting at night and on dark, overcast days. Even on sunshiny days, small window openings may necessitate electric lighting for the most desirable rendering of the interior as well as to light effectively for task performance.

In a room with daylight, the adaptation level

may be very high, especially if a bright sky or snowy surfaces are visible through the window. Parts of the room will then appear gloomy and objects seen against the window will appear in silhouette. This effect may be lessened by increasing the luminance in the interior or by reducing the luminance of the window by using control elements such as building overhangs, vertical fins and similar building elements, opaque and translucent screens, shades, draperies or curtains, and landscaping elements including trees and shrubbery. Shades, draperies, and slat-type blinds are also used for blocking, filtering, or redirecting the daylight.

See Section 7 in the 1981 Reference Volume for additional information on daylight availability and daylighting design.

LUMINAIRES

Lighting equipment for interior living spaces ranges from novelty plug-in lamps to completely engineered package luminous ceilings. Described below are the key categories of luminaires listed by their lighting distribution characteristics. The lighting distribution characteristics of some of these luminaires, such as recessed, ceiling mounted and track mounted luminaires, may be similar and the choice will depend on structural conditions, esthetics and economics. Examples of each type are pictured and some indication is given of features to look for and for suitable applications. To select lighting equipment wisely, the designer must be able to read the various types of literature presented by the manufacturer for this purpose.

Recessed Luminaires

Recessed luminaires are installed in the ceiling, and distribute all of their light downward. They have what is known as a *direct* distribution. However, by the use of various diffusers, lenses, and reflector and louver configurations, and by employing different types of lamps, recessed luminaires can be made to distribute their light in a wide variety of patterns. See Fig. 10-21, upper left. Candlepower distribution curves offer the best means of evaluation.

Asymmetric distribution. While most recessed luminaires used in residential lighting have symmetrical distributions, there is an increasing demand for equipment that is adjustable and which can direct light to walls and other vertical surfaces. A popular type of asymmetric luminaire is the "wall washer." This is designed to direct a broad spread of light at a nearby vertical surface to create nearly uniform illuminance on that surface. While candlepower distribution curve of such a luminaire is helpful in predicting the illuminance on the wall, lux (footcandle) charts are usually provided to simplify the design procedure.

Applications. Fig. 10-22 shows design features, sketches and applications of recessed luminaires having wide, medium, narrow and asymmetric profiles (candlepower distributions). Each unit is listed under its most common profile.

Ceiling-Mounted Luminaires

General Diffusing. General diffusing ceiling-mounted luminaires (see Fig. 10-23a and 10-23b) direct their light in a very wide pattern and are normally used for general illumination because they light walls and ceilings. Their luminance should be carefully considered in relation to room reflectances. Light sources, either incandescent or fluorescent, should not be visible either through or over the luminaire. Although fluorescent lamps are lower in luminance than incandescent, they should always be shielded from direct view. The close-to-ceiling position of luminaires of this type allows them to have higher luminance than similar ones hung lower into the line of sight. For typical residential sizes the average luminance of the total luminous area should not exceed 1700 candelas per square meter (500 footlamberts), except in utility areas. For equipment used in strictly utility spaces luminances as high as 2700 candelas per square meter (800 footlamberts) are acceptable. Within the diffusing element luminance of the brightest 645 square millimeters (one square inch) should not materially exceed twice the average. Luminance ratios between the luminaire and the ceiling should not be more than 20 to 1. Even with glass or plastic of excellent diffusing qualities, spottiness will occur if the lamps are widely spaced or too close to the diffuser.

Diffuse Downlighting. There are many variations of the basic luminaire type, but they are usually characterized by an opaque or partly opaque side or border (see Fig. 10-23c). This restricts the distribution somewhat and limits the effectiveness as a general lighting device. Such luminaires seldom put light on the ceiling,

Fig. 10-22. Recessed Lighting Equipment

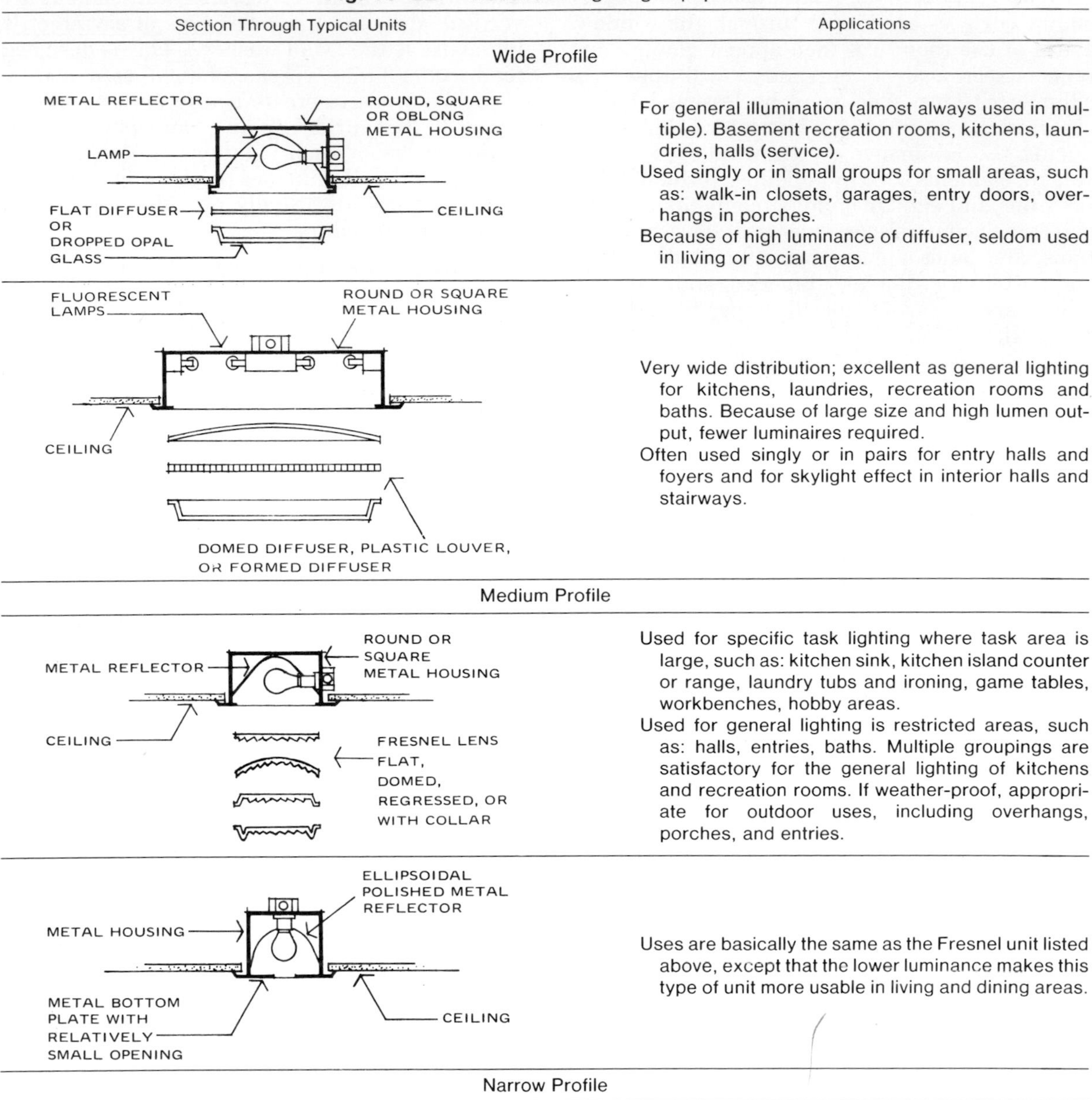

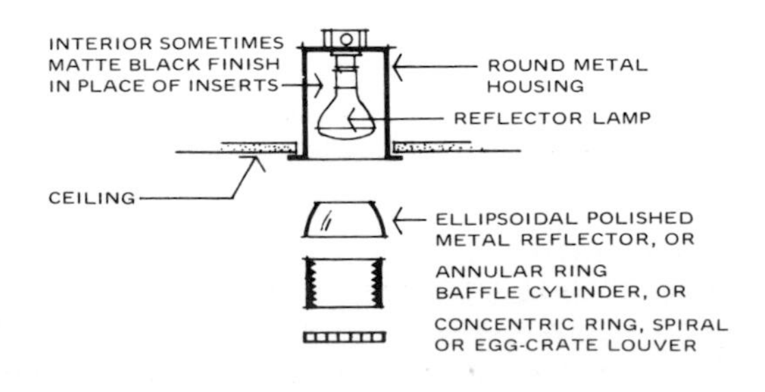

Section Through Typical Units	Applications
Wide Profile	
Metal reflector; round, square or oblong metal housing; flat diffuser or dropped opal glass	For general illumination (almost always used in multiple). Basement recreation rooms, kitchens, laundries, halls (service). Used singly or in small groups for small areas, such as: walk-in closets, garages, entry doors, overhangs in porches. Because of high luminance of diffuser, seldom used in living or social areas.
Fluorescent lamps; round or square metal housing; domed diffuser, plastic louver, or formed diffuser	Very wide distribution; excellent as general lighting for kitchens, laundries, recreation rooms and baths. Because of large size and high lumen output, fewer luminaires required. Often used singly or in pairs for entry halls and foyers and for skylight effect in interior halls and stairways.
Medium Profile	
Metal reflector; round or square metal housing; Fresnel lens	Used for specific task lighting where task area is large, such as: kitchen sink, kitchen island counter or range, laundry tubs and ironing, game tables, workbenches, hobby areas. Used for general lighting is restricted areas, such as: halls, entries, baths. Multiple groupings are satisfactory for the general lighting of kitchens and recreation rooms. If weather-proof, appropriate for outdoor uses, including overhangs, porches, and entries.
Ellipsoidal polished metal reflector; metal housing; metal bottom plate	Uses are basically the same as the Fresnel unit listed above, except that the lower luminance makes this type of unit more usable in living and dining areas.
Narrow Profile	
Round metal housing; reflector lamp	Accent lighting over plants, cocktail tables, etc. Wall lighting—mounted close to textured surfaces, such as: brick, stone, rough wood and fabrics. Task lighting—food preparation areas (may cause specular reflections). In multiple on quite close spacing for general lighting. Most effective when used near perimeter of room so that some light spills onto wall. Dramatic effects for family rooms and formal living areas. Supplementary stair lighting—shadow patterns define treads and risers. Dining tables—provide functional light on dining table to supplement decorative effect from hanging luminaire.

Fig. 10-22. *Continued*

Section Through Typical Units	Applications
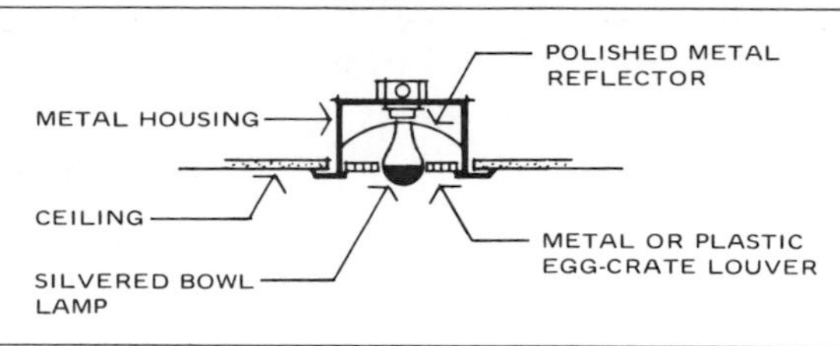 	Same uses as type luminaire above. Low luminance of luminaires very desirable, but larger size sometimes prohibits use in highly styled interiors.
Special Asymmetric Profile	
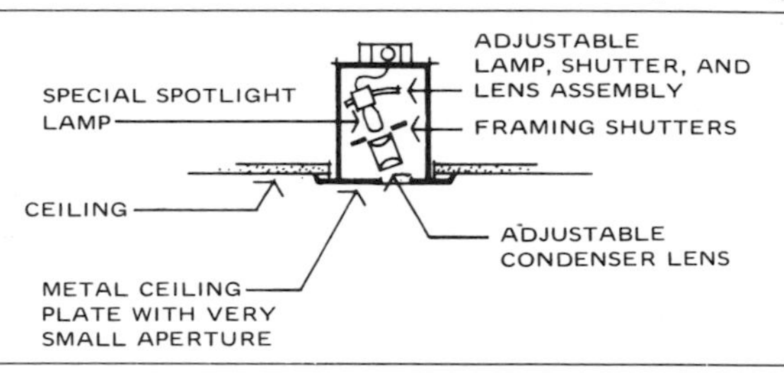 	Adjustable beam can be framed precisely to outline paintings, pictures, maps and niches. When aimed directly down, shutters can frame dining, cocktail, or coffee tables, or other horizontal elements. Special high lumen output lamps are often used in these units. Should have easy access for frequent relamping. May come equipped with top access openings for relamping from above.
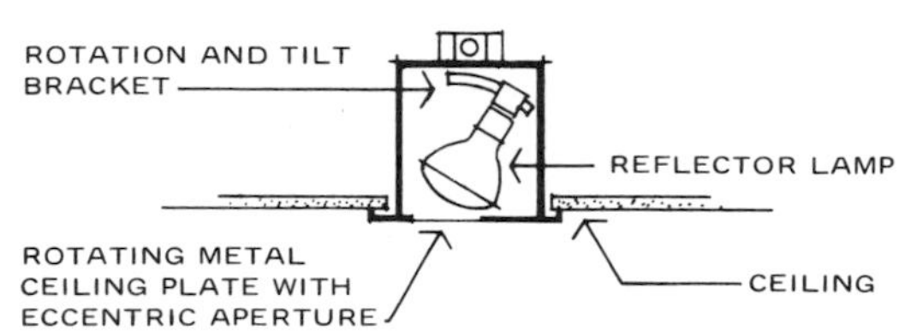 	Useful for gallery or picture lighting and to light sculpture. If scalloping effect is acceptable it can be used for wall lighting and to accent fireplace surfaces. Large size of bottom aperture sometimes makes this unacceptable for highly stylized interiors. Also used for lighting piano music and sewing machines.
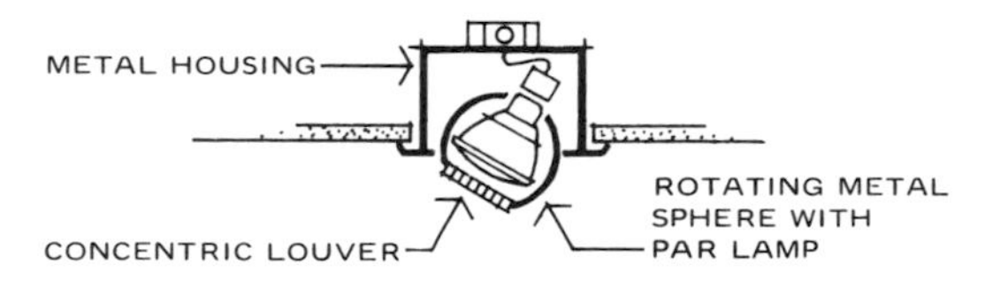 	Same uses as recessed adjustable luminaire as listed above.
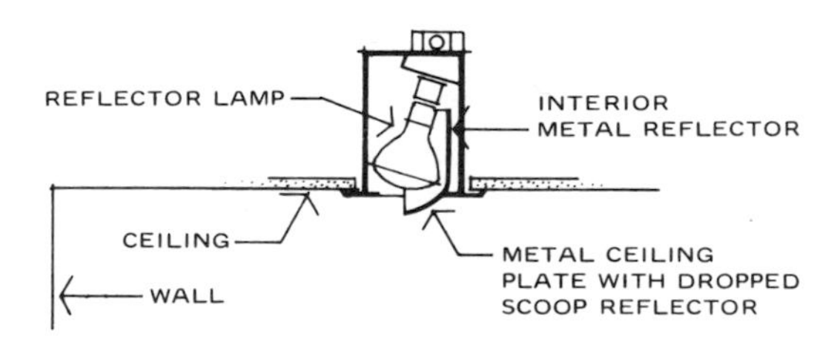 	For uniform illuminance on plane wall surfaces. Effective for lighting murals and for minimizing wall imperfections. Not generally to be used for lighting textured wall surfaces because it directs no grazing light at wall. Spacing of these units is critical—follow manufacturer's recommendations closely.

and in order to light walls effectively they must be located rather close to them. If they are used for general room lighting, they should be limited to no more than 10 square meters [100 square feet] per luminaire and the room finishes should be very light in color.

Downlighting. Downlighting or directional luminaires (see Fig. 10-23d) are designed to be used for accent lighting or supplementary illumination for critical visual tasks such as sewing. If used in multiple for general illumination of larger areas they must be spaced quite close together because of their narrow distribution. At typical room heights one luminaire cannot effectively light an area of more than 2.5 square meters [25 square feet]. Sharp shadows are inherent in the lighting result.

Track Mounted Luminaires

Track and Electrical Feed. Track is generally a linear extruded aluminum housing, containing copper wires to form a continuous electrical raceway. Available in nominal lengths of 0.6, 1.2 and 2.4 meters (2, 4, and 8 feet), it can be joined or cut, or set into a variety of patterns with connectors in the shape of "L's", "T's" or "X's", or with flexible connectors for irregular patterns. It is available in a one, two, three or four circuit mode, for greater capacity and added flexibility in control and switching, and is generally available in several different metallic and/or paint finishes and can often be further accessorized with selections of colored, metallic or wood-toned tape appliques.

Track can be mounted at or near the ceiling surface, recessed into the ceiling plane with special housing or clips, or can be mounted on stems in high ceiling areas (see Fig. 10-24). It may also be used horizontally or vertically on walls and can be hard-wired at one end or anywhere along its length. There is a flexible plug-in-cord kit approved by the Canadian Standards Association for use in Canada.

Circuiting of Track. Single circuit track's electrical distribution system generally has a rating of 20 amps at 120 volts (2400 watts incandescent load). For various lamp wattages and spacings, consult manufacturers and local codes.

Adjustable Track Luminaires. A variety of adjustable track mounted luminaires are available (see Fig. 10-24b) for attachment at any point along the track. These luminaires are in many shapes, styles, colors and metallic finishes housing a large assortment of lamp sizes and shapes. In addition, there are a number of luminaires designed to create special effects for special or decorative applications (see Fig. 10-24c).

Accessories. Accessory equipment is available that allows expanded track lighting application techniques as shown in Fig. 10-24d. Cord-hung pendant, or chain-mounted chandeliers can be suspended anywhere along the tracks length. An "outlet" adapter makes any track luminaire a surface or wall luminaire. The "pin-up" makes any track unit a plug-in-wall lamp. The "display hook" can support mobiles, plants and other decorative accessories. The cord and plug makes installation of track particularly convenient when a wired lighting outlet is not available.

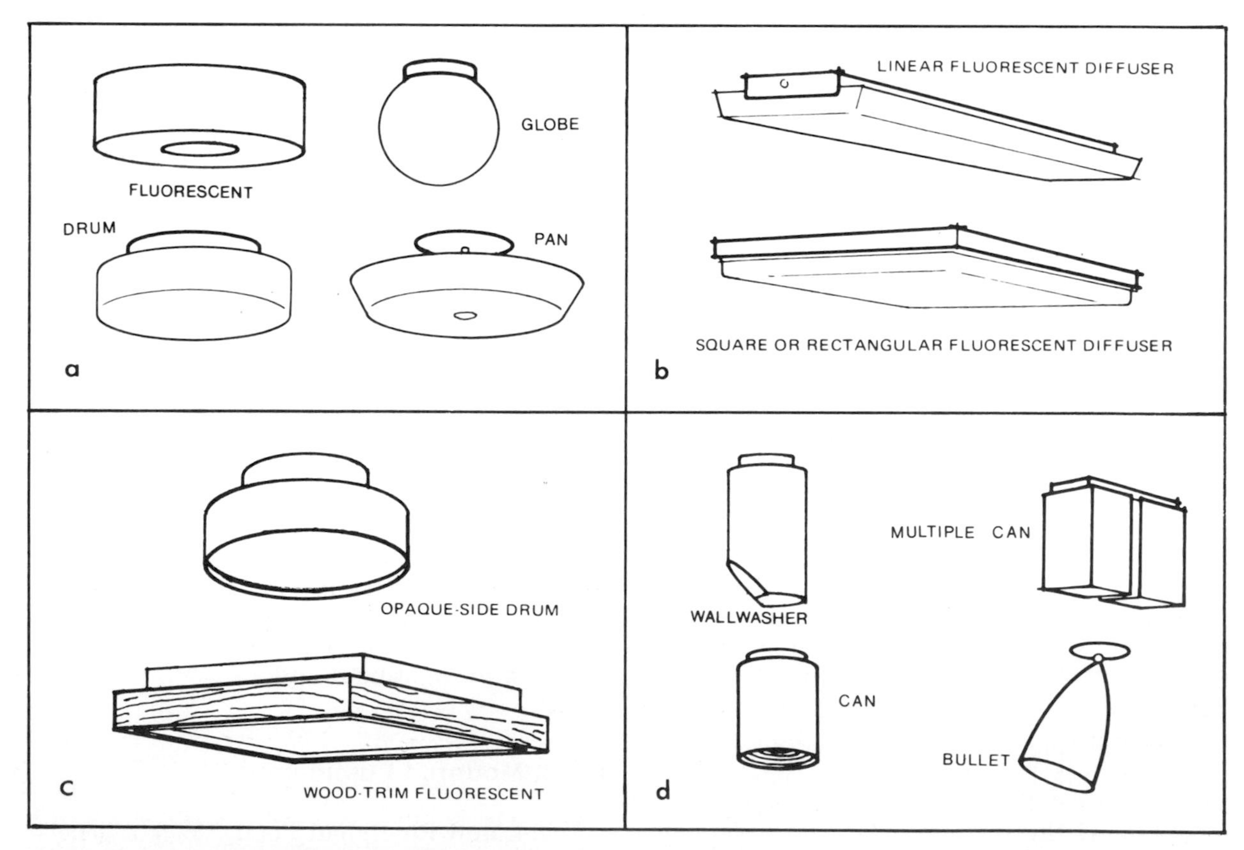

Fig 10-23. Ceiling mounted luminaires: group a, small general diffusing type; group b, large general diffusing type; group c, diffuse downlighting type; group d, downlighting type.

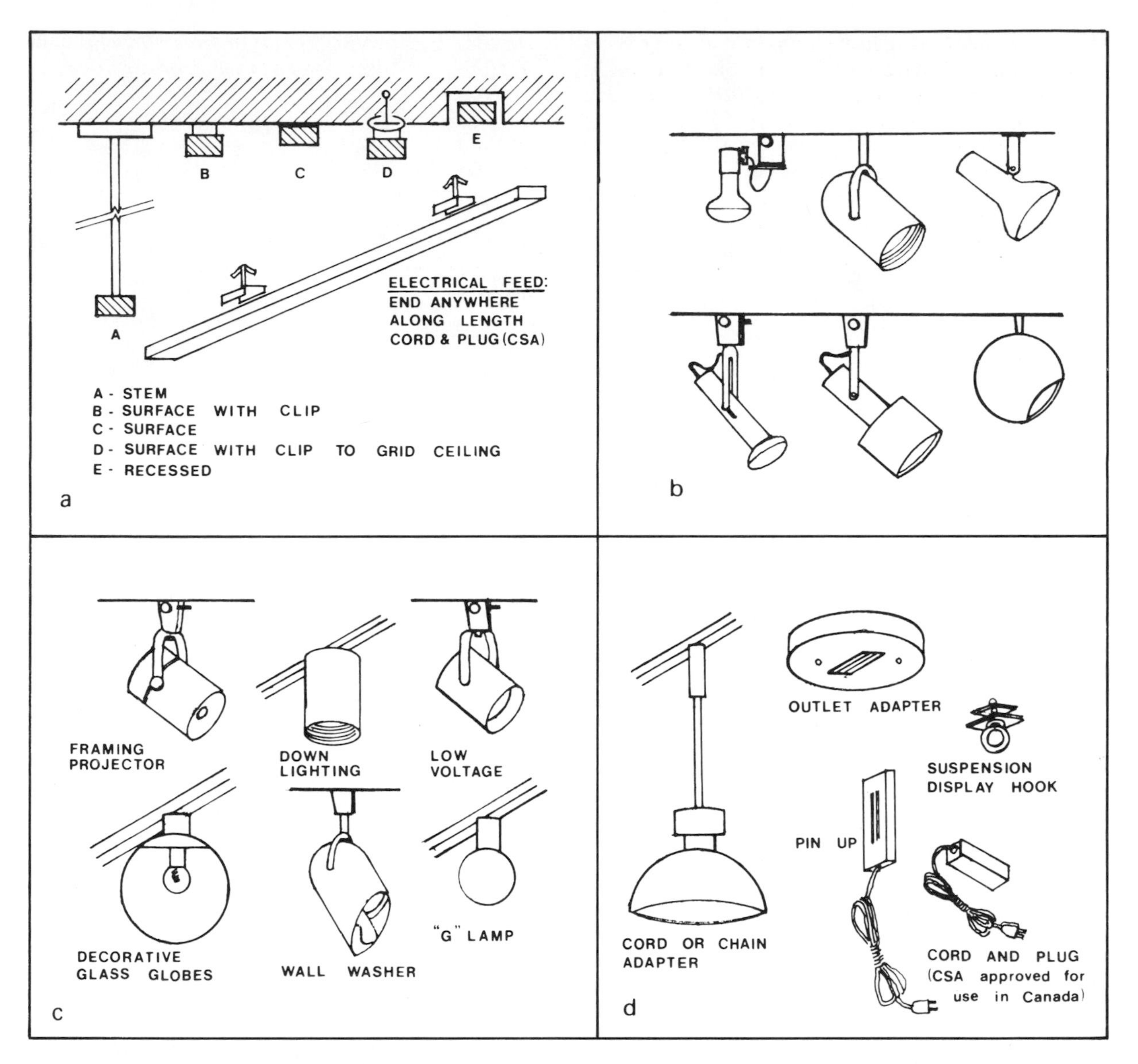

Fig 10-24. Track-mounted luminaires: group a, the track; group b, adjustable luminaires; group c, special effects or application; group d, accessories to expand track lighting techniques.

Pendant Luminaires

General Diffusing. Pendant luminaires with general diffusing characteristics are used for both functional and decorative purposes. See Fig. 10-25. For the latter they are lamped with low-wattage or tinted light bulbs, or provided with dimmer control to keep the luminance low, and hung as points of decorative interest or as "lighted ornaments." When intended for functional lighting and located well above the normal line of sight, they are subject to the same luminance limitations as general diffusing ceiling-mounted luminaires. If hung low enough to be in the line of sight they should be restricted to substantially lower luminances. Fig. 10-26 gives lamping recommendations for this type of equipment. Even relatively low luminances may be annoyingly bright if viewed against dark walls or dark nighttime windows. When luminaires of this kind are used in dining areas the luminance should be rigidly controlled, and additional lighting with a directional component should be installed to illuminate the table top.

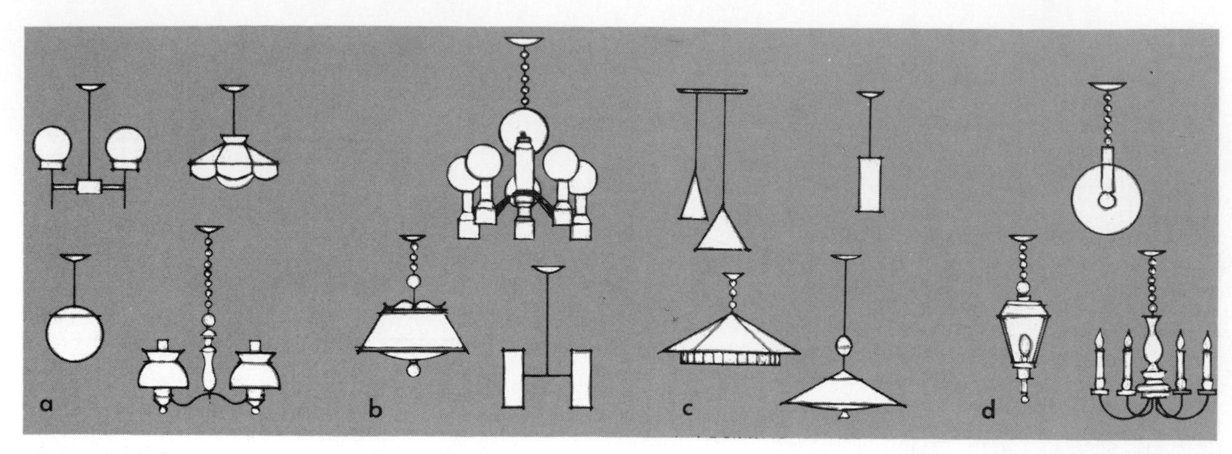

Fig 10-25. Pendant luminaires: group a, general diffusing; group b, direct-indirect; group c, direct downlighting; and group d, exposed lamp.

Direct-Indirect. Direct-indirect luminaires (see Fig. 10-25b) provide generous upward light which contributes to room lighting, as well as functional downlighting. They are characterized by opaque or slightly luminous shades or shields, which often make them suitable for low hanging where they will be in the line of sight, and they are commonly used to light dining tables, game tables, snack bars, etc. When they are hung quite low over coffee tables or planters the downward component can provide accent highlights and in this case top shielding should be considered. Such equipment is often adjustable, and can have added flexibility when suspended from a track. Glass or plastic diffusers should not be visible below the shading element. Bottom shielding or louvering is important when luminaires are mounted in a higher position and viewed from below.

Fig. 10-26. Suggested Lamp Wattages for Visual Comfort for Spherical Uniformly Luminous General Diffusing Luminaires*

Sphere Diameter in Millimeters (Inches)	Mounting Height of Luminaire: 2.1 Meters (7 Feet) or Above — Utility Spaces	2.1 Meters (7 Feet) or Above — Living Areas	1.0 to 1.8 Meters (3.3 to 6 Feet) — Living Areas
150 or less (6 or less)	60W	40W	25W
200 to 250 (8 to 10)	75W	60W	40W
310 to 410 (12 to 16)	150W	100W	60W
460 to 560 (18 to 22)	200W	150W	75W
610 to 710 (24 to 28)	300W	200W	100W

* Assuming typical diffusing material of 50 per cent transmittance and ceilings and walls of recommended reflectances. See Fig. 10-2.

Downlighting. The main function of pendant downlights (see Fig. 10-25c) is to provide accent illumination. They are generally hung at or below eye level. When mounted above eye level, as in stairways, halls, entryways, etc., they require some form of bottom shielding. The lighting distribution is characterized by sharp beam definition and shadow patterns. In general, the larger the luminaire the less sharply defined are the shadows. Almost without exception luminaires of this type need additional general room lighting for visual comfort.

Exposed-Lamp. The main function of exposed-lamp luminaires is to provide decorative highlights, sparkle and accent (see Fig. 10-25d). They are not suitable for general room illumination or for the lighting of specific tasks. Since their use is primarily decorative, lamps should be kept low in wattage or be used on dimmer control. Other room lighting is necessary to insure visual comfort. Because of the high luminances of all incandescent filament lamps, even low-wattage ones and particularly those with clear bulbs, mounting above the normal line of sight is recommended. When luminaires must be suspended low, lamps should be of very low luminance and walls against which they are viewed should be light in color.

Wall-Mounted Luminaires

Basically, wall-mounted luminaires are junior versions of similar luminaires mounted on the ceiling. They have the same distribution characteristics and qualities described in the previous paragraphs. However, their low mounting places them much more in the field of view, and therefore the designer should exercise strict control over luminance. Below is a brief summary of some of the more popular types of wall luminaires and their uses.

Exposed-Lamp. Wall-mounted luminaires with exposed lamps (see Fig. 10-27a) are quite often used for decorative purposes, especially with clear light bulbs. Other room lighting is almost invariably necessary. One type of exposed-lamp unit used for functional purposes as a mirror strip utilizing several low-wattage lamps with diffusing white coating. The comments on the luminance of pendant exposed-lamp luminaires apply also to wall-mounted equipment.

Small Diffuser. Small diffusing wall brackets (see Fig. 10-27b) are often used in pairs, in groups, or in lines for lighting hallways, stairways, doorways and mirrors. In areas like living rooms and family rooms where the luminaires will be in view for long periods of time, the luminance should be very low. Where they will be seen only briefly in passing, as in hallways or doorways or providing functional lighting at a mirror, higher luminances can be tolerated, but other lighting in the room is usually necessary.

Linear. Linear wall brackets (see Fig. 10-27c) are commonly placed over mirrors, under cabinets, and in other locations where functional lighting is required for close-up tasks. They are occasionally used for general lighting in hallways, passages, etc. Good diffusing quality of glass or plastic, or satisfactory opaque shielding is important because of the high lumen output of the lamps employed. Higher luminances can be tolerated when the luminaire is seen briefly, or when it is used for a specific task, as at a bathroom mirror. Well-designed fluorescent brackets, nearly always with opaque shielding, may be used in living areas to light walls, provide general room lighting, or even to supply functional light at desks or seating arrangements placed against the wall. They may be extended to cover the entire length of a wall. In these applications they perform exactly like structural wall brackets.

Directional (Adjustable and Fixed). Directional luminaires having strong downward components are often mounted on the wall for accent and display lighting, over planters, room dividers, and sculpture. Directional luminaires providing upward lighting can be used in groups for general illumination. Both types are most effective when shielding is opaque and the light source cannot be seen from any point in the room. See Fig. 10-27d for directional units.

Portable Lamps

Although all portable lamps are considered by homemakers and designers alike as decorative elements in the furnishing scheme, there are a great many cases where they must be extremely functional as well as good-looking. In many interior-living spaces, portable lamps are relied upon to make a major contribution to the general illumination as well as provide illumination for the most difficult visual tasks. Such lamps must be chosen not only for their decorative suitability, but also for the amount and quality of the light they deliver on the task area and their emission of upward light to add to the general illumination. Moreover, a well-designed lamp can perform effectively only if it is placed correctly in relation to the task and to the eyes of the user.

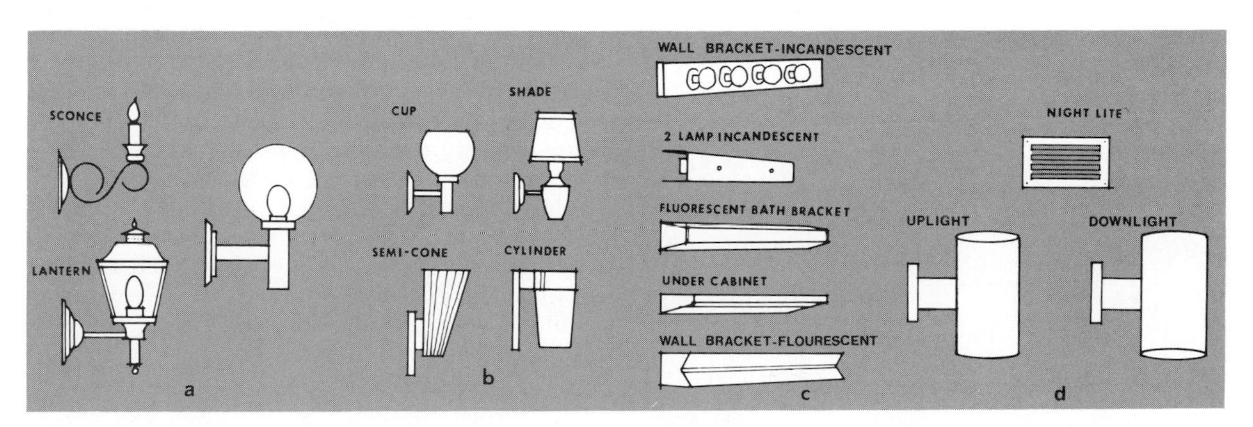

Fig 10-27. Wall luminaires: group a, exposed lamp; group b, small diffuser; group c, linear diffuser; group d, directional (adjustable and fixed).

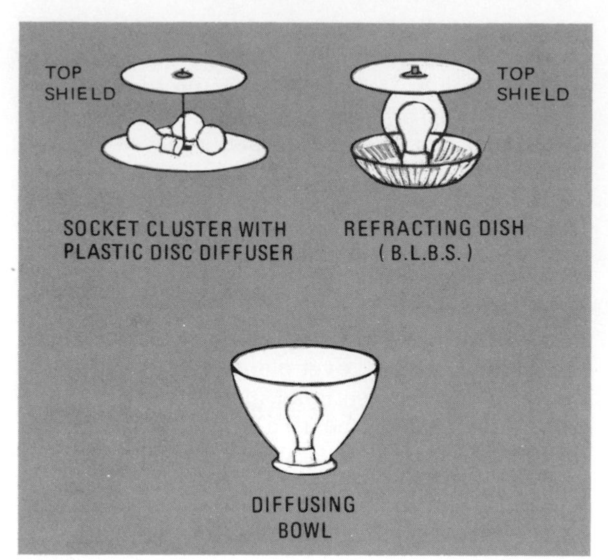

Fig 10-28. Typical under the shade components for portable lamps.

The following paragraphs list some check points that will serve as criteria in the selection of lamps and their proper placement.

Under-the-Shade Components. If the lighting in a room is well balanced and the room has favorable reflectances it is not essential that each portable lamp have a light-control device under the shade; the ordinary inside-frosted or white bulb will provide a reasonable amount of diffusion. However the addition of diffusing, reflecting, refracting, or shielding elements results in improved quality that is highly desirable, particularly if the visual task is to be difficult or prolonged. Lamps for the study desk *must* make provision for excellent lighting quality.[2] Fig. 10-28 illustrates some of the constructions to be found in portable lamps on the market today, taking into account the sizes and types of light bulbs currently available. As new, more powerful, light sources are introduced in portable lamps, the need for control and quality-improving elements will increase.

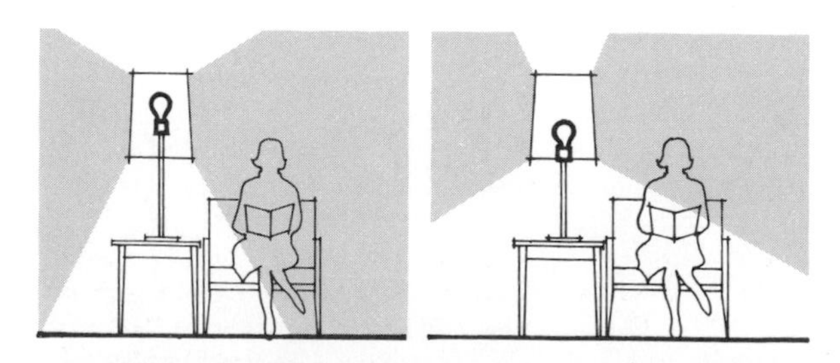

Fig 10-29. Influences of light source position on light distribution of portable lamps. Left, high position. Right, low position.

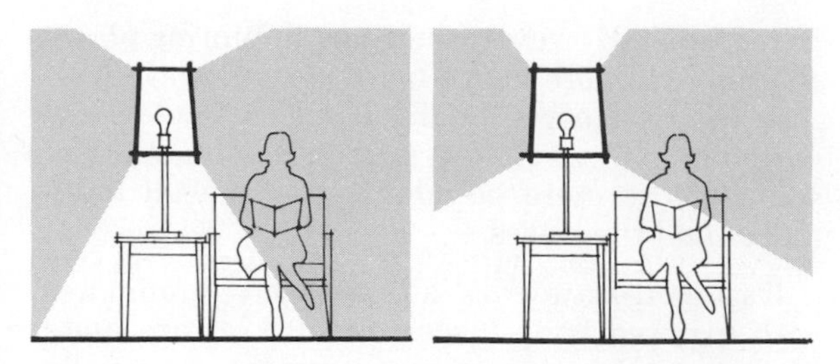

Fig 10-30. Influence of the shade dimensions on light distribution of portable lamps. Shade on the left is 355 millimeters (14 inches) wide. Shade on the right is 410 millimeters (16 inches) wide.

Provision for Varying Light Output. Some means of varying the light output of portable lamps is a desirable design feature, because they are often used for purposes other than specific task illumination. For television viewing, entertaining, or conversation a lower lighting level is often preferred. The utmost control is obtained when a lamp has a built-in full-range dimmer, or when it is used with a separate table-top dimmer. However, high-low and three-way switches for three-way light bulbs or clusters of low-wattage bulbs are suitable means of achieving flexibility.

Position of Light Source. If the light source is close to the bottom edge of the shade, more light will reach the reading plane (see Fig. 10-29). However, the present practice is to center the bulb so that the luminance of the shade is more nearly uniform from top to bottom. A better solution would be to mount one source high in the shade to provide upward light, while placing sources responsible for the task-plane illumination near the lower edge.

Characteristics of Shade. Deep, narrow shades are to be avoided for lamps intended to provide useful illumination; they are double-ended bottle-necks, restricting the spread of both the downward and upward light. Fig. 10-30 illustrates the difference in performance between a 360 millimeter-(14-inch) and a 410 millimeter-(16-inch) shade.

The inner surface of the shade should have a very high reflectance (otherwise much light is lost by absorption), and the shade material should have good diffusing quality and suitable transmittance. In living areas a pleasing effect is achieved when the shades blend smoothly rather than contrast sharply with their backgrounds, in both color and luminance. White or light shades usually should be used against light walls, and darker shades of lower luminance against darker walls.

Translucent shades provide horizontal or cross-lighting, and if all shades have approximately the same luminance the effect is more harmonious. Opaque shades create pools of light above and below, and little light on vertical surfaces. Unless there are other light sources in the room, such as lighted valances or brackets, the effect can be visually uncomfortable. On the other hand, shades of too high transmittance and too little diffusion, such as some made of white fiber glass or laminated plastic, show "hot spots" that are unattractive and distracting. The degree of translucence is vital to visual comfort, and especially important in a room lighted exclusively by portable lamps.

A light meter, preferably one that reads to 5000 lux [500 footcandles], will help to determine suitable shade luminances. The light-sensitive cell is placed in direct contact with the shade and moved from top to bottom to get an average. If the shade is to be seen against walls having 40 to 60 per cent reflectance, the meter should read 1600 (150) or less. With wood-paneled and other walls having reflectances substantially lower than 40 per cent, visual comfort demands much denser shades, and the meter should read no more than 540 (50), preferably less. (Meter reading in footcandles equals luminance of shade in footlamberts).

Lamp Placement for Reading. For reading and other serious visual work the lamp should never be in a position where the bright inner surface of the shade is visible to the user. If the lamp is directly beside the head, as an end table lamp usually is, or a floor lamp at a sofa or chair located against the wall, the lower edge of the shade should be just about at eye level—no lower, to permit as wide a spread as possible, and no higher, to provide proper shielding. Fig. 10-31 illustrates the correct placement. Lamps must be carefully selected for size appropriate to the furniture with which they will be used. In some cases, chair-seat height as well as table height is a determining factor.

Lamps for Dressers and Dressing Tables. Lamps used on dressers and dressing tables are notable exceptions to the general rules for portable lamps. Here, instead of light being directed from the bottom of the shade to a seeing-task, the light useful for make-up must be transmitted through the shade onto the face, and the outside of the shade becomes the light source. White fiber glass, cellulose acetate, and high-transmittance silks or plastics are suitable materials. For a shade with a lower diameter of 230 to 280 millimeters (9 to 11 inches) a lux (footcandle) meter in contact with the surface should show an average reading of 2500 to 4000 [250 to 400].

Centers of the shades should be approximately at cheek level. This usually means that dresser lamps must measure about 560 millimeters (22 inches) from the bottom of the base to the center of the shade, and dressing table lamps 380 millimeters (15 inches) to the center of the shade. In

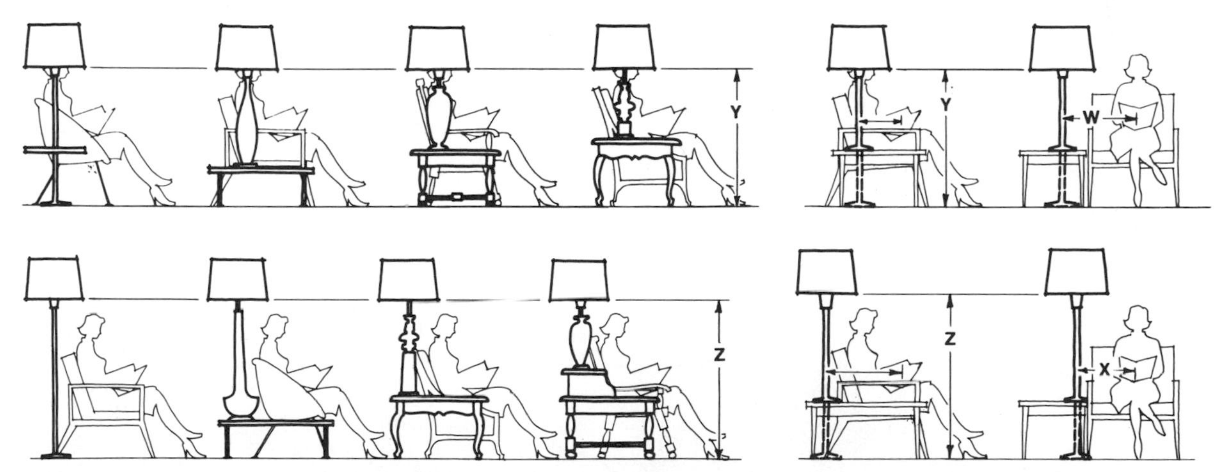

Fig 10-31. Average seated eye level (*Y*) is 970 to 1070 millimeters (38 to 42 inches) above the floor. Lower edge of floor or table lamp shades should be at eye level when lamp is beside user. This is the correct placement for most table lamps, and for floor lamps serving furniture placed against a wall. Floor lamps with built-in tables should have shades no higher than eye level. For user comfort—when floor-lamp height to lower edge of shade or lamp-base-plus-table height is above eye level (*Z*) 1070 to 1250 millimeters (42 to 49 inches), placement should be close to right or left rear corner of chair. This placement is possible only when chairs or sofas are at least 250 to 310 millimeters (10 to 12 inches) from the wall. *W* = 510 millimeters (20 inches). *X* = 380 millimeters (15 inches).

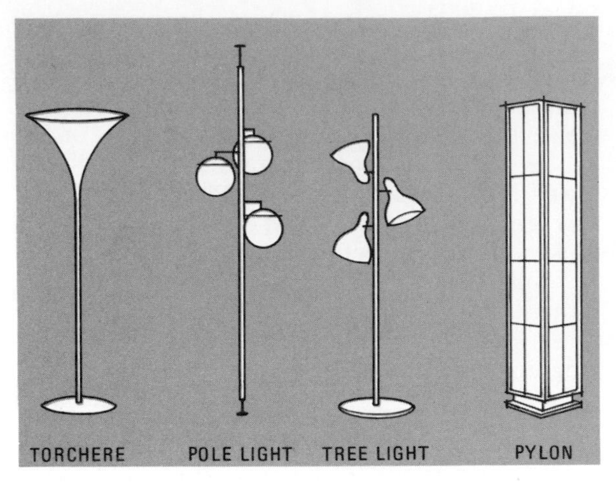

Fig 10-32. Typical portable lamps for general room illumination.

general, dressing table lamps cannot be placed closer than 910 millimeters (36 inches) apart or they will be bothersome visually and will also tend to obscure the mirror.

Lamps for General Room Lighting. Many portable lamps are designed not to provide local illumination but to contribute to general room lighting (see Fig. 10-32). Some are of the completely indirect type, sending all their light to the ceiling to be re-directed over the room. Others are luminous forms, with or without an indirect component; their soft glow can be a most effective element in the lighting pattern of a room. Light sources should never be visible from any seated or standing position within the room, the large luminous surfaces at eye level should not give a reading in excess of 540 (50) on a lux (footcandle) meter placed against them (a luminance of 170 candelas per square meter or 50 footlamberts). The lighting effectiveness of some of the indirect types can be increased by equipping them with reflectorized light sources.

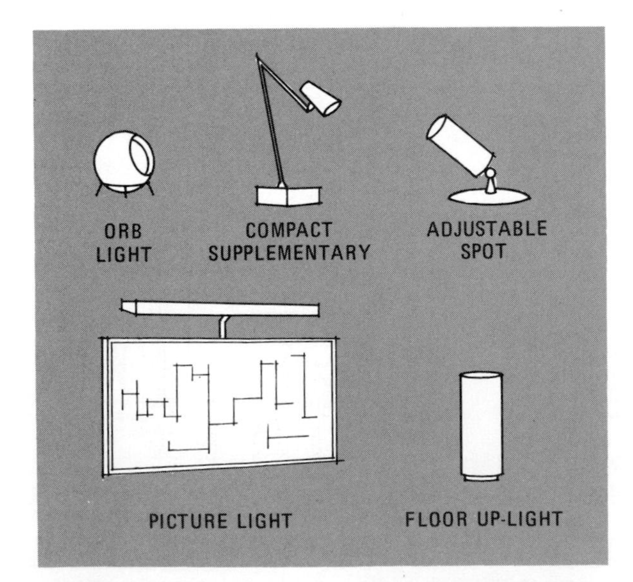

Fig 10-33. Typical portable lamps producing directional lighting.

Lamps for Directional Lighting. Portable lamps providing directional light are made in a wide variety of forms, a few of which are illustrated in Fig. 10-33. Some are used only to supply decorative accent light on a restricted area. Others have excellent application as sources of supplementary illumination for difficult visual tasks. For this purpose they should never be used without other lighting in the room; otherwise the strongly directional light will create sharp shadows, and luminance ratios between the task and the surrounding area will be excessively high.

Structural Lighting Elements

Lighting is frequently built in as part of the structure, or uses structural elements, such as space between joists, as part of the luminaire. Structural lighting includes not only those widely used types illustrated in Fig. 10-34, but also such applications as bookshelves with integral concealed lighting, lighted niches, lighted coffers, artificial skylights, extended soffits, lighted mantels, handrails, china cabinets, luminous ceilings and many others.

The principal advantage of structural lighting is its close correlation with the architecture of the room, offering a very unobtrusive method of providing light. Perhaps its principal disadvantage is the fact that designers have such great freedom. They must make sure that the installation will be architecturally correct, will be simple to construct, and will provide optimum lighting quality. The same basic considerations of lamp concealment, diffuser luminance, direction of light, and light-source color that apply to all types of luminaires apply here also. Structural lighting is usually designed for linear sources, most commonly fluorescent. Dimming devices or alternate switching arrangements may be included to provide more than one lighting effect from the same luminaire. Many manufacturers of light source equipment have prepared excellent technical guides for simplifying the design of

Fig. 10-34. Structural Lighting Elements

LIGHTED CORNICES

Cornices direct all their light downward to give dramatic interest to wall coverings, draperies, murals, etc. May also be used over windows where space above window does not permit valance lighting. Good for low-ceilinged rooms.

VALANCES

Valances are always used at windows, usually with draperies. They provide up-light which reflects off ceiling for general room lighting and down-light for drapery accent. When closer to ceiling than 250 millimeters (10 inches) use closed top to eliminate annoying ceiling brightness.

LIGHTED WALL BRACKETS (HIGH TYPE)

High wall brackets provide both up and down light for general room lighting. Used on interior walls to balance window valance both architecturally and in lighting distribution. Mounting height determined by window or door height.

LIGHTED WALL BRACKETS (LOW TYPE)

Low brackets are used for special wall emphasis or for lighting specific tasks such as sink, range, reading in bed, etc. Mounting height is determined by eye height of users, from both seated and standing positions. Length should relate to nearby furniture groupings and room scale.

LIGHTED SOFFITS

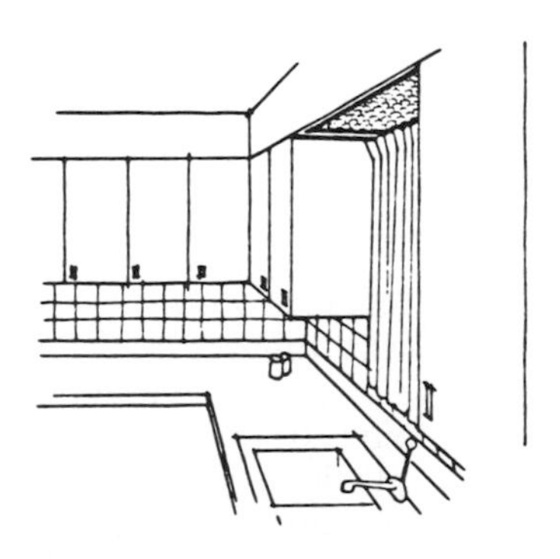

Soffits over work areas are designed to provide higher level of light directly below. Usually they are easily installed in furred-down area over sink in kitchen. Also are excellent for niches over sofas, pianos, built-in desks, etc.

LIGHTED SOFFITS

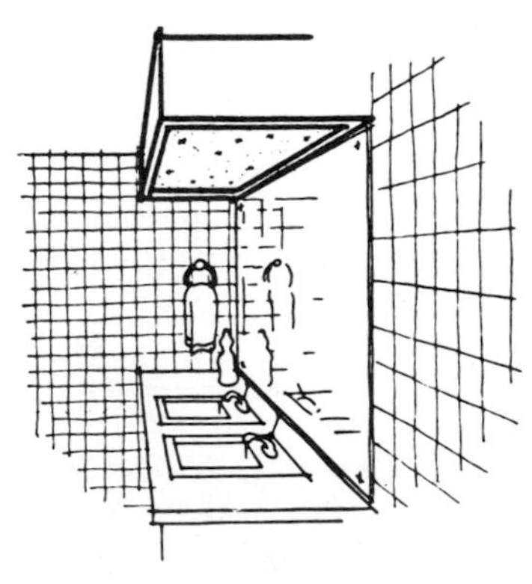

Bath or dressing room soffits are designed to light user's face. They are almost always used with large mirrors and counter-top lavatories. Length usually tied to size of mirror. Add luxury touch with attractively decorated bottom diffuser.

Fig. 10-34. *Continued*

LUMINOUS WALL PANELS

Luminous wall panels create pleasant vistas; are comfortable background for seeing tasks; add luxury touch in dining areas, family rooms and as room dividers. Wide variety of decorative materials available for diffusing covers. Dimming desirable for most efficient use of energy.

LIGHTED CANOPIES

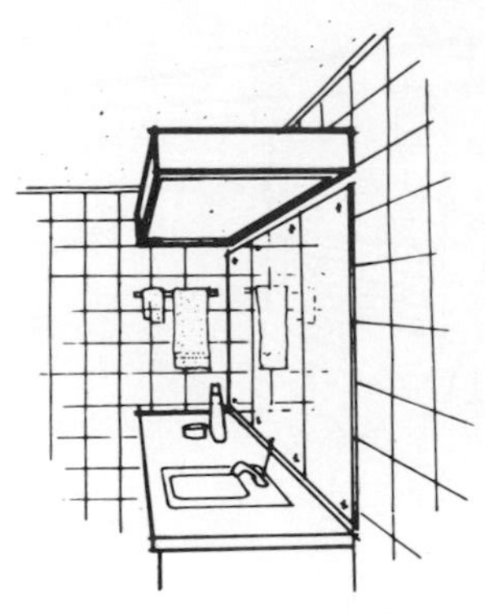

The canopy overhang is most applicable to bath or dressing room. It provides excellent general room illumination as well as light to the user's face.

LIGHTED COVES

Coves direct all light to the ceiling. Should be used only with white or near-white ceilings. Cove lighting is soft and uniform but lacks punch or emphasis. Best used to supplement other lighting. Suitable for high-ceilinged rooms—and for places where ceiling heights abruptly change.

LUMINOUS CEILINGS

Totally luminous ceilings provide skylight effect very suitable for interior rooms or utility spaces, such as kitchens, baths, laundries. With attractive diffuser patterns, more decorative supports, and color accents they become acceptable for many other living spaces such as family rooms, dens, etc. Dimming controls desirable for most efficient use of energy.

good structural lighting. Some manufacturers now offer completely assembled structural lighting "packages."

LIGHTING CONTROLS

Switching has always been a necessity to control electric light, but with the increase in emphasis on better utilization of energy, dimming, too, has almost become a necessity. The designer should make an effort to design the control (switching and dimming) locations with convenience in mind—the more convenient a control is, the more likely a person is to use it.

Controlling Illuminance Levels. Light output of incandescent luminaires or portable lamps can be varied in steps by switching and continuously by dimming, from no output to full output, and from full light output to less than one per cent output for fluorescent luminaires. Therefore, a desired amount of light can be set for the activity being performed. For instance, in a dining room the lighting can be set to full output for the setting of the table. When the meal is served, the lighting can be set to a different level for normal dining or to a lower level when a more intimate and romantic effect is to be achieved. In the living room area, one level of lighting may be set for reading while a second is set for a social gathering.

Decorative Effects with Dimming. Dimming can also be used effectively to create decorative effects. For example, by having dimming capabilities, lighting for a piece of sculpture or a painting can be varied to produce the level desired for viewing a particular work of art. A chandelier which utilizes clear decorative incandescent lamps, can be dimmed to a point where the lamps simulate very closely the color of candlelight and the feeling of a candelit space.

For additional information on dimming see Section 2.

ENERGY CONSIDERATIONS

Decisions early in the lighting design process can clearly affect the energy utilization for any lighting installation. Using the criteria in this Section together with those energy considerations covered in Section 4 can result in residential lighting solutions that are energy efficient.

The homeowner often must work within the limitations of lighting equipment which is already in place and achieve power saving through modification of those light sources and luminaires, unless remodeling or equipment replacement is feasible[4]. If structural changes are made, then the remodeled area may be treated as new construction and the connected power for lighting planned as if for original design.

To determine how much power is being used in an existing lighting system each room should be examined and the connected power totaled. This total should include ballast watts for fluorescent luminaires and the equipment check should include portable lamp types as well as ceiling luminaires.

See Section 4 for a procedure for determining an appropriate lighting power budget for residential spaces.

REFERENCES

1. Residence Lighting Committee of the IES: "Design Criteria for Lighting Interior Living Spaces," *Light. Des. Appl.*, Vol. 10, p. 31, February and p. 35, March, 1980.
2. "IES Lighting Performance Requirements for Table Study Lamps," *Illum. Eng.*, Vol. LX, p. 463, July, 1965.
3. Crouch, C. L. and Kaufman, J. E.: "Illumination Performance for Residential Study Tasks," *Illum. Eng.*, Vol. LX, p. 591, October, 1965.
4. Residence Lighting Committee of the IES: "Energy-Saving Tips for Home Lighting," *Light. Des. Appl.*, Vol. 4, p. 40, April, 1974.
5. Color Committee of the IES: "Color and the Use of Color by the Illuminating Engineer," *Illum. Eng.*, Vol. 57, p. 764, December, 1962.

Theatre, Television and Photographic Lighting

SECTION 11

Even though lighting for theatre, television and film have many similarities, they have many differences. In theatre, one lights a three-dimensional subject for the eye. In television and film, one lights for a camera to transform a two dimensional medium (height and width) into a three-dimensional look, adding depth.

For theatre, television and film, the lighting system design and luminaire choice is based on production plans. The size and complexity of the system is based on production needs, from elementary training facilities to professional facilities. In all facilities, however, the budget usually determines the degree of complexity.

The theatre design requires information concerning types of programs (opera, ballet, drama, variety) which will be produced by resident groups or touring companies. Television design requires information concerning types of productions (variety shows, dramas, news, soap operas, panel shows) which will be produced for network or local broadcasting, or closed circuit or syndication release. The actual illuminance levels for television vary from under 1000 lux [100 footcandles] to several thousand lux [several hundred footcandles], depending upon the type of video camera in use.

In lighting for film, both still and motion picture, the function of the lighting is not only to give the third dimension, but to produce the photochemical changes required to produce the image. Thus, illuminance is determined largely by the type of film, *i.e.*, its sensitivity.

NOTE: References are listed at the end of each section.

LUMINAIRES, LAMPS AND CONTROL SYSTEMS

Luminaires for Theatre, Television and Photographic Lighting

In theatre, television and photographic (film) lighting different types of luminaires are used to produce a quality of light output which falls generally within three basic categories:

1. *Soft light*—diffuse illumination with indefinite margins. Its output produces poorly defined shadows. (Luminaires—scoop, soft-light.)
2. *Key light*—illumination with defined margins. Its output produces defined, but soft-edged shadows. (Luminaire—Fresnel spot-light.)
3. *Hard light*—illumination which produces sharply defined geometrically precise shadows. (Type of luminaire—ellipsoidal spotlight.)

The basic type of luminaires used in theatre, television and film production are described as follows (see Fig. 11-1):

Fresnel Spotlight. The Fresnel spotlight is a luminaire which embodies a lamp, a Fresnel lens and, generally, a spherical reflector behind the lamp. The field and beam angles* can be varied

* Those points on the candlepower distribution curve at which the candlepower is 10 per cent of the maximum candlepower define the field of the lighting unit. The included angle is the *field angle*.

Those points on the candlepower distribution curve at which the candlepower is 50 per cent of the maximum candlepower define the beam of the lighting unit. The included angle is the *beam angle*.

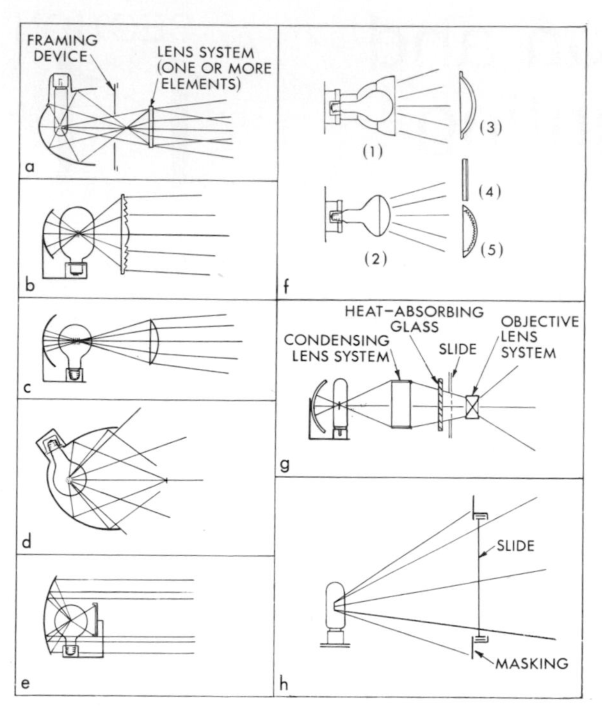

Fig. 11-1. Optical characteristics of stage lighting equipment. a. Ellipsoidal-reflector spotlight. b. Fresnel-lens spotlight. c. Plano-convex lens spotlight. d. Scoop-type floodlight. e. Parabolic-reflector floodlight. f. Striplight: (1) reflector with general service lamp; (2) reflectorized lamp; (3) glass roundel, plain or colored; (4) sheet color medium; (5) spread lens roundel, plain or colored. g. Lens-type scenic slide projector. h. Non-lens type scenic slide projector (Linnebach type).

by changing the distance between the lamp and the lens. This action is called focusing. The distance between the lamp and reflector is determined by the optical designer and does not vary during focusing.

The quality of illumination produced by the Fresnel spotlight tends to be intermediate or hard, and the beam is soft edged. The illumination varies considerably, depending on the optics of the luminaire. Typical luminaires have a beam angle of 10 to 40 degrees, depending on the relative position of the lamp and lens.

Fresnel spotlights are generally equipped with tungsten-halogen incandescent lamps with C-13 or C-13D planar filaments. Many Fresnel spotlights are now available using a compact source metal halide lamp as well. In order to shape the light beam, barn doors* are used as well as snoots.† The light beam may also be colored or diffused with materials placed in its color frame.

* A set of swinging flaps, usually two or four (two-way or four-way) which may be attached to the front of a luminaire in order to control the shape and spread of the light beam.

† A metal tube that can be mounted on the front of a spotlight to control stray light. Also known as a funnel, top hat or high hat.

Fresnel spotlights are manufactured in lens diameters from 76 to 610 millimeters (3 to 24 inches) and in wattages from 75 to 10,000 watts. Also available are Fresnel spotlights that utilize dual filament lamps, *i.e.*, 1250/2500 watts or 2500/5000 watts. The lamps are designed to work with both filaments on (full power) or with either filament on (half power) giving the luminaire even more versatility. In addition, remote operation (pan, tilt, focusing, on/off) is available on some luminaires.

Scoop. The scoop is a floodlight consisting of a lampholder, lamp and reflector with a matte or brushed finish. The lamp and reflector may have a fixed or variable relationship. Scoops are equipped with front clips to hold a color frame containing either color media or diffusion material. The scoop produces illumination having a field angle of 90 to 180 degrees. The quality of the illumination is considered soft, and the shadow sharpness depends primarily on the texture of the reflector. Scoops are available from 300 to 450 millimeters (12 to 18 inches) in diameter and are usually equipped with tungsten-halogen lamps from 500 to 2000 watts. The lamps are usually frosted for further softness of illumination. In general, the larger the diameter of the scoop, the softer the light output.

Non-Lens Luminaires. The non-lens luminaire (primarily used in film location applications) embodies a lamp, reflector and frequently a focus mechanism to change the light output by varying the field and beam angles. The quality of the illumination produced by a non-lens luminaire tends to be intermediate or hard depending on lamp type and reflector finish. External beam control is possible with barn doors; however, the degree of control is somewhat limited. In addition, the luminaires can be equipped with scrims, heat filters, daylight correction filters, or color frames. There are many different types of non-lens luminaires which can be considered as spotlights or floodlights. They utilize tungsten-halogen lamps in the range of 650 to 2000 watts.

PAR-Luminaire. PAR-luminaires embody a PAR lamp, lampholder and housing; the performance of the luminaire depends upon the type of PAR lamp selected.

As the beam pattern of most PAR lamps tends to be oval, the luminaire is designed so that the lamp may be rotated to cover the desired area. The beam of a PAR lamp can be shaped and glare reduced by barn doors and the intensity increased by an intensifier skirt—an exterior reflector that continues the parabolic contour of the internal lamp reflector.

PAR luminaires using 650- or 1000-watt lamps are designed to accommodate either single lamps or groups of lamps in clusters of 3, 6, 9, 12, etc. Luminaires using PAR-38 or PAR-56 lamps may be arranged in linear strips defined as borderlights or striplights.

Striplight. A striplight is a compartmentalized luminaire. Each compartment contains a reflector lamp, or a lamp and reflector, and a color frame. Compartments are arranged in line, and wired on 2, 3 or 4 alternate circuits, each circuit being colored with a suitable color media.

Striplights provide an over-all wash of illumination over a stage. They may also be located at the front of a stage to provide an over-all low illuminance level (termed footlight).

Cyclorama Lights. Cyclorama or cyc lights provide an over-all wash of illumination over the cyclorama curtain for background illumination. There are two types of cyc lights: the striplight, which is a compartmentalized luminaire with lamps on 200- to 300-millimeter (8- to 12-inch) centers, and the "cluster" lights, with units mounted on 1.8- to 2.4-meter (6- to 8-foot) centers.

Striplights, when mounted either from above or on the floor, are limited to a 3.6-meter (12-foot) high wash unless lights are used on both the top and bottom or have an asymmetric reflector to improve the beam distribution. Normally, a single group, either above or floor mounted, will have a falloff of wash greater than 50 per cent from top to bottom or vice versa.

"Cluster" lights, on the other hand, will light a cyclorama up to 9 meters (30 feet) high by lighting just from the top. The falloff of light levels is less than 5 per cent. Since these lights are mounted on 2.4-meter (8-foot) centers, rather than the 300-millimeter (12-inch) centers for striplights, the power savings is substantial. Striplights, for 2-color units, use 3300 watts per meter (1000 watts per foot), whereas the "cluster" lights use 820 watts per meter (250 watts per foot), for the same 2-color coverage.

Ellipsoidal Spotlights. The ellipsoidal spotlight, or pattern spotlight, consists of a lamp and ellipsoidal reflector, mounted in a fixed relationship. The light is focused through the gate of the unit where the beam can be shaped with the use of either push shutters, a gobo or an iris. The shaped beam then goes into the lens system which focuses the beam.

The output of the ellipsoidal spotlight is a very hard edge light with precise beam control. By focusing the lens system, the hard edge can be somewhat softened.

The lens diameter and focal length will determine the throw and coverage of the unit. They are available from 90-millimeter (3½-inch), 400-watt units to 300-millimeter (12-inch), 2000-watt units.

The effective throw of the larger units is about 30 meters (100 feet).

Follow Spot. A follow spot is a special type of spotlight, stand mounted, with a shutter, iris and color changer for color media. Most follow spots today utilize a tungsten-halogen lamp or an arc source and a lens system.

A new lensless tungsten-halogen unit has recently been introduced. For high intensity spotlights the carbon arc is being replaced with compact source metal halide lamps and xenon.[1] All of these arc units produce more output, watt for watt, than the tungsten-halogen units.

All follow spots are designed for the long throws required for their application.

Parabolic Spotlight. The parabolic spotlight consists of a lamp and parabolic specular reflector. Some luminaires have a reflector in front of the lamp to redirect light into the main reflector. Other luminaires are equipped with spill-rings to minimize spill light and glare. In most types of parabolic spotlights the lamp and reflector are adjustable to produce a wide or a narrow beam of light—the closer the lamp to the parabolic reflector the wider the beam. This luminaire produces a hard beam of illumination which cannot be easily controlled, except in part by spill-rings. (A parabolic spotlight is also known as a sun spot or beam projector.)

Soft Light. A soft light luminaire is a well diffused, almost shadow free light source used in special applications. All of the light in these units is reflected off a matte finish reflector before reaching the subject, and due to this design, they are not efficient sources of light.

They are available from 1000 to 8000 watts in size and are used where either shadows or reflections must be minimized.

Arc Light. An arc light luminaire uses a carbon arc as the source of illumination. Such luminaires are manufactured in two basic forms: (1) equipped with a Fresnel lens for general illumination, and (2) equipped with a reflector and optical system for follow spot operation. Carbons available for these luminaires are of various types producing corresponding intensities and color temperatures (see Fig. 11-2 and Fig. 11-3). Carbons and ballast equipment are available for either dc or ac operation.

Fig. 11-2. Arc Lamps Used in Motion Picture Set Lighting

Unit	Beam Divergence (degrees)		Amperes	Arc Volts	Positive Carbons	Negative Carbons
	Minimum	Maximum				
Type 40 Duarc	90	90	40	36	8 mm × 12 inch CC MP studio	7 mm × 9 inch CC MP studio
Type 90	8	44	120	58	13.6 mm × 12 inch MP studio	7/16 × 8½ inch CC MP studio
Type 170	8	48	150	68	16 mm × 20 inch HI MP studio	½ × 8½ inch CC MP studio
Type 450 Brute	12	48	225	73	16 mm × 22 inch Super HI MP studio (white flame) or 16 mm × 22 inch Super HI YF MP studio (yellow flame)	17/32 × 9 inch H.D. Orotip
Titan	12	48	350 white flame	78 white flame	16 mm × 25 inch ULTREX HI WF studio (white flame)	11/16 × 9 inch CC MP studio
					or	
			300 yellow flame	68 yellow flame	16 mm × 25 inch ULTREX HI YF 300 special studio (yellow flame)	

Lamps for Theatre, Television and Photographic Lighting

The most prevalent light sources for theatre stages, television and film production, and professional still photography are tungsten-halogen lamps that have been designed especially for these types of service. Conventional incandescent spotlight lamps continue to be used, but are gradually being supplanted by tungsten-halogen lamps designed to retrofit existing luminaires.

Carbon arcs continue to be used (principally for motion picture production), but have been supplanted to a great extent—originally by multiple-lamp arrays of 650-watt or 1000-watt tungsten-halogen PAR lamps, often with integral daylight filters—by compact source metal halide lamps. These compact-arc lamps are also used to an increasing extent to provide supplementary light of approximate daylight color for television as well as film production, and have been adopted into such specialized equipment as scenic projectors and follow spots. Xenon short-arcs of the types commonly used for motion picture projection are also used in some follow spots and scenic projectors.

Virtually all types of light sources find occasional use in theatre, television and film production. For example: metal halide high intensity discharge lamps are used to light such areas as stadiums and arenas, providing light levels and color quality suitable for televising performances, and have the capability of providing "daylight fill" for movie and television productions. Fluorescent lamps (white and/or colors) on special dimming systems may light cycloramas in television studios or theatre stages, or backings of motion picture sets. Xenon flash tubes, similar to those for photo studios, may be used in a repetitive-flash mode for spectacular theatrical effects. Lasers and light-emitting diodes may also be used for spectacular effects. Standard PAR and R lamps, incandescent sign/decorative and indicator lamps have a variety of common "theatrical" applications.

Extensive data on all types of light sources are found in Section 8 of the 1981 Reference Volume, including lamp operating principles, physical and performance characteristics, and tables of commonly available types.

Tungsten-Halogen and Incandescent Lamps.[2-5] Data for lamps most frequently used in lighting for theatre stages, television and motion picture production, and professional still

Fig. 11-3. Beam Characteristics of Typical Motion Picture Studio Arc Lighting Units

Unit	Lamp	Beam Width (degrees)	Lumens	Approximate Peak Candlepower (candelas)
Type 40—Duarc	115V dc arc (40 amp, two 36V arcs in series)	150	76,500	32,000
Type 90—High-intensity arc spot	115V dc arc (120 amp, 58 arc V)	10	18,700	2,100,000
		18	26,000	550,000
		44	62,500	150,000
Type 170—High-intensity arc spot	115V dc arc (150 amp, 68 arc V)	10	47,000	5,700,000
		18	75,000	2,300,000
		48	130,000	300,000
Type 450—Brute high intensity arc spot	115V dc arc (225 amp, 73 arc V)	12	117,000	10,000,000
		16	159,000	7,300,000
		20	172,000	5,000,000
		30	228,000	2,500,000
		48	260,000	1,00,000

photography are found in Section 8 of the 1981 Reference Volume, Figs. 8-91 and 8-94. In many cases lamps may appear to be mechanically interchangeable with each other, *i.e.*, they have the same base and light source location. However, caution should be exercised. For example, in newer, more compact luminaires designed specifically for tungsten-halogen, the bulbs of some incandescent lamps may be too large to fit. Furthermore, there may be differences in filament configuration that could affect the luminaire optical performance.

Another possible limitation on lamp interchangeability is that some luminaires may not provide adequate heat dissipation for higher-wattage lamps. Lighting equipment manufacturers should be consulted for maximum allowable wattage.

ANSI Codes. The American National Standards Institute (ANSI) assigns three-letter designations for incandescent lamps (including tungsten-halogen lamps) used in photographic, theatre and television lighting applications. The letters are arbitrarily chosen and do not, in themselves, describe the lamps. Neither do they imply tolerances or minimum levels of performance with respect to such non-mechanical characteristics as life, light output, etc. When an ANSI code is assigned to a lamp type at the request of a lamp manufacturer, it may be used by all manufacturers as a commercial ordering code. The assignment of an ANSI code defines some basic parameters of the lamp such as nominal wattage, type of base, size of bulb, light center length, approximate color temperature, etc., so that physical interchangeability of lamps is assured. However, if lighting equipment performance is particularly sensitive to filament size, coil spacing, etc., the ANSI code does not guarantee equal performance for lamps of different manufacture. Furthermore, lamp manufacturers sometimes make improvements in lamp performance characteristics (life, lumens, etc.) beyond the values that are stated on the original ANSI code application.

Manufacturer's Ordering Codes. The manufacturer's ordering code is another type of lamp designation. Often two or more manufacturers will agree on the code designation for a lamp type. Manufacturer's codes are usually descriptive of some of the lamp characteristics, and often include lamp wattage, bulb size and shape. Some manufacturers of lighting equipment supply lamps marked with their own private codes to identify the proper lamps for their equipment.

Low-Noise Construction. Most tungsten-halogen lamps for theatre, television and film application, and many of the incandescent lamps for such applications, have special *low-noise* construction to minimize audible noise generation when operated on ac circuits. The lamp manufacturer should be consulted for information on which lamps have low-noise construction. Lamps, sockets, wiring, etc., tend to generate more audible noise when used with dimmers that distort the normal ac sine wave than with ordinary autotransformer or resistance dimmers (see page 2-35). Generally noise is not generated on dc circuits.

Caution Notices. Caution notices are generally provided with most lamps for stage and studio service. All tungsten-halogen lamps operate with internal pressure above that of the atmosphere; therefore, protection from lamp abrasion and avoidance of over-voltage operation is advised. The use of screening techniques is advised where appropriate to protect people and surroundings in case a lamp shatters.

Electric Discharge Lamps. A wide variety of enclosed high intensity electric arc discharge lamps employing metal halides as the light-generating medium are in common use for general interior and exterior lighting systems. They are extensively described in Section 8 of the 1981 Reference Volume. In general, they are characterized by: (1) relatively high luminous efficacy (up to over 100 lumens/watt), (2) ac operation at line frequencies of 50 to 60 hertz with consequent cyclic variation in light output, (3) a need for auxiliary ballast equipment to provide striking voltage and current regulation, (4) service lives in the thousands or tens of thousands of hours, (5) spectral distributions that depart substantially from a blackbody, containing many spikes and valleys, yet providing generally acceptable color rendering at apparent color temperatures higher than incandescent lamps, (6) warm-up times of a few minutes to reach full output and color stability, and usually a short cool-down period after turn-off before they will restart, (7) arc luminances substantially below the luminance of typical incandescent tungsten-halogen spotlight lamps and (8) relatively large outer glass bulbs with screw bases, enclosing the arc tube. Such lamps are commonly and successfully used in stadiums and arenas from which sporting events, etc., are televised.

During the 1970's, special types of metal halide discharge lamps have evolved with characteristics that are often better-suited to the requirements of lighting for film and television production outside of the studios, and for scenic projectors and follow spots. Two principal groups[6] of lamps are commonly known as HMI[7] (hydrargyrum, medium-arc-length, iodide) and CSI (compact source iodide). Recently two additional types have been added, DMI and CID[8], both in the same group as HMI. These two groups of lamps share the first three characteristics noted above, but differ in the remaining ones.

HMI, DMI and CID lamps are available in a range of sizes—from 200 to 4000 watts (see Fig. 8-128 in the 1981 Reference Volume). HMI and DMI lamps are of double-ended construction, CID are of single-ended construction and both have relatively short arc gaps, such that arc luminance is several times that of typical incandescent spotlight lamps. This feature makes them useful in follow spots and effects projectors. HMI and DMI have an essentially complete spectrum (see Fig. 11-4) with a very high color rendering index (R_a = 90) at approximately 5600 K apparent color temperature (for CID R_a = 85 at approximately 5500 K), making them well-suited as sources for "daylight fill" in movie and television shooting. Effective service lives range from 300 to 2000 hours. After ignition, they require a minute or so to warm up. Most ballast/igniter equipment will re-strike the lamps immediately after turn-off. The cyclic flicker of these ac operated lamps generally presents no problems for television systems, but requires careful coordination of motion picture frame rates and shutter angles to avoid noticeable flicker effects in the projected films. Ballasts are available which provide waveform and/or frequency modifications that will minimize or eliminate flicker in motion pictures.

CSI lamps are available in 400-watt and 1000-watt sizes (see Fig. 8-128 in the 1981 Reference Volume). The latter wattage is also furnished within a PAR-64 sealed glass reflector, over which interchangeable spread lenses may be used. Life ratings range from 500 to 2000 hours. CSI lamps also have relatively high source luminances that permit effective image projection and follow spot applications. Spectral quality departs substantially from a blackbody (see Fig. 11-5), and is photographically intermediate between daylight and incandescent lamp color. The color rendering index (R_a) is 80 at approximately 4200 K. Filtration via auxiliary color media is often used—blue-tinted to blend with daylight, yellow-tinted to blend with tungsten lamps. As with other metal halide lamps, a few minutes warm-up is required; CSI lamps with mogul bi-post bases can be hot re-struck if suitable igniters are included with the ballast. As with HMI lamps, cyclic flicker can present problems for motion picture work unless suitable precautions are taken. Development continues on control equipment that minimizes such problems.

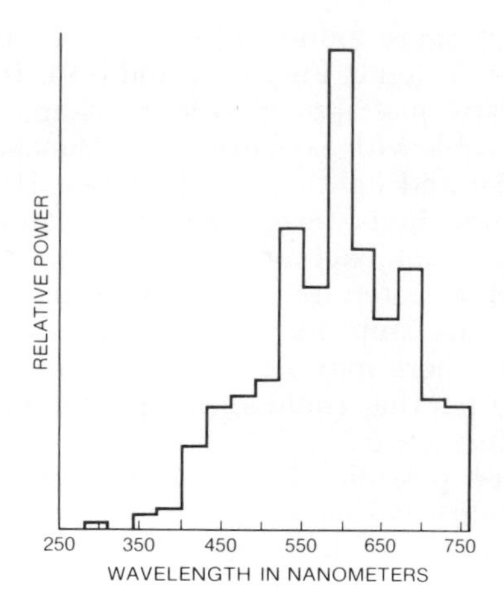

Fig. 11-5. Relative spectral power distribution for a typical CSI lamp.

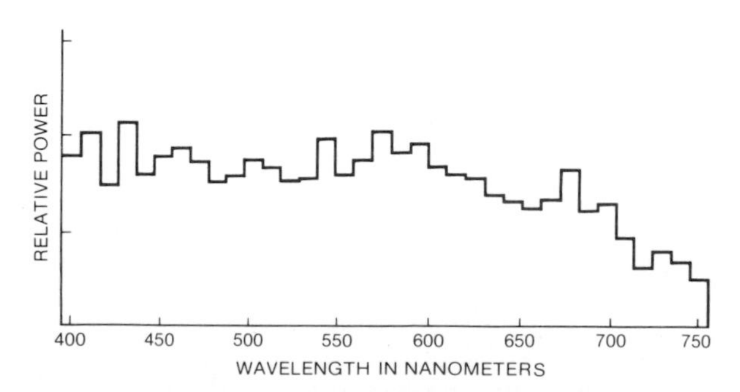

Fig. 11-4. Relative spectral power distribution for a typical HMI lamp.

Control Systems for Theatre and Television Lighting

The design of a lighting control system is based on the artistic and technical needs of projected productions.[9-13] This design is related to the building architecture, luminaire rigging system and density of electrical outlet distribution. Design parameters are expressed in terms of power capability, number of lighting outlets, dimmer bank capacity, interconnection system to assign lighting outlets to dimmers, dimmers to control channels and lighting control facilities.

The performance requirements of a lighting control system are to provide the lighting designers with total flexibility of control over all of the luminaires lighting the set. There must, therefore, be adequate dimmers, non-dim circuits, and control equipment to establish the number of lighting channels required, to assemble these channels into cues, and to switch and fade from one cue to another achieving the desired lighting changes.

Lighting control systems for the theatre and for television differ slightly: theatre lighting control systems have extensive use for memory, require accurately timed faders and be capable of complex simultaneous operations; television lighting control systems require a lower memory capability, less operational features, and benefit from automated facilities to build dimmers into channels.

Lighting control systems divide into two basic categories described as "dimmer-per-outlet" and "power programming". Dimmer per outlet systems provide a dimmer of adequate capacity for every lighting outlet and use a control patch, manual or automated, to assign dimmers into control channels. Power programming systems employ a cord and jack, or slider and bus system to switch individual outlets to larger capacity dimmers.

Dimmer-per-outlet systems provide for maximum flexibility of control and should be considered for lighting installations of more than 300 outlets.

Dimmers. Almost all of the lighting control devices used in theatre and television lighting use SCR dimmers which are manufactured in a range of 2 to 12 kilowatts, or triac dimmers manufactured in a nominal capacity of 2 kilowatts.

For good lighting operation, it is essential that electronic dimmers have stable output, cause no interference to audio and video circuits or to other dimmers, be insensitive to load and have high efficiency. The dimmer curve,[14, 15] that is voltage output relative to control voltage, may be fixed or set to any one of several different curves by adjustment of potentiometers on the dimmer card. This card also provides for optional current and voltage regulation and other specified features.

Dimmers are assembled in portable packs of 6 to 12 dimmers or into dimmer racks custom-built to suit the installation.

All electronic dimmers require some ventilation in order to maintain components within specified operating temperatures. The amount of ventilation required is dependent on the individual manufacturer's recommendations.

A major expenditure in any lighting system is the cost of the electrical installation; therefore, the lighting outlet's power switching system and dimmer bank must be located in close proximity to minimize wiring runs.

Power Switching Systems. There are two basic power programming systems, described as cord patch and slider patch.

The *cord patch* system consists of a plug, cord and load circuit breaker for each lighting outlet, and a group of jacks for each dimmer or non-dim circuit. Luminaires are connected to numbered outlets, which are wired to numbered plugs and cords. These plugs are connected to appropriate dimmer jacks and the circuit breaker may then be energized.

The *slider patch* has a slider assembly and load circuit for each lighting outlet, and a bus for each dimmer or non-dim circuit. Luminaires are connected to numbered outlets which are wired to numbered circuit breakers and sliders. These sliders are positioned to the appropriate dimmer bus and the circuit breaker is switched on. Electrical connection is made from the dimmer bus to the slider contact through slider brushes through a connecting rod to the circuit breaker.

Manual Preset Lighting Control Systems. The basic form of lighting control is the manual preset system which employs a group of manual controllers for each dimmer or control channel. These controllers are arranged in horizontal rows termed presets.

Presets (controllers) are switched to sub-master faders, which are in turn switched to *X* and *Y* master faders for proportional dipless cross fades between presets.

Lighting levels are set as required on individual controllers in each preset, and lighting cues are achieved through sub-master and master controllers.

Memory Lighting Control Systems. Memory lighting control systems are in general hard wire logic control systems and programmed soft wire based systems.

In a hard wire memory system, the dc voltage from the potentiometers is fed into a multiplexer and by timing techniques into the logic section where each channel is sampled and fed into an analog to digital converted (A/D). This information is then inserted into the memory. Later the stored information may be recalled from the memory using a keyboard through a D/A converter, pass through a decoder and be transmitted as control voltage to the dimmers.

Data in a hard wire memory system is generally stored in solid state volatile memories, which are maintained by a battery when ac power is disabled. A magnetic core or permanent type memory system may be used in place of the volatile memory.

An alternative and more sophisticated approach to hard wired memory systems is to employ a microprocessor based system. In this system, the operational program is permanently stored in a "read only memory" (ROM) section which may be varied by the manufacturer to provide additional operational facilities as they are developed.

These systems may incorporate video monitors for displaying cue information and have a floppy disc or cassette for storage of program information.

Computer Controlled Systems. The most recent trend in memory lighting systems is the development of the computer based system. Computer systems consist of a minicomputer, a video terminal, a floppy disc for the operating program and cue storage, an interface board between the computer and the dimmer bank, and finally a hard copy printer to produce cue sheets with appropriate notes. All information is entered into the computer via a keyboard. This information is displayed on a monitor together with appropriate instructions and questions to the operator.

These systems are soft wired based. This means that operational facilities are built into the program, and as additional features are added, the program is revised to provide them. Computer programs are uncommitted and may be written to provide information on dimmer loading, outlet testing, power demand and any other technical data on the lighting control system.

A dimmer-per-outlet system combined with a computer lighting control offers maximum flexibility of operation. The computer will connect dimmers to channels, and program channels into cues; associated equipment will switch and fade from cue to cue as required by the productions.

LIGHTING FOR THEATRES

The two basic categories of theatres are the live-production type and the film or motion picture theatre. The former can be further classified as legitimate, community and school theatres. In the case of motion picture theatres, the two basic classifications are the indoor auditorium-type and the outdoor drive-in theatre.

Lighting requirements for the marquee, lobby and foyer are similar for both live and film auditorium-type theatres, but separate considerations must be given to drive-in theatres. For auditorium-type theatres, it is a case of reducing the lighting level as a patron proceeds from the brightly lighted marquee and street area to the lobby, the foyer and eventually to the auditorium. Basically, the same type of planning is required for the approach to drive-in theatres, remembering that everything must be larger to accommodate the moving vehicle.

At this point, however, the lighting requirements for the various types of theatres change. The so-called houselights, which illuminate the seating or parking area, are quite different in nature. Lighting within the film theatre is minor in scope compared to house and stage lighting required for live productions. Further, in the live theatre a large portion of the on-stage luminaires are designed to be portable so as to accommodate a variety of needs as productions change.

Marquee. Attracting attention is one of the most important factors in the design of theatre exteriors. Here is where much of the selling is done for the current event and for coming attractions. Flashing signs, running borders, color-changing effects, floodlighting and architectural elements are but a few of the many techniques employed. Many present marquee attraction panels are lighted with incandescent filament lamps, fluorescent sign tubing or fluorescent lamps behind diffusing glass or plastic. Opaque or colored letters on a lighted field are generally more effective than luminous letters on a dark field. The principal requirement is uniformity of luminance because spotiness interferes with legibility of letters. The actual luminance is usually determined by the general district luminance. The following values are a guide: low luminance district, 300 to 500 candelas per square meter [90 to 150 footlamberts]; medium, 400 to 700 [120 to 200]; and high, 500 to 1200 [150 to 350].

A studded pattern of small sign lamps is a typical soffit treatment for many theatres. Reflectorized lamps are being employed for this purpose in many theatres to provide illumination on the sidewalk and some degree of luminance at the soffit, as well as effective ceiling pattern. Light here must serve the additional functions of increasing sales and providing security at the box office. A range of 200 to 500 lux [20 to 50 footcandles] is typical. Infrared units for heating and snow melting should be considered in localities where required by climatic conditions. See Section 19 for further discussion.

Lobby. An illuminance of 200 lux [20 footcandles] is desirable in theatre lobbies. The ceiling treatment is often integrated with the marquee soffit. Many lighting treatments are applicable here; some requirements are easy maintenance, designs that retain appearance and architectural suitability, and a pattern of luminances that attracts as well as influences the flow of traffic.

Built-in lighting with fluorescent lamps, spotlighting or transillumination are successful techniques for poster panels. Luminances range from 70 to 350 candelas per square meter [20 to 100 footlamberts], depending on surroundings and brightness competition. An important consideration is to allow sufficient depth so fairly uniform illuminance may be obtained.

Proceeds from refreshment stand sales may be a large share of the total gross income. Built-in case lighting, spotlighting, luminous elements and signs (often in color) help gain attention. Lamps are used in hot food dispensers both for attraction purposes and heat.

Foyer. Usually a restful, subdued atmosphere is desirable in the foyer. Illumination from large, low luminance elements, such as coves, is one good method. Wall lighting and accents on statuary, paintings, posters and plants are important in developing atmosphere. Light must not spill into the auditorium. General illuminance levels of from 50 lux [5 footcandles] for motion picture theatres, to 150 lux [15 footcandles] for live-production theatres are recommended.

Live-Production Theatres

Included among the live-production theatres are the legitimate, community and educational theatres, including public schools, high schools

Fig. 11-6. Two of the most common types of live-production theatre stages: (upper) the traditional proscenium stage in New York State Theatre in Lincoln Center, and (lower) the open stage or "thrust-type."

and colleges. The term live-productions refers to the presence of live actors on-stage as opposed to presentation of a film show. Although there are many varieties of indoor and outdoor live-production theatres, such as amphitheatre, music tent, arena, open stage, etc., the most common are the traditional proscenium and the open stage or "thrust-type." See Fig. 11-6.

The proscenium-type of theatre is composed, typically, of a seating area and a stage area. It may serve not only as a theatre, but also as an assembly and lecture hall, study room, concert or audio-visual aids (including television) area, and it may also house numerous other activities. Considerable attention is being given to the development of speech and theatre arts programs not only in schools but also as a community activity among adult groups. The many uses of the theatre require well planned and properly equipped lighting facilities.

Seating Area. The seating area should be provided with well diffused comfortable illumination. Because of the many purposes the seating area often serves, different illuminance levels are necessary. Basic illumination, well diffused and producing an illuminance of the order of 100 lux [10 footcandles] should always be provided in the seating area. It should be under dimmer control, preferably from several stations (stage lighting control board, projection booth, etc.). However, there should be transfer capabilities so that the lighting will not be accidently turned on during performances. Lighting equipment for the basic illumination may include general downlighting units, coves, sidewall urns, curtain and mural lights, etc. Supplementary illumination is required where the seating area is also used for visual tasks to provide an additional 200 lux [20 footcandles] over the basic level. This supplementary illumination should be directed evenly over the seats and should be controlled separately from the basic illumination. The supplementary illumination should preferably be achieved by a downlighting system, properly shielded to prevent direct rays from becoming distracting. Selected circuits of the supplementary lighting system may also be used as work lights for cleaning, rehearsals, etc. Separate control of certain basic or supplementary circuits for emergency use should be provided at the rear of the seating area and other locations as may be required. This control must be independent of dimmer or switch settings. In accordance with local and national codes, a system of emergency lighting which would include panic lights, exit lights, shielded aisle and step lights, and other required lighting must be provided.

Stage Area.[16] Proper lighting for dramatic presentations extends beyond visibility to the achievement of artistic composition, production of mood effects, and the revelation of forms as three-dimensional. These functions of stage lighting result from the manipulation of various qualities, quantities, colors and directions of light, and these vary from one performance to the next and even continually throughout a single performance. The layout is affected by the amount and kind of use planned for the theatre.

Stage areas to be lighted include stage apron or forestage, acting areas upstage from the proscenium arch, including areas above and below the stage floor, extension stage areas, auxiliary acting areas in the auditorium, foreground scenery and properties, and background scenery and properties.

Basic Lighting Functions. An appreciation of the dramatic potential of lighting begins with an understanding of its basic functions, *i.e.*: to provide the required degree of visibility; an indication of time and place called motivation; a visual emphasis of selected elements called composition; and an over-all atmosphere called mood.

Visibility. This is the most basic function of lighting in the theatre. For the audience to hear and understand in the theatre, they must be able to see.

Motivation. Motivation or naturalism is the term given to the expression of time and place.

Composition. Composition is revealed artistically through the proper use of light and shadow. Warm light and cool light give plasticity and composition to the visual effect. The concept of the production as indicated by the playwright and chosen by the director or producer determines the approach of the designer using light with its natural characteristics.

Mood. Mood or atmosphere, as created by the total visual effect, brings the stage into focus with the meaning of the play. The final visual effect is provided by equipment which has been chosen by the lighting designer because it supplies the desired output of light in terms of intensity, form, color, and movement.

Intensity. Intensity control is extended by the use of various types of luminaires, lamps, mounting positions, color media and, of course, by altering the illuminance through the use of dimmers. Precise consistent dimmer control is essential for establishing various levels of intensity.

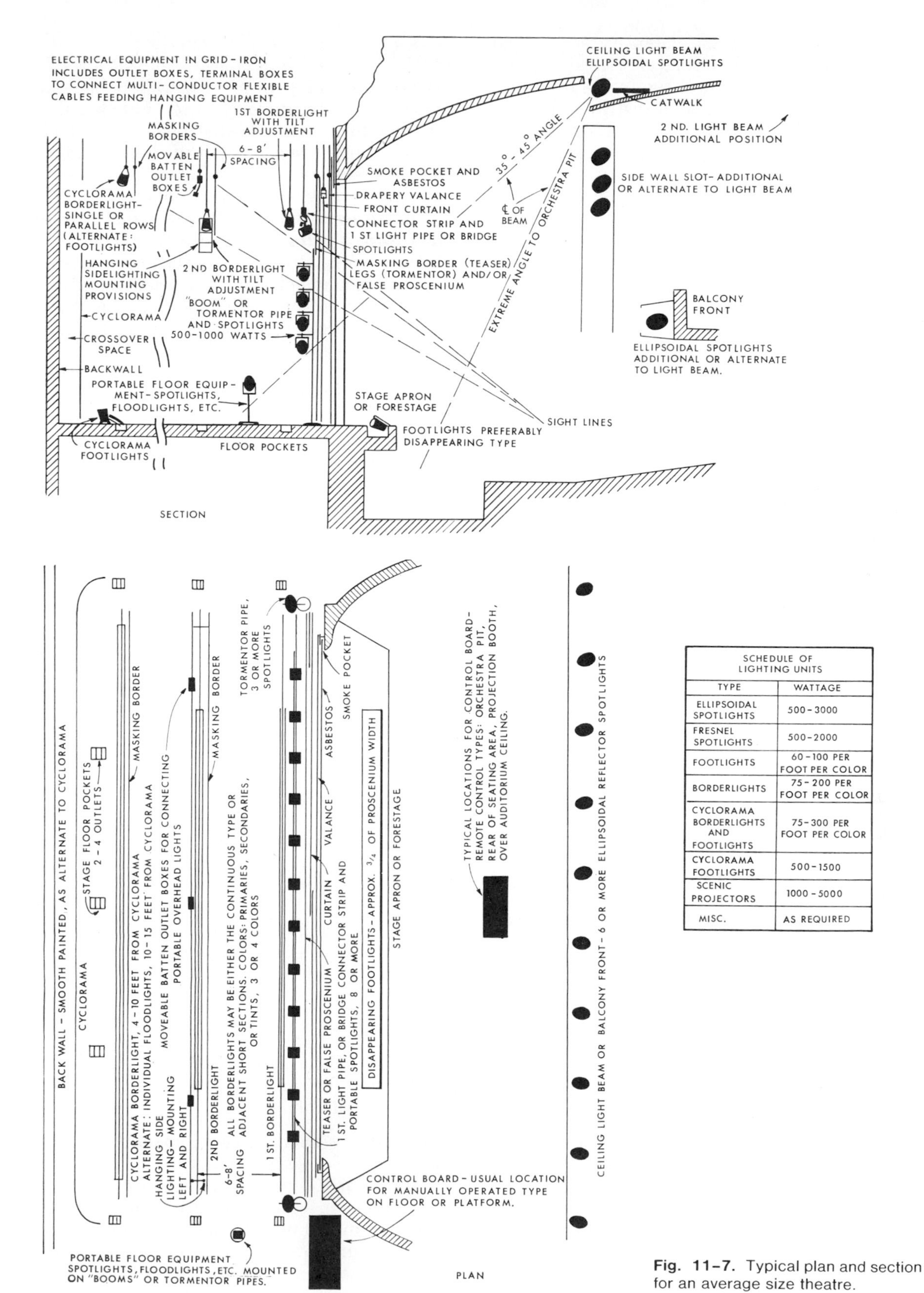

SCHEDULE OF LIGHTING UNITS

TYPE	WATTAGE
ELLIPSOIDAL SPOTLIGHTS	500-3000
FRESNEL SPOTLIGHTS	500-2000
FOOTLIGHTS	60-100 PER FOOT PER COLOR
BORDERLIGHTS	75-200 PER FOOT PER COLOR
CYCLORAMA BORDERLIGHTS AND FOOTLIGHTS	75-300 PER FOOT PER COLOR
CYCLORAMA FOOTLIGHTS	500-1500
SCENIC PROJECTORS	1000-5000
MISC.	AS REQUIRED

Fig. 11-7. Typical plan and section for an average size theatre.

Form. Form, meaning the density, spread and direction of the light rays, calls for a wide variety of types of luminaires and mounting positions. The particular quality of light distribution given by each type of luminaire determines its use.

Color. Color in lighting design is used to accent, enhance, distort and motivate the scene. Color is controlled by means of dimmers plus separate filters in front of each source. A tonal quality can be obtained by the additive mixture of two or more sources.

Movement. Movement consists of a change in one or all of the qualities of light. Aside from a manually-operated follow spot or special remotely-controlled unit, this is accomplished by means of dimming individual units rather than by directional movement.

Lighting Locations.[17] Basic locations for lighting equipment (see Fig. 11-7) may be divided into two groups as follows: (a) locations in front of the proscenium opening (which includes the auditorium ceiling, side walls of auditorium or proscenium, or both, balcony front, follow spot booth, and edge of the stage apron) and (b) locations behind the proscenium opening (such as tormentor pipes for side lights, overhead cyclorama pipes for top lights, cyclorama pit for base lights, and locations employed as needed, such as overhead light pipes, free space at sides and rear of the stage main acting area for floor mounted or hanging equipment, and below stage areas). Though the positions might be fixed, virtually every luminaire is portable and gets shifted around for each production. The focus, direction, intensity and color are generally different for each production.

Locations of Luminaires in Front of the Proscenium Opening.

Luminaires in the Auditorium Ceiling. Stage lighting luminaires in the auditorium ceiling are generally used for the basic purpose of lighting down-stage and apron acting areas. Also, they are sometimes used for the illumination of the proscenium curtains when the angles of throw permit reasonable coverage of the curtains. Each luminaire should produce a clearly defined light beam which can provide an average illuminance of 350 to 500 lux [35 to 50 footcandles] of white light on a vertical plane, with adjustable means for a controlled cut-off, so that the beam can be varied in shape to cover a desired area with little or no spill onto adjacent areas. The ellipsoidal reflector spotlight incorporating framing shutters is an example of such a luminaire, and is the type usually used. See Fig. 11-1. The spotlights are best located behind slots in the ceiling and wall surfaces and are ideally mounted in a continuous slot stretching across the ceiling from side wall to side wall. Whether in one or more rows, spotlights should be angled to project so that beam centers come within 30 to 60 degrees with the horizontal when seen in side elevations, and up to 45 degrees with the stage axis when viewed in plan.

Luminaires in Auditorium and Proscenium Side Walls. Luminaires located on or in the side walls, although not absolutely required, are recommended. They are used mainly as a supplement to the ceiling spotlights, and are of a similar type. Preferably, these luminaires should be recessed in wall slots. They provide lower angles of throw than the ceiling units and offer an excellent opportunity for side lighting with a wide range of different angles of throw.

Luminaires on Balcony Front. Angle lighting from spotlights in the side walls and/or ceiling will provide effective front lighting for most purposes, but there are occasions when the balcony position affords desirable low angles of throw or a soft wash of directional front lighting. Attention must be paid to where the shadows fall. There is a danger that shadows from low angle front spots may fall on the scenery and move as the actor moves. This will cause an unacceptable distraction. Proper access must be provided to these spaces so that lights can readily be put in place, focused, lamped and relamped.

Follow Spot Booth. Follow spots are used to highlight selected performers. A follow spot should be capable of providing a level of at least 2000 lux [200 footcandles] in 2.5-meter (8-foot) diameter area, and should embody means to enable size and shape of the light beam to be varied so that it may be reduced to cover the head of a person only, or be widened to flood a considerable portion of the stage. In addition to the iris diphragm, spread lens and horizontal paired shutters—accessories usually required for this purpose—follow spot equipment often includes either a color wheel or a multi-slide color boomerang that can be operated either from the side or rear of the spotlight, in order to change the color of the light. Light sources for follow spots may be incandescent, arc or compact source metal halide.

Locations Behind the Proscenium Opening.

Overhead Locations. The greatest number of luminaires in any one location upstage of the proscenium will be mounted on the first pipe or bridge immediately upstage of the proscenium. The luminaires for the first pipe or bridge may include spotlights, borderlights and scenic projectors. The majority of the spotlight units are required to produce a soft-edged beam that is widely variable in focus, as for example, is obtained from a Fresnel lens type spotlight. A number of ellipsoidal reflector spotlights are usually mounted in the same row. There should be provisions for mounting additional rows of lights on pipes parallel with the proscenium opening every 2 to 2.5 meters (6 to 8 feet) of stage depth.

Connector Strips for Stage Light Pipes. The basic purpose of a connector strip is to provide a simple and quick method of electrically connecting a number of lighting units, wherever mounted on an associated pipe, by means of a series of outlets. The length of a connector strip mounted on a pipe or bridge should approximate the width of the stage proscenium opening. Outlets should not be spaced closer than 300 millimeters (12 inches) apart and each outlet or group of outlets should be on an individual circuit with a separate neutral. So that any available pipe may be used for luminaires, a more flexible alternative to connector strips fixed to specified light pipes is a number of multiconductor cables, permanently attached to the gridiron, with a box of minimum proportions housing an appropriate number of outlets at the other end. Each cable is long enough so its outlet box can be lowered and attached to either end of a stage pipe.

Borderlights for Stage Light Pipes. A borderlight provides general downlighting across an area of the stage and provides overhead illumination of hanging curtains and scenery. It contributes tonal quality to the over-all lighting effect. Separated control of different sections of a borderlight enables parts of the stage width to be variously accented in brightness and color. Borderlights (other than for cyclorama illumination) should be wired on three or four separate color circuits and should be capable of providing illumination of the whole width of a curtain or flat scenic drop mounted 1.2 meters (4 feet) or more upstage of the borderlight equipment. The illuminance provided by a borderlight in the center of the vertical surface should be not less than 250 lux [25 footcandles] of white light when measured at a point 1.8 meters (6 feet) from the stage floor.

Cyclorama Top Lighting. Cyclorama borderlights must be long enough to illuminate the whole of the visible width of the background, independent of illumination from cyclorama bottom lighting. Cyclorama lighting requires at least double the illuminance provided by other borderlights, and when the cyclorama is an important feature and is to be illuminated with 3- or 4-color lighting in conjunction with deep color filters such as red, green and blue, then the wattage of the associated borderlight equipment may be from 2 to 4 times that of a regular borderlight—sometimes even more. The required lighting level may necessitate two parallel rows of borderlights, using, for example, 250-watt PAR-38 lamps on 150-millimeter (6-inch) centers or 500-watt PAR-56 lamps on 200-millimeter (8-inch) centers in each strip.

Backlighting from an Upstage Pipe. It is desirable to provide for a row of high intensity narrow beam luminaires, as for example, parabolic reflector floodlights or ellipsoidal reflector spotlights, suspended on an upstage pipe, but directed downstage to provide backlighting of artists in the main acting area. For example, there may be one 500- to 750-watt luminaire for every 1.2 to 1.8 meters (4 to 6 feet) of effective stage width.

Mounting for Stage Side Lights. Although other mounting methods may be used where conditions permit, there are two general methods of providing side stage lighting, either on suspended three or four rung ladders or from vertical, floor mounted boomerangs or tormentor pipes.

Special Theatrical Effects. Fluorescent paints, fabrics or other materials responding to long-wave ultraviolet are often used for special theatrical effects. Sources for exciting the fluorescing materials include mercury lamps with filters for absorbing visible radiation, fluorescent "black light" lamps (which also require an auxiliary filter) and integral-filtered fluorescent "black light" lamps. Carbon arc follow spots are sometimes filtered for black light effects. Fluorescent lamps are the most efficient "black light" sources, but the energy cannot be readily focused for long throws or to cover restricted areas. Strobe lights and lasers are used in today's theatre. Great care must be exercised where using lasers.

For greatest dramatic effect, visible light in the scene should be held to a minimum.

Scenic Projection. Increased understanding of the techniques of projection by theatre personnel has refined the basic optical design for lens projection in the live theatre. Presently, the principal refinements involve: increased wattages (up to 10 kW) and the use of compact source metal halide lamps to provide additional scene illumination and increased area coverage; improved slide-making and handling techniques; programmed remote slide changing; extension of wide-angle projection to screen widths up to 1½ times the projection distance; standardization of units to permit easy interchangeability; and the availability of relatively inexpensive and easily mastered remote-control 35mm projectors.

The general availability of the 35mm or 2- x 2-inch slide projector, as well as the increased availability of larger units, has led to much experiment. Multi-sources, wide ranges in the message, wide ranges in composition and build-up of the total picture, multi-screens and mosaic designs, fractured pictures, time and dimming variants, programmed variation, and adjustment of the controllable properties of auxiliary light sources—all combine to make projection an exciting light tool in all theatrical presentations.[18] Picture projectors are sometimes used in theatrical performances.

Touring Concerts, Rock Shows, Etc.

Touring concerts, rock shows, etc. bring their own luminaires, cable, dimmer systems and mounting equipment. They set up towers, trusses, booms, etc., a day before the show; then run through in preparation for the show. The equipment is very specialized and highly sophisticated, and there are companies that specialize in providing for these kinds of shows.

Meeting, Convention and Industrial Show Facilities[19, 20]

Lighting for meeting and conventions requires comfortable illumination and accent lighting for non-theatrical participants. Where free discussion takes place between speakers and audience, the lighting should be free of glare so that prompt recognition of each speaker occurs. Lighting for industrial shows, new product presentations, etc. requires the dramatic lighting of the theatre.

Stage locations may vary considerably from meeting to meeting. A show that uses rear projection may move the stage area 4.5 to 6 meters (15 to 20 feet) forward. Another meeting may require a simple platform with maximum space for an audience seated in school room or conference style. Many meetings use a center area, or theatre-in-the-round arrangement. Other producers find a projected stage along a wall more satisfactory for their presentation.

Many meetings are conducted in multi-purpose spaces that are used for food service, fashion shows, motion pictures, social events and meetings. The ease and speed in which these areas can be changed from one arrangement to another is an important economic operating factor.

Lighting must be coordinated with many other elements. These include wall surface brightness, projection screen location, communications, sound systems, etc. Projection from audio-visual equipment and followspots requires unobstructed views of screens, stages and acting areas. Chandeliers must not be placed in locations that will interfere. Dimmers should be provided.

Theatre-Restaurants, Lounges and Discos

Stage lighting design criteria for theatres and auditoriums is generally applicable for theatre-restaurants, night clubs and lounges. However, theatrical lighting in small areas such as lounges will utilize more compact luminaires. For low ceilings, a basic luminaire is an "inky" with a 76-millimeter (3-inch) Fresnel lens, or adapter accessory with individually adjustable framing shutters, and a 100- to 150-watt lamp. In larger spaces with longer throws, a Fresnel-lens spotlight or a floodlight for 250- to 400-watt lamps and a beamshaper accessory may be used. An alternative would be a small ellipsoidal-reflector framing spot of 650 watts or less, available with wide, medium and narrow beam lens system.

Discos use all of the stage lighting equipment and techniques. Additionally, a tremendous amount of movement in lights is introduced. Mirror shower balls, spinners, rotators, police emergency lights, etc., are used. Control systems include presets, chasers, programmers, etc.

Luminaire Locations. Lights can normally be used at sharper angles, closer to the 30-degree rather than the 45-degree limits. Downlights are used to produce "pools of lights" on dancers and set pieces. Uplights recessed in the stage floor are also used. Side-front, side, and side-back lighting is essential for three-dimensional effects particularly in dance and production numbers. Floor-mounted linear strips are used for horizon effects on cycloramas. These may be of the disappearing type or recessed with expanded metal covers to permit performers to walk over them.

Followspots. Locations for several followspots should be provided for front-side as well as front spotlighting. One or more followspots should be able to cover audience areas. Some performers enter from the audience, and runways are frequently used to bring the chorus closer to the viewers. Side-stages on each side of the main stage are frequently used for bands and stage action and provision should be made for adequate lighting of these areas.

Transparencies. Scrims are frequently used to *hide* the band when playing for a show. The band and performers are frequently *revealed* by bringing up lighting behind the scrim and keeping light off the front of the scrim. These changes may occur on the side or principal center stage.

Special Effects. Mounting devices and no-dim control circuits and receptacles are required for "black lights," projectors, electronic flash, motor-driven color wheels, and dissolves, fog and smoke machines, mirrorballs, etc. Color-organs are used to pulsate lights with music. Plastic covered floors for dancing and entertainment should provide selectable color and pattern effects. In minimum spaces, fluorescent lamps and dimming ballasts are used. If ventilation can be provided, incandescent lamps can be used for special effects.

Controls. Single lights are frequently used. Receptacles should be on individual cross-connect circuits. Permanent grouping should be avoided or minimized if necessary. For more flexibility a greater proportion of dimmers to receptacle circuits are used than in a conventional theatre. Control cross-connects may be used to supplement or in the place of load patching. No-dim controls should be integrated with the dimmer controls, *i.e.*, voltage sensitive relays controlled by the same type potentiometers as used for the dimmer controls.

Motion Picture Theatres

Auditorium-Type Theatres. The objectives of auditorium lighting in the motion picture theatre may be outlined as follows: (a) to create a pleasing distinctive environment; (b) to retain brightness and color contrasts inherent in the picture; (c) to create adequate visibility for safe and convenient circulation at all times; and (d) to provide comfortable viewing conditions.

For general lighting during intermission, 50 lux [5 footcandles] is considered minimum. During the picture, illumination is necessary for safe and convenient circulation of patrons. Illuminance values of 1 to 2 lux [0.1 to 0.2 footcandles] represent good practice. Screen luminance with picture running is in the range of 15 to 20 candelas per square meter [5 to 6 footlamberts] to less than 3 [1]. The need to eliminate stray light on the screen dictates controlled lighting for at least the front section of the auditorium. Downlighting is one of the most effective methods for this purpose. In general, diffusing elements, such as coves, allow too much light to fall on the screen if they provide adequate illumination in the seating area. Diffusing wall brackets, semi-direct luminaires, and luminous elements are generally too bright to be used for supplying illumination during the picture presentation.

Contrast between the screen and its black border is very high (sometimes more than a thousand to one), creating uncomfortable viewing conditions. Areas around the screen can be raised in luminance if they are not treated with distracting decorations. Light for this purpose may be reflected from the screen under special conditions, supplied by supplementary projectors, or by elements behind the screen.

Curtains may be lighted in color with a projector border during intermissions. Adequate spotlighting on the stage is desirable for announcements and special occasions.

Aisle lights should be low in luminance and spaced close enough to give fairly uniform illuminance in the aisle.

Houselights should be dimmer controlled, rather than controlled merely by on/off switches.

Drive-In Theatres. Lighting for safety is a major concern of the drive-in theatre manager because of the interference of pedestrian traffic going from cars to the refreshment stand, and cars entering and leaving without benefit of headlighting. Screen luminance is much lower than in conventional theatres, which imposes greater limitations on tolerable extraneous luminances. Screen orientation with respect to moonlight, twilight and sky haze is a major factor. Unique opportunities for spectacular and decorative effects include lighting flower beds, plants, pools and fountains, which are not within view of the audience watching the screen.

Entrance signs must be legible at long distances to allow enough time for patrons traveling at high speed to slow down and maneuver to enter. Conventional sign lighting methods for regular theatres are used. The entrance drive can be lighted by street lighting equipment, luminous pylons or low-level units. Particular care is needed to control any luminance so that it will not be distracting from the screen area. Light reflected from trees, surrounding structures, or even haze or mist can be disturbing. Floodlighting, luminous elements, lighted murals and patterns outlined in tubing are popular means of drawing attention to the screen tower.

At least 300 lux [30 footcandles] is recommended in the money exchange area at the box office (generally some distance from the entrance). Spotlighting for cashiers in costume and decorative lighting at this location offer opportunities for showmanship. Lighted ramp markers, lighted ushers' wands, and illuminated fences facilitate movement after car headlights are turned out. Speaker stands are available equipped with lights for ground area and for signaling the refreshment vendor.

Intermission lighting should be at least 5 lux [0.5 footcandles], and can be satisfactorily provided from units mounted on poles. Experience has shown that floodlights for this purpose located at top of the screen tower are glaring and lead to complaints by patrons. Ten to twenty lux [one to two footcandles] is recommended for exit lanes for safety of fast-moving, highly congested traffic. Floodlights along the fence aimed with the flow of traffic plus floodlights on a pole at the rear of the area is one solution. At least 20 lux [2 footcandles] is recommended for convergence of theatre traffic with highway traffic.

Refreshment stand interior light must be shielded from the outside. High luminance luminaires are objectionable because patrons are dark-adapted when they come in. Lighting the counter area is usually sufficient. In stands with a viewing room, vertical glass windows require highly-shielded, low-luminance luminaires to prevent reflections. Windows splayed out at the front reflect patrons instead of the ceiling pattern, but the former method can be better if properly handled.

LIGHTING FOR TELEVISION

Television broadcasting requires extensive preplanning of the lighting arrangements, switching, dimming, etc., due to the necessity of continuous dramatic action. An exceptionally high degree of mechanical and electrical flexibility is mandatory. Multiple scenes are arranged for continuous camera switching. Consequently, the lighting for each scene is pre-set without interference with adjoining sets. The quantity and quality of the lighting needed for television production depends upon the type of camera pickup tube used (Image Orthicon, Vidicon, Plumbicon) and the reflectance of the subject. The choice of lenses (size and focal length) is governed by the same optical principles commonly applied to all photographic devices.

Studio Lighting for Monochrome Television

There are two aspects of lighting for television production. The first has to do with controlling the light in terms of quantity, color and distribution to produce a technically good picture. The second has to do with the design of lighting to produce a dramatic and artistic visual effect for broadcasting.

Camera Pickup Tubes. The 5820 image orthicon, the vidicon (SB_2S_3), and the lead oxide target vidicon (PbO) of the Plumbicon* type are the pickup tubes most commonly used in studio telecasting. Not as commonly used, because they are just emerging, and usually employed in color cameras are several other tubes with transfer characteristics identical to that of the lead oxide vidicon, with somewhat greater response in the longer wavelength region (*e.g.*, Saticon**). In general, the lighting techniques recommended for the lead oxide vidicon would apply to these pickup tubes.

Their spectral sensitivity characteristics are shown in Fig. 11-8 and the normal lighting level required is 900 lux [90 footcandles]. The tubes do not lend themselves to a contrast range exceeding 40 to 1. Lens stops are adjusted for proper exposure, usually between *f*4 and *f*11, depending on the lighting used. For example, with an illuminance of 1000 to 1250 lux [100 to 125 footcandles] of incident light a lens stop of *f*8 normally results in a broadcast quality picture

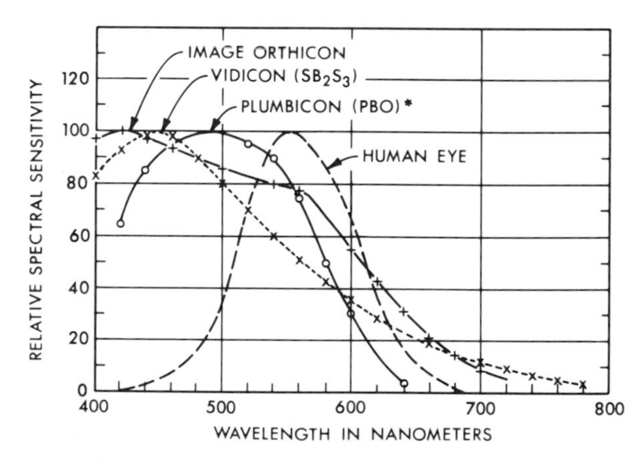

Fig. 11-8. An equal energy relative spectral response curve of an image orthicon, an SB_2S_3 vidicon, a Plumbicon,* and the human eye.

* Registered trademark N. V. Philips of the Netherlands.
** Registered trademark of NHK (Japan Broadcasting System).

(provided the average reflectance of the subject matter equals that of flesh tones, *i.e.*, a 40 to 50 per cent reflectance). White and some dark areas can be used in each picture to aid camera level adjustment. These should be of limited area and have enough reflectance to guarantee the contrast range mentioned above.

Using the Vidicon Camera. Before giving the recommended illuminance at which to operate the vidicon camera, it is necessary to know some of the interlocking facets of high- and low-level lighting. Lower levels (100 to 400 lux [10 to 40 footcandles]) are usually found in existing room lighting since most modern offices and classrooms have a lighting level of over 400 lux [40 footcandles]. The higher levels (1000 lux [100 footcandles] and up), when not available under the existing room lighting, may almost always be obtained by using standard television spotlights and floodlights.

To determine whether to use ordinary room lighting (which may be a combintion of incandescent and fluorescent permanently mounted luminaires) or to use spotlights and floodlights, the telecaster should consider the quality of the resulting picture and the standard set as acceptable for transmission. For instance, if there is no motion in the scene either on the part of objects in the scene or on the part of the camera itself (panning, tilting) and if it does not matter whether the picture is somewhat grainy, then the lower level lighting is adequate. If, however, the scene contains moving objects, or the camera makes a fast pan, and the picture must approach broadcast quality, then a higher level should be used.

The vidicon tube, however, has special characteristics that under certain conditions will deliver an "acceptable" picture with ordinary room lighting as the only illumination. In this case, the picture is far too noisy and otherwise unacceptable for minimum broadcast transmission. If, however, the system is only used to observe and count people entering and leaving a store, for instance, detail and picture quality may not be necessary. For this purpose an acceptable picture may be accomplished with as low a level as 200 lux [20 footcandles].*

The vidicon phenomenon called "image retention" is also given other names, some of which are sticking, trailing and lag. It occurs when the actor moves quickly or the camera pans or tilts rapidly to a new position. It is especially noticeable in low-level lighting and becomes less of a problem as the lighting level is increased.

Studio Lighting for Color Television

Response characteristics of color cameras are critical and experience has shown that the spectral power distribution of the light sources illuminating a scene should be essentially uniform. Where filament lamps are used, current practice indicates that they should remain within a 300 K, normally ± 150 K, range for accurate color rendering. While a camera chain and system can be adjusted to operate with almost any continuous spectrum light source by adjusting the gains of the red, blue and green channels, a wide variation in the spectral characteristics of the light sources within the picture area will cause unpredictable changes in color reproduction.

Incandescent filament lamps have been favored for color studio lighting due to their continuous spectrum, their relatively low cost and long life, and the fact that they are available in a wide range of sizes with approximately the same kelvin temperature. Incandescent filament light source color temperatures of 2900 K to 3350 K are currently used. These lamps are favored because their housing equipment is generally less bulky and of lighter weight, and the smaller filament light sources allow for better optical control of the light beam. Furthermore, they readily permit use of color filters at the light source.

A camera initially balanced for incandescent filament lighting may be used to take scenes outdoors in daylight with the color temperature variations corrected through the use of suitable filters on the camera lens. Variations in the amount of light can be controlled by the camera iris.

Mixed light sources, such as incandescent, high intensity gas discharge and fluorescent, ordinarily should not be used to light a scene simultaneously because the camera will not transmit a good color picture where the spectral power distribution of the light sources vary greatly. However, new developments in fluorescent and high intensity gas discharge light sources are now providing greater flexibility for mixing sources. This can be attributed to discharge lamps being manufactured with much better, and in some cases, continuous spectral power distributions. Some sports arenas have mixed 75 per cent high intensity gas discharge and 25 per cent incandescent sources successfully.

* For further information on light levels and camera stops on 106 special and industrial applications, see Reference 21.

For televising color pictures, lighting levels of 1900 to 3770 lux (175 to 350 footcandles) incident are found satisfactory. Some newer cameras will produce studio equality pictures at a light level as low as 750 lux [75 footcandles]. Many studios, especially those producing color spectaculars, have found it desirable to provide up to 5000 lux [500 footcandles]. This is done to compensate for older model color cameras or to provide greater depth of field.

In those cases employing very low light level, the problems of focus may be complicated by the very narrow depth of field afforded by wide open lenses.

	Illuminance Readings in Lux (Footcandles) at Test Positions					
	1	2	3	4	5	6
A	—	1000 (100)	1000 (100)	1000 (100)	—	—
B	—	—	—	—	1000 (100)	—
C	1250 (125)	1250 (125)	1250 (125)	1250 (125)	—	1000 (100)
D	1250 (125)	1500 (150)	1500 (150)	1250 (125)	—	1000 (100)
E	—	1500 (150)	1500 (150)	1250 (125)	—	1000 (100)
F	1250 (125)	—	1250 (125)	1250 (125)	—	—

Distance	Meters	Feet
V	9.1	30
W	27.4	90
X	30.5	100
Y	45.7	150
Z	91.4	300

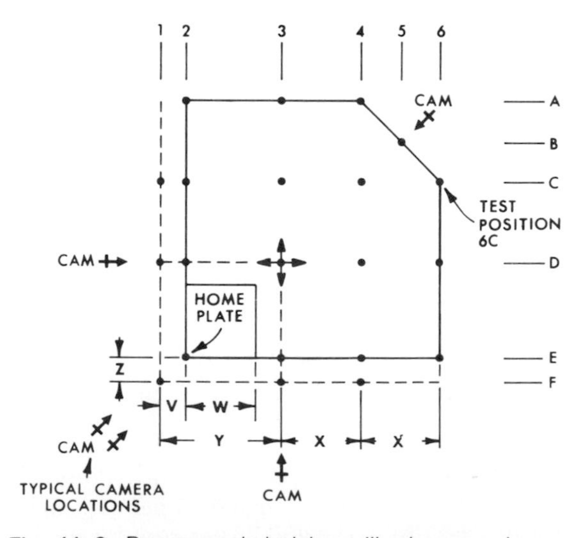

Fig. 11-9. Recommended minimum illuminance and survey test positions for color television pickup of baseball.

Videotape and Kinescope Recordings

The same requirements as given in all the above apply to video and kinescope recording. In addition, the contrast range should average 20 to 1, but under controlled conditions the contrast range can be extended to 30 to 1.

Television Film Production

The significant element in lighting for television film production is the control of subject contrast range. Most television film reproduction systems require a picture luminance range not exceeding 25 to 1. This contrasts with the current theatre projection range of about 100 to 1. It is important, therefore, to limit the luminance range of filmed subjects. A fair amount of such control can be obtained by the introduction of "fill" light to raise the luminance of significant shadow areas. This does not mean that flat lighting is desirable but rather that the lighting ratios used for modeling should be lower than is normally used when making films for theatrical projection.

In shooting color film for television, the same general rules apply. Additionally, it is necessary to bear in mind that color contrast is not enough to insure an acceptable signal. Brightness contrasts essentially constitute the signal seen on a monochrome receiver tuned to a color broadcast. Consequently, it may be necessary to light in such a way that areas having hue differences only are placed at different lighting levels. A poor monochrome picture is particularly conspicuous when scenery is painted in several pastel shades, all having similar reflectances.

Projected Backgrounds

Rear projection is the projection of a scene onto a translucent screen from behind. It is used to simulate background scenery, which may take the form of stationary objects as produced by a slide; moving effects such as clouds, water, etc., as produced by an effects machine; or continuous motion simulating moving trains, motion from an automobile, etc., as produced by a motion picture film.

For realism, projected highlight levels should be between ½ to 1 times live highlight levels. It is desirable to have a projected highlight luminance of 250 candelas per square meter [70 footlamberts] measured facing the screen from the camera side of the screen when the lighting level in the acting area is 1000 lux [100 footcandles].

Chroma Key

The production technique known as chroma key is a special effect that enables any background material to be matted into a scene. In the studio a color camera views the subject against a backdrop of a primary color that has sufficient saturation to produce a full output level in the corresponding channel of the camera. This signal output is used to key a special effects generator so that all information except the wanted subject is matted out of the original studio scene. Information from any other source such as a film chain or video tape recorder can then be inserted in the matted portions of the signal.

The primary color used in the backdrop is chosen on the basis that it is not present in the color of the wanted subject. When human subjects are used, blue is usually the optimum background color because of its absence in flesh tones.

Additional precaution must be taken to avoid the use of the background color in costumes or stage props.

The lighting level on the backdrop must be high enough to produce a full output signal from the camera without excessive noise. However, the same lighting must not be so high that light reflected from the background will appear on the subject and thereby create spurious keying signals.

The color camera used for the chroma key technique must have sufficient video bandwidth in the color channels so that key pulses with good transient edges can be developed.

Field Pickups of Sports Events

The lighting recommendations for sports (see Section 13) are based upon player and spectator visibility needs. When telecast, especially in color, or when filming is involved, additional factors must be considered. These include (1) providing a directional balance to the vertical illumination, (2) providing the quantity of illumination required by zoom lens systems, (3) providing uniformity of illuminance within a 2 to 1 maximum to minimum ratio and (4) providing good color quality.

Layouts for a baseball and football field indicating the normal camera locations are shown in Figs. 11-9 and 11-10. Since both vertical and horizontal illumination is important, a typical test location and vertical illuminance test direction is illustrated. The vertical illuminances should be recorded in their directions as illustrated; however, only two of the four directions need to meet the illuminance requirement. Measurements of horizontal and vertical illuminances in four directions should be made at a height of 900 millimeters (36 inches) above the playing surface.

The latest television systems and motion picture film are sensitive enough to obtain quality pictures at levels of 800 to 1600 lux (75 to 150 footcandles). This corresponds to the normal illuminances required in large sports facilities such as college or major league football stadiums and major league ballparks. The main variations in the normal lighting requirements that are needed for television and film is the need to consider vertical illumination and the reduced uniformity limit.

It is also important to provide illumination in the spectator area adjacent to the playing field for crown shots and for wide angle shots of the playing field. This is best accomplished from behind the spectators to limit glare for both the spectators and cameras' positions in the spectator seating area. Glare is also an important consideration for players, and it can be minimized if the light comes from the sides of the normal direction of play. This is easily identified in a football stadium, but in a baseball park the light direction is taken as being at right angles to a line from home plate to centerfield.

Luminaire or Pole Location. Pole locations as shown for baseball in Section 13 are applicable when designing a lighting installation for color television. However, for football, pole locations will be required in accordance with Fig. 11-11 to provide the illumination for the camera located at the end of the field. Extreme caution should be exercised in the selection and aiming of floodlights so as not to create objectionable glare for players, spectators or side-line cameras.

Measurement of Illuminance

Meters generally used in television production measure incident light and consequently do not take into account the reflectances of the scene components. Since measurements by such meters are not indicative of the scene luminance variations, the operator must keep this constantly in mind when planning the lighting for different sets. However, an incident light meter

	Illuminance Readings in Lux (Footcandles) at Test Positions								
	1	2	3	4	5	6	7	8	9
A	1000 (100)	1000 (100)	1000 (100)	1000 (100)	1000 (100)	1000 (100)	1000 (100)	1000 (100)	1000 (100)
B	1100 (110)	1250 (125)	1350 (135)	1350 (135)	1350 (135)	1350 (135)	1350 (135)	1250 (125)	1100 (110)
C	1250 (125)	(1500 (150)	1500 (150)	1500 (150)	1500 (150)	1500 (150)	1500 (150)	1500 (150)	1250 (125)
D	1100 (110)	1250 (125)	1350 (135)	1350 (135)	1350 (135)	1350 (135)	1350 (135)	1250 (125)	1100 (110)
E	1000 (100)	1000 (100)	1000 (100)	1000 (100)	1000 (100)	1000 (100)	1000 (100)	1000 (100)	1000 (100)

Distance	Meters	Feet
W	6.1	20
X	15.2	50
Y	48.8	160
Z	91.4	300

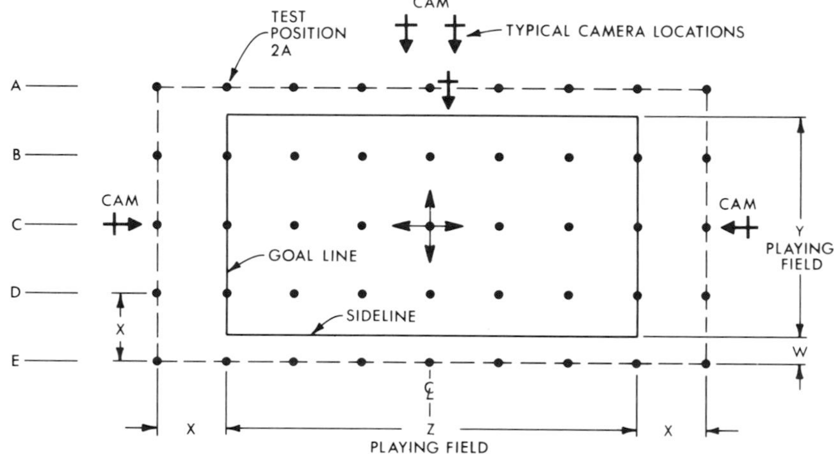

Fig. 11-10. Recommended minimum illuminance levels and survey test position for color television pickup of football.

in the hands of an experienced person can be a very useful tool. The quantity of light is measured with a photocell meter holding the photosensitive element in the scene at approximately camera lens height and at the angle of and facing the lens. Extreme light contrasts can be used upon such measurements. The practical scene contrast range for operational use is 30 to 1.* Color contrast should not be confused with brightness, since two different colors can have the same luminance. Unless careful consideration is given to tonal rendering of colors, a scene that appears beautiful to the eye can appear on a receiver as a flat monotone picture.

Types of Illumination

Base Light or Fill Light. Base light or fill light is usually supplied by floodlights using incandescent lamps, including etched tungsten-halogen which supply a broad source of soft illumination. It is desirable to aim base lights at a 12- to 15-degree angle below horizontal.

Key or Modeling Light. Key or modeling light is usually supplied by Fresnel lens spotlights ranging in lamp size from 500 to 10,000 watts. Fresnels are equipped to hold supplementing masking devides, such as barn doors, snoots and color frames. The barn door fits in front of the lens and is used to limit the bottom, top or sides of the light beam. These units can either be hung or used on floor stands and are generally aimed at a 20- to 40-degree angle below horizontal.

Back Light. Back light is used for separation. Back lights are hung behind a subject and are aimed at approximately a 45-degree angle to light the back of the head and shoulders, and to separate the subject from the background. Back light can be from one-half to equal that of the

* See Videotape and Kinescope Recordings

front light, depending upon the reflectance of the hair and the costume.

Set Light. Set light is used to decorate or help give dimension to scenery. The amount of light necessary is totally dependent on the reflectance of the scenery. It must be kept in mind that flesh (a person's face) reflects 40 to 45 per cent of the light falling upon it. Therefore, the major part of the background must be kept below the reflected light level of the face. A gray scale value of 75 per cent of skin tone is an adequate average.

A graphic representation of the above types of lighting is shown in Fig. 11–12.

There are many other luminaires that can be used to help dramatize a show. A few are: sun spots, ellipsoidal follow spots, pattern projectors and strip lights.

Balancing for Correct Contrast. It is important to have the proper balance among the four types of light discussed above. If, for instance, the set is painted in a color or gray scale value that reflects more light than the flesh of the actor, skin tones may appear darker than desired in the picture. This means that the set light should be reduced. A quick way to do this is to use spun glass diffuser material on the set-lighting units. This material is available in 1- by 4-meter (3- by 12-foot) rolls and is cut to fit the unit. One 0.38-millimeter (0.015-inch) thickness cuts about 20 per cent of the light. Additional thicknesses can be used until the correct contrast is obtained. The problem generated when using spun glass is that the character of the light has been altered from somewhat firm image forming to soft, diffuse, flood, and less directional ambient light. If the front, or fill light, is brighter than the key light the use of spun glass will help too in solving that problem.

Another medium that is sometimes used to balance the lighting is one or more layers of ordinary house window screening and black window screening material has the virtue of not changing the optical characteristics of the light.

If the installation includes dimmers and a cross-connecting system, the various lighting units can be grouped and then dimmed until the desired contrast is obtained. There is no apparent effect in black and white television. In the case of color television, a limitation imposed is that the color temperature decreases 10 K per volt for 120-volt lamps. As stated before differences of 300 K contained in one scene will be perceptible. Further, a very low color temperature contains little blue energy and may introduce unwanted noise in the picture.

Lighting Equipment Installation

The method of supporting the luminaires depends to a great extent on the ceiling height and the intended use of the studio. Where the height is low (3.5 to 5 meters [12 to 16 feet]) a permanent pipe or track grid is usually installed from which the units are hung directly or through available pantographs which permit individual vertical setting. See Fig. 11–13. In any case, the units are capable of complete rotation and tilting. In high-ceiling studios and in television theatres, the lighting units are supported either from fixed pipe or track grids or on counterweighted pipe or track battens (see Fig. 11–13).

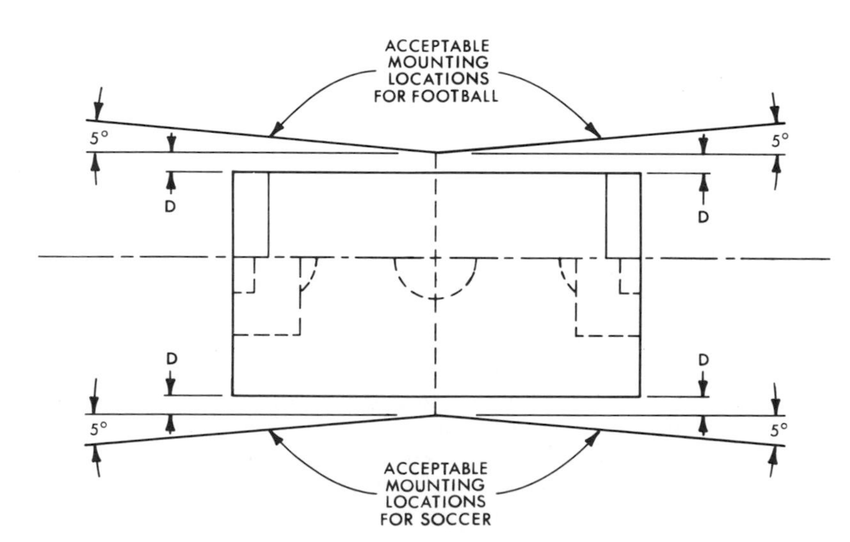

Fig. 11–11. Recommended pole locations for color television pickup of football and soccer. Mounting locations to be in areas defined in Fig. 11–10; minimum mounting height to be in accordance with Fig. 13–25. (See Section 13.) Distance *D* is 4.6 meters (15 feet).

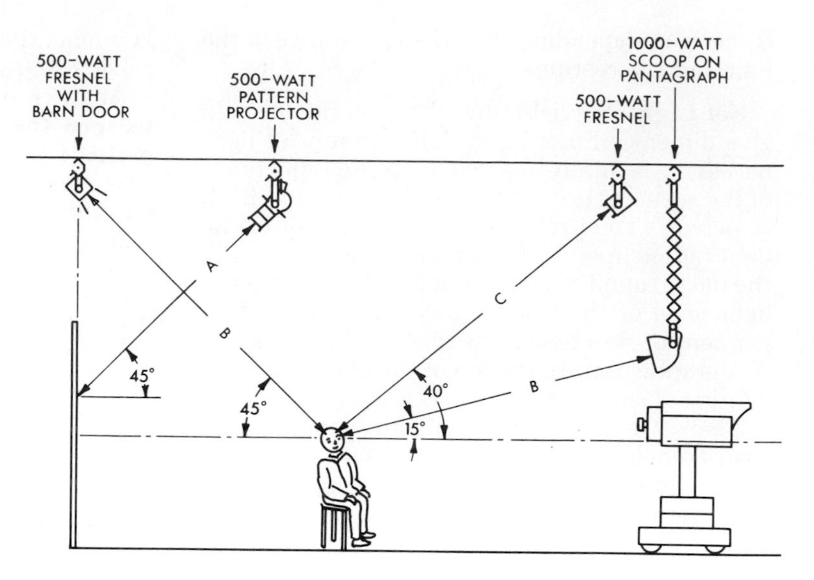

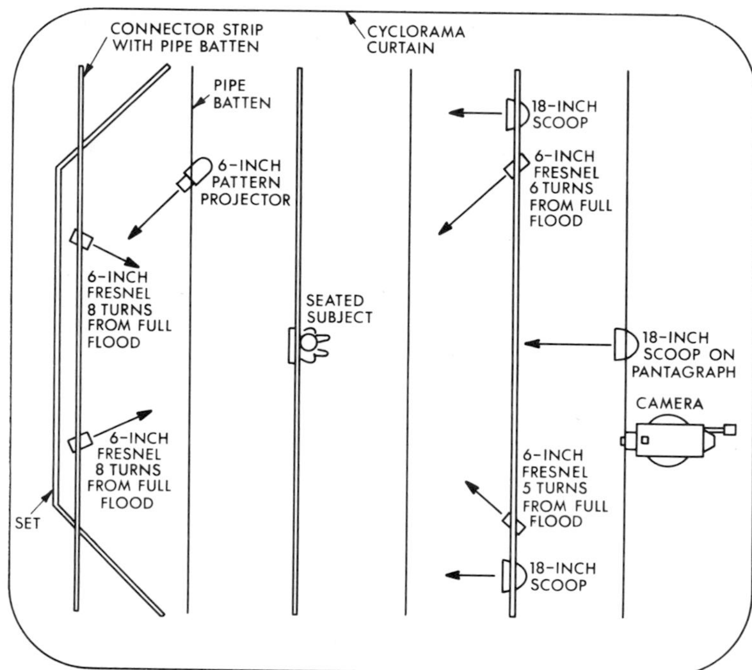

Fig. 11-12. Diagram showing good practice in luminaire location and aiming angles. Dimensions A = 3 meters (10 feet), B = 3.7 meters (12 feet) and C = 4 meters (13 feet).

TV Control Room Lighting

The lighting system for use during control room operating periods should meet the following requirements:

1. *General illumination.* A diffused, evenly distributed illuminance of 50 lux [5 footcandles] should be provided by low intensity sources located so as to avoid any specular reflections in picture monitors, clock faces, windows, control panels, console desks or similar surfaces, as seen from normal positions occupied by operation personnel.

2. *Work illumination.* Localized higher illuminances of 250 ± 50 lux [25 ± 5 footcandles] should be provided on the production consoles, control consoles, switching consoles and announcer's desk. See Fig. 11-14.

3. *Color temperature.* The color temperature of control room operating lights should be 6500 K ± 1000 K in order to provide an approximate match with the color temperature of standard color television monitors.

4. *Emergency lighting.* Emergency power from a separate source should be provided. This would allow activation of emergency lighting units de-

signed to provide safety for personnel in event of failure of the main lighting operating circuits.

The lighting system for use during maintenance of the area should have a level of approximately 250 ± 50 lux [25 ± 5 footcandles]. This system would be independent of the production lighting outlined above and would be used when installing equipment, repairing equipment, moving equipment or for cleaning chores.

Educational Television

In addition to commercial television channels, there are certain channels which have been allocated for educational broadcasting exclusively. Most of these are owned and operated by universities or other educational bodies. Their operations closely parallel the commercial stations in that the same type cameras and lighting are used, even though the budget they have to work with may be only a small fraction of that of their commercial counterparts.

Closed Circuit Television (Cable Television)

A recent trend, which is becoming a major television application, is the use of television on a closed circuit basis. Such telecasts cannot be

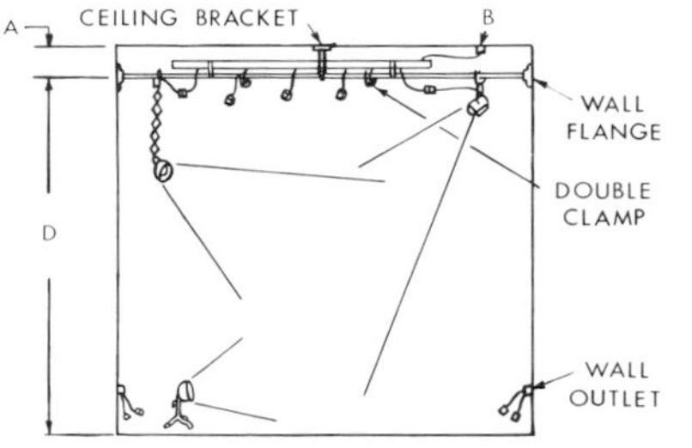

Fig. 11-13. Diagram (left) showing a typical overhead grid system for mounting lighting equipment in a small low-ceiling studio. Typical large studio grid system (photograph below) with raising and lowering capability. Dimensions A = 300 millimeters (12 inches), B = 100- × 100-millimeter (4- × 4-inch) duct, C = 32-millimeter (1.5-inch) I. D. pipe, and D = 3 to 3.7 meters (10 to 12 feet).

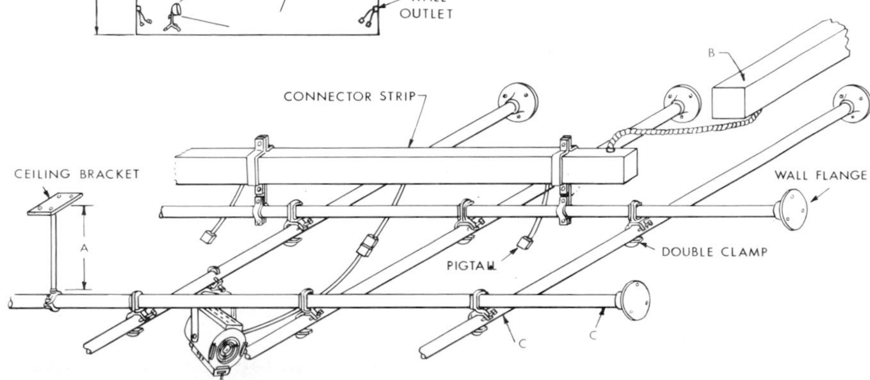

Fig. 11-14. Television control room equipped with localized downlights for the production and control consoles.

picked up by the conventional home receiver but can be received by those who have a set wired-in to the originating circuit.

In closed circuit educational television, a person giving a lecture or classroom demonstration in one room can be viewed in other rooms in the same building or in other buildings on campus. One lecturer can thus contact hundreds or thousands of students at one time. Such installations have been installed in many elementary schools, high schools and colleges.

The principal difference between the usual lighting practices of commercial and closed circuit television is that most closed circuit systems, due to financial limitations, utilize vidicon cameras. The techniques for lighting closed circuit television are identical to those of the commercial counterpart of the particular camera type being utilized as explained in previous sections.

Film Review Rooms for Television

When films are being prepared for television release, it is desirable that they be screened in a room especially designed to simulate the telecine system[22] and hence present the film as the home viewer would see it on a properly adjusted television receiver. The Canadian Broadcasting Corporation[23], European Broadcast Union[24] and the Society of Motion Picture and Television Engineers (SMPTE) are among those that have each defined clear operating standards in this regard. It is generally agreed that a projector fitted with a xenon source should be used, thereby producing a reflected color temperature of approximately 5500 K, largely in recognition of the commonly standardized television monitor white at D_{6500} as well as the motion picture theatrical standard at 5400 ± 400 K.* Likewise, it is generally agreed that a neutral gray screen of 20 to 30 per cent reflectance should be used to simulate the darkened studio monitor condition and thereby establish proper contrast relationships with the light surround. Either a projected or a self-luminous light surround may be used to judge brightness and provide an anchor for eye adaptation. Care must be taken to insure that the surround be of the same color quality as the reflected screen image, although most often only at 10 per cent of the total screen luminance. Reflected screen luminance, measured in the open-gate condition is usually specified to be within the range of 100 to 140 candelas per square meter [30 to 40 footlamberts], assuming a matte screen is employed. The SMPTE Recommended Practice, RP-41 (1974) is in agreement with these values.[25] Another alternative film preview room was outlined earlier by the Eastman Kodak Company in which a highly directional screen is employed in order to allow the use of a conventional classroom type projector equipped with a blue glass correction filter. The boost in screen gain offsets the severe loss in luminance from the filtered tungsten source and provides adequate screen viewing for a limited number of observers.[26]

* Various organizations have specified the reflected color temperature somewhat differently; the value of 5500 ± 500 K would represent the median value only, rather than any recognized standard in and of itself. The principal constraint is that the color quality be measured and maintained at the elevated value and that it not exceed these values, even though there may be some disparity among the specified aim point from one standard to another.

PHOTOGRAPHIC LIGHTING

Photographic lighting is used by amateur photographers, portrait and commercial photographers, industrial photographers, and cinematographers. The needs of these photographers vary drastically, as does the film material and lighting systems. A portrait photographer may use strobe lights to minimize the discomfort for the subject. Strobes and flash bulbs will stop the motion of moving objects. Color photography must have a compatibility between the film spectral requirements and the light source spectral output. In motion picture photography, the stroboscopic effect of fluorescent and high intensity discharge light sources must be minimized with special ballasts or synchronized with the film speed and shutter angle. The lighting requirements are, therefore, many and varied.

Photosensitive Materials

Commonly used photosensitive films and plates include the following:

Panchromatic (sensitive to all colors—black and white image)
Orthochromatic (sensitive to all colors except orange and red—black and white image)
Color (sensitive to all colors—color image)
Infrared (sensitive to red and infrared—black and white image)

Spectral sensitivity curves are given in Fig. 11-15.

For photography, light sources must emit energy in the spectral region in which the photographic material is sensitive. Even in black-and-white photography, color delineation in the form of faithful gray values is required. In black-and-white photography, photographers endeavor to secure a scale of grays in their negatives corresponding to the various brightnesses of the subject; thus, it is necessary that the film and the light source complement each other. Where this is not possible, it is general practice to employ a filter at the camera lens.

For photography in color, the spectral quality of the illumination is even more critical. Color emulsions are "balanced" for use with a particular quality of light. Because most color photography materials are based on three emulsion layers, each sensitive to a relatively narrow spectral band, adjustment by filtering to a light source other than the one for which the material was originally intended calls for precise filter selection.

The quantity of light that a film receives is a function of object luminance, of exposure time and of lens aperture. "Exposure" equals *illuminance at the film × time.* This relationship frequently is referred to as the reciprocity law. Longer or shorter exposure than a particular film is designed for will result in "reciprocity failure." The film manufacturer will suggest adjustments to compensate. Exposure time may be governed by factors such as the necessity for stopping motion, flashlamp and flashtube characteristics, and subject reflectance.

The most common system of expressing the light-gathering ability of a lens is the *f*-system in which the *f*-value of a lens is given as the *focal length ÷ the diameter of the lens opening.* For adjustable-aperature lenses, any particular ap-

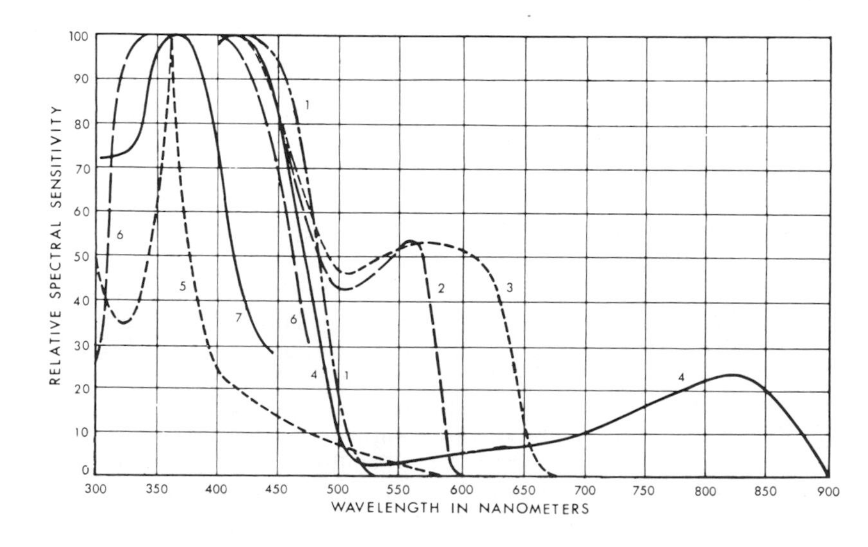

Fig. 11-15. Spectral sensitivity curves for common types of photographic and photoprocess materials:
1. Blue Sensitive
2. Orthochromatic
3. Panchromatic
4. Infrared Sensitive
5. Bichromate Coating
6. Diazotype Paper
7. Blueprint Paper

erture setting is referred to as the "*f*-stop." For a given scene luminance, illuminance at the film is proportional to the inverse of the square of the *f*-value.

There are a number of systems for evaluating the sensitivity (speed) of film and plates.[27] The American National Standards Institute (formerly ASA) has standardized a procedure for determining film speed. For example, a particular film may be said to have an ASA rating of 64.

The following formula, embodying the reciprocity law and the factors of lens aperture and film rating, gives the relationship of the several elements affecting exposure for objects of average reflectance:

$$E = \frac{K \times f^2}{T \times S}$$

where E = illuminance in line with camera on subject being photographed.
f = *f*-value at which the lens aperture is set.
S = speed of film, according to the ASA system.
T = time of exposure (seconds).
K = a constant based on the various elements used. For E in lux, 200 (in footcandles, 20) is a satisfactory value for negatives of average density (ASA).

Instead of basing the exposure on incident illuminances, the average luminance L may be substituted for E if a corresponding change is made in K.

Exposure Meters

The formula above is the basis of exposure meter design and operation, since all exposure meters of the photoelectric-cell type are essentially luminance-measuring devices. A photographic exposure meter, however, is not corrected for the eye sensitivity curve, but reads total energy. The meter consists of a photovoltaic cell, an ammeter of high sensitivity and a calculator. A hood or louver is provided in front of the sensitive cell to limit the acceptance angle to approximately 30 degrees, a rough average of the angle intercepted by the lenses of both still and movie cameras.

The customary method of using a photoelectric exposure meter (luminance type) is to hold it near the camera and point it toward the subject, thereby assuming that the meter "sees" the area being photographed, much as does the camera lens. Frequently a scene may include large areas, such as an open sky or a dark surrounding doorway, that may result in a luminance indication on the meter scale having little relation to the brightness of the subject. An underexposure or overexposure of the subject will result unless the proper precautions are taken. These include holding the meter at such a distance from the subject as to include only the subject.

The design of some meters permits the removal of this hood so that the cell will respond to illumination from an almost 180-degree solid angle when making illuminance measurements. When using a meter of this type, a different method (often called the incident-light method) is used. The meter is held close to the subject but pointed in the general direction of the camera. The meter reading indicates the illuminance on the subject. Meters of this type usually include a provision in the calculator for arriving at the correct shutter speed and lens aperture. If not, the formula given at the left can be applied.

In motion-picture photography the lens aperture forms the only variable, inasmuch as the exposure time is fixed by picture frequency. Exposure meters designed for this work give *f*-numbers for a specific film speed.

Guide Number System. Since it is not practical to employ exposure meters in connection with the use of flash lamps, there has come into general use a system of guide numbers which greatly simplifies the statement and use of exposure information in connection with these sources.

Flash Lamp Photography

The five important elements affecting exposure in flash photography are:

1. Luminance of the subject (affected by light output of flash source used, reflector used and reflectance of subject).
2. Film speed rating.
3. Shutter timing.
4. Distance from the light source to the subject.
5. Lens aperture.

Photographers usually have a particular size or sizes of photographic flash lamp and reflector, a particular type of film, and an established practice as to the shutter speed they prefer to use. Thus, items 1, 2 and 3 are fixed and it is possible to combine them empirically to provide a guide number that is the product of the aperture (*f*-number) and the distance from subject to lamp.

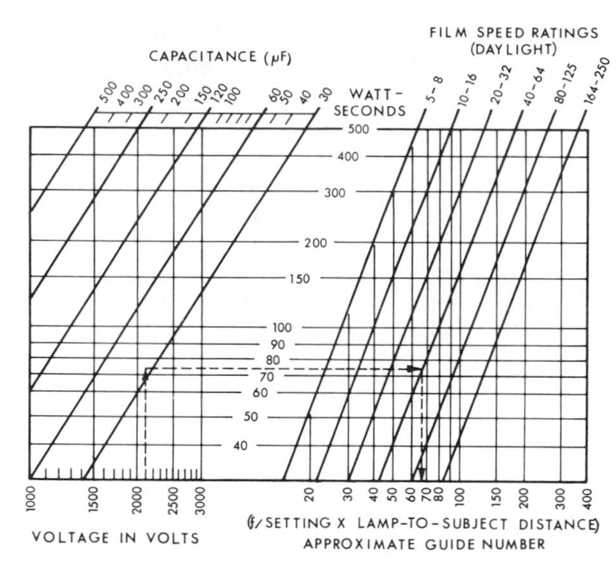

Fig. 11-16. Approximate Guide Numbers for flashtubes operated at various loadings. Film speed ratings are ASA and distances are in feet.

Since these are both second power functions and in inverse relationship, it remains merely to divide the guide number by the distance from lamp to subject to obtain the aperture setting. It becomes a simple matter to remember the guide number applicable to a particular lamp, film and shutter speed. The guide-number system is useful also in conjunction with other light sources, when an exposure meter is not available.

The American National Standards Institute, through its photographic committees, has worked out a standard procedure for obtaining guide numbers. This standard sets up the following formula:*

$$\text{Guide Number} = \sqrt{0.004Lt \times M \times S}$$

where Lt = light output of the flash source, expressed in lumen-seconds

M = reflector factor

S = film speed rating (ASA system).

For "time," "bulb," and exposures to 1/30 second the full lumen-second rating of the flash lamp may be used in the formula as the value of Lt. For higher shutter speeds the rated lumen-seconds of the lamp must be reduced since not all of the light will pass through the shutter during the time it is open. The following are approximate values for typical foil-filled lamps:

* In determining the aperture setting using the Guide Number derived using this formula, the distance is in feet. When distance is in meters, the distance must be converted to feet or a Guide Number used derived from a DIN film speed rating.

Shutter Speed (Seconds)	Percentage of Rated Lumen-Seconds of Lamp
Open, 1/25, 1/30	100
1/50, 1/60	70–80
1/100, 1/125	55–65
1/200, 1/250	35–40
1/400, 1/500	18–23

Reflector factors are controlled by the contour, size and surface finish of the reflector and by the physical size of the flash lamp.

Typical flash lamp values are as follows:

Reflector Diameter (Inches)	Finish	Bulb Diameter (Inches)	M
2	Polished*	15/32	9
3	Polished	27/32	8.5
4–6	Polished	1½	8
4–6	Satin	1½	5
6–7	Polished	1⅞	6
6–7	Polished	2⅜	4
Large studio	Polished	2⅝	6

The film speed ratings are usually supplied with the film or obtainable from the film manufacturers.

Exposures with flashtubes (see Section 8 of 1981 Reference Volume) depend principally on the watt-seconds of output of the power supply unit, but are also influenced by the design of the reflector and the flashtube used. A chart applicable to flashtubes (see Fig. 11-16) gives recommended guide numbers for loadings up to 500 watt-seconds.

* Due to the wide range of reflectors in use for this small diameter flash lamp, this information is intended only as a guide for a good 2-inch polished parabolic reflector.

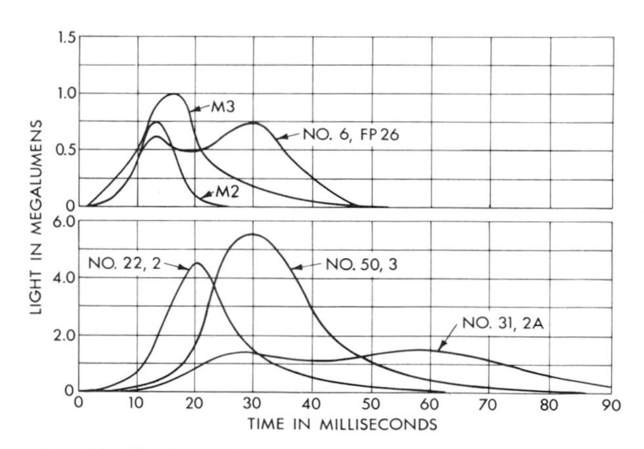

Fig. 11-17. Approximate time-light curves for several photoflash lamps; for further data see Fig. 8-112.

Flash Lamp Synchronizers. For most flash lamp photography, it is desirable to operate the camera shutter at speeds of 1/125 to 1/250 of a second in order to minimize the effect of any ambient illumination and to stop motion. Since approximately 5 milliseconds is required for a shutter of the pre-set type to reach full opening and 17.5 to 20 milliseconds for the lamp to reach peak light output, the synchronizing device must first apply the current to the lamps and then 12.5 to 15 milliseconds later trip the shutter.

Two or three flashlight-type dry cells customarily are used to supply the igniting power. Within two or three milliseconds (0.002 to 0.003 second) a flashlamp filament is heated to a sufficient temperature to ignite the priming material. In the case of lamps filled with shredded aluminum or zirconium wire, the burning primer sends a shower of sparks through this material, initiating its combustion at about 10 milliseconds. The burning metal reaches peak light output at about 17.5 to 20 milliseconds for the smaller and average-size lamps and at 30 milliseconds for the largest size. See Figs. 11-17 and 8-112.

A more recent flash lamp development provides self-contained ignition energy in the form of a cocked spring which ignites the primer when triggered by the camera synchronizer, thus permitting simple batteryless flash lamp systems.

Photographic Lighting Equipment

Reflectors. Still camera lenses ordinarily cover an angle of about 45 degrees; therefore, for lighting equipment placed at or near the camera, reflector beam patterns for complete light utilization should fill an angle of about 45 degrees. However, difficulties caused by inaccurate aiming of the reflector and other variables are minimized by filling a 60-degree cone with reasonably uniform lighting levels. A luminaire with a 60-degree beam angle, of course, provides lower illuminances toward the edges of the scene (slightly less than 1/2 of the center-beam value), but this is seldom objectionable since the point of interest in a picture is generally in the middle and a lower exposure at the edges is not serious.

The shadows and contrasts that help to light a person as normally seen are usually "soft," such as those produced by a light source of appreciable angular size. Large reflectors of 400 to 600 millimeters (16 to 24 inches) in diameter produce more natural modeling and are commonly used in portrait studios or other applications where size is not a handicap. Flash lamp equipment used with hand-carrier cameras necessarily have smaller reflectors which produce somewhat sharper shadows (less natural).

Photographic Lighting Techniques

Successful photography requires a rather narrow range of illuminances so that both the brightest parts (highlights) and the darkest parts (shadows) will be fully rendered in the final print or transparency. This range is much narrower than can be used for vision, particularly in the case of color photography. With typical subject reflectance ranges, the recommended maximum illuminance range with a scene is 10 to 1 for black and white, and 4 to 1 for color film.

A general requirement peculiar to color photography is that the apparent color temperature of all of the light sources used must be the same. The eye readily accepts illumination of mixed color temperature; photographic film does not.

Another general requirement of photographic lighting arises from the monocular vision of the camera. To compensate for the lack of stereo depth, the best lighting on photographic subjects emphasizes their roundness, form and spatial relationship. This is largely a matter of lighting direction, such as lighting from the side or the back.

In photography, two types of illumination are needed to produce a likeness of a subject:

1. *General illumination*, if used alone, produces a negative that is flat and without modeling. Such illumination does not produce prominent shadows, and density differences in the negative are caused for the most part by differences in the reflectance of various portions of the subject. This general, over-all illumination goes by several names, among which are front light, broad light, flat light, camera light and basic light.
2. *Modeling light*, if used alone, produces a negative in which the highlights can be well exposed but the shadows are clear and show no detail at all. Modeling lights are usually highly directional and are used for the express purpose of casting shadows and forming highlights.

Background Luminance. A factor closely related to lighting is background luminance. For ordinary subjects, the background should not be very dark, very light, or too close behind the subject; neither should it be of exactly the same luminance as important parts of the subject, because such a condition would have the effect of merging the subject with the background. The less detail and the fewer the distracting spots in the background, the better.

Lighting Installation Photography

The photography of "existing light" installations does not require special photographic materials or equipment. Attention to focus, proper exposure and composition are necessary; the use of a tripod is recommended.

The finished picture should represent what the actual installation looked like.[28] Because most installations have a luminance range that exceeds the acceptance of the film, various techniques in the taking and finishing are used to compensate or "compress" the luminance of the scene to produce a satisfactory photograph.

In the taking of the picture, exposures may be split, "fill light" may be introduced, lamp shades may be lined and bulbs may be substituted. When fill light is used, great care must be taken not to introduce unnatural shadows into the scene. This may mean bouncing light off of walls or ceiling. In black and white photography, a 10 to 1 ratio of scene illuminance (maximum) is desirable; in color, a 4 to 1 ratio (maximum).

The old rule of exposing for shadows and developing for highlights is still valid for black and white photography. Through experience and testing, the photographer finds the best combination of lighting technique, film and processing that yields an easily printed negative. The printer can also use various methods of compensation such as dodging, flashing and burning-in to overcome the deficiencies of a poorly executed negative.

When using color film, exposure should be for the highlight areas where detail is desired. It is also imperative that all the light sources be similar in color. Cool white fluorescent lamps cannot be combined with incandescent fill light, for example, without a noticeable color mismatch in the photograph. One combination of lamps that has proved generally compatible is incandescent lamps and warm white deluxe fluorescent lamps, as often used in residential photography.

Installations of discharge lamps photographed with color positive material present balancing problems. Because all films are designed to respond to a continuous spectrum and high intensity discharge lamps have discontinuous spectrums, filtration in some degree is required to produce a realistic color balance. This exception is clear mercury which has no red and, therefore, can never be corrected properly.

Lamps and film manufacturers can generally furnish recommendations of suitable filters for balancing the various color-positive films to specific discharge lamps—both fluorescent and high intensity types.

A much simpler procedure is to photograph the installation on unfiltered *color negative* material and work with a professional color laboratory to produce prints or slides with the correct color—the necessary filtration being done in the laboratory.

Darkroom Lighting

In general, any type of darkroom safelight filter must transmit light which will have the least effect on the photographic material and yet will provide most illumination for the eye. Any photographic material will fog if left long enough under safelight illumination.

The placement as well as the size and type of the safelight lamp will depend on the purpose which the light is to serve. The two types of darkroom illumination are (1) general, to supply subdued illumination over the whole room without concentration at any one point and (2) local, to supply higher illumination on some particular point or object. These are combined, dependent upon the size of the room and the type of work.

Because of the varying sensitivities of the different classes of photographic materials, several safelight filters are available, differing in both color and intensity. These have been scientifically prepared by the manufacturers and it is, therefore, never safe to use substitutes. Other materials may appear to the eye to have the same color as a tested safelight filter, but they will frequently have a much greater photographic action. The use of makeshift safelight substitutes is a very fertile source of darkroom troubles.

The following listing indicates available types of safelight filters:

Color	Material Used With
Clear yellow	Contact printing papers
Bright orange	Bromide and other fast papers and lantern slide plates
Greenish-yellow	Better than orange for judging print quality
Orange-red	Ordinary films and plates
Deep red	Orthochromatic films and plates
Green	Panchromatic films and plates
Yellowish-green	X-ray film
Special green	Infrared films and plates only

Film manufacturers should be consulted for more complete information.

Professional Motion Picture Photography

Motion picture set lighting is both a science and a creative art. The objectives in the entertainment type of motion picture photography from the standpoint of light are twofold. First, it is necessary that sufficient light be used to properly expose the film, and second, that the types and sizes of units are available which will give the director of photography or cameraman a maximum of control over illuminances, distribution of light and color temperature.

The entire illusion created by motion pictures is done with light. Directors of photography use light as the painter uses pigment. In order to obtain depth, roundness, smooth skin texture, color and tonal separation, sharp shadows, no shadows, streak-light (as in sunlight coming through a window or doorway), or the effect of a single source when using a hundred or more lights, they must have not only a wide assortment of light sources but all practical devices for the control of such sources.

The basis of motion picture set lighting is the "key-light" which simulates an actual indicated source appearing in the scene (a window or a table lamp, for example), or arbitrarily establishes a source if none is indicated. It is almost always the brightest area of light in the scene and is the reference point which the cinematographer uses to calculate basic exposure.

Customarily, key-light is established at a level which will permit a negative of the desired density for optimum print quality. All other illumination on the set (such as "fill" or back lighting) is arranged for position and level in relation to the key-light. The balance to be achieved between the key-light and the secondary illumination is largely dependent upon the psychological, emotional and dramatic mood required in the scene to be photographed. These are artistic considerations dictated by the script and/or the director's style.

When it is realized that the director of photography is actually making a series of snapshots of approximately 1/48-second exposure; that the characters themselves are moving about; and that the camera is often traveling on a dolly, or swinging through the air on a camera crane, it becomes apparent that motion picture photography is really an art as well as a science, and all lighting equipment and accessories should be designed for maximum operational latitude and versatility even if sacrifices in efficiency are to be made.

Since the cinematographer must deal with mechanical problems as well as artistic intangibles, the palette is somewhat more limited than that of the painter who can take great liberties in the use of basic materials. The cinematographer is restricted by the speed of the film stock, the technical characteristics of lenses, and the shifting spatial relationships of the elements within a given scene. For example, the slower the film, the more the aperture of the lens must be opened—with a proportional decrease in the depth of field. Thus, when using a film stock rated as ASA 64, an illuminance of 1700 lux (160 footcandles) must be available to permit shooting at *f*/2.8. This *f*-stop has its own sharply defined depth of field; however, if the available level is only 430 lux (40 footcandles), the lens aperature must be opened to *f*/1.4, with a resultant decrease in depth of field. Conversely, if the artistic requirement of the scene demands a greater depth of field (*e.g.*, that which is available at *f*/5.6), then the level will have to be raised to 6900 lux (640 footcandles).

Black and white production is complex since is requires that visual effect be achieved by *chiaroscuro*, or black and white shadow patterns. This permits the use of dimmer control of individual luminaires without concern for the effect of variations in color temperature. In color motion picture production, the cinematographer has the additional tool of color, but with the restriction that color temperature be maintained so that it is consistent within each scene. As a result, dimmers can only be used in a restricted fashion.

Professional color film is manufactured to meet either of two basic lighting conditions. "Daylight" balanced emulsions are created for exposure under conditions approximating 5600 K. "Tungsten" balanced emulsions are for exposure with approximately 3200 K. However, it should be understood that there is a reasonable degree of latitude in the color temperature range that each emulsion will reasonably expose.

The acceptable ranges for the color temperature of the exposure lighting cannot be precisely specified. With most professional film, variations may be readily corrected in the laboratory so long as the following limits are not exceeded:

Daylight 5600 ± 600 K

Tungsten 3200 ± 200 K

Since the only "objective" standard is one that must conform to reality, the subjective test of the "skintone" of the human face is usually chosen as the standard when final print corrections are determined.

Light Sources. The use of lighting in "naturally" illuminated situations, requires great care in order to be certain that severe color distortions are not introduced. These may be uncorrectable in the laboratory due to the distorted character of the mixture. Examples might be (1) an interior location "naturally" lighted with fluorescent light and supplemented with incandescent sources, and (2) a natural exterior (daylight) where electric lighting is used as supplemental fill to reduce shadow contrast.

Incandescent filament, including tungsten-halogen types, are available in a wide variety of forms. For motion picture production lighting, see Fig. 8-91. Among the most interesting types developed for film use are the dichroic coated PAR-36 and PAR-64 types which emit a "daylight corrected" light. Carbon arc sources can be used for mixing with 3200 K incandescent light sources, with the use of a yellow-flame carbon and YF-101 filter. See Fig. 11-18.

Daylight sources are used under quite different conditions from those involving the 3200 K sources. When the 3200 K sources are used they usually provide all of the illumination that the cinematographer may require on the stage. Daylight sources are usually used where the predominant light is daylight itself. "Daylight" is a broad term; its color temperature varies widely depending upon the time of day, the proportion of sunlight to skylight, the existence and character of clouds, etc. "Daylight" sources are basically used as supplemental lighting to modify the effect of real daylight with respect to varying color temperature and density of shadows. Daylight sources include:

1. High-wattage carbon arcs. These operate on dc current and require a resistance "grid" (ballast) to provide the proper arc voltage when operating from 120-volt systems. The sizes range from 120-amp arcs to 350-amp arcs.
2. Short-arc, high-pressure xenon sources have recently been used for this application. They are available in sizes up to 4.5 kilowatts and require a somewhat complex power supply and high-voltage start system.

In the past decade, because of increasing studio costs, plus a greater emphasis upon dramatic realism, there has been an increasing trend toward shooting on locations. This trend includes not only the filming of exterior scenes, but those which take place within actual interiors as well. This modification of production procedure, which will surely become more prevalent, demands the development of lighter weight, more highly efficient filming equipment in all categories—most definitely including lighting units.

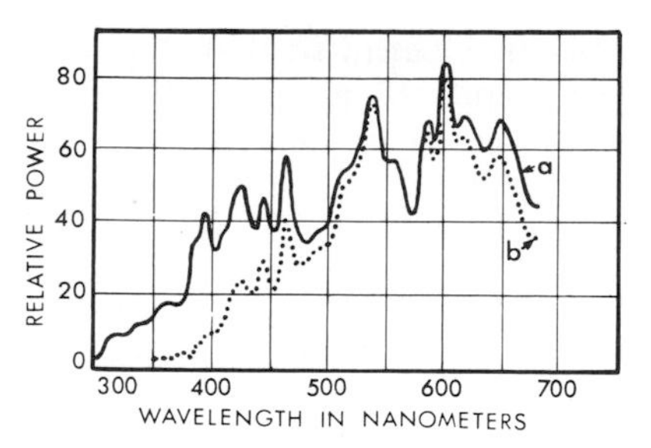

Fig. 11-18. Spectral power distribution curves for yellow flame carbon. a. Without filter. b. With YF-101 filter.

Luminaires. The selection of luminaires is dictated more by the dramatic requirements of the script than by the size of a particular set. In general, motion picture production requires what is called "controlled" lighting and because of this the Fresnel lens spotlight is the most popular type of luminaire. Fresnel lens luminaires offer a controllable beam ranging from an 8- to 10-degree "spot" to a 30- to 45-degree "flood." Because they can be successfully barn-doored, they offer the kind of control that permits the cinematographer to compose his lighting.

The introduction of the compact and the coiled-coil filaments of tungsten-halogen lamps has resulted in the design of a type of luminaire which produces "semi-controlled" illumination. These luminaires are "focusing" but their barn-door control is very limited. They do not use a lens to achieve their focusing ability. This is done by moving the filament with respect to the reflector or vice versa. Since there is direct illumination from the filament itself, in addition to the varying illumination from the functioning of the reflector, it is possible to change the level and coverage of the light, but the shadow tends to be double. This restricts predictable control of the barn-dooring function. The effective lumen output is quite good, since very high collection efficiencies are possible and there is no loss through a lens. In general, these luminaires are very light in weight and very compact. They are available in wattages up to 2kW.

Another important luminaire in film production is the "shadowless" or "soft" light. Since the cinematographer must always control the "scene" it is desirable to be able to illuminate the subject and reduce the shadows (wrinkles on a face, etc.) on the subject and behind the subject. This can be achieved with the uses of large, even sources. Large incandescent lamps have been used in luminaires such as the bulky, generally awkward and inefficient "cone" light. The long straight filaments of certain of the tungsten-halogen lamps have permitted development of more manageable and efficient soft lights, since the filaments are easily hidden.

The compactness of tungsten-halogen lamps has led to the redesign of other luminaires and the creation of new and more convenient special luminaires. The long filament of some of the tungsten-halogen lamps has permitted a positive redesign of the family of luminaires called "broads". These luminaires are excellent for broad, general base illumination and provide a very flat pattern with high efficiency. A miniature version of this type of luminaire has been developed. It provides very flat illumination and can be barn-doored to a very sharp cut-off in the horizontal direction. These small luminaires can be placed in tight areas and permit the cinematographer greater latitude in developing his desired effects.

The development of tungsten-halogen PAR lamps with compact filaments, plus the development of dichroic coatings which can survive the high temperatures, have resulted in an entire line of luminaires for daylight fill use on location.[29] The most popular luminaires of this type utilize either PAR-36 lamps, of about 650 watts each, or PAR-64, 1000-watt units. Since the reflector and the precise positioning of the filament result in illumination of considerable "punch," they can provide powerful daylight-type illumination when used in clusters. The loss of light due to the dichroic filter is partially offset by raising the color temperature of the tungsten source within the lamp to approximately 3400 K. The weight and size per lumen output of these luminaires is very favorable, as compared to the daylight arcs, so that they have become exceptionally popular on location.

PICTURE PROJECTION LIGHTING

Satisfactory picture projection requires not only careful selection of light source and optical elements for projecting the picture but also of the screen and its surroundings in relation to the seating area from which it is to be viewed.[30] The basic requirement is that the picture luminance be of a value such that the proper contrasts of highlights and shadows are achieved at a satisfactory over-all level.[31] Some illumination in the seating area is essential for the convenience of the audience, safety, discipline, etc.[32] See Theatres, page 11-10. However, if light from the seating area is allowed to fall on the screen the desired contrasts are reduced, and the over-all luminance must be increased to restore the proper relations.[33]

The logical place to start in planning projection is the area in which the pictures will be shown. This establishes the luminance level, the type of screen to be used, and the amount of light needed from the projector.

Luminance Levels for Motion Pictures

Luminance levels recommended by the Society of Motion Picture and Television Engineers (SMPTE) are predicated upon the presence of minimum stray light on the screen itself and a practicable balance between characteristics of photographic materials and available light. The weight given these factors, and therefore the determination of the most desirable projection conditions, differs somewhat among the five typical applications of projected pictures: (1) review rooms, (2) theatres, (3) drive-in theatres, (4) auditoriums and (5) classrooms.

Definitions. Screen measurements in motion picture projection are made with the projection light source adjusted to normal operating conditions, with the projector shutter running, and with no film in the gate. This screen luminance level is approximately 10 times the average luminance of the pictures projected from normal films.[33, 34] Generally, the incident light falling upon the projection screen decreases from the center toward the edges; the incident light measured at points on the horizontal center line which are at a distance in from the screen edge equal to 5 per cent of the screen width, is normally 60 to 80 per cent of illuminance at the center.[35] The luminance of the screen from the audience position may parallel the variation of the illuminance, or may vary from it in a more complex pattern as indicated in the description of the properties of screen surfaces.[36–38]

Review Rooms. During the preparation of motion pictures, the producer, the motion picture film laboratory personnel, and others examine the film many times from the original test shots through many stages to the final release print. The films are projected in a specialized theatre known as a "review room." These installations are designed specifically for the inspection of motion pictures and are built to accommodate a small reviewing group of usually 10 to 20 people. The actual picture size may be small or large depending upon the space available, but the viewing conditions are chosen to duplicate as nearly as possible actual theatre viewing from the most desirable seating locations. All of the viewing conditions are capable of precise control and it is generally practical in review rooms to hold these variables to a minimum tolerance.

For review rooms, luminance levels are generally based upon the standard[39] for screen luminance and viewing conditions. This provides for an aim luminance of 55 ± 7 candelas per square meter (16 ± 2 footlamberts) at a color temperature of 5400 K ± 400 K, at the center of the screen.

See page 11-24 for a discussion of review rooms for films made for television release.

Theatres. Facilities constructed primarily for the presentation of projected pictures to an audience are designed for a minimum of stray light, adequate and permanent projection facilities, optimum viewing angles, screen sizes and surfaces adequate for the facility in which they are installed, etc. Under these optimum conditions the audience receives the best in projected pictures.

For primary theatre projection the standard[39] specifies 55 ± 7 candelas per square meter (16 ± 2 footlamberts) at the center of the screen. The Committee on Theatrical Projection Technology of the SMPTE is currently working on the inclusion of other important characteristics in this specification; it is suggested that current reports of the Committee be consulted.

Drive-In Theatres. Viewing conditions are quite different from those of the indoor theatres with perhaps the most significant change being the much smaller angle which the screen subtends at the observer's eye; this in turn changes several psychophysical factors. Permissible screen luminances have been shown by experience to be lower than those for comparable quality projected images in indoor theatres, even though the drive-in projects the same identical standard prints. An applicable luminance standard has not yet been devised. The SMPTE has issued a guide[40] in which the minimum screen luminance in drive-ins is 15 candelas per square meter (4.5 footlamberts).

Auditoriums. These are facilities where provision is made for projecting motion pictures to an audience although this is not the sole use of the facility and compromises may be made in the interest of other functions. Stray light may not be as well controlled, projection facilities are more limited, screen sizes may be smaller, etc. It is suggested, nevertheless, that the theatre conditions be used as a design aim.

Classrooms. Frequently projection in classrooms is handicapped by an irreducibly high stray light level, generally temporary projection equipment and screens, and frequently little control of seating arrangements. Under these conditions optimum pictorial quality can seldom be achieved, although a satisfactory presentation of information to the classroom group is frequently possible. Acceptable luminance levels are 55 ± 7 candelas per square meter (16 ± 2 footlamberts)[41]; consideration should be given to many other factors in the selection of the particular room, its equipment, etc.[42, 43]

Luminances for Slide Projection, Etc.

There has been less study of optimum projection conditions for slides, slide films, opaque projectors, etc., but all proposed standards have agreed on 34 candelas per square meter (10 footlamberts) as a desirable aim luminance. The projection of continuous-tone pictures seems to be governed by the same factors as affect projection of conventional motion pictures; the projection of line drawings, tables, and other high contrast images is less critical of luminance level and less critical of stray light, as discussed in the following section.

Surround Illuminance

Illuminances on surfaces other than the projection screen are controlled for both esthetic and practical reasons. In review rooms illuminance levels must be adequate for limited movement among a small group; in theatres and auditoriums provision must be made for complete seating of the audience in semi-darkness, and for reasonable safety, etc.; in drive-in theatres illumination must permit both automobile and pedestrian traffic; in classrooms illumination should permit note-taking—although the usual problem is the adequate reduction of ambient illumination. In all applications visual comfort is desired.

In the design of ambient lighting the two prime requirements are avoidance of all high luminance sources visible to the audience, and direction of the light so that as little as possible falls onto the screen surface.[33] (With directional screen materials the disadvantages of stray light will depend upon both its level and its angle of incidence upon the screen.) Good pictorial reproduction demands an adequate tone scale in the projected picture; maximum highlight luminance is limited by and cannot exceed screen luminance; minimum shadow luminance must be at least equal to the stray light luminance. The ambient illumination must be so designed and directed as to hold stray light luminance of the screen to as low a value as possible; suggested luminance ratios are listed in Fig. 11-19.

Fig. 11-19. Permissible Luminance Ratios for Projected Pictures: Conventional Screen Luminance[a] vs Screen Luminance Resulting from All Non-Image Luminance[b]

Projection Facility[c] Type					Type of Material	Recommended Ratio[d]
1	2	3	4	5		
×	×	×	×		Motion pictures at optimum.	300:1
		×	×	×	Full scale black and white and color, where pictorial values are important and color differences must be discriminated.	100:1
				×	Color diagrams and continuous tone black and white in high key.	25:1
				×	Simple line material such as text, tables, diagrams, and graphs.	5:1

[a] Measured with no film in the aperture; therefore maximum image high-light luminance will normally be 25 to 60 per cent of the screen luminance.[34]

[b] Measured with the projection lens capped; therefore minimum shadow luminance will approach this screen luminance resulting from all nonprojected illumination.

[c] Type of projection facilities: 1—Review Room; 2—Theatre; 3—Drive-In Theatre; 4—Auditorium; 5—Classroom

[d] Reference 33.

Screen Types

Screens can be classified generally as reflective or translucent, depending upon whether the projected picture is viewed from the same side as the projector or from the opposite side. Reflective types may be either directional or nondirectional depending upon whether or not the luminance changes with viewing angle. Translucent screens generally are of the directional type. Most reflective theatre screens are perforated to permit sound transmission from speakers located behind the screen; these perforations represent about six per cent of the total surface area.

The directional characteristics of a screen surface result from the nature of the reflection or transmission, and produce a variation in luminance as a function of both incidence and viewing angles.[38] For screens showing significant directionality, it is customary to curve the screen surface into a portion of a cylindrical surface, with axis vertical and radius equal to the distance from projector to screen; this minimizes the variations in incidence angles.

The general nature of such reflective patterns is suggested in Fig. 11-20. Off-axis response introduces many complexities which should be studied in detail.[36-38, 44] The number of special screen surfaces is increasing rapidly but may be classified among the following types:

Matte Surface. Such surfaces are practically non-directional. In other words their luminance is substantially the same at all viewing angles. Practical reflective matte screens have surfaces of high reflectance but since the light is distributed through-out a complete hemisphere the maximum attainable luminance is limited. On typical clean, new matte screens an incident illuminance of 10 lux (1 footcandle) produces a luminance of 2.6 to 3 candelas per square meter (0.75 to 0.90 footlamberts). Matte surfaces are recommended where viewers will be distributed over a wide angle in the viewing area. They are applied in theatres, auditoriums and classrooms whenever the screen sizes and available light permit a satisfactorily bright picture. Translucent matte screens have such a low transmittance that they are not of practical importance. Whenever a higher luminance than that available from matte screens is desired, one of the following types of directional screens may be specified if their more restrictive installation requirements can be met.

Semi-Matte Surface. In this type of surface a material is incorporated in the paint or surface finish to give it a slight gloss. This can be done in such a way that it improves the reflectance or "gain" over a certain distribution angle. When carefully done, with central luminance gain held to values not in excess of 1.5 to 2, an over-all improvement in picture luminance can be obtained without excessive "hot-spotting" or glare appearing near that point on the screen where an image of the projector lens would be reflected by a specular reflector.

Metallized Surface. A metallic pigment (usually aluminum) may be incorporated in a paint or otherwise applied to the surface of a suitable supporting material to make a useful reflective screen. Such a surface is mostly very directional but can be made to vary in its reflection characteristics. Because of this versatility these screens can be "tailor-made" to give maximum luminance consistent with acceptable viewing in any given situation. Directional screens of this type will show a "hot-spot". Furthermore, plane screens will appear brighter on the near than on the far side when viewed from side positions when the screen is flat. Such effects can be minimized by curving the screens approximately to the projection radius. Therefore, installations of screens of this type should be made only after careful consideration has been given to all viewing factors, and then in accordance with specifications carefully drawn with the special characteristics of the screen in mind. A metallized screen must be provided whenever projection of 3-D pictures or slides with polarized light is anticipated, because matte screens depolarize the incident polarized light.

Lenticulated Surface. Practical reflective screens have been developed with small uniformly-shaped and spaced lens-elements impressed in the surface. These control the direction of light reflection so that the maximum luminance will be obtained within certain specified viewing angles. Within these angles, moreover, the luminance is generally more uniform than for non-lenticulated types of directional screens. The highest luminance for a given incident illumination is obtained with screens having lenticulated surfaces. Their characteristics are even more specialized than for metallized surfaces and manufacturer's recommendations must be followed in their selection and installation if satisfactory results are to be achieved.

Practical lenticulated translucent screens have been produced so far only in limited sizes; this has restricted their use to specialized applications such as projection under conditions of high ambient light, where the controlled luminance of a small lenticulated screen permits the presentation of acceptable pictures.

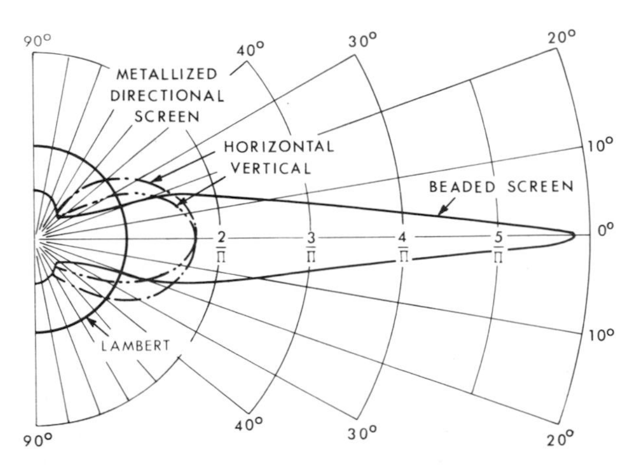

Fig. 11-20. Luminance patterns for typical screens. Goniometric distribution for a nonabsorbing perfect diffuser (Lambert), a beaded screen and a metallized screen showing different horizontal and vertical patterns. The radial scale is graduated in candelas per square meter observed under illuminances of one lux. A matte screen luminance pattern approaches that of the perfect diffuser.

Beaded Surface. This may be either a reflective or translucent screen; the surface is covered with small glass or plastic spheres which reflect or transmit most of the light in the general direction of the projection axis. Such a screen will appear the brightest when viewed along the axis of projection and will darken quite rapidly as the viewing angle increases away from this axis. The property of reflecting a high proportion of the light back to its source offers some control of stray light from sources that are not located in line with the audience. Such screens are useful for homes and small classrooms where the viewing angle requires limited horizontal distribution.

Maximum and Minimum Viewing Distances

Viewing distances are determined by a number of factors including resolution of picture detail,[45] limitations of graininess and sharpness in the projected image, psychophysical impressions of presence and identity within the projected action, limitations of comfortable viewing posture, structural limitations of viewing angle, etc.

Over the past several decades there has been intensive effort in the motion picture industry to increase the horizontal viewing angles so that the viewer can more readily identify himself with the projected story, and simultaneously to reduce the imperfections in the projected image in order to produce a more complete illusion of reality. At the same time the vertical viewing angles have remained relatively constant since they are more effectively limited by design of theatres and auditoriums (especially balcony overhangs), etc. For design purposes, therefore, Fig. 11-21 may be taken as an approximate guide. Improvements in picture sharpness, graininess and resolution eventually may make closer viewing positions possible, but further studies on the psychophysical response are still not conclusive concerning the so-called wide-screen presentation methods.

Limitation of Viewing Angle

The angle (measured from the projection axis) within which an audience may see a presentable picture depends upon both geometric distortions and luminance variations on the screens. Objectionable geometric distortions of the image on a flat screen become apparent at angles beyond approximately 30 degrees from the normal to the screen. Since screen surfaces may be curved cylindrically to obtain better luminance distribution from directional screens or to increase audience participation, geometrically desirable viewing angles may thereby be further limited. In addition, since all of the directional screens show some luminance fall-off as the viewing angle is increased, acceptable maximum values are also set by the point at which screen luminance falls below the minimum tolerable.

Practical viewing angles are determined by both the geometry of the theatre and the goniometric reflection characteristics of the screen surface. It is suggested that these two factors be so adjusted that at the maximum useful viewing angle, the luminance of the center of the screen be no less than half its maximum luminance.

Projected Screen Image Dimensions

The projected screen image size is determined by three factors, namely: projector aperture dimensions; focal length of projection lens; and throw, *i.e.*, distance from lens to screen. In motion picture projection, be it 8, 16, 35- or 70mm, the projector aperture is always smaller than the available printed image on the film to allow for printer aperture variations, height unsteadiness and side-weave. The projected image area through the aperture is further masked on the screen to permit a clean-cut picture outline with a minute "spill" being absorbed by the black velvet masking. In still, slide or filmstrip projection, the screen image borders are the same as determined by the original mounts.

The ratio of width to height of the projected or finally masked screen image is referred to as *aspect ratio* and should be considered as an esthetic rather than a technical measurement.

For many years motion picture projection was limited to 1.33:1 aspect ratio which still prevails for the 8mm format and 16mm and 35mm films for television. Theatrical wide-screen systems from a single piece of film have, in search for the optimum audience appeal, greatly departed from that aspect ratio. In 35mm, two methods are generally adopted. Using linear, non-anamorphic lenses, pictures are presented as "wide screen" productions in 1.66:1, 1.75:1 or 1.85:1 aspect ratios, the last of these being the most commonly used.

Introducing an anamorphic "squeeze" ratio of 2:1 led to establishing an aspect ratio of 2.35:1 for productions justifying more "scope" for stories so told with 35mm film. There is of course an increasing amount of 16mm prints now available with 2:1 "squeezed" image information which will result in presentation aspect ratios of 2.35:1 up to 2.55:1.

The actual screens used in any projection system must of course be oversized but will have approximately the aspect ratio of the system involved. In theatrical installations it is customary to use the largest screen that can be accommodated on the stage. All kinds of aspect ratios are presented with variable screen masking, mostly remote controlled.

Since filmstrips and all professional slides are designed for aspect ratios of 1.33:1 or greater, and the 2 × 2 slides can be positioned either horizontally or vertically, it is necessary to employ a square screen for such showings.

The basic factors to determine the requirements for any linear, *non-anamorphic* projection set-up are related as follows:

Fig. 11-21. Limitations of Viewing Distances and Angles[a]

	Review Rooms[b]	Theatres and Auditoriums[c]	Drive-in Theatres[d]	Classrooms[e]	
				Matte Screens	Beaded Screens[f]
Front row of seats		1.2	2.8		
Minimum viewing distance	3.0			2.7	3.3
Maximum viewing distance	5.0	8.0	14	8.0	8.0
Maximum viewing angle (degrees)	10	30	30	30	20

[a] Distances are measured from the screen surface in multiples of picture height; angles are measured from the screen normal in degrees.
[b] Suggested values; no current standard.
[c] Recommendations: Society of Motion Picture and Television Engineers Theatre Engineering Committee.
[d] Values typical of present practice; no formal recommendations or standards available.
[e] Recommendations: Society of Motion Picture and Television Engineers Committee on Nontheatrical Equipment.[41]
[f] These values should also be used for metallized and lenticular screens unless the manufacturer's data indicate that performance at greater viewing angles is satisfactory.

Fig. 11-22. Projector Aperture Dimensions

Projection Mode	Film Size	Aperture Width Millimeters	Aperture Width Inches	Aperture Height Millimeters	Aperture Height Inches	Projection Optics	Projection Aspect Ratio[a]	Applicable ANSI Standard[b]
Motion Picture Film								
Standard	8mm	4.37	0.172	3.28	0.129	Normal	1.33/1	PH22.20—1975
Standard	Super 8	5.31	0.209	4.00	0.157	Normal	1.33/1	PH22.154—1976
Standard	16mm	9.65	0.380	7.21	0.284	Normal	1.33/1	PH22.8—1969
Anamorphic Print	16mm	9.65	0.380	7.21	0.284	Anamorphic[c]	2.66/1	
Standard (TV)	35mm	20.96	0.825	15.24	0.600	Normal	1.33/1	PH22.58—1954
"Wide Screen"	35mm	20.96	0.825	11.35	0.447	Normal	1.85/1	PH22.58—1954
Anamorphic Print	35mm	21.31	0.839	18.16	0.715	Anamorphic[c]	2.35/1	PH22.106—1971
"Todd AO"								
"Super Panavision"	70mm	48.59	1.913	22.05	0.868	Normal	2.21/1	
"Ultra Panavision"	70mm	48.59	1.913	22.05	0.868	Anamorphic[d]	2.35/1 plus	
Slides, etc.								
2 x 2 Slides[e]								PH3.43—1969
	35mm	23.01	0.906	34.14	1.344	Normal	0.67/1	
	35mm	34.14	1.344	23.01	0.906	Normal	1.48/1	
	Roll 1⅝ x 1⅝ inches	37.97	1.495	37.97	1.495	Normal	1.00/1	
Filmstrips	35mm	23.01	0.906	17.27	0.680	Normal	1.33/1	PH1.24—1955
Lantern Slides	4 x 3¼ inches	76.2	3.00	57.15	2.25	Normal	1.33/1	PH3.43—1969

[a] Aspect ratio is the quotient of projected width divided by projected height. This ratio will be reduced if the vertical projection angle is not 0 degrees.
[b] Present standards or latest revisions thereof as issued by the American National Standards Institute, New York, N.Y.
[c] These anamorphic processes require projection lenses with a horizontal image magnification greater than the vertical magnification by a ratio of 2:1.
[d] In present 70mm anamorphic processes the horizontal magnification is 1.25:1 greater than the vertical.
[e] 35mm slides in the 2 x 2 format will fit in the projector with the 34.14 millimeter (1.344 inch) frame dimension oriented either horizontally or vertically. Horizontal orientation is preferred for professional use.

$$\text{Focal Length} = \frac{\text{Aperture Width} \times \text{Throw}}{\text{Picture Width}}$$

$$\text{Picture Width} = \frac{\text{Aperture Width} \times \text{Throw}}{\text{Focal Length}}$$

$$\text{Throw} = \frac{\text{Focal Length} \times \text{Picture Width}}{\text{Aperture Width}}$$

$$\text{Aspect Ratio} = \frac{\text{Picture Width}}{\text{Picture Height}} \text{ or } \frac{\text{Aperture Width}}{\text{Aperture Height}}$$

$$\text{Picture Height} = \frac{\text{Picture Width}}{\text{Aspect Ratio}}$$

In linear projection the aspect ratio of the projected image is equal to the aspect ratio of the projector aperture or slide mount. This is not the case involving projection of *anamorphic* or "squeezed" pictures. Their requirements for a projection set-up are as follows:

Focal Length

$$= \frac{\text{Aperture Width} \times \text{Squeeze Ratio} \times \text{Throw}}{\text{Picture Width}}$$

Picture Width

$$= \frac{\text{Aperture Width} \times \text{Squeeze Ratio} \times \text{Throw}}{\text{Focal Length}}$$

Throw

$$= \frac{\text{Focal Length} \times \text{Picture Width}}{\text{Aperture Width} \times \text{Squeeze Ratio}}$$

Screen Image
Aspect Ratio

$$= \frac{\text{Picture Width}}{\text{Picture Height}} \text{ or } \frac{\text{Aperture Width} \times \text{Squeeze Ratio}}{\text{Aperture Height}}$$

Due to changing projection and wide screen systems, as well as standards, most tables compiling such data as picture sizes, aspect ratios and available focal length of lenses become easily obsolete or misleading. The preceding formulas permit the calculation of all data required for each individual projection condition. The presently used aperture dimensions are listed in Fig. 11-22. Fig. 11-23 lists picture widths.

Wide Screen and Special Processes

With the expansion of the screen in theatres different production and presentation methods were devised.

Fig. 11-23. Picture Sizes Obtained with Various Focal Length Lenses

Focal Length of Lens (Inches)	Type of Projector	Projector Aperture to Screen Distance (Feet)																		
		5	10	15	20	25	30	35	40	50	70	80	100	150	200	250	300	400	500	600
		Width of Picture (Feet)																		
0.5	8mm	1.7	3.4	5.1																
	Super 8	2.1	4.2	6.3																
1.0	8mm	0.8	1.7	2.6	3.4	4.3	5.1													
	Super 8	1.0	2.1	3.1	4.2	5.2	6.3													
	16mm	1.9	3.8	5.7	7.6	9.5	11.4	13.3	15.2	19.0	26.6	30.4	38.0							
2.0	16mm	0.9	1.9	2.8	3.8	4.7	5.7	6.6	7.6	9.5	13.3	15.2	19.0	28.5	38.0					
	35mm			6.1	8.2	10.2	12.3	14.4	16.4	20.6	28.8	32.9	41.2	61.8	82.5	103.1	123.7	165.0		
	35mm Anamorphic			12.4	16.6	20.8	25.0	29.2	33.4	41.8	58.5	66.9	83.7	125.6	167.5					
2.5	16mm	0.7	1.5	2.2	3.0	3.8	4.5	5.3	6.0	7.6	10.6	12.1	15.2	22.8	30.4	38.0				
	35mm				6.5	8.2	9.8	11.5	13.1	16.4	23.0	26.3	32.9	49.4	65.9	82.5	99.0	132.0	165.0	
	35mm Anamorphic				13.3	16.6	20.0	23.3	26.7	33.4	46.8	53.5	66.9	100.4	134.0	167.5				
3.0	16mm	0.6	1.2	1.9	2.5	3.1	3.8	4.4	5.0	6.3	8.8	10.1	12.6	19.0	25.3	31.6	38.0			
	35mm					6.8	8.2	9.6	10.9	13.7	19.2	21.9	27.4	41.2	54.9	68.7	82.5	110.0	137.5	165.0
	35mm Anamorphic					13.8	16.6	19.4	22.2	27.8	39.0	44.6	55.7	83.7	111.6	139.6	167.5			
	2 x 2 Slide																			
	35mm		4.4	6.6	8.9	11.1	13.4	15.6	17.8	22.3	31.3	35.8	44.8	67.2						
	1⅝ x 1⅝		4.9	7.4	9.8	12.3	14.8	17.3	19.8	24.8	34.8	39.8	49.7	74.7						
3.5	35mm					5.8	7.0	8.2	9.4	11.7	16.4	18.8	23.5	35.3	47.1	58.9	70.7	94.2	117.8	141.4
	35mm Anamorphic					11.8	14.2	16.6	19.0	23.8	33.4	38.2	47.8	71.7	95.7	119.6	143.5	191.4		
4.0	16mm	0.4	0.9	1.4	1.9	2.3	2.8	3.3	3.8	4.7	6.6	7.6	9.5	14.2	19.0	23.7	28.5	38.0		
	35mm						6.1	7.2	8.2	10.2	14.4	16.4	20.6	30.9	41.2	51.5	61.8	82.5	103.1	123.7
	35mm Anamorphic						12.4	14.5	16.6	20.8	29.2	33.4	41.8	62.7	83.7	104.6	125.6	167.5		
	2 x 2 Slide																			
	35mm		3.3	4.9	6.6	8.3	10.0	11.7	13.4	16.7	23.5	26.8	33.5	50.4	67.2					
	1⅝ x 1⅝		3.6	5.5	7.4	9.2	11.1	13.0	14.8	18.6	26.1	29.8	37.3	56.0	74.7					
	Lantern Slide	3.5	7.3	11.0	14.8	18.5	22.3	26.0	29.8	37.3	52.3	59.8	74.8	112.3	149.8					
5.0	35mm							5.7	6.5	8.2	11.5	13.1	16.4	24.7	32.9	41.2	49.4	65.9	82.5	99.0
	35mm Anamorphic							11.6	13.3	16.6	23.3	26.7	33.4	50.2	66.9	83.7	100.4	134.0	167.5	
	2 x 2 Slide																			
	35mm		2.6	3.9	5.3	6.6	8.0	9.3	10.7	13.4	18.7	21.4	26.8	40.3	53.7	67.2				
	1⅝ x 1⅝		2.9	4.4	5.9	7.4	8.9	10.3	11.8	14.8	20.8	23.8	29.8	44.8	59.7	74.7				
	Lantern Slide	2.7	5.7	8.7	11.8	14.8	17.7	20.7	23.7	29.7	41.7	47.7	59.7	89.7	119.8	149.8				
6.0	35mm								5.4	6.8	9.6	10.9	13.7	20.6	27.4	34.3	41.2	54.9	68.7	82.5
	35mm Anamorphic								11.0	13.8	19.4	22.2	27.8	41.8	55.7	69.7	83.7	111.6	139.6	167.5
	2 x 2 Slide																			
	35mm		2.1	3.3	4.4	5.5	6.6	7.7	8.9	11.1	15.6	17.8	22.3	33.5	44.8	56.0	67.2			
	1⅝ x 1⅝		2.4	3.6	4.9	6.1	7.4	8.6	9.8	12.3	17.3	19.8	24.8	37.3	49.7	62.2	74.7			
	Lantern Slide	2.3	4.8	7.3	9.8	12.3	14.8	17.3	19.8	24.8	34.8	39.8	49.8	74.8	99.8	124.8	149.8			
8.0	35mm									5.1	7.2	8.2	10.2	15.4	20.6	25.7	30.9	41.2	51.5	61.8
	35mm Anamorphic									10.3	14.5	16.6	20.8	31.3	41.8	52.2	62.7	83.7	104.6	125.6
	2 x 2 Slide																			
	35mm		1.6	2.4	3.3	4.1	4.9	5.8	6.6	8.3	11.7	13.4	16.7	25.1	33.5	42.0	50.4	67.2		
	1⅝ x 1⅝		1.7	2.7	3.6	4.6	5.5	6.4	7.4	9.2	13.0	14.8	18.6	27.9	37.3	46.6	56.0	74.7		
	Lantern Slide		3.5	5.4	7.2	9.1	11.0	12.9	14.7	18.5	26.0	29.7	37.2	56.0	74.7	93.5	112.3	149.8		
10.0	35mm										5.7	6.5	8.2	12.3	16.4	20.6	24.7	32.9	41.2	49.4
	35mm Anamorphic										11.6	13.3	16.6	25.0	33.4	41.8	50.2	66.9	83.7	100.4
	Lantern Slide		2.8	4.3	5.8	7.3	8.8	10.3	11.8	14.8	20.8	23.8	29.8	44.8	59.8	74.8	89.8	119.8	149.8	
20.0	Lantern Slide				2.8	3.5	4.3	5.0	5.8	7.3	10.3	11.8	14.7	22.3	29.8	37.3	44.8	59.8	74.8	89.8

These processes can be classified into two categories according to film formats used,[46-48] namely:

1. Standard 35mm film plus optical devices for wide screen presentation.
2. 65mm camera negative and 70mm release print methods.

There were several special processes in use utilizing multiple negative and multiple print methods such as in Cinerama with three 35mm prints or Circarama whereby nine camera originals and prints are used for a 360-degree picture presentation method.

The aim of these systems is to obtain a "wrap-around" or "audience participation effect" to enhance the story telling impact. Due to inherent defects in "lacing" projection methods such as Cinerama it is now attempted to utilize a single 70mm print and project it into the deeply curved screen. A more recent addition to these special processes is called "Dimension 150," meaning that a 150-degree field of view can be used and projected to the viewer. Numerous other systems have been published, most of which did not reach commercial application for entertainment purposes.

Projection Booths[49]

For detailed information on the design and facilities recommended for the projection room and the screen presentation in theatres it is suggested that the reports of the Committee on Theatrical Projection Technology of the Society of Motion Picture and Television Engineers be studied in the *Journal of the Society of Motion Picture and Television Engineers.*

Required Light Output of Projectors

In order to determine the required light output of a projector, it is necessary to know the picture size that satisfies the viewing conditions and the average reflectance at the applicable viewing angles of the screen to be used. With this information, the lumens required to meet the luminance recommendations can be calculated by the formula:

$$\Phi_s = \pi^* \times \frac{L_c \times K_d \times A}{L_f}$$

where Φ_s = Luminous flux reaching the screen, in lumens.

L_c = Luminance of the screen at its center in candelas per square meter.

K_d = Distribution weighting factor. This factor will be 1.00 for a screen which is uniformly illuminated; for the more general situation of side-to-center distributions of 60 to 80 per cent in illuminance, factors of 0.72 to 0.86 are suggested.

A = Area of screen in square meters.

L_f = Luminance factor. The ratio of the luminance of the screen to the luminance of a nonabsorbing perfect diffuser receiving the same illumination. This factor may be a function of the angles of illumination and viewing.

* π is omitted when luminance is in footlamberts and area is in square feet.

Classroom Projection. Lumens-at-screen values to satisfy the recommended luminance values for classroom projection are given in Fig. 11-24 for several screen sizes. Only one set of values is given for beaded screens because the luminance differences encountered over the range of viewing positions embrace the recommended luminance range.†

† For comprehensive information see: "Foundation for Effective Audio Visual Projection," Kodak Pamphlet No. S-3, Eastman Kodak Co., Rochester, NY 14650 and Reference 50.

Fig. 11-24. Screen Requirements for Classroom Projection

1. Screen Lumen Requirements

Screen Size		Matte Screen		Beaded Screen
Meters	Feet	Lumens for 17 cd/m² (5 fL)	Lumens for 70 cd/m² (20 fL)	Lumens for 17–70 cd/m² (5–20 fL)
.76 x 1.00	2.5 x 3.33	55	210	45
.91 x 1.22	3 x 4	75	305	65
1.14 x 1.52	3.75 x 5	120	475	105
1.37 x 1.82	4.5 x 6	170	690	150
1.60 x 2.13	5.25 x 7	235	940	205
1.82 x 2.44	6 x 8	305	1225	265
2.06 x 2.74	6.75 x 9	385	1540	340
2.29 x 3.05	7.5 x 10	480	1915	415
2.74 x 3.66	9 x 12	690	2750	600
3.20 x 4.27	10.5 x 14	935	3745	815
3.66 x 4.88	12 x 16	1230	4920	1070

2. Screen Size Requirements[33]

Picture Width		Maximum Audience Size	
Meters	Feet	Matte Screen	Beaded Screen
1.0	3.3	35	20
1.3	4.2	50	30
1.5	5.0	75	45
1.8	5.8	100	60
2.1	7.0	150	90

REFERENCES

1. Hatch, A. J.: "Updating the Follow Spot," *Light. Des. Appl.*, Vol. 4, p. 54, March, 1974.
2. Clark, C. N. and Neubecker, T. F.: "Evolution in Tungsten Lamps for Television and Film Lighting," *J. Soc. Motion Pict. Telev. Eng.*, Vol. 76, p. 347, April, 1967.
3. Levin, R. E.: "New Developments in Tungsten-Halogen Lamps," *Ind. Photogr.*, Vol. 17, p. 38, November, 1968.
4. Lemons, T. M. and Levin, R. E.: "Tungsten-Halogen Replacement Lamps for Standard Incandescent Types," *J. Soc. Motion Pict. Telev. Eng.*, Vol. 77, p. 1194, November, 1968.
5. Lemons, T. M. and Levin, R. E.: "The Rating Problem-Lamps in Luminaires," *J. Soc. Motion Pict. Telev. Eng.*, Vol. 78, p. 1064, December, 1969.
6. Schelling, W. F.: "HID Lamps for Television Remotes," *Light. Des. Appl.*, Vol. 9, p. 2, April, 1979.
7. Lemons, T. M.: "HMI Lamps," *Light. Des. Appl.*, Vol. 8, p. 32, August, 1978.
8. Glickman, R.: "CID: The Latest Compact Source AC ARC Lamp for Film and Television," *Light. Dimens.*, Vol. 4, p. 10, May, 1980.
9. Rubin, J. E. and Crocken, W. E.: "Q-File Random Access Memory Control for Theatre and Television," *J. Illum. Eng. Soc.*, Vol. 1, p. 329, July, 1972.
10. Pincu, T. L.: "Memory-Assisted Dimming," *Light. Des. Appl.*, Vol. 4, p. 50, March, 1974.
11. Pearlman, G.: "Functional Criteria for Memory Lighting Control Systems," *Light. Des. Appl.*, Vol. 9, p. 27, March, 1979.
12. Garrard, M., Ghent, E. and Seawright, J.: "A High-Speed Digital Control System," *Light. Des. Appl.*, Vol. 4, p. 14, March, 1974.
13. Miller, K. H. and Wittman, L. J.: "A Dimmer-Per-Circuit Approach to Stage Lighting," *Light. Des. Appl.*, Vol. 9, p. 29, March, 1979.
14. Shearer, C. W.: "Which Dimmer Curve—Why?," *J. Illum. Eng. Soc.*, Vol. 1, p. 325, July, 1972.
15. Otto, F. B.: "A Curve for Theatrical Dimmers," *Light. Des. Appl.*, Vol. 4, p. 44, March, 1974.
16. Committee on Theatre, Television and Film Lighting of the IES: "Lighting for Theatrical Presentations on Educational and Community Proscenium-Type Stages," *Illum. Eng.*, Vol. 63, p. 327, June, 1968. Bentham, F.: *The Art of Stage Lighting*, Taplinger Publishing Company, New York, 1969. Committee on Theatre, Television and Film Lighting of the IES: "Stage Lighting—A

Guide in the Planning of Theatre and Public Building Auditoriums," To be published.
17. Davis, B.: "Frontlight Positions—An Informal Plea for Diversity," *Light. Des. Appl.*, Vol. 5, p. 62, June, 1975.
18. Tawil, J. N.: "Staging with Light Patterns and Scenic Projections," *Light. Des. Appl.*, Vol. 8, p. 27, January, 1978.
19. Gill, G. and Sorensen, C. E.: "Making Available Light Available," *J. Soc. Motion Pict. Telev. Eng.*, Vol. 75, p. 310, March, 1966. "Elements of a Successful Meeting Area," *Film and Audio-Visual Annu.*, 1966.
20. Moody, J. L.: "'Hanging' a One-Night Stand," *Light. Des. Appl.*, Vol. 9, p. 22, March, 1979. Fiorentino, I.: "Lighting for Mixed Media," *Light. Des. Appl.*, Vol. 4, p. 22, March, 1974.
21. Subcommittee on Lighting for the Vidicon Camera of the Theatre-Television Committee of the IES: "Lighting for the Vidicon Camera," *Illum. Eng.*, Vol. LVIII, p. 387, May, 1963.
22. Quinn, S. F. and Wachholz, E.: "The Design of Film Review Rooms for Color Television," *J. Soc. Motion Pict. Telev. Eng.*, Vol. 80, p. 93, February, 1971.
23. Canadian Telecasting Practices Committee, CTP-1: "Viewing Rooms for Evaluation of 16mm Color Films for Television," *J. Soc. Motion Pict. Telev. Eng.*, Vol. 78, p. 483, June, 1969.
24. European Broadcast Union: "Viewing Conditions for the Appraisal, by Means of Optical Projection, of Colour Films Intended for Television Presentation," Tech 3091-E, September, 1970.
25. Society of Motion Picture and Television Engineers: "Evaluation of Color Films Intended for Television," SMPTE Recommended Practice, RP 41-1974, July, 1974.
26. Eastman Kodak Company: "The Television Film Preview Room," Pamphlet S-1, Motion Picture and Audiovisual Markets Division, Rochester, NY, 1970.
27. Jones, L. A.: "Measurement of Radiant Energy with Photographic Materials," *Measurement of Radiant Energy*, Forsythe, W. E., Editor, McGraw-Hill Book Co., Inc., New York, 1937.
28. Jones, B. F.: "Good Color Slides without Gadgetry," *Illum. Eng.*, Vol. LVIII, p. 116, March, 1963. Ulrich, J. D.: "The Lighting of Lighting," *Light. Des. Appl.*, Vol. 4, p. 33, March, 1974.
29. Levin, R. E. and Lemons, T. M.: "Application of Tungsten Halogen Lamps in Theatrical Luminaires," *Illum. Eng.*, Vol. 64, p. 47, January, 1969.
30. Stote, H. M. (Editor): *The Motion Picture Theatre*, Society of Motion Picture and Television Engineers, New York, 1948.
31. Guth, S. K., Logan, H. L., Lowry, E. M., MacAdam, D. L., Schlanger, B., Hoffberg, W. A. and Spragg, S. D. S.: "Screen Viewing Factors Symposium," *J. Soc. Motion Pict. Telev. Eng.*, Vol. 57, p. 185, September, 1951.
32. Allen, C. J.: "Lighting the School Auditorium and Stage," *Illum. Eng.*, Vol. XLVI, p. 131, March, 1951.
33. Estes,R. L.: "Effects of Stray Light on the Quality of Projected Pictures at Various Levels of Screen Brightness," *J. Soc. Motion Pict. Telev. Eng.*, Vol. 61, p. 257, August, 1953. Eastman Kodak Company: "Legibility Standards for Projected Material," *Kodak Sales Service Pamphlet No. S-4, 1956.* "The Foundation for Effective Audio-Visual Projection," *Kodak Sales Service Pamphlet*, Rochester, New York.
34. Tuttle, C. M.: "Density Measurements of Release Prints," *J. Soc. Motion Pict. Telev. Eng.*, Vol. 26, p. 548, May, 1936.
35. Lozier, W. W.: "Reports on Screen Brightness Committee Theatre Survey," I. *J. Soc. Motion Pict. Telev. Eng.*, Vol. 57, p. 238, September, 1951; and II. *J. Soc. Motion Pict. Telev. Eng.*, Vol. 57, p. 489, November, 1951.
36. Berger, F. B.: "Characteristics of Motion Picture and Television Screens," *J. Soc. Motion Pict. Telev. Eng.*, Vol. 55, p. 131, August, 1950.
37. D'Arcy, E. W. and Lessman, G.: "Objective Evaluation of Projection Screens," *J. Soc. Motion Pict. Telev. Eng.*, Vol. 61, p. 702, December, 1953.
38. Hill, A. J.: "A First Order Theory of Diffuse Reflecting and Transmitting Surfaces," *J. Soc. Motion Pict. Telev. Eng.*, Vol. 61, p. 19, July, 1953.
39. "Screen Luminance and Viewing Conditions for Indoor Theatre Projection of Motion-Picture Prints," PH22.196-1978, American National Standards Institute, New York.
40. "Minimum Screen Luminance for Drive-in Theatres," SMPTE Recommended Practice RP12-1972, *J. Soc. Motion Pict. Telev. Eng.*, Vol. 81, p. 929, December, 1972.
41. Committee on Nontheatrical Equipment of the SMPTE: "Recommended Procedure and Equipment Specifications for Educational 16mm Projection" *J. Soc. Motion Pict. Telev. Eng.*, Vol. 37, p. 22, July, 1941.
42. "Planning Schools for Use of Audio-Visual Materials": No. 1, "Classrooms," July, 1952; No. 2, "Auditoriums," February, 1953; and No. 3, "Audio Visual Instructional Materials Center," January, 1954, National Education Association, Dept. of Audio-Visual Instruction, Washington, DC.
43. Will, Jr., P: "Eyes and Ears in School," *Architect. Rec.*, Vol. 99, p. 66, February, 1946.
44. Vlahos, P.: "Selection and Specification of Rear-Projection Screens," *J. Soc. Motion Pict. Telev. Eng.*, Vol. 70, p. 89, February, 1961.
45. Lowry, E. M.: "Screen Brightness and the Visual Functions," *J. Soc. Motion Pict. Telev. Eng.*, Vol. 26, p. 490, May, 1936.
46. Beyer, W.: "Wide Screen Systems," *Am. Cinematogr.*, p. 44, October, 1960.
47. Beyer, W.: "Wide Screen Production Systems," *Am. Cinematogr.*, p. 296, May, 1962.
48. *Wide Screen Motion Pictures,* Society of Motion Picture and Television Engineers, New York.
49. Beyer, W.: "The Research Council Developments for Better Theatre Projection," *J. Soc. Motion Pict. Telev. Eng.*, Vol. 69, p. 792, November, 1960.
50. School and College Committee of the IES: "Guide for Lighting Audiovisual Areas in Schools," *Illum. Eng.*, Vol. LXI, p. 477, July, 1966.

Outdoor Lighting Applications

SECTION 12

Building exteriors and surrounds are lighted for both utilitarian and decorative purposes. This section is concerned with the *decorative aspects* of floodlighting buildings, monuments, fountains, exhibitions and gardens, and with the *attraction and utilitarian* aspects of service station lighting. For other utilitarian outdoor lighting applications see Section 13 for Sports and Recreational Area Lighting, Section 14 for Roadway Lighting, Section 15 for Aviation Lighting, and Section 17 for Lighting for Advertising.

Floodlighting Design Procedure

The following procedure may be helpful in designing a floodlighting installation:

1. Determine the decorative effect desired.
2. Consider codes, including those relating to power limits, operating times, light trespass, etc., as well as electrical.
3. Determine the location of the floodlights.
4. Determine the illuminance value to be used (See Fig. 12-1 and Fig. 2-2).
5. Select the appropriate equipment (lamps, luminaries and controls).
6. Determine the number of units required and the wattage of the lamps and the controls to be used.
7. Check the uniformity and coverage of lighting (Formulas and tabular detail are given in Section 9 in the 1981 Reference Volume).
8. Prepare operating and maintenance programs.
9. Check compliance with codes.

Building Floodlighting

The floodlighting of stores, shopping centers, offices and other places of business is intended to attract attention to these buildings and to create a favorable impression with passersby. In this sense, the floodlighting of these buildings is often a subtle and dignified, yet highly effective, form of advertising.

Public buildings, churches and monuments are generally lighted as an expression of civic pride, although here, too, the advertising aspect is present if the end result is to attract new people, business, and industry to the community.

Decorative floodlighting is essentially an art rather than a science. While calculations of luminance will generally be necessary, successful floodlighting depends to a large extent on the designer's ability to manipulate brightness relationships, textures and colors. Thus, floodlighting is part of the architectural vocabulary and as such can be utilized to help create a nighttime image of a structure, sculpture or garden, thereby extending the hours of their usefulness.

Principles of Floodlighting Design. The first step in floodlighting design, as listed in the above design procedure, is to establish the effect desired, or more accurately, to investigate the effects possible. The daylighted appearance may be helpful. Daylighting is usually a combination of strongly directional sunlight and diffuse sky light. The color of the former is warm, while that of the latter is cool. Shadows, therefore, are never "black", but simply less bright and bluer. Daylighting varies continually with the time of day and year and with the weather. Floodlighting, by contrast, is highly controllable and can therefore be utilized to present the building in a continuously favorable aspect, or in a character that is not seen during the day. See Fig. 12-2.

The major viewing locations or directions may help determine a floodlighting approach. If the building is to be seen primarily from moving automobiles, these viewing locations will most often be at some considerable distance and generally not head-on. The over-all impression cre-

NOTE: References are listed at the end of each section.

Fig. 12-1. Recommended Illuminances for Floodlighting*

Surface Material	Reflectance in Per Cent	Surround: Bright	Surround: Dark
		Recommended Level in Lux [Footcandles]	
Light marble, white or cream terra cotta, white plaster	70–85	150[15]	50[5]
Concrete, tinted stucco, light gray and buff limestone, buff face brick	45–70	200[20]	100[10]
Medium gray limestone, common tan brick, sandstone	20–45	300[30]	150[15]
Common red brick, brownstone, stained wood shingles, dark gray brick	10–20†	500[50]	200[20]

* See also Fig. 2-2. For poster panels, see Fig. 17-14.

† Buildings or areas of materials having a reflectance of less than 20 per cent usually cannot be floodlighted economically, unless they carry a large amount of high-reflectance trim.

ated by the major elements of the building, especially the upper areas, may be the most important consideration. At close viewing distance the main concern may be with the ground level 1acade, sidewalk and landscaping and with the effect of light on the building material and construction details.

It is usually desirable to locate the main, or key, lighting so that there will be some modeling effect. See Fig. 12-3. If floodlights are aimed on the line of sight from the viewer to the building, the effect will tend to be flat and often uninteresting.

The key lighting need not be a single source or several sources located at a single point. Modeling can be achieved if a series or line of floodlights are aimed in the same direction.

Deep shadows may be softened by low levels of diffuse floodlighting at an angle relative to the key lighting. A cool color may help recall the daylighted appearance.

Where it is desirable to have key lighting from two directions, modeling effects may be achieved by using contrasting tints or saturated colors.

Location of Floodlighting Equipment. Physical limitations imposed by either the relationship of the building to its surrounding or by local regulations may drastically limit the designer in the number of solutions or effects available. In general, four locations may be considered: on the building itself; on adjacent ground (see Fig. 12-4); on poles or ornamental standards (see Fig. 12-5); and on adjacent buildings. To completely express the structure in light, several, or all, of these locations may be required.

Floodlighting located close to the surface to be lighted and aimed at a grazing angle will tend to emphasize the texture of the surface (see Fig. 12-6), especially when viewed from nearby. However, defects in the surface—ripples, dents, misalignment of building panels, etc.—will also tend to be emphasized. Depending on the spacing and nearness to the surface, "scallops" from individual units may also be visible although these are not necessarily objectionable. Mounting the floodlights farther and farther from the surface de-emphasizes or flattens the texture and usually improves brightness uniformity.

Setbacks generally offer ideal opportunities to mount and conceal floodlighting. See Fig. 12-7. It is desirable that each of the setback areas be lighted to a different level or should vary in brightness vertically so as to retain the setback appearance. Increasing thc average luminances

Fig. 12-2. The White House in Washington, D.C., a revered national symbol, is impressive at night as well as by day due to the artful use of floodlighting techniques using luminaires concealed from spectators' view.

Fig. 12-3. Key lighting from a single location in front and to the side reveals character of masonry and produces dramatic light and shadow.

Fig. 12-4. Two examples of ground-mounted floodlighting. High pressure sodium floodlighting, on the left, produces essentially a uniform brightness pattern over the entire surface of a large building. On the right, architectural treatment is reenforced with a single floodlight per bay, producing a regular pattern which predominates when seen from a distance. Masonry texture becomes important for close viewing.

Fig. 12-5. The cool color effect of metal halide lamps, pole-mounted to light the front and ground-mounted to light the sides, is contrasted with the warmth of tungsten-halogen lamps lighting the portico from the ceiling behind the columns.

of the upper setback areas 2 to 4 times over that of the lower areas is generally considered to create an apparent equality of brightness of all the areas and to increase the impression of height.

Direct or reflected glare from floodlights can detract from even the most interesting floodlighting installation and can be a source of annoyance to neighbors. The lamp and reflector should be shielded or louvered so that brightness cannot be seen from any normal viewing locations. Special attention should be given to entrance areas of buildings used at night so as to avoid direct glare which could make steps and curbing hazardous to pedestrians.

Floodlights should be located or shielded so that units do not light adjacent units thereby revealing their presence.

Landscaping, walls or wells are highly useful in minimizing direct glare. In addition, they help conceal the equipment during the day, a most important aspect of the total design.

Reflected glare from specular surfaces such as polished marble, tile, glass and metal trim should also be considered. Sight lines should be checked to insure that reflected images of lamps or reflectors will not detract from the over-all effect. This is especially important in areas where there is considerable pedestrian traffic and the surface is seen from nearby.

Fig. 12-7. Setback-mounted narrow beam floodlights, concealed behind a parapet, light the upper stories of this building.

Interior Floodlighting. Modern construction, utilizing large expanses of glass, as windows or curtain walls, can often be expressed most effectively at night in terms of interior brightnesses. The simplest technique is to operate part of the interior lighting of spaces next to windows. In spaces which are electrically space-conditioned this technique may in some cases also provide a measure of the nighttime heating required in cold weather.

Fig. 12-6. Grazing floodlighting from units concealed in moat emphasizes the texture of the sculptured facade.

Where there is no control of the interior lighting, such as in tenanted office buildings, separate interior lighting may be provided. This could take the form of supplementary wall and ceiling lighting such as coves and valances. Lamps can often be concealed in soffits, in window headers, or in sills. These would light the window reveal and blinds or curtains which are kept closed at night.

A major advantage of the interior approach to floodlighting is that indoor lighting equipment can be utilized and maintenance can be simplified.

Illuminance Values. To serve as a design and calculation guide, illuminance values for building floodlighting are given in Fig. 12-1. Because of the decorative and advertising nature of building floodlighting, these should be considered as guides only. Variation from these values is to be expected depending on the type of building, its location, and the ultimate purpose for floodlighting.

Color in Floodlighting. The values given in Fig. 12-1 are based on "white" light. Where saturated colors are desired, color filters can be applied. Since these produce the wanted color by absorbing or reflecting the unwanted colors, the amount of light transmitted by the filter is sharply reduced. The transmittance of saturated color filters usually falls within the following ranges: amber, 40 to 60 per cent; red, 15 to 20 per cent; green, 5 to 10 per cent; and blue, 3 to 5 per cent. Fig. 12-8 indicates the factors by which incandescent filament lamp wattage must be increased when it is desired to provide equal illuminance in white and color. Relatively less colored light than white light is needed for equal advertising or decorative effect. The second line of Fig. 12-8 gives factors by which clear-bulb incandescent filament lamp wattage should be multiplied in order to achieve an advertising or decorative effect in color comparable to that obtained with a given wattage lamp emitting white light.

Actually even "white" light sources produce a color of light that is characteristic with the source. Incandescent lamps, including the tungsten-halogen types, produce light which is warm in appearance, rich in red, and weak in blues and greens. Most "white" fluorescent lamp colors used in floodlighting appear cooler or bluer than incandescent. Fluorescent lamps are also efficient producers of colored light. The high intensity discharge sources, such as mercury, metal-halide, and high pressure sodium, have individual color characteristics. For example, clear mercury emphasizes blues and greens, deluxe white mercury is similar to cool-white fluorescent, and high pressure sodium is warmer in appearance than incandescent, but is weak in the deep reds and blues. Low pressure sodium is the most efficient producer of monochromatic yellow light. See Section 8 of the 1981 Reference Volume. These color differences can be used to visually separate or accentuate elements of building design, or a combination of sources can be employed to create the desired color.

Maintenance. Although maintenance is important in most lighting systems, the ultimate success of a floodlighting installation often depends upon just how well it is maintained. To this end the functions of relamping, cleaning and re-aiming should be considered as a fundamental part of the total design.

Fig. 12-8. Approximate Factors by Which Clear Bulb Incandescent Filament Lamp Wattage Must Be Multiplied to Compensate for the Absorption of Various Color Filters

Desired Effect Relative to Clear Bulb	Approximate Multiplying Factor			
	Amber	Red	Green	Blue
Equal illuminance	2	6	15	25
Equal advertising or decoration	1.5	2	4	6

Monument and Statue Floodlighting

The design of floodlighting for monuments (see Figs. 12-9 and 12-10) and statues aims at the achievement of a natural lighted appearance. See Fig. 7-26. The relationship of shadows and brightness is of utmost importance. This is particularly true where the human form is concerned. When obelisk-type structures are lighted, depth can be maintained by adjusting the brightness of each face so that there will be a contrast between each face in the field of view of an observer at any one time.

Fig. 12-9. The frigate Constellation, a national historical monument, is lighted by tungsten-halogen floodlights of various beam spreads, mounted on the mooring dolphins and aimed upward.

Fig. 12-10. A natural daylight appearance is achieved at night at the Piazza dei Miracoli in Piza by use of strategically placed metal halide equipped floodlights.

Exposition and Public Garden Floodlighting

In the lighting of spectaculars, such as expositions and public gardens, the primary considerations are to create beauty and to enhance form and design of surface. Lighting is utilized to reveal or conceal, to accent or subdue, to create moods, to control traffic, to provide heat (infrared radiation) and to illuminate and create animation. See Fig. 12-11.

Light can be used as a tool to transform a daytime environment to a completely new setting at night. Buildings and grounds become a fairyland under the spell of properly applied lighting schemes. For example, the ordinary daytime appearance of a group of buildings and exhibits can at night, under the creative genius of the architect and illuminating engineer, become a delight to the visitor, designed to transport him into an environment of specific moods. Rather than creating areas of high brightness for attraction and resultant gaudiness, often the goal is to achieve an elegant atmosphere of mystery. See Fig. 12-12.

When lighting outdoor areas for decoration, it is not necessary to maintain the natural daytime appearance. Originality and novelty are the dominant factors. Beautiful and unusual effects may be achieved by imaginative application of the almost infinite list of light sources and equipment available today. The engineer and architect who may be involved in the over-all planning should maintain extremely close cooperation because the ideas and equipment used are limited only by their mutual imagination.

General Principles. Decorative effect is the primary objective rather than efficacy in terms

Fig. 12-11. The major program area at Expo '70 in Osaka, Japan featured the "Grand Roof," a steel lattice structure lighted from below by 800 metal halide floodlights.

Fig. 12-12. Translucent glass screen walls highlight traditional Japanese architecture at Expo '70 in Osaka. Fluorescent lamps inside the walls are spaced to provide maximum brightness at the front and gradually decreasing brightness down the sides.

of lumens per watt as is usually the case in forms of functional lighting although power budget limitations, where applicable, may influence the final design.

Probably the most important single rule that should be followed in decorative as well as functional type lighting installations is to conceal the light source. (See Fig. 12-13 and Location of Floodlighting Equipment on page 12-2.) Lighting equipment may be shielded from view by trees, shrubs, rocks, buildings, and structures, or shrubbery planted expressly for this purpose. Conventional equipment may be placed in a suitable cowling, which takes on an appearance in keeping with that of the area. For example, an optical assembly might be mounted inside of a hollowed stone or the cowling may be made to simulate a toadstool. In addition, all equipment might be painted as an aid in camouflaging.

Sometimes bright spots appearing immediately in front of lighting equipment mounted on the ground can be virtually as objectionable as viewing the light source itself. Also, spill light on surrounding branches and foliage may create undesirable effects. Therefore, easily adjusted equipment is preferred for garden and architectural lighting. Sometimes special louvers and shields will be required.

Illuminance Values. Every scene at an exposition or public garden has a center of attraction. See Fig. 12-11. The lighting installation should be designed to create higher illuminances at that location with adequate surrounding light to prevent this attraction from appearing detached from the immediate environs.

Color. Considerable caution should be exercised when selecting colors of light sources. For

Fig. 12-13. Integrated landscaping and building lighting with light sources hidden from view.

example, clear mercury sources should not be used to illuminate a reddish brick facade. However, clear mercury sources are excellent for enhancing the green of foliage. During the autumn when leaves become yellow, red, and brown, incandescent filament sources, which are rich in red and yellow, are a better choice than the clear mercury lamp. At the same time, the imaginative designer should not overlook the possibility of using other high intensity discharge sources which efficiently produce a wide range of less subtle colors including yellow, red, green and blue. Furthermore, the entire rainbow of colors is available from fluorescent sources.

In choosing the light sources to illuminate flower beds which contain a variety of delicate natural colors care should be taken and, in some cases, it may be found necessary to experiment before the final choice is made. Generally, the high color rendering fluorescent lamp colors should be considered. Pale blue light on some structures will simulate moonlight; warmer colors may be used in foreground areas to accentuate the effect.

Fountains

The use of water has been an important architectural element for centuries and certainly has been one of the major attractions at world's fairs.[8] Still water acts as a mirror and will reflect clear images of lighted objects. See Fig. 12-12. When churned into spray and foam, water is an excellent diffuse reflector and will absorb incident light and will appear to change color to match that of the light.

A single jet may be made attractive by installing directly beneath it, in a suitable water-tight enclosure, a narrow beam of light projecting along the stream. In the case of more complex displays, the effects which may be obtained by varying water flow, number of in-service jets, and colors of light are unlimited.

Floodlight Types. Attempts have been made to illuminate sprays and jets of water from floodlights operating in air as well as under water. Although the initial investment is usually more, the results are generally more desirable when underwater locations are used.

When wide area water effects are to be lighted such as water curtains, sprays and jet rings, wider beam spread floodlights should be used. Narrow beams are needed to adequately illuminate high projected streams.

The cascade formed by water spilling over the lip from an inner basin can also be effectively lighted by installing colored lamps under the lip. The lip should be designed to churn the curtain of water thereby making it turbulent and diffuse in appearance. Lamp spacing around the basin should not exceed the distance back from the waterfall. For optimum uniformity in apparent brightness, fluorescent strips may be preferred in this application.

Size and Location of Fountains. As a practical point, the radius of the outer pool is usually not less than the maximum water projection height. In extremely windy locations the radius may be increased to two times the maximum water projection, so as to not continually wet the landscape and/or spectators.

If a fountain is to be illuminated the actual location is best selected where there is a minimum of ambient light, especially when colored light is to be used. Any environmental light tends to wash out the relatively low brightness produced by underwater color floodlighting either by reducing its contrast with the surroundings or by adapting the spectator's eyes to a higher level.

Christmas Lighting

Christmas is the one season of the year when everyone is especially conscious of decorative

Fig. 12-14. An entrance is lighted with 7-watt C-7½ lamps in foil covered stylized tree on door and 10-watt C-9½ lamps in festooning and hard board tree in front of entrance. Two 150-watt PAR lamps are used, a green one behind the tree and a clear spot out front. This produces a shadow tree in green.

Fig. 12-15. Christmas trees on the Ellipse in back of the White House—lighted with D27 clear flasher lamps and with 10S11 sign lamps.

lighting. See Figs. 12-14 and 12-15. Several publications, appealing especially to the homemaker, offer helpful hints on the many uses of miniature lamps at Christmas time. Other applications include the use of miniature lamps for other festive occasions and for colorful patio and garden lighting. For the latter application low-voltage miniature equipment is available. It should be noted that not all string sets and devices have Underwriters Laboratories, Inc. approval.

Service Station Lighting

The design of service stations tends toward a specific architectural style, such as colonial, ranchtype structures, etc., with landscaping, fencing, walkways, and other treatments utilizing outdoor lighting. Light sources for such installations should be considered in the initial design.

The basic objectives of service station lighting are:

1. To aid the rapid identification of the station and its product.
2. To facilitate safe entrance and exit into and out of the station.
3. To provide adequate light on the pump island and its adjacent area to permit the attendant or customer to easily perform tasks (see Fig. 2-2).
4. To provide a well lighted building interior and exterior.
5. To provide an over-all installation that is attractive to the prospective user.

Additional lighting objectives are:

A. To provide driveway lighting between the approach and island to complete the traffic pattern.
B. To illuminate the auxiliary service areas.
C. To illuminate additional driveway areas.
D. To illuminate parking areas.

See Fig. 12-16 for a correlation of service station areas and the above objectives.

Identification. The first objective to be considered in designing service station lighting is the identification of the product, services rendered, and facilities available. This is achieved by adequately illuminating the product identification sign and the principal vertical surfaces of the station and canopies.

Signs. Location, legibility and brightness are the primary factors to be considered in sign lighting. Signs should be located to provide maximum advertising and identification value from all possible traffic angles. Illuminances for both internally and externally lighted signs are given in Section 17, on outdoor signs.

When utilizing a sign pole for the mounting of approach or driveway lighting units, caution should be taken so that the lighting units will not distract from the advertising appeal of the sign.

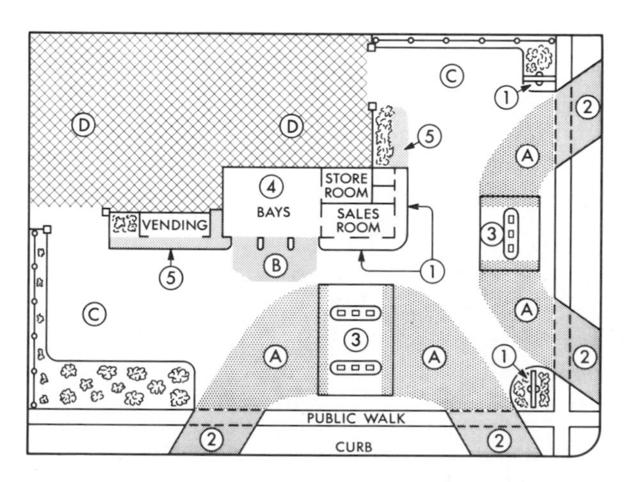

Fig. 12-16. Service station plot plan defining areas for lighting. Numbers and letters refer to text above.

The use of a minimum pole height of 4.5 meters (15 feet) should permit the mounting of a properly designed approach light below the sign where there is less tendency to distract from the sign itself. When a sign pole is located 3 meters (10 feet) or more from the approach curb, use of a separate light pole should be considered.

Building Lighting. Lighting the prominent faces of the station building and canopies serves to advertise the station's presence and indicate that the station is open and ready for business. Soffit type lighting is a commonly used method for the lighting of the building face. The internal lighting of spires, cupolas, canopy edges, etc., as well as the accenting of the building's architectural features with appropriate floodlighting also is used. See Building Floodlighting, page 12-1.

Approach Lighting. Adequate lighting of the approaches from the edge of the highway or street curb to the station property is a vital part of good service station lighting. The limits to the approaches should be readily identified to permit the motorist to see them well in advance. While some luminaire brightness may be desirable from an attraction viewpoint, extreme care should be taken to insure that the type of luminaire or the aiming does not cause glare that will interfere with the vision of passing motorists or those entering or leaving the station. A minimum clearance of 4.3 meters (14 feet) should be contemplated where luminaries are designed to extend out over the approaches.

Pump Island Area. The pump island is a primary point of sale and its lighting should reflect its commercial importance. Sufficient illuminance should be provided in this area not only to enable the operator to perform his tasks quickly and accurately, but also to prominently illuminate the product dispenser and the merchandise displays in the area. As part of its evaluation, the types of pumps to be used (lighted or unlighted) should be considered, as well as the use of island canopies.

Building Interior. The building interior is the office and major work area of the station. The salesroom, which is used for merchandise displays, limited stock supplies and sales transactions should be illuminated in accordance with the requirements of a general office area. Areas for automatically vended foods, beverages, etc., also may be involved. Lighting in the service bay areas should be adequate to permit efficient performance of the various services rendered. Auxiliary lighting for merchandise displays also may be required. Well lighted rest rooms are essential to encourage cleanliness and acceptance by the public. Special lighting consideration should be given to the walkways and approaches to the rest rooms as well. For totally self service stations the office (cashier's location) should be lighted to be the focal point of the station.

Driveway. Limited driveway lighting (sections marked *A* in Fig. 12-16) completes the path-of-light from entrance to island to exit and adds to the active business like appearance of the station. Depending on the magnitude of the operation and services available, extended driveway lighting (C) plus auxiliary lighting (B) may be required to provide sufficient visibility for outdoor service operations such as replacing headlamps, servicing batteries, changing tires or installing chains. Where there is a patio or housing for vending machines within the driveway area, there should be adequate lighting to promote cleanliness and instill public confidence.

The Surround. The over-all objective in lighting a service station is to make it stand out by contrast from the areas around it, to attract the attention of the motorist in need of its services. In achieving these objectives, care must be taken that the type of equipment used, mounting heights, etc. do not create an annoyance, through glare and spill light, to surrounding neighbors. The requirements of applicable codes and ordinances must be considered in the selection of equipment as well as the economic factors of maintenance, operating costs and flexibility for future expansion.

REFERENCES

1. Benson, B. S., Jr.: "The Application of Floodlighting to Buildings," *Illum. Eng.*, Vol. XLIX, p. 367, August, 1954.
2. Eshelby, R. A.: "Outdoor Decorative Lighting Techniques," *Illum. Eng.*, Vol. L, p. 155, April, 1955.
3. Nightingale, F. B.: *Lighting as an Art*, Knight Publishing Co., Skyforest, California, 1962.
4. Schuler, S.: *Outdoor Lighting*, D. Van Nostrand Co., New York, 1962.
5. Nightingale, F. B.: *Garden Lighting*, Knight Publishing Co., Skyforest, California, 1958.
6. Gladstone, B.: *The Complete Book of Garden & Outdoor Lighting*, Harthside Press, Inc., New York, 1956.
7. Travis, B. A. and Faucett, R. E.: "Outdoor Lighting at Seattle World's Fair—Century 21," *Illum. Eng.*, Vol. LVII, p. 651, October, 1962.
8. Hamel, J. S., Langer, R. A., and Culter, C. M.: "New York World's Fair Lighting," *Illum. Eng.*, Vol. LIX, p. 644, October, 1964.
9. "Expo 67 and Its Lighting," *Illum. Eng.*, Vol. LXII, p. 453, August, 1967.
10. "Expo 70," *Illum. Eng.*, Vol. 65, p. 458, August, 1970.
11. Faucett, R. E.: "Engineering of Lighted Fountains," *Arch. Rec.*, Vol. 126, p. 218, July, 1959.

Sports and Recreational Areas[1, 2]

SECTION 13

The lighting of areas for sports and recreational activities, especially those located outdoors, involve problems not encountered in other fields of lighting.

SEEING PROBLEMS IN SPORTS

The objective of a sports lighting installation is to provide a luminous environment appropriate for the playing and viewing of the sport by controlling the brightness of the playing object and its background to the extent that the object will be visible regardless of its size, location, path and velocity, for any normal viewing position of spectator or player, in an energy efficient manner. In a majority of sports, this objective is achieved by the illumination of vertical rather than horizontal surfaces.

Objects To Be Seen

Dimensions and reflectances of typical objects to be seen in sports activities vary over a wide range resulting in different lighting requirements. For example, objects may vary in size from a man to a golf ball, and in reflectance from 1 per cent (a hockey puck) to 80 per cent (a golf ball). The combination of these sizes, reflectances, and how they are viewed in contrast with their backgrounds will influence the lighting design.

NOTE: References are listed at the end of each section.

Background Brightness

In many sports, the normal background against which an object must be viewed by a player comprises all surfaces or space above, below and on all sides of the player's position.

Because a ball or other object may move rapidly through the field of view, the background brightness, which is seldom uniform, will vary rapidly. For example, outdoors in daylight a baseball may be viewed against the relatively dark shaded grandstand area at one instant and in the next be silhouetted against the sky or sun. A football may be viewed against dark green grass, white snow, clear sky, a mottled pattern of spectators or a player's jersey.

With illumination from a few high candlepower soures concentrated on an outdoor playing field and filling the space above to a limited altitude only, most of the background area is relatively dark. Great care should be taken to be sure that, in addition to providing relatively uniform illuminance, the sources are placed so that a ball will seldom be viewed against any bright portions of the sources.

One of the most effective ways to reduce the effects of glare is to keep the luminance of the surround to a reasonable level. This can be done very effectively for indoor sports by the proper selection of reflectances and finishes for walls and ceilings. Control of the surround luminance is much more difficult in outdoor locations; however, a great deal can be done in this regard. Adequate light in the stands for safety considerations, light colored fences, together with provisions for providing some illumination on the ground immediately around the playing field, aid considerably in improving the surround conditions.

Observer Location

In designing lighting for sports, careful consideration should be given to the requirements and comfort of each of the three observer groups; players, officials and spectators, whose orientation with respect to the object of play differs. The normal fields of view of each group differ also, and in the case of player and official there may be no fixed location. The probable variation in location and field of view will be different for each sport.

In providing adequate illumination of proper quality for one group, no glare should be introduced into the field of view of the other two groups, if at all possible.

QUALITY AND QUANTITY OF ILLUMINATION

Diffuse illumination, such as that provided by an overcast sky on an outdoor playing field during the day, or that provided by an indirect lighting system in an interior with high reflectance ceilings, walls and floors, is considered to be of excellent quality for sports. Indoors, the design problems are quite similar to those encountered in other interior areas. Outdoors, the problem is more difficult, and it has been necessary to develop locations and mounting heights for concentrated sources which will provide satisfactory visibility of the object. See Fig. 13-1.

Number and Location of Sources

The shape and surface characteristics of the object to be seen and its probable orientation with respect to the observer are important factors in establishing satisfactory luminaire mounting heights and locations. Fortunately, balls with diffuse surface reflectance are the most common objects to be viewed. A point source located in such a position that its central axis forms an angle of not more than 30 degrees with the observer's line of sight (apex at the ball) will for practical purposes illuminate the entire side of the ball facing the observer. See Fig. 13-1a. If the angle is increased to 90 degrees, the ball will remain lighted over half its visible surface, as shown in Fig. 13-1b. Fig. 13-1c shows the same tennis ball lighted from above by two point sources that form angles of 85 degrees with the line of sight and from the ground by reflected light. It appears that the arrangement shown in Fig. 13-1c will provide illumination satisfactory for more observer locations than will the others. However, in the practical case, the specific source locations selected will be those which offer the best compromise between desired illumination diffusion and minimum glare in the majority of observer locations.

Uniformity of Illuminance. It is necessary that illuminances by fairly uniform (with no sharp changes in level) at points throughout the entire space above the playing area through which the object may travel, since a fast-moving object passing quickly from a light to a dark space will appear to accelerate. This occurs when there is inadequate overlap of floodlight beams. Such a condition distorts the players' judgment of the object's trajectory. In terms of horizontal illuminance, for acceptable uniformity (for sports in which play is skillful, the visual task is severe or there are likely to be spectators) the ratio of *maximum to minimum illuminances should not exceed three to one within a given area.*

Glare

The reduction or elimination of objectionable glare is one of the principal quality objectives in sports lighting. A floodlight is inherently a glare source, and whenever possible, it is imperative to diminish the effects of glare by locating the luminaires away from the normal lines of sight. The angle between the luminaire and the normal line of sight is affected both by the luminaire location in a horizontal plane and by the mounting height.

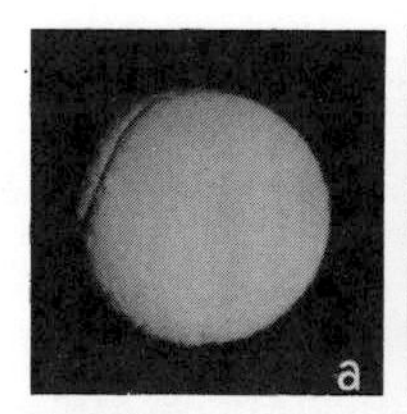

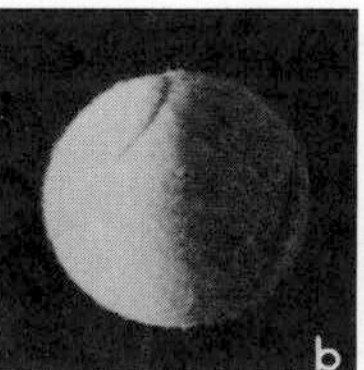

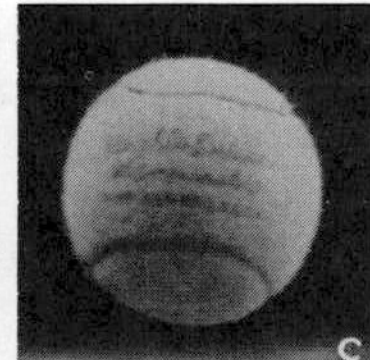

Fig. 13-1. Appearance of a tennis ball lighted in different ways: (a) by single source 30 degrees to right of line of sight; (b) by single source 90 degrees to left of line of sight; (c) by light from two sources above and at 85 degrees from line of sight, and by light reflected by the ground.

To remove all of the luminaires in an installation from the normal lines of sight of each observer group, however, is not always feasible. For many sports, the luminaire locations must reflect a compromise between reducing glare and providing light from the proper direction. Physical obstructions may also require changes from recommended luminaire locations. In all cases, a careful evaluation should be made of the possible lines of sight of both players and spectators.

In those instances where glare cannot be avoided in positioning the luminaires, consideration should be given to the type of lamp, the use of glare shields, lenses or some form of louvering to reduce the luminaire luminance. These means can also be used to reduce the amount of spill light into surrounding areas, if the sports installation, for example, is located in the heart of a residential zone. When shields are used, particularly the louver type, additional floodlights may be required to compensate for the reduced efficiency of the floodlights.

For unidirectional sports, such as bowling, racing, handball, archery, golf, etc., it is desirable and possible to provide much higher vertical illuminance from one direction as well as to locate the luminaires so that they are almost completely removed or shielded from the normal field of view.

The aiming of floodlights, even with correct luminaire locations and mounting heights, determines, to a large extent, whether the uniformity, direction and candlepower toward the eye are satisfactory. For information concerning the correct fundamentals of aiming floodlights, see page 13–18.

Illuminance Values

It is important that illuminances be sufficient for comfortable and accurate seeing, first, to enable the players to perform their visual task, and second, to enable the spectators and the audience viewing the event on television[3] to follow the course of the play. See also Section 11.

In those sports where large numbers of spectators are expected, such as in large football and baseball stadiums, the illuminance is determined by the amount required for the spectators in the row of seats farthest removed from the playing area to follow the course of the play. This condition may require several times the amount of light found satisfactory to the players. The illuminance suggested for sports (see Fig. 2–2, page 2–16) are those which are currently considered values of good practice, taking into consideration both players and spectators.

The illuminances in Fig. 2–2 are in most cases stated as horizontal lux (footcandles) in service. It is recognized that the vertical component of the illumination on the playing area is important in most sports. This is particularly true in the "aerial" games, where both players and spectators rely, to a considerable degree, on the vertical illumination on or near the playing area, and in some cases well above the playing area. The vertical components of illuminance have usually been found adequate where the horizontal illuminances meets the values in the table, except where noted otherwise, and the lighting equipment is positioned at mounting heights and locations conforming to accepted good practice, as covered later.

Daylighting for Sports

Daylight usually provides adequate illuminance to permit satisfactory participation in morning and afternoon outdoor contests, even on cloudy and overcast days. In the design of gymnasiums and other interiors used for daytime athletics, the daylighting principles set forth in Section 7 of the 1981 Reference Volume should be applied.

Windows. Windows in a gymnasium are not considered desirable, particularly behind the baskets of the basketball court and at either end of an indoor tennis court, as they are unsatisfactory in the illuminances that they provide, and are a potential blinding glare source. Skylights should be screened to prevent breakage. Since screening may have a very low transmittance, the utilization factor for screened apertures will be low (15 to 60 per cent, depending on screen transmittance characteristics).

LIGHT SOURCES

There are many light sources available for sports lighting applications. Each type has certain advantages and disadvantages and the proper selection will depend upon the particular requirements of the installation being considered, the economics and perhaps some personal preference of the system designer or owner.

A comparison of the major characteristics of the various light sources are shown in Fig. 13-2 which provides guidance for choosing the light source for a particular application. This chart must be tempered by the actual application for the light source which might alter the rating shown. For example, in the case of a low ceiling interior lighting installation, the "low" rating given in the lumen output per lamp and the degree of light control ratings for the 40-watt fluorescent source, might be considered most advantageous to obtain a low source luminance and a wide beam spread. More detailed information on the various light sources can be found in Section 8 of the 1981 Reference Volume.

Incandescent Lamps

The chief advantages of incandescent lighting are its low initial cost, good color rendering properties, and optical control capabilities. Disadvantages are shorter lamp life and lower efficacy (lumens per watt) as compared to the other light sources.

Fluorescent Lamps

Advantages of fluorescent lighting are its high luminous efficacy, long lamp life and good color rendering properties. Its inherently lower luminance offers a further advantage for some applications, especially those in which mounting heights are relatively low and short projection distances are acceptable. Such applications can include tennis, bowling, curling, trampolines, spectator stands and a variety of indoor sports.

High Intensity Discharge Lamps

The family of high intensity discharge lamps include mercury lamps, metal halide lamps and high pressure sodium lamps. Although each of these lamp types has its own specific characteristics, they have the following characteristics in common: long lamp life and high luminous efficacy when compared to incandescent lamps, a time delay and slow build up of light output when the lighting system is first energized or when there is a power interruption.

Metal halide lamps, because of their high efficacy and good color rendering properties are most often used for sports lighting applications. The requirements for color television broadcasting of sports events are best met by metal halide lamps, when cost of operation and energy consumption are considered.

High pressure sodium lamps have the highest efficacy of this group. Where energy consumption is of greater importance than color rendering, these are being applied to some sports lighting applications.

Mercury lamps with phosphor coatings provide good color rendering and long life, but are gradually giving way to the higher efficacy and better beam control possibilities of the metal halide lamps.

Special Design Considerations

In selecting a lighting system for use in a sports or multi-purpose recreational area, several im-

Fig. 13-2. Comparative Characteristics of Light Sources for Sports Lighting Purposes (There are Four Ratings for Each Characteristic—High, Good, Fair and Low)*

	Lumen Output Per Lamp	Efficacy	Life Expectancy	Color Acceptability	Degree of Light Control	Maintenance of Lumen Output
Incandescent	Low	Low	Low	High	High	Good
Tungsten-Halogen	Fair	Low	Low	High	High	High
Mercury	Good	Fair	High	Low	Good	Good
Phosphor Mercury	Good	Fair	High	Fair to Good	Fair	Fair
Metal Halide	High	Good	Fair	Good to High	Good	Fair
High Pressure Sodium	High	High	Fair	Fair	Good	Good
40-Watt Fluorescent	Good	Good	Good	Good to High	Low	Good
High-Output Fluorescent	Fair	Good	Good	Good to High	Low	Good
1500-mA Fluorescent	Good	Good	Fair	Good to High	Low	Fair

* See also Special Design Considerations.

portant factors should be given special consideration:
1. Over-voltage operation of incandescent and tungsten halogen lamps often can be used to economic advantage in sports lighting. This is especially important if a lighting system is used for only a few hundred hours or less each year. With over-voltage operation, the lamps will deliver more lumens per watt but their average life will be reduced. In general, operation at 10 per cent above rated lamp voltage is recommended if the lamps are in use for less than 200 hours per year, and 5 per cent if annual use is 200 to 500 hours per year. If annual use exceeds 500 hours, lamp operation at rated voltage is recommended. The approximate increases in light output, lumens per watt, watts and the reduction in average lamp life are shown in Fig. 8-12 of the 1981 Reference Volume.
2. High intensity discharge and fluorescent light sources, when operated singly on alternating current circuits, may produce a flicker on rapidly moving objects. This condition, called stroboscopic effect, can be minimized by connecting lamps or luminaires on alternate phases of a three-phase supply, or by employing two-lamp lead-lag or series sequence start ballasts where available.
3. When a quiet surround is an important factor in multi-purpose areas, consideration should be given to the possibly objectionable disturbance resulting from ballast "hum" produced by high intensity discharge and fluorescent systems. Remote mounting of ballasting equipment may be desirable.
4. High intensity discharge lamps have a time delay and slow buildup of light output when the lighting system is first energized or when there is a power interruption. Because of this time delay characteristic, it may be desirable to include an incandescent lighting system to provide emergency stand-by illumination, particularly over spectator areas.
5. When fluorescent lamps are used outdoors, they may need protection from the wind and low temperature in order to maintain light output and starting ability. Special lamp designs are available for operation under such conditions. See Section 8 of the 1981 Reference Volume.

LIGHTING FOR INDOOR SPORTS

The walls and ceilings of interiors used for sports provide a means for controlling background brightnesses, assist in diffusing the available light, and make possible a variety of convenient lighting equipment arrangements. The design and calculation procedures for interiors used for sports are outlined in Section 9 of the 1981 Reference Volume. However, in addition to luminaire mounting height, spacing and lumen output, and illuminance uniformity on a horizontal reference plane, which are important factors in all installation plans, it is necessary in designing sports lighting to consider the following factors:
1. Observers have no fixed visual axis or field of view. During the course of the game, the ceiling and luminaires may frequently be included in the visual field.
2. The object of regard will have no fixed location, and may be viewed at floor level, near the ceiling, or at any level in between.
3. It is particularly important for observers to be able to estimate accurately object velocity and trajectory.

The location of sport play can be divided into general areas used for more than one sport and areas designed for a particular sport. The lighting system must meet the varied or particular requirements for the sport, or sports, played in the given area.

The sports, themselves, may be generally divided into two classes: sports which are aerial in part or whole, and sports which are at or close to ground level.

Aerial Sports

Badminton, basketball, handball, jai-alai, squash, tennis and volleyball are considered aerial sports. The type of action encountered during normal participation in these activities is such that the ceiling may be in the observer's field of view during a large portion of the playing period. In planning general lighting installations for these sports, therefore, every effort should be made to select, locate and shield the light sources to avoid introducing glare into the observer's field of view. For these sports in particular, adequate overlapping of the luminaire beam patterns is imperative to insure proper vertical illumination over the entire height of the playing area.

Low Level Sports

Archery, billiards, bowling, fencing, curling, hockey, shuffleboard, skating, rifle and pistol ranges, swimming, boxing and wrestling, and other sports in which observers in the normal course of play do not look upward are called low level sports. General lighting may be planned more easily for these sports than for aerial sports, since luminaire luminance is less critical.

General Areas

General areas used for sports would be field houses, gymnasiums, community center halls and other multi-purpose areas. The sports normal to such areas are badminton, basketball, volleyball, fencing, shuffleboard and other similar sports. The criteria used for designing the lighting system for such areas can best be demonstrated by the design criteria for a gymnasium.

Gymnasiums

The modern gymnasium has become a multipurpose as well as a multi-sport area, serving a variety of needs of the student body and community in general, including such activities as assemblies, concerts and dances. Where the creation of a mood or atmosphere is the objective, this may be done by the use of portable or temporary auxiliary lighting equipment. It is possible, however, and at times desirable, to incorporate dimming, switching or other means of illumination control into the over-all scheme of design.

For most high school applications, a general lighting system providing a level of 300 lux [30 footcandles] will answer the majority of requirements. If, however, exhibitions and matches are played, it is recommended that this level be increased to a minimum of 500 lux [50 footcandles]. Principles of good design would indicate that this could probably be most easily achieved by designing for the higher level and providing switching or dimming means for the reduction in level. Fig. 13-3 is a layout for the basketball area lighting for a gymnasium. Fig. 13-4 shows installations meeting current recommendations.

To prevent breakage, it may be necessary to protect luminaires with wire guards or grids. Any reduction of the light output of the unit by the addition of such a device should be taken into consideration and compensated for in the initial system design. Ceiling heights of gymnasiums may vary, but minimum recommended mounting height for luminaire mount is 6.7 meters (22 feet).

The position of luminaires and windows in a gymnasium can present serious problems. Fig. 13-5 demonstrates the hazards of improperly located luminaires and unshielded fenestration.

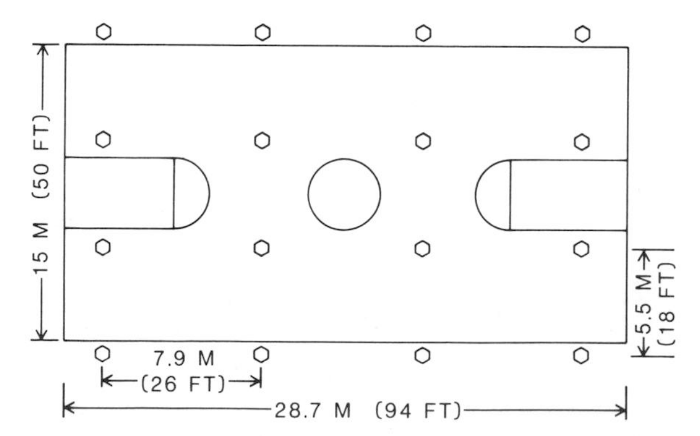

Fig. 13-3. Lighting layout for an indoor basketball court. Minimum luminaire mounting height should be 6.7 meters (22 feet).

Field Houses

The field house and the gymnasium closely resemble each other as far as sports activities are concerned. The field house may, however, be larger in dimension and serve a somewhat wider range of sports. Among these are indoor track and field events, skating, and such outdoor sports as may be driven indoors by inclement weather. Portable floors and seating facilities are in common use. General lighting levels and methods dictated by particular sports will meet the needs for the participants, but may require considerable increases to meet the needs of the spectators. The resultant lighting system design should, therefore, meet the requirements for the anticipated activities in the field house as well as provide for the spectators. This could include consideration for aerial and low level sports, versatile control or individual systems for the various sports, and increased illuminance levels and beam control to meet the needs of the spectators.

Specialized Areas

Lighting layouts which illustrate the adaption of the previously stated principles to certain specialized indoor sports areas are shown in Fig. 13-6 through 13-15. It is important to recognize that these layouts are not the only acceptable method which can be used for lighting a particular sports area. Other types of luminaires, light sources, and in some instances, luminaire locations may

Fig. 13-4. A gymnasium lighted with suspended HID luminaires (left) and one with recessed HID luminaires (right).

Fig. 13-5. Caution should be exercised in positioning luminaires relative to critical surfaces such as glass basketball backboards (upper left), to avoid blinding, reflected glare. Windows behind glass backboards in gymnasiums (lower left) can produce direct glare, and unshielded windows (upper right) and skylights are a potential source of both direct and reflected glare. (Note how reflections on the floor veil the floor markings.) Windows can produce unwanted veiling reflection on swimming pool surfaces (lower right) reducing the ability of life guards to see beneath the surface or even some objects on the surface.

be used satisfactorily. These layouts merely show one or more ways in which the lighting objective has been accomplished. For recommended illuminance values see Fig. 2-2, page 2-16.

Badminton. Badminton is an aerial sport, and required ceiling heights of 7.6 meters (25 feet) minimum and upwards to 12.2 meters (40 feet) desirable. A brown or green color is recommended for the walls and ceiling to provide good contrast for the white shuttle. A dark finish is also recommended for the floor. To minimize glare, a well controlled lighting system mounted along the sideline, or an indirect system, is recommended. See Fig. 13-6.

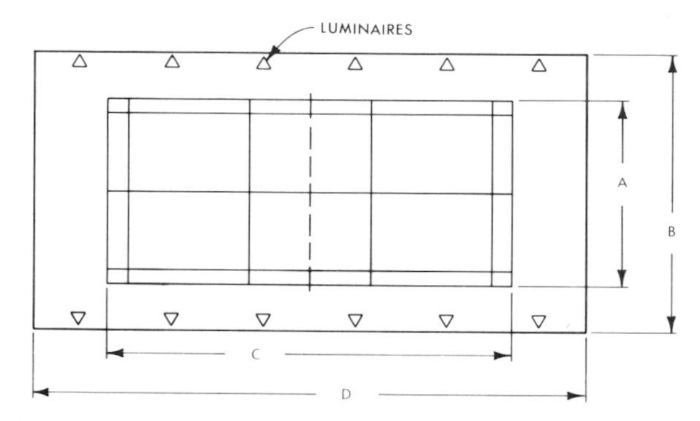

Fig. 13-6. Recommended lighting layout for badminton. Lamp size and luminaire quantities for each class of play are dependent upon the specific room characteristics and luminaires used. Luminaires should be semi-direct and carefully shielded. A = 6.1 (20), B = 9.1 (30), C = 13.4 (44) and D = 18.3 meters (60 feet).

Billiards. It is desirable to have a layout of the location of the tables before establishing the location of luminaires, so that they can be placed over the tables, to provide the best lighting possible and create the fewest number of shadows. Luminous ceilings or other general lighting systems can be utilized. Billiard tables are approximately 1.5 by 2.7 meters (5 by 9 feet) in size and are usually located so that they are 1.5 meters (five feet) apart, side by side and at least 1.8 meters (six feet) from the adjoining wall. The minimum recommended mounting height of the ceiling and light source is 2.3 meters (7.5 feet). The preferred height is 3 to 3.7 meters (10 to 12 feet). The ceiling should be a light color with a reflectance of 75 to 85 per cent. Recommended illuminances might be substantially increased for public attraction or business considerations.

Bowling. Lighting for bowling is often governed more by public attraction and increased business considerations than any other factor. Bowling is considered a low-level sport which is divided into three areas—the approaches, the lanes, and the pins. General illumination methods are utilized in the approach area. This area often includes seating for spectators as well as participants with lighting utilized to create a pleasant atmosphere. The lighting of the lanes should be well shielded for the bowler and the

Fig. 13-7. Typical lighting arrangements for bowling. The ceiling luminaires should be completely shielded from the view of the bowler. To avoid severe luminance differences and to make maximum use of reflected light, the ceiling should be maintained at a reflectance of 70 per cent or better. All dimensions X in the elevation are 3.7 meters (12 feet).

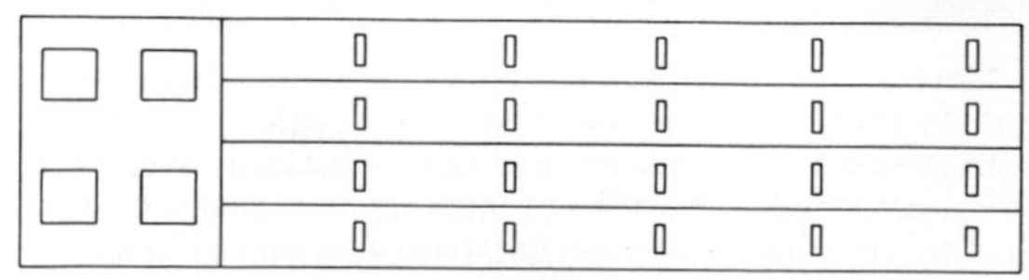

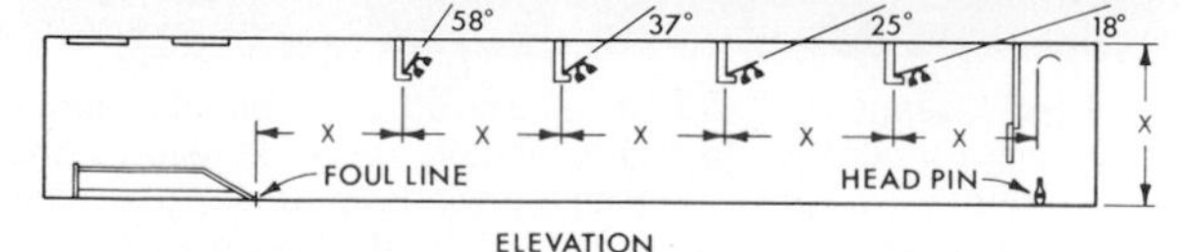

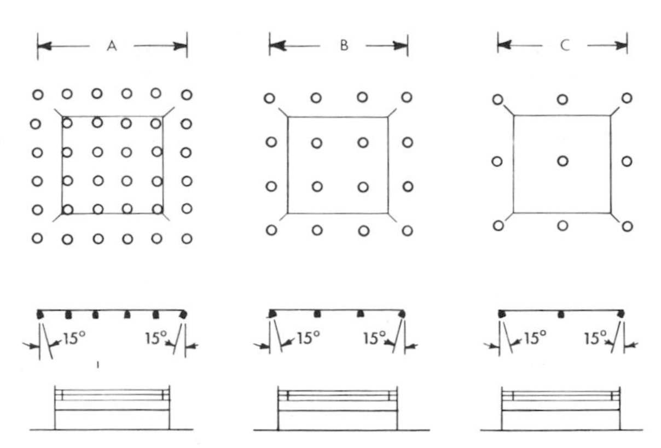

Fig. 13-8. Recommended layouts for indoor boxing or wrestling rings. Luminaires are direct with concentrating distribution from 6.1-meter (20-foot) mounting height. A = 9.1 (30), B = 8.2 (27) and C = 7.9 meters (26 feet).

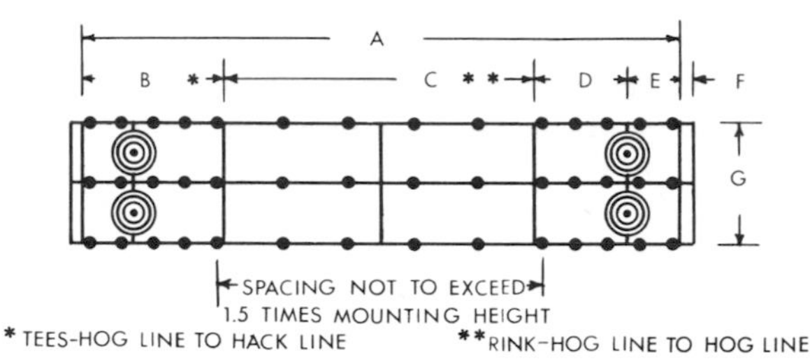

Fig. 13-9. Recommended lighting layout for curling. Lamp size and luminaire quantities for each class of play are dependent upon the specific room characteristics and luminaires used. A = 42 (138), B = 10 (33), C = 21.9 (72), D = 6.4 (21), E = 3.7 (12), F = 1.2 (4) and G = 8.7 meters (28.5 feet).

shielding is often an architectural element of the structure. This ceiling area should be finished with a high reflectance, non-gloss, light paint which maintains a 70 to 85 per cent reflectance. The illumination of the pins is so directed as to provide high illuminance on the vertical as seen by the bowler. A typical layout is shown in Fig. 13-7.

Boxing and Wrestling. The recommendations for illumination are governed by the requirements of the spectators which completely outweight the requirements of the participants. A recommended layout is shown in Fig. 13-8.

Curling. The indoor curling rink is classified as a low-level installation system. Direct or semi-direct luminaires, with wide spread distribution, mounted between rinks provide the best method of illumination. The minimum mounting height of the luminaires is 3.7 meters (12 feet) and the ceiling and wall finishes should have a reflectance of over 60 per cent to provide good luminance ratios. Fig. 13-9 shows a recommended layout for curling.

Handball, Racquetball and Squash. The handball, racquetball and squash court with its white walls and ceiling presents definite luminaire beam control problems for these aerial sports. The wall and ceiling finish should be a white, non-glossy paint with a reflectance of 75 to 85 per cent. The luminaires should be recess-mounted in the ceiling with a carefully shielded spread distribution. In these areas, adequate pro-

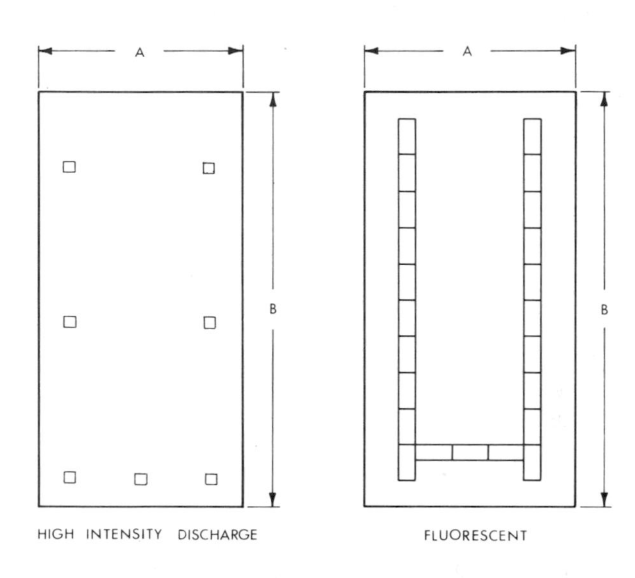

Fig. 13-10. Recommended lighting layouts for squash, four-wall handball and racquetball. Lamp size and luminaire quantities for each class of play are dependent upon the specific room characteristics and luminaires employed. Tournament class of play is illustrated here. A = 6.1 (20) and B = 12.2 meters (40 feet).

tection of the luminaires from possible breakage through the use of guards or impact-resistant covers is vitally important. See Fig. 13-10.

Hockey Rink. Lighting for indoor hockey rinks requires extreme care in the selection and location of luminaires. Not only should direct glare from the luminaires be considered, but the possible loss of visibility due to reflected glare from the ice is of equal importance. Care should be exercised so that no shadows from the boards and nets cause difficulty in following the course of play. All luminaires should be mounted above the line of sight of the spectator in the most elevated seat at the greatest distance from the playing area. This provides an uninterrupted view of the playing area, minimizes possible direct glare to the spectators, and improves general appearance. See Fig. 13-11.

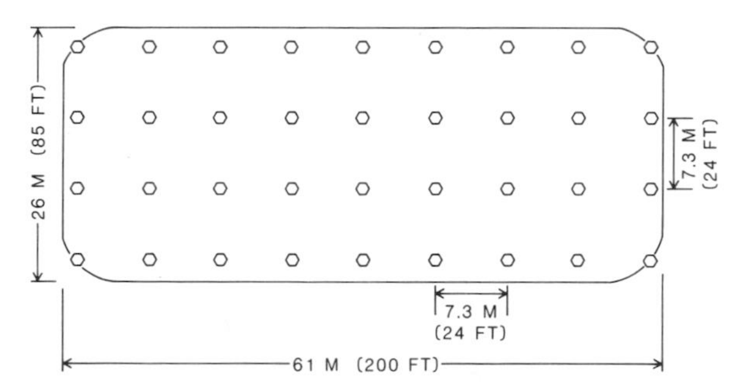

Fig. 13-11. Typical lighting layout of direct high intensity discharge luminaires at an indoor hockey arena.

Jai-Alai Court. Due to extreme speed of the ball in play (over 240 kilometers (150 miles) per hour), careful consideration must be given to illuminances, shielding of luminaires, and surface texture of paint on the court walls and floor. Glare is to be avoided at all costs. Play is fast and serious accidents are not uncommon. Colors recommended are: grass green for the frontis, lateral and rebote; off-white for the floor; and dark red for the foul stripes. Luminaires should be mounted above the top screen for physical protection. Viewing is done from the open side of the court, again through a protective screen. It is good practice to provide for dimming in the audience area at the start of play.

Shooting (Archery, Pistol and Rifle Ranges). Indoor archery, pistol and rifle ranges present similar illumination problems. Major emphasis is placed upon the illuminance at the target and the distance from the firing line to the target. In the case of the indoor pistol and rifle range, which has a 15-meter (50-foot) distance, the recommended vertical illuminance on the target meets the requirements for the distance from the firing line to the targets and the size of the targets. In the case of archery, distances of 18 to 50 meters (60 to 150 feet) are normal

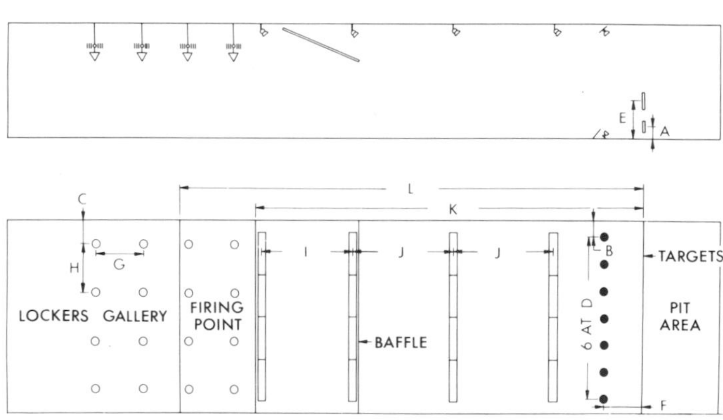

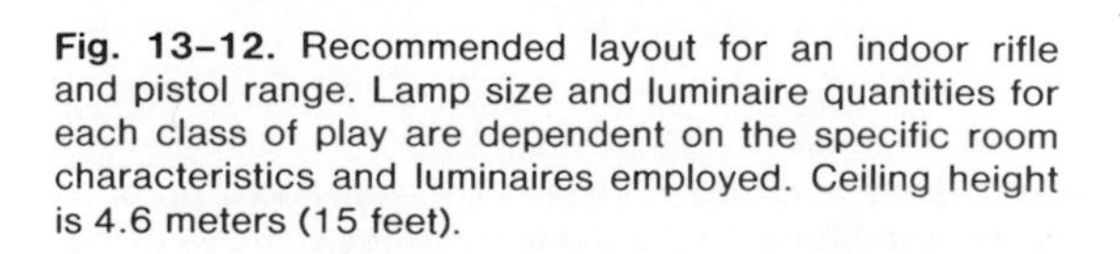

Fig. 13-12. Recommended layout for an indoor rifle and pistol range. Lamp size and luminaire quantities for each class of play are dependent on the specific room characteristics and luminaires employed. Ceiling height is 4.6 meters (15 feet).

Distance	Meters	Feet	Distance	Meters	Feet
A	0.5	1.5	G	1.8	6
B	0.6	2	H	2.0	6.5
C	0.9	3	I	3.7	12
D	1.1	3.5	J	4.1	13.5
E	1.5	4.8	K	15.2	50
F	1.5	5	L	18.3	60

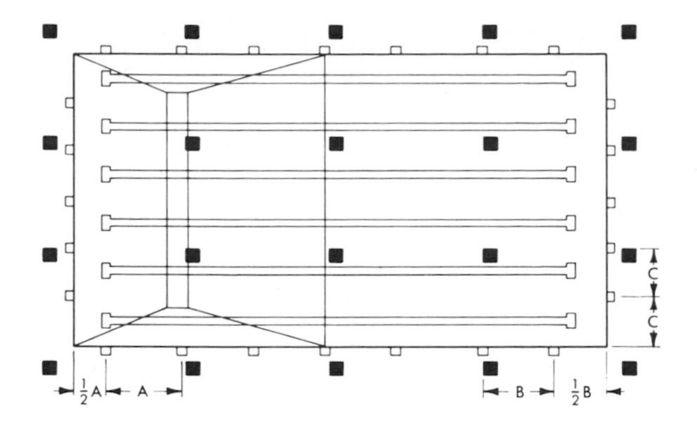

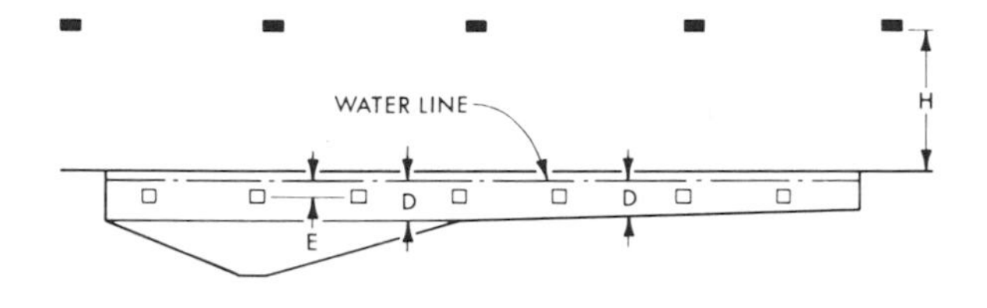

UNDERWATER*

Location of Pool	Lamp Lumens Per Square Meter [Foot] of Pool Surface (width × length)
Outdoors	600 [60]
Indoors	1000 [100]

Dimensions

Lamp Lumens	A Maximum in Meters (Feet)†	B Maximum in Meters (Feet)†	E in Millimeters (Inches) Below Water Line	
			Minimum	Maximum
3750 to 8000	2.4 (8)	3 (10)	300 (12)	380 (15)
9900 to 33,000	3.7 (12)	4.6 (15)	460 (18)	610 (24)

* *C* dimension is equal to the swimming lane width to minimize glare and accidental damage.

† Where *D* is over 1.5 meters (5 feet).

Above lighting uses especially designed floodlights not covered by NEMA Classification or Type. Two systems are used—wet niche and dry niche. The former uses submersible units, while in the latter the casings or niche linings are cast in the pool walls with the floodlights behind them. Use minimum number of floodlights that will satisfy distribution and lumens per square meter [foot]. At the ends of the pool, the *C* dimension can be doubled or units eliminated especially at the shallow end or for narrow pools.

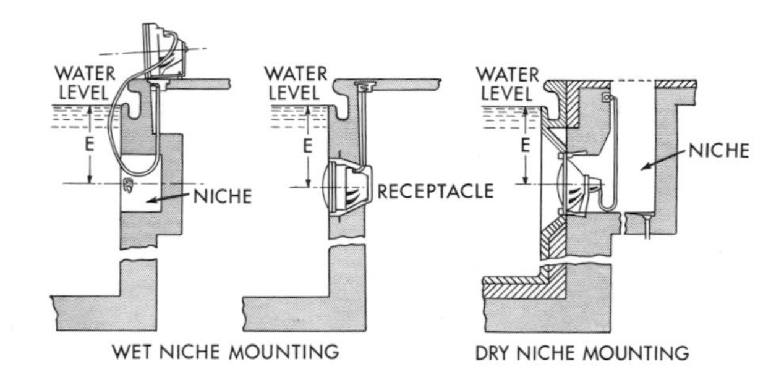

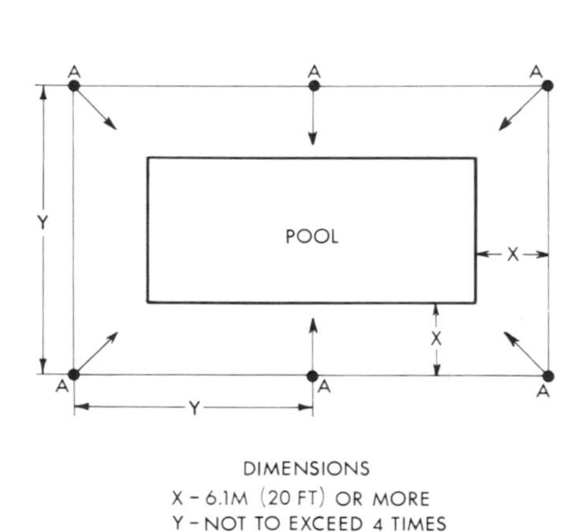

Fig. 13-13. Lighting recommendations for swimming pools. Locate lighting equipment or life guards' positions so as to minimize direct and reflected glare. For overhead lighting in indoor and outdoor pools mounting height should be 6.1 meters (20 feet) minimum. For outdoor pools use floodlights type 5 or 6, class GP (see footnote page 13-13), and underwater lighting and layout shown.

between the firing line and target. The recommended illuminance on the vertical target again considers the distance and the size of the target. The typical layout for shooting ranges is shown by the example in Fig. 13-12 which illustrates the standard pistol and rifle range.

Swimming. The lighting of swimming pools is multi-fold. It is to light: (1) the water surface; (2) the floor of the pool; and (3) the deck area around the pool adequately, and for the safety of the persons using the pool. Underwater luminaires should be so located to give complete illumination to all underwater areas. Refer to the National Electrical Code and applicable local codes for specific placement of luminaires.

For underwater lighting, luminaires should be properly located in the pool walls to provide adequate illuminances throughout the pool, but should not be placed in line with a swimming lane where competitive swimmers would make a turn and possibly kick the light during the turn. It is therefore quite important that the luminaires should be located between the lanes so that it would not in any way interfere with competitive swimming.

The overhead lighting of the indoor pools can be executed in a way similar to lighting any indoor space with proper spacing and location throughout the ceiling. In the event there is crawl space above the ceiling, it is desirable to select luminaires that can be relamped from above. In the event the luminaires must be relamped from below, it would seem desirable to locate them over the deck rather than over the water and aim some of them toward the water. This will eliminate the need for servicing the overhead luminaires from a pool location. Fig. 13-13 illustrates the recommended practice for swimming pools.

Indoor Tennis Courts. The area under consideration for indoor play approximates 15 by 37 meters (50 by 120 feet) per court. Suggested interior finishes are: ceilings and upper walls, light, non-glossy 80 to 90 per cent reflectance; walls, lower 3.7 meters (12 feet), dark, non-glossy, maximum reflectance 60 per cent (usually dark green); court surfaces, porous or nonporous with low reflectance, typically 25 per cent.

The luminaires for direct lighting should provide a minimum of 20 per cent up-light. The direct light from the luminaires should be controlled with baffles, louvers or other shielding techniques to reduce the possibility of glare that would distract the players. The baffles should provide cutoff at 45 degrees in the direction of play.

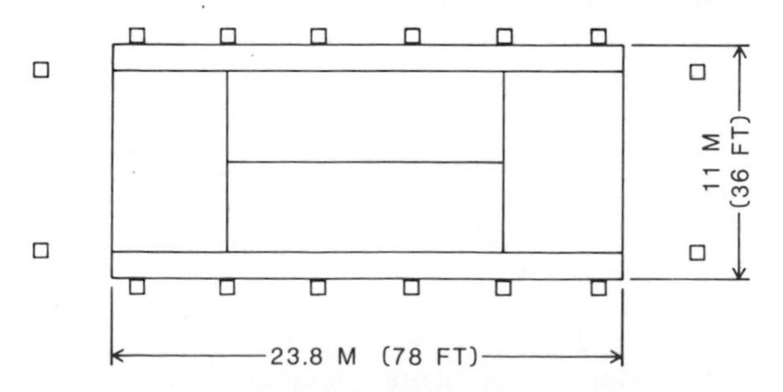

Fig. 13-14. Lighting layout for indoor tennis court. Minimum luminaire mounting height is 6.7 meters (22 feet).

For indirect lighting, the luminaires should provide a wide beam spread so that there is a high degree of beam overlap, producing "even" illuminance on the ceiling (reflecting surface).

Figure 13-14 shows a typical layout for an indoor lighting system with these special considerations:

(1) The choice of lamps and luminaires is critical and careful consideration should be given to their selection.
(2) Luminaires in the back court may be tilted slightly toward the back of the enclosure so that the light source or its reflected image cannot be seen from the opposite court.
(3) The number of squares in Fig. 13-14 do not necessarily indicate quantities.
(4) Where courts are more than 5.5 meters (18 feet) apart, two rows of luminaires are necessary.
(5) If guards on indirect luminaires are visible, they may be finished flat black to reduce reflected glare.
(6) Indirect luminaires should not be placed closer than 1.5 meters (5 feet) below the ceiling to avoid hot spots.
(7) Direct downlighting is not recommended because of the glare a player may encounter when serving or hitting an overhead ball.
(8) Skylights may be considered as an energy saving device. However, they should be located over the space between the courts. One should consider the added initial costs, possible increased heating and cooling costs, possible leaks, possible condensation and loss value when covered with snow or on dark days. The sun's rays need to be diffused. A full lighting system is needed for night play.

Indirect Lighting

Many aerial sports require the upper walls and ceiling to be finished with a high reflectance semi-gloss white paint. This area is illuminated by the upward component of the semi-direct type luminaires and becomes an added factor in the overall quality of illumination. One method of increasing the quality of illumination is to utilize a totally indirect lighting system. See Fig. 13-15. Such systems may be less efficient than a semi-direct system, but often provide other benefits and at times present the only adequate method of obtaining satisfactory results.

Inflatable structures are finding a wide usage in the sports field. This is especially true for skating rinks, swimming pools and tennis courts. These inflatable structures normally mount on a

Fig. 13-15. Indirect lighting of a tennis center.

foundation that varies between ground level and a wall of up to 2.4-meter (8-foot) height. The structures cannot support overhead items such as luminaires and, therefore, indirect lighting answers the lighting need. The design of such structures normally employs the use of materials which provide a high reflectance matte surface. It is very important that such material is utilized to eliminate glare caused by specular reflections and to prevent increasing the lighting load as the surface reflectance decreases.

A major consideration in the design of an indirect lighting system is the uniformity of illuminance over the entire surface. Hot spots around the luminaire's location can be as distracting as a direct view of the luminaire or light source by the participant. The number of luminaire locations need only be governed by the uniformity which can be achieved and architectural or surface elements which could create deep shadows on the surface being illuminated. These could be as distracting as a black cloud might be in an otherwise clear blue sky.

LIGHTING FOR OUTDOOR SPORTS

In the following discussion, where various "classes" of sports are indicated, the classifications follow league ratings where they exist. In general, these ratings are indicative of the skill and speed of play to be expected, and correlate closely with the relative number of spectators regularly accommodated. This latter factor determines the maximum distance at which a spectator may be observing the playing area, and consequently has a direct bearing on the angular size of the object to be seen and, therefore, on the quantity of the light required. Figs. 13-16 through 13-24 present data for layouts considered good practice.* For recommended illuminances see Fig. 2-2, page 2-16.

Baseball

Baseball presents a severe, though not prolonged, seeing task. The ball is small, moves rapidly, and is viewed at varying distances against variable background brightness. The necessity for concentration is intermittent. The large number of possible observer locations and the movement of the players also introduce difficulties. See layout shown in Fig. 13-16.

In providing adequate and uniform illuminance for baseball, it is standard practice to consider the infield as including a 9.1-meter (30-foot) strip outside all baselines and to consider the outfield as including a 9.1-meter (30-foot) strip outside each foul line.

The floodlights should be aimed so that the beam overlap will provide lighting from two directions at almost every outfield point and from four directions over most of the infield.

Junior League Baseball

This classification of baseball includes such leagues as Pony, Colt, Khoury, Little, Teen-Age, etc. In general, the standard baseball principles apply here also. However, an auxiliary strip outside the baselines and foul lines equal to one-third the length of the baseline is recommended in each instance to be lighted to the same illuminance as the adjacent playing area. See Fig. 13-17.

* For other sports see reference 1. For floodlight types (see NEMA FA 1-1973):

Type	Beam Spread (degrees)
1	10 up to 18
2	18 up to 29
3	29 up to 46
4	46 up to 70
5	70 up to 100
6	100 up to 130
7	130 and up

For floodlight classes: HD = heavy duty, GP = general purpose or enclosed ground-area, O = ground-area open, and OI = ground-area with reflector insert (see Section 1 of 1981 Reference Volume).

Class of Baseball (regulation)	Floodlights		Minimum Mounting Height to Bottom Crossarm	
	Type	Class	Meters	Feet
Major league	3, 4 or 5	GP	37	120
AAA or AA	3, 4 or 5	GP	33.5	110
A and B	3, 4 or 5	GP	27	90
C and D	3, 4 or 5	GP	21	70
Semi-professional and municipal	3, 4 or 5	GP	21	70
	4, 5 or 6	OI		
Recreational	3, 4 or 5	GP	21	70
	4, 5 or 6	OI		

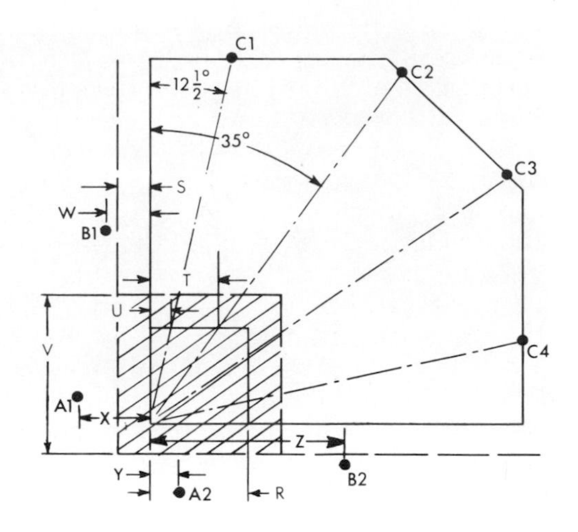

Distance	Meters	Feet
R	27	90
S	9	30
T	19	63
U	6	20
V	45.5	150
W	9–18	30–60
X	12–24	40–80
Y	6–9	20–30
Z	40–55	130–180

Fig. 13-16. Recommended lighting layouts for baseball fields. These layouts are based on the following total playing area including a strip 9 meters (30 feet) wide outside each foul line—12,310 square meters (132,500 square feet). Infield area (shaded), 2090 square meters (22,500 square feet). Outfield area, 10,220 square meters (110,000 square feet).

Combination Sports Field

The combination layout is never as satisfactory as two individual lighting systems. Nevertheless, athletic fields are laid out for daytime seasonable playing of several sports, usually for a two- or three-game combination of baseball, softball, football and soccer. Lighting such a combination field for night play requires special attention, since the lighting requirements for each individual sport must be considered in developing the final lighting plan. The final design will be largely affected by the relative location of the several fields, and the limiting restrictions which each specific arrangement may impose.

Sometimes baseball and softball are played with the same home plate and foul line locations. In these cases, baseball pole locations and mounting heights can be made entirely satisfactory for softball lighting by means of a system (switching or other) that will permit lighting only as many floodlights as are necessary and properly aimed to cover the softball area.

A number of equipment locations is possible when overlapping baseball or softball and football fields. Mounting height on a pole should be the greatest height recommended for any sport served by that particular pole. See Fig. 13-18.

Football, Soccer and Rugby

Football, soccer and rugby are a combination of aerial and ground play requiring adequate lighting from ground level to approximately 15 meters (50 feet) above ground. The symmetrical field utilized for these sports affords easy provision for good lighting. Pole locations may prove the biggest problem because of existing facilities. See layout Fig. 13-19.

Golf

The lighting of a golf course for night play involves problems not generally encountered in other sports. The area involved is many times larger than the average sports area. Although the sport is basically unidirectional in nature, the frequent orientation of fairways in direct opposition to each other, and the extreme variations in terrain make the selection of pole locations, beam types and luminaire orientation a much more critical problem than in most sports areas.

The tee should be lighted so that neither a right- nor left-hand player will shadow his ball. High vertical illuminance values down the fairway should be provided to permit the player to follow the small sphere for the full length of that area while it is traveling at a speed of 160 kilometers (100 miles) per hour or more and to locate it after it has come to rest.

Each green should be lighted from at least two directions to minimize harsh shadows. Care should be taken in the selection and aiming of the floodlights so that glare from the units does not handicap either the player or those on adjacent fairways. See Fig. 13–20. Special consideration must be given areas not covered by the general lighting system. Some will present physical hazards or require special accent. Examples: sand traps, water hazards, bridges, steep grades, roughs, areas adjacent to greens, pathways, etc.

Platform Tennis

Platform tennis is sometimes called "paddle," but it is not to be confused with paddle tennis. It is played on a court 9 meters by 18 meters (30 feet by 60 feet), usually raised off the ground and surrounded by wire fencing 3 meters (10 feet) high. One plays with a paddle somewhat larger and heavier than a ping pong paddle. Balls striking the screening are in play until they bounce a second time. Balls are heavier than tennis balls and are made of solid sponge rubber. It is similar to tennis but only one serve is permitted. Most often it is played as a doubles game and usually in winter. See Fig. 13–21.

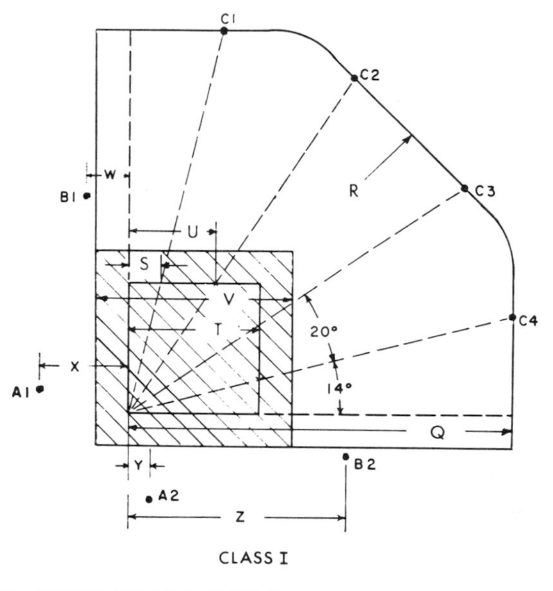

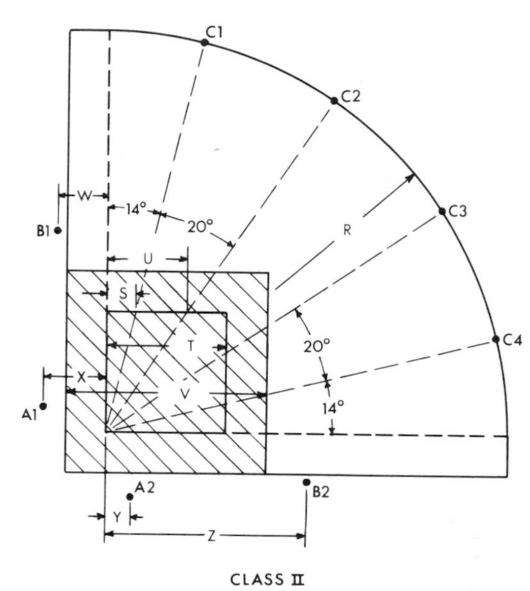

Class of Baseball	Floodlights		Area in Square Meters (Square Feet)		Minimum Mounting Height to Bottom Floodlight in Meters (Feet)	
	Type	Class	Infield	Outfield	Poles A and B	Poles C
I	3, 4 or 5	GP	930 (10,000)	2290 (24,700)	12 (40)	15 (50)
	4, 5 or 6	O or OI				

Class of Baseball	Floodlights		Area in Square Meters (Square Feet)		Minimum Mounting Height to Bottom Floodlight in Meters (Feet)	
	Type	Class	Infield	Outfield	Poles A and B	Poles C
II	3, 4 or 5	GP	1450 (15,625)	4330 (46,600)	15 (50)	18 (60)
	4, 5 or 6	OI				

Distance		Q	R	S	T	U	V	W	X	Y	Z
I	Meters	55	56	4.5	18	12	30	6–9	9–15	1.5–4.5	27–34
	Feet	180	185	15	60	40	100	20–30	30–50	5–15	90–110
II	Meters		76	5.7	23	15.4	38	7.3–14	11–20	3–7.6	34–44
	Feet		250	18.7	75	50.6	125	24–45	35–65	10–25	110–145

Fig. 13–17. Lighting recommendations for Junior League Baseball. (a) Class I—baselines 18 meters (60 feet) or less. (b) Class II—baselines 18 meters (60 feet) and up to 23 meters (75 feet).

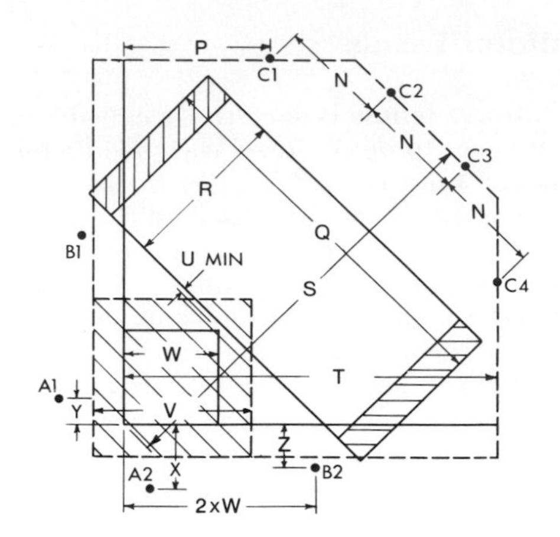

Sport (class)	Floodlights		Minimum Mounting Height to Bottom Floodlight Crossarm in Meters (Feet)	
	Type	Class	Poles A1, A2, B1, B2, C1 and C4	Poles C2 and C3
Baseball (semiprofessional and municipal)	3, 4 or 5	GP	21 (70)	27 (90)
Football	4, 5 or 6	OI		

Fig. 13-18. Recommended lighting layout for a combination baseball and football field where baseball is the primary sport. The combination layout is not as satisfactory for either sport as individual layouts. Re-aiming for each sport will increase the effectiveness. The layouts are based on the following: Total playing area, including a strip 9 meters (30 feet) outside each foul line—12,310 square meters (132,500 square feet). Infield area (baseball)—2090 square meters (22,500 square feet). Outfield area—10,220 square meters (110,000 square feet).

Distance	Meters	Feet
N	30	100
P	41	135
Q	110	360
R	49	160
S	122	400
T	107	350
U	1.5	5
V	45.5	150
W	27	90
X	12-24	40-80
Y	6-9	20-30
Z	12	40

Classification*	Distance—Nearest Sideline to Floodlight Poles		Spectator Seating Capacity	Number of Poles	Floodlights	
	Meters	Feet			Type	Class
I	Over 43	Over 140	Over 30,000	6	1 or 2	GP
	30-43	100-140		6	2 or 3	GP
II	23-30	75-100	10,000-30,000	6	3	GP
	15-23	50-75		8	3, 4	GP
III	9-15	30-50	5000-10,000	8	4	GP
IV	4.5-9	15-30	5000	10	5	GP
					6	OI
					6	O
V	4.5-9	15-30	No fixed seating facilities	10	5	GP
					6	OI
					6	O

Fig. 13-19. Recommended lighting layouts for football fields. Any of the six pole plans at the right or any intermediate longitudinal spacings are considered good practice with local field conditions dictating exact pole locations.

Distance	Meters	Feet
R	4.5-9	15-30
S	9-23	30-75
T	23	75
U	26	85
V	30	100
W	37	120
X	45.5	150
Y	55	180
Z	23 or over	75 or over

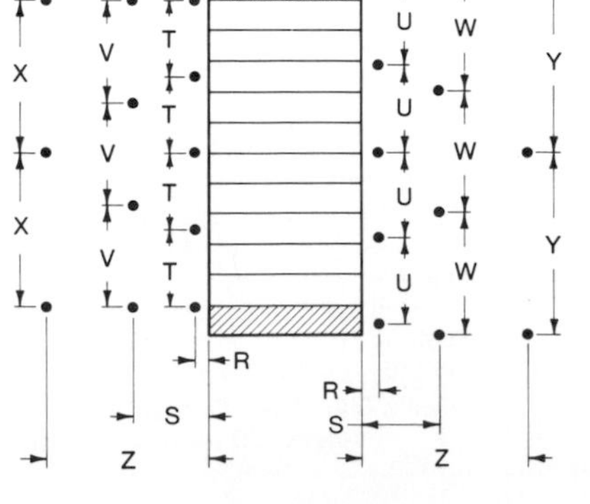

* It is generally conceded that distance between the spectators and the play is the first consideration in determining the class and lighting requirements. However, the potential seating capacity of the stands should also be considered. For minimum mounting height see Fig. 13-25.

Skiing

Lighting of a slope for night skiing in many respects resembles the lighting of a golf course for night play. Both are large-area unidirectional sports. The directional quality of the light is even more important in the case of skiing due to the downhill speed often attained on the more precipitous areas. Variations in terrain and slope make pole locations, beam types and luminaire orientation critical.

Due to the large physical areas involved and the accompanying economic factors, illuminances must be kept to a minimum commensurate with safety. Levels of ten to twenty lux (one to two footcandles), horizontal, have proved adequate provided reasonable distribution of light is maintained. Dark areas or spots are to be avoided at all costs. In the case of unusual terrain features, moguls, etc., more frequent pole spacing is required than on straight slopes. See Fig. 13-22 for typical layouts.

Softball

Softball also follows the same general principles as baseball. Fields may vary in outfield distance from 50 to 85 meters (160 to 280 feet). See layout Fig. 13-23. Dimensions for slow-pitch softball are essentially the same as standard softball. See layout and footnotes relative to slow-pitch softball.

1. Each green should be lighted from at least two directions to minimize harsh shadows.
2. Pole locations should be confined to the 40° cross-hatched zone indicated in front of the green.
3. Pole spacing should be equal to or less than 3 times mounting height.
4. The maximum horizontal illuminance measured at any place on the green area should not be greater than 3 times the minimum measured at any other place on the green area.
5. Care should be taken in placement of luminaire poles around green so as to neither obstruct the approaching drive nor create objectionable glare in the eyes of the approaching golfer, as well as the eyes of golfers on other fairways.

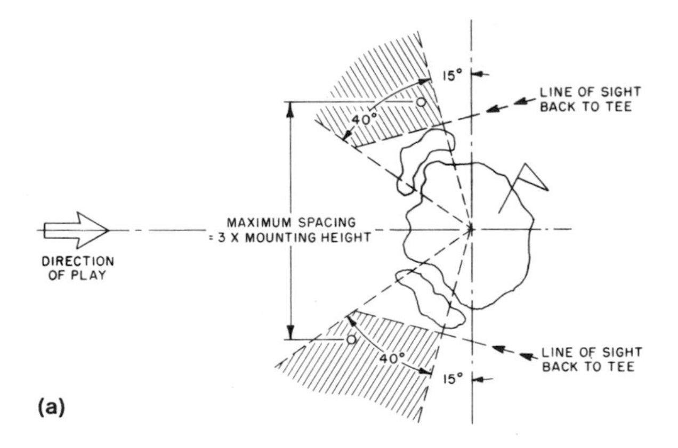

1. Use one pole located a minimum of 1.5 meters (5 feet) behind back edge of tee. Extremely wide tees may require more than one floodlight location.
2. Floodlight mounting height above tee should be equal to or greater than one half the width of the tee but in no case less than 9 meters (30 feet). Good practice indicates higher mounting heights for deep tees.

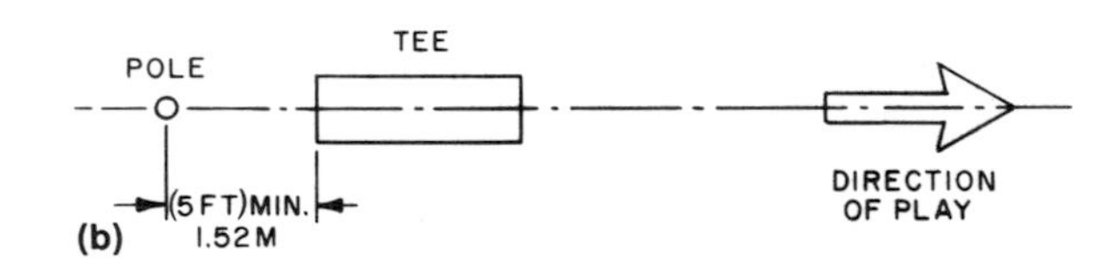

1. Vertical planes should be considered to:
   (a) Extend the full width of the fairway at the point in question,
   (b) Be perpendicular to the centerline of the fairway,
   (c) Extend from fairway centerline elevation to a point 15 meters (50 feet) above the fairway centerline.
2. Vertical planes should be considered to be at points midway between fairway poles.
3. The first vertical plane should be considered to be no less than 30 meters (100 feet) from the tee pole.
4. The ratio of average to minimum illuminance at any point in the plane under consideration should be no more than 7 to 1.
5. Minimum mounting height should be 11 meters (35 feet) above the pole base; however, it may be necessary to adjust this if unusual terrain features exist.
6. Spacing between poles should be coordinated with photometric characteristics of floodlight employed, terrain existing at site and other lighting design criteria.

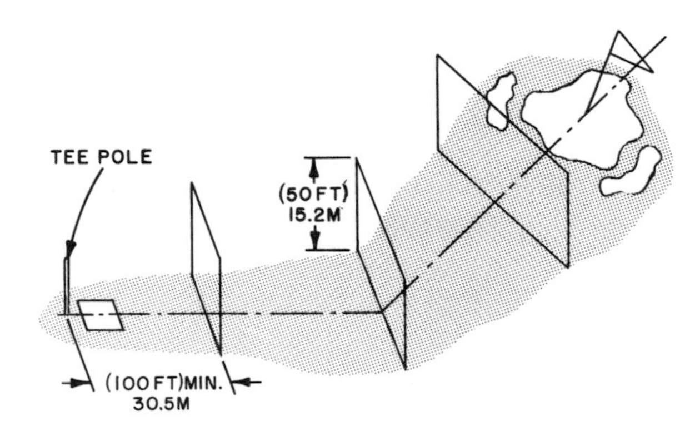

Fig. 13-20. Recommended lighting layouts for (a) golf course greens, (b) tees and (c) fairways.

Tennis

Being a fast aerial sport, outdoor tennis is confined to a smaller area than sports like baseball and football. In order to produce the best quality, luminaire beam control and luminaire location must be considered. Typical lighting systems for various classes of play are illustrated in Fig. 13-24.

Luminaires located at the back court lines should provide beam control that will cut off objectionable glare in the opposite court. This can be done by directional aiming and shielding of the light source. The mounting height of uncontrolled luminaires should be considerably greater than those providing beam control and light source shielding.

Continuous rows of luminaires along the sideline, should be controlled with baffles, louvers or other shielding techniques to reduce the possibility of glare that would distract the players. The baffles should provide cut-off at 45 degrees in the direction of play.

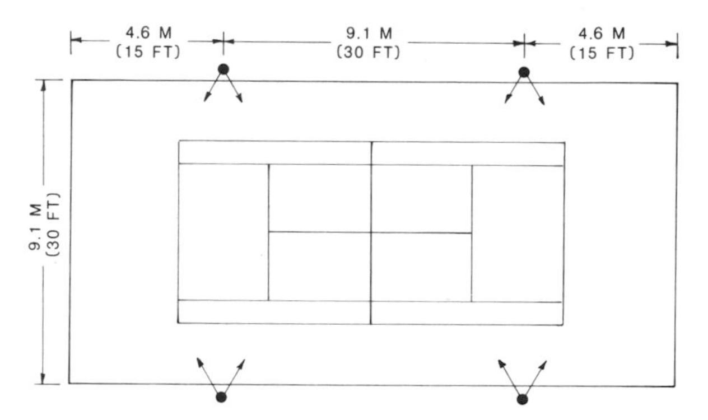

Fig. 13-21. Layout for platform tennis.

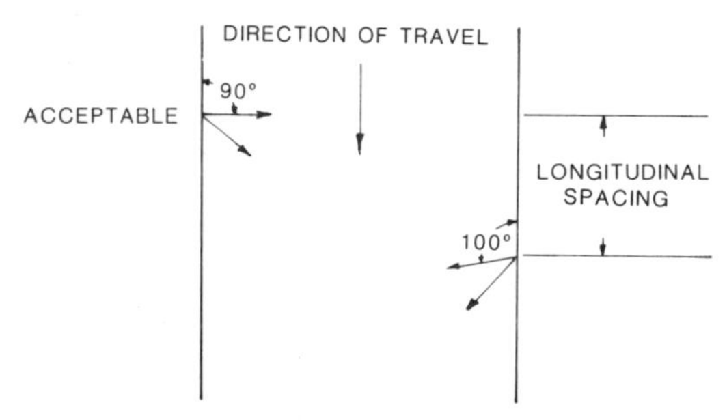

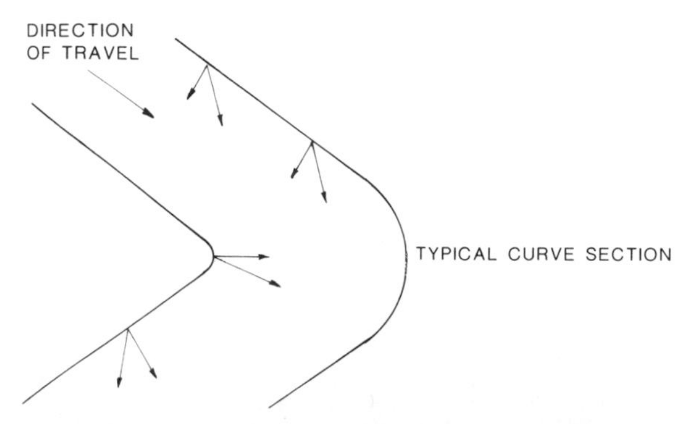

Fig. 13-22. Pole locations and luminaires aiming directions for typical straight and curved ski slope sections. Longitudinal spacing of poles not to be greater than 4 times effective mounting height.

Quantity and Location of Poles

Positions of poles are generally dictated by the existing facilities. Care should be taken to insure that luminaires are displaced from the normal lines of sight of players and spectators. Recommended pole locations should be used when designing new installations.

Mounting Height

The angle between the horizontal playing surface and a line drawn through the lowest mounted floodlight and a point one-third the distance across the playing field should not be less than 30 degrees. At the same time, a minimum of 6 meters (20 feet) for ground sports and 9 meters (30 feet) for aerial sports should be observed. This can be expressed in a formula described in Fig. 13-25.

Aiming of Floodlights

In any sports lighting project, proper aiming of the luminaires upon installation is vitally important in order that the user may secure the full benefits of the quality that the manufacturer has built into the equipment, and of the layout that the engineer has provided. Each luminaire must be carefully directed to its appropriate point on the playing field if the lighting system is to provide both the horizontal and vertical uniformity and the freedom from objectionable glare for which the installation was designed. To facilitate the actual aiming process, an aiming or "spotting" diagram, prepared in advance, is generally employed.

Computer-oriented point calculation methods make it possible to predetermine accurately the illuminance distribution provided by any given

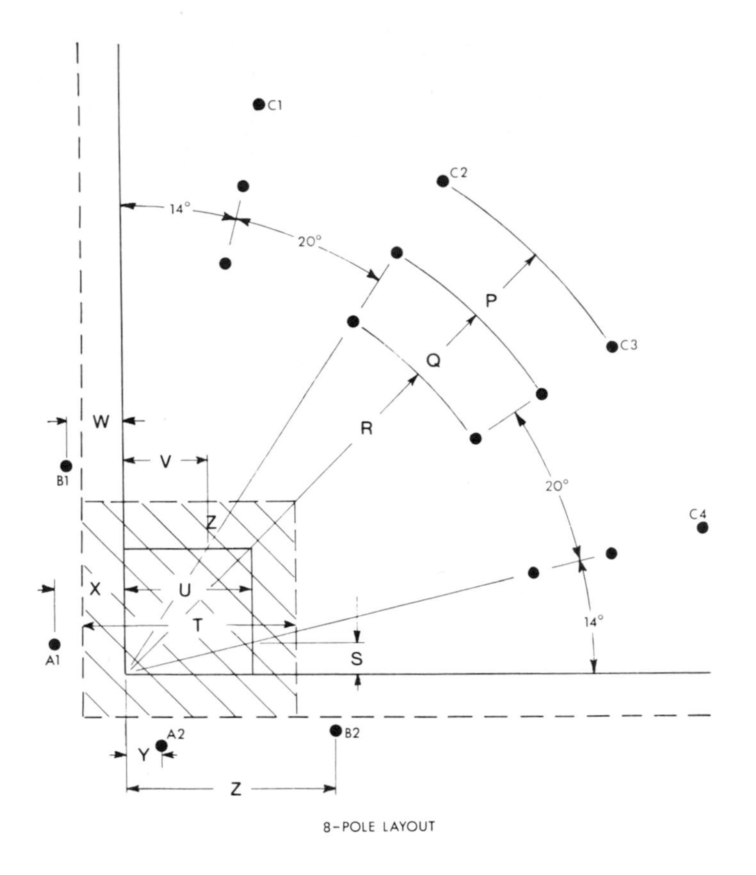

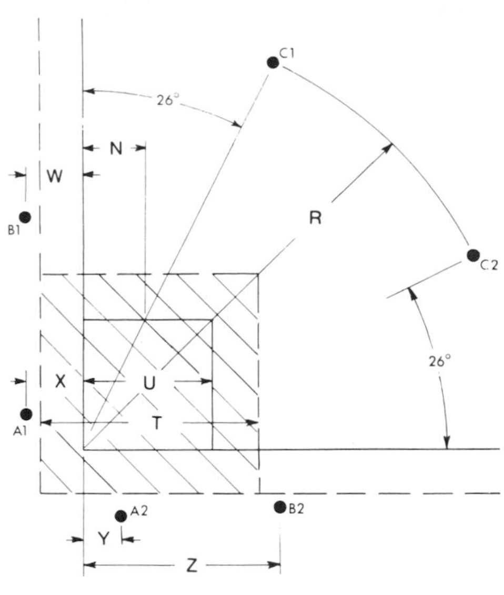

6-POLE LAYOUT

Distance	Meters	Feet
N	8.5	28
P	85	280
Q	73	240
R	61	200
S	4.5	15
T	30	100
U	18	60
V	12	40
W	6-9	20-30
X	7.6-15	25-50
Y	1.5-4.5	5-15
Z	27-33.5	90-110

Class	Outfield Distance	Floodlights		Minimum Mounting Height to Bottom Floodlight Crossarm			
				A and B Poles		C Poles	
		Type	Class	Meters	Feet	Meters	Feet
		8-Pole Layout					
Professional and championship	P	3, 4, or 5	GP	15	50	18	60
	Q			15	50	17	55
Semi-professional	P	3, 4, or 5	GP	12	40	17	55
	Q	4, 5, or 6	OI	12	40	15	50
Industrial league	P	3, 4, or 5	GP	11	35	15	50
	Q	4, 5, or 6	OI	11	35	14	45
	R	6	O	11	35	12	40
		6-Pole Layout					
Recreational	R	5	GP	11	35	12	40
		4, 5, or 6	OI				
		6	O				

Poles: 6 for recreational, 8 for other classes.
Note: Supplementary corner poles may be installed to carry overhead wire around boundary rather than across playing area.
For slow-pitch softball—tournament class is same as industrial league, recreational class same as recreational above.

Fig. 13-23. Recommended lighting layouts for softball. These layouts are based on the following total playing area including a strip 6 meters (20 feet) wide outside each foul line: Infield area—930 square meters (10,000 square feet); Outfield area [61 meters (200 feet)]—2770 square meters (29,820 square feet); Outfield area [73 meters (240 feet)]—4200 square meters (45,240 square feet); Outfield area [85 meters (280 feet)]—5870 square meters (63,200 square feet).

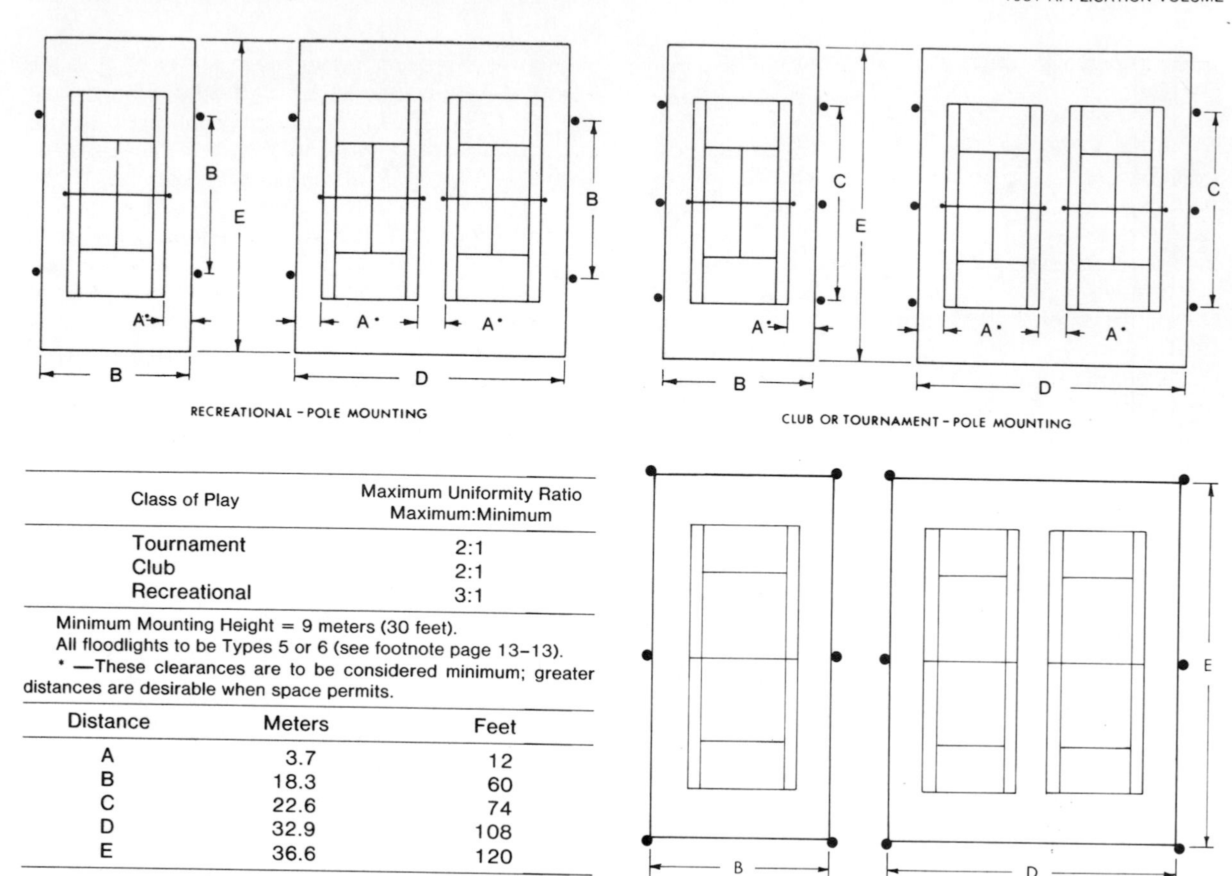

Class of Play	Maximum Uniformity Ratio Maximum:Minimum
Tournament	2:1
Club	2:1
Recreational	3:1

Minimum Mounting Height = 9 meters (30 feet).
All floodlights to be Types 5 or 6 (see footnote page 13-13).
* —These clearances are to be considered minimum; greater distances are desirable when space permits.

Distance	Meters	Feet
A	3.7	12
B	18.3	60
C	22.6	74
D	32.9	108
E	36.6	120

Fig. 13-24. Recommended lighting layouts for single and double outdoor tennis courts.

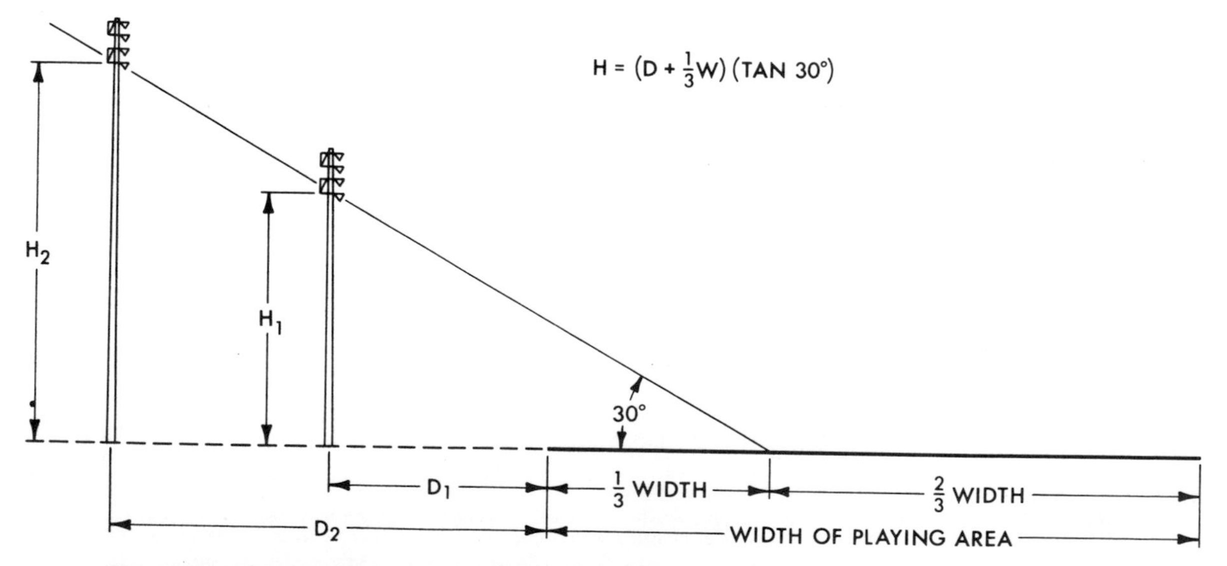

Fig. 13-25. For adequate mounting heights, a line drawn from a point one third the distance across the playing field to the lowest mounted floodlight should form an angle with the horizontal of not less than 30 degrees. In addition, minimum height for ground sports should not be less than 6 meters (20 feet); for aerial sports not less than 9 meters (30 feet).

aiming pattern. It is general practice to base spotting or aiming diagrams for certain sports employing a symmetric field (such as football, or for minor sports where relatively few floodlights are employed) on scale plots of the floodlight beam spread and the area to be lighted, previous calculations, and practical experience with similar installations. The procedure is as follows:

From an end elevation view, similar to that shown for a football field in Fig. 13-26, the vertical aiming of the floodlight beam axes can be determined to obtain approximately uniform horizontal illuminance across the field together with sufficient "spill," "direct filament," or "beam-edge" light in the space above to provide adequate illuminance to a height of approximately 15 meters (50 feet) above the field. In this connection, care should be taken to minimize the amount of light from the upper portion of the floodlight beams falling in the opposite stands. A limited number of point calculations based on a single group of floodlights aimed along a line perpendicular to the axis of the pole crossarms can be made without excessive difficulty to check the graphical appraisal of the proper vertical aiming angle, and such calculations will increase the accuracy of the aiming diagram; particularly if more than one row of floodlights from a single

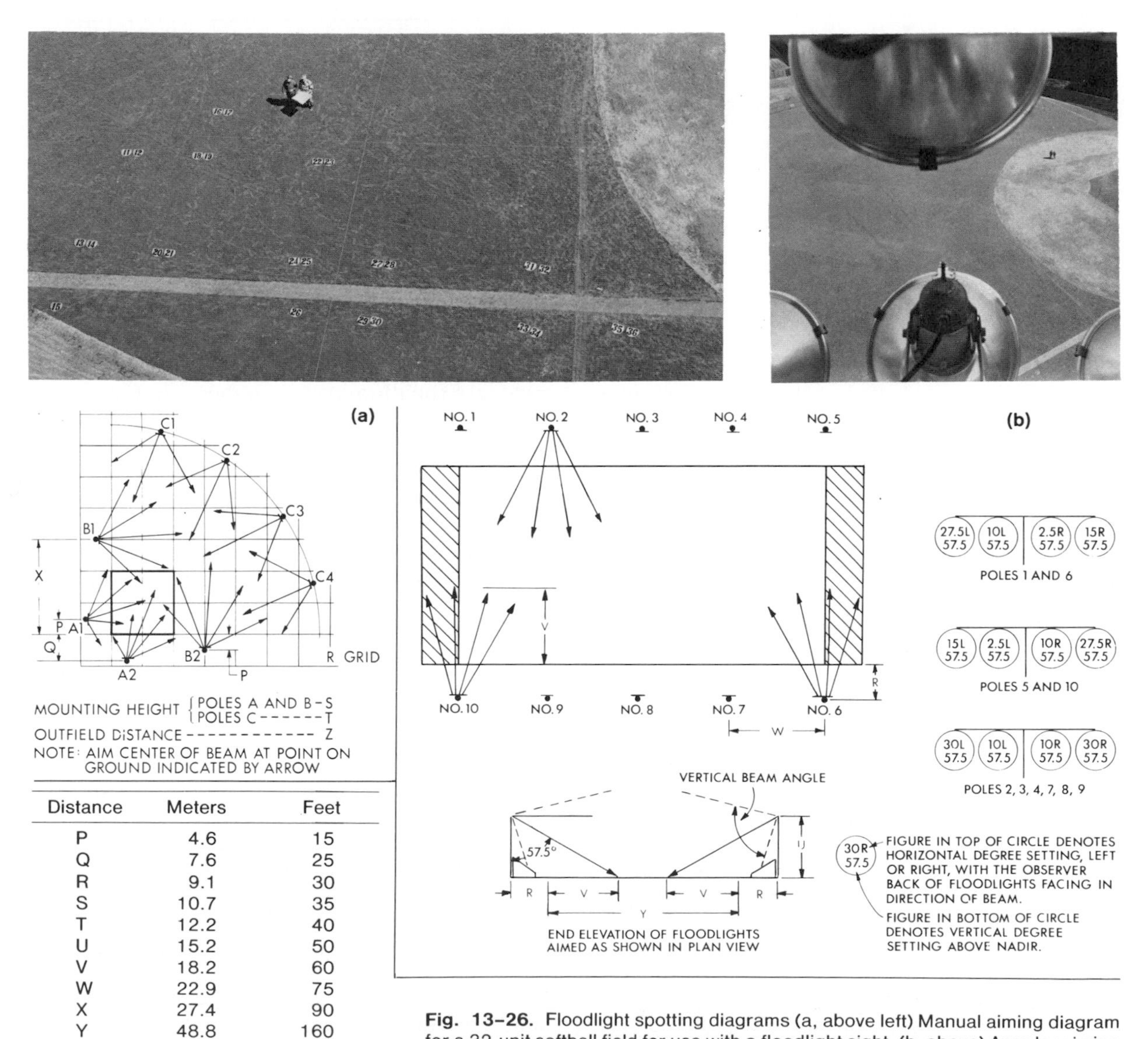

Distance	Meters	Feet
P	4.6	15
Q	7.6	25
R	9.1	30
S	10.7	35
T	12.2	40
U	15.2	50
V	18.2	60
W	22.9	75
X	27.4	90
Y	48.8	160
Z	61	200

Fig. 13-26. Floodlight spotting diagrams (a, above left) Manual aiming diagram for a 32-unit softball field for use with a floodlight sight. (b, above) Angular aiming diagram for a 40-unit football field.

pole is required. The plain view of the field makes it possible to plan horizontal aiming of the floodlights to provide approximately uniform horizontal illuminance in the longitudinal direction of the field.

It will be noted in Fig. 13-26b that relatively wide beam floodlights (Type 5) are used because the poles are close to the playing field. It will also be noted that the upper parts of the beams of the two sets of floodlights indicated fall in the opposite stands. However, since these are the wide beam type, the candlepower in the upper portions of the beam (more than 16 degrees from the beam center) will be low, and the spill brightness from them will be well within comfortable limits when evaluated with respect to the relatively high luminance of the field itself.

For installations differing appreciably from the standard recommendations or involving a large number of floodlights, it is desirable to obtain an aiming diagram prepared by the manufacturer of the lighting equipment.

Field Methods of Adjusting Floodlights

There are several ways to put spotting or aiming information to use in making an installation. First, and most accurate, is manual aiming of the floodlight beam centers at predetermined spots on the playing field area. This may be accomplished by using built-in beam sights, or by placing necessary beam sights against the floodlights parallel to their optical axes. Then by gridding both the aiming diagram and the playing area into sections, perhaps 9 meters (30 feet) square, markers can be placed at the aiming points on the field designated on the drawing, and the sights aimed at those points.

A second aiming method for directing the floodlights is to calculate or determine graphically from the aiming diagram, the vertical and horizontal angular settings of each floodlight. Most floodlights are equipped with degree scales which may then be set to those angles. The accuracy of this system may be less than that of the one described above unless the poles are set accurately and the crossarms carefully leveled and aligned. A difference of only a few degrees may move the beam center 6 meters (20 feet) or more on the field.

A third aiming method which may be used successfully with practice is to stand an observer on the field, short of the aiming point (so the line from the floodlight to the aiming point passes approximately through the observer's eyes), and observe the floodlight, preferably through binoculars. As the floodlight is moved by an assistant, the observer then estimates the position in which the lamp filament (or concentric reflector rings) appears exactly centered in the floodlight aperture. An alternate observation method that may be used with the narrowbeam type (specular reflector) floodlight is to light the lamp and, with smoked glasses on (preferably with binoculars), estimate when the entire reflector appears uniformly bright and at a maximum brightness. The latter methods are inherently less accurate than the first method but can be satisfactory when relatively large numbers of medium or wide beam floodlights are directed into the same general area.

ENERGY CONSIDERATIONS

Today's technology in luminaire, light source and control design provides the tools which, if properly applied, can make a significant contribution toward effective energy management. The following considerations and ideas deal specifically with energy management of sports and recreational areas. For additional information see Section 4.

1. Determine the activity to be carried on in the area. Use the illuminance as shown in Fig. 2-2 as a design basis. Then design the lighting for the activity. To conserve space many recreational areas are designed for a variety of activities—each with its own lighting requirement. For example, one area might include recreational baseball, Class V football and industrial softball, with the fields overlapping. Design lighting for the sport requiring the most light and include switching for turning off units not needed for the other sports. Due to placement of the fields it may be necessary to add poles and luminaires to adequately cover the other fields.
2. Luminaire efficiency is not necessarily the criterion for luminaire selection. The ability of the luminaire to control the light output so that it is effective on the area and not wasted off the area is the real measure. Use a luminaire which delivers the most "utilized" light.
3. Use of high intensity discharge lamps is on the increase in recreational area lighting. Mercury, metal halide and high pressure sodium lamps have long life and high luminous efficacy compared to incandescent lamps. Their use should be seriously considered.
4. Enough switching flexibility should be designed into the system so that if part of the area

is not being used, the luminaires for that area can be turned off. Or, if the area is to be used for only a specific time, it should be turned on and off automatically. Controls could be manually operated switches, time clocks, photoelectric controls, coin meters or the like.

5. Most arenas are designed for multiple uses. Basketball, hockey, public skating, ice shows, and other sports and non-sports activities may take place in the same arena. Each activity has its own recommendations. Various control schemes, including switching and dimming can provide the lighting flexibility required and reduce the energy used at the same time.

6. A relamping schedule should be established and luminaires cleaned periodically. Luminaires should be kept in good working condition with parts replaced as needed. To assist in this program a list should be provided covering operation, maintenance and replacement parts so that the system can be kept in design condition.

DESIGN FACTORS

Details of floodlighting calculations pertinent to sports lighting can be found in Section 9 of the 1981 Reference Volume and in reference 1.

REFERENCES

1. Committee on Sports and Recreational Areas of the IES: "Current Recommended Practice for Sports Lighting," *Illum. Eng.*, Vol. 64, p. 457, July 1969.*
2. Committee on Sports and Recreational Areas of the IES: "Tennis Court Lighting—A Revision to the IES Sports Lighting Practice," *J. Illum. Eng. Soc.*, Vol. 4, p. 292, July, 1975.
3. Committee on Sports and Recreational Areas and the Committee on Theatre, Television and Film Lighting of the IES: "Interim Report—Design Criteria for Lighting of Sports Events for Color Television Broadcasting," *Illum. Eng.*, Vol. 64, p. 191, March, 1969.

* Contains a very comprehensive list of sports lighting references.

Roadway Lighting

SECTION 14

The principal purpose for fixed lighting of public ways for both vehicles and pedestrians is to create a nighttime environment conducive to quick, accurate and comfortable seeing for the user of the facility. These factors, if attained, combine to improve traffic safety, achieve efficient traffic movement, and promote the general use of the facility during darkness and under a wide variety of weather conditions.

In the case of vehicles, and as a supplement to vehicular headlight illumination, fixed lighting can enable the motorist to see details more distinctly, locate them with greater certainty and with sufficient time to react safely toward roadway and traffic conditions present on or near the roadway facility. Pedestrians must be able to see with sufficient detail to readily negotiate the pedestrian facility and recognize the presence of other pedestrians, vehicles and objects in their vicinity. When proper application of fixed lighting principles and techniques are used, the visibility provided on these public ways can provide economic and social benefits to the public including:

(a) Reduction in night accidents and attendant human misery and economic loss;
(b) Prevention of crime and aid to police protection;
(c) Facilitation of flow of traffic;
(d) Promotion of business and industry during night hours; and
(e) Inspiration for community spirit and growth.

This section considers only fixed roadway lighting. See Section 16 for information on vehicle headlighting.

NOTE: References at the end of each section.

Factors Which Influence Seeing and Visibility

Most aspects of traffic safety involve visibility. The fundamental factors which directly influence visibility are:

1. The brightness of an object on or near the roadway.
2. The general brightness of the background of the roadway.
3. The size of an object and its identifying detail.
4. The contrast between an object and its surroundings.
5. The ratio of pavement luminance to the surroundings as seen by the observer.
6. The time available for seeing the object.
7. Glare.
8. The visual capabilities of the pedestrian and/or motorist (including windshield conditions, eyeglass conditions and/or physical conditions of the individual.)

Adequate visibility on public ways at night results from lighting (both fixed and vehicular) which provides adequate luminance contrasts with good uniformity, together with reasonable freedom from glare.

Method of Discernment

Discernment by Silhouette. An object is discerned by silhouette when the general luminance level of all or a substantial part of the object is lower or higher than the luminance of its background. This method of discernment predominates in the observation of distant objects on lighted roadways. Silhouette discernment depends upon the pavement surface reflectance.

Discernment by Surface Detail. When an object is seen by virtue of variations in brightness or color over its own surface, without regard to its contrast with its background, it is discerned by surface detail.

Glare in Roadway Lighting

The common term "glare", as it affects human vision, is subdivided into two components which are not completely independent but which are discussed separately below. These are:
1. *Disability Glare* (which may not be apparent to the observer). It acts to reduce the ability to see or spot an object. It is sometimes also referred to as "blinding glare" or "veiling glare."
2. *Discomfort Glare.* It produces a sensation of ocular discomfort but does not affect the visual acuity or the ability to discern an object.

While both forms of glare reactions are caused by the same light flux, the many factors involved in roadway lighting such as source size, displacement angle of the source, illuminance at the eye, adaptation level, surround luminance, exposure time, and motion do not affect both forms of glare in the same manner, nor to the same degree. The only two factors common to both forms of glare are illuminance at the eye and the angle of flux entrance into the eye. Even these factors have varying effects on the two forms of glare. It is generally true that when Disability Glare is reduced, it follows that there will also be a reduction in Discomfort Glare, but not necessarily in the same relative amount. On the contrary, it is entirely possible to reduce the Discomfort Glare of a system but at the same time increase the Disability Glare.

It is impossible to eliminate Disability Glare completely since the pavement, surrounding buildings, and the objects which are viewed have a definite luminance which projects some light flux into the eye.

The amount of Disability Glare can be calculated and measured. Unfortunately, such is not the case with Discomfort Glare, because it must be evaluated subjectively. Different people vary considerably in their appraisals of the borderline between comfort and discomfort.

Too often judgment is passed on the anticipated effectiveness of a roadway lighting system as a result of casual visual static observations. It should be recognized that in such instances an appraisal can only be made of the Discomfort Glare effect since the eyes may not be conscious of and cannot evaluate Disability Glare as such. Often this becomes misleading, especially when observing a single luminaire at close range rather than a system of several luminaires at different distances. It should also be recognized that comparing a single large area luminaire with a smaller one, especially at short distance, can be misleading as to the anticipated glare effects of an entire system.

ROADWAY, WALKWAY, BIKEWAY, AND AREA CLASSIFICATIONS[1]

Roadway, Walkway and Bikeway Classifications

Freeway. A divided major roadway with full control of access and with no crossings at grade. This definition applies to toll as well as nontoll roads.

Expressway. A divided major roadway for through traffic with partial control of access and generally with interchanges at major crossroads. Expressways for noncommercial traffic within parks and park-like areas are generally known as parkways.

Major. The part of the roadway system that serves as the principal network for through traffic flow. The routes connect areas of principal traffic generation and important rural highways entering the city.

Collector. The distributor and collector roadways serving traffic between major and local roadways. These are roadways used mainly for traffic movements within residential, commercial and industrial areas.

Local. Roadways used primarily for direct access to residential, commercial, industrial, or other abutting property. They do not include roadways carrying through traffic. Long local roadways will generally be divided into short sections by collector roadway systems.

Alleys. Narrow public ways within a block, generally used for vehicular access to the rear of abutting properties.

Sidewalks. Paved or otherwise improved areas for pedestrian use, located within public street rights-of-way which also contain roadways for vehicular traffic.

Pedestrian Way. A public walk for pedestrian traffic not necessarily within the right-of-way for a vehicular traffic roadway. Included are skywalks (pedestrian overpasses), subwalks (pedestrian tunnels), walkways giving access to parks or block interiors and midblock street crossings.

Isolated Interchange. A grade separated roadway crossing which is not part of a continuously lighted system, with one or more ramp connections with the crossroad.

Isolated Intersection. The general area where two or more noncontinuously lighted roadways join or cross at the same level. This area includes the roadway and roadside facilities for traffic movement in that area. A special type is the channelized intersection in which traffic is directed into definite paths by islands with raised curbing.

Bikeway. A public street, highway or separate path, identified as part of a bicycle travel network. Bikeways may consist of the following:
1. *Type-A Bikeway.* A strip within or adjacent to a public roadway or shoulder, marked for bicycle travel;
2. *Type-B Bikeway.* An improved strip identified for public bicycle travel and located away from a roadway or its adjacent sidewalk system.

Area Classifications

Commercial. That portion of a municipality in a business development where ordinarily there are large numbers of pedestrians during business hours. This definition applies to densely developed business areas outside, as well as within, the central part of a municipality. The area contains land use which attracts a relatively heavy volume of nighttime vehicular and/or pedestrian traffic on a frequent basis.

Intermediate. That portion of a municipality often characterized by a moderately heavy nighttime pedestrian activity such as in blocks having libraries, community recreation centers, large apartment buildings or neighborhood retail stores.

Residential. A residential development, or a mixture of residential and commercial establishments, characterized by a few pedestrians at night. This definition includes areas with single family homes, town houses and/or small apartment buildings.

Rural. Open land with little or no commercial or residential development.

CLASSIFICATION OF LUMINAIRE LIGHT DISTRIBUTIONS[1]

Proper distribution of the light flux from luminaires is one of the essential factors in efficient roadway lighting. The light emanating from the luminaires is directionally controlled and proportioned in accordance with the requirements for seeing and visibility. Light distributions are generally designed for a typical range of conditions which include luminaire mounting height, transverse (overhang) location of the luminaires, longitudinal spacing of luminaires, widths of roadway to be effectively lighted, arrangement of luminaires, percentage of lamp light directed toward the pavement and adjacent areas, and maintained efficiency of the system.

Several methods have been devised for showing the light distribution pattern from a luminaire. (See Figs. 14-1 through 14-5). For practical operating reasons the range in luminaire mounting heights may be kept constant. Therefore, it becomes necessary to have several different light distributions in order to light effectively different roadway widths, using various luminaire spacing distances at a particular luminaire mounting height. All luminaires can be classified according to their *lateral* and *vertical* distribution patterns. Different *lateral* distributions are available for different *street width-to-mounting height ratios.* Different *vertical* distributions are available for different *spacing-to-mounting height ratios.* Distributions with higher vertical angles of maximum candlepower emission are necessary to obtain the required uniformity of

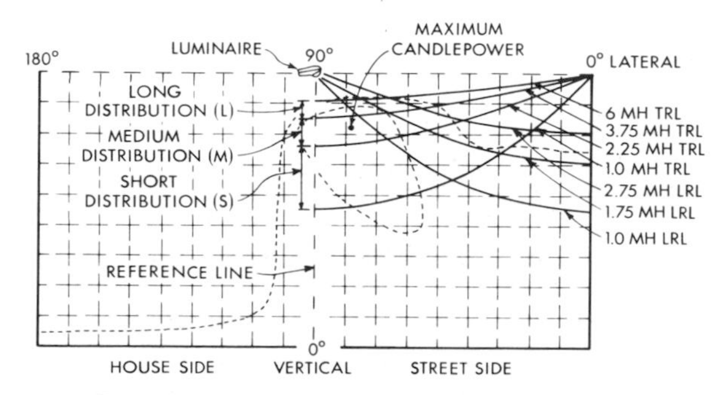

Fig. 14-1. Recommended vertical light distribution boundaries on a rectangular coordinate grid (representation of a sphere). Dashed lines are isocandela traces.

illuminance where longer luminaire spacings are used (as on residential and light traffic roadways). These higher vertical emission angles produce a more favorable pavement luminance which may be desired for silhouette seeing, where traffic volume is relatively light. Distributions with lower vertical angles of maximum candlepower emission are used in order to reduce system glare. This becomes more important when using high lumen output lamps. The lower the emission angle, the closer the luminaire spacing must be to obtain required illuminance uniformity. Therefore, to achieve specific illuminance results it becomes necessary as a part of any lighting system design to consider and to check the uniformity of illuminance by checking ratios of average illuminance to minimum illuminance.

Luminaire light distribution may be classified in respect to three criteria:

1. Vertical light distribution.
2. Lateral light distribution.
3. Control of light distribution above maximum candlepower.

Classification of light distribution should be made on the basis of an isocandela diagram which, on its rectangular coordinate grid, has superimposed a series of *longitudinal roadway lines* (LRL) in multiples of *mounting height* (MH) and a series of *transverse roadway lines* (TRL) in multiples of mounting height. The relationship of LRL and TRL to an actual street and the representation of such a web are shown in Figs. 14-1 through 14-5. The minimum information which should appear on such an isocandela diagram for classification is as follows:

(a) LRL lines of 1.0 MH, 1.75 MH and 2.75 MH;
(b) TRL lines of 1.0 MH, 2.25 MH, 3.75 MH, 6.0 MH and 8.0 MH;
(c) Maximum candlepower location and half maximum candlepower trace; and
(d) Candlepower lines equal to the numerical values of 2½, 5, 10 and 20 per cent of the rated bare lamp lumens.

Vertical Light Distributions. Vertical light distributions are divided into three groups: short (S), medium (M) and long (L). See Figs. 14-1 and 14-4.

Short Distribution. A luminaire is classified as having a short light distribution when its maximum candlepower point lies in the "S" zone of the grid which is from the 1.0-MH TRL to less than the 2.25-MH TRL. See Figs. 14-1 and 14-2 (Note 2A).

Medium Distribution. A luminaire is classified as having a medium light distribution when its maximum candlepower point lies in the "M" zone of the grid which is from the 2.25-MH TRL to less than the 3.75-MH TRL. See Figs. 14-1 and 14-2 (Note 2B).

Long Distribution. A luminaire is classified as having a long light distribution when its maximum candlepower point lies in the "L" zone of the web which is from the 3.75-MH TRL to less than the 6.0-MH TRL. See Figs. 14-1 and 14-2 (Note 2C).

Lateral Light Distributions. Lateral light distributions (see Figs. 14-3 and 14-4) are divided into two groups based on the location of the luminaire as related to the area to be lighted. Each group may be subdivided into divisions with regard to the width of the area to be lighted in terms of the MH ratio. Only the segments of the half maximum candlepower isocandela trace which fall within the longitudinal distribution range, as determined by the point of maximum candlepower (short, medium or long), are used for the purpose of establishing the luminaire distribution width classification.

Luminaires At or Near Center of Area. The

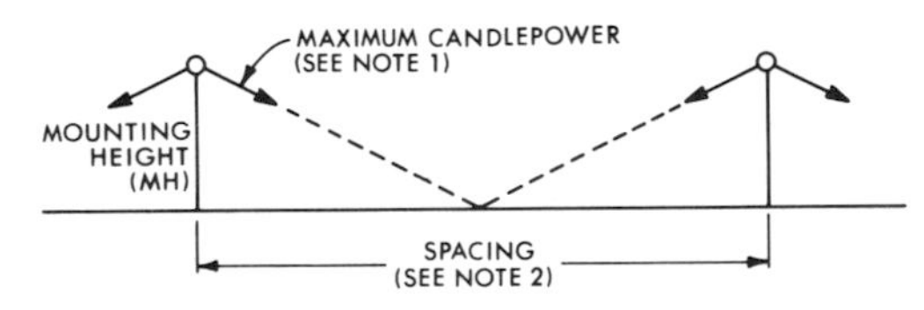

Note 1: Maximum candlepower beams from adjacent luminaires should at least meet on the road surface.
Note 2. Maximum luminaire spacing generally is less than:
"A"—Short Distribution—4.5 MH;
"B"—Medium Distribution—7.5 MH; and
"C"—Long Distribution—12.0 MH

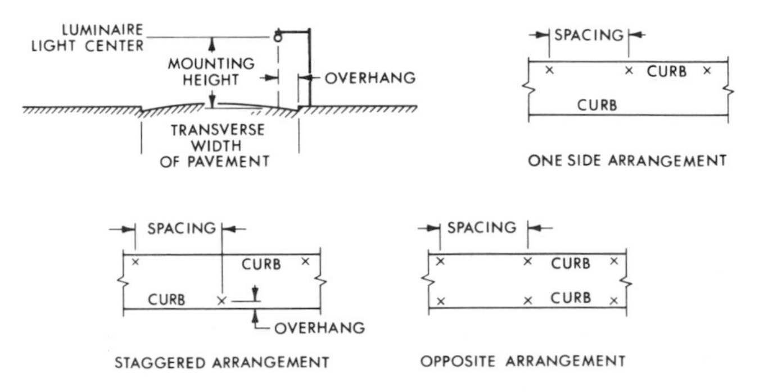

Fig. 14-2. Typical lighting layouts showing spacing-to-mounting height relationships and terminology with respect to luminaire arrangement and spacing.

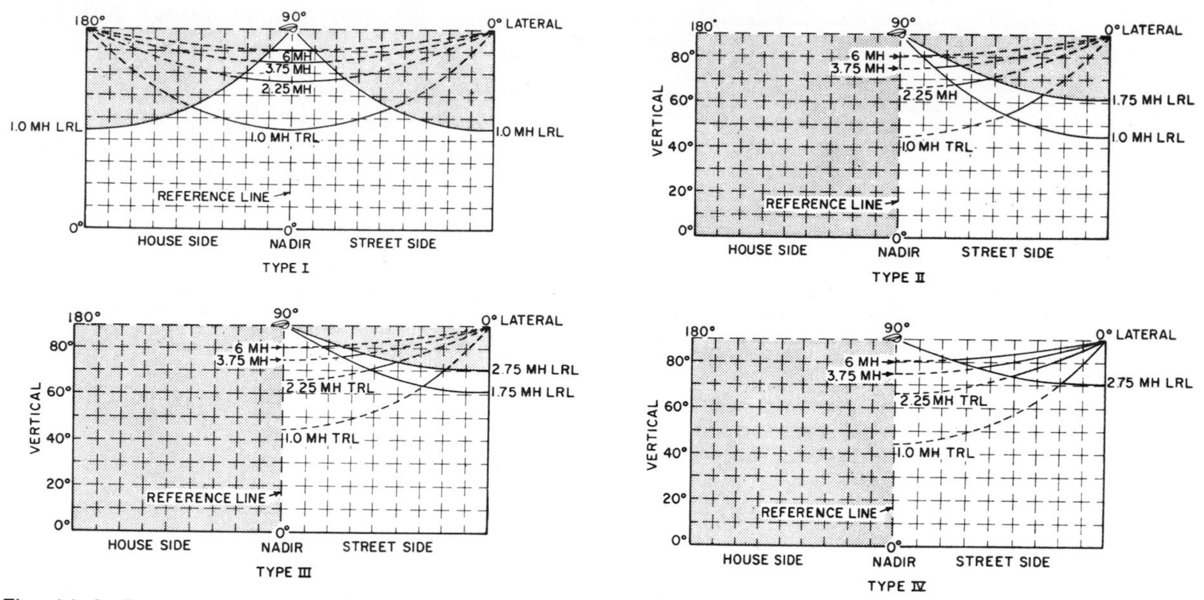

Fig. 14-3. Recommended lateral light distribution boundaries on a rectangular coordinate grid (representation of a sphere).

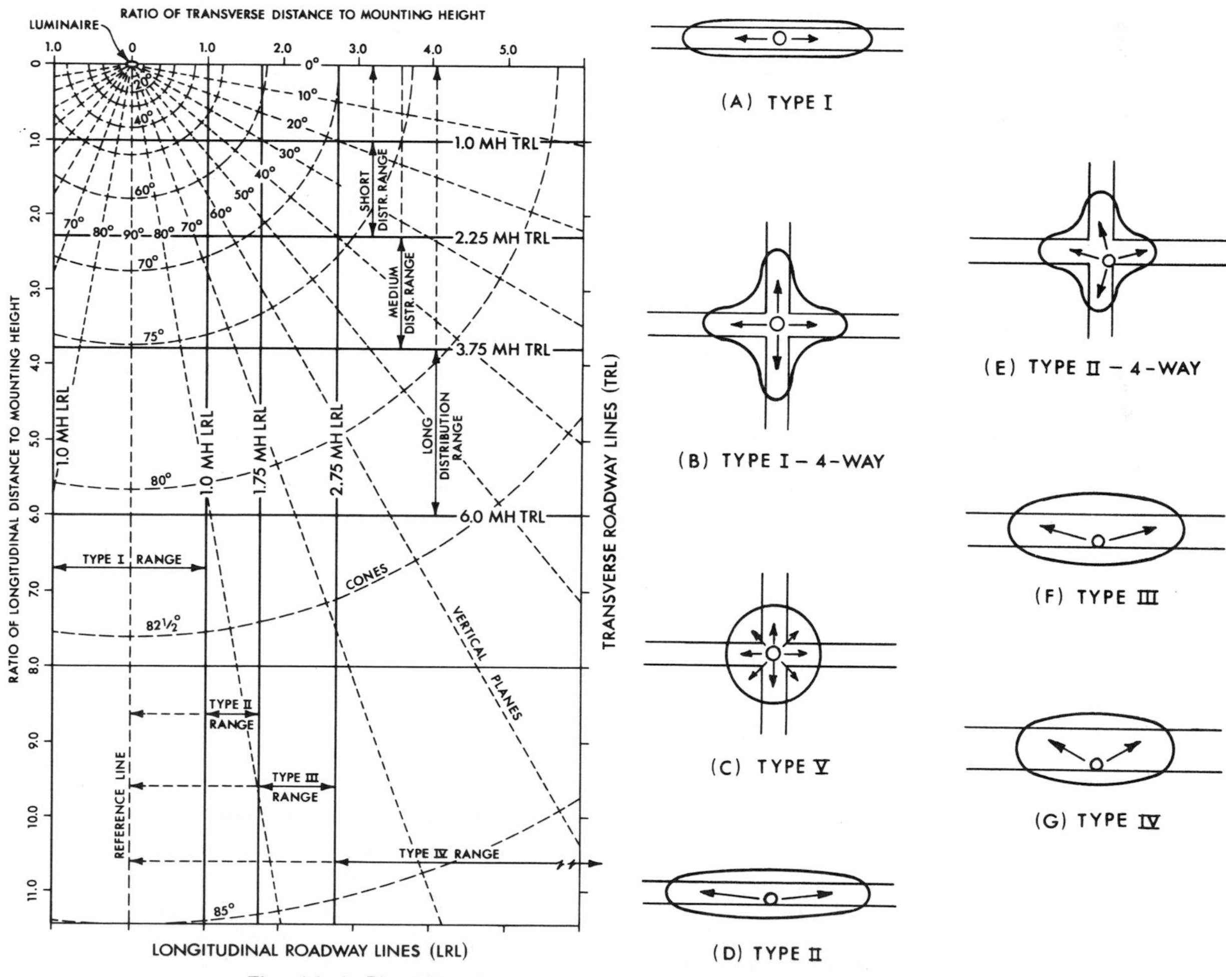

Fig. 14-4. Plan view of roadway coverage for different types of luminaires.

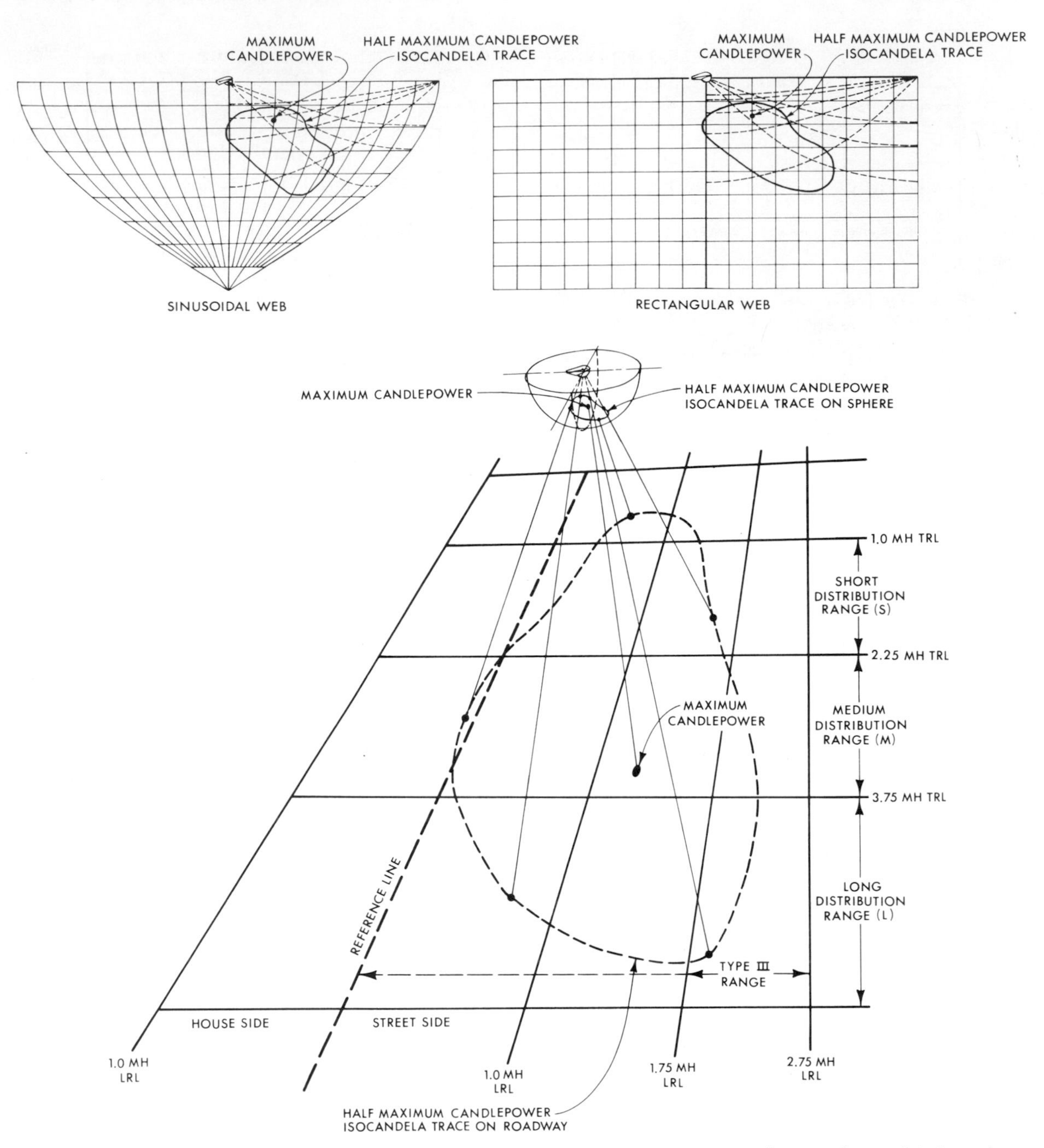

Fig. 14-5. Diagram showing projection of maximum candlepower and half maximum candlepower isocandela trace from a luminaire having a Type III—Medium distribution, on the imaginary sphere and the roadway. Sinusoidal web and rectangular web representation of sphere are also shown, with maximum candlepower and half maximum candlepower isocandela trace.

group of lateral width classifications which deals with luminaires intended to be mounted at or near the center of the area to be lighted has similar light distributions on both the "house side" and the "street side" of the reference line.

Type I. A distribution is classified as Type I when its half maximum candlepower isocandela trace lies within the Type I width range on both sides of the reference line which is bounded by 1.0-MH house side LRL and 1.0-MH street side LRL within the longitudinal distribution range (short, medium or long distribution) where the point of maximum candlepower falls. (See Fig. 14-4A.)

Type I Four-Way. A distribution is classified as a Type I four-way when it has four beams of the width as defined for Type I above. (See Fig. 14-4B.)

Type V. A distribution is classified as Type V when the distribution has a circular symmetry of candlepower distribution which is essentially the same at all lateral angles around the luminaire. (See Fig. 14-4C.)

Luminaires Near Side of Area. The group of lateral width classifications which deals with luminaires intended to be mounted near the side of the area of the lighted vary as to the width of distribution range on the street side of the reference line. The house side segment of the half maximum candlepower isocandela trace within the longitudinal range in which the point of maximum candlepower falls (short, medium or long) may or may not cross the reference line. In general it is preferable that the half maximum candlepower isocandela trace remains near the reference line. The variable width on the street side is as defined.

Type II. A distribution is classified as Type II when the street side segment of the half maximum candlepower isocandela trace within the longitudinal range in which the point of maximum candlepower falls (short, medium or long) does not cross the 1.75-MH street side LRL. (See Fig. 14-4D.)

Type II Four-Way. A distribution is classified as a Type II four-way when it has four beams each of the width on the street side as defined for Type II above. (See Fig. 14-4E.)

Type III. A distribution is classified as Type III when the street side segment of the half maximum candlepower isocandela trace within the longitudinal range in which the point of maximum candlepower falls (short, medium or long) lies partly or entirely beyond the 1.75-MH street side LRL, but does not cross the 2.75-MH street side LRL. (See Fig. 14-4F.)

Type IV. A distribution is classified as Type IV when the street side segment of the half maximum candlepower isocandela trace within the longitudinal range in which the point of maximum candlepower falls (short, medium or long) lies partly or entirely beyond the 2.75-MH street side LRL. (See Fig. 14-4G.)

Control of Distribution Above Maximum Candlepower. Although the pavement luminance generally increases when increasing the vertical angle of light flux emission, it should be emphasized that the disability and discomfort glare also increase. However, since the respective rates of increase and decrease of these factors are not the same, design compromises become necessary in order to achieve balanced performance. Therefore, varying degrees of control of candlepower in the upper portion of the beam above maximum candlepower are required. This control of the candlepower distribution is divided into three categories.

Cutoff. A luminaire light distribution is designated as cutoff when the candlepower per 1000 lamp lumens does not numerically exceed 25 (2½ per cent) at an angle of 90 degrees above nadir (horizontal), and 100 (10 per cent) at a vertical angle of 80 degrees above nadir. This applies to any lateral angle around the luminaire.

Semicutoff. A luminaire light distribution is designated as semicutoff when the candlepower per 1000 lamp lumens does not numerically exceed 50 (5 per cent) at an angle of 90 degrees above nadir (horizontal), and 200 (20 per cent) at a vertical angle of 80 degrees above nadir. This applies to any lateral angle around the luminaire.

Noncutoff. The category when there is no candlepower limitation in the zone above maximum candlepower.

Variations and Comments. With the variations in roadway width, type of surface, luminaire mounting height, and spacing which may be found in actual practice, there can be a large number of "ideal" lateral distributions. For practical applications, however, a few types of lateral distribution patterns may be preferable to many complex arrangements. This simplification of distribution types will be more easily understood and consequently there will be greater assurance of proper installation and more reliable maintenance.

When luminaires are tilted upward it raises the angle of the street side light distribution. Features such as cutoff or width classification may be changed appreciably. When the tilt is planned the luminaire should be photometered and the light distribution classified in the position in which it will be installed.

Types I, II, III and IV lateral light distributions should vary across transverse roadway lines other than that including the maximum candlepower in order to provide adequate coverage of the rectangular roadway area involved. The width of the lateral angle of distribution required to cover adequately a typical width of roadway varies with the vertical angle or length of distribution as shown by the TRL (transverse roadway line). For a TRL 4.5 MH, the lateral angle of distribution for roadway coverage is obviously narrower than that required for a TRL 3.0 MH or for a TRL 2.0 MH.

For typical roadway conditions it is desirable to approach very closely the light distributions prescribed. Purposeful variations from these distributions are permissible when such variations become necessary. Several examples of these purposeful variations are:

1. Linear source luminaires which provide broad Type I or Type II distributions and which project the maximum candlepower lower than specified.
2. Directional lighting for one-way streets and divided highways, where the light projected in the direction of traffic is substantially reduced in the high vertical angle.
3. Linear source luminaires parallel to the street to obtain reduced glare and increased utilization.
4. Luminaires mounted at low mounting heights.
5. Types IV and V luminaire distributions with extra upward light for illuminating building fronts.
6. "Offset mounting" style luminaires designed to be located at a lateral distance from the area to be lighted.

For "high mast" installations involving multiple luminaires on one structure or support, the entire group of luminaires may be considered as a single composite luminaire for purposes of determining distribution type, cutoff classification, maximum candlepower, etc. Photometric data may be supplied in this form.

Other purposeful variations from the distributions specified may be found advantageous from time to time for special applications.

Fig. 14-6. Recommendations for Roadway Average Maintained Illuminance on the Horizontal

Vehicular Roadway Classification*	Urban**					
	Commercial		Intermediate		Residential	
	Lux	Foot-candles	Lux	Foot-candles	Lux	Foot-candles
Freeway†	6	0.6	6	0.6	6	0.6
Expressway†	15	1.4	13	1.2	11	1.0
Major	22	2.0	15	1.4	11	1.0
Collector	13	1.2	10	0.9	6	0.6
Local	10	0.9	6	0.6	4	0.4
Alleys	6	0.6	4	0.4	4	0.4

Note: The recommended illuminance values shown are meaningful only when designed in conjunction with other elements. The most critical elements as described in this section are as follows:

(a) Light depreciation
(b) Quality
(c) Uniformity
(d) Luminaire mounting heights
(e) Spacing
(f) Transverse location of luminaires
(g) Luminaire selection
(h) Traffic conflict areas
(i) Border areas
(j) Transition lighting
(k) Alleys
(l) Roadway lighting layouts

* See page 14-2.
** See page 14-3.
† Both mainline and ramps.

LIGHTING DESIGN[1]

The design of a lighting system involves many variables which include visibility factors, economic and esthetic considerations, as well as equipment and material usage for optimal effectiveness. The design process follows these major steps:

1. Determination from roadway or walkway classification, and adjacent land use (area classification), of the quantity of light desired in average illuminance on the horizontal. (See Figs. 14-6 and 14-7.)
2. Formulation of a tentative concept as to luminaire location and mounting height relative to the roadway to be lighted. Lamps in general use today in roadway lighting are shown in Figs. 8-88, 8-97, 8-98, 8-120, and 8-126 in the 1981 Reference Volume.
3. Selection of a luminaire light distribution type to be used. (See Fig. 14-8.)
4. Detailed calculations using several tentative light source types and sizes, luminaires, mounting heights and maintenance conditions to determine spacings, luminaire locations and illuminance levels achieved (average and minimum).
5. Comparative calculations on several possible systems to determine relative factors of uniformity, economics and glare control.
6. Selection of final design or re-entry of design process at any step above.

It is important that roadway lighting be planned on the basis of traffic information which includes the factors necessary to provide traffic safety and pedestrian security. Some of the factors applicable to the specific problem which are to be carefully evaluated are:

(a) Type of lane use development (Area Classification) abutting the roadway or walkway;
(b) Type of route (Roadway, Walkway or Bikeway Classification);
(c) Traffic accident experience;

Fig. 14-7. Recommendations for Average Maintained Illuminance on the Horizontal for Pedestrian Ways

Walkway and Bikeway Classification*	Minimum Average Levels		Average Levels for Special Pedestrian Security**			
			Mounting Heights 3 to 5 meters [9 to 15 feet]		Mounting Heights 5 to 10 meters [15 to 30 feet]	
	Lux	Footcandles	Lux	Footcandles	Lux	Footcandles
Sidewalks (roadside) and Type A bikeways						
Commercial areas	10	0.9	22	2.0	43	4.0
Intermediate areas	6	0.6	11	1.0	22	2.0
Residential areas	2	0.2	4	0.4	9	0.8
Walkways distant from roadways and Type B bikeways						
Park walkways and bikeways	5	0.5	6	0.6	11	1.0
Pedestrian tunnels	43	4.0	54	5.0	—	—
Pedestrian overpasses	3	0.3	4	0.4	—	—
Pedestrian stairways	6	0.6	9	0.8	—	—

Note: Crosswalks traversing roadways in the middle of long blocks and at street intersections should be provided with additional illumination producing from 1.5 to 2 times the roadway illuminance level recommended in Fig. 14-6.

* See pages 14-2 and 14-3.

** For pedestrian identification at a distance.

(d) Street crime experience and security requirements; and
(e) Roadway geometric features.

Illuminance Requirements

The recommended illuminance values are given in Figs. 14-6 and 14-7. They represent the lowest average levels which are currently considered appropriate for the kinds of roadways or walkways in the various areas. Numerous installations have been made at higher values. Furthermore, the recommendations assume design of proper uniformity and use of applicable types of luminaire light distributions, lamp sizes, mounting heights, spacings and transverse locations. These values do not represent initial illuminance, but should be in-service values of systems designed with proper light loss factors. These elements are reviewed in subsequent paragraphs.

Fig. 14-6 values are for urban areas. Research has not yet defined lighting values for roadways in rural areas. The levels specified in the "Residential" column for the appropriate roadway are often used when rural areas are lighted.

Light Depreciation

The recommended values in Figs. 14-6 and 14-7 represent average illuminance when the luminaires are at their lowest output. This condition occurs just prior to lamp replacement and luminaire washing. It is impossible to attempt the design of a lighting system without knowing in advance the light losses to be expected. Since illuminance values depreciate by as much as 50 per cent or more between relamping and luminaire washing cycles, it is imperative to use lamp lumen depreciation (LLD) and luminaire dirt depreciation (LDD) factors which are valid and based on realistic judgment. See page 14-26 for the use of light loss factors in roadway calculations.

Fig. 14-8. Guide for Luminaire Lateral Light Type and Placement

Side of the Roadway Mounting			Center of the Roadway Mounting		
One Side or Staggered	Staggered or Opposite	Grade* Intersection	Single Roadway	Twin Roadways (median mounting)	Grade* Intersections
Width up to 1.5 MH	Width beyond 1.5 MH	Width up to 1.5 MH	Width up to 2.0 MH	Width up to 1.5 MH (each pavement)	Width up to 2.0 MH
Types II-III-IV	Types III and IV	Type II 4-way	Type I	Types II and III	Types I 4-way and V

Note: In all cases suggested maximum longitudinal spacings and associated vertical distribution classifications are: Short distribution = 4.5 MH, Medium distribution = 7.5 MH, and Long distribution = 12.0 MH.

* Local street intersection.

Quality

Quality of lighting relates to the relative ability of the light available to provide the contrast differences in the visual scene in such a manner that people may make quick, accurate and comfortable recognition of the cues required for the seeing task. The quality of lighting of installation "A" is higher than that of installation "B" if, with the same average illuminance level, visual recognition of typical tasks is faster, easier and/or done more comfortably under installation "A."

Many factors are interrelated to produce a high quality of lighting. The following factors are involved, but quantitative values and relative importance cannot be given:

1. Disability glare and discomfort glare should be minimized.
2. Reflected glare will conceal some contrast differences.
3. Pavement luminance if increased will improve contrast situations.
4. Light on vertical surfaces is desirable.
5. Uniformity of horizontal and vertical illuminance, as well as uniformity of pavement luminance and other background areas, affects quality.

It should be recognized that in many instances, changes intended to optimize one factor relating to quality will adversely affect another and the resultant total quality of the installation may be degraded.

Uniformity

The illuminance values in Figs. 14-6 and 14-7 are minimum and provide effective visibility only when combined with uniformity, or relatively even illumination spread on the pavement and sidewalks.

Uniformity may be expressed in several ways. The average-level-to-minimum point method uses the average illuminance on the roadway design area divided by the lowest value at any point in the area. Under this method, the average-to-minimum ratio should not exceed 3 to 1 for any roadway in Fig. 14-6 excepting local residential streets, which may have a ratio as high as 6 to 1.

Luminaire Mounting Height

Mounting heights of luminaires have, in general, increased substantially during the past several decades. The advent of modern, more efficient and larger size (lumen output) lamps has been the basic reason. Engineers have increased mounting heights in order to obtain economic and esthetic gains in addition to increased illuminance uniformity when utilizing the newer large lamps. Mounting heights of 12 meters [40 feet] and higher are used along roadways and the cluster mounting of luminaires is used at interchanges. The advent of suitable servicing equipment and lowering devices has made this practical.

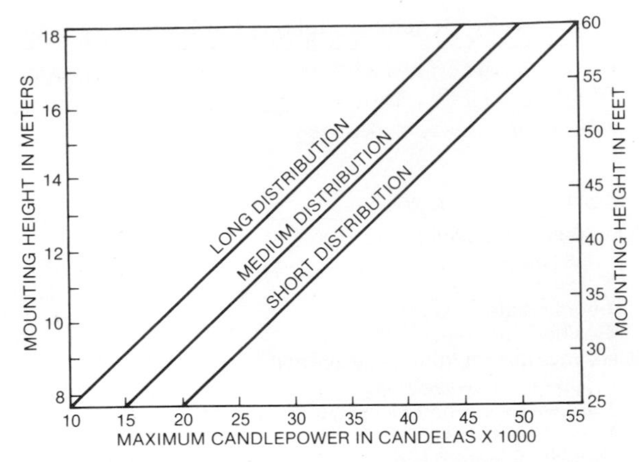

Fig. 14-9. Minimum luminaire mounting heights based on current practice.

During this same period there has been a trend to lower mounting heights in some cases. In general, this has been due to esthetic considerations. An example is the use of pole top mounted luminaires in residential areas.

When designing a system, mounting height should be considered in conjunction with spacing and lateral positioning of the luminaires as well as the luminaire type and distribution. (See Fig. 14-9.) Uniformity and illuminance levels should be maintained as recommended regardless of the mounting height selected.

Increased mounting height may, but will not necessarily, reduce discomfort glare and disability veiling luminance. It increases the angle between the luminaires and the line of sight to the roadway; however, luminaire light distributions and candlepower also are significant factors. Glare is dependent on the flux reaching the observer's eyes from all luminaires in the scene.

High Mast Lighting. Multilevel interchanges or roadway sections consisting of multiple traffic lanes may be advantageously illuminated by the use of high mast type units where high intensity sources are suspended in clusters at heights of 25 to 55 meters [80 to 180 feet]. Such lighting design offers advantages regarding traffic safety due to the reduction in the number of luminaire supports. The high mast units also offer greater flexibility in luminaire support location.

Luminaire Spacing

The spacing of luminaires is often influenced by the location of utility poles, block lengths, property lines, roadway configurations and the terrain features. It is generally a more economical practice to use larger lamps at reasonable spacings and mounting heights than to use smaller lamps at more frequent intervals with lower mounting heights. This is usually in the interest of good lighting provided the spacing-to-mounting height ratio is within the range of light distribution for which the luminaire is designed. The desired ratio of lowest illuminance at any point on the pavement to the average illuminance should be maintained. The disregarding of luminaire light distribution characteristics and the exceeding of maximum spacing-to-mounting height ratios can cause loss of visibility of objects between luminaires.

Transverse Location of Luminaires

Types II, III and IV luminaires are intended (unless designated as offset luminaires) to be mounted over or near the edge of the roadway. Types I and V are generally designed to be mounted over or near the center of the area to be lighted. Usually, luminaire overhang exceeding 0.25 mounting height does not contribute to visibility and often increases system glare and cost.

Optimum luminaire location is best determined by reference to the photometric data showing illumination distribution and utilization. Other factors that must be considered are:

1. Access to luminaires for servicing.
2. Vehicle-pole collision probabilities.
3. System glare aspects.
4. The visibility (both day and night) of traffic signs and signals.
5. Esthetic appearance.
6. Trees.

Luminaire Selection

Luminaire light distribution classifications are intended to serve as a means of selecting a luminaire which is a good candidate for further calculations to determine if it is optically and economically suitable for lighting a particular roadway from the proposed mounting height and mounting location. Within the boundaries of the light distribution system described above, a multitude of combinations are possible. Fig. 14-8, however, tabulates preferred lateral light distributions and maximum recommended longitudinal spacings for various geometric factors encountered in common practice.

The fact that a luminaire is assigned a particular classification does not assure the designer that it will produce the recommended quantity and quality of illumination for the roadway configuration and mountings shown in Figs. 14-6 and 14-7. The relative amount and control of light in areas other than the cone of maximum candlepower are equally important in producing good visibility in the final system and are not considered in the classification system.

There should be no hesitation in using a luminaire of a particular classification on roadway configurations other than those recommended in Fig. 14-8 if adequate calculations confirm that it meets performance requirements.

Traffic Conflict Areas

The illuminances in Fig. 14-6 are for roadway sections which are approximately straight and nearly level. Intersecting, converging or diverging roadway areas require higher illuminances. The illuminance within these areas should at least be equal to the *sum* of the values recommended for each roadway which forms the intersection. Such areas include ramp divergences or connections with streets or freeway mainlines. Very high volume driveway connections to public streets and midblock pedestrian crosswalks should be illuminated at least to a 50 per cent higher level than the average route value.

Border Areas

There is value in illuminating areas beyond the roadway proper provided it is appropriate to the environment and not objectionable to the adjacent property use. It is desirable to widen the narrow visual field into the peripheral zone in order to reveal objects and enhance eye adaptation. It also improves depth perception and perspective thus facilitating the judgment of speed, distance, etc. Such illumination should normally diminish gradually away from the road.

Border areas and median strips often are landscaped, attractive areas. Both their daytime and nighttime esthetic appearance often can be enhanced by roadway lighting. This should be considered at the time of system design and is a factor to be considered in planning the illumination and selecting the hardware and street furniture.

Where extraneous light intrudes on adjacent properties, shielding or other suitable means may be used to contain the light within the desired space.

Transition Lighting

It is good practice to gradually decrease brightness in the driver's field of view when emerging from a lighted section of roadway. This may be accomplished by extending the lighting system in each exit direction using approximately the same spacing and mounting height but graduating the size of the lamp used. A recommended procedure is to utilize the design value for the roadway as the calculation base. Using the design speed of the roadway, the reduced lighting-level sectors should be illuminated for a 15-second continuous exposure to the sector illuminance level of one-half of the preceding higher lighted sector, but the average illuminance in the terminal sector should not be less than 2.7 lux (0.25 footcandle) nor more than 5.4 lux (0.5 footcandle).

Alleys

Experience has proved that well lighted alleys reduce the criminal's opportunity to operate and hide under cover of darkness. Alleys should be adequately lighted to facilitate police patrolling from sidewalks and cross streets, especially in commercial areas. Generally, such lighting also meets vehicular traffic needs.

Pedestrian Walkways and Bikeways

Proper and adequate lighting of walkway and bikeway areas is essential to the safe and comfortable usage of such areas by pedestrians (herein assumed to include cyclists). In the great majority of instances, walkways and bikeways are located adjacent to illuminated roadways and no specific or separate lighting is provided for such pedestrian ways other than the incidental lighting afforded from the house side distribution of roadway luminaires. All too often, such incidental lighting does not produce the proper quality or illuminance for the comfort and safety of pedestrians unless the designer reviews the walkway or bikeway lighting segment and makes modifications to the roadway lighting system to correct any deficiencies. It is recommended that all initial roadway lighting designs be checked for conformance to illuminance requirements prescribed for pedestrian ways adjacent to roadways and that revisions and/or additions be made when and as necessary to achieve complete conformity. Photometric data provided by the supplier of the roadway luminaires can be used for the checking and designing of sidewalk or roadside bikeway illumination as well as for roadway illumination. Where the primary goal of a particular lighting system is to provide adequate illumination for sidewalks and Type A (roadside) bikeways, initial designs should give greater attention to such pedestrian ways with subsequent checks made for conformance to roadway lighting requirements.

Illuminating those walkways and bikeways (Type B) which are not associated with or are substantially distant from roadways permits a greater freedom of system and luminaire design. The designer should exercise good judgment in providing a quality of light that is particularly suitable for these pedestrian ways and in accordance with recommendations.

The recommended values of walkway and bikeway illuminance located in various types of areas are listed in Fig. 14-7. These represent average maintained illuminances on horizontal surfaces and should be considered as minimum, particularly where security and/or pedestrian identification at a distance is important. Visual identification of other pedestrians and objects along walkways is dependent to a great degree on vertical surface illumination. Therefore, different values are shown in the table for two common ranges of mounting heights. This variation in horizontal levels will compensate in part for the need for the vertical component of lighting. Recommended illuminance levels for mounting heights at or near 5 meters [15 feet] are intended to be the average of the levels given for 3 to 5 meters [9 to 15 feet] and 5 to 10 meters [15 to 30 feet] ranges. In selecting luminaires for mounting heights over 10 meters [30 feet] (lighting levels not shown), due care should be exercised to avoid conditions of illumination which will not permit easy identification of distant pedestrians.

To provide well-illuminated surrounds for such pedestrian ways as walkways and bikeways (Type B) through parks, it is further recommended that the area bordering these pedestrian ways for a width of 2.5 meters [8 feet] on each side be lighted to levels of at least one-third (⅓) that suggested for the walkway or bikeway. This is also applicable to similar marginal areas such as depressed entrances to building basements, gaps between building fronts, and dense shrubbery, and other locations where pedestrian safety is of utmost concern.

The average-to-minimum uniformity ratio in illuminating pedestrian ways where special pedestrian security is not essential should not exceed 4 to 1, except for residential sidewalks and Type A bikeways in residential areas, where a ratio of 10 to 1 is acceptable. For added personal security, pedestrian identification at a distance is important. Such identification is dependent to a great degree on illumination of vertical surfaces, which is related to mounting height. Where special pedestrian security is deemed desirable, the uniformity ratio should not exceed 5 to 1 for any walkway or bikeway.

Situations Requiring Special Consideration

Roadways have many areas where the problems of vision and maneuvering of motor vehicles are complex, such as grade intersections, abrupt curves, underpasses, converging traffic lanes, diverging traffic lanes and various types of complicated traffic interchanges. The design of roadway lighting for these areas demands special consideration.

When all of these areas are analyzed it becomes apparent that there are the following three basic factors which are fundamentally different from those encountered on normal straight roadway areas:

1. Motor vehicle operators are burdened with increased visual and mental tasks upon approaching and negotiating these areas.
2. Silhouette seeing cannot be provided in many cases due to the vehicle locations, pedestrian locations, obstructions and the general geometry of the roadway.
3. Adequate vehicle headlight illumination often cannot be provided. This is due to the geometry of roadways, lack of stopping room within headlight distances at speeds above 55 kilometers per hour [35 miles per hour], and the fact that vehicle headlight illumination follows rather than leads the progress of a vehicle in the negotiating of turns.

The lighting of such areas, at first glance, appears to be a very complicated problem. It becomes apparent upon analysis, however, that all such areas consist of several basic types of situations or a combination of these. The basic situations are treated individually in the following paragraphs.

Grade Intersections, Balanced Heavy Traffic. See Fig. 14-10. These intersections may have unrestricted traffic flow on both roadways, restriction by means of stop signs on one or both of the roadways, control of the traffic by signal lights, control of the traffic by police officers, or other means. Some intersections are complicated by pedestrian traffic as well as vehicular traffic. The lighting problem on all of these, however, is fundamentally the same. The illuminance in these areas should be the summation of the levels of the intersecting roads. Refer to Fig. 14-6 for appropriate levels.

Luminaires should be located so that illumination will be provided on vehicles and pedestrians in the intersection area, on the pedestrian walkways and on the adjacent roadway areas.

Fig. 14-10b shows a larger, more complex grade intersection. The lighting problems and techniques are similar to the small intersections. The size, however, may make the use of more and larger luminaires mandatory. Refer to Figs. 14-6 and 14-8 for proper illuminance values and guide for luminaire distribution type and placement.

Curves and Hills. See Fig. 14-11. The visual problems in motor vehicle operation increase on curves and hills. In general, gradual large radius curves and gently sloping grades are lighted satisfactorily if treated as straight level roadway surfaces. Sharper radius curves and steeper grades, especially those having crowns at the crest of hills, warrant closer spacing of luminaires in order to provide more uniform pavement luminance and illuminance. See Figs. 14-11e and 4-11f.

The geometry of abrupt curves, such as those found on traffic interchanges and many roadway areas, requires careful analysis. Headlight illumination is not effective in these situations and silhouette seeing cannot be provided in some instances. Luminaires should be located to provide ample illumination on vehicles, road curbings and berms, guard rails, etc. Many vehicle operators may be unfamiliar with these areas

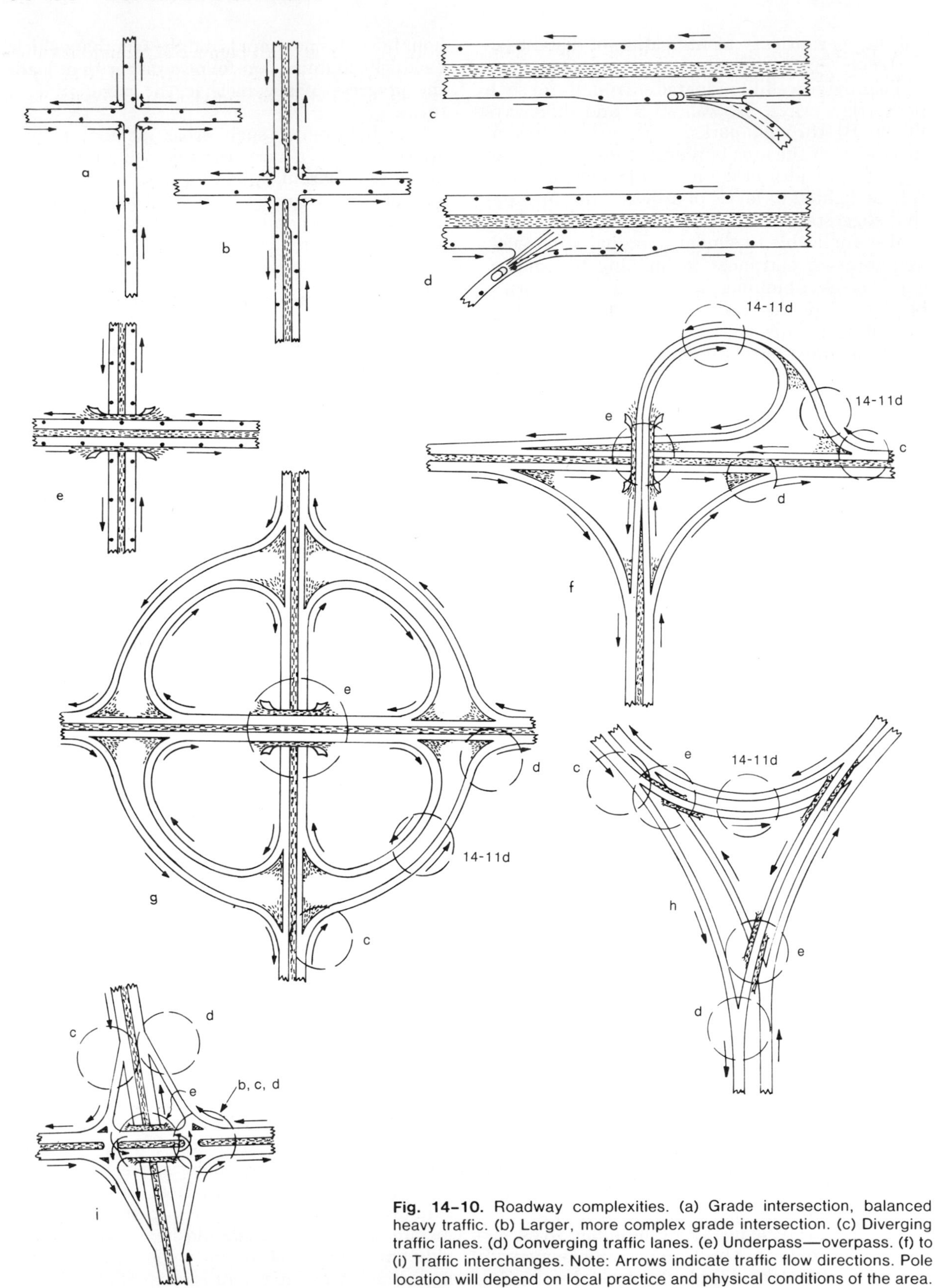

Fig. 14-10. Roadway complexities. (a) Grade intersection, balanced heavy traffic. (b) Larger, more complex grade intersection. (c) Diverging traffic lanes. (d) Converging traffic lanes. (e) Underpass—overpass. (f) to (i) Traffic interchanges. Note: Arrows indicate traffic flow directions. Pole location will depend on local practice and physical conditions of the area.

and illumination on the surround greatly helps their discernment of the roadway path. See Figs. 14-11c and d.

Proper horizontal orientation of luminaire supports and poles on curves is important to assure balanced distribution of the light flux on the pavement. See Fig. 14-11a.

When luminaires are located on grade inclines, it is desirable to orient the luminaire so that the light beams strike the pavement equidistant from the luminaire. This assures maximum uniformity of light distribution and keeps glare to a minimum. See Fig. 14-11b.

Underpass—Overpass. See Fig. 14-10e. Short underpasses such as those encountered where a roadway goes beneath a two- or four-lane roadway can generally be illuminated satisfactorily with standard luminaires if they are properly positioned. Luminaires on the lower roadway should be positioned so that the pavement illumination from those on either side of the overpass will overlap well underneath the structure and so that the horizontal illuminance levels recommended in Fig. 14-6 are provided. Care should be taken that the illuminance uniformity does not fall below the minimum values recommended. These luminaires should also provide adequate illuminance on the abutments of the overpass.

Longer underpasses, where such overlapping of the illumination from the street luminaires cannot be accomplished, require special treatment. Longer underpasses also greatly reduce the entrance of daylight, warranting illumination during the daytime. This is justified to very high levels.

Converging Traffic Lanes. See Fig. 14-10d. Converging traffic lanes frequently have all the problems of abrupt curves, plus the problem of direct illumination on vehicles in the adjacent traffic lanes. Here, automobile headlight illumination is ineffective and silhouette seeing cannot be provided for many of the situations. It is also essential to provide good direct side illumination on the vehicles entering the main traffic lanes. Refer to Fig. 14-6 for proper illuminance levels.

Diverging Traffic Lanes. See Figs. 14-10c. Diverging traffic lanes warrant extremely careful consideration because these are areas where motorists are most frequently confused. Luminaires

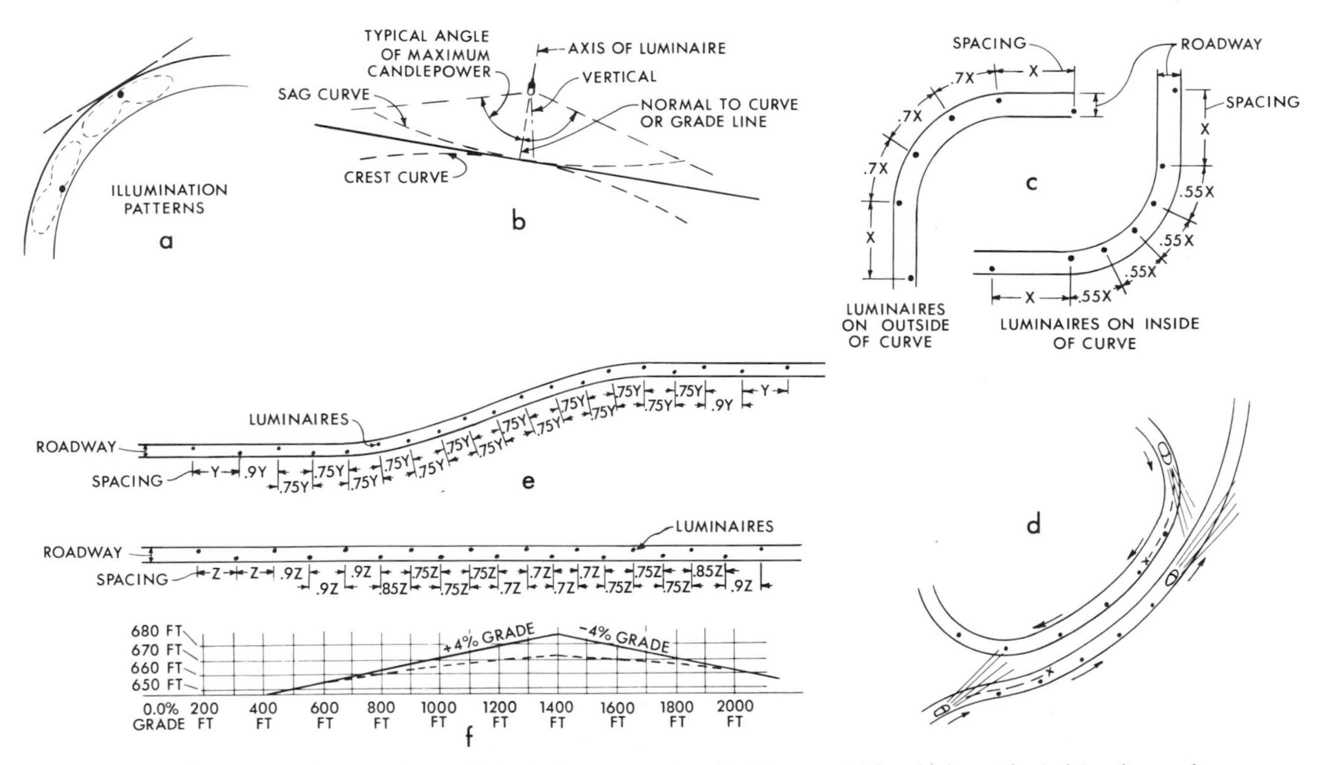

Fig. 14-11. Typical lighting layouts for horizontal curves and vertical curves. (a) Luminaires oriented to place reference plane perpendicular to radius of curvature. (b) Luminaire mounting on hill (vertical curves and grade). (c) Short radius curves (horizontal). (d) Vehicle illumination limitations. (e) Horizontal curve, radius 305 meters (1000 feet), super elevation 0.06 foot per foot. (f) 380-meter (1250-foot) vertical curve with four per cent grade and 230-meter (750-foot) sight distance. (In this illustration, 100 ft = 30m)

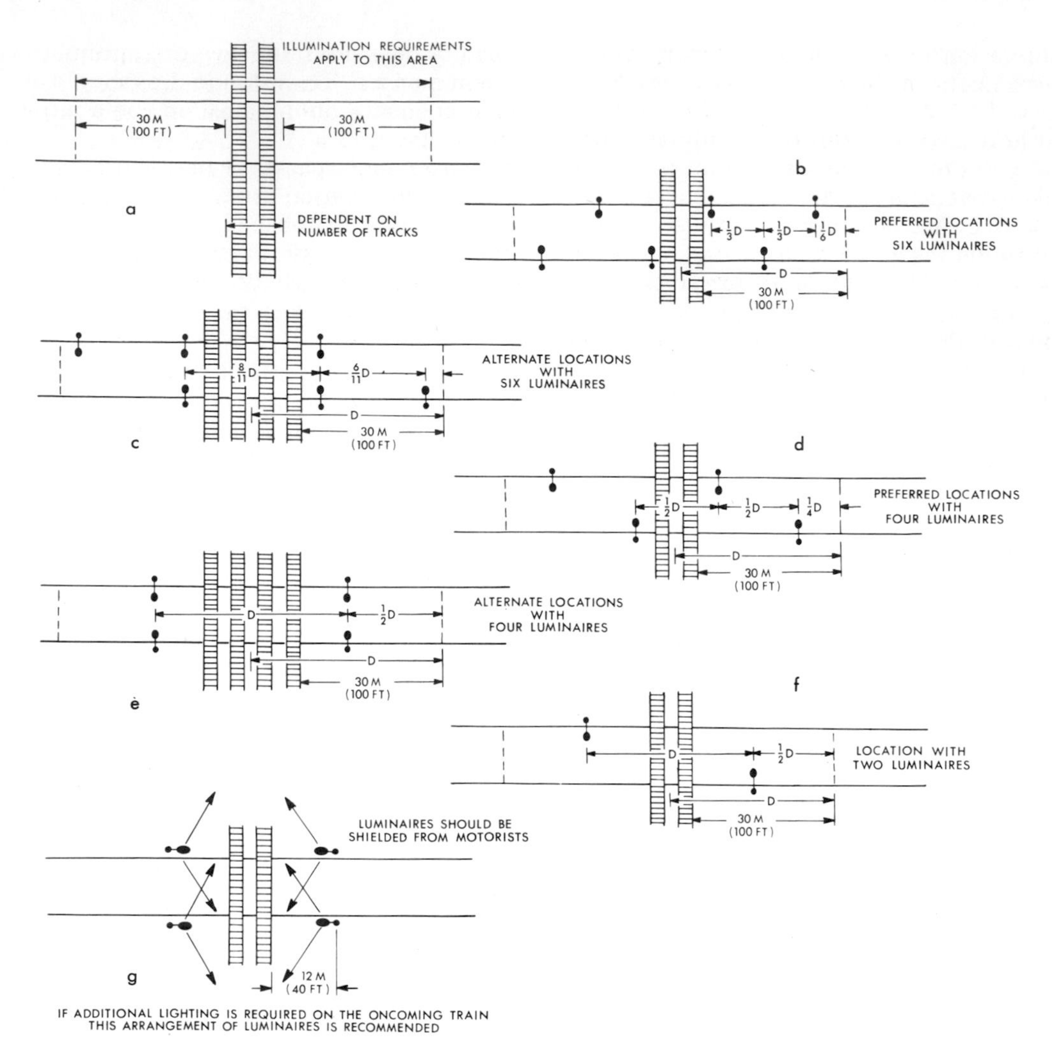

Fig. 14-12. Railroad grade crossings.

should be placed to provide illumination on curbs, abutments, guard rails and vehicles in the area of traffic divergence. Lighting also should be provided in the deceleration zone. Diverging roadways frequently have all the problems of abrupt curves and should be treated accordingly. Refer to Fig. 14-6 for proper illuminance levels.

Interchanges—High Speed, High Traffic Density Roadways. See Figs. 14-10f, g, h and i. Interchanges at first glance appear to be complex lighting problems. Analysis, however, shows that they are comprised of one or more of the basic problems that are dealt with in previous paragraphs and may be treated accordingly.

When designing lighting for interchanges the regular roadway lighting system will usually provide sufficient surround illumination in the field of view to reveal all the complexes and features of the entire scene, and allow drivers to know at all times where they are and where they are going. An inadequately lighted interchange with too few luminaires may lead to confusion for the driver, by giving misleading clues due to the random placement of the luminaires. The preceding statement does not apply to high mast lighting.

When continuous illumination for the entire interchange area cannot be provided it may be desirable to illuminate intersections, points of access and egress, curves, hills and similar areas of geometric and traffic complexities. In these cases illumination should be extended beyond the critical areas and graduated downward in level as outlined below. Two fundamental reasons for this are:

1. The eyes of the driver, adapted to the level of the illuminated area, need a graduated reduction in illuminance level upon leaving the lighted area to maintain vision during the period of dark adaptation.

2. Traffic merging into a major roadway from an access road is often slow in accelerating to the rate of speed on the major roadway. Illumination along this area for a distance beyond the access point extends visibility and facilitates the acceleration and merging process.

It is good practice to gradually reduce the illuminance level in these departure zones as outlined under Transition Lighting on page 14-12.

It is preferable to accomplish the reduced illuminance level in departure zone sectors by using light sources of lower lumen output and maintaining, insofar as practicable, luminaire spacings and configurations similar to those used on the interchange area. This maintains a similar geometry to help avoid confusion and, more importantly, reduces luminaire luminance and disability veiling luminance in graduated steps to assist in eye adaptation to the lower levels produced by fixed lighting or by vehicle headlights.

Railroad Grade Crossings. Railroad grade crossings should be adequately lighted to permit identification of existence of a crossing, presence or absence of a train in the crossing, and recognition of unlighted objects or vehicles at or near the railroad crossing.

Grade crossings are normally identified by means of identification signs with the message on a vertical face, and/or markings on the pavement surface. Illumination direction and illuminance level should permit visual recognition of such signs and markings. Minor variation of the basic lighting layouts shown in Fig. 14-12 may be desirable, depending on the exact locations of such signs or markings.

General principles to be followed in selecting and locating equipment are as follows:

1. Illuminance level over track area, starting 30 meters [100 feet] before the crossing and ending 30 meters [100 feet] beyond the crossing, should be in accordance with Fig. 14-6 or twice the level of an adjacent area of the same roadway, whichever is higher, but never less than 11 lux (1.0 footcandle). See Fig. 14-12a.
2. Luminaire classification in accordance with the width of roadway. See Lateral Light Distributions, page 14-4.
3. Pole location (Figs. 14-12b through f) should provide illuminance uniformity as outlined under Uniformity on page 14-10.

Light of a cautionary color may be used; however, distinctive color sources depend for effectiveness upon recognition by the observer of the meaning of that color.

Trees. See Figs. 14-13 to 14-15. Roadway lighting and tree foliage need not conflict. Judicious pruning will permit effective lighting, reduce system glare and generally improve the appearance of the street. Design compromises, involving deviations from preferred system layouts with respect to luminaire spacings, mounting height and transverse locations, may be necessary. Any such deviation can generally be compensated for by resorting to center suspension or lower mounting heights and closer spacings, with

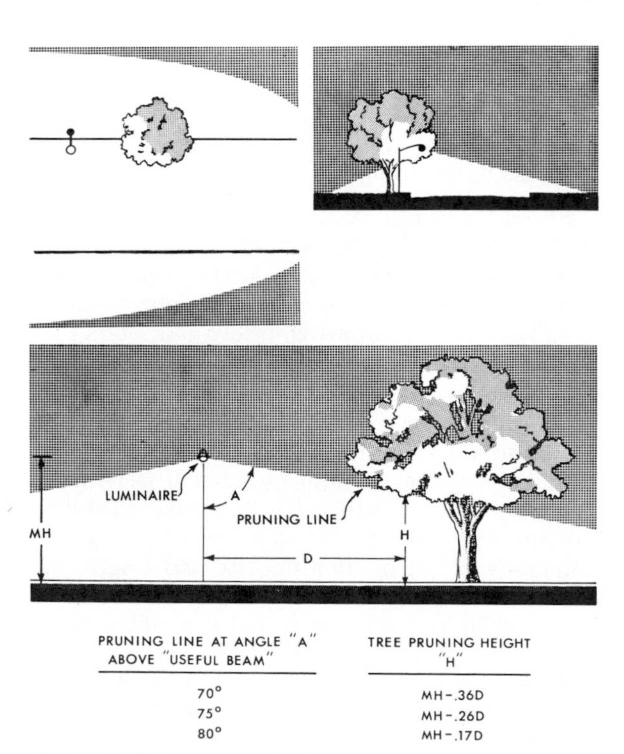

PRUNING LINE AT ANGLE "A" ABOVE "USEFUL BEAM"	TREE PRUNING HEIGHT "H"
70°	MH-.36D
75°	MH-.26D
80°	MH-.17D

Fig. 14-13. Recommended tree pruning to minimize conflict with roadway lighting.

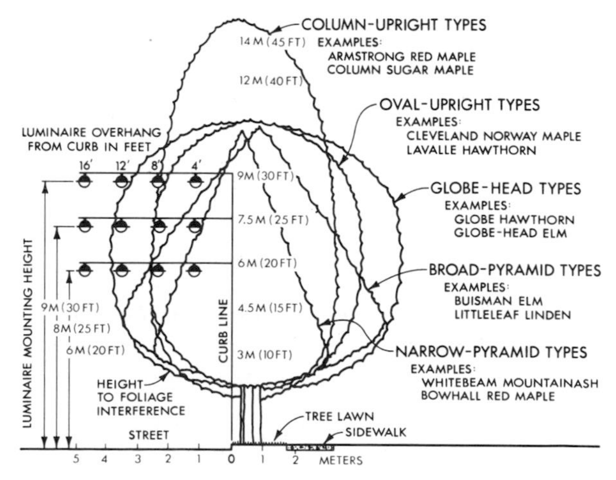

Fig. 14-14. Height to foliage interference for different types of trees and luminaire overhang from curb. Tree examples by E. H. Scanlon.

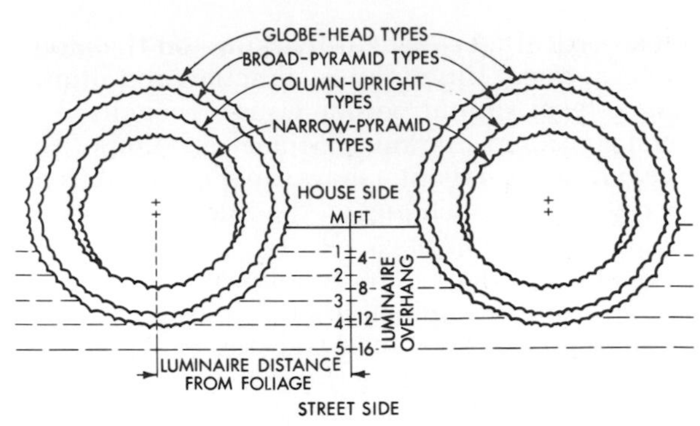

Fig. 14-15. Longitudinal and transverse location of luminaires as related to different types of trees.

smaller lamp sizes and, if necessary, lower angle of maximum candlepower. Also, irregular spacing of individual luminaires up to 20 per cent of average spacing can be tolerated, providing no two consecutive luminaire locations are involved. Transverse deviation of an individual luminaire should only be made where there is no other reasonable compromise.

Although foliage interference mostly affects roadway illumination, there may be instances on local traffic residential streets where it can also affect important sidewalk illumination. Generally this problem can be solved by either altering the luminaire positions, by pruning, or a combination of both methods.

Tunnels[2]

The basic objective of tunnel illumination is adequate and comfortable visibility both by day and at night for the users of the tunnel. Some of the main factors involved in the illumination design are:

1. A minimum of "black-hole effect" at the tunnel entrance. (In daylight hours when the motorist approaches the tunnel entrance, an object in the tunnel cannot be seen if its luminance and that of its immediate surroundings is much lower than the luminance to which the eye of the motorist is still adapted at that moment.)
2. The design of an adequate number of tunnel zones of sufficient length and different illuminances to provide eye adaptation from bright sunlight to the minimum level of tunnel lighting used.
3. The location and alignment of light sources.
4. A minimum of glare.
5. A minimum of flicker effect of alternate light and dark areas.
6. Adequate reflectance in tunnel linings.
7. Pleasing color contrast in tunnel linings.
8. Proper signing and signal lighting.

For lighting purposes every artificial or natural covering of a road, irrespective of the length and nature of the covering, is considered a tunnel. Although the term underpass is in common usage, there is no agreement on the definition of an underpass. Hence, the term tunnel is used exclusively.

Recommended illuminances for tunnels are given in average horizontal values on the roadway at minimum conditions. With wall reflectance of 70 per cent or higher, and the use of luminaires which illuminate the roadway and walls, the horizontal illuminance recommended will usually produce satisfactory visibility. It is considered that under these conditions, adequate horizontal illumination will also provide adequate vertical illumination.

In vehicular tunnels, the design speed may be as high as 110 kilometers [70 miles] per hour, and in heavy traffic, the vehicular spacings may be as close as 15 meters [50 feet] or less. Much of the driver's attention is directed to the pavement markings and the curbing defining the traffic lanes, as well as to other vehicles ahead in his lane and to the vehicles in adjacent lanes.

Uniform illuminance on horizontal and vertical surfaces is necessary for good visibility within tunnels. In the lanes of tunnels which are adjacent to the walls, luminance of the vertical surface is of major importance. At the tunnel entrances a high wall luminance is particularly valuable in reducing the "black-hole effect."

Short Tunnels. A short tunnel is one where, in the absence of traffic, the exit is clearly visible from a point ahead of the entrance portal. For lighting purposes, the length of a short tunnel is usually limited to approximately 45 meters [150 feet]. Some tunnels up to 120 meters [400 feet] long may be classified as short if they are straight, level and have a high width and/or height to length ratio.

In most cases no lighting system is required inside short tunnels for adequate driver visibility. Daytime penetration from each end, plus the silhouette effect of the opposite end brightness, generally assures satisfactory visibility. Tunnels between 20 and 45 meters [75 and 150 feet] in length may require lighting if daylighting is restricted due to roadway depression or the proximity of tall buildings in urban areas. In this case the tunnels should be treated as long tunnels and entrance lighting provided.

For tunnels approximately 20 meters [75 feet] or less in length, the proper positioning and mounting height of pole-mounted street lights adjacent to each end of these tunnels, usually provides satisfactory nighttime lighting. When higher mounting heights are used for these street lights, the poles must be located farther away from the tunnel to get the light into the tunnel entrance. This may mean that the uniformity ratio is exceeded and some tunnel lighting may be necessary.

On roadways over 15 meters [50 feet] in width, or where the tunnel has a center row of columns, it may be desirable to position pole-mounted units on opposite sides of the roadways at each end. Wherever a shoulder-mounted pole can be eliminated by attaching luminaires to a necessary bridge structure, it should be done for the sake of safety.

For tunnels from approximately 20 to 45 meters [75 to 150 feet] in length, a nighttime lighting system is usually required to provide a maintained level of approximately two times, but not over three times, that recommended in Fig. 14-6 for the connecting open roadway.

Long Tunnels. In straight tunnels, generally over 45 meters [150 feet] in length, where the exit brightness takes up too small a part of the driver's field of vision to serve as an effective background for silhouette discernment, the structure is classified as long for lighting purposes. All tunnels where the exits cannot be seen from points ahead of the entrance portals, are also classified as long.

After the Open Road Zone the following main zones are considered:

1. For day conditions
(a) The Grid Zone, which is optional.
(b) The Main Tunnel Entrance Zone.
(c) The Central Tunnel Zone.
If the Grid Zone is not used, two or more illuminances are required in the Main Tunnel Entrance Zone, with the first section at a high illuminance.
2. For night conditions the tunnel is considered as a single zone for the entire tunnel length.

Grids used may be of many different designs but are usually constructed to permit passage of snow and rain but not direct sunlight. The average grid is designed to reduce the daylight on the roadway by about 70 per cent. This can be reduced further at the end of the Grid Zone by the use of black paint on the grid, and/or by decreasing their width and increasing the width of the solid section on each side of the grids.

Daytime Lighting. For daytime lighting two simple principles are recommended to produce comfort and safety for the motorist:

1. Illuminance values under daytime conditions in adjacent zones of long tunnels should not exceed a reduction ratio of 10 to 1.
2. The time that the motorist requires in passing from the highest to the lowest illuminance, should be 4 seconds or more.

The lighting in the Main Tunnel Entrance Zone under daytime conditions is a major lighting design problem. It is generally assumed that the luminance near a tunnel entrance can be regarded as the luminance to which the driver is adapted. This adaptation luminance is determined by the average luminance of the visual field at this location. As the driver approaches the tunnel entrance, the central part of the driver's visual field is mostly focused on the darker tunnel entrance. As this becomes larger on the approach to the tunnel, the adaptation luminance of the driver is greatly reduced.

Considering as an example a case where there is no grid and the driver's adaptation level is conservatively assumed to be that produced by an average of 50 kilolux [5000 footcandles], the illuminance in the Main Tunnel Entrance Zone just after the Main Tunnel Portal should therefore be a minimum of 5000 lux [500 footcandles] horizontal maintained, according to the 10 to 1 maximum reduction ratio for adjacent zones. If the Main Tunnel Portal has a large opening which gradually slopes down to the main tunnel cross section, this will help in the transition from full daylight to the tunnel lighting.

With no grid used, the Main Tunnel Entrance Zone is divided into two or more zones for lighting. If only two zones are used in the example considered the second zone should have at least a 500 lux [50 footcandle] horizontal maintained level and each of the zones would have a minimum travel time of 2 seconds.

The illuminance for the Central Tunnel Zone of all long tunnels should be at least 50 lux [5 footcandles] horizontal maintained, which is between two and three times the recommended level for *nighttime* lighting of the open road at each end of the long tunnel. For tunnels up to 600 meters [2000 feet] in length it may be desirable to use up to 100 lux [10 footcandles] in that zone.

To facilitate the transition from high daylight levels to practical levels of lighting at the Main Tunnel Entrance, a grid is sometimes installed over the roadway ahead of the Main Tunnel Portal. This use of a Grid Zone aids adaptation to high and low values of daylight and appears to be a good method of combating the black-hole effect. The Grid Zone, if used, becomes part of the tunnel.

Since the average grid is designed to reduce the illuminance on the road at a point soon after the entrance to the Grid Zone by 70 per cent, in the example above, the 50 kilolux [5000 footcandle] level on the Open Road would be reduced to 15 kilolux [1500 footcandles] in the Grid Zone. By use of solid sections of increasing width at the side of the grid, the illuminance on the road can be further reduced to 5000 lux [500 footcandles] at the end of the Grid Zone. Black paint may also be used at the last part of the Grid to cut down on the light transmission.

With a reduction to 5000 lux [500 footcandles] at the end of the Grid Zone, the Main Tunnel Entrance Zone should have a level of 500 lux [50 footcandles] and the Central Zone 50 lux [5 footcandles] minimum maintained. In the example given, the length of the Grid Zone and the Main Tunnel Entrance Zone should each have a minimum travel time of 2 seconds.

The length of the Grid Section should be *increased* as compared to the example given, if there is considerable direct sunlight penetration into the entrance of the grids. The Grid Section may be *decreased*:

1. For long straight tunnels from 45 to 300 meters [150 to 1000 feet] in length and made in proportion to the tunnel length.
2. For entrance from a depressed roadway with close vertical side walls since this results in a reduced lighting level.
3. When one or more short tunnels for cross streets over the depressed roadway are comparatively near the long tunnel.
4. Where large buildings or trees shade the tunnel entrance.

Nighttime Lighting. At nighttime the full length of long tunnels is considered as a single zone for lighting purposes, and the illuminance in the entire zone should be between two and three times the level of lighting on the open roadway approaches as listed in Fig. 14-6.

Adjacent Areas. Special lighting is normally not used at tunnel exits as the motorist's eyes adjust quickly to the increasing luminance as the exit is approached; however, when a tunnel has two-way traffic, both portals must be treated as entrances.

All tunnels, urban and non-urban, which have nighttime lighting, should have at least 150 meters [500 feet] of roadway lighting before each tunnel entrance and after each tunnel exit in accordance with the recommended levels in Fig. 14-6.

Fig. 14-16. Recommended Illuminance Levels for Roadway Safety Rest Areas*

Rest Area	Illuminances		Uniformity Ratio
	Lux	Footcandles	
Entrance and Exit			
Access Lanes	3.2 to 6.5	.3 to .6	6:1 to 3:1*
Gores	6.5	.6	3:1
Interior Roadways	6.5	.6	3:1
Parking Areas	11	1.0	3:1
Activity Areas			
Major	11	1.0	3:1
Minor	5.4	.5	6:1

* The illuminance values recommended represent the condition just prior to cleaning and/or group relamping as calculated and planned in the design procedure.

Safety Rest Areas[3]

Safety rest areas on limited access highways are an important feature to the through motorist and there is a general agreement that for the public to obtain maximum benefits from the construction of a safety rest area it must be available to the traveling public 24 hours a day, and be considered safe to enter and stay for short periods of time without fear for their safety and security. To obtain this condition, these areas must be adequately lighted for nighttime use.

In designing a lighting system for a safety rest area, geographical location, topographical location, motorist's comfort and safety, landscaping and architectural treatment, and pedestrian appearance must be considered. An important benefit to be derived from proper lighting is the ease of policing areas during nighttime hours.

One of the prime design considerations is the motorist's visibility while traveling along an unlighted main highway or in the rest area, *i.e.*, the motorist should not be disturbed by glare or spill light from luminaires placed adjacent to the roadway within the rest areas. While traversing the entire length of the adjacent rest area, the motorist should be able to discern any vehicle leaving the rest area, as well as the traffic moving along the main roadway.

The over-all design of the lighting is divided into general areas as follows:

1. Entrance and exit.
2. Interior roadways.
3. Parking areas.
4. Activity areas.

These have been defined for separate consideration as each is to be used for a specific and different purpose.

The illuminance values recommended in Fig. 14-16 represent the condition just prior to cleaning and/or group relamping as calculated and planned in the design procedure.

Entrance and Exit. Entrance and exit areas are defined as the deceleration and acceleration lanes adjacent to the main roadway, leading to and from the gore (approach nose) area. The entrance and exit lanes should be lighted so that the driver entering or leaving the rest area can safely make the transition from the main roadway to the rest area and vice versa. At the same time, the driver electing to continue along the main roadway would be able to do so without having his vision impaired by luminaire brightness or spill light. The driver should likewise be able to discern vehicles leaving or entering the roadway.

The assumption is often made that access lanes to safety rest areas should be lighted in the same manner as ramps at interchanges; however, interchange ramps are usually designed for higher speeds and traffic densities. Safety rest area entrance and exits with relatively low traffic density are deceleration and acceleration lanes leading to and from low speed roadways within the area, and thus must be designed with this purpose in mind.

It is recommended that the illumination along the deceleration lane may vary, but the maximum illuminance should occur at the gore point between the deceleration lane and the beginning of the interior roadways. This is based on the use of from three to five luminaire locations along the length of the change lanes.

Similarly, a higher level should occur at the exit gore (merging end) and may decrease at a nominal distance to a place where a motorist can be considered able to merge into the through traffic lanes. The motorist on the through lanes must be able to see an exiting vehicle, make a proper decision and adjust to the traffic flow. The lower illuminance levels are desirable to allow for the motorist's eye adaptation from the illuminated area to the unlighted roadway.

It is recommended that luminaires used in these areas be a Type II short cutoff to confine the main light to the deceleration and acceleration lanes, and to restrict high angle brightness. Luminaires used may be of lower wattage than those normally used in order to achieve good uniformity and low luminance.

In the event that the main roadway is continuously lighted beyond the confines of the rest area, deceleration and acceleration lanes should be lighted to a level equal to that of the main roadway.

Interior Roadways. Interior roadways are those between the entrance gore point and the parking areas, and vice versa to the exit. In some instances they can be extremely long or hardly exist at all.

As these roadways are interior and off the mainline roadway, the designer may desire to use the same type of equipment used on the entrance and exit lanes, or select another type. The designer should keep in mind that there may be an added maintenance problem when several different types of luminaires and lamps are required within one area.

Parking Areas. Illumination of both automobile and truck parking areas should be designed so that the motorist can distinguish features of the area—as well as discern pedestrians moving about the area—while still within the vehicle. The area should be lighted so that the motorists can read information-signs which direct them to various parts of the area.

Careful attention should be provided to special areas, such as, handicap ramps, sanitary disposal stations and other items which may require special detailing. This may be obtained by placing a luminaire in close proximity to a particular area for maximum illuminance.

Consideration might be given to providing a different color of illumination for improved color rendition or area designation—still using the levels in Fig. 14-16. Should this be the case, it is recommended that the entire area be so lighted for maintenance purposes.

Activity Areas. The activity areas are those areas designed for pedestrian use. The use of light sources which render skin tones favorably is recommended in these areas.

Main activity areas are those which include such structures as comfort stations, information centers, etc., as well as the walkways to and from these locations and to the parking area.

Minor activity areas are those which include picnic tables, dog walks, etc., and their associated walkways and facilities. Generally, the lighting levels in the main activity areas will be higher than those in the minor areas.

Area floodlighting may be provided for architectural or other purposes, should the designer so desire. Caution should be observed that stray light is not directed toward or reflected on the main roadway to endanger the passing motorist—not only from the design standpoint, but in the final setting of the luminaires. There should be adequate illumination so that persons using the facilities may view the surroundings and tasks without difficulty, providing a sense of security.

Maintenance. Safety rest areas are frequently in remote, isolated areas, and require more rigid maintenance and supervision than areas normally visible to residents of the locality. It must be remembered that these areas have been designated *safety rest areas*; therefore, any system of maintenance must be considered on the basis of optimum operation for the safety of the nighttime traveler while driving as well as walking in these areas.

EXTERNALLY ILLUMINATED ROADWAY SIGNS

A roadway sign that is properly designed for the daytime can be adequately lighted to convey the same message at night.

Color and shape of roadway signs convey information to the motorist as evidenced by the green and blue background on interstate signs, and the shapes of interstate and U.S. route shields. To maintain integrity of messages conveyed by sign color, nighttime illumination must not degrade the color sensation from that when viewed under daylight conditions.

Sign Color Standards. Sign color standards have been internationally established by usage and agreement. Colors have been assigned meanings in the United States by the Department of Transportation.* Roadway sign colors cover the entire range of the visible spectrum. The problem is to adequately illuminate the sign face while closely reproducing color values.

Lamp Selection. Selection of a light source based on *color rendition*, *efficacy* and *lamp life* results in the following order of lamp desirability:

1. *High desirability*
(a) Fluorescent
(b) Mercury, deluxe
(c) Metal halide, clear
(d) Metal halide, phosphor coated
2. *Medium desirability*
(a) Incandescent filament
(b) Mercury, color improved
(c) Tungsten-halogen
3. *Low desirability*
(a) Mercury clear
(b) High pressure sodium
(c) Low pressure sodium

* *Manual on Uniform Traffic Control Devices.*

Ambient Luminance. Ambient luminance is the background brightness against which the sign will be viewed by the driver of a vehicle. There is no approved method for determining ambient luminance, nor is there agreement on a definition of area classification. The following definitions of *high*, *medium* and *low* ambient luminance are assigned for purposes of this discussion:

1. *Low.* Rural areas without illumination or with very low illuminance levels. This would include background of mountains, fields, trees, shrubbery, grass and rural interchanges.
2. *Medium.* Illuminated freeways and freeway interchanges or interchanges with small commercial developments and minimum street lighting levels.
3. *High.* Central business district areas with high street lighting levels and brightly lighted advertising signs. A freeway through or adjacent to a downtown area would illustrate high ambient luminance.

Luminance of Externally Illuminated Signs. Luminance will determine how well the signs will attract attention among competing distractions of other roadway lighting and surrounding advertising signs. The difference in reflectance between message and background will determine how quickly the message is read by drivers. Illuminance uniformity over the sign face will add to the effectiveness. Five elements important in producing an effective roadway sign are:

1. Ambient illuminance.
2. Sign illuminance above ambient.
3. Reflectance of message and background.
4. Ratio of message to background illuminance.
5. Uniformity ratio of over-all sign illuminance.

Recommended minimum maintained illuminances of externally illuminated roadway signs are shown in Fig. 14-17.

Uniformity. Uniformity of sign illuminance is

Fig. 14-17. Recommended Illuminances for Externally Illuminated Roadway Signs

Ambient Luminance	Illuminance		Luminance*	
	Lux	Footcandles	Candelas/Square Meter	Footlamberts
Low	100	10	24	7
Medium	200	20	48	14
High	400	40	96	28

* Maintained reflectance of 70 per cent for white sign letters.

expressed as a ratio of maximum-to-minimum levels on the sign face—that area enclosed within the border line. A uniformity ratio of 6 to 1 is recommended as an acceptable ratio. Lower ratios providing more even illuminance will produce a more pleasing appearance and a more legible sign.

Location of Luminaires. Luminaires may be located at either the top or bottom of the sign, remotely from ground level, or from adjacent lighting standards or poles. A study of the surround at any given installation may dictate which location is best suited for that installation. Most highway authorities recommend locating luminaires at the bottom of the sign.

General considerations for the source location are as follows:

1. *Top mounted sources*
(a) Snow will not collect on the cover.
(b) Luminaires will not hide the message.
(c) Glare of source is partly shielded from opposing traffic.
(d) Reflected source may produce glare.
(e) Luminaires may produce daytime shadows.

2. *Bottom mounted sources*
(a) Reflected source will not produce glare.
(b) Luminaires may hide message from some viewing angles.
(c) Cover will collect snow and dirt.
(d) May cause direct glare to traffic approaching from opposite direction by spill light under sign—unless shielded.
(e) No daytime shadows produced by luminaire.

3. *Ground or remote located sources*
(a) Requires a carefully controlled light source.
(b) May produce glare if not properly shielded.
(c) Location is prone to vandalism.

PARKING FACILITIES LIGHTING[5]

Objectives. From the standpoints of traffic safety; protection against assault, theft and vandalism; convenience and comfort to the user; and in many instances, for business attraction, adequate parking facility lighting is vital in today's motorized society. A well-lighted parking facility provides safety and security for the public.

Illumination Requirements. The illumination requirements of a parking facility depend on the type of usage the facility receives. For open area parking three levels of activity have been established and are defined as *High*, *Medium* and *Low*. These levels reflect both traffic density and intensity and are described by the following examples:

High Activity:
- Major league athletic events
- Major cultural or civic events
- Major regional shopping centers

Medium Activity:
- Fast food facilities
- Area shopping centers
- Hospital parking areas
- Transportation parking (airports, etc.)
- Cultural, civic or recreational events
- Residential complex parking

Low Activity:
- Local merchant parking
- Industrial employee parking
- Educational facility parking

Illuminance values shown in Fig. 14-18 are maintained average values for vehicular areas with uniformity ratios as shown. Also shown are minimum safety values for areas used by pedestrians. If additional illumination is needed for pedestrian security and identification at a distance, use appropriate maintained average values in the final columns.

In open parking facilities exits, entrances, loading zones, pedestrian crossings and collector lanes should be given special consideration to permit ready identification and to aid in providing safety.

The entrance area of a covered parking facility is defined as the portal or physical entrance to the covered portion of the parking structure and 15 meters [50 feet] beyond the edge of the covering into the structure.

Control of Light and Brightness. For all parking facilities care should be taken in aiming lighting equipment to limit spill light and excessive glare to adjacent private and public property.

The type of source used for parking facility lighting will depend upon the color of light desired, light control capabilities, maintenance characteristics and economics.

Poles should be spaced to permit sufficient overlapping of adjacent beams to minimize harsh shadows and avoid dark pockets throughout the area. The lighting design should provide uniform illuminance at the facility as prescribed by the uniformity ratio of average illuminance to minimum illuminance.

Energy Management. From the standpoint of energy management it may be desirable to reduce the lighting levels in certain parking facilities during periods of reduced activity. For example, during peak traffic periods at stadiums

Fig. 14-18. Recommended Maintained Illuminances for Open and Covered Parking Facilities

Open Parking Facilities								
Level of Activity	For Vehicular Traffic			For Pedestrian Safety		For Pedestrian Security		
	Lux*	Footcandles*	Uniformity Ratio	Lux**	Footcandles**	Lux*	Footcandles*	Uniformity Ratio
Low activity	5	0.5	4:1	2	0.2	9	0.8	5:1
Medium activity	11	1	3:1	6	0.6	22	2	5:1
High activity	22	2	3:1	10	0.9	43	4	5:1

Covered Parking Facilities				
Areas	Day		Night	
	Lux***	Footcandles***	Lux*	Footcandles*
General parking and pedestrian areas	54	5	54	5
Ramps and corners	110	10	54	5
Entrance areas	540	50	54	5
Stairways and lobbys (refer to Fig. 2-2)				

* Average on pavement
** Minimum on pavement
*** Average on payment—sum of electric lighting and daylight

the "High" activity lighting levels may be required, but while the game is being played or during hours of reduced activity the "Medium" or "Low" activity lighting levels may be adequate.

ROADWAY ILLUMINATION DATA AND CALCULATIONS

The following is an example of a simple and straightforward calculation procedure to determine average illuminance and illuminance at a specific point on a roadway. For a detailed treatment of the subject, including calculations for high-mast and pedestrian walkway lighting, the reader is referred to Reference 1.

Determination of Average Illuminance

The average illuminance over a large pavement area in terms of lux (footcandles) may be calculated by means of a "utilization curve" of the type shown in Fig. 14-19.

Utilization Curves. Utilization curves, available for various types of luminaires, afford a practical method for the determination of average illuminance over the roadway surface where lamp size, mounting heights, width of roadway, overhang and spacing between luminaires are known or assumed. Conversely, the desired spacing or any other unknown factor may readily be determined if the other factors are given.

The Coefficient of Utilization, as shown in Fig. 14-19, is the percentage of rated lamp lumens which will fall on either of two strip-like areas of infinite length, one extending in front of the luminaire (street side), and the other behind the luminaire (house side), when the luminaire is level and oriented over the roadway in a manner equivalent to that in which it was tested. Since roadway width is expressed in terms of a ratio of luminaire mounting height to roadway width, the term has no dimensions.

Light Loss Factors. There are a number of causes of light loss. They are listed on page 4-21. For each cause, a factor can be determined. All individual factors can be multiplied together to obtain one total light loss factor. Some factors, usually due to less than ideal operating conditions, exist initially and continue through the life of the installation. They may, however, have too little effect to justify correction or be too costly to correct. The significant light loss factors in roadway calculations are:

Lamp Lumen Depreciation. Information about lamp lumen depreciation is available from manufacturers' tables and graphs for lumen depreciation and mortality of the chosen lamp. Rated average life should be determined for the specific hours per start; it should be known when burnouts will begin in the lamp life cycle. From these facts, a practical group relamping cycle will be established and then, based on the hours elapsed to lamp removal, the specific lamp lumen depreciation (LLD) factor can be determined.

Consult manufacturers' data or Section 8 of the 1981 Reference Volume.

Luminaire Dirt Depreciation. The accumulation of dirt on luminaires results in a loss in light output, and therefore a loss on the roadway. This loss is known as the luminaire dirt depreciation (LDD) factor and is determined as follows:
1. Determine the dirt category (very clean, clean, moderate, dirty or very dirty)
2. From the appropriate dirt condition curve in Fig. 14-20 and the proper elapsed time in years of the planned cleaning cycle, the LDD factor is found.

Luminaire Ambient Temperature. The effect of ambient temperature on the output of some luminaires may be considerable. Each particular lamp-luminaire combination has its own distinctive characterstic of light output versus ambient temperature. To apply a factor for light loss due to ambient temperature, the designer needs to know the highest and lowest temperatures expected and to have data showing if there are variations in light output with changes in ambient temperature for the specific luminaire to be used.

Voltage to Luminaire. In-service voltage is difficult to predict, but high or low voltage at the luminaire will affect the output of most luminaires.

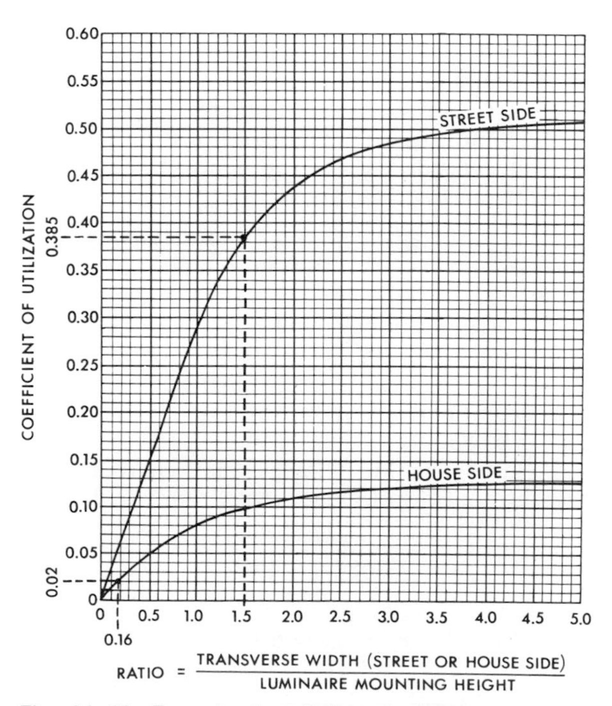

Fig. 14-19. Example of coefficient of utilization curves for luminaire providing Type III-M light distribution.

Ballast Factor.* If the ballast factor of the ballast used in a luminaire (fluorescent or high intensity discharge) differs from that of the ballast used in the actual photometry of the luminaire, the light output will differ by the same amount. The manufacturer should be consulted for necessary factors.

Luminaire Component Depreciation.
1. Luminaire surface depreciation results from adverse changes in metal, paint and plastic components which result in reduced light output.
2. Because of the complex relationship between the light-controlling elements of luminaires using more than one type of material it is difficult to predict losses due to deterioration of materials. Also, for luminaires with one type of surface, the losses will be affected by the type of atmosphere in the installation. No factors are available at present.

Change in Physical Surround. As much as possible should be known about future changes that may affect any of the above roadway conditions, such as whether trees or border areas will be added, or nearby buildings constructed or demolished.

Burnouts. Unreplaced burned-out lamps will vary in quantity, depending on the kinds of lamps and the relamping program used. Manufacturers' mortality statistics should be consulted for the performance of each lamp type to determine the number to burn out before the time of planned replacement is reached. Practically, quantity of lamp burnouts is determined by the quality of the lighting services program incorporated in the initial design procedure and by the quality of the physical performance of the program.

Total Light Loss Factor. The total light loss factor is simply the product of multiplying all the contributing factors described above. Where factors are not known, or applicable, they are omitted. At this point, if it is found that the total light loss factor is excessive it may be desirable to reselect the luminaire and/or lamp, or modify the cleaning and/or maintenance schedule.

Formulas for Computation. The basic formula for determination of average horizontal illuminance is as follows:

* See "IES Approved Method for Determining Luminaire-Lamp-Ballast Combination Operating Factors for High Intensity Discharge Luminaires," *Illum. Eng.* Vol. 65, p. 718, December, 1970.

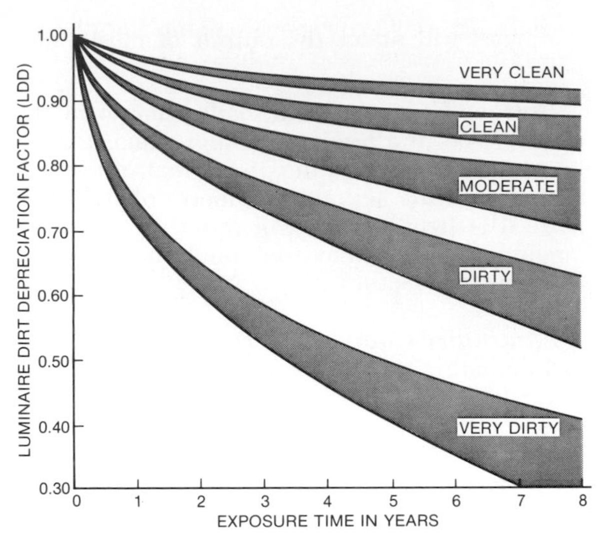

SELECT THE APPROPRIATE CURVE IN ACCORDANCE WITH THE TYPE OF AMBIENT AS DESCRIBED BY THE FOLLOWING EXAMPLES:

VERY CLEAN—No nearby smoke or dust generating activities and a low ambient contaminant level. Light traffic. Generally limited to residential or rural areas. The ambient particulate level is no more than 150 micrograms per cubic meter.

CLEAN—No nearby smoke or dust generating activities. Moderate to heavy traffic. The ambient particulate level is no more than 300 micrograms per cubic meter.

MODERATE—Moderate smoke or dust generating activities nearby. The ambient particulate level is no more than 600 micrograms per cubic meter.

DIRTY—Smoke or dust plumes generated by nearby activities may occasionally envelope the luminaires.

VERY DIRTY—As above but the luminaires are commonly enveloped by smoke or dust plumes.

Fig. 14-20. Chart for estimating roadway luminaire dirt depreciation factors for enclosed and gasketed luminaires.

Average Illuminance

$$= \frac{\text{Lamp Lumens} \times \text{Coefficient of Utilization}}{\text{Pavement Area per Luminaire}}$$

This formula is usually expanded as follows:

Average Illuminance =

$$\frac{(\text{Lamp Lumens}) \times (\text{Coefficient of Utilization})}{(\text{Spacing Between Luminaires})^{**} \times (\text{Width of Roadway})}$$

It can be seen that with this expression of the formula, it is possible to solve horizontal average illuminance, or spacing, or lamp lumens as desired. A further modification of this formula is necessary to determine the average illuminance on the roadway when the illuminating source is at its dirtiest condition (see Light Loss Factors above). For such a calculation, the formula is expressed as follows:

Average Illuminance =

$$\frac{(\text{Lamp Lumens}) \times \left(\begin{matrix}\text{Coefficient of}\\ \text{Utilization}\end{matrix}\right) \times (\text{Light Loss Factor})^{*}}{(\text{Spacing Between Luminaires})^{*} \times (\text{Width of Roadway})}$$

** This is the longitudinal distance between luminaires if spaced in staggered or one-side arrangement. This distance is one-half the longitudinal distance between luminaires if luminaires are arranged in opposite spacing.

* This value may be experimentally determined or estimated if not known.

Typical Computations. To illustrate the use of a utilization curve, Fig. 14-19, a typical calculation is provided as follows:

Given: Roadway with layout as in Fig. 14-21.
Staggered luminaire spacing, 37 m (120 ft)
Roadway width curb-to-curb (pavement), 15 m (50 ft)
Luminaire mounting height, 9 m (30 ft)
Luminaire overhang, 1.5 m (5 ft)
Light loss factor, (0.6)
Lamp with 20,000 lumens initial.

Required: To calculate the minimum average illuminance for the above roadway.

Solution: For average illuminance:

1. Determine the coefficient of utilization for the "street side" of the luminaire: Ratio (street side) (from Fig. 14-21)

$$\frac{15 - 1.5}{9} = \frac{13.5}{9} = 1.50$$

$$\left(\frac{50 - 5}{30} = \frac{45}{30} = 1.50\right)$$

Coefficient of utilization from Fig. 14-19 for ratio 1.50 is 0.385.

2. To determine "house side" coefficient of utilization: Ratio (house side) (from Fig. 14-21)

$$\frac{1.5^{**}}{9} \cong 0.16$$

$$\left(\frac{5^{**}}{30} = 0.16\right)$$

** Use actual overhang distance from curb to point below luminaire.

Coefficient of utilization from Fig. 14–19 for ratio 0.16 is 0.02.

3. Total coefficient for "street side" plus "house side" is 0.405.

4. To determine average illuminance on roadway use the formula from above, giving:

$$\text{Average Illuminance} = \frac{20{,}000 \times .405 \times 0.6}{37 \times 15}$$

$$= 8.7 \text{ lux}$$

$$\text{Average Illuminance} = \frac{20{,}000 \times 0.405 \times 0.6}{120 \times 50}$$

$$= 0.8 \text{ footcandle}$$

Determination of the Illuminance at a Specific Point

The determination of the horizontal illuminance at a specific point may be determined from an "isolux" ("isofootcandle") curve, Fig. 14-22, or by means of the classical point method of calculation.

Isolux (Isofootcandle) Diagram. An isolux diagram is a graphical representation of points of equal illuminance connected by a continuous line. These lines may show values on a horizontal plane from a single unit having a definite mounting height, or they may show a composite picture of the illuminance from a number of sources arranged in any manner or at any mounting height. They are useful in the study of uniformity of the illuminance and in the determination of the illuminance at any specific point. In order to make these curves applicable to all conditions, they are computed for a given mounting height, but horizontal distances are expressed in ratios of the actual distance to the mounting height. Correction factors for other mounting heights are usually given in the tabulation alongside the isolux curves.

Typical Computations. To illustrate the use of the isolux (isofootcandle) diagram, a typical calculation is as follows:

Given: Same as given for typical computations for average illuminance.

Required: To determine the illuminance at point "A" on Fig. 14–21 which is the total of contributions from luminaires 1, 2 and 3.

Solution using an isofootcandle diagram:

1. The location of point "A" with respect to a point on the pavement directly under the luminaire is dimensioned in transverse and longitudinal multiples of the mounting height. Assume that the luminaire distribution provides isofootcandle lines (horizontal footcandles) as shown in Fig. 14–22. Point "A" is then located on this diagram for its position with respect to each luminaire.

2. To determine the contribution of each luminaire to point "A":

a. Luminaires Numbers 1 and 3:

Locate point "A"—Transverse 5 feet (1.5 meters) to "house side":

$$\frac{5}{30} = 0.16 \text{ times mounting height}$$

Longitudinal 120 feet (37 meters) along pavement

$$\frac{120}{30} = 4.0 \text{ times mounting height}$$

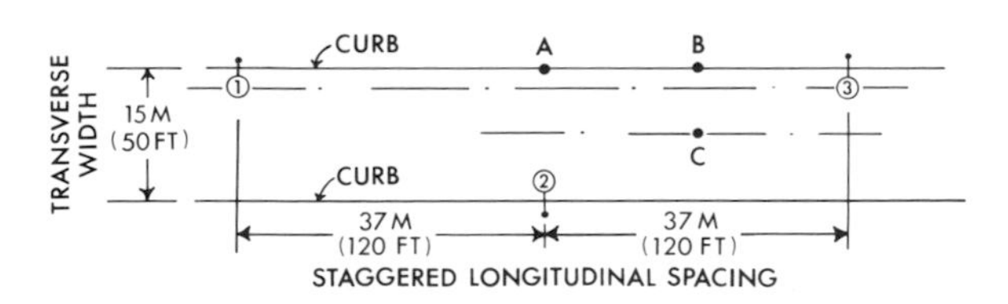

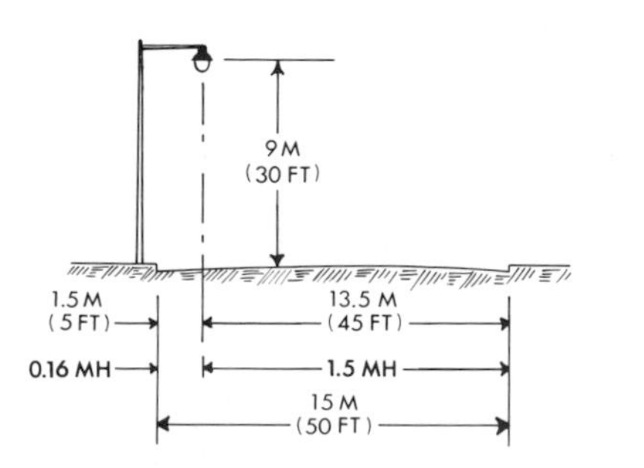

Fig. 14–21. Layout of luminaire and roadway assumed for typical computation.

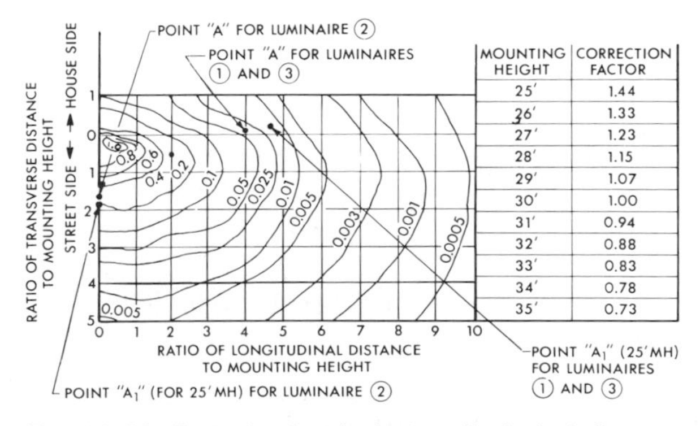

MOUNTING HEIGHT	CORRECTION FACTOR
25'	1.44
26'	1.33
27'	1.23
28'	1.15
29'	1.07
30'	1.00
31'	0.94
32'	0.88
33'	0.83
34'	0.78
35'	0.73

Fig. 14–22. Example of an isofootcandle (isolux) diagram of horizontal footcandles on pavement surface for a luminaire providing a Type III-M light distribution, per 1000 lumens of lamp output times 10.

At point "A" for these luminaires, the estimated footcandle value from Fig. 14-22 isofootcandle diagram is 0.04 footcandle. This is from each luminaire, Numbers 1 and 3. Both luminaires together provide 0.08 footcandle.

Were an isolux diagram used, the value read at this point would be 0.43 lux. Using the standard footcandle/lux conversion factor of 10.76 also results in an illuminance value of 0.43 lux, or a combined value of 0.86 lux.

b. Luminaire Number 2: Locate point "A"—Transverse 45 feet 13.5 meters to "street side":

$$\frac{45}{30} = 1.5 \text{ times mounting height}$$

Longitudinal location is 0, directly across from the luminaire. At point "A" for this luminaire, the estimated footcandle value from Fig. 14-22 is 0.3 footcandle.

3. The total at point "A" from the three luminaires is 0.08 plus 0.3 = 0.38 footcandle. The value of 0.38 footcandle is based on 1000 lamp lumens times 10 and clean luminaires with a lamp producing rated output. The initial footcandle level, therefore, is 0.38 × 2 = 0.76 footcandles. If it is desired to express the footcandle level in terms of the footcandles when the illuminating source is at its lowest output and when the luminaire is in its dirtiest condition, this can be done by utilizing the procedure covered above (0.76 × 0.6 = 0.46 footcandles).

4. To use these data for a mounting height other than the one for which the isofootcandle curves are made, it is necessary to find the correct new location on the diagram as well as apply a correction factor to the footcandle value at this new location. The following procedure may be used.

a. Compute new transverse and longitudinal distance-to-mounting height, and locate points on the diagram as outlined in the following calculation:

Example for 25-foot (7.5-meter) mounting height;

Luminaire Numbers 1 and 3—Point "A_1";
Transverse 5 feet (1.5 meters) to "house side":

$$\frac{5}{25} = 0.2 \text{ MH}$$

Longitudinal 120 feet (37 meters) along pavement:

$$\frac{120}{25} = 4.8 \text{ MH}$$

Point "A_1" is located on the isofootcandle diagram, Fig. 14-22, with these new dimensions.

b. Obtain estimated footcandle values at the new locations and multiply these values by the correction factor for the new mounting height.

Footcandle value estimated at point "A_1," Fig. 14-22, is 0.015 footcandle. This is multiplied by the correction factor for 25 feet which is 1.44.

0.015 × 1.44 = 0.0216 footcandle from each luminaire, Numbers 1 and 3. Both luminaires provide 0.043 footcandle.

Luminaire Number 2—Point "A_1."
Transverse 45 feet to "street side":

$$\frac{45}{25} = 1.8 \text{ MH}$$

Longitudinal location is still 0, directly across from luminaire. The estimated footcandles from Fig. 14-22 is 0.2 footcandle. This is multiplied by the correction factor, 1.44.

$$0.2 \times 1.44 = 0.288 \text{ footcandle.}$$

The total at point "A_1" is

$$0.043 + 0.288 = 0.331 \text{ footcandle.}$$

As before this must then be multiplied by the ratio of the actual lamp lumens to the lamp lumens of the isofootcandle diagram (20,000/10,000 = 2) for the initial footcandle level.

Uniformity Ratios

The illuminance uniformity requirements (see page 14-10) should be determined by computing the ratio:

$$\frac{\text{Minimum Horizontal Illuminance}}{\text{Average Horizontal Illuminance}}$$

It can also be expressed as the ratio:

$$\frac{\text{Average Horizontal Illuminance}}{\text{Minimum Horizontal Illuminance}}$$

Sufficient number of specification points over the roadway should be checked, as outlined above, to ascertain accurately the location and value of the minimum point. If the values at points "A," "B," and "C" as shown in Fig. 14-21, are first determined, the approximate location of the minimum point may be located or its location will become more apparent.

The average illuminance on the roadway pavement should be computed as shown above, taking care to use the same lamp lumen output and other conditions as used in determining the minimum illuminance value.

REFERENCES

1. *American National Standard Practice for Roadway Lighting, ANSI/IES RP-8, 1977,* American National Standards Institute, New York. (Committee on Roadway Lighting of the IES: "American National Standard Practice for Roadway Lighting," *J. Illum. Eng. Soc.*, Vol. 7, No. 1, October, 1977.)
2. Subcommittee on Lighting of Tunnels and Underpasses of the Roadway Lighting Committee: "Lighting of Tunnels," *J. Illum. Eng. Soc.*, Vol. 1, No. 3, April, 1972.
3. Subcommittee on Rest Areas of the IES Roadway Lighting Committee: "Lighting Roadway Safety Rest Areas," *J. Illum. Eng. Soc.*, Vol. 4, No. 1, October, 1974.
4. Subcommittee on Roadway Sign Lighting of the IES Roadway Lighting Committee: "Roadway Sign Illumination," *J. Illum. Eng. Soc.*, Vol. 4, No. 1, October, 1974.
5. Subcommittee on Off Roadway Facilities of the IES Roadway Lighting Committee: "Lighting of Parking Facilities." To be published.
6. VanDusen, Jr., H.A.: "Roadway Lighting System Design," *J. Illum. Eng. Soc.*, Vol. 3, p. 115, January, 1974.
7. Clark, F.: "The Case for Step-by-Step Procedures for Calculations in Roadway Lighting Design," *Illum. Eng.*, Vol. 65, p. 637, November, 1970.

Aviation Lighting

SECTION
15

Aviation lighting falls into two principle categories: aircraft lighting—that on or inside the aircraft; and aviation ground lighting—that on the landing facility or airport.

Aircraft lighting, categorized by location as exterior lighting, crew station lighting and passenger interior lighting, covers a broad spectrum of lighting equipment ranging from that used for visual collision avoidance to instrument lighting to decorative lighting in the passenger cabins.

Aviation ground lighting provides for the illumination of aprons, hardstands, parking areas, taxiways and runways and in addition includes a good deal of signal lighting. This lighting conveys information to pilots by means of color, location, flash characteristics and/or pattern of lights rather than by illumination of areas or objects. To provide additional visual guidance to pilots, other visual aids such as reflectors, markings, etc. are also used at airports.

Standardization

In no field of lighting is standarization more important than in aviation lighting, both in ground lighting and in aircraft lighting. The interstate and international scope of operations make imperative the establishment of standards of color and pattern of all the visual aid systems essential to the safe operation of aircraft. These regulations and standards are originated by government agencies and departments and by the military and, in addition, by international organizations such as the ICAO (International Civil Aviation Organization) and NATO (North Atlantic Treaty Organization). In most instances in this Section, the relevant regulation or regulating body is mentioned but before planning or modifying any aviation lighting or visual aid project the appropriate agencies or standards should be consulted first.

NOTE: References are listed at the end of each section.

AIRCRAFT LIGHTING

Aircraft lighting equipment, particularly that used on the exterior surfaces of the airplanes, must perform satisfactorily over a wide range of environmental conditions including extremes of temperature, pressure and vibration, not normally found in most lighting applications.

For the most part, luminaires are designed to perform specific functions when installed in specific locations in specific types of airplanes. Therefore, there is little opportunity to achieve any significant standardization in luminaires for aircraft lighting. It is more likely that lighting functions will be standardized and performed by luminaires which will vary due to space configuration and environmental requirements to fit in the specific applications. Types of luminaires performing the same functions will vary widely when designed for different categories of aircraft.

Aircraft Categories. Broad categories of aircraft are: general aviation, commercial and military.

General Aviation Category. General aviation includes all branches of aviation other than airline (commercial) or military. It runs a gamut from small private, single-engine airplanes to sophisticated multi-engine turbine-powered airplanes as used by business. These include those aircraft used for air taxi and charter operations.

The minimum lighting requirements for these general aviation aircraft are found in Part 23 of the Federal Aviation Regulations (FAR's) for fixed wing airplanes and Part 27 for helicopters. Part 91 (Operating Requirements) requires the use of position or navigation lights and anti-collision lights for night operations. Position lights (red and green wing tip lights and white tail light) are so named because they are intended to indicate to an observer, the position of the aircraft. Landing lights are required on aircraft for some commercial operations.

Commercial Category. Commercial or airline aircraft must include lights meeting FAR Part 25 for a fixed wing and Part 29 for helicopters—transport types. Operating requirements for air carrier (airline) operations are found in Part 121 of the FAR's. These include minimum lighting requirements for and use of position, anticollision, landing, instrument, indicator, wing ice floodlights and emergency egress lighting and lighted no-smoking, fasten-seat-belt signs. One large wide bodied air transport aircraft has more than 560 individual lights of 55 different types.

Military Category. Military aircraft also consist of a wide variety of types of aircraft ranging from relatively simple, small trainers to high performance fighters, bombers, large transport aircraft and helicopters. While military aircraft have many of the same lighting requirements as general aviation or commercial, they will frequently require special lights for specific military functions such as formation flying, in-flight refueling operations and approach lights for operations onto aircraft carrier decks. There are differences in requirements for intensity distribution for both position and anticollision lights among the FAR's, Navy specifications, Air Force and Army specifications. Navy exterior lighting requirements are generally covered in Specification MIL-L-006730. Navy interior lighting requirements may be found in military specification MIL-L-18276. Requirements for lighting of Air Force and Army airplanes are described in Specification MIL-L-6503H. Other specifications, usually referenced in these documents have been issued for various detail lighting requirements.

Some international standardization is attempted by use of documents published by the International Civil Aviation Organization (ICAO) for civil aviation and the North Atlantic Treaty Organization (NATO) for military aviation.

Aircraft Electrical Systems

Most new aircraft are now equipped with either 28 volt dc or 120/208-volt, 400-hertz, 3-phase electrical systems. Until recently all small single-engine airplanes were equipped with 14-volt dc systems. Many of the more standard lamps are designed to operate on either 14 or 28 volts. In many cases lamps for aircraft using the 400-hertz systems may be lower voltage, 6 to 10 volts, and operated from individual transformers. The use of low voltage incandescent filaments can produce increased strength for resistance to shock and vibration as well as more compact size for improved optical efficiencies.

Exterior Lighting of Aircraft

Position Lights. For many years, aircraft have been equipped with a navigation or position light system consisting of red lights on the left wing tip, green light on the right wing tip and white lights on the tail of the aircraft or more recently on the trailing edge of each wing tip. These lights are basic to all systems and required for night operations. Intensities of red and green position lights in the forward direction on different aircraft range from 40 candelas (minimum required by FAR's) to more than 300 candelas in some cases. Commercial aircraft are commonly equipped with dual position lights for redundancy. Location of the white tail light on the trailing edge of the wing tips, or in some cases on the outboard trailing edges of the horizontal stabilizer, permits light locations for easier maintenance and results in these lights providing more attitude information when viewed from the rear.

Anti-collision Lights. Anti-collision lights are required on all aircraft flying at night and may be either red or white lights. Current Federal Aviation Regulations require that they produce a minimum of 400 candelas of effective intensity near the horizontal planes, reducing to 20 candelas at vertical angles of 75 degrees up or down. Condenser discharge flashtubes (strobes) are commonly used for both red and white anti-collision lights. Many aircraft are equipped with anti-collision light systems which include red flashing lights located top and bottom of the fuselage and supplemented with white flashing lights on the wing tips. Strobe lights are operated at energy levels of 15 to 100 joules per flash and flash rates of 50 to 80 flashes per minute. Effective intensities range from 100 to more than 4000 candelas. Location of white high intensity flashing lights on the wing tips results in significantly fewer problems due to reflections or backscatter which may interfere with the crew's vision from the cockpit.

Use of high intensity white lighting is re-

Fig. 15-1. Taxiing light mounted on nosewheel strut.

stricted during taxiing operations at the airport. Lights on the wing tip or tail of one aircraft may be only a few meters (feet) from the pilot of another aircraft. It has become common practice to operate the red anti-collision lights to indicate that turbine engines are operating while an airplane is stationary on the ground.

Landing Lights. Landing lights are commonly equipped with 200-millimeter (eight-inch) diameter (PAR-64), 600-watt, or 150-millimeter (5 ¾-inch) diameter (PAR-46), 450-watt sealed reflector type lamps. Large aircraft will commonly be equipped with four of these lamps in a landing light system with one in each outboard wing section in retractable units and one in a fixed light in each inboard wing root. On some aircraft fixed lighting is located on the landing gear struts. The use of landing lights in the daytime to improve the conspicuity of the aircraft has become quite common, particularly when aircraft are within 16 kilometers (ten miles) of a tower-controlled airport or lower than 3000 meters (10,000 feet) above sea level. If only landing-gear mounted or retractable landing lights were available, it would be necessary to slow the airplane down to gear extension speed. In order to avoid this, some airplanes are now being equipped with fixed recognition lights in the wing tips or wing leading edges of some aircraft just for identification purposes. These lights will usually produce a relatively narrow beam with an intensity of approximately 70,000 candelas.

Helicopters are equipped with search and landing lights which are controlled in both elevation and azimuth by operating a 4-way switch commonly located on the collective pitch control.

Taxi Lights. Taxi lights normally have a wide horizontal and narrow vertical beam and are commonly mounted on landing gear struts. In some cases, they are mounted on the moveable section of a nose wheel strut as shown in Fig. 15-1. At other times, there are fixed taxi lights located in wing root cavities to illuminate area outboard and ahead of the aircraft to identify taxiway turnoff areas.

Auxiliary Lights. All large aircraft are equipped with *fixed floodlights* arranged so that the crew may visually inspect wing leading edge surfaces and nacelles at night to determine whether or not ice build-up is a problem. These lights are usually located in the side of the fuselage and sometime in the outboard sides of the engine nacelles. Another type of exterior floodlight which has become popular are lights arranged to floodlight the sides of the vertical fin. These are commonly called "logo" lights and may be located in the surface of the horizontal stabilizer or in some cases in units on the trailing edge of the wing tips aimed at the vertical fin.

Electroluminescent lamp strips have become very common for use as *formation lights*. Strips, usually about 50 millimeters (two inches) wide and up to one meter (three feet) long are placed on the sides of the fuselage, the vertical fin and on the top and bottom surfaces of the wing tips. Modern fighter aircraft may use upwards of 0.45 square meters (700 square inches) of green electroluminescent lamps for this purpose. The luminance of these lights are continuously controlled from 0 to about 70 candelas per square meter (20 footlamberts).

For inflight refueling, tanker airplanes are equipped with a variety of incandescent floodlights and signal lights supplemented with electroluminescent strip lights. The lights are designed to make it possible for the pilot of the refueling airplane to maintain the proper location with respect to the tanker and by use of various types of signal lights to receive pertinent information on the progress of the refueling process. The most recent tanker airplane is equipped with a special anti-collision or rendezvous light which consists of dual red and white flashtubes coded in color and flashing sequence providing a signal which unmistakably identifies the aircraft as a tanker airplane.

Carrier based aircraft are normally equipped with red, green and amber lights which can be seen in the daytime during landing operations by the landing signal officer aboard the carrier. These lights will indicate to the landing signal officer the approach attitude of the aircraft.

Crew Station Lighting

In sophisticated aircraft, crew station lighting has become rather complex. The use of luminous devices such as cathode ray tubes have contributed to the over-all lighting problems of the cockpit areas. Most multi-engine aircraft are equipped with instruments which have the lighting designed into the instrument as an integral part of the instrument.

Switch panels are normally plastic panels about 5 millimeters (3/16 inch) thick and which are lighted with small embedded subminiature or electroluminescent lamps. See Fig. 15-3.

The color of the instrument and panel lighting is red where the mission of the aircraft may require the pilot to achieve a high degree of dark adaptation. Many Air Force airplanes are equipped with blue-white instrument lighting. The blue filters were originally added to permit operation at very low levels without the extreme yellow color associated with operation of incandescent lamps at very low color temperatures. Commercial airplanes usually use white unfiltered light from long-life 5-volt subminiature incandescent lamps. See Fig. 15-2.

Most aircraft are also equipped with instrument floodlighting systems in addition to integral instrument and panel lighting to provide low level background lighting in order to avoid any autokinetic effect which may be produced when lighted markings are viewed against a perfectly black background. Also, aircraft are commonly equipped with "thunderstorm" floodlights capable of providing a higher level of about 1000 lux (100 footcandles). These lights insure the continued visibility of instruments in case dark adaptation is suddenly destroyed by bright lights such as a lightning flash.

Another important part of crew station lighting is the indicator or annunciator lights. Normally, requirements are established that annun-

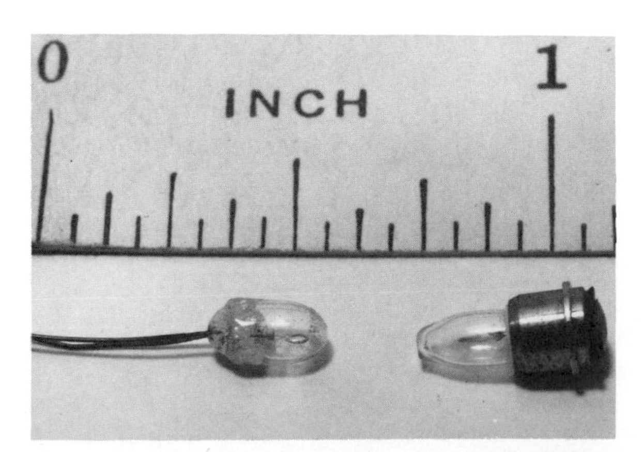

Fig. 15-2. Subminiature lamps (T-1, 5.0 volts) for instrument and control panel lighting.

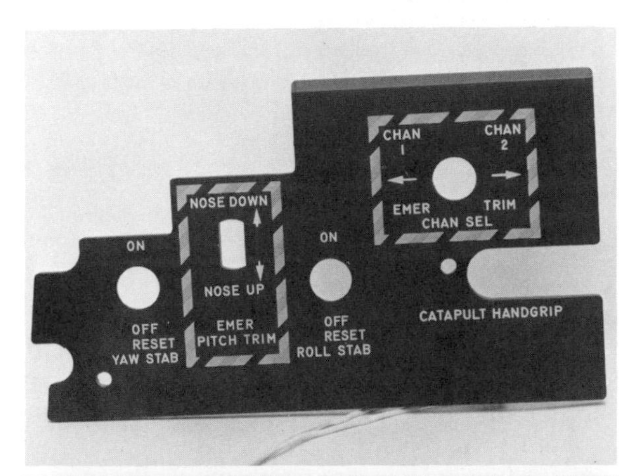

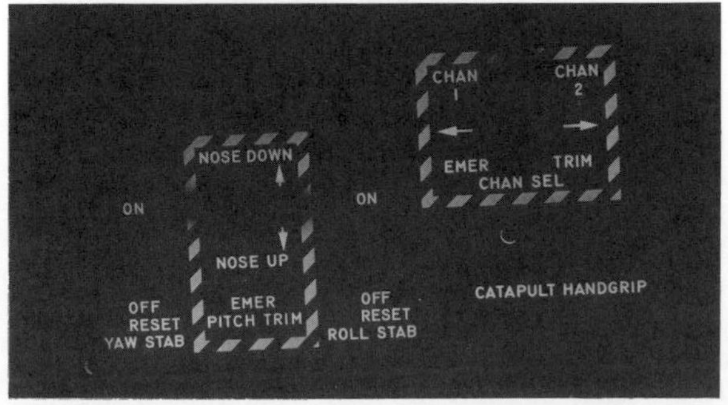

Fig. 15-3. Electroluminescent control panel. Top is unlighted daylight view. Bottom is lighted night view.

Fig. 15-4. A portion of a modern commercial airplane showing some of the passenger cabin lighting. (1) Individual passenger reading lights. (2) Incandescent down lights. (3) Pendant sign: No Smoking—Fasten Seat Belt. (4) Fluorescent indirect luminaire operated from inverter ballasts, fully dimmable. (5) Information sign. (6) Emergency exit signs operated from separate battery power supplies. (7) Fluorescent window reveal lights, fully dimmable.

ciator lights be unreadable in bright sunlight unless they are energized. Further, they must be perfectly readable with incident illumination of 100 kilolux (10,000 footcandles). Currently available annunciator lights do fully meet the requirements for sunlight readability.

Most airplanes are also equipped with map lights which are easily controlled by the pilot to provide a desired level of very controlled illumination on navigational charts.

Passenger Interior Lighting

Passenger interior lighting varies from one aircraft to another since considerable use is made of lighting in conjunction with the decorative designs in the cabin. Fig. 15-4 illustrates some of the cabin lighting in a wide-body jet aircraft.

Individual controlled passenger reading lights are an important part of the cabin lighting system. These lights provide a level of 250 to 300 lux (25 to 30 footcandles) on the reading plane. They must provide carefully controlled beam patterns to avoid interference with adjacent passengers. The distance from the light may vary from less than 760 millimeters (30 inches) to as much as 2 meters (80 inches).

A variety of lighted signs are used in the passenger interiors of airplanes. These include such information as no smoking, fasten seat belts, return to seat, etc. In some cases, large signs are located on the forward bulkheads of compartments and in other areas, smaller signs are located in passenger service units for each row of seats. Frequently, the design of the sign is part of the decorative scheme in the airplane interior. The use of symbols is becoming more common.

Generally, passenger cabins will be equipped with general lighting which may be fluorescent, incandescent or combinations of both. Use of valance lights over window reveals on either side of the cabin is very common. Also, many airplanes are equipped with strip lights which floodlight the ceilings of the cabins from locations above the bag racks. Both incandescent and fluorescent lamps are used for this application.

Cabin lights other than reading lights and signs are dimmable by the cabin crew either continuously or in steps. Both hot and cold cathode fluorescent lamps are in use.

Emergency Egress Lighting. Emergency egress lighting has become an important part of safety equipment in modern airline aircraft. Lights are installed in the ceiling to provide illuminances of more than 0.5 lux (0.05 footcandles) at seat arm-rest level. Other lights are installed to illuminate the exit areas. Federal Aviation Regulations specify minimum lighting requirements for area lighting as well as the size and luminance and contrast of signs marking exits and directing passengers to these exits.

All lights and signs are powered by battery power supplies which activate manually or automatically on failure of the aircraft electrical power system.

Areas outside the airplane such as escape slides and over-wing exit paths are also lighted.

AVIATION GROUND LIGHTING

Airport/Runway Classification

The type of ground lighting and other visual aids required at an airport is usually based upon the visibility conditions under which operations are conducted and the types of electronic navigational aids available at the airport to support such operations.

Visibility conditions are categorized based upon the distance from which a pilot can detect a 10,000-candela light source (Runway Visual Range) and the decision height above the runway at which a missed approach is initiated if the required visual reference has not been established (Decision Height). See Fig. 15-5.

Depending on the types of navigational aids available, runways are classified as Visual or Instrument runways. A visual runway is a runway intended solely for the operation of aircraft using visual approach procedures. An instrument runway is a runway having an instrument approach procedure utilizing electronic navigational facilities. If these facilities provide only horizontal guidance or area type of navigational information the runway may be defined as a non-precision runway. If the runway has Instrument Landing Systems (ILS) or a Precision Approach Radar (PAR), the runway may be defined as a precision instrument runway.

Taxiway Guidance Systems

Visual taxi guidance between the runways and the destination on the airport is provided by centerline lights, edge lights, reflectors, signs and markings.

Fig. 15-5. Airport/Runway Visibility Condition Categories

Category	Runway Visual Range		Decision Height	
	Meters	Feet	Meters	Feet
I	720	2400	60	200
II	360	1200	30	100
IIIA	210	700	—	—
IIIB	45	150	—	—
IIIC	0	0	—	—

Taxiway Centerline Lighting. A taxiway centerline lighting system is designed to facilitate aircraft ground traffic movements.

Application. Centerline lights are provided on taxiways intended for use in runway visual range conditions less than 360 meters (1200 feet) and are recommended for all airports with runways having precision approach procedures, particularly at high traffic density airports.

Characteristics. Taxiway centerline lights are steady burning lights of variable intensity green. The luminous intensity and distribution are adequate for the conditions of visibility and ambient light in which use of the taxiway is intended. Fig. 15-6 shows a typical photometric centerline light characteristic.

Location. Centerline lights on a straight section of a taxiway are spaced at longitudinal intervals of not more than 30 meters (100 feet) except that:

1. On short, straight sections, intervals less than 30 meters (100 feet) are provided.
2. On a taxiway intended for use in runway visual range conditions of less than a value of the order of 360 meters (1200 feet) the longitudinal spacing is not to exceed 15 meters (50 feet).

On a taxiway intended for use in runway visual range conditions of less than a value of the order of 360 meters (1200 feet) the lights on a curve are not to exceed a spacing of 15 meters (50 feet) and on a curve of less than 360 meters (1200 feet)

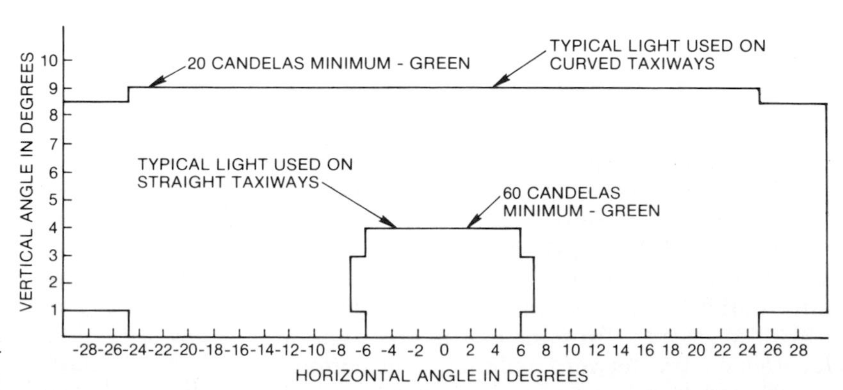

Fig. 15-6. Typical photometric characteristics for taxiway centerline lights.

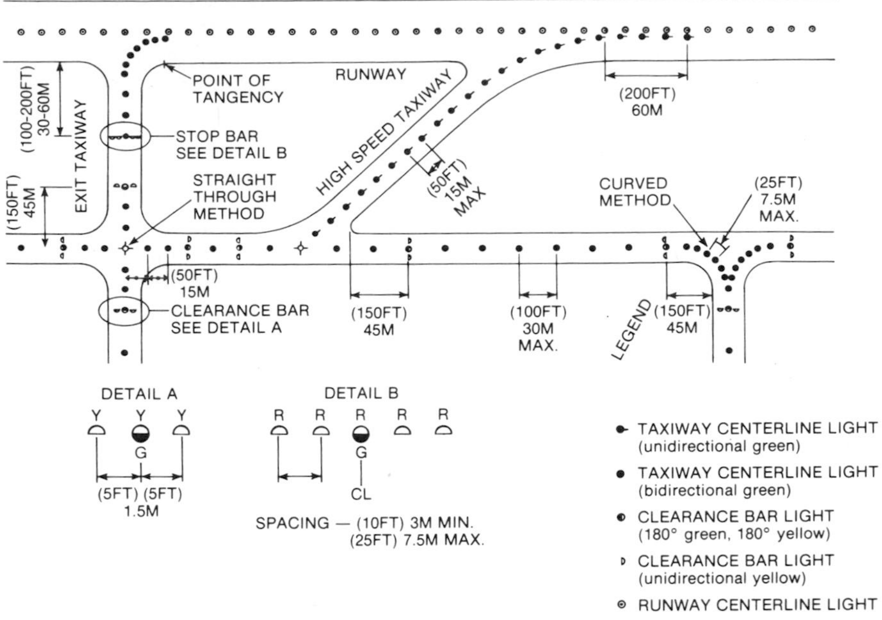

Fig. 15-7. Typical centerline lighting configurations.

radius, the lights are spaced at intervals of not greater than 7.5 meters (25 feet). This spacing may extend for 60 meters (200 feet) before and after the curve.

Unidirectional centerline lights on a high speed exit taxiway commence at a point at least 60 meters (200 feet) before the beginning of the taxiway centerline curve and continue beyond the end of the curve to a point on the centerline of a taxiway where the aircraft can be expected to reach normal taxi speed. The lights on that portion parallel to the runway centerline are either on the runway centerline or if runway centerline lights are provided, offset from the runway centerline up to 900 millimeters (36 inches).

At taxiway intersections, centerline lights may be installed around the curves at the spacing indicated above or by using the straight thru method (see Fig. 15-7).

Taxiway Edge Lights. A taxiway edge lighting system is a configuration of lights which define the lateral limits of the usable taxiing area.

Application. Taxiway edge lights are provided along taxiways intended for use at night and along taxiways not provided with centerline lights. Taxiway edge lights need not be provided where, considering the nature of the operations, adequate guidance can be achieved by other means, such as reflectors.

Characteristics. Taxiway edge lights are steady burning blue lights which may be of variable intensity. The lights show up to at least 30 degrees above the horizontal and at all angles of azimuth necessary to provide guidance to a pilot taxiing in either direction. Lights may be shielded so that they cannot be seen at angles of azimuth in which they may be confused with other lights.

Location. Taxiway edge lights on a straight section of a taxiway are spaced at uniform longitudinal intervals of not more than 60 meters (200 feet). The lights on a curve are spaced at intervals less than 60 meters (200 feet) so that a clear indication of the line of the curve is provided. The lights are to be located not more than 3 meters (10 feet) from the edge of the taxiway.

Stop Bars.

Application. Stop bars are provided at a taxiway-runway intersection or other taxi-holding positions when it is desired to provide traffic control by lights. The use of stop bars requires control by air traffic services. Stop bars should be provided at taxi-holding positions used in conjunction with a precision approach runway category III.

Characteristics. Stop bars consist of unidirectional lights, showing red in the direction of approach to the intersection or taxi-holding position, spaced at intervals of 3 meters (10 feet) across the entire width of taxiway. The intensity and beam spreads of stop bar lights should not be less than that of taxiway lights provided.

Location. Stop bars are located across the taxiway at the point where it is desired that traffic stop. See Fig. 15-7.

Clearance Bars.

Application. A clearance bar is provided at a taxiway intersection where it is desired to define a specific aircraft holding position. See Fig. 15-7.

Characteristics. Clearance bars consist of at least three fixed unidirectional lights showing yellow in the direction of approach to the intersection with a light distribution similar to taxiway lights provided. The lights are installed symmetrically about the taxiway centerline with individual lights spaced 1.5 meters (5 feet) apart.

Location. Clearance bars are normally located at a point between 30 meters (100 feet) to 60 meters (200 feet) from the near edge of the intersecting taxiway or runway.

Taxiway Centerline Reflectors. Centerline reflectors are installed to improve the conspicuity of the taxiway centerline during the hours of darkness.

Application. Centerline reflectors may be provided on taxiways not equipped with centerline lights. In snow areas, special reflectors that are compatible with snow removal equipment are preferred. Centerline reflectors normally do not replace lights or painted markings; but rather, supplement those visual aids.

Characteristics. Taxiway centerline reflectors are bidirectional green reflectors capable of withstanding normal operations of aircraft and maintenance vehicles. The reflectors are small units with a maximum height of 15.9 millimeters (5/8 inch), and designed for cementing on the taxiway surface. The reflectors should have a wide divergence angle of light return to meet most operational requirements.

Location. Centerline reflectors on straight sections and large areas of taxiway, and on high speed exits are spaced at longitudinal intervals of not more than 15 meters (50 feet). On tight taxiway curves (120 meters (400 feet) or less radius) reflectors are spaced at 7.5-meter (25-foot) intervals. The layout of centerline reflectors is similar to the layout described for centerline lights.

Taxiway Edge Reflectors. Elevated edge reflectors are installed to improve the conspicuity of taxiway edges during the hours of darkness.

Application. Edge reflectors may be provided along unlighted taxiways and to supplement centerline lighting systems.

Characteristics. Elevated taxiway edge reflectors are generally cylindrical in shape and blue in color, with 360-degree reflective bands. They may be flexible or rigid (with frangible mounts), and are constructed to withstand aircraft blast.

Location. The spacing of elevated edge reflectors is not greater than that specified for taxiway edge lights.

Taxi Guidance Signs. The purpose of the taxi sign system is to give guidance within the aircraft operating area. Signs also substantially aid the traffic controller by simplifying instructions for taxiing clearances, routing and holding aircraft. The classes and types of signs comprising a sign system will vary according to the operational needs of that airport.

Application. A taxi guidance sign system consists of four basic classes of signs:

1. *Destination signs* to aid pilots in their inbound and outbound taxiing routes.
2. *Intersection signs* to warn pilots of a taxiway or runway intersection.
3. *Exit signs* to identify an exit from a runway.
4. *Special purpose signs* to alert pilots that they are entering special condition areas.

Characteristics. Signs are lightweight and frangibly mounted. Retaining chains are used to prevent signs, which have broken from their mountings, from blowing away.

Two types of signs installed on airports are:

1. *Mandatory signs,* which, if ignored by a pilot in an aircraft on the ground, could cause a hazard involving an aircraft landing or taking off. Mandatory signs have white letters on a red background and are illuminated. *Runway* intersection and special purpose signs are mandatory signs.
2. *Information signs* enable pilots to determine location or route and, if ignored, could cause a hazard to taxiing aircraft. Information signs are yellow with black letters (or vice versa) and generally are illuminated. Signs which are not illuminated are retroreflective. Destination and *taxiway* intersection signs are information signs.

Location. All signs are located as near as possible to the edge of the pavement, but at sufficient distance to be clear of jet engine pods and engine blast effects.

1. *Destination signs* are installed on the far side of the intersection and in the most visible location, usually the left side of the taxiway.
2. *Intersection signs* are installed on the near side of the intersection on the side toward which traffic normally turns. If traffic turns both left and right, the sign is installed on the left.
3. *Exit signs.* A runway exit is usually marked by a sign on the near side of the exit with taxiway identification and a properly oriented arrow, depicting the exact angle.

4. *Special purpose signs* are located so as best to identify special condition areas such as Category II and ILS clearance areas.

Taxiway Markings. Taxiway markings are provided to meet operational needs and are applied in an aviation yellow surface color meeting the chromaticity and luminance factors given in Appendix 1, Annex 14, Aerodromes, a publication of the International Civil Aviation Organization (ICAO).

Application.

1. *Taxiway Centerline Markings* are applied on paved taxiways.
2. *Taxiway Edge Markings* are applied when the edge of the full strength pavement of a taxiway is not readily apparent.
3. *Taxiway Holding Line Markings* are applied at those places on a taxiway where it is desirable to hold aircraft clear of an active runway.
4. *Critical Area Hold Line Marking* is applied to keep aircraft clear of electronic sensitive and obstacle free areas.
5. *Taxiway Identification Markings* are applied where difficulty is encountered in locating taxiway identification signs.

Runway Guidance Systems

Runway visual aids are installed to provide guidance during landings and takeoffs. These visual aids consist of reflectors, signs and painted markings as well as edge, centerline, touchdown zone, slope and approach lights.

Approach Lighting Systems. An approach lighting system is a configuration of lights usually disposed symmetrically about the extended runway centerline. The system starts at the beginning of the usable landing runway surface (threshold), and extends horizontally out toward the aircraft approaching for a landing. Approach lights provide visual information on runway alignment, roll guidance and horizon reference during the last stages of approach to a landing.

Application. The selection of a particular approach lighting system is based on the worst-case meteorological visibility conditions permissible for operations conducted on the particular runway.

Characteristics. An approach lighting system is comprised of white and/or colored steady-burning lights, white sequentially flashing lights, or a combination of both. Steady-burning lights may be unidirectional or omnidirectional, and are usually of variable intensity, depending on the system which they comprise. The sequentially flashing lights may be unidirectional or omnidirectional; have a definite flash duration, intensity and spatial orientation; and may be of variable intensity and flash sequentially, starting with the unit furthest from the threshold, inward toward the threshold at the rate of either one or two flashes per light per second, depending on the system which they comprise. Directional lights have both azimuth and elevation orientations, using the extended runway centerline and horizontal as reference. Most systems incorporate the use of frangible or low-mass structures to hold the individual lights in proper orientation; these structures are designed to minimize damage to landing aircraft, if the structures are inadvertently struck.

Fig. 15-8. Runway Approach Light Applications

Runway Usage		Approach Light System
Instrument Systems	Categories II & III	High Intensity Approach Light System with Sequenced Flashers (ALSF-2)
	Category I*	Modified Calvert System (High Intensity)
		Alpha System (High Intensity)
		Medium Intensity Approach Light System with Runway Alignment Indicator Lights (MALSR)
	Other Instrument not Categories I, II, or III*	Omnidirectional Approach Light System (ODALS)
		Medium Intensity Approach Light System (MALS)
		Medium Intensity Approach Light System with Sequenced Flashers (MALSF)
Visual Flight Only*		Non-Instrument Approach System (Low Intensity)

* The use of an individual approach light system for this runway usage is subject to approval by the local authority having jurisdiction.

Approach lighting systems are listed in Fig. 15-8 and are described below.

1. *Approach Light System for Category II and III Runways.* The approach light system for Category II and III operations uses white, green and red unidirectional steady-burning lights, and white unidirectional sequenced flashers, all of variable intensity. The photometrics of a typical steady-burning white light are shown in Fig. 15-9. The photometrics for the white flashing light is of the same shape as shown in Fig. 15-9, but the intensity is 20,000 maximum and 8000 minimum effective candelas within the area shown. The flash rate is twice per second per light. The system layout is shown on Fig. 15-11.

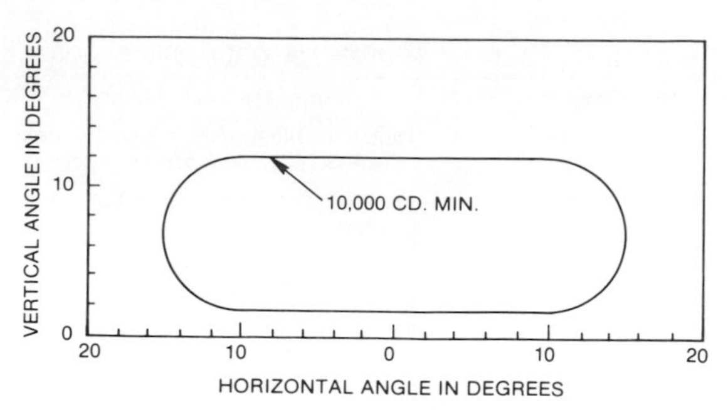

Fig. 15-9. Photometric characteristics of a typical white approach light (steadyburning).

2. *Approach Light Systems for Category I Operations.* There are three configurations for instrument approach Category I operations, all three of which are termed centerline crossbar systems. Two of these systems are classed as high intensity systems, and the third as a medium intensity system.
   a. *Modified Calvert System.* The modified Calvert system uses white, red and green steady-burning lights of variable intensity, and white unidirectional sequenced flashers, which may be of variable intensity. The white steady-burning lights have the following minimum intensity characteristics: ± 7.5 degrees divergency in the horizontal plane at 20,000 candelas, ± 12.5 degrees divergency in the horizontal plane at 5000 candelas; beam spread in the vertical plane may be up to 30 per cent less than that of the horizontal plane.

   The requirements for the sequenced flashers is a minimum effective intensity of 10,000 candelas with a beam spread of not less than 20 degrees horizontally and 10 degrees vertically, with a flash rate of one or two flashes per light per second. The system layout is shown in Fig. 15-12.
   b. *Alpha System.* The Alpha system (also locally known as ALSF-1, High Intensity Approach Light System with Sequenced Flashers) differs only slightly from the ALSF-2 (see Fig. 15-11). The lights have the same photometric characteristics as the ALSF-2 system.
   c. *Medium Intensity Category I System.* The Medium Intensity Approach Light System with Runway Alignment Indicator Lights (MALSR) uses steady-burning unidirectional white and green lights and white sequenced flashers which are usually unidirectional, and all lights are of variable intensity. The white steady-burning lights are grouped in bars of five lights. Fig. 15-10 shows the MALSR layout.
3. *Approach Lights for Other Instrument Runways.* Of the three systems in this category, two systems may be thought of as simplified versions of the MALSR system. The other system is unique in that it uses only omnidirectional sequenced flashers.
   a. *Omnidirectional Approach Light System (ODALS).* The omnidirectional approach light system consists of seven white omnidirectional flashing lights of variable intensity and with a flash rate of one flash per second per light. The two lights on the threshold flash in unison. The layout is shown in Fig. 15-13.

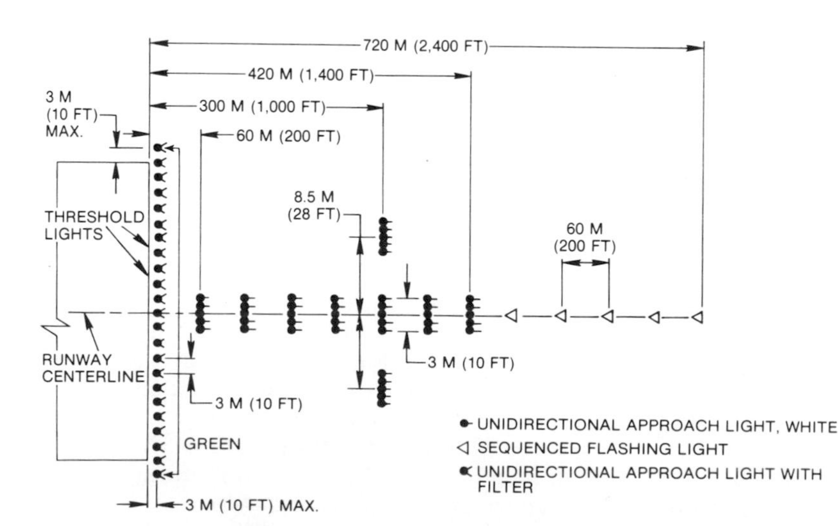

Fig. 15-10. Medium intensity approach light system with runway alignment indicator lights (MALSR).

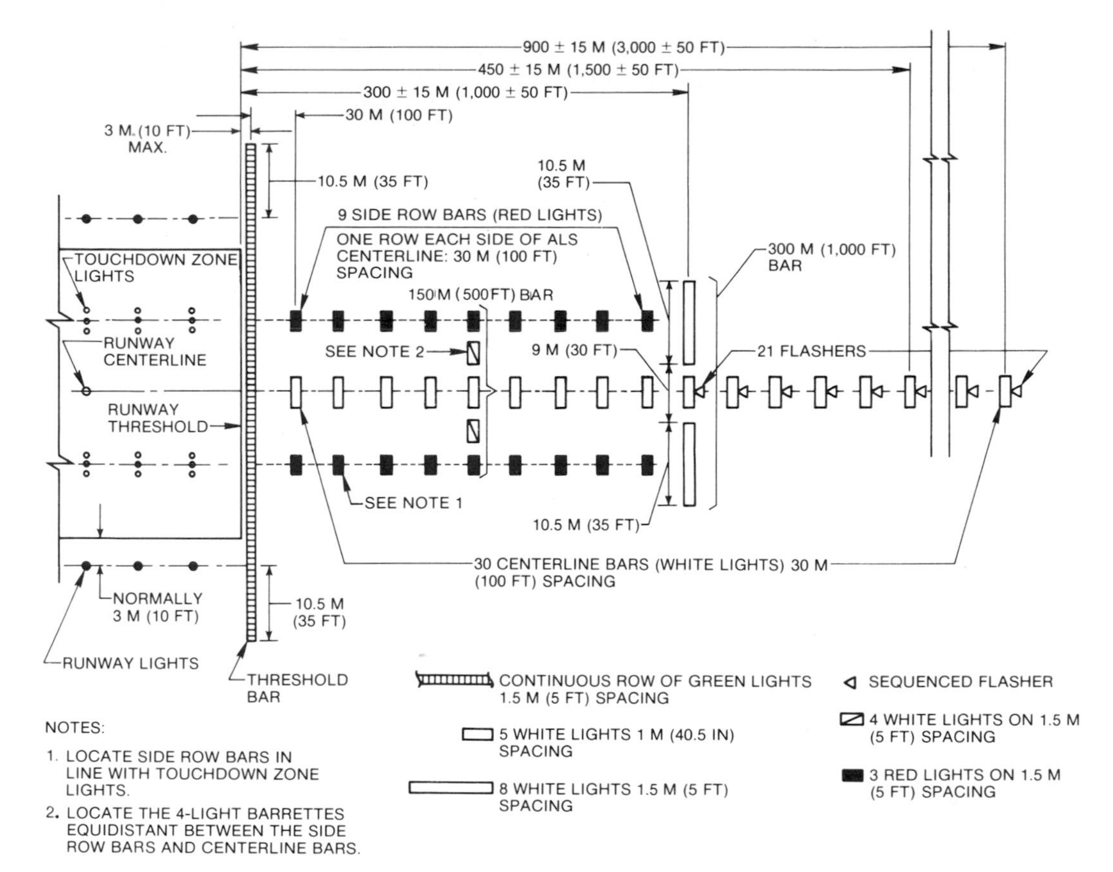

Fig. 15-11. Approach light system with sequenced flashers: ALSF-2.

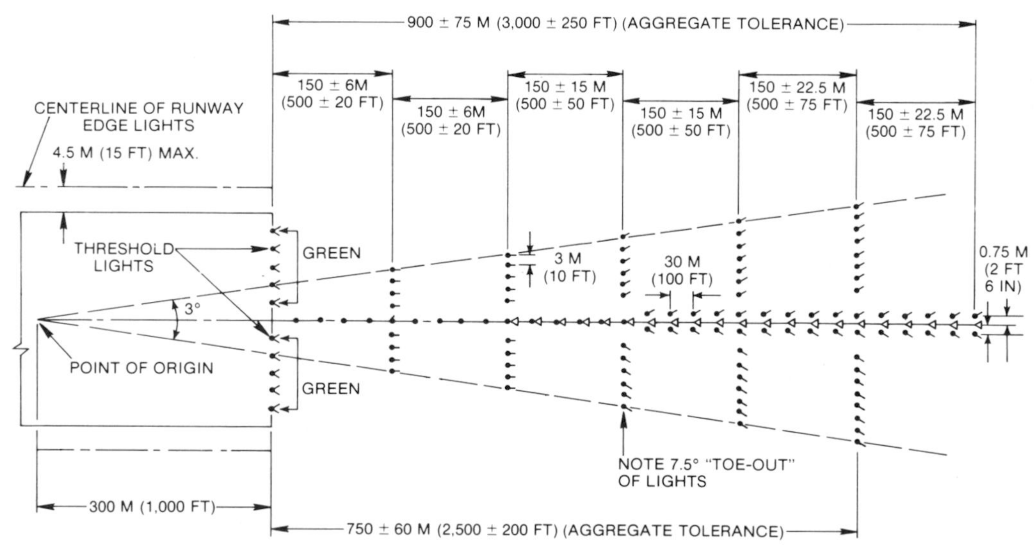

Fig. 15-12. Modified Calvert system.

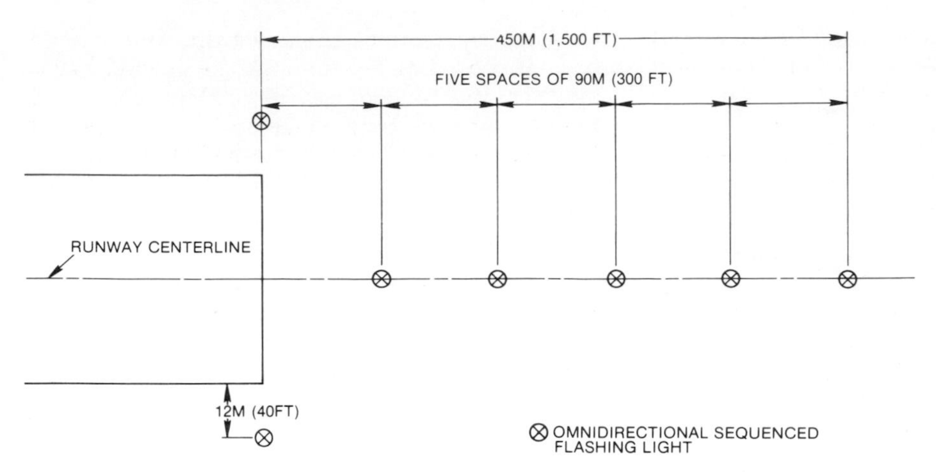

Fig. 15-13. Omnidirectional approach light system (ODALS).

b. *Medium Intensity Approach Light System (MALS).* The medium intensity approach light system (MALS) is identical to the first 420 meters (1400 feet) of the MALSR shown in Fig. 15-12 and may be considered a building block of a MALSR for phased development.

c. *Medium Intensity Approach Light System with Sequenced Flashers (MALSF).* The MALSF consists of a MALS with three sequenced flashers at the three outer light bar locations. These flashers are added to the MALS at locations where high ambient background lighting, or other reasons, requires these lights to assist pilots in making an earlier identification of the system. These lights flash in sequence toward the threshold at a rate of twice per second per light, and are the same as the flashers used in the MALSR.

4. *Non-Instrument Approach System.* The non-instrument approach light system is also a centerline cross-bar system. It consists of a single or double omnidirectional steady-burning light, either aviation yellow or aviation red, and of single intensity. The system layout is shown in Fig. 15-14.

Visual Approach Slope Indicator Systems (VASIS). The VASIS is a configuration of lights which furnish the pilot with visual approach slope information to provide safe descent guidance.

Application. The VASIS is intended for day and night use during Visual Flight Rule (VFR) weather conditions. While several different forms of VASIS are in use throughout the world, the Red/White system is the most commonly encountered and can be considered the standard used by nations on the North American continent. As such it will be the only system considered here.

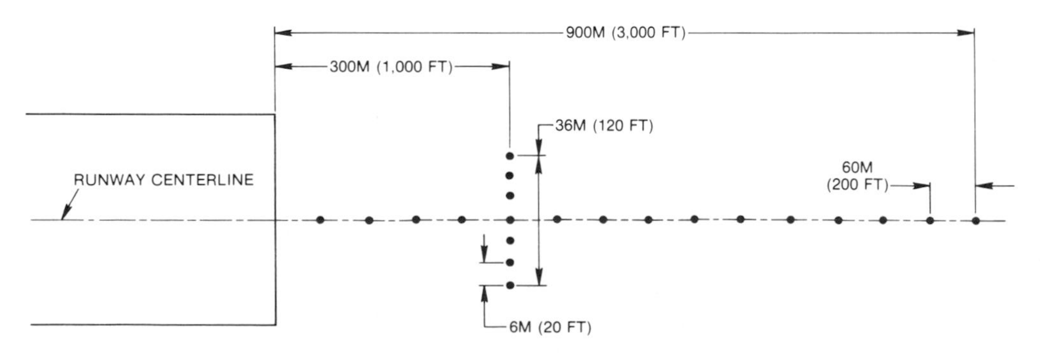

Fig. 15-14. Non-instrument approach system.

Fig. 15-15. Configurations and Usage for the Various Types of Red/White (VASIS) Visual Approach Slope Indicator Systems

Type	Configuration	Usage
VASI-16	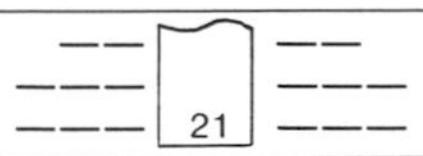 21	Wide Bodied Jet Operations at Major International Airports
VASI-12	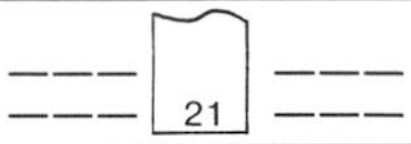 21	Conventional Jet Operations at Major International Airports
VASI-6	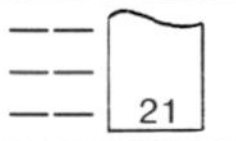 21	Wide Bodied Jet Operations at Domestic Airports
VASI-4	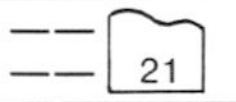 21	Conventional Jet Operations at Domestic Airports
VASI-2	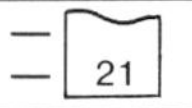 21	Smaller Airports Having no Turbojet Operations

Configurations. The configurations and usage for the various types of Red/White VASIS are as shown in Fig. 15-15.

Characteristics. In the basic 2-Bar System (VASIS-2, -4, and -12), each light box unit projects a split beam of white (above aiming line) and red (below aiming line) light. When the aircraft is on the proper glide path the pilot will observe the downwind light bar to be white in color while the upwind bar appears to be red. If the approach is too high, both light bars will appear white. An approach at too low an angle will cause both bars of the system to be seen as red.

The basic 3-bar VASI System (VASIS-6 and -16) adds an additional bar to the conventional 2-Bar System so as to form a second "On-course" Red/White corridor, at a higher angle, above the "On-course" signal of the 2-Bar System. Conventional aircraft fly the normal Red/White signal for the lower corridor, which is formed by the down-wind and middle bars of the system, disregarding the color signal of the added third (or up-wind) bar of the system. Wide-bodied jet aircraft, having higher wheel-to-eye heights, fly the higher angle "On-course" signal formed by the middle and up-wind bars of the system, disregarding the color signal of the downwind bar. The approach path formed by the added bar is approximately ¼ degree higher than the usual 3 degree approach path of the 2-Bar System, so as to provide the necessary higher threshold clearance for wide-bodied jet operations.

Location. Light box units are located as follows: for the basic 2-Bar System the downwind bars are 30 to 245 meters (100 to 800 feet) from the landing threshold and the upwind bars 155 to 305 meters (500 to 1000 feet) beyond; for the basic 3-Bar System the downwind and middle bars are the same as the downwind and upwind bars of the 2-Bar System but the upwind bar is another 155 to 305 meters (500 to 1000 feet) beyond; and in both systems the first row of boxes is spaced 15 to 18 meters (50 to 60 feet) to the side of the runway with the other rows spaced 4.8 to 5.2 meters (16 to 17 feet) farther away from the runway.

Runway Identifier Lights (RIL). Runway identifier lights (also known as runway end identifier lights, REIL) are a pair of white flashing lights located near the threshold of a runway. They provide rapid and positive identification of the threshold.

Application. Runway identifier lights are usually most effective in identifying runways surrounded by a preponderance of other lighting, or lacking contrast with surrounding terrain. They are usually installed on visual or instrument runways other than Categories I, II, or III.

Characteristics. Runway identifier lights are two simultaneously flashing white lights, and may be either unidirectional or omnidirectional. Intensity may be variable. The flash rate is one flash per second for the omnidirectional and two flashes per second for the unidirectional RIL. The omnidirectional RIL provide excellent circling guidance, while the unidirectional RIL provide only limited circling guidance.

Location. A typical RIL layout is two lights located 12 to 22.5 meters (40 to 75 feet) on both sides of the runway, 3 meters (10 feet) ahead of the runway threshold, maximum, and turned 15 degrees away from the runway centerline.

Runway Edge Lights. A runway edge light system is a configuration of lights used to outline the lateral and longitudinal limits of usable landing or take-off area during periods of darkness and restricted visibility conditions both day and night.

Application. The selection of a particular lighting system is based on the type of operations conducted on the particular runway. (a) Low Intensity Runway Lights (LIRL) are installed on runways at Visual Flight Rule (VFR) airports. (b) Medium Intensity Runway Lights (MIRL) may be installed on runways having a non-precision Instrument Flight Rule (IFR) procedure for either circling or straight-in approaches. (c) High Intensity Runway Lights (HIRL) are installed on runways having precision IFR procedures and for runways utilizing Runway Visual Range (RVR) instrumentation.

Characteristics. Runway edge lights are steady burning lights having variable intensity white color, except that:

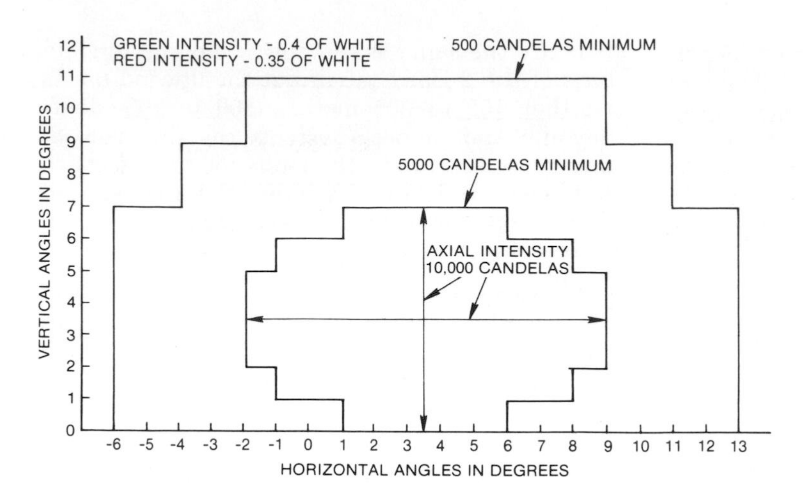

Fig. 15-16. Photometric characteristics of typical high intensity runway edge, threshold and end lights.

1. In the case of a displaced threshold, the lights between the beginning of the runway and the displaced threshold show a red color in the aircraft landing approach direction; and
2. A section of the lights 600 meter (2000 feet), or one-half of the runway length, whichever is less, at the opposite end of the runway from which the straight instrument approach is made, the color of the lights may be yellow. Runway threshold lights are fixed unidirectional lights showing a green color in the direction of approach to the runway. The intensity and beam spread of the lights are designed for the conditions of visibility and ambient light in which the runway is intended to be used.

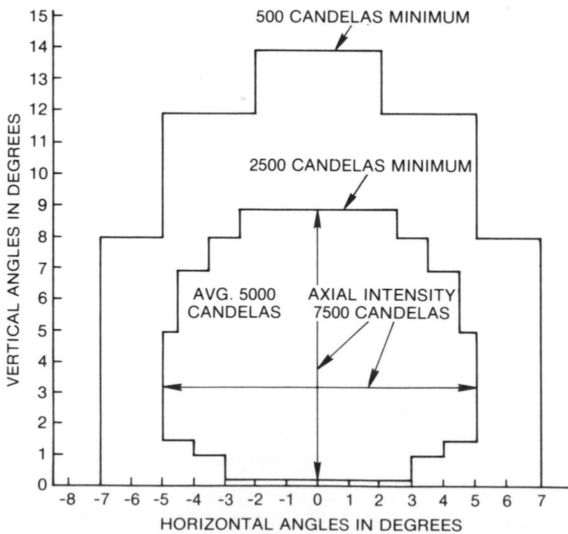

Fig. 15-17. Photometric characteristics of typical high intensity runway centerline lights.

Runway end lights are fixed unidirectional lights showing a red color in the direction of the runway. The intensity and beam spread of the lights are designed for the conditions of visibility and ambient light in which the runway is intended to be used.

The runway threshold and end lights, except where a displaced threshold might exist, are usually combined in one luminaire. Typical high intensity edge, threshold and end light photometric characteristics are shown in Fig. 15-16.

Location. The runway edge lights are located on a line not more than 3 meters (10 feet) from the edge of the full strength pavement which is designated for runway use. The longitudinal spacing of the lights should not exceed 60 meters (200 feet) and be located such that a line between light units on opposite sides of the runway is perpendicular to the runway centerline.

The combination of threshold and runway end lights are located on a line perpendicular to the extended runway centerline not less than 0.6 meters (2 feet) nor more than 3 meters (10 feet) outboard from the designated threshold of the runway. The designated threshold is the end of the pavement (surface) useful for aircraft operations. The lights are installed in two groups located symmetrically about the extended runway centerline. For instrument runways each group of lights contains not less than four lights; for other runways, not less than three lights. In either case, the outermost light in each group is located in line with the runway edge lights. The other lights in each group are located on 3-meter (10-foot) centers toward the extended runway centerline.

For a displaced threshold the threshold lights are located outboard from the runway and are separate and distinct from the runway end lights.

The innermost light of each group is located in line with the line of the runway edge lights, and the remaining lights are located outward, away from the runway on 3-meter (10-foot) centers on a line perpendicular to the runway centerline. As the displaced runway area is usable for specific operations (takeoff, rollout, taxiing) runway edge lights are installed to delineate the outline of this area.

Runway Centerline Lights. Runway centerline lights are intended to provide after-touchdown rollout and also for takeoff guidance.

Application. Runway centerline lights are provided on instrument runway, Category II and III and on runways used for takeoff in visibility less than 480 meters (1600 feet) Runway Visual Range (RVR). These lights are recommended on instrument approach runway, Category I.

Characteristics. Runway centerline lights are steady burning having variable intensity white color except that for the last 900 meters (3000 feet) from the runway end; alternate red and white from 900 meters (3000 feet) to 300 meters (1000 feet) from the runway end, and red from the 300 meters (1000 feet) to the runway end. Runway centerline lights are not visible on approach in the area that may be displaced from landing on the full strength runway due to navigational hazards. Photometrics of typical runway centerline lights are shown in Fig. 15-17.

Location. Runway centerline lights are located along the centerline of the runway at a uniform longitudinal spacing of 15 meters (50 feet).

Runway Touchdown Zone Lights. The runway touchdown zone lights are landing aids located in the area that an aircraft would normally land on a runway.

Application. These lights are provided in the touchdown zone of an instrument runway.

Characteristics. Touchdown zone lights are steady burning lights having variable intensity white color. Photometrics of typical touchdown zone lights are shown in Fig. 15-18.

Location. Touchdown zone lights extend for 900 meters (3000 feet) from the landing threshold.

Runway Centerline Reflectors. Centerline reflectors are installed to improve the conspicuity of the runway centerline during the hours of darkness.

Application. Centerline reflectors may be provided on runways not equipped with centerline lights. In snow areas, special reflectors that are compatible with snow removal equipment are desired. Centerline reflectors normally do not

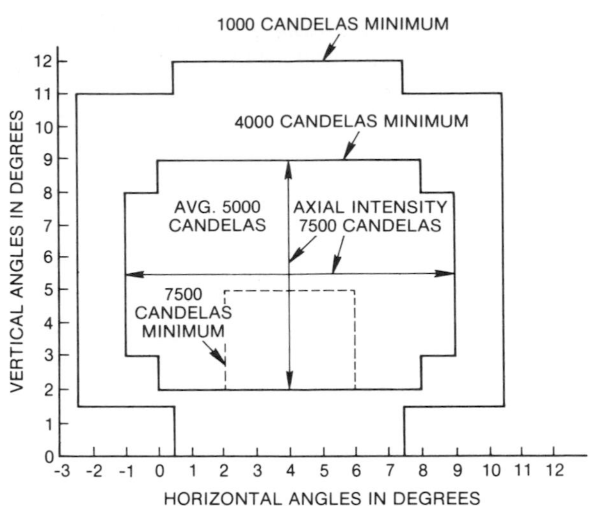

Fig. 15-18. Photometric characteristics of typical high intensity runway touchdown zone light.

replace centerline lights or painted markings; but rather, supplement these visual aids.

Characteristics. Runway centerline reflectors are bi-directional white, or red and white reflectors capable of withstanding normal operations of aircraft and maintenance vehicles. The reflectors are small units with a maximum height of 15.9 millimeters (5/8 inch) and designed for cementing on the runway surface. The reflectors should have a wide divergence angle of light return to meet most operational requirements.

The color coding of the reflectors is identical to the color coding used for runway centerline lighting. For displaced thresholds, blanked-out markers are used in the direction of aircraft approach in pavements denied for landings.

Location. Runway centerline reflectors are spaced at longitudinal intervals of not more than 15 meters (50 feet).

Runway Markings. Runway markings are provided to enhance the safety and efficiency of aircraft operations and are normally white in color. They vary with the operational requirements of the runway.

Runway Distance Markers. Used primarily on military airfields and civil airports where military aircraft operate, Runway Distance Markers are to provide the pilot of an aircraft with an indication of the remaining operational length of the runway during takeoff or landing.

Application. Markers consist of pairs of signs (one on each side of the runway) at each 300 meters (1000 feet) showing the number of such distances remaining from that point to the end of the runway.

Characteristics. Markers are painted with a flat black background and white numbers. Markers are illuminated and frangibly mounted.

Location. Markers are placed at 300 meters (1000 feet) intervals down both sides of the runway and at a uniform distance from the runway edge (15 meters (50 feet) or more). A marker is omitted where its position falls on an intersecting runway or taxiway.

Barrier Engagement Markers. Hookcable or Barrier Engagement Markers are of a similar configuration to Distance Markers except with a large illuminated orange circle instead of numbers. They are used to mark the position of arresting gear.

Gate and Parking Guidance

The gate and parking system is designed to give pilots visual lead-in guidance to gate and parking positions and it is capable of providing final alignment and stop indications.

Application. A visual docking or parking guidance system is installed when it is necessary to position an aircraft by visual aids on an aircraft stand.

Characteristics. The system provides azimuth and stopping guidance. The azimuth guidance unit provides left and right guidance so that the pilot is informed of the position of the aircraft in relation to the longitudinal guidance line. The stopping position indicator shows the stopping position for the aircraft for which the stand is intended and also provides closing rate information near the stopping position.

Location. The azimuth guidance unit is so located that its signals are visible from the cockpit of an aircraft and aligned for use by the pilot occupying the left seat. The stopping position indicator is so located that its signals are visible from the cockpit of an aircraft and should be preferably usable by the pilot and the copilot. It is colocated with the azimuth guidance unit.

Apron/Ramp Floodlighting and Ground Lighting/Marking

Apron/ramp lighting is provided at night or during low visibility conditions. This lighting may include both apron/ramp taxi route guidance and general area floodlighting.

Application. When operations are conducted in visibility conditions of 360 meters (1200 feet) RVR or less, the most effective route guidance within apron/ramp areas is provided with taxiway centerline lights. With visibilities greater than 360 meters (1200 feet) RVR, taxiway centerline lighting is also recommended for this purpose, especially if complex routes must be delineated. The edges or boundaries of apron/ramp areas are usually delineated by blue elevated edge lights. Route guidance within apron/ramp areas may also be provided by centerline reflectors, elevated edge reflectors and painted markings, as used on taxiways. Floodlighting is provided to perform the various functions of loading and unloading passengers and cargo, as well as aircraft turn-around maintenance.

Characteristics. Centerline lights, elevated edge lights, reflectors and painted markings used to delineate apron/ramp routes and boundaries have the same characteristics as those specified for use on taxiways.

The color rendering of the apron/ramp floodlights is very important and should be such that the colors used for aircraft marking connected with routing, servicing and for surface and obstacle marking, can be correctly identified.

The average illuminance of the floodlighting system should be as follows:

1. *Aircraft stand:* Horizontal illuminance–20 lux (2 footcandles) with a uniformity ratio (average to minimum) of about 4 to 1.
Vertical Illuminance–20 lux (2 footcandles) at a height of about 2 meters (6 feet) above the apron in relevant directions.
2. *Other apron areas:* Horizontal illuminance–50 per cent of the illuminance on the aircraft stands.

Location. Equipment is located as follows:

For centerline guidance, centerline lighting, reflectors, and/or painted markings are installed along all apron/ramp routes using the same criteria as for taxiways.

For boundary and edge guidance, edge lights and/or reflectors spaced not over 30 meters (100 feet) apart are installed along the boundaries or edges of the apron/ramp areas so that aircraft and supporting vehicles maneuvering within the apron will not wander off usable pavement.

For floodlighting, apron floodlighting is located so as to provide adequate illuminances on all apron service areas, with a minimum of glare to pilots of aircraft in flight and on the ground, airport controllers and personnel on the apron. The aiming arrangement of the floodlights is to be in such a way that an aircraft stand receives light from two or more directions to minimize shadows.

Airport Beacons

An airport beacon is installed on or adjacent to the airport to aid the pilot in visually locating the airport after dark.

Application. An airport beacon is provided at each airport intended for use at night except when, in special circumstances, a beacon is regarded as unnecessary having considered the requirements of the air traffic using the airport, the conspicuity of the airport features in relation to its surroundings and the other aids useful in locating the airport.

Characteristics. The airport beacons show alternating white and green flashes. The frequency of the flashes are from 12 to 30 per minute and preferably not less than 20 per minute. The light from the beacon shows at all angles of azimuth. The minimum effective intensity of white flash at various elevations should be as shown in Fig. 15-19. The green flash intensity should be at least 20 per cent of those in Fig. 15-19. See page 3-25 in the 1981 Reference Volume for the method of obtaining the effective intensity of a light flashed at a known rate from measurements of the average intensity.

Location. The beacon should be located as follows:

1. Minimum distance:

a. For airports with runways 960 meters (3200 feet) or less in length, the beacon should be located at least 105 meters (350 feet) from the nearest runway centerline.

b. For airports with runways over 960 meters (3200 feet) in length, the beacon should be located at least 225 meters (750 feet) from the runway centerline.

2. Maximum distance:

For all airports, the beacon should not be located more than 1500 meters (5000 feet) from the nearest point of the usable landing areas, except in cases where surrounding terrain will unduly restrict the visibility of the beacon. In these cases, the distance may be increased to a maximum of 3.22 kilometers (2 miles) from the usable landing area provided that airport itself is readily identifiable from the beacon location.

Fig. 15-19. Airport Beacon Minimum Effective Intensity of White Flashes at Various Elevations

Elevation Angle (degrees)	Minimum Effective Intensity of White Flash (candelas)
1 to 2	25,000
2 to 8	50,000
8 to 10	25,000
10 to 15	5,000
15 to 20	1,000

Wind Direction Indicators

An airport is equipped with one or more wind direction indicator(s).

Location. If only one wind indicator is provided, it is located near the center of the runway complex. It should be visible to aircraft in flight or on the movement area and installed in such a manner as to be free from the effects of air disturbances caused by nearby objects. If more than one wind indicator is provided, they are located at each runway end, in an area 45 to 90 meters (150 to 300 feet) from runway edge and 150 to 450 meters (500 to 1500 feet) down the runway from runway threshold. The wind indicator should not be located in any ILS glide or localizer critical area.

Characteristics. The wind direction indicator is in the form of a truncated cone made of fabric. The cone is at least 2.5 meters (8 feet) in length but not more than 4 meters (12 feet). The diameter at the larger end may range from 450 to 900 millimeters (18 to 36 inches). A framework is provided for a portion of the length of the cone to permit viewing in calm wind conditions. The color is international orange, or orange and white bands with five alternate bands. The first and last bands are orange. The wind indicator is capable of indicating the correct wind direction at a speed of 3.5 kilometers (2.2 miles) per hour and fully stream at a speed of 10.5 kilometers (6.5 miles) per hour. The wind indicator is lighted for night use and mounted on lightweight frangible tower when installed near a runway.

Electrical Circuits for Lighting

Except for smaller airports which usually use multiple (parallel) circuits, the power to approach, runway, taxiway and other field lighting equipment is distributed underground by series lighting circuits.

Application. Electric power is usually made available for the following facilities:

1. Lighting to enable air traffic services personnel to carry out their duties.
2. All obstruction lights which are essential to ensure safe operations.
3. Meterological equipment.
4. Security lighting.
5. Field lighting (all aids listed in this Section).

Power Source Requirements and Characteristics. For air traffic safety, in addition to normal power supply, an emergency standby power supply is necessary. In the event of a failure of a normal power supply, automatic transfer to the emergency standby takes place. The maximum time from the normal power supply to emergency standby is as shown in Fig. 15-20. The type, size and rating of the emergency power source required is dependent upon the minimum power necessary to operate the essential visual aids to permit safe landing, takeoff and ground movement of aircraft under specific visibility conditions.

Location. The field lighting power system center is centrally located with reference to the loads so that power losses due to long distribution distance may be minimized.

Series Lighting Circuit Equipment Function(s) Constant Current Regulator. In order to maintain required brightness to suit airport visibility conditions, the power to the series field lighting circuit is provided by a constant current regulator. This regulator is designed to produce constant current output for variations in field circuit loads and for variations in voltage of the ac power source. Each regulator can be set locally or remotely to produce one, three or five different constant current outputs, corresponding to one, three or five steps of brightness.

Field Lighting Control Panel. The control panel normally is located in a control tower and allows the air traffic controller to select and control the approach, runway and taxiway systems as well as other essential lighting, such as wind cone direction indicators and obstruction lights, and power to visual navigational aids.

Isolating Transformers. The primary function of the isolating transformer is to provide continuity in a series lighting circuit so that a single lamp failure does not interrupt the primary circuit. The second important function of the isolating transformer is to provide electrical isolation of the lamp in the luminaire from the high voltage circuit for safety purposes.

Fig. 15-20. Emergency Power Switch-Over Time for Airport Lighting

Runway	Aids Requiring Power	Maximum Switch-Over Time (seconds)
Non-Instrument	VASIS*	No Requirement
	Runway edge	
	Obstruction*	
Instrument Approach	Approach lighting system	15
	VASIS*	15
	Obstruction*	15
Precision Approach Category I	Approach lighting system	15
	Runway edge	15
	Essential taxiway	15
	Obstruction*	15
Precision Approach Category II and III	Approach lighting system	1
	Runway edge	1
	Runway centerline	1
	Runway touchdown zone	1
	Essential taxiway	15
	Obstruction*	15

* VASIS and obstruction lights should be supplied with secondary power when the competent authority considers their operation essential to the safety of flight operations.

HELIPORTS

Heliport lighting and marking provides visual operational guidance to helicopter operators for landing and takeoff from a helipad. Heliports consist of one or more helipads.

Marking for helipads should clearly identify the area as a facility for the arrival and departure of helicopters. Numerous types of heliport marking have been proposed.

Perimeter lights form a square around the landing area made up of equally spaced yellow lights with an odd number of lights per side. Each side has a minimum of five lights. A typical layout is shown in Fig. 15-21. Perimeter lights are omnidirectional, yellow in color, may be semi-flush or elevated, and are provided with intensity control. The beam peak is between 7 degrees and 9 degrees of elevation with a minimum of 40 candelas between 3 and 15 degrees of elevation, 15 candelas between 15 and 25 degrees of elevation and a minimum of 5 candelas between 25 and 90 degrees of elevation.

A heliport beacon is a beacon used to identify the location of the heliport. It is located within 0.8 kilometers (a half mile) of the helipad and flashes a color code green-yellow-white at a rate of between 30 and 60 flashes per minute.

Helipad floodlighting is used to illuminate the helipad landing area and is used to augment the guidance of perimeter lights. It also aids in depth perception. Helipad floodlights are white in color with a wide beam spread and no upward component.

Helipad landing direction lights are provided to indicate the final direction of approach to the landing pad. Landing direction lights are yellow in color normally elevated and seven in number. A typical installation is shown in Fig. 15-21. While there may be more than one set of landing direction lights only one is illuminated at one time. The landing direction lights have the same characteristics as the perimeter lights.

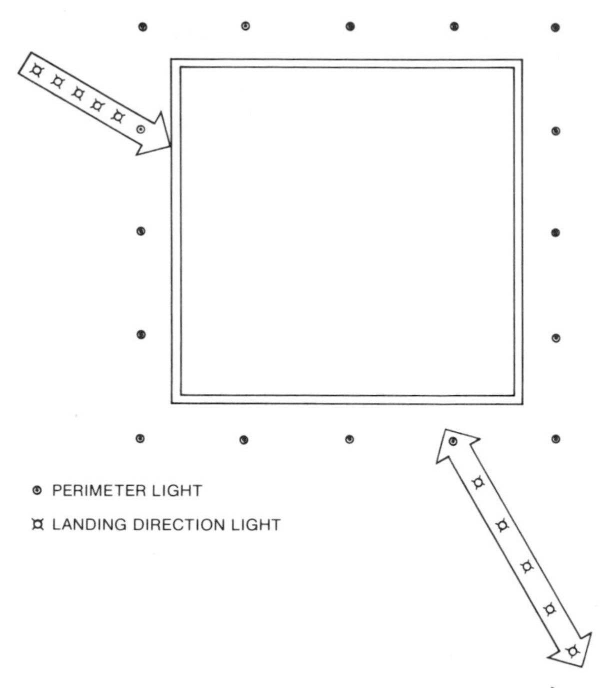

Fig. 15-21. A typical layout for helipad lighting.

Helipad approach lights provide identification of the preferred approach to the helipad in the initial stages. Helipad approach lights are unidirectional elevated units, white in color, with the peak beam adjustable between 3 degrees elevation and 12 degrees elevation and an intensity twice that of the perimeter lights.

OBSTRUCTION IDENTIFICATION

Any object which penetrates an established set of imaginary surfaces or which exceeds a height of 60 meters (200 feet) at the airport site *may* be required to be marked and/or lighted so that the obstruction will be conspicuous both day and night. However, whether an obstruction is to be marked and/or lighted, and the extent thereof, is determined as part of an aeronautical study conducted by the proper authority.

Marking. The purpose of marking a structure is to warn of its presence during daylight hours.

Characteristics. Two types of markings are used:

1. Painting. Objects with essentially unbroken surfaces are painted to show alternately aviation surface orange and aviation white.
2. Markers. Markers should be used to mark obstructions when it has been determined that it is impracticable to mark such obstructions by painting. Two types of markers are used: spherical markers not less than 500 millimeters (20 inches) in diameter and colored aviation surface orange, and colored flags. The flags may be of checkerboard pattern, aviation orange or two triangular sections.

Installation.

1. A painted pattern.
2. Spherical markers are normally displayed on overhead wires at equal intervals.
3. When it has been determined that the use of paint or spherical markers is technically impracticable, flag markers should be displayed around the obstruction, on top of the obstruction, or around its highest edge.

Lighting. The purpose of lighting a structure is to warn of its presence during both day and night conditions.

Characteristics. Three commonly used obstruction lighting systems are:

1. Aviation red obstruction lights, consisting of flashing beacons and steady burning lights, are used for night operations and during periods of limited day meteorological visibility.
2. High intensity white obstruction lights consisting of variable intensity, flashing, white light are used for both day and night operations.
3. Dual lighting consisting of aviation red obstruction lights for night operations and high intensity white lights for day operations.

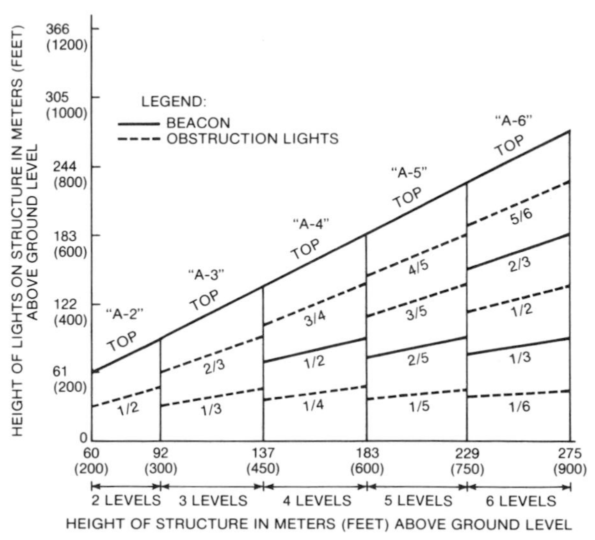

Fig. 15-22. Standards for lighting an obstruction of more than 60 meters (200 feet) and less than 275 meters (900 feet).

Installation. Fig. 15-22 contains typical standards for lighting an obstruction more than 60 meters (200 feet) and less than 275 meters (900 feet) with aviation red obstruction lights.

AIRPORT PARKING AREAS

The lighting of automobile parking areas and roadways in and around an airport must provide visibility for control tower operators and pilots and promote safe and efficient movement of motor vehicles and pedestrians.

The lighting must not interfere with nighttime visibility of the control tower operators and incoming pilots. At night, control tower operators work in semi-darkness. Their eyes must be dark-adapted to enable them to see aircraft maneuvering in the air and on the ground. Any appreciable amount of brightness in their fields of vision will greatly reduce their ability to see. The same is true for incoming pilots. But, in addition to their ability to see under dark adaptation, the nearby roadway and parking area luminaires should not be visible above the horizontal to avoid confusion between the pattern that the luminaires may form and the pattern of runway marker lights.

For the parking area, where the public park their own cars, the major considerations in providing illumination are to eliminate accidents, make it easier to locate parking spaces and to locate cars on return, and to discourage petty larceny and criminal assault.

Illuminance Recommendations. It is considered good practice to provide a minimum of at least 10 lux [one footcandle] with a maximum of 20 lux [two footcandles] average maintained in service, with a uniformity ratio of not greater than 3 to 1 between the average and minimum point anywhere on the roadway, and 4 to 1 in the parking area.

Focal points in parking areas, such as entrances and exits, should have an average illuminance of at least twice that of the general parking area. Also, points of heavy pedestrian crossings should have that same increased illuminance.

Special Requirements. It is desirable to limit the ambient light on the control tower windows to no more than one lux [0.1 footcandle] of ambient light. Tests at eight lux [0.8 footcandle] and three lux [0.3 footcandle] proved to be objectionable. Lowering this to one lux [0.1 footcandle] did satisfy the control tower operators. The illumination (vertical) on the control tower windows is ambient light accumulated from all sources, including reflected light from paved surfaces and direct high-angle light from luminaires. In areas that are subject to snowfall, the reflected light may become a significant factor.

Although it is difficult to control reflected light and glare, some factors that contribute to reflected light can be controlled to some degree, such as: the location of lighting equipment with relation to the location of the control tower, and the candlepower and angle at which the main beam is directed in relation to the control tower. The reflected angle should not be in the direction of the tower.

Direct light or stray light falling on the tower may be controlled in several ways: by using luminaires or floodlights which have positive optical control such that no direct or stray light is emitted above the horizontal, and by proper location of these sources with relation to the tower and the height of the tower.

The view of the runways and runway approaches, taxiways, and ramp areas should be considered when locating luminaires, so that during darkness they provide minimum obstruction as glare sources. The lighting poles used must be designed so that specified obstruction clearance requirements are met.

REFERENCES

1. Aviation Committee of the IES: "IES Guide for Calculating the Effective Intensity of Flashing Signal Lights," *Illum. Eng.*, Vol. LIX, p. **747**, November, 1964.
2. Aviation Committee of the IES: "Recommended Practice for Airport Parking Area Lighting," *Illum. Eng.*, Vol. 63, p. 590, November, 1968.

Transportation Lighting

SECTION
16

The general principles established for good interior and exterior lighting apply also in the transportation field. However, limited supply of electric power and the special basic characteristics of this power in automobiles, buses and railway cars often make it a more difficult and expensive problem to provide interior illumination of recommended quantity and quality. Also, the tremendous length of roadways and railways, the fact that they are used intermittently, and their exposure to a wide variety of weather conditions are important factors which complicate the lighting problem.

The following requirements are of special importance in the road and rail transportation fields. They apply to the lighting of practically all types of vehicles but are more essential in some than in others.

1. Adequate illuminance on the task plane. Where the task tends to vibrate, a relatively higher level is needed than for a fixed seeing task.
2. Adequate illuminance at special locations to permit safe entrance, exit and movement of passengers.
3. Lighting for fare and ticket collection.
4. Adequate vision of operator, particularly freedom from reflections or glare spots in his field of view.
5. Minimum glare and luminance ratios for passengers' seeing comfort.
6. Cheerful and attractive appearance of the vehicle inside and also as seen from without.

Most vehicles are long and narrow and have low ceilings and large window areas. Also, the desires of the passengers and the loading of the conveyance vary. These factors make the lighting problems difficult. On the other hand, the position of the occupants and the operator is generally known and fixed so that good lighting results are obtainable through a careful study of all phases of the problem.

NOTE: references are listed at the end of each section.

In the transportation field, direct current power supplies are predominant, a wide variety of voltages are involved, and power is often limited. For these reasons, one of the principal problems is to adapt lamps and lighting equipment to the power available and particularly to provide suitable power for fluorescent lighting, which is discussed beginning on page 16-11.

AUTOMOBILE LIGHTING

Most automobiles depend on a 12-volt wet storage battery kept charged by a generator (dc or rectified ac) driven by the car engine. In addition to the lighting, the electrical system must supply power for starting, ignition, heater blowers, radios, air conditioners, power seats, power windows, defrosters and other special equipment. A single wire grounded wiring system is commonly used. With the increased use of structural plastics, which are non-conducting, two-wire systems are sometimes used in part.

Exterior Lighting

Two categories of exterior lighting equipment are commonly used on motor vehicles—lighting units to see by and lighting units to be seen. The first group comprises headlights, back up lights, cornering lights, fog lights, spot lights, etc. The second group includes a variety of signal and position or presence lights whose function is to convey information between drivers of vehicles.

Headlighting.* The most difficult and important illuminating engineering problem in the au-

* This discussion primarily concerns headlighting and signal lighting in United States of America (USA). The paragraphs that follow, entitled *Canadian, Mexican and European Regulations* discuss differences.

Fig. 16-1. Lamps Used in the Four-Lamp Headlighting System

Lamp Type	No. of Filaments	Filament Position	Provides
1 (round) or 1a (rectangular) Mounted inner or lower position	1	At focus	Primary part of upper beam
2 (round) or 2a (rectangular) Mounted outer or upper position	2	At focus Below focus	Lower beam Part of upper beam

tomobile field is the headlighting system, for it is not easy to provide good road lighting without creating glare for an approaching driver.

The simplest headlighting system in use comprises two identical two-filament sealed-beam units, each of which provides an upper and a lower beam. A toe-board mounted foot-operated switch or a steering column mounted hand operated switch permits the use of either the lower beams or the upper beams. Both 178-millimeter (7-inch) round and 142- by 200-millimeter (5-⅝- by 7¾-inch) rectangular units are in use. The beam patterns of the two different shapes are essentially identical.

A four-lamp (often called dual) headlighting system is also used. It consists either of four 146-millimeter (5-¾-inch) diameter or four 100- by 165-millimeter (4- by 6-½-inch) rectangular sealed beam units of two types with one of each type mounted on either side on the front of the vehicle. Further details are summarized in Fig. 16-1. A foot or hand operated switch permits the use of the two lower beam filaments only, in traffic and passing situations, or the four upper beam filaments for country driving.

The sealed beam unit construction consists of a lens and reflector hermetically sealed together with a filament or filaments enclosed within. They have a standardized two or three blade connector and are usually filled with an inert gas under some pressure. Those units utilizing tungsten-halogen light sources have a small tubular envelope inside the unit which encloses the tungsten filament and is filled with one of the halogens under pressure. Each sealed beam unit has three bosses on the lens face. These are used in conjunction with a mechanical aiming device to provide correct aim without the necessity of a darkroom or aiming screen.

The total maximum luminous intensity of all the high beam lamps on a vehicle must not exceed 150,000 candelas (raised only recently (1978) from 75,000 candelas). It is expected that units providing this total intensity will be available in both the round and rectangular four lamp systems.

The increase in seeing distance is not nearly in proportion to the increase in intensity. Higher intensity also means more glare and the necessity to change to lower beam at greater distances.

Signal Lighting. Signal lamps include tail lights, stop lights, turn signals, front position light, parking lights, side marker lights, and reflex reflectors. Signal indications must be unmistakable. The number of indications or messages conveyed must be kept to a minimum to avoid confusion and to be easily understood by the public.[8] The following indications will cover most situations:

1. Indications of the presence of a vehicle proceeding in a normal manner and its direction of travel.
2. Indication that the brakes are being applied and the vehicle is stopping or is stopped.
3. Indication that the vehicle is disabled or obstructing a traffic lane.
4. Indication that the driver intends to change the direction of travel.

Effective signal indications may be accomplished by the following means:

Color of Signal Indication. For automotive use the colors available are uncolored (white), yellow, red, green and blue. The usually accepted meaning of these colors is: uncolored—indication of the presence of the vehicle, commonly considered as approaching; yellow—indication that caution is needed in approaching the vehicle displaying the yellow light; and red—indication that the vehicle displaying the red light is a traffic hazard and that the approaching vehicle may have to stop.

Fire and police vehicles use many combinations of these colors with blue in combination with red and/or white becoming common for police vehicles. Very high intensity gaseous discharge or rotating colored spotlights add much attention-getting quality to these lights.

Alternate flashing yellow or red lights are used on school buses to alert motorists to the presence of children and indicate in many states that the motorist must stop until the lights are turned off.

Lamps in specific configurations and locations are used on trucks, trailers and other vehicles over 2 meters (6 feet 8 inches) wide to indicate wide and possibly slow-moving vehicles.

Intensity of Signal Light. Intensity of the signal light has a definite bearing on its conspicuity. In giving signal indications, different intensities may be used, provided the ratio of the intensities is sufficiently great. In the contemplated use on automobiles, the signal indication must be recognized unmistakably when first seen. Consider the situation where a car comes over the brow of a hill or around a sharp curve, and suddenly the driver sees a vehicle ahead of his car. The driver must know at once whether this vehicle is proceeding normally or is stopping or stopped. A change of intensity, as used on many vehicles today, is useless, because the ratio of the two intensities is too small. For example, a driver not in sight of the car ahead when the intensity changed and therefore not seeing the change in intensity occur, has no sure way of knowing whether he is looking at a bright tail light or a dim stop light—information which may be of vital importance.

Difference in intensity, to be a reliable means of conveying different indications, must be from units having a ratio of intensities not less than 5 to 1. Larger ratios are desirable.

Pattern of Display. Patterns of display, such as a single unit for one indication, two units, one above the other, for another indication, or three units in a triangular pattern for still another indication, are not good signal patterns since they are not distinguishable at greater distances.

Method of Display. The signal light may be displayed as a steady burning light or as a flashing light. Two units may be displayed by alternate or simultaneous flashing.

Canadian, Mexican and European Regulations

Canadian Regulations. Canadian automotive lighting regulations are almost identical to those in the USA except that Canada permits the use of either European headlamp beam patterns or those of the USA sealed beam units.

Canadian federal regulations as established by Transport Canada (formerly Department of Transportation) control the equipment on new vehicles built in or imported into Canada. The individual provinces, however, regulate the use of lighting equipment. As a result many variations of lighting devices are permitted to be sold for aftermarket installation.

Mexican Regulations. Mexico has no federal regulations regarding automotive lighting equipment. It is expected, however, that all new vehicles have basic lighting equipment, *i.e.*, headlamps and parking, tail, stop, license, and turn signal lights. As a result, not all vehicles have hazard warning, backup, and side marker lights. The Mexican government is presently developing a proposal to require these in the future.

The Department of Transportation and Police Departments do have subjective performance regulations for lighting equipment on vehicles purchased for their use. Vehicles complying with either USA or European regulations will meet these requirements.

European Regulations. Most European countries follow the recommendations of the United Nations committee WP29 or the regulations of the Common Market. These are either identical or very similar but differ from USA regulations as follows:

1. European headlamp lower beams have a sharp top-of-beam cut-off with less glare light above the horizontal than the USA beam. Seeing ability is roughly comparable to the USA beam, each having advantages in certain areas.
2. The European upper beam system maximum is 225,000 candelas while the USA system maximum is 150,000 candelas. Upper beam patterns of the two systems are quite similar.
3. Sealed beam headlamp units are not required in Europe and lens and reflectors are of many shapes and sizes. In addition to clear lenses, selective yellow (a very pale yellow) is permitted or, in some countries, required. Several different standardized lamps are used. Tungsten-halogen lamps are in most common use.
4. In a few north European countries lower beams or a pair of special front running lamps must be used in the daytime during winter months as an aid to being seen.
5. European regulations for rear position (tail) lamps, front position (parking) lamps, stop lamps, and turn signal lamps generally have lower candela requirements than do USA regulations. Rear turn signals must be yellow and front turn signals must be white (colorless) whereas in the USA rear turn signals may be either red or yellow while front turn signals must be yellow. Side marker lamps and hazard warning lamps are not required. Rear fog lamps are required in a few countries. These are manually switched, steady burning, red, rear lamps about the intensity of a stop lamp. Trucks must have triangular reflex reflectors but do not use clearance or identification lamps as in the USA. Special large reflective panels are required in some countries on long vehicles.

Interior Illumination

The average person does not expect to read or write continuously in a passenger automobile either while driving or while the vehicle is parked. Standards of luminance and illuminance have not been established for interior illumination for passenger cars; however, installations should be planned to provide illumination for casual inspection of road maps and other printed matter, and for safety in getting out. The installation should be in harmony with the style of the car interior. Lamps employed range from 1½ to 21 candelas and should be shielded to prevent direct glare. In addition, all lamps should be located and/or shielded to prevent reflections in the windshield from obscuring the driver's view of the road, if the lamps should accidentally be turned on while the car is in motion.

Panel-board or instrument lighting for automobiles should be designed for utilitarian requirements with decorative considerations given second place. The average driver uses the various instruments for reference rather than continuous viewing; nevertheless they should be easily and quickly readable. It is essential that lighting units and instrument faces be so placed that they are not reflected from the windshield or shiny trim surfaces into the driver's eyes. Provision for dimming panel lights is desirable to avoid excessive brightness and interference with the driver's view of the road at night. Illumination is provided by: small lamps recessed behind glass, plastic or other light-transmitting materials; similar lamps used for edge lighting of recessed or raised numerals; lamps located at the top or bottom or in front of the panel faces (direct illumination); or ultraviolet excitation of fluorescent numerals and pointers. Electroluminescent lamps have also been used.

SPECIFICATIONS FOR EXTERIOR LIGHTING OF MOTOR VEHICLES

The mass production methods characteristic of the automotive industry encourage extensive standardization and through the cooperation of the Society of Automotive Engineers (SAE), the Illuminating Engineering Society, safety engineers, and state motor vehicle administrators, standards have been developed over a period of years covering the characteristics and procedures for testing automotive lighting equipment. The US Department of Transportation has issued standards for automotive lighting equipment, in general following the SAE standards with some variations and additions. The SAE standards are published annually in the *SAE Handbook* (also available in reprint) and are reviewed at least once every five years and continued or revised. The standards outline specifications and tests for the various lighting devices covering such details as photometry, color, vibration, moisture, dust and corrosion.

Since National Highway Traffic Safety Administration (NHTSA) regulations change from time to time, that agency should be contacted for the latest information before proceeding with new designs. It should be remembered that NHTSA frequently specifies an older SAE standard (as indicated by an earlier suffix letter) than that published in the latest *SAE Handbook*.

The American Association of Motor Vehicle Administrators has an *equipment approval program* with participation by most states and Canadian provinces, for not only lighting devices, but other safety equipment as well.

REFLEX DEVICES IN TRANSPORTATION LIGHTING

Retro-reflecting devices or reflex reflectors are important in transportation lighting, signaling and for directional guides. They have been standard equipment on the rear of automobiles since about 1935. Other applications include: reflector flares for highway emergency markers; railroad switch signals; clearance markers for commercial vehicles; luminous warnings and direction signs; delineators of highway contours; marine buoys; contact markers for airplane landing strips; bicycle front, rear and side markers; belts and markers for traffic officers; luminous paving strips; and luminous advertising display signs.

Principles of Operation

A reflex reflector is a device that turns light back toward its source. There are several specific types of these devices; however, the principle of operation, namely the production of brightness in the direction of the source, as shown in Fig. 16-2, is the same for all types. The greater the accuracy of the design of the reflex reflector, the narrower will be the cone of reflected visual brightness. The narrower the cone, the brighter will be the signal. Two optical systems in use are the triple reflector (corner of a cube) and the lens-mirror device. The triple reflector is most commonly used in automotive devices.

Triple Reflectors

The triple reflector makes use of the principle of reflection from plane surfaces where the angle of incidence is such that total reflection takes

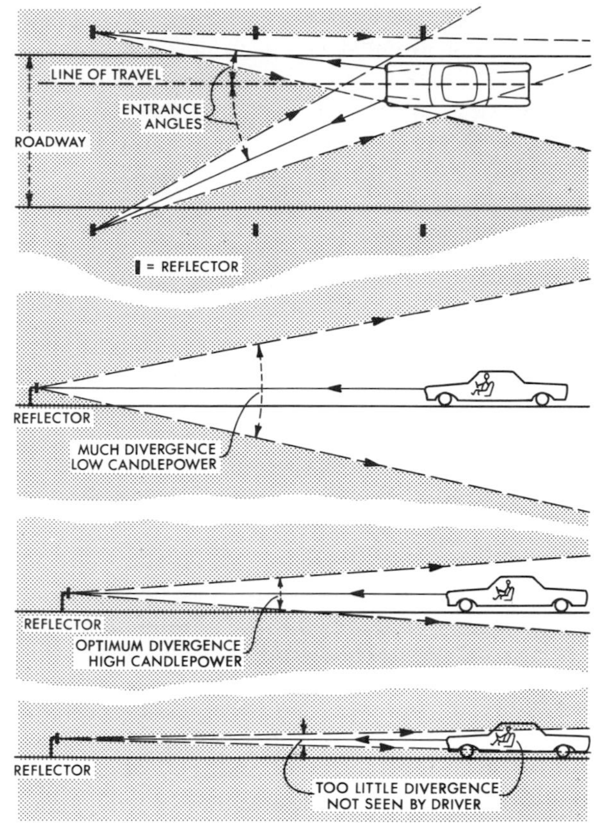

Fig. 16-2. Effect of the divergence of reflex devices on their angular coverage and intensity.

place. Three optically flat surfaces arranged mutually at right angles, as the inside corner of a cube, form a system such that any ray of light which has been successively internally reflected from the three surfaces will be reflected back upon the source.

A plaque of transparent glass or plastic with a continuous pattern of small adjacent cube corners molded into the back as shown in Fig. 16-3 is a commonly used form of reflex reflector. Acrylic or polycarbonate plastics are the most commonly used plastics and are more adaptable than glass to accurate shaping of the prisms, resulting in greater intensity of the return beam. Economy of manufacture, lightness, and shatter resistance are other advantages. A 76-millimeter (three-inch) plastic reflex can be readily seen from up to 300 meters [1000 feet] from an automobile with headlights set for country driving.

Lens-Mirror Reflex

The lens-mirror button consists of a short focal length lens and mirror combination designed to some extent with respect to chromatic and spherical aberration such that the lens focuses the light source upon the mirror, the mirror and lens returning the reflection in the direction of the source. See Fig. 16-4. An aggregation of small lenses pressed into a plaque with a mirrored backing formed into concave surfaces properly designed will produce a very satisfactory reflex reflector.

Another device that produces a wide spread of light but lower luminance is the spherical transparent glass bead embedded in a diffuse reflecting material such as white or aluminum paint. This type of reflex is used largely in signboards, center stripes on highway pavements, etc.

Maintenance and Construction

For good maintenance, all of the reflecting and transmitting surfaces should be kept clean and where possible free from moisture. The construction should be such that the rear surfaces are sealed and waterproofed, either as a part of the

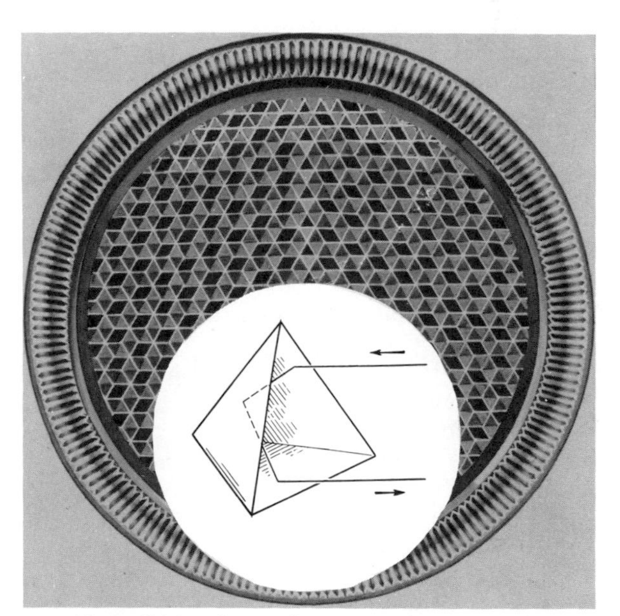

Fig. 16-3. Triple mirror reflectors comprise aggregates of concave cube corners.

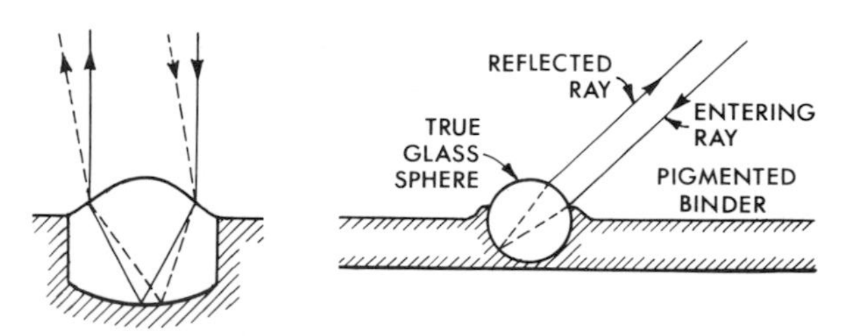

Fig. 16-4. Light paths in button and spherical ball type lens-mirror reflexes.

reflex reflector, or open to a sealed compartment such as in the case of a reflex reflector formed as a part of a lamp lens. Moisture on a totally reflecting surface will lessen the reflection. If moisture should accumulate on a dusty surface, optical contact takes place and light would pass through the surface rather than be reflected. Any roughening or etching of the transmitting surface will also tend to reduce the efficiency.

PUBLIC CONVEYANCE LIGHTING-ROAD AND RAIL[11]

The material that follows covers the lighting of interiors of public passenger road and rail vehicles. The general principles established for lighting of fixed interiors as covered in earlier sections apply to the lighting of public conveyance interiors. This has been made feasible through the development of solid state devices and the availability of new light sources and luminaires.

Illuminance Values. In modern road and rail conveyances, passenger seeing tasks vary widely; from seeing to board or exit, deposit fares or have fares collected, find seat accommodations, read and write, view advertising cards, to residental type tasks on inter-city trains such as found in dining cars, lounges and washrooms.

Illuminances for the specific tasks found in road and rail conveyances listed in Fig. 2-2 represent values that have been found to be satisfactory in practice. In general, they are consistent with similar tasks in other land applications and have been tempered only to recognize the factors of adaptation and comfort where the exterior surround is in darkness and when passengers must move from lighted to unlighted areas.

The illuminance values in Fig. 2-2 are maintained values which are to be provided on the visual tasks regardless of their location or plane. To insure the availability of these values, the original lighting design should make allowance for the decrease in light output caused by luminaire dirt accumulations and depreciation of other interior surfaces such as walls, ceilings, floors and upholstery and by depreciation in light source output.

Fig. 16-5. Recommended Luminance Ratios for Public Conveyances

To achieve a comfortable balance of luminances, it is desirable to limit luminance ratios between areas of appreciable size from normal viewing points as follows:

Areas Involved	Acceptable Limit
Between task and adjacent surroundings	1 to 1/5
Between task and more remote darker surfaces	1 to 1/10
Between task and more remote lighter surfaces	1 to 10
Between luminaire or windows and surfaces adjacent to them	20 to 1
Anywhere within the normal visual field	40 to 1

Fig. 16-6. Suggested Reflectance Values for Surfaces in Public Conveyance Interiors

Surface	Reflectance
Ceilings	60–90%
Walls	
Upper	35–90%
Lower	35–60%
Floors	15–35%
Upholstery	15–35%
Furniture	25–45%

Quality of Illumination. High luminaire and window luminances and high luminance ratios will produce uncomfortable seeing conditions (see Fig. 16-5), and prolonged exposure will generally result in eye fatigue. To avoid discomfort from high luminances, which may reach the eyes directly or indirectly through reflections from shiny surfaces or from improperly shielded light sources, lighting equipment should be located as far from the line of sight as possible. Glossy reflecting surfaces should be covered or modified to reduce glare, and luminaires and windows should have proper brightness control. See Sections 6 and 7 in the 1981 Reference Volume.

In passenger spaces, the average luminance of the total luminous area of a luminaire should not exceed 1700 candelas per square meter [500 footlamberts]. Within a diffuser area, the maximum luminance (the brightest 645 square millimeter area [square inch]) should not materially exceed *twice* the average. In strictly utility spaces luminances as high as 2700 candelas per square meter [800 footlamberts] are acceptable. In either case, the luminance ratio between the luminaire and the ceiling should not be more than 20 to 1 as shown in Fig. 16-5.

Luminances within the remainder of the environment also should be balanced in accordance with Fig. 16-5. The use of matte finishes of reflectances recommended in Fig. 16-6 will help to achieve this balance.

Colors of objects appear to change with the surface finish of the object. It should be noted that matte finishes will reflect diffuse light and give an object a consistent color appearance while glossy surfaces may lose their color when viewed at a direction near the specular angle.

Finishes such as velvet and deep pile carpeting appear darker than smooth surface materials such as vinyl or a plastic laminate of the same color.

Choice of Light Source. Incandescent and fluorescent lamps are normally used in the lighting of public conveyance interiors. Incandescent lamps are usually considered where (1) operating hours are short, (2) a high degree of light control is necessary, (3) interior surfaces are warm in color and (4) general illuminance values are low. Fluorescent lamps are considered where (1) operating hours are long, (2) general illuminance values are higher, (3) the linear shape is desired and (4) surfaces are cool in color.

Road Conveyances—City, Inter-City, and School

General lighting is provided for passenger movement along aisles and for seat selection. City buses leave general lighting on at all times because of the frequent movement of passengers when boarding and exiting the bus. Inter-city buses usually turn general lighting off when on the road to allow the driver's eyes to adjust to the outside luminances. For school bus lighting, the general principles and requirements are the same as those for city buses and inter-city buses; however, higher levels of general lighting are provided for surveillance of active children, whether the bus is moving or stopped, being boarded or exited. Lighting should be directed or shielded in such a way that the driver's vision is not impaired by reflections or glare spots in the field of view.

Boarding and Exiting. The seeing tasks are the steps, ground or platform area. The steps and/or an area extending 1.2 meters [4 feet] outside the door should be illuminated to allow passengers to see the base of steps, curbs or platform area. The plane of the task is horizontal and on steps, curbs or platforms, with the viewing distance varied according to the height of the individual.

Illumination (See Fig. 2-2 for values) should be provided at the center of each step and 450 millimeters [18 inches] from the bottom step on the ground, centered on the doorway.

Fare Collection. The seeing task is one of identifying money or tickets and depositing fares, and, therefore, the fare box and the immediate area around the box should be illuminated. The plane of the task is horizontal and at the top of the fare box.

Illumination should be provided at the top of the fare box on a horizontal plane.

Aisle Lighting. The task is one of observing the floor area for obstacles and the seat area for accommodation, generally while the vehicle is moving. The plane of the task for walking is at floor level and for seat selection at the back of seats.

Illumination should be provided on a horizontal plane at the center line of the aisle floor and for seat selection, on a horizontal plane at the top center of each seat back.

Advertising Cards. The seeing task is viewing opaque or back-lighted translucent advertising cards placed at the top of side walls. The plane of the task is vertical to 45 degrees.

Illumination on opaque cards should be provided on the face of the card. Luminance of back-lighted advertising cards should be measured at the face of the luminaire without the card over the diffuser. Within the diffuser area, the maximum luminance should not materially exceed *twice* the average.

Reading. The seeing task is generally one of reading newspapers, magazines and books and is generally the most difficult seeing task encountered in conveyances. The task plane, which is at 45 degrees, should be as free from reflection as possible.

For a seated passenger, the illumination should be provided on a plane 430 millimeters [17 inches] above the front edge of the seat on a 45-degree angle. For the standing passenger, measurement should be taken 1420 millimeters [56 inches] above the floor on a 45-degree plane at the edge of the aisle seat as if a passenger were reading facing the seat.

Typical Lighting Methods.

City Buses. For boarding and exiting, luminaires should be designed, located and arranged so as to minimize the casting of shadows, prevent glare for both passenger and driver, maintain a uniform illuminance and remain permanently in adjustment. Typical luminaire locations are the ceiling, over the steps or at the side of the step well.

For fare collection luminaires can be ceiling mounted or a local light may be used.

Fig. 16-7 illustrates one approach for general lighting for aisle seat selection, advertising cards and reading.

Inter-City Buses. For boarding and exiting,

Fig. 16–7. City bus. Fluorescent luminaires mounted end-to-end over the passenger seats provide diffuse semi-direct illumination. Dome lights are incandescent battery powered for emergency lighting.

lighting of step wells and the outside ground area is the same as for city buses.

Fig. 16–8 is an example of luminaires located for reading. Illumination for fare collection, aisle lights and seat selection is provided by an indirect system. A low-level night light system is often desirable so that the driver may clearly see throughout the entire bus. At minor stops, a special aisle lighting system of moderate level is turned on to facilitate passenger movement without disturbing those who are sleeping.

Special luminaires are often provided for the baggage compartment above the seats.

Fig. 16–8. Inter-city bus. Individually controlled beam-type incandescent lights are mounted in the base of the baggage rack to provide each passenger with a reading light. An indirect lighting system is built into the edge of the baggage rack for general interior illumination.

Rail Conveyances—Rapid Transit

Rapid transit is used here to indicate intra-city rail service such as subway or surface railways. It is generally characterized by fast travel conditions ranging from a few minutes to an hour. Standing passengers are a normal condition of this service. The seeing tasks and their lighting are the same as those described for Road Conveyances, except that the illumination for boarding and exiting should be provided on the tread of the entrance and platform areas.

Typical Lighting Methods. For boarding and exiting, provisions should be made for illumination of the threshold and the steps of the car. In addition, car illumination should supplement platform illumination for at least 1.2 meters [4 feet] from the car body at the location of the doors. In general, fare collection is made before entering the boarding platform; however, where rapid transit cars have fare collection facilities, it is necessary to provide illumination by an internal light in the fare box and/or a ceiling light.

The use of incandescent sources for aisle, seat selection, advertising cards and reading has been discarded in favor of fluorescent luminaires. This is due, primarily, to the efficiency, long life and general diffused type of illumination which is available from fluorescent luminaires. The normal type installations use a row of luminaires down the center of a car, or a row of luminaires on each side of a car center as shown in Fig. 16–9, or as a cornice on each side of the car. In some cars a third row of fluorescent luminaires is added between the outside rows, often with an air distributor combined for an integrated ceiling.

Where transverse seats are installed in the car, the lighting can be accomplished by use of transverse luminaires. They can be mounted above each seat provided there are no difficulties in connection with the car structural elements and air distribution system.

Some back-lighted advertising card luminaires are designed to provide adequate lighting for both standees and seated passengers (Fig. 16-10). Supplementary center strip luminaires are also needed to provide illumination. In cars using opaque cards, illumination can be supplied from two rows of continuous fluorescent luminaires over the passenger seats as in Fig. 16-9.

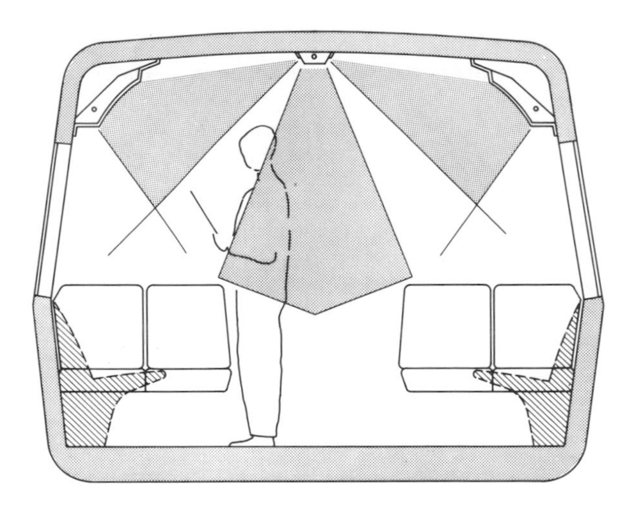

Fig. 16-10. Back-lighted car card luminaires with supplementary center strip lighting.

Rail Conveyances-Inter-City and Commuter

Seeing tasks on inter-city and commuter trains are much the same as on other public conveyances. The only differences are those additional tasks in the facilities provided for passenger comfort during long distance travel, such as food preparation areas, diners and washrooms. Because of the shorter travel distances and less travel time involved, some tasks described here do not apply to the commuter train.

Fig. 16-9. Rapid transit. Continuous rows of 40-watt fluorescent luminaires along each side provide diffuse illumination.

Boarding and Exiting. The seeing tasks are the platform, steps and vestibule floor, and all are horizontal at floor level. The viewing angle is practically vertical for the individual to observe conditions of tread surface and platform alignment with the car floor.

Illumination should be provided at the center of the vestibule floor and at the longitudinal center line of the steps.

Fare or Ticket Collection. There are no special lighting requirements for fare or ticket collection since this normally takes place while passengers are seated.

Aisle Lighting and Seat Selection. The seeing task is one of observing the floor area for obstacles and the seat area for accommodations. The plane of the task is at floor level for walking and at the top center of each seat back for seat selection.

Illumination should be provided on a horizontal plane at the center line at floor level. For seat selection, illuminance should be measured on a horizontal plane at the top center of each seat back.

Advertising cards. The seeing task is the same as for road conveyances except that some cards may be placed on bulkheads at the end of passenger compartments.

Reading. The seeing task is one of reading magazines, newspapers, books or business correspondence for an extended period of time. For the plane of the task and measurement of illuminances, see Road Conveyances.

Food Preparation. The seeing task is one of preparing foods and beverages and for cleaning the area. The plane of the task is horizontal on work counters 910 millimeters (36 inches) from the floor.

Illumination should be provided on the work counters from the front to the back edge.

Dining. The seeing tasks are eating and drinking, and menu and check reading. The plane of the task is horizontal to 45 degrees from the horizontal 200 millimeters (8 inches) in from the front edge of the table at each seat position.

Illumination should be provided on a plane 45 degrees up facing the seated diner, 200 millimeters (8 inches) from the front edge of the table or counter.

Lounge. In addition to relaxing and conversing, there are seeing tasks such as reading and card playing. For reading, the task plane is the same as for Road Conveyances. For card playing the task is horizontal at the playing surface.

General illumination should be provided on a horizontal plane 760 millimeters (30 inches) from the floor.

Washrooms. The chief tasks are shaving or makeup. Because the apparent distance of the face or figure as viewed in the mirror is twice its actual distance from the mirror, and because the details to be seen in shaving and critical inspection are usually small and of low contrast with background, the visual task may be a severe one. The task area in standing position consists of two 150- by 220-millimeters (6- by 8⅝-inch) planes at right angles with each other, converging at a point 410 millimeters (16 inches) out from the mirror, and centered vertically 1550 millimeters (61 inches) above the floor. See page 10-5. They represent the front and sides of the face. A third plane 300 millimeters (12 inches) square, its front edge also 410 millimeters (16 inches) out from the mirror, is tilted up 25 degrees above the horizontal and represents the top of the head.

Recommended illuminances should be provided 1550 millimeters (61 inches) above the floor and 410 millimeters (16 inches) from the mirror with the plane facing in the direction of the light source.

Typical Lighting Methods. The design of the interior lighting system of inter-city and commuter coaches are much the same. The main differences are the additional facilities on inter-city trains such as washrooms and lounges that provide passenger comfort during long distance travel.

All of the general requirements of good lighting such as elimination of direct and reflected glare or deep shadows and excessive luminance ratios that will interfere with good vision apply to railroad cars. A railway passenger car has certain fundamental limitations which affect the design of lighting systems. The length, height and width of the car are fixed by track gauge and clearances. The inherent physical characteristics of the car with its maximum utilization of space, air ducts, wire ways, structural members and limitations of power supply, also may complicate the lighting design.

Fig. 16-11. Inter-city railway passenger coach. A single row of fluorescent luminaires mounted at the center of the ceiling provides general lighting. Individual reading lights are located under the baggage rack.

Boarding and Exiting. Vestibule and platform lighting should be designed so that passengers board and detrain safely. Both track and vestibule levels are generally provided with a luminaire over each trap door, illuminating vestibule and step areas. Supplementary lighting may be provided by step lights located in the step well to increase illumination over steps, and/or a leading light located adjacent to steps, beamed to give increased illumination in front of the steps.

Fare or Ticket Collection. See Inter-City Buses.

Aisle and Seat Selection. General lighting can be provided indirectly where the luminaire is mounted in the extreme corner of the baggage rack and outer wall of the car, by ceiling mounted luminaires which direct the light to the desired areas as in Figs. 16-11 and 16-12 and by luminous ceilings. Individual reading lights are desirable as shown in Fig. 16-11 since the baggage rack casts shadows on the window seats.

Reading. Illumination for reading and writing can be accomplished by individually controlled reading lights mounted either in the aisle edge of the baggage rack, or at the joint line of the baggage rack and the outer wall of the car. In the case of the closed type of baggage rack, these lights are mounted directly above the seats.

Food Preparation. Higher illuminance values are given in these areas for the utmost efficiency of operating personnel and to insure good appearance of food and its proper inspection. Quality of lighting is important, especially in regard to color. In addition to general lighting, usually supplied by ceiling mounted luminaires, supplementary lighting should be provided over all work areas. The service window between the kitchen and pantry should be well lighted to facilitate food inspection. Refrigeration cabinets should have at least one lamp per compartment operated by automatic door switches.

Dining Area Lighting. The functions of the lighting system in dining areas are to enhance the appeal of the interior decorations, table settings and food, and to assist in providing a comfortable, pleasant atmosphere for the diners. Quality and color of light under these circumstances are of more importance than quantity which should, however, be adequate for safety and convenience. The design of the lighting system should be governed by the over-all decor of the car and the effects desired. Direct incandescent downlighting on a table provides sparkle to the silverware and glassware that cannot be obtained from diffuse illumination, and is an important aid in the stimulation of eye appeal.

Lounge. General illumination should be such as to meet the requirements for relaxing or card playing and to provide sufficient illumination to the upper walls and ceiling for the elimination of high luminance ratios. Supplementary lighting units may be required to furnish a higher level for prolonged reading and typing or processing business forms. Fig. 16-13 shows a typical lighting design for a lounge car with supplementary individual downlights at each chair.

Fig. 16-12. Commuter railway passenger coach. Two rows of fluorescent luminaires are ceiling mounted over the seats to provide general lighting for reading.

Sleeping Car (Pullman). The bedroom compartment for all intents and purposes can be considered the passenger's travelling apartment and should afford some home environments. It should lend to the comfort, convenience and beauty of the accommodations. As in any good lighting installation, the ability to perform the visual task is the primary consideration. The visual tasks in any sleeping accommodation are similar to those in the home.

The arrangement of berths and provisions for upper storage in the daytime severely limits the ceiling area available for luminaires. Structural members, air ducts and conduit runs also have a limiting effect on useable ceiling area. In general, the location and maximum size of a luminaire in the ceiling is fixed by the conditions outlined above.

Berth lighting units should be designed to provide suitable illumination for reading in bed. Here again, as in the case of the ceiling luminaires, freedom of design and optimum use of materials are limited by the physical characteristics of the application. Berth lighting units should be so designed as to provide a concentrated beam of light for the reading task and a component for general illumination to relieve excessive luminance ratios.

Washrooms and Toilet Sections. The lighting design should provide general illumination and mirror lighting from luminaires on ceiling, side walls or in combination. A luminaire generally located in the ceiling should be provided in the toilet section.

LAMPS AND ELECTRIC POWER SYSTEMS FOR TRANSPORTATION LIGHTING

Lamps

Both incandescent filament and fluorescent lamps are used for public conveyance lighting,

Fig. 16-13. Railroad parlor car. Two continuous rows of 40-watt fluorescent lamps are concealed by the parcel rack on each side of the car. Supplementary lighting is provided by locally controlled fluorescent lamps centered above each chair.

but there is a trend toward use of more fluorescent lighting in the newer units. The high efficacy, long tubular shape, and low luminance of the fluorescent lamp make it well suited to vehicular lighting. However, few public conveyances have the normal 60-hertz 118-volt alternating current for which most fluorescent lamps and accessories are designed; consequently, special electrical systems have been devised to facilitate the use of fluorescent lamps in this field.

Multiple incandescent filament lamps suitable for transportation are described in Section 8 in the 1981 Reference Volume in Fig. 8-90, series lamps in 8-96, and fluorescent in 8-114 through 8-117.

Electric Systems for Operating Fluorescent Lamps in Public Conveyances

Fluorescent lamps were primarily developed for ac operation and are generally more efficient and satisfactory when operated on ac, but a dc power supply for lighting has been the accepted standard in the transportation field until recent years. Certain sizes of fluorescent lamps may be operated directly on the dc available on the vehicle or ac may be generated from the available dc supply by means of various types of conversion units.

With dc operation of fluorescent lamps, a resistance type ballast is used to control the current.[12] Ballast loss should be included when determining the total power required for the lighting. Also, in lamps over 600 millimeters [24 inches] in length, the direction of the current flow through the lamps should be periodically reversed to prevent the reduction in light output at the positive end of the lamp caused by the gradual drift of the mercury to the negative end. The useful lamp life on dc burning is reduced to approximately 80 per cent of that on ac burning. Also, special provisions are needed to assure dependable lamp starting at low line voltage and at lower ranges of ambient temperature.

The use of power conversion equipment to convert from dc to ac is common practice. See Fig. 16-14. The conversion may be to 60 hertz; however, the trend is toward higher frequencies in order to gain over-all efficiency and reduced weight in the auxiliary equipment.

There are two common methods of conversion in use to produce ac from the basic dc power source available:

Rotary Machines. Such devices as rotary

converters, motor alternators, booster inverters, etc., are being used on dc voltages to generate various ac output voltages. Gasoline-electric or diesel electric[13] equipment is avilable for mounting beneath a car and for a "head-end" ac power system.

Inverter Systems. These systems produce ac energy usually of a high frequency ranging from 400 hertz to 25 kilohertz. The lower frequencies are sometimes used in applications where the noise of the vehicle overcomes the audible frequency hum. Optimum frequencies are those above the hearing range. These frequencies also provide minimum size of equipment and maximum over-all system efficiency.

Any standard fluorescent lamp can be used to advantage on high frequency power; however, the ballast and all component equipment should be designed for good lamp performance, taking into consideration ample open circuit voltage and correct operating watts for the fluorescent lamp involved.

MARINE LIGHTING[14]

The objectives of shipboard lighting are to provide for the safety and well being of the passengers and crew, to provide adequate illumination for the various tasks encountered aboard ship, and to provide a home-like environment with comfortable, well-lighted staterooms and public spaces.

Interior Lighting

Lighting in the interior of a ship involves spaces similar to many of those encountered on land. However, due to low ceiling heights, space limitations, and the requirements for watertight, guarded or explosion-proof luminaires in certain locations, there are some differences necessary in the design and application of luminaires from those encountered on land.

The principles set forth in such Sections as 5, 6, 7, 9, 11 and 13, having to do with lighting practices in land installations, will be applicable aboard ship. Illuminances for the tasks in the spaces normally encountered on board ship are found in Fig. 2-2. The values listed represent maintained values, measured without daylight or light supplied by supplementary units such as mirror lights, berth lights, gauge lights, etc., but

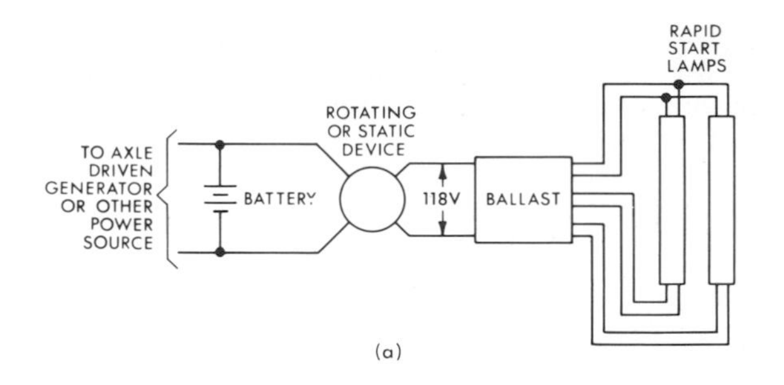

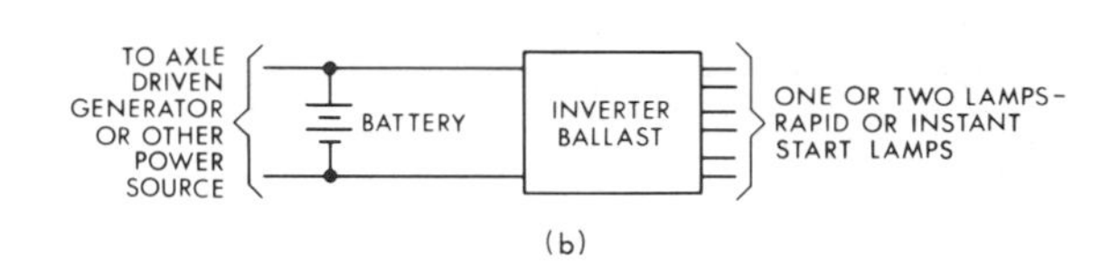

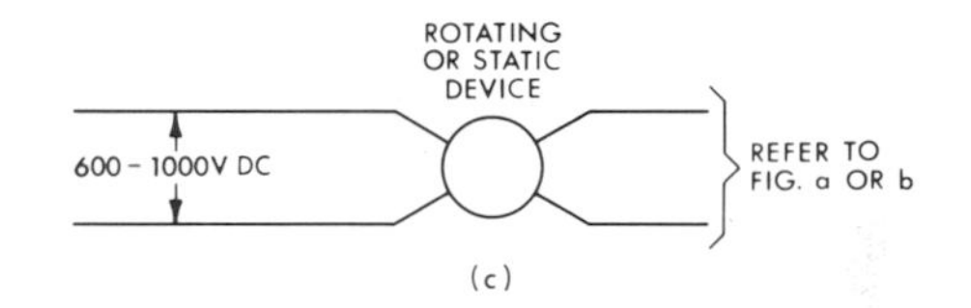

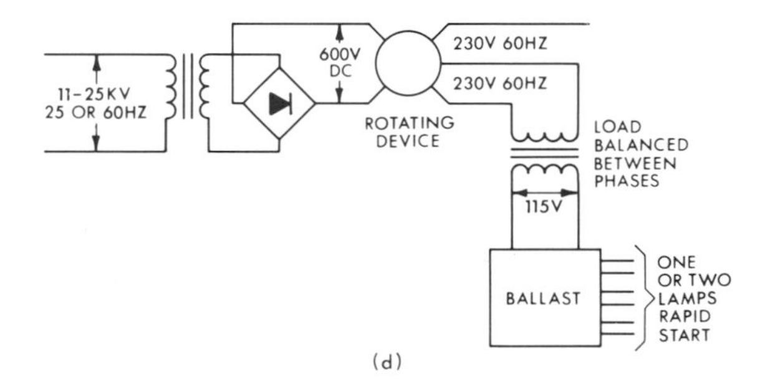

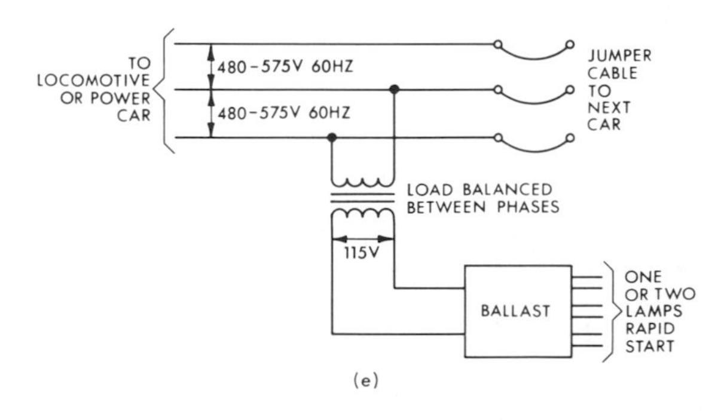

Fig. 16-14. Typical circuits for operating fluorescent lamps in the transportation field: (a) circuit for converting available dc to 60 Hz ac; (b) circuit using available dc; (c) circuit for converting high voltage dc; (d) circuit for converting high voltage ac; and (e) circuit used with power source from locomotive.

include all units normally contributing to the general illumination such as wall brackets, floor lamps and table lamps. They are based on the safety, utilitarian and decorative characteristics of the areas concerned.

Regulatory Body Requirements and Other Standards

The requirements for safety to life for ships of United States registry are established by the US Coast Guard. Regulations regarding ships' electrical plant and lighting are contained in *US Coast Guard Electrical Engineering Regulations.*[15] These regulations reflect the lighting requirements published by the International Conference on Safety of Life at Sea. Detailed requirements for the construction of marine-type luminaires are contained in Underwriters Laboratories "Standard for Marine-Type Electric Lighting Fixtures"[16] and "Standards for Electrical Lighting Fixtures for use in Hazardous Locations."[17]

Luminaires which have been examined and found to comply with the Underwriters Laboratories standards are identified in two different ways:

1. *Underwriters Laboratories service label*—provided for most luminaires, the exception being general utilitarian deck and bulkhead luminaires. Labeled luminaires are identified by Underwriters Laboratories label service symbol together with the designation "Marine-Type Electric Fixture" or "Marine-Type Recessed Electric Fixture." The labels also include other applicable luminaire information, such as "Inside-Type," "Inside Dripproof-Type," "Outside-Type (Fresh Water)," or "Outside-Type (Salt Water)."
2. *Underwriters Laboratories re-examination service*—used for unwired general utilitarian type luminaires for deck and bulkhead mounting where exposed to the weather or other wet or damp locations. These luminaires are listed by catalog number in Underwriters Laboratories "Electrical Construction Materials List,"[18] but are not labeled. Listed luminaires are characterized by having junction boxes, globes and frequently guards. Special luminaires that are not within the scope of Underwriters Laboratories "Standard for Marine-Type Electric Lighting Fixtures" or luminaires of such limited use that inspection by Underwriters Laboratories would be uneconomical may be given special consideration by the US Coast Guard.

The American Bureau of Shipping establishes requirements for construction of vessels for certifying them as being eligible for insurance issued by members of marine insurance underwriters.[19] The Committee on Marine Transportation of the Institute of Electrical and Electronics Engineers (IEEE) has published a "Recommended Practice for Electric Installations on Shipboard"[20] which serves as a guide for the equipment of merchant vessels with an electric plant system.

Seeing Tasks in Marine Lighting

Living and Public Spaces. Stateroom, dining, recreation, office, and medical spaces for passengers, officers and crew should have general illumination as well as local lighting where seeing tasks are involved. Direct- or indirect-type luminaires can be used to give illumination without glare, and to allow safe movement throughout these areas, as well as sufficient light for cleaning and maintenance purposes. Supplementary lighting should be provided in all areas where reading, serving, dressing and make-up are intended. A means for controlling the general lighting should be conveniently located at each main entry, as well as local control for supplementary lighting. Figs. 16-15, 16-16 and 16-17 show typical living and public spaces.

Passageways, stair foyers and stairs should be uniformly illuminated by luminaires installed overhead, in the corner between the bulkhead and the overhead, or in other suitable locations. Care should be exercised in locating the luminaires to illuminate cross passages, stairs, room numbers, etc. Some of the lamps in these areas should be supplied from the emergency lighting system(s) to provide for safety and escape in an emergency.

Passenger entrances are the first spaces on board the ship seen by the passenger, and therefore, particular attention should be paid to the quality of illumination provided. Emergency lighting should be provided to enable escape.

Passengers' and officers' dining room, passengers' lounges, libraries, smoking rooms, cocktail lounges, bars, and ballrooms are all public spaces and the principles of lighting involved are the same as for similar land installations (see Figs. 16-16 and 16-17), but emergency lighting should be provided to enable escape.

Ship's Offices and Navigation Spaces. During nighttime operation of the ship, the *wheelhouse* must be in complete darkness, except for necessary instrument lights. General

Fig. 16-15. Stateroom. General illumination is from fluorescent valance and incandescent downlights; supplementary lighting is provided by incandescent portable lamps.

Fig. 16-16. Officers' dining room. General lighting is provided by recessed fluorescent lighting. A warm tone is provided by incandescent lamps in the bracket above the wide mirror.

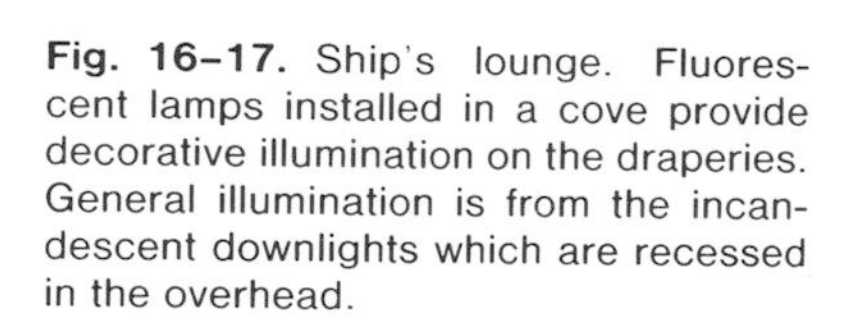

Fig. 16-17. Ship's lounge. Fluorescent lamps installed in a cove provide decorative illumination on the draperies. General illumination is from the incandescent downlights which are recessed in the overhead.

Fig. 16-18. Two rows of two-lamp fluorescent luminaires alternating with 60-watt incandescent units provide general illumination in this galley.

lighting should be provided for cleaning and maintenance purposes only. Passageway lights, which would interfere with navigation when a door to the wheelhouse is opened, should be controlled by a door-operated switch. In order not to interfere with dark adaptation of the eyes of the personnel on duty, instruments should be illuminated by red light of wavelengths not less than 590 nanometers and luminance not greater than about 0.3 candela per square meter [0.1 footlambert].

General lighting in the *chart room* should be provided by ceiling luminaires. Adjustable chart table lamps, fitted with red filters, and a lamp for illumination of the chronometers should be installed and controlled by a momentary contact switch. The chart table lamp should be supplied from the emergency lighting system.

Service and Operating Spaces. Careful attention should be paid to the quality of illumination in *galleys*, *shops* and *other service spaces*. Reflected glare and high luminance ratios should be avoided. The general illumination should be provided by ceiling luminaires with a concentration over benches and working surfaces; supplementary detail lighting should be provided at griddles and range tops, sewing machines, and other points where specific tasks are performed. See Fig. 16-18.

Machinery spaces, steering gear rooms and other ship operating spaces should be provided with general illumination by means of luminaires installed so that piping and other interferences do not obstruct light. The general lighting should be supplemented by detail lighting of gauges, switchboards and other points where meter reading or specific functions require higher illuminances. See Fig. 16-19.

Exterior Lighting

General Deck Lighting. Outside lighting of decks should be by means of watertight deck and bulkhead luminaires. Protective guards should be provided except in certain passenger areas such as passengers' promenades where it is desirable to achieve a certain decorative effect. In certain areas, such as those below, illuminances in excess of that required for safe night passage should be provided.

Cargo Handling. Modern ships are usually fitted with permanent lighting installations in the cargo holds, and receptacles are normally located at the hatches for use with portable cargo clusters. In addition to the foregoing, floodlighting of the deck area in the vicinity of cargo hatches is recommended. This may be accom-

Fig. 16-19. Operating deck console and control panel lighting is provided by a fluorescent general lighting system in this installation. Supplementary lighting is also used at the control panel.

Fig. 16-20. Tanker with high-tower deck lighting using floodlights housing two high intensity discharge 400-watt lamps.

plished by the use of floodlights suitably located on the ship's structure. At least two such floodlights should be provided at each hatch. See Fig. 16-20.

Recreation Areas. Where illumination is desired for nighttime sports activities, floodlights suitably located to illuminate the game areas are recommended.

Lifeboat and Life Raft Launching. Incandescent floodlights should be provided at each lifeboat and life raft for launching. The location of the light should be such that it can be trained on the boat or the boat gear during launching operations. These floodlights should be permanently connected to the emergency system.

Stack Lighting. Where stack floodlighting is desired, floodlights equipped with focusing adjustment are recommended. They should be provided in sufficient quantity and located to insure an even light distribution on the entire stack, except when special effects are desired.

Navigation Lights. Requirements for running anchor and signal lights should conform to the "Rules of the Road" and should be installed to conform to the governing rules for the waters in which the ship is to navigate.[21] A running light indicator panel, equipped with both visible and audible indication of light failure, should be provided for control of the side, masthead, range and stern lights. In general, all navigation and other special lights used for navigational purposes should be energized from the emergency lighting system.

Searchlights. Where a navigational searchlight is desired, it should be located on top of the wheelhouse and operable from within the wheelhouse. A portable signaling searchlight should be provided for operation from either bridge wing. The light should be either battery operated or energized from the emergency lighting system. Signaling may be effected by keying the lamp current or by movement of shutters fitted to the searchlight.

Naval Vessels

Naval vessels are designed for specific military purposes to a set of military characteristics. One of these is "Habitability", of which the lighting installation is a very important factor. The US Navy Bureau of Ships has made an exhaustive study of naval shipboard lighting and has established and standardized on a very efficient group of luminaires to insure meeting the exacting tasks encountered on board and to provide as pleasant an environment as is possible.

Aids to Marine Navigation

Most lighted aids to marine navigation utilize incandescent filament light sources; however, increasing use is being made of gaseous discharge light sources in new and in modernized aids to navigation.

The light beams projected by lighthouse beacons may be produced by one of the following types of apparatus: (a) flashing incandescent filament lamps, flash tubes, or flashing xenon short-arcs mounted in fixed cylindrical Fresnel lens assemblies; and (b) steady burning incandescent filament lamps, mercury lamps, or mercury-xenon short-arcs mounted in rotating assemblies of Fresnel flash panels or in rotating assemblies of reflectors. The effect of both types of apparatus is to produce a distinctive flashing characteristic which permits ready sighting and identification of the lighthouse by reference to appropriate navigational charts on a Light List.

Some lighthouses and other lighted aids to marine navigation (minor battery operated shore lights) are characterized by steady burning or flashing lights projecting distinctive colors over various sectors of azimuth. This effect is accomplished by use of one of the two basic types of apparatus described above together with suitable opaque screens and color filters.

Major lighthouses comprise but one of several classes of lighted aids to marine navigation. By far, the bulk of money, maintenance time and material is applied by the United States Coast Guard to more than 15,000 minor and secondary lighted aids to navigation, consisting of battery powered lighted buoys and shore mounted marine navigational beacons.

Complete details of design and application of lighting equipment for aids to marine navigation are contained in US Coast Guard, *Ocean Engineering Report No. 37 (CG-250-37)*; "Visual Signalling: Theory and Application to Aids to Navigation," 1970.

RAILWAY GUIDANCE SYSTEMS

Railway train operating personnel receive guidance through lighting units of three main categories:

1. Exterior lights on the train, including headlights and marker lights.
2. Interior cab signals.
3. Wayside signals.

Voice communications by wire and radio serve various crew coordinating functions and yard movements; however, actual train movement into any main line segment of the rail system is directed by a signal light indication.

Exterior Lights on Trains

Locomotive headlights are classified either as road service giving 240-meter (800-foot) object visibility, or as switching service for 90-meter (300-foot) object visibility, as governed by the regulations of the Federal Railroad Administration, Department of Transportation. New headlight equipment consists of two all glass sealed beam lamps mounted in a single housing, each lamp projecting a beam about 5 degrees wide of 300,000 candelas. The pair of lamps exceeds the performance and reliability of the old single reflector headlight. Shielding is provided to minimize veiling glare from stray light illuminating atmospheric particles in the line of sight. In addition to lighting possible obstructions on the right of way the headlight also activates colored reflectorized markers at switch locations. The rear lights on trains consist of two red markers spaced horizontally apart.

Locomotive Cab Signals

By suitable track circuits and electric receiving equipment, automatic signal lights inside the cab can be made to show signal aspects corresponding to those of the wayside signals. This is useful in times of poor visibility due to atmospheric conditions or other obstructions. Changes of wayside signal aspects in advance of the train may be displayed promptly in the cab thus expediting response to the change. When cab signals are supplemented with speed control, the engineer is required to limit his speed to that prescribed by the cab signal to prevent automatic brake application.

Rapid Transit Cab Signals

In rapid transit systems speed commands are continuously transmitted, through the rails, precisely and exclusively to the train intended. Onboard, the cab signal displays the commands, and the overspeed control system compares the actual train speed with the maximum speed allowed by the cab signal. See Fig. 16-21. If the actual speed exceeds the limit displayed, the system warns the motorman, audibly, that a brake application is required. If he fails to take action immediately, the control system automatically stops the train.

Wayside Signals

The movement and speed of trains into each segment of trackage is permitted only by the adjacent signal indication, with some advance information provided by the range of the signal beam and the preceding signal.

Wayside Signal Range. The beam intensity and range considerations for a lighted signal are based upon the estimated safe visual range by day in clear weather. For red and green signals it is common to use the formula: $D = \sqrt{186\ I}$, where D is the range in meters and I the intensity in candelas or $D = \sqrt{2000\ I}$, where D is the range in feet and I the intensity in candelas of the same type signal when equipped with colorless optical parts.

Fig. 16-21. In this type of rapid transit cab signal, lighted segments on the speedometer show the motorman the highest speed permitted. The six windows below the speedometer light to show yellow, green, and red aspects and speed limits.

Yellow and lunar white lenses provide a somewhat longer range, but blue only about one-third the distance *D*.

By the use of these formulas and the candlepower distribution curve of a signal beam, it is possible to lay out a chart or plan that shows the ground area over which a particular signal will be within visible range in clear weather. This signal range plan can be superimposed over a track plan to see whether the signal will have visibility over a particular track approach. See Fig. 16-22. Signal manufacturing companies have prepared range charts for their various signal units embodying the large variety of horizontal beam spreading and deflecting auxiliary lenses available.

A horizontal deflecting or spreading prismatic element may be chosen to provide visibility along a curved track approach. A vertical deflecting prismatic element is necessary to enable an engineer at very close range to see a signal high overhead or to see a dwarf signal close to the ground.

External Light Interference. By making the front surface of lenses and roundels convex, rather than flat, it is possible to scatter most of the external light reflected from the front surface. Also, so that daylight will produce negligible interference with the signal under most conditions, hoods or visors are always used. Occasionally flat auxiliary roundels are inclined at selected angles. The incorporation of reflectors in the optics of a signal involves particularly careful analysis to prevent reflected external light. A typical deep parabolic reflector, as used in ordinary spotlights, could flash false indications from external light if used in signals.

Color. Train operating personnel are selected as having normal color vision. The colors used by railroads in North America are with very few exceptions governed by the Signal Manual Part 136 of Association of American Railroads. See Fig. 16-23. These color specifications contain basic definitions for the colors to be displayed in service and the tolerances for color-limit filters to be used to inspect signal glassware. The primary standard filters controlling these inspection filters are deposited with the National Bureau of Standards in Washington, D.C.

The Association of American Railroads currently specifies five colors: red, yellow, green, blue and lunar white; however, only red, yellow and green are used for long range color signals. Both the red and yellow are somewhat deeper colors than those used in street traffic signals. Blue performance is limited by the very low blue emission of incandescent signal lamps and consequent transmission of about 2 per cent by blue glass. Lunar white is the term applied to white light as filtered by a bluish glass which raises the apparent color temperature or whiteness. The lunar white aspect from an incandescent signal lamp will appear about 4000 to 5000 K and from a kerosene unit about 3000 to 4000 K. Purple is no longer recommended as a signal color because of low filter transmission and because of variable impression made upon different observers.

As is commonly understood, red is associated with the most restrictive signal indications, green

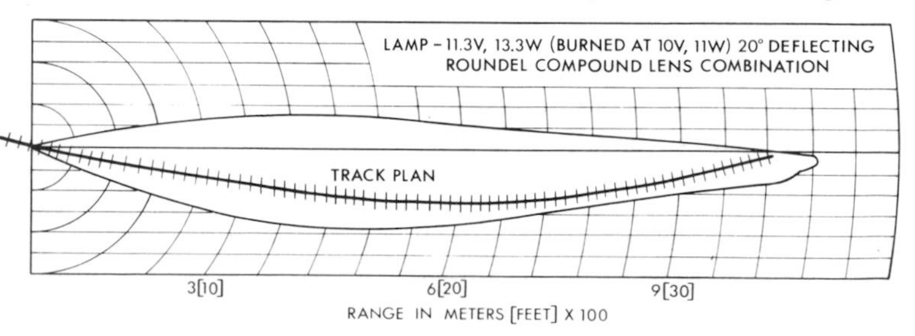

Fig. 16-22. Range chart for searchlight-type signal-unit with part of a track plan superimposed to show range of useful coverage.

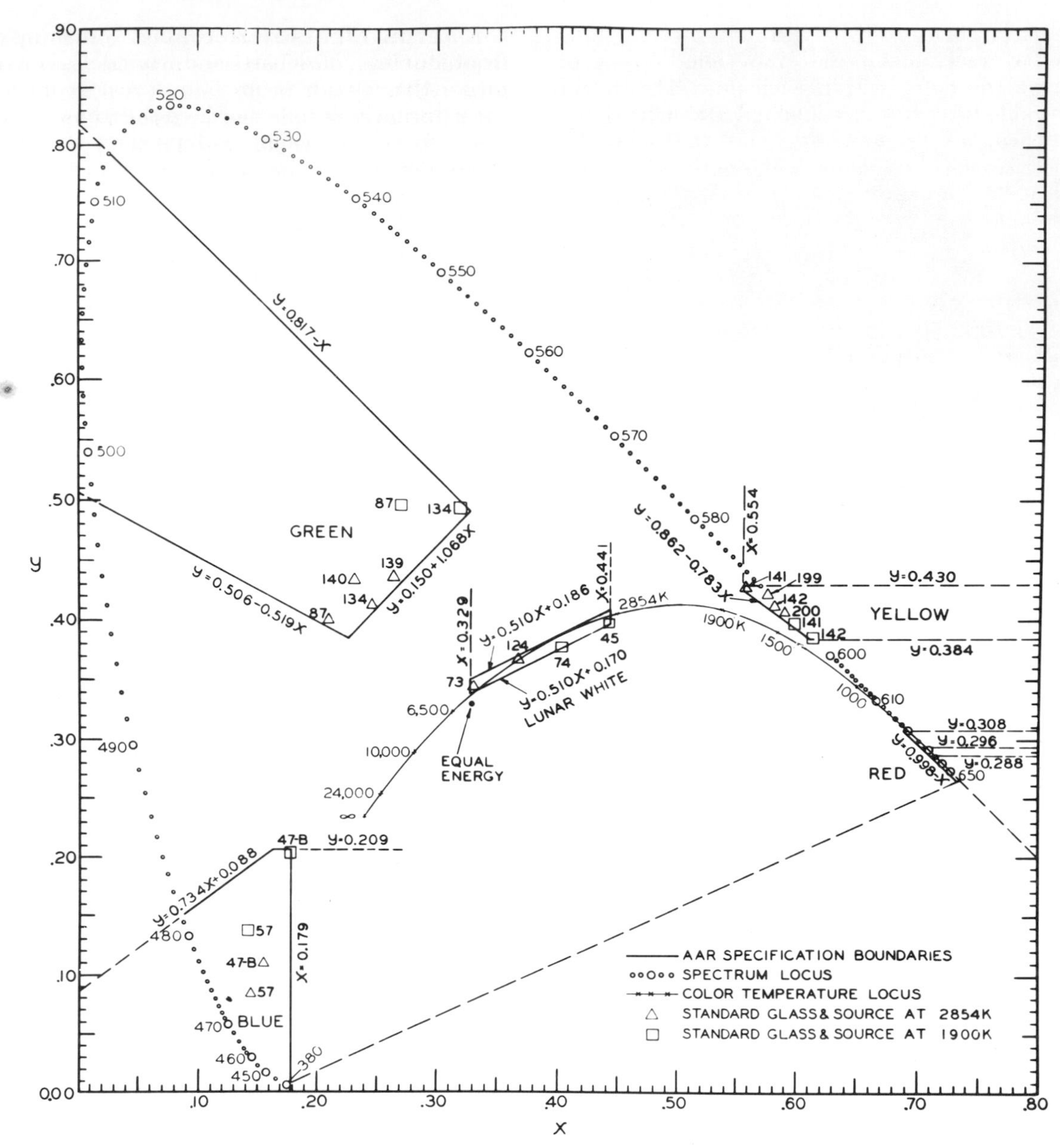

Fig. 16-23. Railway signal color specifications plotted on CIE chromaticity diagram.

with the least restrictive, and yellow intermediate. For the specific meaning of the signal aspects, many of which involve two or more lights shown together, see *American Railway Signalling Principles and Practices*, Chapter II, published by the Association of American Railroads.

Wayside Signal Types. Modern signal units and their arrangement on a mast all have some feature suggestive of the early semaphore unit which had a blade for day viewing and associated color disks that swing in front of a lamp for night viewing. The three types of signals which depend entirely upon light are described below as: position-light signal, color-light signal and color-position-light signal. See Fig. 16-24. All utilize large targets or black backgrounds to permit relatively low wattage lamps to show up without blending into the sky. Lamps are usually of 18 to 40 watts with very small filaments that must be precisely located at the focal point by a prefocus base. Each signal unit is accurately aimed using a sighting device during installation or is adjusted by radio instructions from a viewer down the track.

Color-Light Signal. A color-light signal may involve separate lights with colored lenses for each color, or the signal lighting units may have internal mechanism and moveable filters to change the color within each unit. This latter moveable filter unit is called a searchlight signal and permits three units on a mast to display the widest variety of color combinations, for instance, "red over yellow over yellow," or "green over yellow over red," etc.

Position-Light Signal. The position-light signal is a type of wayside signal which does not depend upon color discrimination by the engineer. In this type, a number of lamps (maximum nine) are mounted on a circular target—eight lights arranged in a circle, one in the center. By operating three lamps at a time, the aspect of the signal may be a vertical, a horizontal or a diagonal row. Each of the target lights is aligned by its own projector system in the direction of the approaching train. Yellow lenses are normally used to achieve distinctiveness from other non-signal lights.

Color-Position-Light Signal. The color-position-light signal is a type which utilizes a combination of the principles of the color-light and the position-light systems. Here also there are several lights on a target. These may be lighted in pairs: vertical pair (green), horizontal pair (red), right and left diagonal pairs (yellow and lunar white, respectively).

Power Sources for Signals. The lighted aspect displayed is controlled by relays at the signal, actuated by coded impulses in the track circuit; however, power for the lamp comes from storage batteries at the location for complete dependability. The batteries may be used alone or as standby for ac service. To provide long lamp life and reduce the probability of a dark signal, lamps are usually burned at 90 per cent of rated voltage.

Control Panels for Signal System

The movement of trains and sections of track system thus occupied are represented on control panels at "interlocking" or at "centralized traffic control" centers. On such a panel the operator has before him push buttons that operate relays for switches and signals along a portion of the rail line which may be a local yard or several hundred kilometers (miles) of track. Lights indicate the response of the switches and signals.

A track diagram for the territory is studded with indicator lights which show when a train occupies certain sections of track along the line or what route has been established.

Railroad-Highway Grade Crossing Lights

Warnings for highway traffic at grade crossings is provided by train actuated flashing red lights. Generally four pairs of horizontally spaced alternate flashing red lights of 18 to 25 volts are used. Track circuits which sense the presence of an

a

b

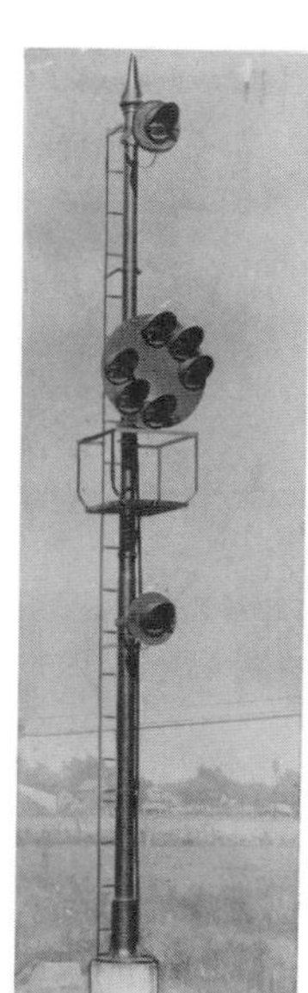
c

d

Fig. 16–24. a. Position-light signal. b. Color-light signal. c. Color-position light signal. d. Searchlight-type of color-light signals.

approaching train control the lights. The light beam width is usually 30 degrees and provides usable visibility of 300 meters [1000 feet]. The lights normally operate from ac with dc standby.

REFERENCES

1. Roper, V. J. and Howard, E. A.: "Seeing with Motor Car Headlamps," *Trans. Illum. Eng. Soc.*, Vol. XXXIII, p. 417, May, 1938. Roper, V. J., and Scott, K. D.: "Silhouette Seeing with Motor Car Headlamps," *Trans. Illum. Eng. Soc.*, Vol. XXXIV, p. 1073, November, 1939. Roper, V. J. and Meese, G. E.: "Seeing Against Headlamp Glare," *Illum. Eng.*,. Vol. XLVII, p. 129, March, 1952.
2. Land, E. H., Hunt, J. H., and Roper, V. J.: "The Polarized Headlight System," *Highway Research Board Bulletin No. 11*, National Research Council, Washington, D.C., 1948.
3. Davis, D. D., Ryder, F. A., and Boelter, L. M. K.: "Measurements of Highway Illumination by Automobile Headlamps under Actual Operating Conditions," *Trans. Illum. Eng. Soc.*, Vol. XXXIV, p. 761, July, 1939.
4. de Boer, J. B. and Vermeulen, D.: "On Measuring the Visibility with Motorcar Headlighting," *Proc. Int. Comm. Illum.*, 1951; de Boer, J. B. and Vermeulen, D.: "Motorcar Headlights," *Philips Tech. Rev.*, Vol. 12, p. 305, May, 1951.
5. Bone, E. P.: "Automobile Glare and Highway Visibility Measurements," *Highway Research Board Bulletin No. 34,* National Research Council, Washington, D.C., 1951.
6. Boelter, L. M. K. and Ryder, F. A.: "Notes on the Behavior of a Beam of Light in Fog," *Illum. Eng.*, Vol. XXXV, p. 223, March, 1940.
7. Finch, D. M.: "Lighting Design for Night Driving," *Illum. Eng.*, Vol. XLV, p. 371, June, 1950.
8. Committee on Motor Vehicle (Exterior) Lighting of the IES: "Lighting Study Project Report on Motor Vehicle (Exterior) Lighting," *Illum. Eng.*, Vol. LIX, p. 660, October, 1964.
9. "Tests for Motor Vehicle Lighting Devices and Components," *SAE Handbook*, p. 271, 1971.
10. Spencer, D. E. and Levin, R. E.: "Guidance in Fog on Turnpikes," *Illum. Eng.*, Vol. LXI, p. 251, April, 1966.
11. Committee on Interior Lighting for Public Conveyances of the IES: "Interior Lighting of Public Conveyances—Road and Rail," *J. Illum. Eng. Soc.*, Vol. 3, p. 381, July, 1974.
12. Brady, C. I., Slauer, R. G., and Wylie, R. R.: "Fluorescent Lamps for High Voltage Direct Current Operation," *Illum. Eng.*, Vol. XLIII, p. 50, January, 1948.
13. "Uses of Electric Power," *Westinghouse Eng.*, Vol. 8, p. 2, January, 1948.
14. Subcommittee on Marine Transportation of the Committee on Interior Lighting for Public Conveyances of the IES: "Recommended Practice for Marine Lighting," *J. Illum. Eng. Soc.*, Vol. 3, p. 397, July 1974.
15. Publication No. CG-259, 46CFR, Parts 110 to 113, US Coast Guard, Washington, D.C., latest issue.
16. "Standard for Marine-Type Electric Lighting Fixtures," Publication No. UL-595, Underwriters Laboratories, Inc., Chicago, Illinois, latest issue.
17. "Standard for Electric Lighting Fixtures for Use in Hazardous Locations," Publication No. UL-844, Underwriters Laboratories, Inc., Chicago, Illinois, latest issue.
18. "Electrical Construction Materials List," Underwriters Laboratories, Inc., Chicago, Illinois, latest issue.
19. *Rules for Building and Classing Steel Vessels for Service on Rivers and Intracoastal Waterways*, American Bureau of Shipping, R. R. Donnelley and Sons, Co., Lakeside Press, Chicago, 1963.
20. "Recommended Practice for Electrical Installation on Shipboard," Publication No. 45, Institute of Electrical and Electronics Engineers, New York, New York, latest issue.
21. *Rules of the Road:* "International—Inland," Publication No. CG-169, latest issue; "Great Lakes," Publication No. CG-172, latest issue; "Western Rivers," Publication No. CG-184, latest issue, US Coast Guard, Washington, D.C.
22. Campbell, J. H.: "New Parameters for High Frequency Lighting Systems," *Illum. Eng.*, Vol. LV, p. 247, May, 1960.
23. *Signal Manual, Specification No. 69,* Part 136, April, 1960. Association of American Railroads, Washington, D.C. 20036.
24. "Equivalent Indications for Semaphore, Color Light Position and Light and Color Position Light Signal Aspects," *Proc. of Signal Section*, Vol. XLII, No. 2. Association of American Railroads, Washington, D.C. 20036.
25. Gage, H. P.: "Practical Considerations in the Selection of Standards for Signal Glass in the United States," *Proc. Int. Congr. on Illum.*, p. 834, 1928.
26. Gibson, K. S. and Haupt, G. W.: "Standardization of the Luminous-Transmission Scale Used in the Specification of Railroad Signal Glasses," *J. Res. Nat. Bur. Stand.*, Research Paper RP 1688, January, 1946.

Lighting For Advertising

SECTION 17

Through the knowledge and use of new materials, new light sources and advanced techniques a wide variety of signs can be produced for today's needs. Illuminated advertising signs—exposed lamp, luminous tube, luminous element, etc.—while differing in several respects from other forms of advertising, definitely tie in with any over-all promotional activity. Signs, whether small identification or large spectacular, mass viewed signs (see Fig. 17-1) can quickly gain the observers' attention through the combined use of size, color and motion.

Sign Characteristics

Electric signs may be classified by illumination method:
1. Luminous letter signs (illuminated letters, nonilluminated background) such as exposed lamp signs, exposed luminous tube signs, raised glass or plastic letter signs, etc.
2. Luminous background signs (illuminated background, silhouette) such as translucent plastic or glass faced signs with interior light sources such as fluorescent lamps, incandescent lamps or luminous tubing.
3. Floodlighted signs, such as painted bulletins and poster panels.
Signs may also be classified by the application function, such as: single-faced luminous elements, window or roof; double-faced projecting, horizontal or vertical, etc.

Size. Physical location, desired legibility range, and brightness determine the minimum letter height required for legibility. However, to attain advertising effectiveness, letter heights of twice minimum height for legibility generally are employed. Vertical columns of letters, though usually an aid in increasing the apparent size of a sign, are more difficult to read than horizontal arrangements.

NOTE: References are listed at the end of each section.

Brightness. Letter or background brightness and contrast between letter and background are factors influencing the legibility of a letter and the rapidity with which it is recognized. Contrast between the average sign brightness and that of its surround determines, in a large measure, the manner in which the sign stands out. Brightness and contrast attract attention.

Location and Position. The advertising value of a sign depends on the greatest possible number of persons seeing it. This is a function of its location.

Distinctiveness. One of the elements of a good electric sign is that it possesses distinctiveness and individuality. It should create a pleasing, favorable impression, should have public appeal and should be remembered easily.

Motion. Motion increases the attracting power and memory value of a sign. It capitalizes on the instinctive trait of people to be aware of and to give heed to moving things.

Color. Color is an important factor in legibility. Often color is incorporated in a sign because it provides contrast. It may aid in attracting attention. It may add distinctiveness.

EXPOSED LAMP SIGNS

Exposed Incandescent Filament Lamp Signs

These signs are constructed so that the lamps are exposed to direct view. This type is well suited for applications where long viewing distances are involved, as well as for small, high brightness signs. Motion and color can be incorporated in such signs as shown in Fig. 17-1.

Fig. 17-1. A spectacular sign located in a highly competitive advertising area where close to a half million people pass daily. Computer programmed red, blue, green and white incandescent filament 30- and 60-watt R-20 lamps provide a colorful, animated display for day and night viewing.

Legibility. Legibility is primarily a function of letter size and form or design, lamp spacing and brightness, and also contrast between letter or design and background.

Block letters possess greater legibility than ornamental styles, script and special forms, although the latter types may be used to gain distinctiveness. Wide, extended letters are more legible than tall, thin letters.

Reflectors. When wide-angle viewing is relatively unimportant, reflectors for increasing the directional candlepower or reflectorized lamps may be used with advantage. In this way, the brightness at the viewed angle can be greatly increased, even to the point of being effective during daylight hours, or lower wattage lamps may be used for the same advertising effectiveness.

Letter Size. The letter height employed on an exposed lamp sign usually is greater than the minimum height necessary to gain recognition. For purposes of advertising and quick reading, it is common practice to provide exposed lamp signs with letter heights that are 50 to 100 per cent greater than those necessary for legibility.

For simple block letters (width equal to three-fifths the height) with a *single row of lamps*, in typical locations, the minimum height for legibility is given by the formula:

$$H_r = \frac{D}{500}$$

where H_r = minimum vertical height of letter, for recognition, from top lamp to bottom lamp in meters (feet).

D = maximum distance at which letter is legible to a majority of people in meters (feet).

For letters with *multiple rows of lamps*, the height should be increased by three times W, the distance between outside rows of lamps in a stroke:

$$H_r = \frac{D}{500} + 3W$$

Letter width, height and stroke and lamp spacing are illustrated in Fig. 17-2.

Lamp Spacing. The proper spacing between lamps to obtain an apparently continuous line of light is determined by the minimum viewing distance. Spacing may be estimated by the following formula:

$$s = \frac{D_{min}}{1500}$$

where s = spacing between centerlines of lamps in meters (feet).

D_{min} = minimum viewing distance in meters (feet).

In very bright locations the above spacing should be decreased by 25 to 35 per cent.

To produce a "smooth" line of light the above spacing should be decreased by 50 per cent. At

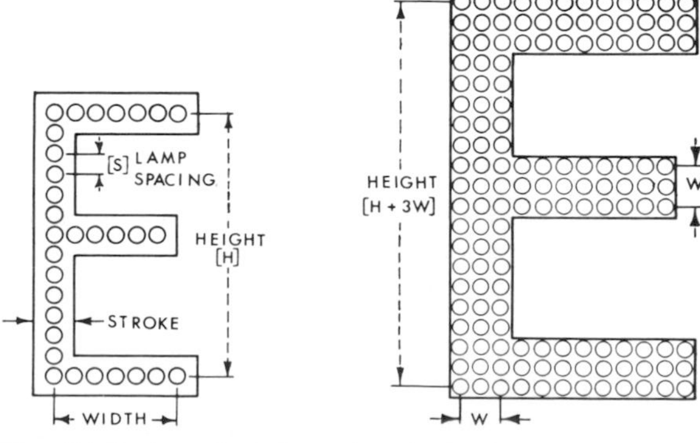

Fig. 17-2. Important dimensions in the design of exposed lamp letters.

Fig. 17-3. Lamp Wattages for Various District Brightnesses

District Brightness	Typical Sign Lamp Wattage
Low	6, 10, 11
Medium	10, 11, 15, 25
Bright	25, 40

Fig. 17-4. Relative Wattage of Clear and Transparent Colored Incandescent Filament Lamps Required for Approximately Equal Advertising Value

Color	clear	yellow	orange	red	green	blue
Wattage	10–11	10–11	15	25	25	40

viewing distances of less than 150 meters [500 feet], a smooth line of light will generally not be possible because low wattage, medium based lamps (6-, 11- or 15-watt S-14) require spacings of 50 to 60 millimeters [2 to 2½ inches] to permit easy maintenance.

Lamp Wattage Rating. The incandescent lamp wattage employed depends upon the general brightness of surroundings, and background, as the sign is viewed. Consequently, a roof sign, even if located in a brightly lighted district in the business center of a city, might at night always be viewed against a dark sky. Such a sign would require the same lamps called for in low brightness areas.

Fig. 17-3 indicates the typical lamp wattages found in signs in various areas, classified according to district brightness.

If incandescent lamps with colored bulbs or clear bulbs with colored accessories are employed, lower letter brightness will result than when equal wattage lamps with clear bulbs are used alone.

Usually, however, less colored light is necessary to create equal advertising effectiveness. It is, therefore, not necessary to increase wattage in direct proportion to the output of the color lamps. This is taken into account in Fig. 17-4.

Both transparent and ceramic coatings are used to color bulbs. In general, the transparent coatings have higher transmittances than the ceramic, thus appear brighter. In addition, the filament is visible for added "glitter" at near viewing distances.

Note that signs lighted in cool colors, blue and green for example, will generally be less legible than those with clear or warm colored lamps, because cool colors appear to swell or "irradiate" more than warm.

Lamp Types. For exposed lamp signs located where rain or snow could fall on relatively hot glass, vacuum-type incandescent lamps are recommended. They are available in 6-, 10-, 11-, 15-, 25- and 40-watt ratings in both clear and colored bulbs.

For high speed motion effects, a 20-watt gas-filled clear lamp is available. The filament heats and cools very rapidly producing a clean, sharp, on-off action. It is used for scintillation effects, running borders and traveling message signs, and wherever afterglow is undesirable.

Channels. Incandescent lamps are often set into channels. This improves legibility of the sign when viewed at an angle, and increases contrast by reducing background spill light. It does not prevent the strokes of the letters from merging together when viewed at a distance since this phenomenon occurs in the eye.

It is very desirable to employ electrically grounded metal channels to separate incandescent filament lamps and luminous tubing when combined in a sign. Without the channel the electric field generated by the tubing causes the filament to vibrate, thereby reducing the life of the lamp.

Daytime Effective Exposed Lamp Signs

Exposed high-candlepower light sources can be used to create electric signs that have as much or more advertising value during the day as do conventional exposed lamp signs at night. Since traffic is generally greater during daylight hours, greatly increased readership usually results with the cost per advertising impression remaining comparable to night viewed lamp signs. The technique is adaptable to signs ranging in size from small store signs to sports scoreboards (see Fig. 17-5) to major spectaculars (see Fig. 17-1).

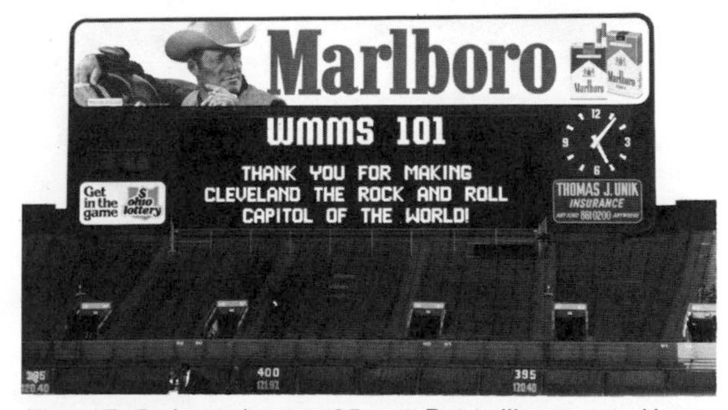

Fig. 17-5. Incandescent 25-watt R-14 silicon coated lamps are used here for a daytime effective scoreboard and animation panel. Lamps are dimmed for night viewing.

Fig. 17-6. Daytime Attraction Power of Several Lamp Types (Viewed Perpendicular to Plane of Sign)

Lamp	Spacing		Distance (meters [feet])		
	millimeters	inches	330 [1100]	750 [2500]	1300 [4300]
25-watt PAR-38 Flood	150	6	F	P	NR
	305	12	P	NR	NR
	460	18	—	—	NR
75-watt PAR-38 Flood	150	6	E	E	G
	305	12	G	G	F
	460	18	—	—	NR
150-watt PAR-38 Flood 75-watt PAR-38 Spot	150	6	*	E	E
	305	12	*	E	G
	460	18	—	—	F
150-watt PAR-38 Spot	150	6	*	E	E
	305	12	*	E	E
	460	18	—	—	E

—Spacing inadequate.
*—Brighter than normally necessary.
E—Excellent.
G—Good.
F—Fair.
P—Poor.
NR—Not recommended

Letter height and lamp spacing in a daytime lamp sign depend primarily on lamp candlepower and on maximum and minimum viewing distances.[1]

For the greatest majority of daytime sign applications, 75-watt PAR-38 floodlamps on 150-millimeter [6-inch] centers will adequately meet advertising and identification needs. Higher candlepower sources should be used with discretion since there is a real possibility of making the sign too bright for comfort. A guide for choosing applicable lamp size is given in Fig. 17-6.

Minimum letter height for legibility is the same as for nighttime exposed lamp signs.

The designer should recognize that a daytime sign utilizing PAR-type lamps is a highly directional display, that is the sign's luminance is a function of the candlepower distribution of the lamp. With a PAR flood, for example, the luminance is reduced to 10 per cent of maximum when viewed 30 degrees off axis. This characteristic also provides automatic dimming of the sign as the motorist drives toward the sign, and under the beam.

Nighttime Viewing. A sign of sufficient brightness to compete successfully with daylight will, in most cases, require dimming at night in order to prevent loss of legibility due to irradiation and the possibility of excessive glare. The need for dimming appears to occur for the 75-watt PAR-38 flood at about 250 lux [25 footcandles] daylight illumination, vertically, on the back of the sign. Except for the highest candlepower lamps, dimming by reducing line voltage 50 per cent has generally proved satisfactory. This can be accomplished by placing the primary windings of the supply transformers in series. Greater dimming, especially for very high candlepower sources, through multiple tap or variable transformers may be required. Continuously variable brightness depending on sky brightness may be accomplished by regulating a dimming system with a photocell.

Luminous Tube Signs

Luminous tube signs (see Fig. 17-7) are constructed of gas-filled glass tubing which, when subjected to high voltage, becomes luminescent in a color characteristic of the particular gas used, the gas and the color of the tubing combined, or of the fluorescent phosphors coating the inner wall.

Color. Fluorescent tubing may be made to emit almost any desired color by mixing different

Fig. 17-7. Exposed luminous tube signage (neon) on the interior wall of a small bank provides a distinctive and recognizable identification for pedestrians and for motorists driving on an adjacent major highway. The silhouetted facade sign provides further identification when interior view is obstructed by pedestrians and cars.

Fig. 17-8. Luminous sign for attraction both day (left) and night (right).

phosphors. Most colors have a higher lumen output per watt rating than the gaseous tubing without a fluorescent coating. Color produced by any one of the gases may be modified by using colored glass tubing, which will transmit only certain colors.

Effective Range. The range of effectiveness for advertising purposes of tube signs is approximately that of exposed incandescent lamp signs of the same size, color and luminance—75 meters [250 feet] to over 3 kilometers [several miles].

Legibility. For block letters of a width equal to three-fifths of their height, the minimum letter height that will be legible to most people is approximately the same as that for exposed lamp signs.

Tubing Sizes. Standard sizes of tubing for signs range from 9 to 15 millimeters, outside diameter, but larger tubing is available.

Transformers. Several forms of high-leakage-reactance type transformers are manufactured to supply the high voltage necessary to start and operate sign tubing. This voltage is of the order of 5000 to 15,000 volts. After a tube sign is lighted, one-third to one-half of the starting voltage is necessary to keep it operating. The usual range of operating current for tube signs is between 10 and 50 milliamperes.

LUMINOUS ELEMENT SIGNS

A luminous element sign can be created by transilluminating or backlighting a plastic or glass panel which may be either integrally pigmented, externally painted or opaqued. The pigmentation or paint film diffuses the light, providing uniform brightness over the desired portion of the sign face. See Fig. 17-8. A wide variety of colors is possible with both the integrally pigmented media and the available translucent lacquers.

Design Data for Luminous Elements

Proper lighting is important in assuring the best attraction value and readability in these signs, but other points must be considered along with the lighting.

Contrast. High contrast, either in color or brightness, between message and background panels should be provided. Opaque letters on a light background are generally preferred for commercial signs because of the attraction value. Where communication is of prime importance, such as in illuminated traffic signs, very light letters on dark backgrounds are generally specified.

Light Sources. Selection of the light source is based on the brightness required, size and shape of the sign, desired color effects, flashing or dimming requirements, environmental temperature conditions, and service access requirements. Line sources of light such as fluorescent lamps, luminous tubing or custom sign tubing provide efficient illumination. Special diffusing screens should be used with spot sources, such as incandescent or high intensity discharge lamps, to prevent appearance of "hot spots" of brightness on sign faces.

Sign Brightness. Adequate sign brightness should be provided but it is important that it should not be "overdone." This depends primarily upon the desired sign brightness, use and the environment. See Fig. 17-9.

Calculating Number of Lamps Needed. When white, yellow or ivory backgrounds are used, a formula for estimating the spacing of the lighting elements in millimeters (inches) is as follows:

$$\text{Spacing Between Lamps (Equal on Centers)} = \frac{K \times \tau \times \Phi/l}{L}$$

where K = 250 when l is in meters and 9.4 when l is in feet (a constant for the combined interreflectance characteristics of the sign enclosure).

l length of lamp in meters (feet).

L = luminance required (from Fig. 17-9).

τ = transmittance of the media (from manufacturer's literature or measurement).

Φ/l = the lumens per meter (foot) of lamp to be used. This is obtained by dividing the manufacturer's initial lumen output for the given lamp by the number of meters (feet) of length.

This spacing is based on providing a clearance between the lighting elements and the sign face material equal to the center-to-center spacing figure just obtained. However, with both internally pigmented media and with lacquer coatings, it is possible to obtain satisfactory diffusion in many cases where the clearance between the lighting elements and the sign face is less than the center-to-center spacing of the lighting elements. There are minimums, in all cases, where light element streaking will appear on the sign face if the clearance-to-spacing ratio is too low. Actual clearance between lamps and sign face media can be determined in a test mock-up where experience or published data are lacking.

Adaptation of Formulas for Point Light Sources. The formulas above may be used for point sources by substituting the initial lamp lumen output for Φ/l and by taking the square root of the value obtained multiplied by 1000 (12). This gives spacing that is equal in both directions. The clearance between the surface of the lamps and the sign face media should be not less than that derived from the following formula:

Minimum Clearance

$$\text{(in millimeters)} = 12.5\sqrt{\text{Wattage of Lamp}}$$

$$\text{(in inches)} = 0.5\sqrt{\text{Wattage of Lamp}}$$

This is to prevent excessive overheating of the sign face media by direct radiation from the lamp. Having determined the necessary spacing, the number of lamps can be readily calculated.

Obscuring Lamp Sockets. Where fluorescent lamps are to be used, the dimensions of the sign should be such that the lamp sockets are located just beyond the translucent face area. This will prevent shadows at the edges directly over the lamp sockets. Series arrangement of tubes in large signs using more than one tube per row requires an overlapping of the tubes of at least 76 millimeters [3 inches] to prevent shadows similar to those at the sockets.

All internal sign components such as structural framing, sheet metal backgrounds and ballasts should be coated with at least two coats of high reflectance white paint to derive maximum lighting efficiency and prevent shadows.

Venting. Provision for venting and air circulation may be necessary depending upon the environmental temperature conditions of the sign. This is primarily to maintain the efficiency

Fig. 17-9. Recommended Luminous Background Sign Luminances

Range of Sign Luminance		Potential Areas of Application
candelas/square meter	footlamberts	
70 to 350	20 to 100	Lighted facades and fascia signs
250 to 500	75 to 150	Bright fascia signs as in shopping centers
450 to 700	125 to 200	"Low" brightness areas where signs are relatively isolated or have dark surrounds
700 to 1000	200 to 300	Average commercial sign such as for gas station identification
1000 to 1400	300 to 400	High rise signs and signs in areas of high sign competition
1400 to 1700	400 to 500	For emergency traffic control conditions where communication is critical

Fig. 17-10. Maximum Spacing of Three Dimensional Letters

Observation Angle (degrees from plane of sign)	Minimum Spacing*
5	12 D
10	6 D
15	4 D
20	3 D

* D = depth of formed or fabricated letter.

and prolong the life of the lighting elements and supporting equipment (ballasts, etc). In signs with lamps on very close centers, forced ventilation may be necessary to prevent overheating of the sign face media.

Readability. Readability of a luminous panel sign depends to a great extent on four factors:
1. The size and proportions of letters as well as letter design configuration.
2. Letter spacing.
3. Color and brightness contrast between letter and background.
4. Brightness of a sign face.

Size and Proportions of Letters. With dark letters and light background colors, the following formula may be used to determine the minimum, letter height:

$$H = \frac{D}{600}$$

where H = letter height in meters (feet), and
D = maximum distance of legibility in meters (feet).

The proportions of a letter designed for maximum readability are—width = 60 per cent of height, and stroke = 15 per cent of the height in a sans serif block Gothic style.

Spacing of Letters. The above width and stroke proportions are effective in preventing blending of letter lines where maximum distance of legibility is required. Considerable license is possible with spacing of painted letters; however, for maximum distance legibility, the spacing between letters should be balanced between 15 per cent of letter height and visual equalization of "white masses such as between a "W" and an "A", etc.

Letters and illustrations, or insignia, of bold silhouette rather than fine detail are preferable for long distance legibility. Fine detail and stylized script letters find their greatest use where they will normally be observed from relatively close quarters, such as in downtown shopping areas and in shopping malls.

Three dimensional formed and fabricated letters should be spaced on the basis of their depth of forming and the minimum acuteness of angle of observation. See Fig. 17-10.

The ratios given in Fig. 17-10 are also useful in the design of letters to be formed or fabricated where their principal legibility will be a function of the acuteness of angle of observation. For example, based on a minimum observation angle of 10 degrees and a proposed depth of forming of 50 millimeters [2 inches], the minimum opening in a letter like an "O" should be 6 × 50 = 300 millimeters [6 × 2 = 12 inches]. Hence, letter designs for acute observation angles will be "extended" and the average letter width will be as much as the letter height, or more, to meet the requirements of acute observation angles.

Luminance and Readability of Sign Face. The brightness of the sign face has a significant influence on the readability of the sign. A sign which is too bright can suffer loss of readability from the halo effect around the letters. Insufficient lighting will reduce distance of legibility. The recommendations for sign luminance in given districts, shown in Fig. 17-9, are within safe limits. In some cases, where high background brightness is required, elimination of this halo effect is achieved by applying a 13 millimeter [½-inch] wide stripe of black opaque paint around the out-line of the letter. This applies particularly to those signs employing flat cut-out, formed or fabricated letters attached to the light colored backgrounds. In signs using dark backgrounds with light letters, "debossing," or forming depressed areas, rather than the conventional raised letter areas, eliminates halation.

Luminous Building Fronts or Facades

The same basic data for design of luminous elements applies, in general, to luminous portions of building fronts. However, the illuminances need not be designed for greater than 350 candelas per square meter [100 footlamberts] of surface luminance. In an area of low level environmental lighting, 85 candelas per square meter [25 footlamberts] of surface luminance will be found adequate.

Building Fascia or Belt Signs

Service access requirements in relatively long fascia or "belt" signs often call for use of a single row of lamps along the top or bottom edges of the enclosures. This low level light input system produces a surface luminance in the 70 to 350 candelas per square meter [20 to 100 footlambert] range. A double row raises the surface luminance to the 250 to 500 candelas per square meter [75 to 175 footlambert] range.

Matters of major design importance in systems for obtaining uniform light distribution or even lighting of the fascia surface, are:

1. The depth of the sign cabinet is the major factor in obtaining uniform light distribtion.
2. The use of specially shaped sign enclosures with sloping, parabolic or elliptical contoured backs does not improve the light distribution over the straight back sign cabinet.
3. It is possible to improve the lighting uniformity with special reflectors at the light source. (A parabolic reflector troffer will permit shallower depth of sign cabinets for given heights.)
4. The luminance uniformity ratio of the sign face media, as shown in the following formula:

$$\frac{\text{Luminance Near Light Source}}{\text{Luminance Farthest Away}} = \text{Uniformity Ratio}$$

(A ratio of 1 would be optimum. A ratio of 2 may be tolerated in some installations but should be considered the maximum allowable. Ratios of 1.3 to 1.5 will be satisfactory for a majority of installations.)

5. Very high output fluorescent lamps are as efficient in obtaining uniform light distribution as aperture lamps.

Examples of Fascia Signs.

Up to 1.2 Meters (4 Feet) in Height: Satisfactory luminance uniformity, depending upon the depth of the sign cabinet can be produced with a straight backed sign box, interior painted with high reflectance white paint, equipped with one or two lines of fluorescent lamps (1500 mA) shielded with a baffle plate and light output controlled with a thin gauge sheet metal egg crate grid. See Fig. 17-11*a*. Surface luminance values versus uniformity ratios are tabulated in Fig. 17-12.

Up to 1.8 Meters (6 Feet) in Height: A straight sign cabinet with either baffles and egg crate grid (see Fig. 17-11*a*), or with a parabolic reflector lighting troffer (see Fig. 17-11*b*), may be used depending upon the required luminance and available depth of the cabinet.

In Fig. 17-12, two 1500 mA fluorescent lamps produce fair luminances for a fascia sign, but at the expense of a uniformity ratio close to the maximum. By using the parabolic troffer system, a more satisfactory uniformity ratio can be achieved as longs as the depth of the sign cabinet is from 460 to 610 millimeters (18 to 24 inches) deep. The lower luminances with the troffer are due to the use of a single tube.

Up to 2.4 Meters (8 Feet) in Height: A straight sign cabinet with a parabolic troffer (see Fig. 17-11*b*) is required to obtain reasonably uniform lighting distribution. The data supplied in Fig. 17-12 will be helpful in both the design and lighting of "belt" or fascia signs.

Luminous Fascia Colors Other than White. The information above is confined to the results using 3.2-millimeter (.125-inch) thick integrally pigmented sign face media having a transmittance of 40 per cent. Other whites and other colors with lower transmittance values will produce surface luminance values below those shown. When using colors other than white, it is necessary to apply a spray coating of white paint to the inside surface in order to obtain comparable light distribution qualities.

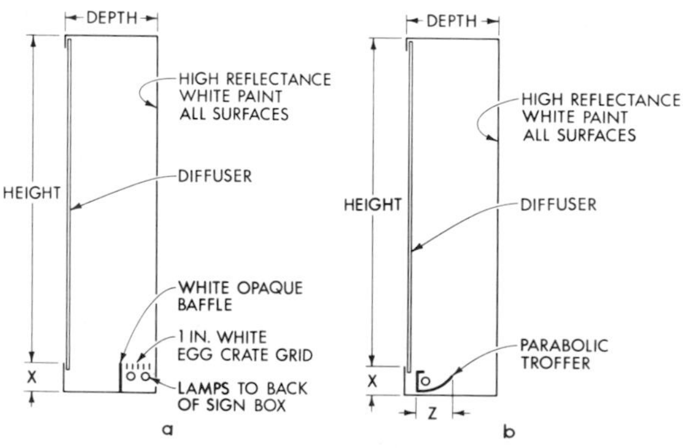

Fig. 17-11. Typical construction of fascia signs (a) 1.2 or 1.8 meters (4 or 6 feet) in height, (b) 2.4 meters (8 feet) in height. Diffuser is 3.2 millimeters (0.125 inch) thickness. Dimension X = 200 (8) and Z = 230 millimeters (9 inches).

FLOODLIGHTED SIGNS

Lighting Poster Panels, Bulletin Boards and Vertical Surface Signs

The most important factors which contribute to the conspicuity of an illuminated sign are its area and its brightness. However, there exist several relatively complex factors affecting legibility of signs, many of which are psychological as well as physical. See page 17-1 under Sign Characteristics.

Fig. 17-12. Luminance of Most Distant Area In Candelas/Square Meter (Footlamberts) Versus Uniformity Ratio[a]

Type of Lights	Depth of Sign Box in Millimeters (Inches)					
	200 (8)	230 (9)	300 (12)	450 (18)	500 (20)	600 (24)
	a. For Fascia Signs up to 1.2 Meters (4 Feet) in Height					
One row of lamps	220 (65)/3.0	—	40 (70)/1.9	220 (65)/1.3	—	—
Two rows of lamps	340 (100)/3.5	—	450 (130)/2.1	450 (130)/1.4	—	—
	b. For Fascia Signs Up to 1.8 Meters (6 Feet) in Height					
Two 1500 mA lamps[b]	—	—	210 (60)/4.0	270 (80)/1.9	310 (90)/1.8	330 (95)/1.6
Parabolic troffer[c]	—	150 (45)/2.8	160 (48)/2.0	170 (50)/1.4	190 (55)/1.3	210 (60)/1.2
	c. For Fascia Signs Up to 2.4 Meters (8 Feet) in Height					
Two 1500 mA lamps[b]	—	170 (50)/9.0	—	260 (75)/3.2	—	170 (50)/1.8
Parabolic troffer[c]	—	120 (35)/2.9	—	140 (40)/1.1	—	70 (20)/1.5

[a] All data based on 40 per cent transmittance for sign face media. [b] See Fig. 17-11*a*. [c] See Fig. 17-11*b*.

Standardized sign dimensions are shown in Fig. 17-13.

General Guides for Floodlighting Signs. The following is a list of recommendations to be considered in designing the floodlighting of signs.

1. The brightness of the sign panel should be sufficient to stand out in contrast with its surroundings. Fig. 17-14 lists recommended illuminances.
2. The luminance should be sufficiently uniform to provide equal legibility over the message area. A maximum to minimum ratio of 4 to 1 is desirable.
3. The lighting should cause neither direct nor objectionable reflected glare at the normal viewing positions.
4. The lighting equipment should not obstruct the reading of the sign from normal viewing position, nor produce daytime shadows objectionably reducing the legibility of the sign.
5. The lighting equipment should require a minimum of maintenance and provide low cost annual operation.
6. The maintenance of the system should be adequate to achieve the designed illuminance level.

Location of Lighting Equipment. Some of the factors to be considered when determining whether luminaires should be mounted across the top or bottom of a floodlighted sign are:

A. *For top mounted units*
  1. Cover may collect less dirt, snow, etc.
  2. Luminaires will not hide message.
  3. Reflected glare more apparent.
  4. Luminaires may produce daytime shadows.
  5. Sign itself usually shields direct view of lamps from opposing traffic.
  6. Luminaires may be more difficult to service.

B. *For bottom mounted units*
  1. Cover may collect more dirt, snow, etc.
  2. Luminaires may hide message from some viewing angles.
  3. Reflected glare is minimized.
  4. No daytime shadows.
  5. Shielding may be necessary to hide direct view of lamps from opposing traffic.
  6. Luminaires may be easier to service—as from a catwalk.

Light Sources for Floodlighted Signs

There is no single type of source that can be described as "best" for sign floodlighting. Most lamps, regardless of type, can be used with dif-

Fig. 17-13. Basic Structure Dimensions of the Outdoor Advertising Association of America, Inc.

Display	Outside Dimensions (Uniformly Lighted Area)	
	Meters	Feet
Poster Panel	3.7 × 7.5	12.25 × 24.50
Painted Bulletin	4.25 × 14.5	14 × 48

Fig. 17-14. Recommended Illuminances for Poster Panels, Bulletin Boards and Other Advertising Signs

Average Reflectance of Advertising Copy	Recommended Illuminance in Lux [Footcandles]	
	Bright Surrounds	Dark Surrounds
Low	1000 [100]	500 [50]
High	500 [50]	200 [20]

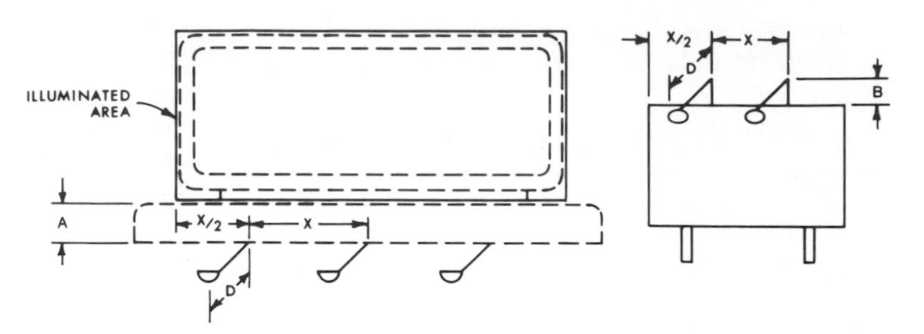

Fig. 17-15. Typical floodlighted sign installations and illuminance results. Dimension A = 0.76 to 1.0 meter (2½ to 3½ feet) and *B* = 0.76 meter (2½ feet).

d = Length measured from copy face to luminaire hub.

x = Space between luminaires $\left(\frac{\text{Length of illuminated area}}{\text{Number of luminaires}}\right)$

x (max.) = 2½d

(a) For Metal Halide Luminaires

Display	Number of Luminaires	d Meters (Feet)	x Meters (Feet)	Approximate Maintained Average Lux (Footcandles)*
Poster Panel	3	1.2 (4)	2.5 (8.2)	550 (55)
Painted Bulletin	6	1.45 (4.75)	2.4 (8)	500 (50)
	4	1.45 (4.75)	3.7 (12)	350 (35)
	3	1.45 (4.75)	4.9 (16)	250 (25)

* Based on 400-watt metal halide lamp, Light loss factor = .71.

(b) For Deluxe-Mercury Luminaires

Display	Number of Luminaires	d Meters (Feet)	x Meters (Feet)	Approximate Maintained Average Lux (Footcandles)*
Poster Panel	4	1.2 (4)	1.9 (6.1)	550 (55)
Painted Bulletin	8	1.45 (4.75)	1.8 (6)	500 (50)
	6	1.45 (4.75)	2.4 (8)	350 (35)
	4	1.45 (4.75)	3.7 (12)	250 (25)

* Based on 400-watt fluorescent-mercury lamp, Light loss factor = .77. If same quantities of 250-watt lamps are used, multiply illuminance by .52. If same quantities of 175-watt lamps are used, multiply illuminance by .37.

(c) For Tungsten-Halogen Luminaires

Display	1500-Watt			500-Watt		
	Number of Luminaires	x Meters (Feet)	Approximate Maintained Average Lux (Footcandles)*	Number of Luminaires	x Meters (Feet)	Approximate Maintained Average Lux (Footcandles)*
Poster Panel	3	2.5 (8.2)	1150 (115)	4	1.9 (6.1)	500 (50)
Painted Bulletin	6	2.4 (8)	1000 (100)	8	1.8 (6)	550 (55)
	3	4.9 (16)	500 (50)	4	3.7 (12)	300 (30)

* Light loss factor = .89. Distance from sign to luminaire = 2.4 meters/(8 feet), aiming point is at a distance down from top of sign equal to ⅓ height of sign.

ferent reflector and lens combinations to realize various beam patterns which may be required. Therefore, choices of light source for a given sign are generally made for reasons of initial cost, operating cost (including maintenance cost), end result desired, color or novelty. Section 8 in 1981 Reference Volume contains a detailed discussion of available light sources, but there are some general guidelines which should be followed, based on the inherent characteristics of the various lamps. Below are characteristics to be considered in selecting the light source. Implied comparisons are relative to the other sources and are based on equal illuminance on the sign.

Incandescent. Advantages include good color rendering, small size permits accurate beam control, and very good operation in cold weather. Disadvantages are relatively low output (lumens per watt), short lamp life and higher operating costs.

Tungsten-Halogen. Advantages include good color rendering, high lumen maintenance (output remains practically constant throughout life) and very good operation in cold weather. Disadvantages are relatively low output (lumens per watt), medium lamp life and higher operating costs.

Mercury. Advantages include long lamp life, high output (lumens per watt) and low operating cost. Disadvantages are high initial cost, fair beam control (due to larger source), and color rendering capability below that of incandescent or tungsten-halogen—(although usually acceptable).

Metal Halide. Advantages are good lamp life, exceptionally high output (lumens per watt), good color rendering capability and low operating cost. Disadvantage is high initial cost.

Fluorescent. Advantages are moderately long life, high output (lumens per watt) and low operating cost. Disadvantages are high initial cost, very little beam control due to large source, and variable output due to changing temperatures.

Lighting Systems for Floodlighted Signs

Concurrent with consideration of a particular light source, there should be an evaluation of the other elements such as lamp housing, mounting arrangements, and auxiliary equipment if required. Primarily, due to improved lamp performance, metal halide lamp systems have made significant inroads in this area. The majority of existing signs, however, are still lighted by fluorescent equipment. Incandescent lamps, with their associated housings are, in general, less expensive to install initially, since they require no auxiliaries (such as ballasts), but their use has dwindled considerably, in favor of the more efficient sources.

Application Data. Regardless of which system is used, the most economical floodlighting system is one which utilizes the fewest number of floodlights containing the highest wattage lamps. It is easier to install, control and maintain. It also uses less power for the same illuminance than a system using more but smaller units. However, illuminance uniformity and appearance may require the selection of a system utilizing a larger number of smaller units. It may be necessary to draw a careful balance between the two extremes. Regardless of which type of source is selected, the beam patterns should be overlapped so that any given area receives light from at least two units. This requirement is usually satisfied if the acceptable uniformity ratio is achieved.

Metal Halide Systems. The 400-watt metal halide lamp is the size most frequently used for sign lighting. Fig. 17-15 includes information on positioning a typical luminaire for both top and bottom mounting and shows the maintained illuminance obtained with specific luminaire location data. Fig. 17-16 is a view of a typical sign.

Mercury Systems. Deluxe-mercury lamps in the 175-, 250-, and 400-watt sizes have many applications in sign lighting where their color rendering capability is acceptable. The life of these lamps is unusualy long, and they are well-suited to locations where maintenance is infrequent, or where the signs are relatively inaccessible. Luminaire location is usually determined in the same manner as for metal halide systems. Maintained illuminance values obtained on typical signs are shown in Fig. 17-15.

Tungsten-Halogen Systems. Both 500- and 1500-watt tungsten-halogen lamps are used for sign lighting. See Fig. 17-15 for typical location details for bottom or top mounted units and the maintained illuminance obtained on typical signs.

Fluorescent Systems. General rules for applying fluorescent units are shown in Fig. 17-17. Aiming angle θ is usually 45 degrees. One important consideration is that fluorescent lamps are temperature and wind velocity sensitive, and most photometric data are obtained at an ambient temperature of 25°C (77°F). See Fig. 17-18 for typical lumen output data for fluorescent floodlights in varying ambient temperatures. Illuminance values obtained on typical signs are shown in Fig. 17-19. For higher illuminance values, twin units are used or units are added at the bottom of the sign and aimed upward. See Fig. 17-20 for typical fluorescent floodlights mounted in tandem.

Fig. 17-16. View of lighted sign using metal halide lamps.

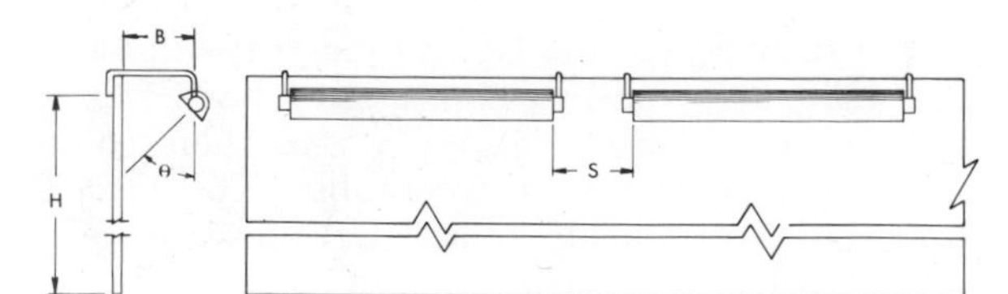

Fig. 17-17. General rules for applying fluorescent floodlights.*

1. S should not exceed B.
2. Overhang B should not be less than approximately 0.4H.
3. If H exceeds 4.5 meters (15 feet), floodlights are recommended across both bottom and top of area in order to assure an acceptable uniformity ratio.
4. B should be as large as practical.

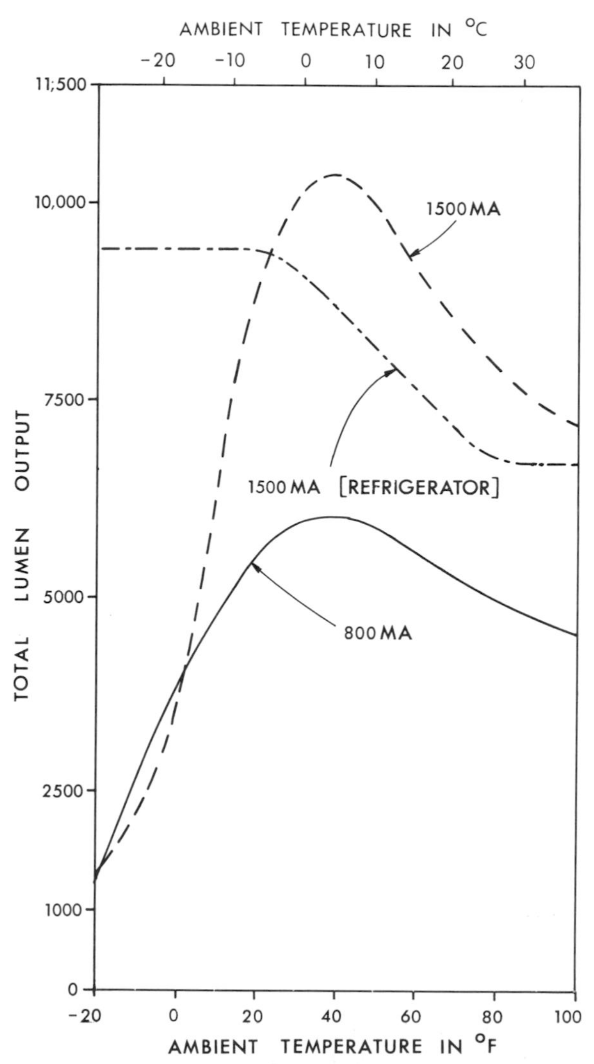

Fig. 17-18. Lumen output versus temperature characteristics for typical fluorescent floodlights.

* These "Rules of Thumb" are based upon the use of a white enameled reflector which produces a symmetrical distribution in a plane perpendicular to the lamp. B and θ may vary with specific reflector types.

Fig. 17-19. Illuminance Obtained on Vertical Surfaces From Enclosed Fluorescent Floodlights[a] (See Drawing in Fig. 17-17)

Display	Number of Lamps[b]	Average Illuminance Maintained in Service in Lux (Footcandles)[c] (Approximate MA—Rating)	
		800	1500
Poster Panel	2-units	140 (13)	250 (23)
	3-units	220 (20)	360 (33)
Painted Bulletin	3-units	120 (11)	190 (18)
	4-units	140 (13)	250 (23)
	5-units	180 (17)	310 (29)

[a] Calculations of illuminance based upon $B = 1.5$ meters (5 feet).
[b] See Fig. 17-17 for general overhangs B and aiming angles θ.
[c] Assuming a light loss factor of 0.65. Uniformity ratio max/min should be less than 4/1. 96T12 fluorescent lamps at 10 °C (50 °F).

Fig. 17-20. Typical fluorescent floodlights mounted in tandem to illuminate an advertising sign.

REFERENCES

1. Hart, A. L.: "Some Factors that Influence the Design of Daytime Effective Exposed Lamp Signs," *Illum. Eng.*, Vol. LI, p. 677, October, 1956.
2. O'Day, R.: "Fluorescent Lighting," *Outdoor Advertising Association News*, February, 1956.
3. Peek, S. C. and Keenan, J. P.: "Outdoor Applications of New Reflector Contour Designs for Higher Output Fluorescent Lamps," *Illum. Eng.* Vol. LIV, p. 77, February, 1959.
4. Baird, N. F. and Schmitz, R. B.: "Effective Use of Colored Lamps on a Computerized, Animated Sign," *Light. Des. Appl.*, Vol. 8, p. 38, October, 1978.

Underwater Lighting *

SECTION 18

Divers, manned submersible vehicles, unmanned instrument platforms and permanent bottom installations use underwater light for a variety of tasks. Underwater television as well as motion picture and still photography are dependent on light.

Seeing ranges in water vary from a few centimeters (inches) in harbors and bays to a few hundred meters (feet) in unusually clear parts of the oceans. The underwater world is generally blue-green due to the color filtering effects of water. The great external pressure of water at depth and the corrosive effects of sea water also provide unusual challenges to the designer of lighting equipment or lighting systems for underwater applications.

Terms and Definitions

The terms and definitions given here are peculiar to underwater lighting. The definitions here generally follow the recommendations of the Committee on Ocean Optics of the International Association for the Physical Sciences of the Ocean (IAPSO). Neither the notation nor the units are universal, but they are widely used and are essentially those recommended by IAPSO. Additional terms and definitions useful in underwater lighting calculations are found in Section 1 in the 1981 Reference Volume.

It should be noted that in the past some authors have used terms such as "absorption" or "extinction" to mean "beam attenuation" as it is defined here, and that sometimes decade values for the beam attenuation, absorption or scattering coefficients have been given, *i.e.*, coefficients smaller by the factor $1/\log_e 10$. To further confuse matters either the term "extinction coefficient" or "vertical extinction coefficient" has sometimes been used in lieu of "diffuse attenuation coefficient."

Absorption Coefficient: the ratio of radiant flux lost through absorption (dF_a) (in an infinitesimally thin layer of medium normal to the beam) to the incident flux (F)*, divided by the thickness of the layer (dx).

$$a = -(1/F)dF_a/dx \qquad \text{unit: meter}^{-1}$$

Volume Scattering Function: the radiant intensity ($dI(\theta)$) from a volume element (dV) in a given direction (θ) per unit of irradiance (E) of a beam incident on the volume per unit volume. (θ is ordinarily measured from the forward direction.)

$$\beta(\theta) = (1/E)dI(\theta)/dV \qquad \text{unit: meter}^{-1}$$

(Total) Scattering Coefficient: the ratio of radiant flux lost through scattering (in an infinitesimally thin layer of the medium normal to the beam) (dF_s) to the incident flux (F), divided by the thickness of the layer (dx); equivalent to the integral of the volume scattering function over all directions.

$$b = -(1/F)dF_s/dx = \int_0^{4\pi} \beta(\theta)d\omega$$

$$= 2\pi \int_0^{\pi} \beta(\theta) \sin \theta d\theta \qquad \text{unit: meter}^{-1}$$

Beam Attenuation Coefficient: the sum of the absorption coefficient (a) and the scattering coefficient (b).

$$c = a + b \qquad \text{unit: meter}^{-1}$$

Note: Sometimes the symbol α is used instead of c. Hence, a transmissometer is often called an "alphameter".

Diffuse Attenuation Coefficient for Irradiance: the ratio of irradiance lost through ab-

* Swimming pool lighting is discussed in Section 13. Note: References are listed at the end of each section.

* IAPSO symbol. IES symbol is Φ.

sorption and scattering (in an infinitesimally thin horizontal layer of the medium) (dE) to the incident irradiance (E), divided by the thickness of the layer (dz).

$$K = -(1/E)dE/dz \qquad \text{unit: meter}^{-1}$$

Filtering Properties of Sea Water

Different wavelengths of light are absorbed by different amounts in water. Sea water that is both deep and clear has a spectrum that closely resembles distilled water with peak transmission in the blue near 480 nanometers.

The absorption curves for other bodies of water differ greatly from this theoretical or ideal distribution due to the presence of silt, decomposition of plant and animal material, pollution and living organisms. A common source of change is plankton in the water; plankton absorb short wavelengths of energy much more than long wavelengths. Therefore, the peak of the transmission curve in a body of water moves to the yellow-green (510- to 570-nanometer) portion of the spectrum. The amount of shift depends on the density of the plankton.

Some typical spectral transmittance curves for different types of water are shown in Fig. 18-1.[1] The shapes of the spectral transmittance curves result primarily from absorption, but the curves include losses from both absorption and scattering processes. Since scattering in natural water is nearly independent of wavelength, it affects the level of the curves and not the shape.

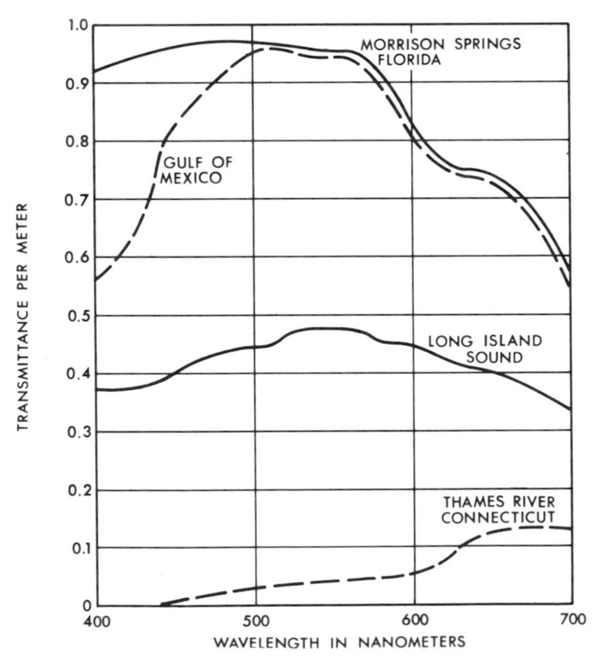

Fig. 18-1. Spectral transmittance of one meter of various bodies of water.

The curve for Morrison Springs in Florida is essentially the same as distilled water and has a maximum transmittance of over 90 per cent at 480 nanometers. The only difference between the samples from Morrison Springs and the Gulf of Mexico is a lower transmittance in the violet and blue portions, presumably due largely to plankton. The Long Island Sound water shows less transmittance throughout the spectrum, with the greatest loss in the blue and blue-green portions. Therefore, it can be said, that typical transmission curves are more symmetrical than that of distilled water. For coastal or moderately shallow water the spectrum of interest lies between about 480 to 500 nanometers.

Waters in heavily polluted areas such as the Thames River in Connecticut[1] transmit very little light at all and the shape of the curve has been completely transformed, with the greatest transmittance in the long wavelengths.

These measured differences are for a one-meter water distance. Since transmittance is related to distance in water by a power function, the wavelength absorption becomes extreme as the distance the light travels increases. Therefore, the relative visibility of different colors can be expected to vary considerably in different types of water and at different distances.

Seeing Distance in Water

Assuming sufficient light, seeing distance in water is limited by scattered light. The loss of contrast and limited seeing distance in water is like that encountered in fog, that is, a diver using artificial lighting has a seeing problem similar to that of a person driving a car in dense fog.

At short distances man or camera can see acceptably well in water with almost any lighting arrangement. At longer underwater distances the quality of underwater images received by the sensor (eye or camera) located near the light source is seriously degraded by light scattered within the volume of water which is common to both the illuminating and sensing systems; *i.e.*, that part of the line of sight which is lighted by the source. Successful long range underwater viewing requires elimination of scattering within the common volume.

Three simple techniques are used to suppress common volume scattering.

Fig. 18-2. Values of One Attenuation Length for Several Bodies of Water[2]

Location	Attenuation Length (meters)
Caribbean	8
Pacific N. Equatorial Current	12
Pacific Countercurrent	12
Pacific Equatorial Divergence	10
Pacific S. Equatorial Current	9
Gulf of Panama	6
Galapagos Islands	4

1. A light shield (septum) can be used to shield most of the line of sight from direct light from the lamp.
2. The light can be offset to the side. No part of the line of sight is close to the lamps; therefore, the inverse-square law and attenuation by absorption and scattering operate to avoid intense lighting of the water close to the sensor. Offset of the lamp may be to the rear or to the front as well as to the side. Offset to the rear offers good illuminance uniformity for large fields of view, but requires increased light source intensity to compensate for the longer distance between light source and object. Offset to the front (source between sensor and object) is effective and efficient, but requires more complex light control to achieve illuminance uniformity.
3. A third method for reducing common volume scattering is provided by crossed polarizers. Both the lamp and the sensor must have polarizers and the object must depolarize in the process of reflection in order to be visible. Absorption of

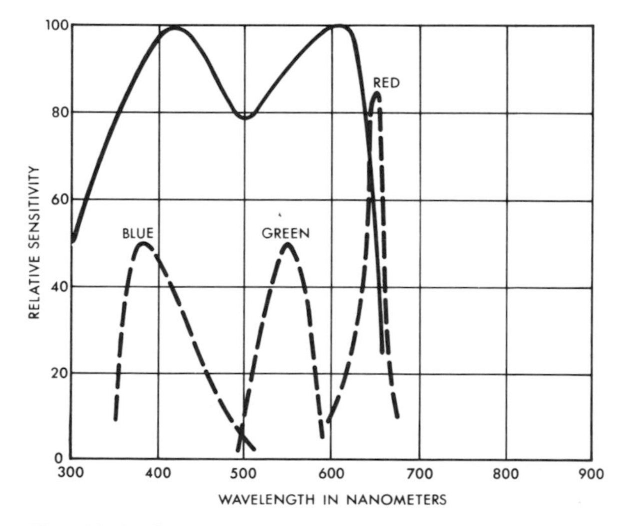

Fig. 18-3. Spectral response of black and white panchromatic negative film (solid line) and of a typical color reversal film (broken line).

light by the polarizers requires an increase in camera integration time, lens speed or sensitivity, or increased lamp output.

To simplify quantitative discussion of seeing distance in water, the concept of the *attenuation length* is often used. One attenuation length is defined as the reciprocal of the beam attenuation coefficient and is the distance in which the radiant flux is reduced to 1/e or about 37 per cent of its initial value. The attenuation length is analogous to the mean free path in physics and the time constant in electronics. Typical values of one attenuation length are shown in Fig. 18-2.[2]

Most underwater photography and viewing is done at distances of less than one attenuation length, because images can be seen with acceptable contrast through such distances in water even with the light source very close to the camera. With natural lighting and horizontal viewing a swimmer can detect a dark object at a maximum of about 4 attenuation lengths and a light object at about 4 to 5 attenuation lengths.[2] At 3 to 4 attenuation lengths good images can be obtained with the aid of the polarizers and septum techniques. Good images can be obtained at 6 attenuation lengths with the light moved near the object in combination with the septum and polarizers. With an ideally arranged combination of these aids and high contrast film developing, usable photographs have been obtained at a distance of 12 attenuation lengths. This distance corresponds to 120 meters (394 feet) in average clear ocean water (one attenuation length equaling 10 meters).

There is a limit to the underwater viewing range which becomes apparent at slightly greater distances. Even with unlimited sensor integration time, self-luminous objects, and silhouetted objects in otherwise unlighted water, objects disappear in their own forward scattered light beyond approximately 15 attenuation lengths. Contrast enhancement extends this limit only slightly.

It should be noted that extremely high attenuation of light exists at such long distances. The laboratory experiments from which the above numbers were derived were performed with stationary objects, moderate sensitivity film (ASA 160), an *f*/2 lens, and a DVY lamp (see Fig. 8-91 in the 1981 Reference Volume). At distances of 1 or 2 attenuation lengths the exposure times were small fractions of a second, but at 5 or 6 attenuation lengths several minutes were required. A time exposure for several hours was needed in order to obtain a picture at 12 attenuation lengths. Such sensor integration times are

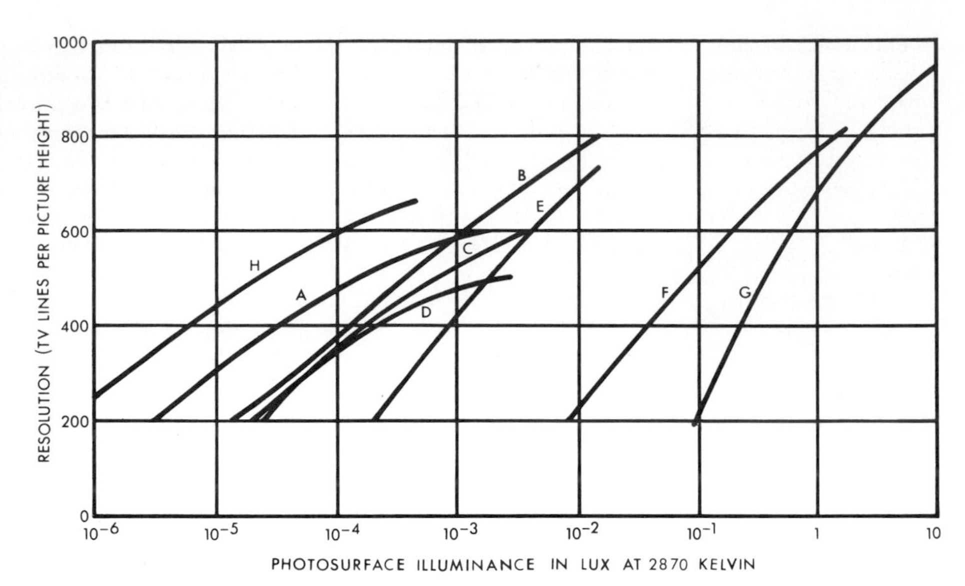

Fig. 18-4. Resolution versus illuminance for various television sensors. Scene contrast is 100 per cent for all curves. A. Forty millimeter image format Secondary Electron Conduction (SEC) tube with 40 millimeter intensifier. B. Image Orthicon. C. Twenty-five millimeter Electron Bombarded Silicon Target SEC tube. D. Twenty-five millimeter SEC tube with 25 millimeter intensifier. E. Forty millimeter SEC tube. F. Broadcast Vidicon. G. Vidicon. H. Twenty-five millimeter Electron Bombarded Silicon Target tube with 25 millimeter intensifier.

not usually feasible, but they give an idea of the amount of light (or sensor sensitivity) required for reviewing at scatter limited ranges in water.

Two other techniques to suppress scatter can be used where their complexity and cost is justifiable.

1. A synchronously scanned narrow beam source and sensor system can reduce the common volume to a minimum.
2. A system using a pulse of light of a few nanoseconds duration (a meter's distance in water) with a sensor which is gated to see only the pulse can also minimize the common volume. Lasers are capable of such pulse lengths.

Such systems are described in articles in the referenced bibliography.[3]

Sensor Characteristics

A wide variety of sensors is available for underwater use. The choice of sensor type will depend on the over-all system requirements and will have a bearing on the characteristics required of the associated underwater lighting equipment.

These sensors may be divided into two general categories: imaging, as in the case of television or film; and non-imaging, for use in instrumentation applications such as photometry and attenuation or scattering measurement.

Photographic Films. Black and white films used in underwater photography have ASA ratings ranging from 12 to 400, and with special development, ASA ratings up to 3200 can be obtained. Negative and positive color films are available with ASA ratings from 25 to 500. The spectral response of typical film are shown in Fig. 18-3.

Television Sensors. Typical characteristics of some television imaging sensors of interest to the underwater lighting system designer are shown in Figs. 18-4 through 18-7.

The Vidicon is the least sensitive as indicated by Fig. 18-4, but is small and requires relatively simple camera circuitry. The range of spectral response is shown in Fig. 18-5. These devices exhibit lag on the order of 20 to 25 per cent (3rd field) and can be damaged by overexposure.

Solid state Vidicons employing lead oxide or

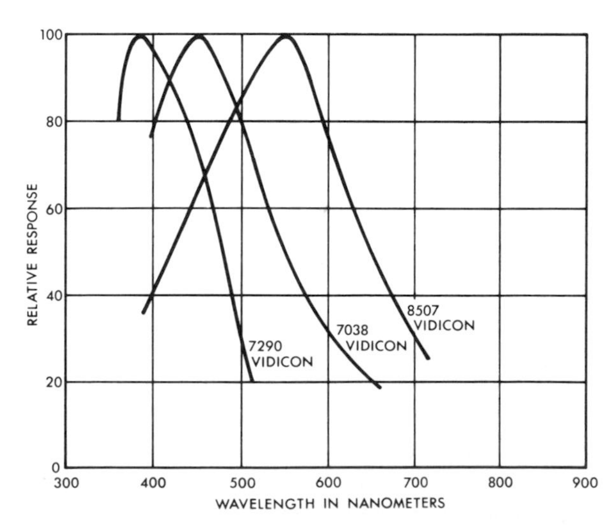

Fig. 18-5. Relative spectral response of Vidicons.

silicon diode array targets are similar to the Vidicon in size, are generally more sensitive and provide higher signal to noise ratios. Residual image or lag is less. The silicon diode target type is highly resistant to effects of overexposure. Both types have a limited dynamic range of light input. Typical absolute spectral response curves of these devices are shown in Fig. 18-6, together with curves for Vidicons used in underwater applications.

The Image Orthicon and Image Isocon are much larger than the Vidicons. These devices are sensitive and capable of high resolution. They exhibit appreciable loss in sensitivity with relative scene motion and require more complex circuitry than the Vidicon types.

The Secondary Electron Conduction or SEC Camera Tube is intermediate in size and sensitivity. It has low lag (10 per cent or less) and is least affected by relative scene motion. Later versions of this type of sensor incorporate burn resistant targets. The Electron Bombarded Silicon Target Camera Tubes (SIT-Silicon Intensified Target, EBS-Electron Bombarded Silicon, etc.) have highly burn resistant diode array targets. These tubes are intermediate in size, are more sensitive as shown in Fig. 18-4, and provide a higher signal to noise ratio than SEC Camera Tubes.

The Orthicon, Isocon, the Electron Bombarded Silicon Camera Tubes and the SEC Camera Tubes use semi-transparent photocathodes having absolute spectral response curves as shown in Fig. 18-7. Any of these devices may be fiber-optics coupled to an intensifier, with increased sensitivity and a change in absolute spectral sensitivity to that of the intensifier used.

Non-Imaging Sensors. Photodiodes and photomultipliers available for use in underwater instrumentation generally employ semi-transparent photocathodes with a spectral response similar to those shown in Fig. 18-7. Characteristics of other types of photosensitive devices are described in Section 4 in the 1981 Reference Volume. Because of the extreme variations possible in the spectral transmittance of water, it is often necessary to match the sensor spectral response to the system application or to make a number of narrowband measurements.

Light Sources for Underwater Use

Light sources used in currently available underwater lighting systems are shown in Fig. 18-8. Incandescent sources are used where instantaneous starting, simplicity and small size are

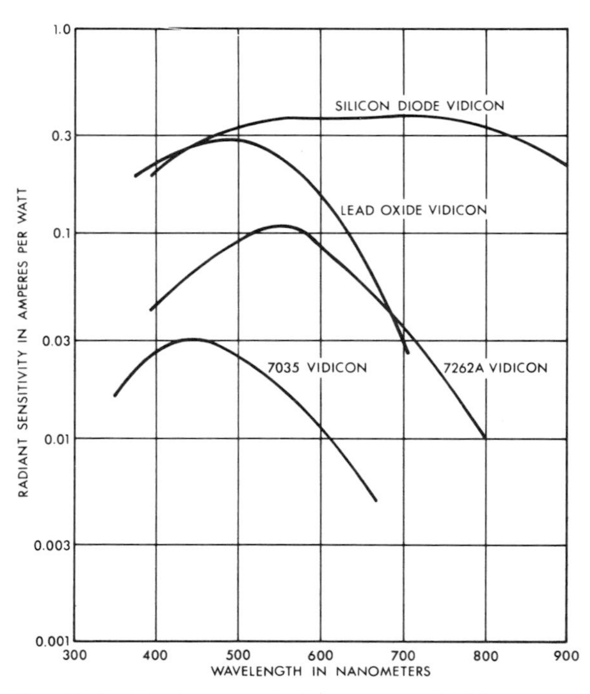

Fig. 18-6. Absolute spectral response of Vidicons and solid state Vidicons.

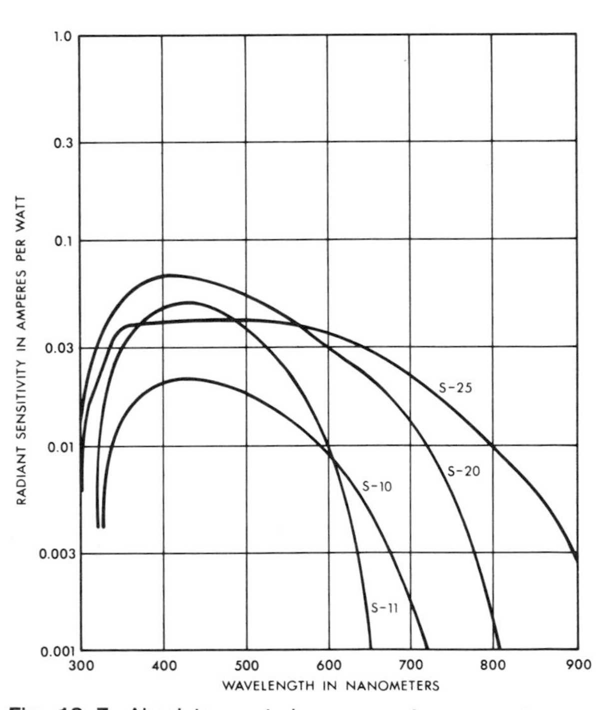

Fig. 18-7. Absolute spectral response of some semitransparent photocathodes.

Fig. 18-8. Currently Available Underwater Lighting Systems (For additional light source data, see 1981 Reference Volume, Figs. 8-91, 8-120, and 8-130.)

Light Source	Power (watts)	Spectral Power Distribution	Recommended Uses
Incandescent (Tungsten-Halogen)	75-1000	2800-3400 K Color Temperature @ 1000-25 hours life. Fig. 8-2 in 1981 Reference Volume	Vision, photography, television
Mercury	175 250	Fig. 8-21	Television, vision
Metal Halide Thallium	150 250 400	Fig. 8-21	Television, vision
Metal Halide Dysprosium Thallium	400	Fig. 8-21	Vision, photography, television
High Pressure Sodium	400	Fig. 8-21	Vision, photography, television
Xenon Flashtube	100-1000 watt-seconds	Fig. 8-66	Photography, slow scan television

requirements. Arc discharge sources are used where their higher efficacy in converting electrical power into radiant flux and longer life are requirements. Typical underwater lighting units are shown in Fig. 18-9.

In choosing light sources for underwater applications, one should consider the filtering properties of the water. Red light is rapidly absorbed by the water. Only at short ranges is it practical to attempt the careful color balance achieved routinely by color photography in air.

Although not yet available in underwater lighting equipment, other light sources applicable to underwater systems, and now found in one of a kind custom built sytems, include Scandium Sodium Metal Halide, Mercury Short Arc and Xenon Short Arc. These have been found to be effective for visual, photographic and television uses.

Electrical characteristics and power supplies for the various light sources are discussed in Section 8 in the 1981 Reference Volume. Also,

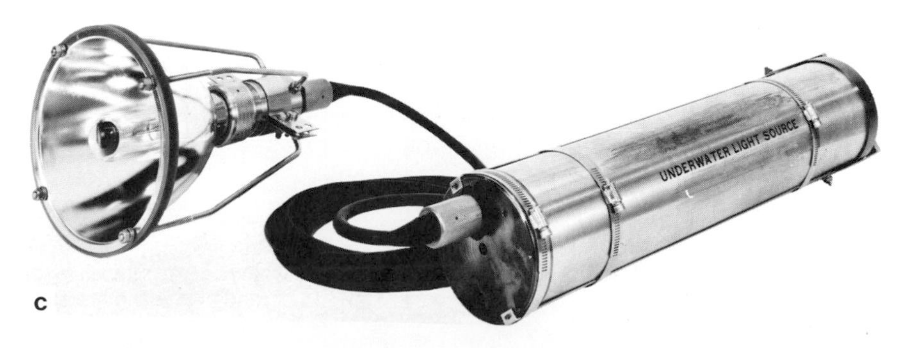

Fig. 18-9. Typical underwater lighting units: (a) incandescent, (b) arc discharge, and (c) pulsed xenon.

there are auxiliary circuits available from underwaer lighting equipment manufacturers which are specially designed for the unusual requirements of various underwater lighting applications. Information peculiar to the design of pressure housings for underwater lights has been presented.[4]

Lasers in underwater systems are discussed in several articles listed in the referenced bibliography.[3]

Underwater Lighting Calculations

The amount of light needed for an underwater task is difficult to predict with any accuracy. Even if underwater light were accurately predictable, wide variations exist in the color filtering and contrast reducing properties of water from place to place, and from day to night and season to season in the same place. For these reasons, the calculations described here are simple approximations, but they are often sufficient for engineering purposes.

Daylighting Calculations. The illuminance due to daylighting on a surface parallel to the water surface can be calculated by

$$E = E_0 e^{-Kd}$$

where E_0 is the illuminance at the surface of the water, K is the diffuse attenuation coefficient for irradiance, and d is the depth from the water surface to the illuminated surface. An observer looking at this surface would see an apparent luminance

$$L = E\rho e^{-cr_2}$$

where ρ is the reflectance of the illuminated surface, c is the beam attenuation coefficient and r_2 is the distance from the illuminated surface to the underwater observer.

Extensive studies of daylight in water have been made.[2]

Artificial Lighting Calculations. The luminance observed by an underwater viewer or camera is

$$L = I\rho\tau/r_1^2$$

where I is the source luminous intensity, ρ is object reflectance, τ is water transmittance (two way path) and r_1 is the source to object distance. The water transmittance can be approximated by

$$\tau = e^{-c(r_1+r_2)}$$

where c is the beam attenuation coefficient of the water, and r_2 is the object to sensor distance. If $r_1 + r_2$ equals one meter, this equation also defines the relation between c and τ for Fig. 18-1. Two factors must be considered in evaluating the validity of using these equations in this simple form: spectral effects (*i.e.*, c, I, ρ, and sensor response are functions of wavelength) and scattering.

A semi-empirical formula has been developed[2] to allow approximate calculation of the amount of scattered light that illuminates the object. This forward scattered component is not included in the simple expression given above for τ, but it can become quite large. The proportion of scattered light in the object illumination increases with range. At one attenuation length, the unscattered object illuminance (calculated by the simple expression given above) is nearly equal to the illuminance due to scattering. At four attenuation lengths, the scattered illumination is about ten times unscattered illumination. The increase in calculated object luminance due to scattering is partially offset by losses due to spectral filtering (see Fig. 18-1).

The value of the beam attenuation coefficient, c, used in approximate calculations is usually the minimum value with respect to wavelength (maximum value in the transmittance curves of Fig. 18-1). For photometric calculations where the photopic eye is the sensor, Fig. 18-1 shows that this choice is a reasonable approximation, because the curves are nearly flat in the region of greatest eye response. When the sensor has a spectral response much different from that of the eye or when more accurate predictions are needed, the radiance observed by the underwater sensor at each wavelength can be calculated as

$$L_\lambda = \rho_\lambda \tau_\lambda I_\lambda / r_1^2$$

where I_λ is the spectral radiant intensity of the source. The expression for τ_λ is the same as for τ except that the attenuation coefficient c is now c_λ, a function of wavelength. The response of the underwater sensor is

$$S = \int_{\lambda_1}^{\lambda_2} S_\lambda E_\lambda d\lambda$$

where S_λ is the spectral response of the sensor (see Figs. 18-3, 18-5, 18-6 and 18-7) and E_λ is the irradiance of the sensor due to L_λ, E_λ and L_λ are related by the usual imaging equation:

$$E_\lambda = \pi t_\lambda L_\lambda / 4(f\text{-number})^2$$

where t_λ is the lens spectral transmittance. Further discussion of spectral calculations can be found in the references.[5-7]

As one might expect from inspection of Fig. 18-1, the light loss due to greater absorption in the blue and red in water is usually smaller than the gain due to scattered light. This means that the simple expressions for L and τ will usually give conservative predictions; *i.e.*, they will predict the need for more light from the source than is actually required. Additional discussion of underwater lighting calculations can be found in a book by Mertens.[8] A review of recent tests and formulas for engineering calculations of underwater lighting is contained in a volume by Duntley.[9]

Measurement Techniques and Instrumentation

Performance of underwater light sources has been tested according to standard measuring techniques for light sources used in air. These data are usually supplied on the information sheets, but can be misleading because, as pointed out above, light absorption is relative to the wavelength. The photometric units often given do not apply when these light sources are used in water.

There are no established methods of instrumentation to evaluate a light source in the water environment. A number of research programs by light source manufacturers are underway but standardization is distant.

REFERENCES

1. Kinney, J. A. S., Luria, S. M. and Weitzman, D. O.: "Visibility of Colors Underwater," *J. Opt. Soc. Am.*, Vol. 57, June, 1967.
2. Duntley, S. Q.: "Light in the Sea", *J. Opt. Soc. Am.*, Vol. 53, February, 1963.
3. *Bibliography on Underwater Photography and Photogrammetry*, Kodak Pamphlet No. P-124, Eastman Kodak Company, Rochester, New York, 1972.
4. Stachiw, J. D. and Gray, K. O.: *Light Housings for Deep Submergence Applications, Parts I and II*, Naval Civil Engineering Laboratory, Report TR-532, 1967, and TR-559, 1968.
5. *Source-Detector Spectral Matching Factors*, Technical Note No. 100, October, 1966, ITT Industrial Laboratories, Fort Wayne, Indiana.
6. Biberman, L. M.: "Apples, Oranges and Unlumens", Opt. Soc. Am., Long Abstract 1967, Spring Meeting ThG 11-1.
7. Moon, P.: *The Scientific Basis of Illuminating Engineering*, McGraw-Hill Book Co., New York, 1936.
8. Mertens, L. E.: *In-Water Photography*, John Wiley & Sons, Inc., New York, 1970.
9. Duntley, S. Q.: *Underwater Lighting by Submerged Lasers and Incandescent Sources*, SIO Ref. 71-1, University of California, San Diego, Scripps Institution of Oceanography, Visibility Laboratory, 1971.
10. Smith, R. C. and Tyler, J. E.: "Transmission of Solar Radiation into Natural Waters," *Photochem. Photobiol. Rev.*, Vol. 1, 1976, Plenum Press, New York.
11. Austin, R. W., Petzold, T. J., Smith, R. C. and Tyler, J. E.: *Handbook of Underwater Optical Measurements*, National Oceanographic and Atmospheric Administration, Washington, DC, 1980.

Nonvisual Effects of Radiant Energy

SECTION 19

Life has evolved under the influence of radiation from the sun and, as a result, man, animals and plants have developed a variety of complex physiological responses to solar radiation, and its daily and seasonal variation, for a variety of biological functions. The spectral power distribution of solar radiation at sea level is shown in Fig. 19-1.

The study of the interaction of biological systems to radiant energy in the ultraviolet, visible and infrared portions of the electromagnetic spectrum is known as *photobiology*. Photobiological responses result from the absorption of radiation by particular molecules in the living organism. The absorbed radiation produces excited states in these molecules which may lead to photochemical reactions of biological consequence. The distinguishing feature of photochemical reactions is that the energy of activation is provided by non-ionizing photons which cause reactions to occur at low or physiological temperatures.

Human vision is discussed in Section 3 of the 1981 Reference Volume. Extra-visual photobiological effects occurring in humans, animals, microorganisms and plants, and non-biological effect on matter are covered in this section.

EFFECTS ON HUMANS AND ANIMALS (PHOTOBIOLOGY)

The effects of solar radiation on humans and animals include such wide ranging phenomena as damage to ocular tissues, and such skin effects as production of Vitamin D_3, erythema, pigmentation, sensitized and allergic reactions, toxicity, and tumor and cancer formation. These effects are generally attributed to exposure to energy in the ultraviolet spectral regions.

NOTE: References are listed at the end of each section.

A variety of diseases are and have been treated with ultraviolet energy alone or in combination with sensitizing drugs. Since the beginning of recorded history psoralens and exposure to solar ultraviolet radiation has been the therapy for vitiligo. Lupus vulgaris was shown to be cured with either ultraviolet from sunlight or carbon arcs by the turn of the century. Psoriasis is now being alleviated with the same therapy which has been applied to vitiligo, using sources of more constant ultraviolet output than from the sun.

Light in the visible region, particularly in the blue (400 to 500 nanometers), is used in the phototherapy of jaundiced infants.

Other extra-visual effects include circadian rhythm and neuroendocrine responses. These are responses that are mediated through the visual system but are responses separate from vision.

Effects on the Eye[1-14]

For the discussion here the radiant energy spectrum is divided into three components: ultraviolet, 200 to 400 nanometers; visible and near infrared, 400 to 1400 nanometers; and infrared, 1400 nanometers to 1 millimeter. Fig. 19-2 summarizes in abbreviated form the over-all effects of radiation as a function of each wavelength band, and indicates that ultraviolet effects include the adverse effects of erythema and photosensitization upon the skin and photokeratitis of the eye, as well as phototherapy and other beneficial effects. The visible and near-infrared effects are confined for the most part to the retina, while the lens is a strong absorber of wavelengths shorter than 400 nanometers. Infrared of wavelengths greater than 1400 nanometers mainly affect the cornea and the lens.

Ultraviolet Radiation Effects. The biological effects of ultraviolet radiation depend upon both the absorption by and the photochemical

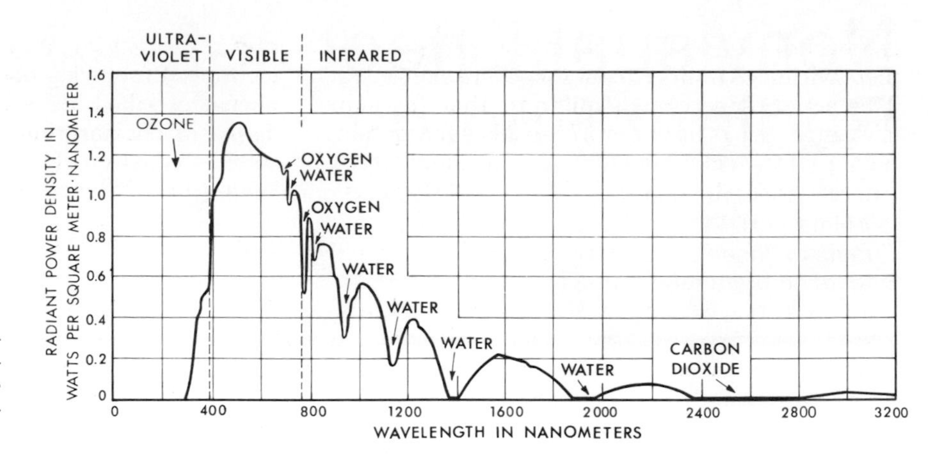

Fig. 19–1. Spectral distribution of solar radiant power density at sea level, showing the ozone, oxygen, water, and carbon dioxide absorption bands.

sensitivity of the various components of the ocular media as a function of wavelength. Fig. 19–3 shows that for wavelengths less than 320 nanometers, nearly all of the radiation is absorbed in the cornea. Between 320 and 400 nanometers, a large proportion of the ultraviolet radiation is absorbed in the lens.

Corenal photokeratitis is a quite painful, but not necessarily serious, inflammation involving the epithelial (outermost) layer of the cornea. The period of latency between exposure and first effects varies from 2 to 8 hours, depending upon the radiant exposure. For moderate exposures the effects are more frightening than serious. The symptoms include conjunctivitis accompanied by an erythema of the surrounding skin and eyelids. There is a sensation of "sand" in the eyes, lacrimation, photophobia and twitching of

Fig. 19–2. Physiological Effects or Applications of Ultraviolet, Visible and Near-Infrared and Infrared Radiation

Effects or Applications	Ultraviolet (200–400 nanometers)	Visible, Near-Infrared (400–1,400 nanometers)	Infrared (over 1,400 nanometers)
Skin	Erythema Carcinogenesis Aging Drug Photosensitivity Melanoma*	Burns Drug Photosensitivity	Burns
Eye			
Cornea	Photokeratitis		Burns, Shocks
Lens	Cataracts (Immediate and long term) Coloration Sclerosis	Near-Infrared Cataracts	Infrared Cataracts
Retina	Aphakics*	Thermal Lesion Photochemical Lesion Shock Lesion Solar Retinitis Macula Degeneration* Loss of Visual Acuity*	
Phototherapy	Psoriasis Herpes Simplex Dentistry	Retinal Detachment Diabetic Retinopathy Bilirubinemia Glaucoma Removal of Port Wine Stains and Tattoos Surgery	
Benefits	Vitamin D Protective Pigmentation	Biological Rhythms Hormonal Activity Behavior*	Radiant Heating

* Extent of effects unknown at this time.

the eyelids. Recovery is rapid and usually complete within 48 hours except for severe exposures. The action spectrum, similar to that for skin erythema, peaks at about 270 to 280 nanometers, falling off to negligible values at 320 nanometers. Recent research, however, shows a shift in the peak to 270 nanometers.

Lenticular effects of ultraviolet radiation have been recently undergoing extensive investigation. The lens undergoes a number of changes with aging, including a yellowing coloration, an increasing proportion of insoluble proteins, sclerosis with loss of accommodation, and cataract, and there is a growing body of evidence, mostly epidemiological, to implicate the ultraviolet radiation in these changes. For example, cataract extractions are significantly greater in India than in Western Europe. Part of the difference may be due to diet and genetic factors, but most authorities believe that exposure to sunlight plays an important role. There is no hard evidence at the present time to link lenticular changes with chronic exposure to sunlight, but some investigators are coming to believe that all wavelengths below 400 nanometers should be excluded, where possible, from the eye, especially since these wavelengths can be easily eliminated by inexpensive sunglasses and contribute nothing to visual perception.

Retinal effects of ultraviolet radiation are problematical, depending on whether enough photons reach the retina to produce an observable effect. Under normal circumstances, the ocular media (cornea, aqueous, lens and vitreous) protect the retina from ultraviolet exposure. In the near ultraviolet (320 to 400 nanometers) the lens effectively shields the retina.

Visible and Near Infrared Radiation Effects. Retinal burns resulting in a loss of vision (scotoma) following observation of the sun have been described throughout history. The incidence of chorioretinal injuries from man-made sources is extremely rare and is no doubt far less than the incidence of eclipse blindness. Until recently it was felt that chorioretinal burns would not occur from exposure to light in industrial operations. Indeed, this is still largely true since the normal aversion to high brightness light sources (the blink reflex and movement of the eyes away from the source) provide adequate protection unless the exposure is hazardous within the duration of the blink reflex. However, the recent revolution in optical technology forged principally by the invention of the laser, has meant a great increase in the use of high-intensity, high-radiance optical radiation sources. Many such sources have output parameters significantly different from those encountered in the past and may present serious chorioretinal burn hazards. In industry, besides lasers, one may encounter sources of continuous optical radiation, such as: compact arc lamps (as in solar simulators), tungsten-halogen lamps, gas and vapor discharge tubes, electric welding units, and sources of pulsed optical radiation, such as flash lamps used in laser research and photochemical investigations, exploding wires, etc. These sources may be of concern when adequate protective measures are not being taken. While ultraviolet radiation from most of these devices must be considered—and may be the principal concern—the potential for chorioretinal injury should not be overlooked. The evaluation of potential retinal hazards from extended sources can be more involved than for collimated sources such as intrabeam laser exposure.

To place chorioretinal injury data in perspec-

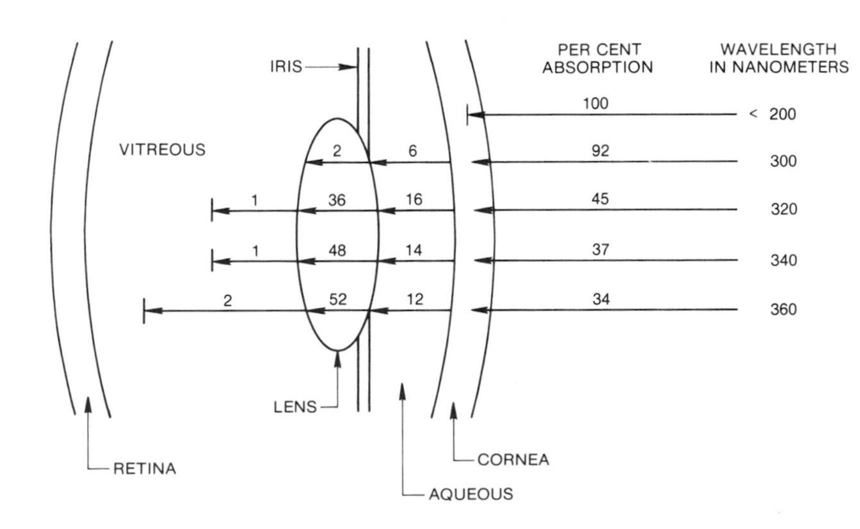

Fig. 19-3. Per cent of energy on surface of cornea absorbed by various layers, after data by Boettner, E.A., and Walter, J.R., MRL-TDR-62.31.

tive, Fig. 19-4 shows the retinal irradiance for many continuous wave sources.[11] It is re-emphasized that several orders of magnitude in radiance or luminance exist between sources which cause chorioretinal burns and those levels to which individuals are continuously exposed. The retinal irradiances shown are only approximate and assure minimal pupil sizes and some squinting for all the very high luminance sources, except the xenon searchlight for which a 7-millimeter pupil was assumed (so as to apply to searchlight turn-on at night).

Light entering the cornea passes through the anterior chamber or aqueous, the lens, the vitreous humor and impinges on the retina (See Fig. 3-1 in the 1981 Reference Volume). This pathlength is the ocular media (OM). Transmittance through the OM is shown in Fig. 3-2 in the 1981 Reference Volume. Examination of those curves shows that between 400 and 1400 nanometers the retina is vulnerable to radiation effects. Between these wavelengths the retina is by far the most sensitive tissue of the body.

In the retina light passes through five layers of neural cells before encountering the photoreceptor cells (the rods and cones) which are the transducers converting absorbed photons of light into electrical impulses to be sent to the brain via the optic nerve tract. Just behind the rods and cones is a single layer of heavily pigmented cells, the pigment epithelium, which absorbs a large portion of the light passing through the neural retina. The pigment epithelium acts like a dark curtain to absorb and prevent backscatter for those photons which are not absorbed in the outer segments of the rods and cones. The neural retina is almost transparent to light. The pigment epithelium is about 10 micrometers in thickness, while the choroid—a layer of blood vessels—behind it ranges from 100 to 200 micrometers. Most of the energy in the form of light which reaches the retina is converted to heat by the pigment epithelium and the choroid. The heat then can cause thermal damage to the retina. Also, chemical damage could occur depending on the total energy of the exposure or the energy per pulse and the time interval between pulses.

Over the past decade, a number of researchers have demonstrated damage to the retina from long term photic insult. For example, when rats and mice are subjected to cool white fluorescent lighting for extended periods of time (weeks to months) they become blind. Histological examination reveals that the photoreceptors in the retinae of these animals have vanished. Similar effects, though not so drastic, have been achieved in laboratory conditions in other species includ-

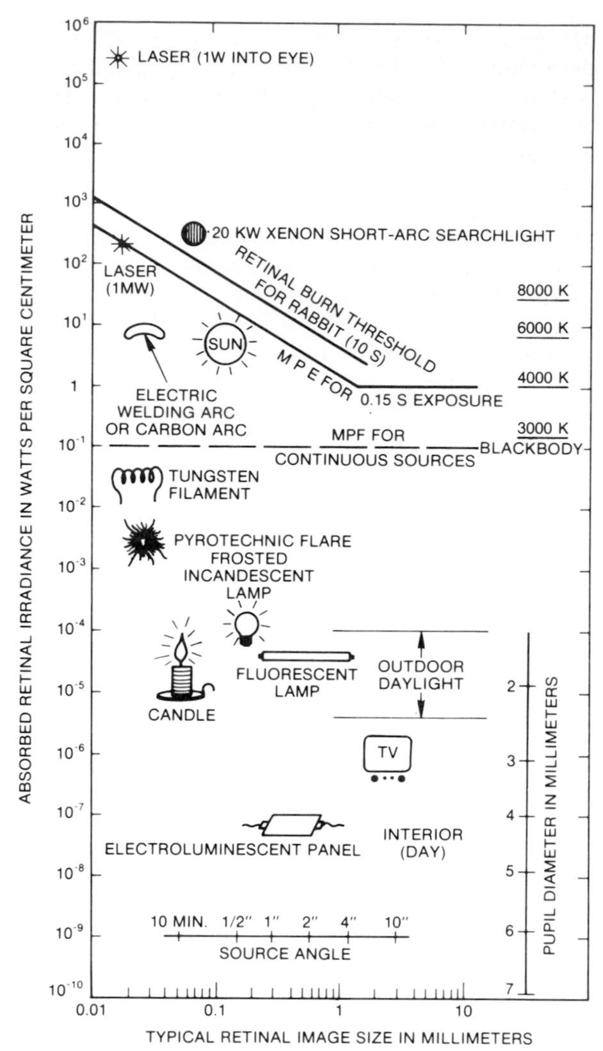

Fig. 19-4. The eye is exposed to light sources having radiances varying from about $10^4 W \cdot cm^{-2} \cdot sr^{-1}$ to $10^{-6} W \cdot cm^{-2} \cdot sr^{-1}$ and less. The resulting retinal irradiances vary from about 200 $W \cdot cm^{-2}$ down to $10^{-7} W \cdot cm^{-2}$ and even lower; retinal irradiances are shown for typical image sizes for several sources. A minimal pupil size was assumed for intense sources, except for searchlight. The retinal burn threshold for a 10-second exposure of the rabbit retina is shown as upper solid line. The maximum permissible exposure (MPE) applied by the U.S. Army Environmental Hygiene Agency in evaluating momentary viewing of continuous wave light sources is shown as lower solid line. Approximate pupil sizes are shown at lower left based upon exposure of most of the retina to light of the given irradiance.

ing pigs and monkeys. Fortunately, humans do not stare for hours directly into banks of lamps.

Retinal Photocoagulation. A therapeutic effect of 400- to 1400-nanometer radiation on the retina involves photocoagulation techniques to repair retinal detachment by both incoherent and coherent (laser) radiation. The original coagulation process involving the "welding" of the detached retina to the sclera was accomplished with incoherent white light from the xenon lamp coagulator. The coagulator has been superseded in many ophthalmological clinics by ruby and argon laser coagulators. Today, the photocoagulation technique in ophthalmology has been applied to the treatment of diabetic retinopathy, glaucoma and many other pathologies involving the eye.

Research over the past decade has demonstrated that in the range of wavelengths between 400 and 1400 nanometers, there are at least three different mechanisms leading to retinal damage. These are:

1. Mechanical or "shock-wave" damage from picosecond and nanosecond pulses of mode-locked or Q-switched lasers.
2. Thermal damage for pulse durations extending from microseconds to seconds. Except for minor variations in transmittance through the ocular media and variations of absorptance in the pigmented epithelium and choroid, thermal damage is not strongly wavelength dependent.
3. Photochemical damage from exposure to short wavelengths in the visible spectrum for time durations and power densities on the retina which preclude thermal effects. Photochemical damage is wavelength dependent.

There is no abrupt transition from one type of damage to the other in terms of exposure time and wavelength. For example, a Nd: YAG laser emitting a pulse train of Q-switched pulses at 1064 nanometers wavelength may produce a combination of shock wave and thermal damage depending on the pulse width in nanoseconds and the time interval between pulses, whereas an acoustically modulated pulse train from an argon ion laser emitting 10 microsecond pulses of 488 nanometers radiation might produce a combination of thermal and photochemical damage.

Infrared Radiation Effects. Wavelengths greater than 1400 nanometers do not involve the retina, but can produce effects on the eye leading to corneal and lenticular damage as well as thermal damage to the skin. Infrared cataract has been reported in the literature for a long time, but there is little or no recent quantitative data to substantiate the clinical observations. The general consensus is that long term exposure to infrared radiation produces an elevated temperature in the lens which, over a period of years, leads to denaturation of the lens proteins with consequent opacification. Some authorities believe that infrared radiation is absorbed by the pigmented iris and converted to heat which is conducted to the lens, rather than by direct absorption of radiation in the lens. Infrared cataract is reported to occur among glassblowers, steel puddlers and others who undergo long term occupational exposure to infrared radiation. Present industrial practices have virtually eliminated this effect in workers of today.

Skin Effects of Radiation[15-23]

The only known benefit of ultraviolet radiation of skin is the production of vitamin D from precursor chemicals which are formed in the skin (see below). Known harmful effects include sunburn and skin cancer and morphologic alterations (wrinkling, irregularity, altered pigmentation, thinning and thickening of skin) which appear as "premature aging." Delayed tanning and increased thickening of skin is a protective response initiated by ultraviolet radiation. The function of immediate tanning is uncertain.

Optical Properties of the Skin. The reflectance of skin for wavelengths shorter than 320 nanometers is low, regardless of skin color; however, from 320 to 750 nanometers reflectance is dependent upon skin pigmentation. The transmission of ultraviolet radiation through the skin depends on wavelength, skin color (melanin content) and skin thickness. In general, transmission increases with increasing wavelength from 280 to 1200 nanometers. Typically for Caucasians, transmittance at the base of the top layer of skin (stratum corneum) is 34 per cent at 300 nanometers and 80 per cent at 400 nanometers.[15] Transmission decreases with increasing melanin content of the skin and with increasing skin thickness. See Fig. 19-5 and 19-6.[23]

The genetically determined skin color is the result of a number of factors, but the primary factor is melanin—its quantity, granule size and distribution all affect skin color. Ultraviolet darkens the basic skin color by affecting the melanin. The immediate *tanning* that occurs with exposure to ultraviolet radiation of wavelengths longer than 300 nanometers and extending into the visible blue region is the *darkening* of existing melanin. Delayed tanning results from ultraviolet stimulation of the melanin producing cells (the melanocytes) to produce additional melanin.[18] Pigmentation from this process begins immediately at the subcellular level and gross changes in skin color can be observed three days after an ultraviolet irradiation and reaches a peak about 1 to 3 weeks later. Fading requires months, as melanin is lost during the normal shedding process.

Melanin protects against ultraviolet damage by reducing ultraviolet transmission through absorption and by scattering which increases effective path length.

Erythema. The delayed reddening (erythema) of the skin caused by exposure to ultraviolet radiation is a widely observed phenomenon. The spectral efficiency of this process, particularly for sunlight wavelengths from 290 to 320 nanometers, has been well studied and there is general agreement as to that action spectrum (see Fig. 19–7 and Fig. 19–12).

The erythema action spectrum of wavelengths shorter than 300 nanometers varies considerably between observers because of the differences in the degree of erythema taken as the endpoint criterion, and in the time of observation after irradiation.[19-21]

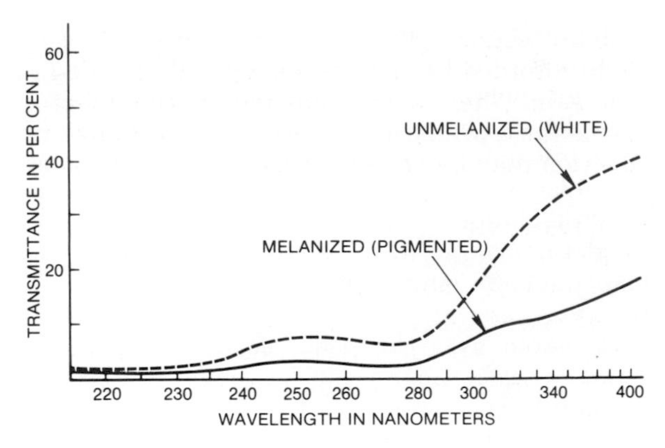

Fig. 19-6. Light-transmission spectra of human epidermis of equal thickness (12 micrometers) obtained from the infrascapular region of two subjects: a fair-skinned Caucasian and a pigmented (medium color) Negro.

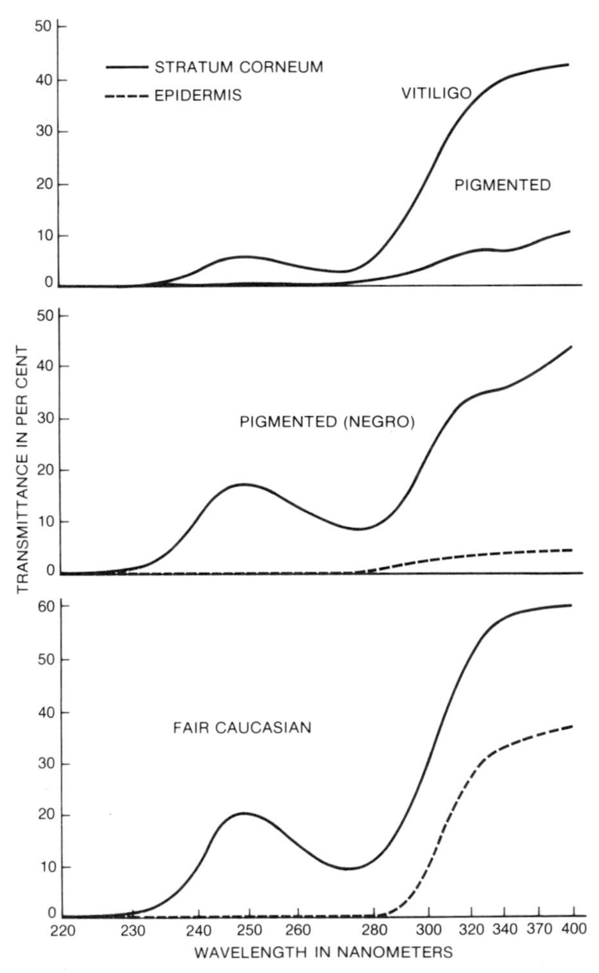

Fig. 19-5. Light-transmission spectra of stratum corneum and epidermis from the scapular region of three subjects: (1) a pigmented Negro with vitiligo (pigmented and nonpigmented stratum corneum); (2) a pigmented Negro (stratum corneum and whole epidermis); and (3) a fair-skinned Caucasian.

Erythema is a component of inflammation of the skin and results from increased blood volume in superficial cutaneous blood vessels. The skin may also be warm and tender.

Approximately 18 millijoules per square centimeter of energy at most effective wavelength (297 nanometers) will cause a barely perceptible reddening in 50 per cent of all Caucasians. This amount of effective energy requires about 12 minutes under an overhead sun and the thin stratospheric ozone layer found in the tropics. When the sun is 20 degrees from the zenith and the ozone layer thickness is about 3.2 millimeters, 20 minutes is typically required for the same degree of reddening.

Exposure to ultraviolet radiation may result in an immediate erythema, particularly at high irradiance levels. This may fade a few minutes after irradiation ceases, and may reappear in from 1 to 3 hours. The greater the dose, the faster the reappearance, and the longer the persistence of erythema.

If the erythema is severe enough, skin peeling (desquamation) will begin about 10 days after irradiation. This rapid sloughing off of the top skin layer results from the increased proliferation of skin cells during recovery after ultraviolet damage. Desquamation counteracts the skin's protective response to ultraviolet radiation for the epidermis does not thicken and the rapid sloughing carries away some of the ultraviolet radiation stimulated melanin granules.

Photoprotection in its common usage refers to

the protection against the detrimental effects of light afforded by sunscreens topically applied to the skin. These sunscreens reduce the effect of ultraviolet, primarily by absorption, although reflection may be of some consequence. Considerable progress has been achieved in recent years in developing sunscreens which are effective and relatively resistant to being washed away by sweating or swimming. Paraaminobenzoic acid in an alcohol base has proven quite effective in preventing sunburn. Other materials in use include benzophenone, cinoxate and sublisobenzone.

Skin Cancer. The three varieties of skin cancer are basal cell, squamous cell and malignant melanoma. The frequency of occurrence is in the order stated, basal cell cancer being the most common; its percentage of the total frequency increasing as the latitude increases. Both basal and squamous cell cancer correlate with solar ultraviolet exposure. These cancers have a very high cure rate if treated promptly. Melanomas are considerably rarer, have a poorer cure rate and have a poorer correlation with ultraviolet exposure.

Effects on Vitamin D and Calcium Metabolism[24-28]

Ultraviolet radiation plays an important role in the production of vitamin D in the skin. This vitamin is essential for normal intestinal absorption of calcium and phosphorus from the diet, and for normal mineralization of bone. Vitamin D deficiency leads to a deficiency of calcium and phosphorus in the bones so that they bend, fracture or become painful. Vitamin D poisoning, on the other hand, leads to excessive absorption of calcium and phosphorus from the diet, and a toxic effect on the skeleton. There is a resultant increase in the calcium concentration in the blood, and precipitation of calcium-phosphate deposits in vital organs, with permanent damage or even death. Vitamin D poisoning also causes increased excretion of calcium in the urine, which can produce kidney stones or bladder stones. When vitamin D poisoning is very mild, the increased urinary calcium excretion may be the only medically important abnormality.

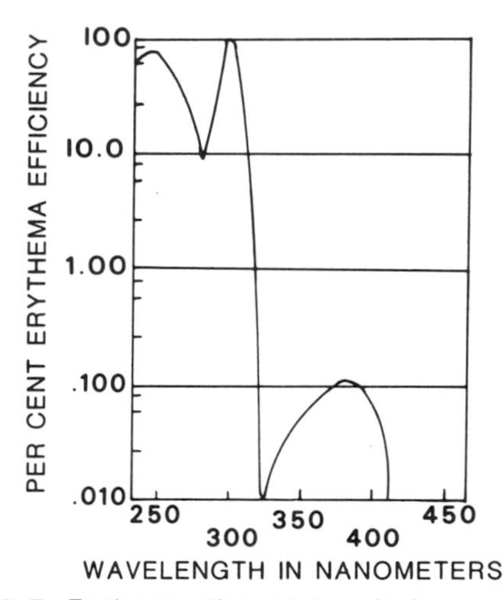

Fig. 19-7. Erythema action spectrum for human skin. Note logarithmic vertical axis. (Modified from Hausser, K. W. and Vahle, W., Sunburn and Suntanning—In *The Biologic Effects of Ultraviolet Radiation with Emphasis on the Skin*. F. Urbach, Editor, Pergamon Press, Oxford, 1969, pp 3-21).

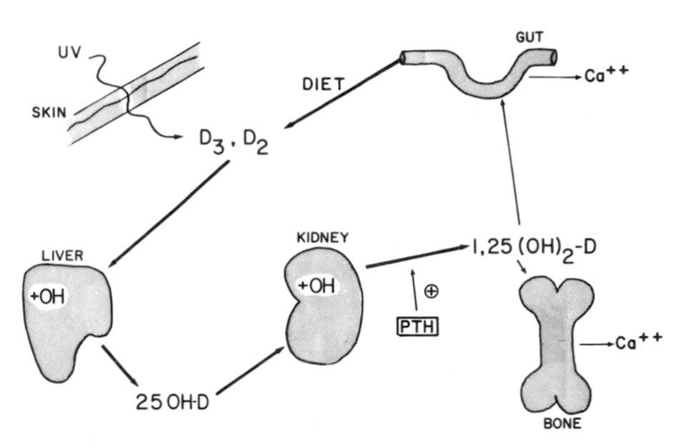

Fig. 19-8. Activation sequence for vitamin D. After synthesis in the skin or absorption from the diet, vitamin D is first converted to 25-hydroxy vitamin D (25-OHD) in the liver and then converted in the kidneys to 1,25-dihydroxy vitamin D (1,25-$(OH)_2$ D), the active D-metabolite which influences intestinal calcium absorption and bone calcium release. The amount of 1,25-dihydroxy vitamin D formed in the kidneys is metabolically regulated by parathyroid hormone (PTH) and a variety of other factors to meet the body's varying needs for calcium and/or phosphorus.

Vitamin D Metabolism. Vitamin D formed in the skin (or absorbed from the diet) is inert until carried by the blood to other organs for activation. Activation involves two sequential chemical changes in the vitamin D molecule (see Fig. 19-8). Only then is the vitamin D fully activated and able to exert its characteristic effects on the intestines and bones.

The activation of vitamin D is regulated to meet the body's needs for calcium and phosphorus. A deficiency of calcium or phosphorus leads to increased production of 1,25-$(OH)_2$D. Conversely, a surfeit of calcium leads to decreased production of 1,25-$(OH)_2$D (see Fig. 19-8). This controlled activation also smooths out the effects of variations in vitamin D intake in the diet, or variations in vitamin D synthesis in the skin. If the supply of 25-OHD (Fig. 19-6) or its parent vitamin D becomes low, so that calcium or phosphorus deficiency begins to develop, the body increases the efficiency of vitamin D activation, assuring maximal effectiveness of whatever vitamin D or 25-OHD is available. On the other hand, when vitamin D excess threatens, the beginning accumulation of excess calcium rapidly halts further activation of the vitamin.

Vitamin D Photoproduction in Skin. Vitamin D is produced photochemically in the skin from 7-dehydrocholesterol, an intermediate in cholesterol biosynthesis. Ultraviolet radiation absorbed by the 7-dehydrocholesterol molecule rapidly converts it to pre-vitamin D. See Fig. 19-9. Pre-vitamin D then isomerizes to vitamin D. This second step does not involve light absorption and occurs slowly over some hours. Vitamin D is thus formed and released into the circulating blood slowly after ultraviolet irradiation of skin, even though the initial photochemical reaction is rapid. The biochemistry of these reactions is currently being investigated in detail in the skin.

Studies in animals and humans clearly show that ultraviolet radiation in the range 250 to 315 nanometers is effective in producing vitamin D in the skin. The action spectrum for this effect has never been directly determined, but indirect studies in animals suggest there is a peak of effectiveness near 297 nanometers. The layer of the skin in which the photochemical reaction occurs is also unknown, and postulated effects of skin thickness and skin pigmentation on the reaction are largely conjectures. The concentration of 7-dehydrocholesterol in human skin is so high that only the availability of ultraviolet irradiation limits the amount of vitamin D biosynthesized there. Consequently, blood levels of 25-OHD in healthy adults are linearly related to the availability of ultraviolet radiation of wavelengths shorter than 315 nanometers, and are higher in summer than in winter, higher at higher altitudes and more tropical latitudes, and higher in people who work outdoors than in those who work indoors.

Light vs. Diet as Sources of Vitamin D. The total amount of vitamin D endogenously synthesized in the skin has not yet been determined in humans. Natural foods contain little vitamin D, with the exception of certain seafoods. To reduce the dependence on environmental radiation, dairy products, cereals and certain other foods are fortified with vitamin D in the United States (USA) and some European countries. In some areas of the USA, the fortification of dairy products is accomplished with vitamin D_2, which is biologically similar to but chemically distinct from the vitamin D_3 produced in the skin. Measurements of blood 25-OHD_2 levels in healthy adults in those areas indicate that dietary vitamin D_2 provides half their total vitamin D supply. The remaining one-half comes from endogenously synthesized vitamin D_3 (plus the small amounts of vitamin D_3 naturally present in foods or added as a fortification to non-dairy foods in these areas).

The incremental vitamin D provided by diet fortification has virtually eliminated childhood vitamin D deficiency in areas where it was formerly a common and serious public health problem, such as countries in high latitudes. Since many adults do not eat vitamin D-fortified foods, they remain relatively dependent on cutaneous synthesis to meet their requirements for vitamin D. The current recommended dietary allowance for vitamin D in adults in the United States is 10 micrograms per day, or 400 International Units. The total amount of vitamin D endogenously synthesized in the skin has not yet been determined, but presumably determines how necessary such dietary vitamin D supplements are in adults. The exact frequency of vitamin D deficiency in USA adults is thought to be quite low (with the possible exception of housebound or institutionalized adults) but this has not been investigated very thoroughly.

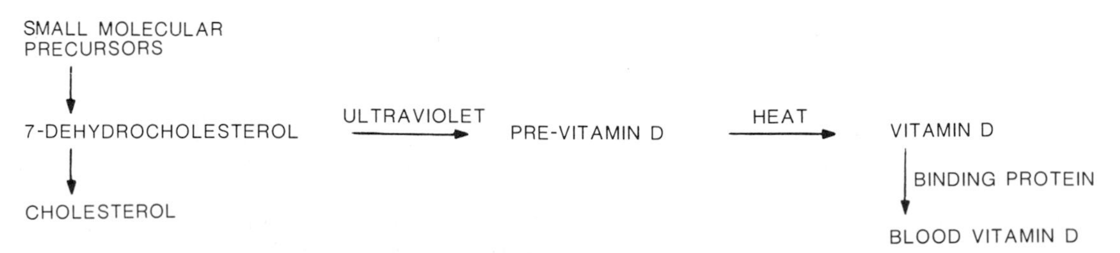

Fig. 19-9. Steps in the photoproduction of vitamin D in the blood.

Heavy exposures to ultraviolet radiation or very large dietary vitamin D supplements result in very high blood levels of vitamin D and 25-OHD, and increased excretion of calcium in the urine. However, the body's regulatory mechanisms generally prevent severe vitamin D poisoning, by shutting down the activation process in the kidneys and thereby preventing the accumulation of poisonous levels of fully active 1,25-$(OH)_2D$. To overwhelm this regulatory system usually requires ingestion of over 1000 micrograms (40,000 International Units) of vitamin D daily for some time. Such severe vitamin D overdosage does not occur from sun exposure in healthy people eating normal diets, but may occur in people with certain rare diseases in which the body's regulatory mechanisms are disturbed.

Biological Rhythms[29-38]

Changes in biological activity which repeat on a regular, cyclic basis exist in organisms throughout the plant and animal kingdoms. They are known as biological rhythms and occur at both multicellular and cellular levels. A variety of biological rhythms are known.[29-32] They vary in period, or length of time necessary to complete one cycle of activity, and in phase, *i.e.*, that particular part of the cycle when the activity takes place. For example, some biological rhythms run on a yearly schedule such as the sprouting of plant seeds, the blooming of flowers, the hibernation of animals, or the migration and sexual development of birds and mammals. Some rhythms occur on a monthly or semi-monthly basis, *e.g.*, breeding cycles in fish and the menstrual cycle in human females.

Circadian Rhythms. Many biological rhythms occur daily and are based on a day-night, light-dark cycle produced by the rotation of the earth on its axis relative to the sun. They have a period of approximately 24 hours and are called circadian rhythms. For example, certain plants, such as bean seedlings, raise their leaves perpendicular to illumination from the sun during the daytime and lower them to the sides of their stems at night. This effect is known as the *sleep-movement* rhythm. In animals and humans, daily fluctuations exist in physical activity, sleep, eating, body temperature, the rates of glandular secretion of certain hormones, and other metabolic processes. Light plays a role in influencing or controlling some of these rhythms. It is not always clear, however, which circadian rhythms are actually induced by light and which are simply synchronized by light.[31]

Many circadian rhythms appear to continue even when the organisms are placed under constant conditions where they are shielded from environmental light-dark cycles. Although the rhythms persist for several days, their period usually becomes slightly longer or shorter than 24 hours.[29] For example, the sleep-movement rhythm in bean seedlings persists for several days under conditions of continuous, dim illumination of visible light. Under such conditions the period of their rhythm increases from 24 to about 27 hours.[32]

Since circadian rhythms persist even in the absence of natural day-night changes in the environment, organisms are believed to possess their own *timing system*, which *sets* the schedule of the rhythms. The popular term for this timing system is "biological clock."[29-31] Little is known about the exact nature or location of biological clocks in organisms. Some scientists believe that organisms have their own built-in timing systems, and others suggest that the clocks are set by the earth's electromagnetic or gravitational forces.[29]

Altering the Rhythms. The schedule of most circadian rhythms in plants and animals can be changed by modifying the cycles of environmental light. For example, mice, which are normally active at night, quickly reset their time of activity when they are placed under a different, artificial visible light-dark cycle in the laboratory. Likewise, plants reverse the phase of their sleep-movement rhythms in a laboratory where they are exposed to light at night and darkness during the day. They raise their leaves during times of illumination and lower them in the dark. This synchronization of an organism's biological rhythms with a new environmental light-dark cycle is called *entrainment.* Other environmental factors may entrain rhythms as well; in fact, some rhythms are insensitive to light cycles. Of those which do respond to light, relatively little is known about what illuminance levels or wavelengths are needed to entrain them. Studies on entraining the daily rhythm in body temperature of rats showed that green light was most effective and that wavelengths in the ultraviolet, blue, yellow and red were less effective.[31, 33] Additional investigation demonstrated that the resynchronization effect was associated with absorption of the green light by rhodopsin, a photosensitive pigment in the retina.[31]

The period of some circadian rhythms can be

increased by exposing organisms in the laboratory to unusually long cycles (*e.g.*, 14 hours of light followed by 14 hours of darkness) or decreased by exposing them to short cycles (*e.g.*, 9 hours of light and 9 hours of dark). Such imposed, unnatural cycles do not seem to have a lasting effect on the rhythm; when the organisms are placed under constant darkness, the period of their rhythms quickly returns to about 24 hours.[29] When extreme cycles are imposed, such as 5 hours of light and 5 hours of darkness, the organisms ignore such cycles and continue to display their natural circadian rhythms.

There are a number of human body processes which change rhythmically, some of which are associated with the 24-hour day-night cycle.[31] Cells, tissues, organs and nervous system are all coordinated with each other and may be sensitive to environmental light-dark cycles.

Human body temperature varies about one °C [2°F] during a 24-hour period; it is lowest during the night rest period and is highest during the day. Certain compounds in the blood plasma vary in amount with a 24-hour rhythm. For example, the concentration of cortisol, a hormone produced by the adrenal cortex, rises during the night (reaching a maximum on awakening) and falls gradually during the day. Studies have shown that blindness upsets this rhythm.[31] The urinary excretion of various compounds such as magnesium, sodium, potassium, and calcium also varies according to a circadium rhythm. This variation, however may relate to feeding patterns as well as to lighting.

The control of photoperiodicity is utilized in the commercial production of decorative plants (see page 19-25) and livestock including a variety of animals[37] and poultry.[38]

Human Biological Rhythms. It is well known that many travelers experience "jet-lag" for several days after being flown east or west where they are subjected to a new day-night cycle. Such shifts in the sleep-waking and eating schedule cause numerous physiological rhythms in the body to be desynchronized from the old schedule and resynchronized to the new. These readjustments in body rhythms probably contribute to the feeling of discomfort.

Environmental lighting levels and cycles influence maturation and responses of gonads in mammals. It has been shown that the absence of light influences the time of onset of puberty. In a clinical study it was observed that sexual maturation (age at which the first menstrual period occurred) of a group of about 300 blind girls occurred several months earlier than in a control group of normal girls.[34]

Some effects of environmental light on humans are produced *directly* by photochemical reactions in the skin, including the stimulation of vitamin D, skin tanning and burning. Skin aging and cancer production are associated directly with exposure to daylight. Those effects of light associated with generating or entraining biological rhythms are *indirect*. There is evidence that such indirect effects are produced via photoreceptors in the retina of the eye.[31] These photoreceptors respond by generating nerve impulses which are transmitted to neuroendocrine glands and organs in the body, which in turn produce metabolic changes through secretion of hormones or other substances. Of particular importance is the light-induced production of hormones by the pineal, an endocrine gland in mammals. Animal and human studies have focused on the indirect effects of light on the synthesis of melatonin, a major pineal hormone.[35] Melatonin has a number of effects on the body including sleep induction, inhibition of ovulation, and modification of hormone secretion from other organs including the pituitary gland, the gonads and the adrenals. The illuminance, wavelength, and timing influences on glandular and metabolic changes in the body have not yet been identified; however, they are presently under study.[35, 36]

It is an important question whether present electrically lighted environments are biologically adequate for humans, and whether or not some might be potentially stressful or harmful to normal or light sensitive individuals. For example, the continuous illumination found in many hospital nurseries may affect circadian rhythms and consequently, the development of newborns. Further investigations on environmental lighting schedules, threshold levels and spectra and their effects on biological rhythms might be beneficial to man on earth and may be important for the success of future space exploration.

Phototherapy of Neonatal Hyperbilirubinemia[39]

Hyperbilirubinemia in neonates is more commonly known as jaundice of the newborn. It is estimated that 50 per cent of all infants develop at least mild jaundice during the first week of life and that about 7 to 10 per cent of neonates have hyperbilirubinemia of sufficient severity to require phototherapy.

Jaundice is the symptom and not the disease. It results from the accumulation of the yellow pigment, bilirubin, due to the infant's inability to rid itself of bilirubin as rapidly as it is produced.

Bilirubin is chemically a tetrapyrrole and is derived principally from the degradation of hemoglobin. In the normal course of events, bilirubin is transported in the blood by binding to albumin. When the bound bilirubin reaches the liver it is then conjugated and excreted into the liver, thence into the bile, to the gall bladder and then to the intestine where it is excreted in the feces. A small amount of bilirubin is excreted in the urine. Infants with hyperbilirubinemia have an inability to bind and excrete bilirubin in the normal manner.

Normal bilirubin concentration in newborns rarely exceeds 1.5 to 2.0 milligrams per deciliter, but when concentrations rise to 5.0 to 6.0 milligrams per deciliter, a quantity of the pigment is transferred to the skin to produce the yellowish skin color known as jaundice. This visible indication is the symptom of hyperbilirubinemia. Upon detection, the condition is monitored by measurement of blood plasma bilirubin level.

If the plasma concentration reaches 10 milliliters per deciliter, or above in low birth weight infants or 15 milligrams per deciliter or more in full-term infants, there is the danger of bilirubin concentration exceeding albumin-binding capacity (bound transport) and free, unconjugated bilirubin circulating in the plasma, reaching the brain—and thus producing bilirubin encephalopathy and irreversible damage from toxic-injury to brain cells, a condition known as *kernicterus*. Kernicterus leads to the development of learning impairment, cerebral palsy, deafness, other serious neurologic injury and may be fatal. Phototherapy is used to prevent a dangerous rise in plasma bilirubin to levels that may cause bilirubin toxicity to the central nervous system. The mechanisms of bilirubin photodegradation *in vivo* are not yet completely understood, and the action spectrum is not completely known.

Phototherapy of neonatal jaundice is an alternative therapy to exchange transfusion. It is administered as whole-body (commonly ventral) irradiation usually by banks of 8 to 10 20-watt fluoresent lamps with some portion of the radiation from lamp emission between the visible spectral region of 400 to 490 nanometers. An infant is placed beneath the phototherapy light unit at a distance of 400 to 500 millimeters [15 to 20 inches] and is kept unclothed (undiapered) with both eyes shielded by an appropriate opaque eye mask to protect the retina. Bilirubin

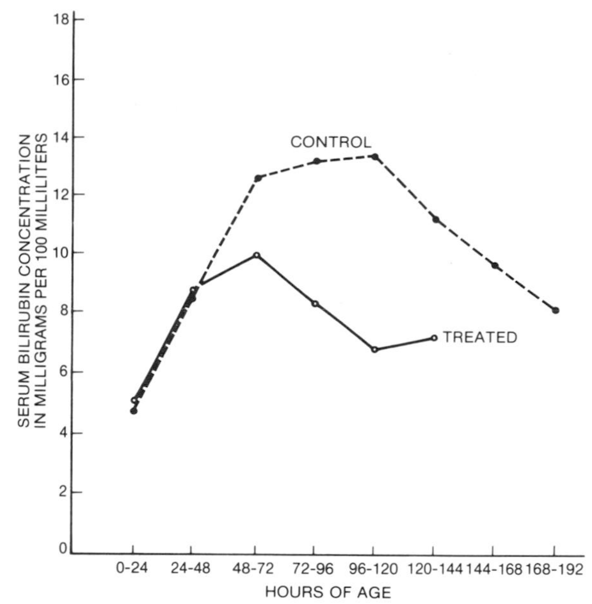

Fig. 19-10. The effect of phototherapy of neonatal hyperbilirubinemia upon the mean serum bilirubin concentrations of 32 infants compared with that of 33 hyperbilirubinemic infants who received no treatment. (From Sisson, 1973a.)

has an absorption peak *in vitro* at about 445 to 450 nanometers and is photochemically altered *in vitro* or *in vivo* by irradiation in this region. The photochemically produced products are relatively non-toxic and are excreted *in vivo*.

Most clinical phototherapy regimens are conducted by constant exposure of infants. Exposure at an irradiance of 0.9 microwatts per square centimeter in the 400- to 490-nanometer range will cause a visual fading of the yellow skin color within 2 to 3 hours, accompanied with a reduction of plasma bilirubin. The irradiance used ranges up to 2.1 milliwatts per square centimeter in the 420- to 490-nanometer region. The effect of phototherapy on plasma bilirubin concentration is shown in Fig. 19-10.

While phototherapy of hyperbilirubinemia has been used for 20 years, optimum irradiance and methods of applying the irradiance have not yet been established. Boosting irradiance to five times the irradiance level of 2.1 milliwatts per square centimeter (420 to 490 nanometers) does not increase the breakdown of bilirubin *in vivo* nor its rate of decline in blood plasma.

Phototherapy is an effective modality in correcting hyperbilirubinemia in the newborn. In contrast to the alternative therapy, exchange transfusion, it is noninvasive and there is little if any risk of mortality. Nevertheless, several side

Fig. 19-11. Side Effects of Phototherapy

Determined *in vitro*
1. Albumin denaturation
2. Diminished riboflavine levels
3. G-6-PD activity loss
4. Glutathione reductase activity loss
5. Mutagenesis in cell cultures

Determined *in vivo* (animals)
1. Retinal damage
2. Increased liver glycogen in rats
3. Retarded gonadal growth (not function) in rats

Determined *in vivo* (human infants)
1. Excess body heat from thermal output of lamps
2. Temporary growth retardation
3. Increased insensible water loss
4. Transient hemolysis (uncommon)
5. Loose, discolored stools
6. Transient skin rash
7. Reduction of whole blood riboflavine
8. Alteration of tryptophan—kynurenine metabolism
9. Alteration of biologic rhythms
10. Physical hazards from inappropriate phototherapy—unit construction
11. Increase in gut transit time
12. Increase in respiration
13. Increase in peripheral blood flow
14. Decrease in circulating platelets

effects have been observed and these are listed in Fig. 19-11.

Equipment Design and Measurement. Some essentials for phototherapy light unit design and radiation measurement are listed below:

Lamp Requirements:
1. Emit sufficient irradiance on the infant skin in the 400- to 490-nanometer spectral region to lower plasma bilirubin levels.
2. Provide sufficient illuminance and color rendering for monitoring and detecting changes in skin color.
3. Does not produce excess infrared irradiation to cause stress of excessive water loss from the infant.

Phototherapy Light Unit Requirements:
1. Utilizes a protective acrylic shield between lamps and infant to absorb ultraviolet radiation from lamps and to protect infant should lamps not be properly installed or be loosened and fall from sockets.
2. Utilizes a heat mirror with those lamps producing a large proportion of radiant output in the infrared to reduce heat load, stress and water loss of the infant.
3. Adequate ventilation is required to prevent build-up due to heat of radiation, conduction and convection and to provide required operative temperature particularly for adequate life and output of fluorescent lamps.
4. The optical design is one that directs the visible light from lamps uniformly over the infant's skin.

Measurement of Radiation:
1. To insure that the irradiance is uniformly distributed and of a sufficient level to perform the phototherapy, measurement instrumentation is required.
2. The instrument used can be one which accurately measures irradiance (milliwatts per square centimeter or microwatts per square centimeter) only in the 400- to 490-nanometer spectral region with optical blocking of all other spectral regions.
3. If an accurate spectral power distribution of the light source is available, an alternative method can be used to estimate the irradiance. The alternative method is to use an illuminance meter and to calculate the irradiance from the lux (footcandle) level by using a conversion factor of microwatts per square centimeter (400 to 490 nanometers) per lux (footcandle). This method is not as precise as *2* above and the conversion factor will differ with each light source having a different spectral power distribution. It is a practical method that can be used to determine level and uniformity of irradiance and to determine when lamps need replacing.

Photochemotherapy[40-49]

Photochemotherapy defines the combination of nonionizing electromagnetic radiation and drug to bring about a beneficial effect. Usually, in the doses used, neither the drug alone nor the radiation alone has any significant biologic activity; it is only the combination of drug and radiation that is therapeutic. PUVA (psoralen and UV-A) is a term used to describe oral administration of psoralen and subsequent exposure to UV-A (longwave ultraviolet, 320 to 400 nanometers). PUVA has proven to be effective in treating psoriasis, vitiligo (a skin disorder with absent pigment cells), certain forms of severe eczema, a malignant disorder called mycosis fungoides and a growing list of other skin disorders.

Psoralens are tricyclic, furocoumarin-like naturally occurring chemicals, some of which can be photoactivated by UV-A. In living cell systems, absorption of energy from photons within the 320- to 400-nanometer waveband (with a broad peak at 340 to 360 nanometers) results in thymine-psoralen photoproducts and transient inhibition of DNA synthesis. When certain psoralens are delivered to the skin either by direct application or by oral route, subsequent exposure to UV-A may result in redness and tanning, which are delayed in onset, occurring hours to days after exposure.

The redness or skin inflammation from PUVA can be severe and is the limiting factor during treatment. The occurrence and degree of redness, however, is related to dose of both the drug and UV-A and is predictable. The redness which results from PUVA differs from sunburn in its time course. PUVA redness may be absent or just beginning at 12 to 24 hours after ultraviolet exposure (when sunburn redness is normally at its peak) and peaks at 48 to 72 hours or later. Because skin diseases can be treated at psoralen-UV-A dose-exposures which are less than the dose-exposure causing severe redness, careful dosimetry permits safe PUVA treatments. The pigmentation which results from PUVA appears histologically and morphologically similar to true melanogenesis (delayed tanning). Pigmentation maximizes about 5 to 10 days after PUVA exposure and lasts weeks to months.

Psoriasis is a genetically determined hyperproliferative epidermal disorder. Until its cause or basic mediators are known, the most effective therapeutic agents must be those which have cytotoxic effects. Many such agents are effective but have the potential of cytotoxic effects on other than cutaneous organ systems. Since PUVA effects require UV-A which penetrates into the skin but does not reach internal organs, PUVA offers the potential for combining the ease of systemic administration with the safety of limiting biologic effects to irradiated skin.

Repeated PUVA exposures cause disappearance of lesions of psoriasis in most patients. Ten to thirty treatments given twice weekly usually are adequate to achieve clearing. Weekly maintenance treatments keep most psoriatics free of evidence of their disease. Although no rebound exacerbation of psoriasis lesions has been seen after discontinuation of therapy, psoriasis recurs weeks to months after PUVA therapy ceases. Patients with recurrent psoriasis respond again to subsequent PUVA therapy. The scalp, body folds, and other areas not exposed to UV-A do not respond to therapy.

Two hours after ingestion of 0.6 milligram per kilogram of 8-methoxypsoralen, patients are exposed to UV-A. Initial UV-A exposure (1.0 to 5.0 joules per square centimeter) depends on degree of melanization and on sunburn history. Exposure must be increased as tanning occurs because the pigmented skin diminishes UV-A penetration to the deeper levels of skin. Ideal exposure sources are those which have high radiant output of UV-A, capability to radiate the entire body surface, little UV-B (280 to 320 nanometers) and infrared, and relatively uniform irradiance at all sites within the radiation chamber. Safety devices and reliable methods of measuring and delivering exact exposure are essential.[49]

The sun can be used as a PUVA radiation source but carries the disadvantage of varying and unpredictable ultraviolet irradiance and spectral distribution at the earth's surface. In tanned or pigmented patients, long exposure times may be required. For example, exposure to both front and back of the body may each be two to three times that needed for a single total-body treatment in an experimental photochemotherapy system. Some patients, however, are willing to tolerate the heat and boredom of sun exposure in order to have the advantage of home treatment. Intense sun, clear skies, metering devices, careful instruction and intelligent, cooperative and motivated patients may make sun PUVA therapy a more reasonable option than hospital or office based treatment.

Exposure to high irradiance of UV-A for prolonged periods of time causes cataract and skin cancer in laboratory animals. These effects are augmented by psoralens. Exposures used in these studies are much greater than therapeutic exposures. Observations in animal systems indicate that the extent of skin cancer induction varies with dose and route of psoralen administration and ultraviolet exposure. Both basal cell and squamous cell carcinomas have been seen in patients treated with PUVA. The incidence of these tumors is highest in patients with a prior history of exposure to ionizing radiation or a previous cutaneous carcinoma. These findings suggest that the potential risk of PUVA-related cutaneous carcinogenesis should be carefully weighed against the potential benefit of this therapy. Special care must be taken in treating patients with prior histories of cutaneous carcinoma or exposure to ionizing radiation.

Experimental animal studies indicate that 8-methoxypsoralen also sensitizes the eyes (cornea and lens) of certain species to UV-A exposure. It is not yet known how this sensitization relates to the use of psoralens in photochemotherapy of humans: although humans have used 8-methoxypsoralen therapeutically for decades, no cataracts attributable to PUVA have been reported. However, it seems wise to limit the use of psoralen photochemotherapy to those with significant skin disease and to use adequate UV-A eye protection during the course of therapy. After ingesting psoralens, patients should protect their eyes at least the remainder of that day.

Physicians must be aware of these theoretical concerns and must carefully observe patients for signs of accelerated actinic damage. Sunglasses which are opaque to UV-A decrease total UV-A exposure to the lens and should be worn on treatment days.

Environmental Lighting Safety Criteria[50-55]

Many exposure limits for optical radiation have been proposed in the literature; however, the only widely accepted standards are for the ultraviolet spectral region. Even these standards have provoked controversy. At present there is movement toward the development of both human exposure limits and product performance standards.[52] For further information refer to such agencies and organizations as the Army Environmental Hygiene Agency, American Conference of Governmental Industrial Hygienists (ACGHI), National Institute of Occupational Safety and Health (NIOSH), Bureau of Radiological Health (BRH) and American National Standards Institute (ANSI).

EFFECTS ON MICROORGANISMS[56-61]

Germicidal (Bactericidal) Ultraviolet

Electromagnetic radiation in the wavelength range between 180 and 700 nanometers is capable of killing many species of bacteria, molds, yeasts and viruses. The germicidal effectiveness of the different wavelength regions may vary by several orders of magnitude, but the ultraviolet wavelengths are generally the most effective for bactericidal purposes.

The bacterium most widely used for the study of bactericidal effects is *Escherichia coli*; these studies have shown the most effective wavelength range to be between 200 and 300 nanometers, corresponding to the peak of the absorption of deoxyribonucleic acid, DNA. The absorption of the ultraviolet radiation by the DNA molecule produces mutations and/or cell death. The relative effectiveness of different wavelengths of radiation in killing a common strain of *E. coli* is shown in Fig. 19-12.

Germicidal (Bactericidal) Lamps. The most practical method of generating germicidal radiation is by passage of an electric discharge through low pressure mercury vapor enclosed in a special glass tube which transmits short ultraviolet radiation. About 95 per cent of the energy is radiated in the 253.7-nanometer band which is very near to the greatest lethal effectiveness. The germicidal output of various commercial lamps is given in Fig. 8-122 in the 1981 Reference Volume.

Hot cathode germicidal lamps are similar in physical dimensions and electrical characteristics to the standard preheat 8-watt, 15-watt and 30-watt fluorescent lamps and they operate on the same auxiliaries. Slimline germicidal lamps are instant-start lamps capable of operating at several current densities within their design range, 120 to 420 milliamperes, depending on the ballast with which they are used. Cold cathode germicidal lamps are instant-start lamps with the cylindrical cold cathode type of electrode. They are made in many sizes and operate from a transformer.

Fig. 19-12. Erythemal and Germicidal (Bactericidal) Efficiency of Ultraviolet Radiation

Wavelength (nanometers)	Erythemal Efficiency	Tentative Bactericidal Efficiency
*235.3	....	0.35
240.0	0.56	....
*244.6	0.57	0.58
*248.2	0.57	0.70
250.0	0.57	....
*253.7	0.55	0.85
*257.6	0.49	0.94
260.0	0.42	....
265.0	....	1.00
*265.4	0.25	0.99
*267.5	0.20	0.98
*270.0	0.14	0.95
*275.3	0.07	0.81
*280.4	0.06	0.68
285.0	0.09	....
*285.7	0.10	0.55
*289.4	0.25	0.46
290.0	0.31	....
*292.5	0.70	0.38
295.0	0.98	....
*296.7	1.00	0.27
300.0	0.83	....
*302.2	0.55	0.13
305.0	0.33	....
310.0	0.11	....
*313.0	0.03	0.01
315.0	0.01	....
320.0	0.005	....
325.0	0.003	....
330.0	0.000	....

* Emission lines in the mercury spectrum; other values interpolated.

The life of the hot cathode and slimline germicidal lamps is governed by the electrode life and frequency of starts. (The effective life of hot cathode germicidal lamps is sometimes limited by the transmission of the bulb. This is particularly true when operated at cold temperatures.) The electrodes of cold cathode lamps are not affected by number of starts, and useful life is determined entirely by the transmission of the bulb. All germicidal lamps should be checked periodically for ultraviolet output to make sure that their germicidal effectiveness is being maintained.

The majority of germicidal lamps operate most efficiently in still air at room temperature, and ultraviolet output is measured at an ambient temperature of 25 °C (70 °F). Temperatures either higher or lower than this optimum value decrease the output of the lamp. Lamps operating in a room at 4 °C (40 °F) produce only about two-thirds to three-fourths as much ultraviolet as at 27 °C (80 °F). Cooling the lamp by passing air currents over it or by submerging it in liquid likewise lowers its output.

Slimline germicidal lamps operated at from 300 to 420 milliamperes and certain preheat germicidal lamps operated at 600 milliamperes are designed exceptions to this general rule. At these high-current loadings the lamp temperature is above the normal value for optimum operation; therefore, cooling of the bulb does not have the same adverse effect as with other lamps. Thus these lamps are well suited for air-conditioning duct applications.

In addition to the energy at 253.7 nanometers, some germicidal lamps generate a controlled amount of 184.9-nanometer radiation which produces ozone in the air. See Fig. 19-13. Since ozone is highly toxic, its environmental concentrations have been limited by the Occupational Safety and Health Administration (OSHA) regulatory mandate to 0.1 part per million or 0.2 milligrams per cubic meter, and care should be taken when choosing germicidal lamps to meet the requirements of these regulations.

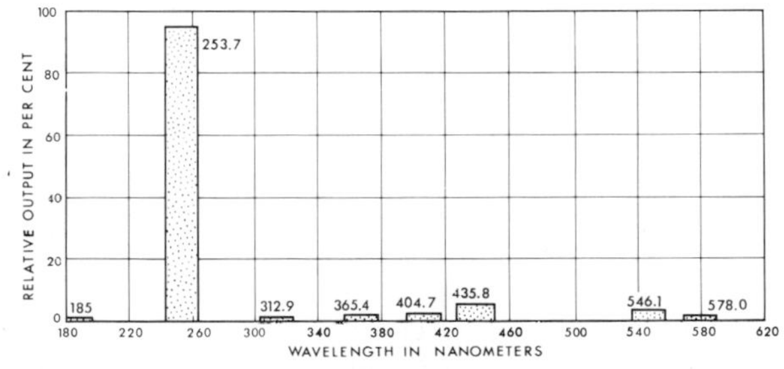

Fig. 19-13. Relative spectral distribution of energy emitted by ozone producing germicidal lamps.

Exposure Time. The ability of radiation to kill an organism is dependent on many parameters, including the complexity of the organism, the ability to repair damage induced by the absorption of ultraviolet radiation, its specific sensitivity, and the presence of a sufficiently high radiant exposure. Fig. 19-14 lists the exposure (joules per square meter) to the 253.7-nanometer ultraviolet radiation required to inhibit or prevent colony formation in 90 per cent of the population of a wide variety of microorganisms.

Photoreactivation. It has been observed that the survival of ultraviolet-irradiated bacteria could be greatly enhanced if the cells were subsequently exposed to an intense source of blue light. Researchers have demonstrated the existence of a photoreactivating enzyme, and established its basic properties in repair of damaged DNA.

The enzyme combines in the dark with cyclobutyl pyrimidine dimers in ultraviolet-irradiated DNA to form an enzyme-substrate complex. The complex is activated by the absorption of energy between 320 and 410 nanometers, the cyclobutyl pyrimidine dimers are converted to monomeric pyrimidines, and the enzyme is released.

Under certain experimental conditions, as much as 80 per cent of the lethal damage induced in bacteria by low energy ultraviolet radiation at 254 nanometers can be photoreactivated, thus indicating the importance of cyclobutyl pyrimidine dimers as lethal lesions. Photoreactivating enzymes have been found in a wide range of species from the simplest living cells, to the skin and white blood cells of man.

Germicidal Effectiveness. The germicidal effectiveness is proportional to the product of intensity times time from one microsecond to a few hours. A nonlinear relationship exists between ultraviolet exposure and germicidal efficacy. If a specific ultraviolet exposure produces a 90 per cent kill of bacteria, doubling the exposure can only produce a 90 per cent kill of the residual 10 per cent for a final 99 per cent. In reverse, a 50 per cent decrease in intensity or exposure results only in a decrease in germicidal efficacy from 99 to 90 per cent.

Precautions. Exposure to germicidal ultraviolet radiation can produce eye injury and skin erythema, and has produced skin cancer in laboratory animals. The American Medical Associ-

Fig. 19-14. Incident Radiation at 253.7 Nanometers Necessary to Inhibit Colony Formation in 90 Per Cent of the Organisms

Organism		Exposure (joules per square meter)
Bacillus anthracis		45.2
S. enteritidis		40.0
B. megatherium sp. (veg.)		13.0
B. megatherium sp. (spores)		27.3
B. paratyphosus		32.0
B. subtilis		71.0
B. subtilis spores		120.0
Corynebacterium diphtheriae		33.7
Eberthella typhosa		21.4
Escherichia coli		30.0
Micrococcus candidus		60.5
Micrococcus sphaeroides		100.0
Neisseria catarrhalis		44.0
Phytomonas tumefaciens		44.0
Proteus vulgaris		26.4
Pseudomonas aeruginosa		55.0
Pseudomonas fluorescens		35.0
S. typhimurium		80.0
Sarcina lutea		197.0
Seratia marcescens		24.2
Dysentery bacilli		22.0
Shigella paradysenteriae		16.3
Spirillum rubrum		44.0
Staphylococcus albus		18.4
Staphylococcus aureus		26.0
Streptococcus hemolyticus		21.6
Streptococcus lactis		61.5
Streptococcus viridans		20.0
Yeast		
Saccharomyces ellipsoideus		60.0
Saccharomyces sp.		80.0
Saccharomyces cerevisiae		60.0
Brewers' yeast		33.0
Bakers' yeast		39.0
Common yeast cake		60.0
Mold Spores	**Color**	
Penicillium roqueforti	Green	130.0
Penicillium expansum	Olive	130.0
Penicillium digitatum	Olive	440.0
Aspergillus glaucus	Bluish green	440.0
Aspergillus flavus	Yellowish green	600.0
Aspergillus niger	Black	1320.0
Rhizopus nigricans	Black	1110.0
Mucor racemosus A	White gray	170.0
Mucor racemosus B	White gray	170.0
Oospora lactis	White	50.0

ation has set a limit of 0.1 microwatt per square centimeter for continuous exposure and 0.5 for 7 hours per day. This conservative limitation can be extrapolated to 22 microwatts per square centimeter for 10 minutes, and a limited 2.5 minutes exposure to 90 microwatts per square centimeter at one meter from a G30T8 lamp, and to 6 to 7 seconds exposure at the distance required to read the caution label on the same lamp while in operation. See Fig. 8-122 in 1981 Reference Volume.

Based on the potential for producing threshold keratitis, the National Institute of Occupational Safety and Health has proposed a value of half of the above as a safe industrial exposure to the eye.

Eye protection is essential for all who are exposed to the direct or reflected radiation from lamps emitting ultraviolet, especially shortwave ultraviolet radiation. Ordinary window or plate glass or goggles that exclude radiations of wavelength shorter than 340 nanometers usually are sufficient protection. However, if the radiation is intense, or is to be stared at for some time, special goggles should be used. Failure to protect the eyes can result in temporary but painful inflammation of the conjunctiva, cornea, and iris; photophobia; blepharospasm; and ciliary neuralgia. Skin protection is achieved by wearing clothing that is opaque to germicidal radiation.

Applications

Air Disinfection in Rooms. In occupied rooms, irradiation should be confined to the area above the heads of occupants as shown in Fig. 19-15. Louvered equipment should be used where ceilings are less than 2.7 meters (9 feet) high to avoid localized high concentration of flux which may be reflected down onto occupants. An average irradiation of 20 to 25 microwatts per square centimeter is effective for slow circulation of the upper air, and will maintain a freedom from respiratory disease organisms comparable with outdoor air.

The upper air disinfection as practiced in hospitals, schools and offices is effective in providing air relatively free of bacteria at the breathing level of room occupants. Personnel movement, body heat and winter heating methods create convection currents through the whole cross section of a room sufficient to provide "sanitary ventilation" of 1 to 2 air changes per minute. This has been compared to a removal of the room ceiling for access to outdoor air. There may be "ultraviolet barriers" down across doors to provide irradiation sufficient to disinfect air at usual movement through the doors. In all cases the ceilings and upper walls should have a germicidal ultraviolet reflectance as low as the 5 per cent, characteristic of most oil and some water-base paints. See Fig. 19-16. "White coat" plaster or gypsum product surfaced wallboard and

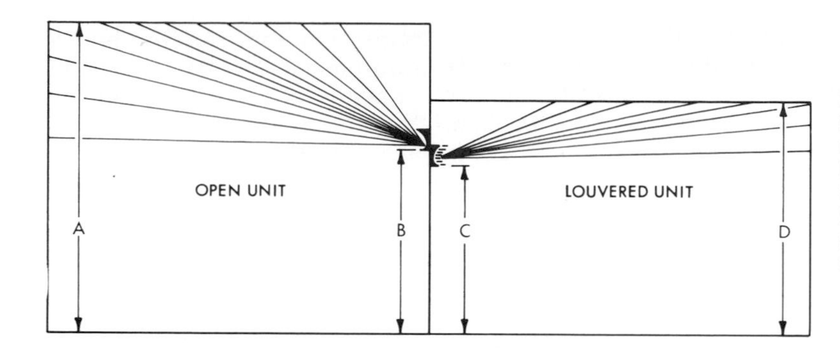

Fig. 19-15. Germicidal lamps for air disinfection in occupied rooms: (a) open unit used in rooms over 2.7 meters (9 feet) in height; (b) louvered unit used where ceilings are less than 2.7 meters (9 feet). Dimensions: A, 3.7 meters (12 feet); B, 2.1 meters (7 feet); C, 2 meters (6½ feet); D, 2.7 meters (9 feet).

Fig. 19-16. Reflectance of Various Materials for Energy of Wavelengths in the Region of 253.7 Nanometers

Material	Reflectance (per cent)
Aluminum	
Untreated surface	40-60
Treated surface	60-89
Sputtered on glass	75-85
Paints	55-75
Stainless steel	25-30
Tin plate	25-30
Magnesium oxide	75-88
Calcium carbonate	70-80
New plaster	55-60
White baked enamels	5-10
White oil paints	5-10
White water paints	10-35
Zinc oxide paints	4-5

acoustical tile may have higher germicidal reflectance and should always be painted. See Fig. 19-17. These precautions are especially important in the hospital infants' ward and somewhat less so in the schoolroom.

In hospital operating rooms, especially where such prolonged surgery as heart, brain and lung operations is done, a combination of upper air and vertical barriers at 25 microwatts per square centimeter is used with head, eye and ear protection in addition to the usual face mask.

Air Duct Installations. It is possible to provide a sufficiently high level of ultraviolet radiation for a 90 to 99 per cent kill of most bacteria in the very short exposure times of duct air at

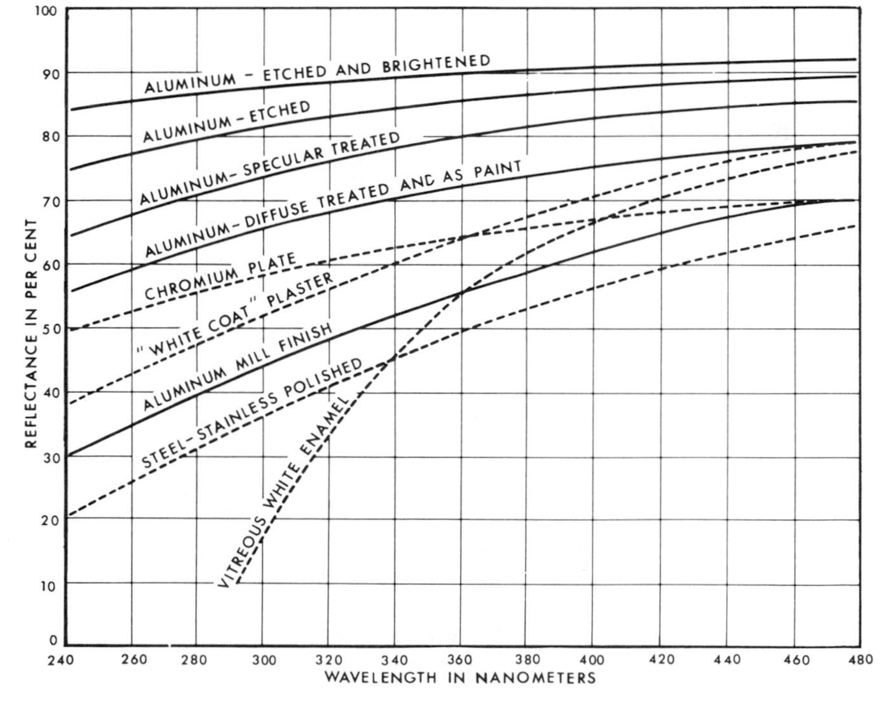

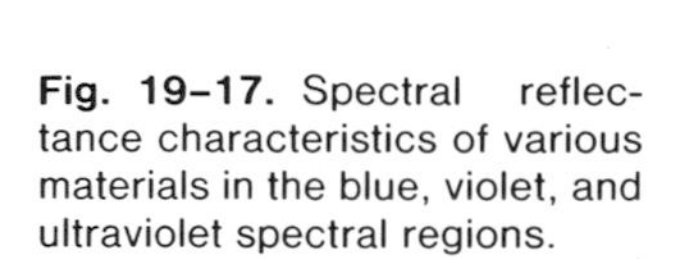
Fig. 19-17. Spectral reflectance characteristics of various materials in the blue, violet, and ultraviolet spectral regions.

usual air velocities. The limitation of the method is that it can only make the duct air equivalent to good outdoor air and its value is in the treatment of recirculated air and contaminated outdoor air in hospitals and pharmaceutical and food processing plants. Duct installations are especially valuable where central air heating and ventilating systems recirculate air through all of the otherwise isolated areas of an institution. Germicidal lamps, often especially designed for satisfactory ultraviolet output even when cooled by high velocity duct air, are installed on doors in the sides of ducts, or inserted across their axis, depending upon the size and shape of the duct and access for servicing. Where possible the best location is across the duct to secure longer travel of the energy before absorption on the duct walls and to promote turbulence to offset the variation in ultraviolet level throughout the irradiated part of the duct.

Liquid Disinfection. Ultraviolet disinfection of water is used where it is essential that there is no residual substance or taste.

Water disinfection methods are similar to those for air except that allowance should be made for some ultraviolet absorption by traces of such natural chemical contaminants as iron compounds and for an 8 to 10 fold greater exposure for wet than for dry bacteria. The product of these two factors calls for water disinfection ultraviolet exposures of 12 to 17 joules per cubic meter. Such exposures are secured by slow gravity flow of water through shallow tanks under banks of lamps, or by lamps in isolating jackets of ultraviolet transmitting glass, immersed directly in the water.

Liquids of high absorptance, such as fruit juices, milk, blood, serums and vaccines, are disinfected with "film spreaders" ranging from high speed centrifugal devices and films adhering to the surfaces of rotating cylinders to gravity flow down screens and inclined planes. The centrifugal devices spread a film of a liquid down to a thinness of the order of its molecular size.

Granular Material Disinfection. The surfaces of granular materials such as sugar are disinfected on traveling belts of vibrating conveyors designed to agitate the material during travel under banks of closely spaced germicidal lamps. During a transit which defines the exposure time, all the particles are brought to the surface frequently and long enough to provide an effective exposure of their surfaces to the irradiation. In the case of sugar, thermoduric bacteria survive the vacuum evaporator temperatures of sugar-syrup concentration and, rejected by the sugar crystals during formation, remain in the final film of dilute syrup left on the crystal surfaces. Ordinarily harmless, they may cause serious spoilage in canned foods and beverages.

Product Protection and Sanitation. Product protection and sanitation is by both air disinfection and surface irradiation as with sugar. In this field, however, the usefulness of germicidal ultraviolet radiation is generally limited to the prevention of contamination during processing instead of a disinfection of an otherwise final product. Germicidal lamps in concentrating reflectors are used to disinfect any air which might contaminate a product during processing and packaging, as during the travel of bottles from washing through filling to capping. They serve to replace or supplement heat in processes where sterilization by heat might be destructive. Where sufficient irradiation to kill mold spores may be impractical, the vegetative growth of mold itself can be prevented by continuous irradiation at the levels lethal to ordinary bacteria as in bakeries and breweries, as on the surfaces of liquid sugar, syrup, fruit juices, and beverages during tank storage. Intensive irradiation of container surfaces can supplement or replace washing between usages.

EFFECTS ON INSECTS

Insect Responses

The increasing popularity of outdoor living, drive-in businesses, and outdoor recreational establishments has been accompanied by intensified insect problems caused particularly by nocturnal insects attracted to light. Similar problems are encountered at lighted farmsteads, animal pens, feed lots, processing plants, and industries operating at night in lighted facilities. Many of these problems can be prevented, or greatly reduced, if the responses of the insect pests involved are considered when designing and planning the use of facilities.

The insect nuisance problems associated with lighting have four distinct but related aspects: (a) the existing insect population in the surrounding vicinity, (b) the attractiveness to insects of the activity carried on in the lighted area, (c) the attractiveness of the lighting system used, and (d) the suitability of the desired insect-free area to sustain insect life. The circumstances of each situation are different, and usually little can be done about the insect population in the surrounding vicinity.

A knowledge of the insects in relation to their normal habitats and the activities to be carried on in the desired insect-free areas usually helps to anticipate problems. Preventive maintenance can be planned based on the known behavioral patterns of the expected insects. Insects likely to cause problems can be broadly categorized as follows:

Insects not attracted by light. Insects not necessarily attracted to light are indigenous to enclosures, buildings, etc., and most live continuously within the desired insect-free area. Included are cockroaches, ants, and flour beetles.

Diurnal insects. These insects normally live outside the desired insect-free area and are active during the day. They are attracted to the area for food, shelter, and/or breeding sites. Examples are house flies, pomance or fruit flies, and honey bees.

Nocturnal insects. This group usually feeds and lives beyond the desired insect-free area and is attracted there by electromagnetic radiation. These insects are the true problem associated with lighting systems. Most nocturnal insects, such as moths, leafhoppers, mayflies, caddisflies, various beetles, midges and mosquitoes, are capable of flight and are active primarily at night.

Research and experience have shown that light-caused insect problems can be greatly reduced with properly designed lighting systems and with proper management of the area to be insect-free.

Phototaxis is the term applied to insect visual response which causes the insect to be attracted to a source of electromagnetic radiation. Insects which are attracted to a radiation source are said to be photopositive or to exhibit positive phototaxis.

The spectral region most attractive to a wide range of insect species, especially nocturnal species, is in the near ultraviolet (310 to 380 nanometers in wavelength). Other species are known to respond to energy in the visible and infrared, as well. A list of some common species attracted to traps equipped with a "black light" (near ultraviolet) lamp can be found in reference 65.

Lighting System Attraction

One means of reducing the insect nuisance is to select light sources having low insect attractiveness. In practice, this involves maximum use of yellow-red light and the reduction of ultraviolet and blue wavelengths.

Mercury lamps generate ultraviolet radiation and filtering to remove the ultraviolet, either by the lamp envelope or a refractor, may also reduce the light output. If filtering is extended to remove some visible wavelengths, to further reduce attraction, the loss will be even greater.

High pressure sodium lamps have a higher lumen per watt output than mercury lamps and with about one-third as much associated insect nuisance as a comparable mercury lamp system. Thus, mercury lamps should be avoided for area lighting either inside or outside buildings where night-flying insects are a potential problem.

If, for example, light-attractive lamps must be used for color rendition reasons, they can be shaded so that all their radiant output is confined to the area to be illuminated. If lamps must remain visible from outside, consideration should be given to using refractors, filters and shields made of glass or plastic material to filter out the ultraviolet radiation. If lamps emitting blue light and near ultraviolet energy are used for lighting, they should not be directly visible at distances beyond a few meters from the illuminated area.

Highly attracting small mercury lamps and exposed incandescent lamps should not be located directly over entrances or work areas, inside or outside of buildings, for insects will be attracted, will annoy people and animals using the entrances, and will gain access to the interior when a door is opened. This problem can be greatly reduced by using yellow insect incandescent or sodium lamps. If fluorescent lamps are to be used gold fluorescent lamps will be less attracting than the "white" type. At the least, an attracting lamp can be shaded so that its radiant output is directed downward and confined to the immediate area. Any type of lamp used to light an entrance or work area should be located a short distance away, with the light directed toward the area to be illuminated.

As insects are also attracted to reflected radiant energy, care should be taken to avoid using a surface or a paint with a high reflectance for visible or ultraviolet radiation.

Decoy Lamps and Insect Traps. In addition to careful lamp selection and shielding designs, the number of night-flying insects within a desired insect-free area also can be reduced by placing attracting "black light" (BL) fluorescent or mercury decoy lamps at a 30- to 60-meter (100- to 200-feet) distance around the perimeter of the area to intercept those insects trying to enter. The location, number, and design of the decoys will depend upon the area to be protected. In addition, collecting-killing devices (traps) can be used to capture those insects attempting to enter, or those having already gained entrance to the lighted area.

Insect traps commonly contain "black light" or other lamps to attract photopositive insects and a means for killing or trapping the insect. One of the common killing mechanisms is an electric grid which electrocutes the insects attracted to the trap. Various designs are available for commercial, industrial and residential use. The placement and number of traps to use will vary with the individual situation and the species of insect involved. Specialists in this field are usually required to determine best placement of traps for solving specific insect problems.

If grid traps are used, they should not be placed so that electrocuted insects fall or are blown into working and food-processing areas. System designs and installations should be in compliance with the *National Electrical Code.*

In agriculture, insect traps of various designs have been used for survey purposes to detect insects in a crop area, to predict the need for pesticide application, and to evaluate effects of insecticide measures. Survey traps used over large areas can be used to determine migration of insects and to predict potential infestation. The trap design usually includes a "black light" or other fluorescent lamp and various means of trapping insects. Designs are altered for specific insect species.

Studies with "black light" insect traps in large tobacco or tomato growing areas of over 293 square kilometers (113 square miles) have shown that 1 or more traps per square kilometer or 3 or more traps per square mile reduce tobacco and tomato hornworm populations. Indications are that insect traps are best used in conjunction with insecticides. The use of such traps would generally reduce the number of insecticide applications in a growing season.

EFFECTS ON PLANTS

Plant Responses

The known light responses of plants are many. Light provides the energy necessary for the conversion of carbon dioxide and water by chlorophyll-containing plants into carbohydrates in the process known as *photosynthesis.* These carbohydrates are essential foods and are the substrate for proteins, fats and vitamins required for the survival of all other living organisms. Oxygen, formed as a by-product, is the source of the atmospheric oxygen used in plant and animal respiration. In addition, most fossil fuel resources

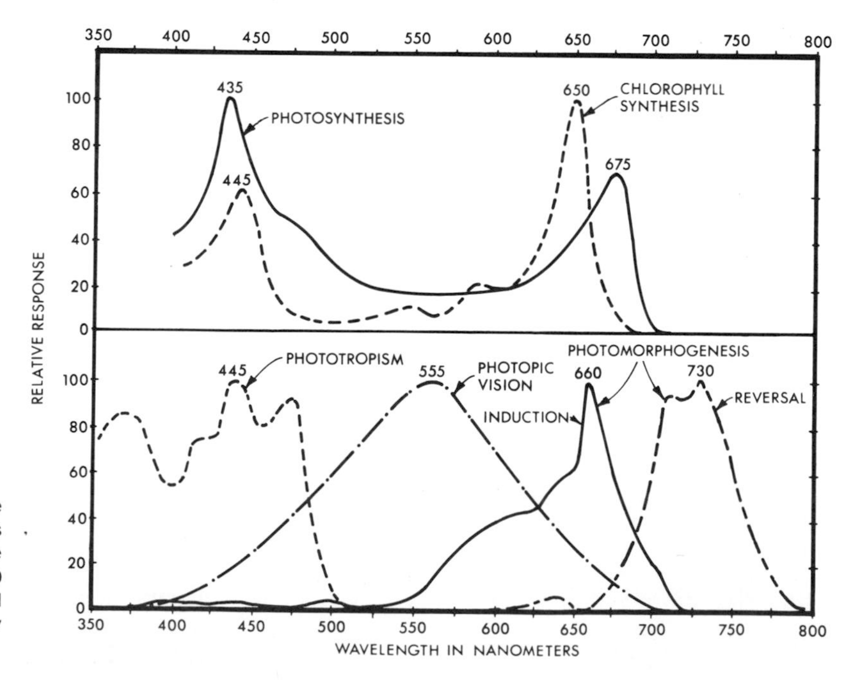

Fig. 19-18. The action spectra of the five major photoresponses of plants show the utilization of energy in three spectral regions (400 to 500 nm, 600 to 700 nm, 700 to 800 nm), compared to photopic vision which utilizes energy in the 380 to 780 nm spectral region.

were derived from photosynthetic processes of a past geological period.

Light is also essential for the formation of such important plant pigments as chlorophyll, carotenoids, xanthophylls, anthocyanins and phytochrome. Light is effective in the opening of stomates, setting internal biological clocks, and in modifying such factors as: plant size and shape; leaf size, movement, shape and color; internodal length; flower production, size and shape; petal movement; and fruit yield, size, shape and color.

The reverse of photosynthesis is *respiration* which does not require either light or chlorophyll but does require food, enzymes and oxygen. Respiration is continuous whereas photosynthesis occurs only in light under normal conditions.

The carbohydrate formed in photosynthesis is oxidized to carbon dioxide, water and energy in respiration. At moderate to high irradiance levels, photosynthesis exceeds the respiration process so the net effect is the production of oxygen from leaves during the light period. If the irradiance level is diminished to the point where the carbohydrate produced is equal to that used in respiration the apparent photosynthetic rate is zero and there is no diffusion of gas from the leaf pores (stomata). This phenomenon is called the compensation point. Plants lighted at the compensation point cannot long survive because stored carbohydrate is used during the dark period. When the carbohydrate reserves are gone the plant will succumb. This is an important concept in the maintenance of plants in interior environments.

In addition to photosynthesis, there are three other major photoresponses of plants: chlorophyll synthesis, phototropism and photomorphogenesis. *Chlorophyll synthesis* is the synthesis of chlorophyll which is formed from the reduction by light of its precursor and photoreceptor, protochlorophyll. *Phototropism* is a light induced growth movement of a plant organ, generally controlled by carotenoid photoreceptors. *Photomorphogenesis* is light controlled growth, development and differentiation of a plant from responses initiated largely by the photoreceptor, phytochrome. The action spectra of these photoresponses are shown in comparison to that of photopic vision in Fig. 19-18.

Photomorphogenic responses of plants are controlled by the blue-green, biliprotein, phytochrome. Like chlorophyll, it has a chromophore that absorbs radiant energy and undergoes excitation, but unlike chlorophyll, the excitation energy is not transferred but is used to change its molecular structure.

The photomorphogenic responses include: photoperiodism, seed germination, stem elongation, and anthocyanin pigment formation. *Photoperiodism* is the response to the relative lengths of light and dark periods.

Light induced movements of plants are phototropism, photonasty and phototaxis. *Phototropism* is the bending of an organ toward or away from the direction of the source of light. *Photonasty* is the movement of plant organs due to changes in illuminance such as the closing of flowers at night and opening during the day. *Phototaxis* is the movement of the whole organism in response to light and is restricted to sex cells of aquatic plants or unicellular aquatic plants.

Limiting Factors for Growth. In addition to light which provides the energy for plants, other requirements must be available in optimum amounts for rapid photosynthesis and growth. These requirements are water, nutrients (inorganic salts), suitable temperature and carbon dioxide. Lack of any one of these requirements will place the plant in stress and will limit or halt growth. The relationship among carbon dioxide, temperature and illuminance is shown in Fig. 19-19 where it can be seen that the photosynthetic rate is increased by an increase in illuminance when the temperature and carbon dioxide levels are increased. These principles are important in the application of light to accelerate plant growth.

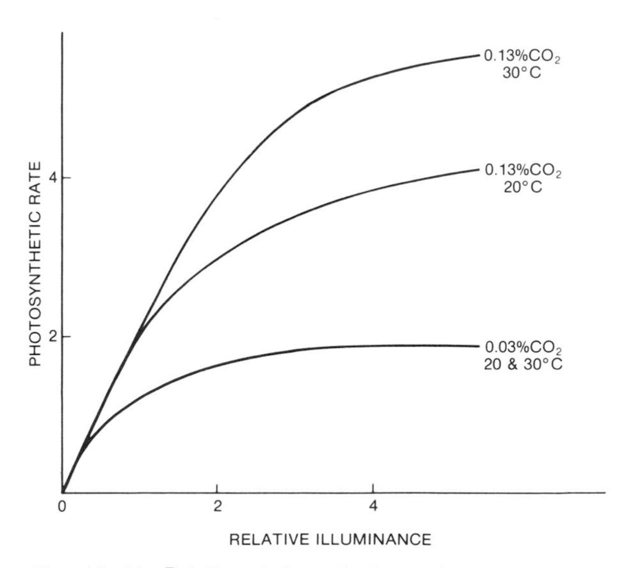

Fig. 19-19. Relative photosynthetic rate in relation to illuminance, temperature and carbon dioxide concentration.

Horticultural Lighting

Horticultural lighting is the application of light sources for the control of growth, flowering or maintenance of plants in interior and outdoor environments. Light sources may be used to duplicate the energy and photoperiod of daylight (sunlight and sky light) for research in a controlled environment. It is more common to use light sources to provide a sufficient irradiance in controlled environment chambers and in other interiors to supplement the level of daylight and to manipulate the photoperiod in greenhouses and other interior environments.

Light Sources. Electric light sources which emit sufficient energy over the entire 300 to 800-nanometer spectral region are effective in photosynthesis and other photoresponses of plants. Light sources which have been made to emit light in limited spectral regions by filtering or other means have been used for special photoresponse purposes. In experimental work, for example, lamps which emit light in the 500- to 580-nanometer region have been used as "safe lights" because this energy is in a low response wavelength region. In contrast, lamps emitting energy in the blue (400 to 500 nanometers), the red (600 to 700 nanometers) or far-red (700 to 800 nanometers) are used in studying various photoresponses which absorb strongly the energy emitted in these spectral regions.

Experimental work in horticultural science uses many types of light sources including carbon arc, incandescent (including tungsten-halogen), fluorescent, mercury, xenon, low and high pressure sodium, and metal halide lamps. Various combinations of lamp types are sometimes used; the most common combinations being fluorescent and incandescent. Other combinations include: high pressure sodium plus metal halide, high pressure sodium plus mercury, low pressure sodium plus incandescent, low pressure sodium plus fluorescent plus incandescent, and metal halide plus incandescent. Low pressure sodium lamps used alone in a controlled environment inhibits the synthesis of chlorophyll in certain species and thus normal chlorophyll synthesis requires that low pressure sodium lamps be combined with fluorescent or incandescent lamps.

It has been generally determined that the type of light source, or combination of types, is more efficient in producing plant material (fresh and dry weight) when the greater portion of the energy, emitted throughout the 300- to 800-nanometer spectral region, is in the 580- to 800-nanometer region. For example, mercury lamps with a phosphor coating which emits in the red (600 to 700 nanometers) spectral region, have been found to be more effective in producing plant material than clear mercury lamps or lamps with a phosphor coating which emits throughout the 400- to 700-nanometer region. Of the various standard fluorescent lamps, the warm white lamp has been found to be most efficient in the synthesis of dry matter in plants. Generally the ability of fluorescent lamps to synthesize dry matter in plants is enhanced with the addition of energy from incandescent lamps. This phenomenon is due to additional energy from the incandescent lamps in the red (600 to 700 nanometers) and especially the energy in the far-red (700 to 800 nanometers) where most "white" fluorescent lamps are relatively deficient. Incandescent lamps which emit a greater percentage of their output in these regions have found wide usage in the control of flowering of horticultural crops. However, many variations in plant responses to light sources are apparently caused by species differences.

Plant growth lamps are presently fluorescent lamps which have been designed exclusively for plant growth and responses. The emission spectra of plant growth lamps are tailored so that nearly all of the energy emitted is that capable of being absorbed by plant photoreceptors for utilization in plant growth. Two types of lamps are manufactured. One type produces emission maxima at 450 and 660 nanometers, closely matching the chlorophyll synthesis curve, but with little emission beyond 700 nanometers. This lamp has generally found wide usage in residential lighting for the growth and color enhancement of house plants requiring relatively low energy levels, especially African violets and gloxinias. The other lamp type has high emission in the blue and red spectral regions but differs from the previously described plant growth lamp and other fluorescent lamps because of its relatively high emission in the 700 to 800 nanometer spectral region. With its far-red emission, the relative growth of several plant species is enhanced and its use generally precludes the use of supplementary incandescent light. This lamp has found use in the growth and flowering of plants which are normally grown in full sunlight, as are many commercial florist and vegetable crops. The spectral energy distribution of these two lamps is compared in Fig. 19-20.

Another approach in lamp spectral design is the simulation of the daylight spectrum in a fluorescent lamp. This lamp simulates daylight at 5500 K. Such a spectral energy distribution is useful in the comparison of plant performance in controlled environments compared to growth in the natural environment.

Radiant Energy Measurement. Horticultural lighting is not illumination. Horticultural light does not conform to the definition of light, as given in Section 1 of the 1981 Reference Volume, *i.e.*, radiant energy that is capable of exciting the retina and producing a visual sensation. Horticultural lighting is plant evaluated radiant energy as illustrated in Fig. 19-18. Because horticultural lighting is not illumination, the lux or footcandle as a unit of radiant energy measurement is limited in its value, especially in comparing plant responses with light from sources having different spectral power distribution characteristics. In horticultural science, illuminance measurements are either discussed in terms of the spectral emission of the light source or are converted to absolute units by calibrating the illuminance meter for the power density in microwatts per square centimeter for each lux or footcandle and for each light source with a different spectral emission. A better alternative is to use a thermopile, bolometer, calibrated silicon detector, or spectroradiometer which is linear with wavelength and energy level. A plant growth photometer has been developed to measure plant growth power density in absolute units (microwatts per square centimeter) in the three spectral regions of major plant response (400 to 500, 600 to 700, and 700 to 800 nanometers). See Fig. 19-21.

Attempts have been made to define photosynthetically active radiation more precisely. In the past it has generally been defined as the energy flux in the wavelengths between 400 and 700 nanometers or 380 and 710 nanometers. Subsequent work on photosynthetically active radiation was based on the photosynthetic rates of an "average leaf" in a similar way that the spectral luminous efficiency curve is based on the "average eye." Recent work has shown that quantum flux in the 400- to 700-nanometer region is a better measure of photosynthetically active radiation than energy flux or the "average leaf," and a meter has been designed for quantum flux measurement in microeinsteins per square meter and per second.[90]

Horticultural Applications

In horticulture there are two general uses for lighting: Photosynthesis and Photoperiodism. In lighting for photosynthesis, light is applied to plants to sustain in part or in total the photosynthetic processes necessary for desired growth. In lighting for photoperiodism, light from various sources is applied to plants to sustain in part or in total the photoperiod necessary to produce a desired flowering response. For many plants, the quantity of light required for photosynthesis can range from 10 to 100 times greater than that required for photoperiodic lighting.

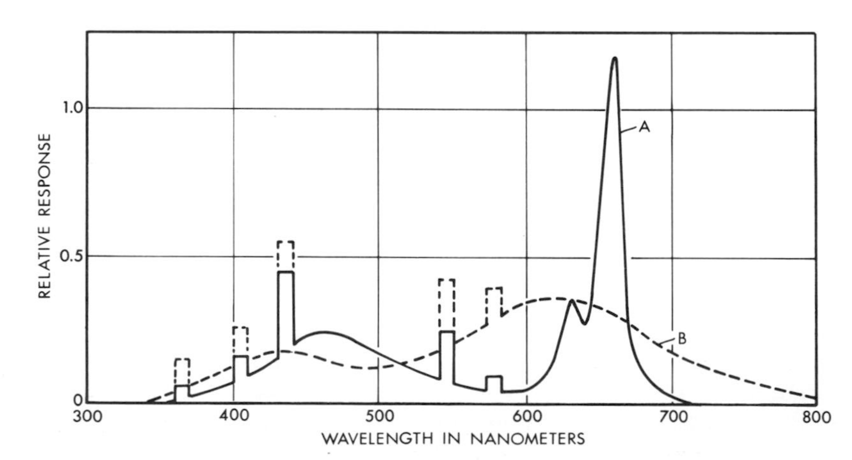

Fig. 19-20. Fluorescent plant growth lamps have been designed with emission spectra tailored so that nearly all of the energy emitted is capable of being absorbed and utilized by the plant photoreceptors for utilization in plant growth. Lamp A has emission maxima at 450 nm and 660 nm with little emission beyond 700 nm, compared to Lamp B which has its major emission also in the blue and red regions as well as emission beyond 700 nm.

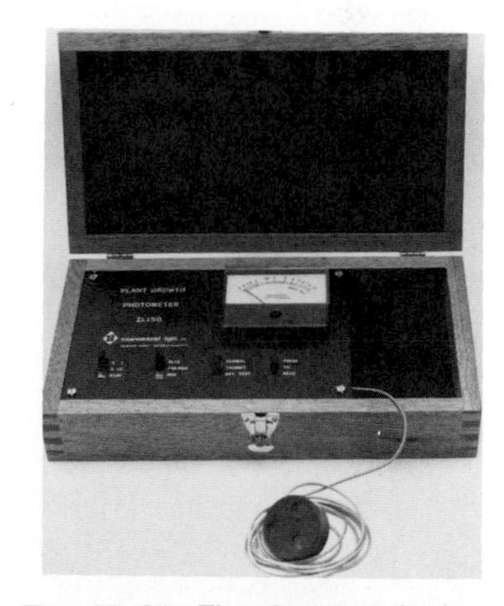

Fig. 19-21. The plant growth photometer is designed to measure radiant energy in absolute units (μW/cm^2) in the three wavelength regions of maximum plant responses (400 to 500, 600 to 700, 700 to 800 nm).

Photosynthetic Lighting. Photosynthetic lighting is used in the greenhouse to supplement daylight during periods of diminished sunlight in winter months, for the growth of out-of-season crops. See Figs. 19-22 and 19-23. This supplementary lighting can be much less than full sunlight—the level being determined by the requirements of the particular plant species.

The different applications of lighting for photosynthesis are as follows:

1. Day Length Extension—Lighting in the greenhouse before sunrise or after sundown to extend the light period.

Fig. 19-22. Over-bench photosynthetic lighting in the greenhouse is a means of increasing crop production efficiency and of timing crops for a market advantage. Luminaires should provide maximum incident light on plants, minimum shading of sunlight, and minimum interference with greenhouse routine. Such luminaires using high pressure sodium lamps are shown for lighting roses (8 to 10 kilolux [800 to 1000 footcandles]).

Fig. 19-23. Under-bench photosynthetic lighting enables the grower to double the growing area within a greenhouse by "stacking" plants.

2. Dark Day—Lighting in the greenhouse on dark, overcast days for the total light period.
3. Night—Lighting in the middle of the dark period when natural carbon dioxide concentrations are high.
4. Underbench—Lighting under benches can significantly increase the growing area within a greenhouse.
5. Growth Room—All lighting is provided by lighting equipment. Such rooms may range from controlled environmental rooms for research to nonproductive areas where seedlings, cuttings, and plants with a low-light requirement can be economically grown. See Figs. 19-24 and 19-30.

In the broadest sense, growth room lighting could also include lighting for plants in all interior spaces, including indoor gardening in homes and commercial areas.

Currently attention is being given to the development of commercial production of salad crops, particularly lettuce, in growth rooms to compete with conventional crop production. Such production of crops is called controlled environment agriculture. Hydroponic or soiless culture of plants is used with this new agricultural venture.

General lighting requirements for the above lighting applications are shown in Fig. 19-25. Specific requirements depend upon plant species and varietal requirements. Particular consideration should be given to photoperiodic plants.

Although definite advantages can be obtained in the greenhouse with photosynthetic lighting alone, it is evident from experimental work that the benefits in the regulation of plant growth are obtained when other interacting factors such as temperature, water, nutrients and carbon dioxide are considered. The greatest growth response is obtained when these factors are present at optimum levels. In research, such optimum levels can be achieved without great concern for costs, but the commercial grower is faced with supplying practical levels of these factors which the particular crop can economically support.

Plant Growth Chambers. Plant growth climatology chambers are now used extensively in agricultural experiment stations, educational institutions, and by industrial research laboratories for the growth of plants under controlled environmental conditions (see Fig. 19-24). The environmental conditions which are controlled and monitored include light, temperature, nutrients, carbon dioxide and oxygen. A research facility which consists of several plant growth chambers (controlled environment rooms) is called a *phytotron.* For most scientific work, the light level is

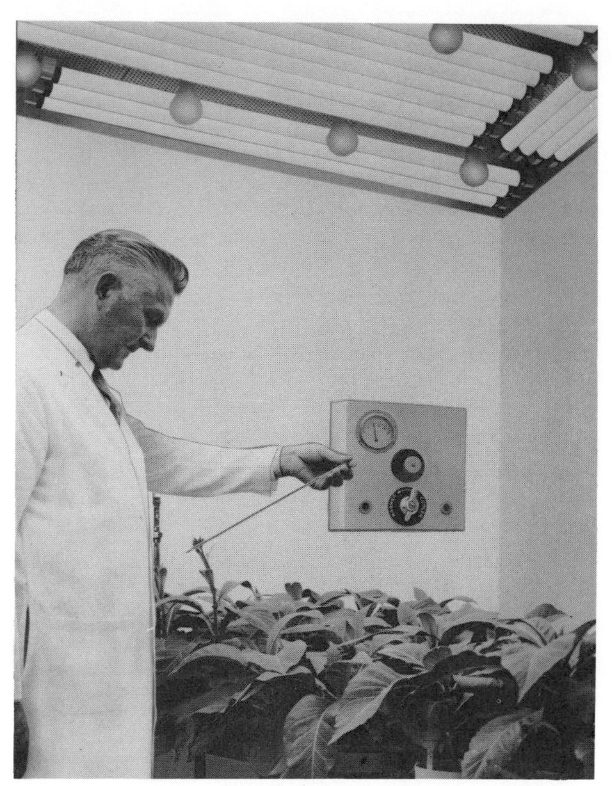

Fig. 19-24. Plant growth chamber 1.5 × 3 × 2.4 meters (5 × 10 × 8 feet), equipped with closely-spaced 1500 mA fluorescent lamps mounted on a perforated white ceiling through which lamp heat is exhausted. Also, 100-watt, 2500-hour life incandescent lamps at uniform spacings provide far-red component. Glossy-white thermal-fabricated walls absorb radiant heat from lamps and reflect light downward to plant area. Initial illuminance was approximately 51 kilolux [5100 footcandles] at the bench. Circuitry of lamp groups provides steps in illuminance.

measured in absolute units. However, growth chamber manufacturers rate the light level within growth chambers in lux or footcandles.

For illuminances up to 50 kilolux [5000 footcandles], 1500-mA 2.4-meter (8-foot) T-12 or T-17 fluorescent (white or cool white) lamps are closely spaced and mounted to a white perforated ceiling through which lamp heat is exhausted. Spaced at uniform intervals between the fluorescent lamps are 100-watt, 2500-hour-life incandescent lamps which provide the far-red component necessary for cuttings and seedlings. For higher illuminances, reflector-type T-10 1500-mA fluorescent lamps permit closer spacing, or high intensity discharge (HID) lamps can be used.

Studies have shown that the combination of plant growth lamps (one type A and one type B of Fig. 19-20) has produced more foilage and earlier fruit set than the combination of cool white fluorescent and incandescent lamps.

For closer climatology control, some growth chamber walls are made of glossy-white, thermal fabricated material which absorbs the infrared energy received from the lamps or specular material that reflects a high percentage of the light downard to plant growth areas. Uniformly spaced groups of lamps are separately circuited to provide several steps of illuminance, and at the same time maintain uniform light distribution. To compensate for early light output depreciation of fluorescent lamps, the system should be designed to produce initially 25 per cent higher illuminance than the maximum level. Circuitry control is used to regulate levels as lamps age. Light meters designed to measure higher illuminances are used periodically to check levels.

Most growth chamber fluorescent or HID lighting systems are operated with ballasts located remotely outside of the growing area. This arrangement divorces ballast heat from the chamber and reduces the air-conditioning load factor.

In chambers where experiments are performed with isotopes to study plant mutation, a shielding medium of lead is built into the chamber walls to prevent transmission of ionizing radiation.

Photoperiodic Lighting. In the United States today the greatest use of lighting in horticulture is in the use of photoperiodism to control the out-of-season flowering of certain species of economic plants which require specific ratios of light to dark periods for flowering. Such plants will remain vegetative rather than flower until these requirements are met. Therefore, plants are classified as to the relative length of light period to dark period needed to set flower buds and to bloom. This knowledge is used to bring plants into bloom when there is a particular market advantage. In Florida and California the flowering of several hundred acres of chrysanthemums in the field is controlled by this type of lighting. See Fig. 19-26.

During winter months it is essential to extend the day length to promote the flowering of long-day (short-night) plants and to inhibit the flowering of short-day (long-night) plants. It is also essential that the grower be able to shorten the day length to promote the flowering of short-day plants and to inhibit the flowering of long-day plants. During summer months, the grower must apply an opaque cloth or plastic covering over the plants for part of the day to simulate a short day. It is essential that the material used be opaque and that the plants be exposed to no light at this time because very low levels are effective

Fig. 19-25. The Requirements for Applications of Photosynthetic and Photoperiodic Lighting

Object of Lighting	Applications	Time Applied	Total Effective Light Period (hours)	Range of Irradiance (mW/m^2)	Light Sources	Luminaires
I. Photosynthetic						
A. Supplementary-Greenhouse						
1. Daylength extension	a. Seed germination, seedlings, cuttings, bulb forcing	4 to 10 hours before sunrise and/or after sunset	a. 12 to continuous	a. 3000–12,000	Fluorescent, high pressure sodium, low pressure sodium, metal halide, mercury and fluorescent-mercury lamps of various wattages with and without internal reflectors and used with or without 10 to 30 per cent of installed watts of incandescent or high pressure sodium plus metal halide	Moisture-resistant luminaires of industrial or custom made designs with mountings fixed or adjustable providing minimum interference with greenhouse routine and uniform light distribution
	b. Mature plants		b. 10 to continuous	b. 8000–50,000		
2. Dark day	as above	Total light period	as above	as above		
3. Night	as above	4 to 6 hours in middle of dark period	16	as above		
4. Underbench	as above	Total light period	10 to continuous	as above	Fluorescent lamps (plant growth lamps)	Moisture-resistant, direct reflector units with mounting for uniform light distribution
B. Growth room						
1. Professional horticulture	Seed germination, seedlings, cuttings, bulb forcing	Total light period	12 to continuous	6000–55,000	Fluorescent lamps or low pressure sodium with or without 10 to 30 per cent incandescent	Industrial direct reflector luminaires which are moisture resistant and are mounted in a shelf arrangement
2. Amateur horticulture	Seed germination, seedlings, cuttings, bulb forcing, mature plants, etc.	Total light period	10 to continuous	as above	Fluorescent lamps (plant growth lamps) with and without incandescent, incandescent and mercury reflector types	General purpose or special luminaires
3. Experimental horticulture	All types of plant responses	Total light period	0 to continuous	0–400,000	Many types used to fit the requirements of tests. Generally fluorescent with 10 to 30 per cent incandescent or HID combinations	Custom or specially built with minimum spacing for maximum light output of lamps with uniform light distribution
4. Interiorscape	Maintenance of mature plants in interiors	Total light period	12 to 16	4000–40,000	Fluorescent, incandescent and HID	General purpose or special luminaires

II. Photoperiodic						
A. Supplementary						
1. Daylength extension (not as effective as 2.)	Long day effect to prevent flowering of short day plants and induce flowering of long day plants	4 to 8 hours before sunrise and/or after sunset	14 to 16	200-2000	Fluorescent, fluorescent-mercury and incandescent lamps	As for photosynthetic supplementary lighting
2. Night break						
a. Continuous	a. as above	a. 2 to 5 hours in middle of dark period	a. 14 to 16	a. 200-2000	a. as above	a. as above
b. Cyclic	b. as above	b. 1 to 4 seconds per minute, 1 to 4 or 10 to 30 minutes per hour as in a.	b. 14 to 15	b. 400-2000	b. Mostly incandescent, or fluorescent lamps with flashing ballasts	b. as above

Fig. 19-26. Photoperiodic lighting enables the grower in southern climates to control the flowering of chrysanthemums in the field. This installation utilized 100-watt incandescent lamps on 3.7- by 4.3-meter (12- by 14-foot) centers.

Fig. 19-27. Over-bed or over-bench photoperiodic lighting in the greenhouse used in conjunction with opaque coverings enables the grower to grow both short-day and long-day plants the year round and to time flowering for the best market period.

in this response. The use of lighting and opaque covering permits the growth and flowering of both long-day and short-day plants the year round.

Long-day responses for both short-day and long-day plants are usually obtained by irradiating plants 4 to 8 hours before sunrise or after sunset or by the more effective 2- to 5-hour light period in the middle of the dark period (called "night break"). See Fig. 19-25.

The most important economic group of photoperiodic plants are short-day plants, chrysanthemums and poinsettias. Such short-day plants remain vegetative with a continuous light period of greater than 12 hours or by a "night break". When flowering is desired, the photoperiod is shortened to about 10 hours and the "night break" is discontinued. By providing long-day plants such as China aster and Shasta daisy with a continuous 16- to 18-hour day with supplementary light, they can be brought into flower, while continuous short days will cause them to remain in the non-flowering or vegetative state.

Incandescent and fluorescent lighting are used for photoperiodic lighting. Clear incandescent in industrial reflector luminaires or reflector incandescent lamps are commonly used in the field (see Fig. 19-26) or in the greenhouse (see Fig. 19-27). Incandescent lamps may produce greater internodal elongation in some plants and this may be undesirable.

Energy savings may be achieved by changing the continuous 2- to 5-hour night break into a series of light/dark cycles. By using a special time switching device, incandescent lamps and fluorescent lamps with a flashing ballast can be cycled on for 1 to 4 seconds per minute, 1 to 4 or 10 to 30 minutes per hour. For most practical applications, the cycle is 20 minutes of continuous light per hour. Use of HID lamps is not practical because life is reduced by frequent starts.

It is often desired to use the higher illuminances required for photosynthetic lighting for photoperiodic lighting; however, these distinctly different lighting applications should not be confused. (See Fig. 19-18.)

Home Hobby Applications. With the aid of lighting and the available lighting equipment for indoor plant culture (see Figs. 19-28 and 19-29), flowering and foliage plants can be taken off the window sill to a place in the room where they can be grown and displayed to the best advantage. Some luminaires are equipped with trays to hold moisture to raise humidity about plants and with timers to turn lights on and off automatically.

Some amateurs, unsatisfied with the number and types of plants which can be grown for decorative purposes, have set up basement gardens of varying sizes in which plants are grown from seed, cuttings, and bulbs as shown in Fig. 19-30. A wide variety of flowering and foliage plants, including all plants of the "house plant" category, have been successfully grown under lights.

Fluorescent lamps in T-8 and T-12 sizes and with typical loadings have been accepted for this type of horticulture. For some plant species, however, it is often considered desirable to include some incandescent light along with fluorescent light to enhance growth and flowering. Incandescent and mercury reflector lamps are used to supplement ambient room light for plants.

Fig. 19-28. Attractive equipment for the home have changed the culture and types of plants grown. Such lighting has replaced "window sill" culture and enables the grower to display and grow plants anywhere in the house.

Commercial and Institutional Applications. Interiorscaping, as interior design with live plants and trees has been called,[70] is now commonplace in lobbies, offices, shopping malls, airport waiting rooms, banks, country clubs, restaurants, entryways of condominiums and apartments, and atriums of large hotels, government, commercial and industrial buildings. Other terms have been used to describe interiorscaping, such as office landscaping or interior landscaping.

Before any plants are chosen or plans are made for interiorscaping, the interior designer, plant specialist or architect must first determine if the environment is suitable—light is commonly the limiting factor:

1. Is there sufficient light (250 lux [25 footcandles] minimum)?
2. Is the temperature range tolerable: day, 18 to 35 °C (65 to 95 °F); night, 10 to 18 °C (50 to 65 °F)?
3. Is the humidity range tolerable (25 to 50 per cent)?
4. Is the light period sufficient (12 to 16 hours)?

If the illuminance is below 250 lux [25 footcandles] and the building owner is unwilling to provide additional light, it is best not to use live plants. If the temperature, humidity and light period are below the minimum requirements

Fig. 19-29. Portable carts made of tubing, with two 40-watt fluorescent lamps in a special reflector mounted over each tray, makes a convenient rack for growth of African violets, gloxinias, and similar house plants.

listed above for any extended period, live plants will not survive. When the environment is not suitable for even plant maintenance, then plants will have to be replaced as they succumb to the environment.

Illuminance Needed. The illuminance in the space will determine the species of plants that will survive. If the illuminance (daylight or electric) is not sufficient for live plants, then supplemental electric lighting can be used, considering that other factors are favorable. The amount of supplemental light is determined by the plant species having the highest light requirement.

The general lighting and supplemental lighting is additive; for example, if the species with the highest light requirement needs 1000 lux [100 footcandles], and the general lighting is 250 lux [25 footcandles], then 750 lux [75 footcandles] of supplemental light will be required. The general and supplemental levels can be measured with an illuminance meter. The levels required for the common plant species used in interiorscaping are shown in Fig. 19-31.

Lighting Acclimatized Plants

Professional plant specialists who work with the plants for interiorscaping make a special effort to use acclimatized plants. Acclimatized plants are plants that have been conditioned for use in the low humidity and low illuminance indoor environments. These plants are taken from greenhouses that often receive full sunlight of 50 to 100 kilolux [5,000 to 10,000 footcandles], to a greenhouse with heavy shade. Here they remain for two months or more before being used for interiorscaping. Also, the watering frequency is reduced to condition the plants for indoor use. Such acclimatization prevents the shock that frequently results in rapid defoliation if plants are taken without conditioning from the bright greenhouse to the interiorscape.

Common Plants and Light Levels

For purposes here, plants are categorized as trees, floor plants and table or desk plants. The lowest illuminances in Fig. 19-31 are the minimum for maintenance. The highest are more satisfactory for good plant condition. The illuminances are for acclimatized plants receiving 14 hours of light per day.

Light Sources. Electric lights may provide part or all of the light for indoor plantings. Various light sources, including daylight, may be combined to provide the lighting for mainte-

Fig. 19-30. Both amateur and professional growers have found value in basement gardens. They allow the amateur to increase the size of his hobby and enable the professional grower to utilize unproductive space for rooting of cuttings and growth of seedlings.

Fig. 19-31. Recommended Illuminances for Acclimatized Plants (14 Hours of Light Per Day)

A. Trees 1.5 to 3 Meters (5 to 10 Feet) Tall

Tree	Illuminances	
	Lux	Footcandles
Araucaria excelsa (Norfolk Island Pine)	above 2000	above 200
Eriobotrya japonica (Chinese Loquator, Japan Plum)	above 2000	above 200
Ficus benjamina 'Exotica' (Weeping Java Fig)	750-2000	75-200
Ficus lyrata (Fiddleleaf Fig)	750-2000	75-200
Ficus retusa nitida (Indian Laurel)	750-2000	75-200
Ligustrum lucidum (Waxleaf)	750-2000	75-200

B. Floor plants 0.6 to 1.8 Meters (2 to 6 Feet) Tall

Plant	Illuminances	
	Lux	Footcandles
Brassaia actinophylla (Schefflera)	750-2000	75-200
Chamaedorea elegans 'bella' (Neanthe Bella Palm)	250-750	25-75
Chamaedorea erumpens (Bamboo Palm)	250-750	25-75
Chamaerops humilis (European Fanpalm)	above 2000	above 200
Dieffenbachia amoena (Giant Dumb Cane)	750-2000	75-200
Dizygotheca elegantissima (False Aralia)	above 2000	above 200
Dracaena deremensis 'Janet Craig' (Green Drasena)	750-200	75-200
Dracaena fragrans massangeana (Corn Plant)	250-750	25-75
Dracaena marginata (Dwarf Dragon Tree)	750-2000	75-200
Ficus elastica 'Decora' (Rubber Plant)	750-2000	75-200
Ficus philippinensis (Philippine Fig)	750-2000	75-200
Howeia forsteriana (Kentia Palm)	250-750	25-75
Philodendron x evansii (Selfheading Philodendron)	750-2000	75-200
Phoenix roebelenii (Pigmy Date Palm)	750-2000	75-200
Pittosporum tobira (Mock Orange)	above 2000	above 200
Podocarpus macrophylla Maki (Podocarpus)	above 2000	above 200
Polyscias guilfoylei (Parsley Aralia)	750-2000	75-200
Rhapis exclesa (Lady Palm)	750-2000	75-200
Yucca elephantipes (Palm-Lily)	above 2000	above 200

Fig. 19-31. Continued

C. Table or desk plants

Plant	Illuminances	
	Lux	Footcandles
Aechmea fasciata (Bromeliad)	750-2000	75-200
Aglaonema commutatum (Variegated Chinese Evergreen)	250-750	25-75
Agalonema 'Pseudobacteatum' (Golden Aglaonema)	250-750	25-75
Aglaonema roebelinii (Peuter Plant)	250-750	25-75
Asparagus sprengeri (Asparagus Fern)	750-2000	75-200
Ciccus antarctiva (Kangaroo Vine)	above 2000	above 200
Cissus rhombifolia (Grape Ivy)	750-2000	75-200
Citrus mitis (Calamondin)	above 2000	above 200
Dieffenbachia 'Exoctica' (Dumb Cane)	750-2000	75-200
Dracaena deremensis 'Warneckei' (White Striped Dracaena)	750-2000	75-200
Dracaena fragrans massangeana (Corn Plant)	250-750	25-75
Hoya carnosa (Wax plant)	750-2000	75-200
Maranta leuconeura (Prayer Plant)	750-2000	75-200
Nephrolepsis exaltata bostoniensis (Boston Fern)	750-2000	75-200
Peperomia caperata (Emerald Ripple)	250-750	25-75
Philodendron oxycardium (cordatum) (Common Philodendron)	250-750	25-75
Spathiphyllum 'Mauna Loa' (White Flag)	750-2000	75-200

nance of plants. Any white light source may be used to provide this light, including incandescent, self-ballasted mercury, mercury (phosphor coated), "white" or sunlight simulating fluorescent, or metal halide (with or without phosphor coating). These lamps are listed in increasing order of their efficacy. Since this is dual purpose lighting for plants and people, the light source has to be sufficiently suitable for both.

Regardless of source used, the visual comfort criteria appropriate to the space should be considered. Moreover, the lighting system should preferably be on a separate electrical circuit with an electrical timer to provide a consistent daylength for the plants.

Aquarium and Terrarium Lighting. Aquaria and terraria are also found in the home, office and school for hobby, decorative and educational purposes.

Aquarium lighting serves both a functional and an ornamental purpose when plants are part of the aquarium environment. Through the process of photosynthesis, lighted aquarium plants increase the oxygen level essential for fish respiration and at the same time reduce the carbon dioxide level, preventing the buildup of carbonic acid which can be harmful to fish. The light also illuminates both the fish and the aquarium. Dramatic colors of both fish and plants are observed when special fluorescent plant growth lamps are used because of the high red and blue emission of such lamps. Colors are more natural with fluorescent lamps which produce the daylight spectrum.

Both fluorescent (T-5, T-8, T-12) and incandescent lamps are used, with a preference for fluorescent because they produce more light and less heat per watt. Lighting requirements for aquaria usually range from 0.25 to 0.5 lamp watt per liter (one to two lamp watts per gallon) of tank capacity.

Terrarium lighting usually requires both fluorescent and incandescent lighting. Fluorescent light is applied at about 200 lamp watts per square meter (20 lamp watts per square foot) for the plant life while an incandescent lamp is used to light a portion of the terrarium to simulate the infrared of sunlight for the animal life (lizards, frogs, etc.) usually found in such environments.

EFFECTS ON MATERIALS

Fading and Bleaching[91-99]

Fading and bleaching of colored textiles and other materials upon exposure to light and other radiant energy is of special interest because of the higher illuminances now employed in merchandising. Consequently, a knowledge of some of the factors involved is important. Some of these (not necessarily arranged in order of importance) are as follows:

1. Illuminance.
2. Duration of exposure.
3. Spectral distribution of the radiation.
4. Moisture.
5. Temperature of the material.
6. Chemical composition of dye or other colorant.
7. Saturation of dye (tints versus saturated colors).
8. Composition and weave of fabric.
9. Intermittency of exposure.
10. Chemical fumes in the atmosphere.

While many researches on fading and colorfastness have been carried out and results published, especially in textile journals, most of them are deficient in data on the illuminances involved. In general, the primary purpose has been improvements in dyes and dyeing methods. Such tests have involved exposures to daylight in various geographical regions and to standardized types of arc lamps, so-called "fading lamps." The National Bureau of Standards has developed standardized methods of conducting such tests and is prepared to standarize carbon arcs in terms of their standards.

One publication[91] presenting a lengthy review, with extensive bibliography, of researches in this field summarizes the subject in this way:

The rate at which a dye fades is governed by seven factors:

1. The photochemistry of the dye molecule.
2. The physical state of the dye.
3. The chemistry of the substrate.
4. The fine structure of the substrate.
5. The presence of foreign substances.
6. The atmosphere.
7. The illumination.

From publication in technical journals over the past years certain generalized conclusions have been derived. However, in view of the fact that the tests have been limited to an infinitesimal percentage of the dyes and textiles in general use, it must be realized that such conclusions have limited application and that many exceptions will be found. With this reservation in mind, the *average* results for several hundred specimens of colored textiles will be discussed.

The illuminance and duration of exposure to any particular light source are obviously the most important factors. Two researches[95, 96] indicate an approximate reciprocity relationship between time and illuminance in the production of fading, *i.e.*, the fading is dependent upon the product of these two factors and is substantially unaffected by variations in both as long as the product is unchanged. A third study[97] disagrees with this conclusion, indicating that the relationship varies from direct reciprocity at higher illuminances.

The spectral distribution of the radiant energy used affects the rate of fading. It has been found that ultraviolet energy of wavelengths shorter than 300 nanometers[93] (not present in energy radiated by most light sources) may cause very rapid fading—and other forms of product deterioration—in some cases. Energy in the region from 300 to 400 nanometers is present in the radiation from most electric light sources in common use but to a much smaller amount per lumen than in daylight. This spectral region apparently produces more fading per unit of energy than an equal amount in the visible spectrum. Absorption of the near ultraviolet in sunlight by filters which absorb very little energy in the visible spectrum has been found to reduce the fading somewhat,[94] but not by as large an amount as is sometimes suggested. In sunlight it has been found that fading is produced by energy through the whole region shorter than the orange-red (approximately 600 nanometers).

Because of the greater amount of energy in the near ultraviolet and blue, per lumen of daylight, than in the light from tungsten and fluorescent lamps, it is logical to expect a somewhat higher rate of fading in daylight, and this is found to be the case. Fig. 19-32[92] shows the results of comparative tests of 108 colored ribbons under daylight, tungsten filament, and daylight fluorescent lamps. While daylight fluorescent lamps with since-obsoleted phosphors were used in that test, all evidence at hand would indicate that results would not be greatly different if lamp colors most commonly used at present had been used. The specimens were exposed to sunlight and skylight through 3-millimeter (⅛-inch) window glass between 9:00 a.m. and 3:30 p.m. on clear days in midsummer. The exposures in lux-hours (footcandle-hours) were automatically recorded throughout the test.

Fig. 19-32. Distribution of 108 Colored Textiles with Respect to Exposure-Ratios (Lux-Hours or Footcandle-Hours) Required to Produce Equal Amounts of Fading with Three Light Sources

Exposure Relative to Daylight	Number of Specimens Equally Faded	
	Incandescent	Daylight Fluorescent
Below 0.5	4	4
0.51 to 0.75	12	9
0.76 to 1.00	12	12
1.01 to 1.25	10	14
1.26 to 1.50	15	13
1.51 to 1.75	14	17
1.76 to 2.00	8	10
2.01 to 2.25	14	14
2.26 to 2.50	2	1
2.51 to 3.00	2	3
3.00 to 4.00	9	6
Greater than 4	6	5
	108	108

Average ratio
Incandescent filament light 1.81
Daylight fluorescent 1.68
For 110 specimens the *average* exposure ratio, daylight fluorescent to incandescent-filament light, was 0.99.

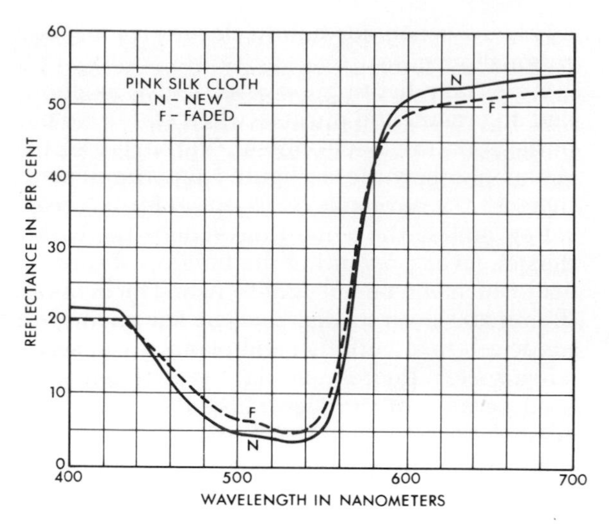

Fig. 19-33. Spectral reflectance of a specimen of pink silk before and after exposure sufficient to cause moderate fading.

Fig. 19-33 shows spectral reflectance curves for new and slightly faded specimens of pink silk cloth. It will be noted that the spectral changes indicate bleaching in regions of maximum absorption and darkening in regions of minimum absorption. These changes are typical of many specimens tested.

Fading appears to be a photochemical process requiring oxygen and is inhibited or greatly reduced in a vacuum.

An increase in moisture content can cause a very large increase in fading, on cellulose in particular, but has less effect on wool.[91]

Temperature appears to have little effect on fading rate of silk and cotton at temperatures below 50 °C (120 °F), but the rate is approximately twice as great at 65 °C (150 °F) as at 30 °C (85 °F).

It is often found that a light tint is more fugitive than a higher concentration of the same dye.

The permissible exposure before some fading may be expected is of major importance in the merchandising field. Tests of approximately 100 textile specimens performed about 1940 showed that half of them showed some fading after exposures of approximately 500 kilolux-hours (50,000 footcandle-hours) to incandescent lamps. A more recent research[97] showed that it required approximately ten times that exposure to produce a minimum perceptible fading with incandescent and fluorescent lamps on more than 100 commercial fabrics. Obviously, improvements in dyes have greatly improved their light-fastness, and fading of merchandise in display cases is not the critical problem which it was when the fluorescent lamp first became an important factor in lighting, leading to usage of greatly increased illuminances.

Fading of merchandise is most readily apparent under conditions where an area receiving a high illuminance is adjacent to areas not exposed. Typical examples are folded neckties and socks stacked on shelves in display cases, the folded edges approaching the lamp closely. In order to reduce the hazard of fading, goods displayed on shelves near the lamps might be rotated to lower shelves on a 1-week to 10-day cycle.

In many modern grocery stores, especially those of the self-service type, packaged meats are displayed in refrigerated cases with relatively high illuminances. Fresh meats show no appreciable color change due to light within any reasonable display period, although unwrapped meats may show changes due to dehydration. Many of the processed meats (veal and pork loaves, bologna, etc.) receive their red color from a curing process using salt or sodium nitrate. Through some reaction of light and air, the processed meats will return to their original or grayed color, and this "fading" takes place very rapidly with some meats. Some, especially veal loaf and bologna, will show perceptible color change in 1500 to 2000 lux-hours [150 to 200 footcandle-hours]. Since the illuminance in some of these cases may be as high as 1000 lux [100 footcandles] it is found that undesirable changes may occur in 1½ to 2 hours. The most susceptible meats should be placed as far away from the lamps as possible.

Research has shown that lighting levels from 500 lux to 2000 lux [50 footcandles to 200 footcandles] have approximately the same effect on shelf life of frozen meats.[98] Depending on the degree of original muscle pigmentation, frozen meat was considered saleable for three to six days; however, above 2000 lux [200 footcandles] shelf life was considerably reduced. Differences in spectral power distribution of the light sources resulted in no apparent or statistically significant differences in rate of color degradation.

Cigars displayed in cases illuminated by fluorescent lamps within the cases may be somewhat bleached by the light, but the exposures required are very long. In a test using seven brands of cigars it was found that an exposure of approximately 400 kilolux-hours [40,000 footcandle-hours] produced a just-noticeable change, and that this exposure could be doubled before the color change reached an objectionable degree. Exposures of this magnitude are much greater than any to be expected in normal merchandising of cigars.

The germicidal (bactericidal) lamp, producing high energy at 253.7 nanometers, has been used as a potent source for accelerated fading.[93] However, these lamps should not be used for accelerated fading tests because no relationship has been found between fading by germicidal lamps and fading by sunlight or commonly used artificial light sources.

Luminescence and Luminescent Materials

The emission of light resulting from causes other than thermal stimulation (incandescence) is known as luminescence. Some of the more important luminescence terminology are discussed in Section 2 and defined in Section 1 of the 1981 Reference Volume.

Photoluminescence occurs to a practical degree in many hundreds of materials when they are exposed to radiation from long wavelengths in the visible through the ultraviolet to the x-ray and gamma-ray regions. The most important practical application of photoluminescent materials (lamp phosphors) is in light sources where mercury ultraviolet is the exciting radiation. These phosphors are oxygen dominated inorganic crystalline materials. Other materials such as the zinc and cadmium sulfides and a wide variety of organic compounds excited by the near ultraviolet (approximately 360 nanometers) are used extensively to achieve spectacular theatrical effects and in various signs and instrument dials. So-called "optical bleaches" are fluorescent organics used as whiteners in laundered items such as shirts, sheets, etc.; they are excited by the near ultraviolet and radiation in daylight to fluoresce a bright blue, thus compensating for the natural yellow-white appearance of the unimpregnated cloth. Super bright orange and red organic dyes which are fluorescent under near ultraviolet excitation are widely used as identification and warning markers—for example in high speed aircraft to aid in rapid visual acquisition to avoid collision. Fluorescent paint, ink and dyed fabrics are available in many colors, including red, orange, yellow, blue and a white that appears blue under ultraviolet. Because these materials transform ultraviolet, violet, and even blue energy into light, as well as reflect incident light, their brightness under daylight is striking. This is true because of the ultraviolet energy in daylight, which, after striking the materials, returns to the eye as light in addition to the daylight reflected by the materials and gives some fluorescent materials an apparent reflectance (under daylight) of 110 per cent or more; that is, they send back more light of a given color than strikes them.

These colored fluorescent materials are especially useful on signal flags and signal panels since they can be identified at greater distances than those with nonfluorescent surfaces. The increased range over which the fluorescent flags can be identified is most apparent during the half-light conditions of dawn and twilight. Organic fluorescent dyed materials are at times used to produce spectacular signs, such as used on streetcars or buses. This use of fluorescent paints and dyes is commercially applicable wherever the long-distance identification of objects is important throughout the hours of daylight. Such materials are also used to produce very colorful clothing.

Other photoluminescence applications include x-ray and γ-ray stimulable crystals which find extensive use in scintillation counters—used for detecting the exciting radiation itself. Chemical analyses are often based upon the use of the characteristic luminescence of certain activator ions in known host media.

Cathodoluminescent materials find their most important application in television screens and in scientific instrumentation such as oscilloscopes, electron microscopes, image intensifiers and radar screens. Here zinc and cadmium sulfides and oxygen dominated phosphors such as the silicates, phosphates and tungstates are used. An improvement in color television screens has resulted from the development of a rare earth (europium) activated deep red phosphate phosphor.

Luminescence attending chemical reactions has been observed widely in both organic and inorganic systems. One of the most interesting is the reaction between the naturally occurring chemicals luciferin, luciferase and adenosine triphosphate (ATP) as it takes place in the firefly.

Ion, sound, friction and electric field excitation of phosphors remain essentially phenomena of relatively little practical application except that the latter has found use in read-out devices.

Phosphorescent Materials. Phosphorescent materials, excited by ultraviolet energy, daylight, or light from electric lamps, have been shown to have a high brightness of afterglow for periods of from 6 to 9 hours, and some of these have a noticeable brightness for as long as 24 hours after the exciting source has been removed.

Phosphorescent materials, generally combinations of zinc, calcium, cadmium and strontium sulfides, can be incorporated into adhesive tapes (plastic over-coatings), paints, and certain molded plastics. Because of the tendency of many plastics either to transmit moisture—which decomposes the sulfide—or to react directly with the phosphor, care should be exercised in the choice of a plastic to carry the phosphorescent powders. Both vinyl and polystyrene plastics have been found well suited to this application.

Phosphorescent materials are suitable only for applications where exposure to light prior to use is possible. While some can be used in spots where a visible brightness is necessary for from 6 to 9 hours, only a few of the many phosphorescent compounds have this degree of persistence. Those manufactured from zinc sulfide have high initial brightness after the light source has been removed, but their useful brightness period does not extend beyond 20 to 30 minutes. Before refinements in the processing of calcium and strontium phosphors were made in 1944, the useful brightness of these types did not extend beyond from 2 to 3 hours after activation. However, now that long-persistence phosphors are available phosphorescent materials are, in many applications, suitable for night-long use.

Brightness reduction (decay) rates are hastened by high temperatures. At very low temperatures (60 kelvin) luminescence may be completely arrested to be released later upon warming.

Radioactive Excitation. This is simply excitation by electrons, ions (atoms, nuclear fragments), or gamma rays—singly or in combination—resulting from the fission or radioactive decay of certain elements. For example, radium emits alpha particles which can excite luminescence when they strike a suitable phosphor. Krypton-85 excites by emission of beta rays or high energy electrons. The sulfide phosphors not only emit light when exposed to ultraviolet energy or light, but also exhibit this property under bombardment by the rays from radioactive materials. Thus, by compounding a mixture of such a radioluminescent material, *e.g.*, zinc sulfide, and a small amount of radioactive material, a self-luminous mixture can be produced. Such a radioactive luminous compound will continue to emit light without the help of external excitation for a very long time (several years) in practical applications. Radioactive-luminous materials have been used for many years on watch and clock dials, and on the faces of other instruments that must be read in the dark. They are the only type of commercially available luminous materials that maintain self-luminosity over long periods of time. The power source was formerly some salt of radium or more frequently the lower-priced mesothorium. Later strontium-90, and more recently polonium, which has certain advantages with regard to cost and safety, are also being used. The bombardment of the fluorescent materials by the radiations from the exciting radioactive materials causes a decomposition of the fluorescent materials, which, of course, limits their life. A good-quality material will be useful for a few years and will maintain a relatively constant brightness during this period. The actual life of a radioactive-luminous paint is controlled to a great extent by its initial brightness, which is varied by changing the concentration of radioactive material in the mixture. Increased brightness means increased radioactive content, and more rapid decomposition of the glowing salt.

Because of the expense of the radioactive substances used to activate this material, radium-luminous paint seldom is used in large quantities or to cover large areas.

EFFECTS OF INFRARED ENERGY[100-109]

Heat may be transferred from one body to another by conduction, convection, or radiation, or by a combination of these processes. Infrared heating involves energy transfer primarily by radiation, although some convection heating may exist simultaneously due to natural or forced air movement.

Transfer of energy or heat occurs whenever any amount of radiant energy emitted by one body is absorbed by another. However, it is the electromagnetic spectrum wavelengths longer than those of visible energy and shorter than those of radar waves that are utilized for radiant heating (770 to 100,000 nanometers). Energy absorption of white, pastel colors, and translucent materials is best obtained by using wavelength emissions longer than 2500 nanometers, whereas the majority of dark pigmented and oxide-coated materials will readily absorb the full range of emissions—visible as well as infrared. Water vapor, steam and other gasses absorb infrared in characteristic bands throughout the spectrum.

Glass and quartz materials effectively transmit infrared energy only out to about 5000 nanometers.

Sources of Infrared Energy

Many sources for producing infrared energy are now available. These can be classified generally as point, line and area sources. Their temperatures, spectral power distribution, and life characteristics vary widely, although source selection generally is not critical unless the products to be heated are selective as to wavelength penetration or absorption as in the case of many translucent plastics.

Maximum design flexibility and economy for industrial installations are generally obtained by using the tungsten filament quartz lamps, alloy resistor quartz tubes, or rod type metal sheath heaters in air or water cooled external reflectors. These are available in power ratings up to 24 watts per millimeter (600 watts per inch) of length, and in sizes from 9.5-millimeters (3/8-inch) diameter by 200 millimeters (8 inches) long to 19-millimeters (3/4-inch) diameter by 1600 millimeters (63 inches) long. Additionally, a variety of screw base lamps, with and without internal reflectors, are available for special applications. All are listed in 1981 Reference Volume, Section 8, Fig. 8–110. Precise voltage ratings as used for lighting service lamps need *not* be followed in using these infrared sources, often employed at voltages as low as 50 per cent of manufacturers' rating. Most of the tungsten filament heaters are designed for a color temperature of 2500 K and a life rating of 5000 hours operated at rated voltage. Metal rod heaters and quartz tubes using coiled alloy resistors are usually designed for 790 °C (1450 °F) operating temperature and a life span of approximately 10,000 hours.

Tungsten filament heaters provide instant on-off response from a power source, and their radiant energy efficiency at 86 per cent of power input makes them a preferred infrared source. Other heaters have thermal inertia varying from about 1 minute for quartz tubes to 4 or 5 minutes for metal sheath heaters. Operating efficiencies are substantially influenced by the design and maintenance of external reflector systems and to a lesser extent by air temperature and velocity within the heating zone. Over-all efficiencies of 35 to 60 per cent are readily obtained in well designed systems where long holding time at designed product temperature is not required. All of the quartz heat sources can accept high thermal shock. However, metal heaters are best qualified for applications subject to mechanical shock and vibration. A variety of porcelain holders and terminals are available for these sources. Specular reflectors of anodized aluminum, gold or rhodium are recommended to direct the radiant energy to product surfaces as desired.

By comparison, gas infrared systems require far heavier and more costly construction to comply with insurance safety standards. Their operating efficiencies must take into account energy loss in the combustion flue products plus other design factors as to eventual energy utilization.

Advantages

The ease with which electric infrared heating can be controlled quickly and reliably with modern radiation detectors and power control devices has greatly advanced its acceptance and range of use.

The ability to focus heat energy on the object to be heated and the ability to turn the source on and off quickly minimizes the heating of the surrounding atmosphere and structure and can reduce the total energy needed.

Heating the object directly eliminates the need for a closed oven, preheating, heavy insulating structures, and the high heat losses to the structure and to the air when opened.

Electric infrared can be used outdoors, in a vacuum, or in other hostile environments when conventional heating systems cannot.

Designed and applied with reasonable care and proper controls, electric infrared can reduce heating energy requirements and costs.

Processes that once required great quantities of fossil fuels and hours in convection ovens, or periods of several minutes in early infrared systems, are often handled in seconds with present day coordination of materials, chemistry and infrared application. The absence of combustion fuel hazard with need for handling large volumes of air, and the space saving afforded for high volume, clean, quality results, keep this form of electric heat in the foreground of manufacturing technology.

Product Heating with Infrared

Infrared radiant energy may be used for any heating application where the principal product surfaces can be arranged for exposure to the heat sources.[100] Modern conveyorized methods of material handling have greatly accelerated use, with heat sources arranged in banks or tunnels. Typical applications include:

1. Drying and baking of paints, varnishes, enamels, adhesives, printers' ink and other protective coatings.
2. Preheating of thermoplastic materials for forming and tacking operation, molds.
3. Heating of metal parts for shrink fit assembly, forming, thermal aging, brazing, radiation testing, and conditioning surfaces for application of adhesives and welding.
4. Dehydrating of textiles, paper, leather, meat, vegetables, pottery ware and sand molds.
5. Spot and localized heating for any desired objective.

Rapid rates of heating can be provided in relatively cold surroundings by controlling the amount of radiant energy, absorption characteristics of the exposed surfaces, and rate of heat loss to the surroundings.[101] Highly reflective enclosures, with or without thermal insulation, are commonly employed to assure maximum energy utilization. Limited amounts of air movement are often essential in portions or all of the heating cycle, to avoid temperature stratification and assure removal of water or solvent vapors.[102] Product temperature control is normally provided by the exposure time to infrared energy, or by the wattage of heaters employed per unit area of facing tunnel area. With modern linear heaters, power densities of 5 to 130 kilowatts per square meter (0.5 to 12 kilowatts per square foot) can accommodate high automation speeds.

Where precise temperatures are needed, the design condition may then be modified by voltage or current input controls to add flexibility for a variety of product conditions, handling speeds, chemical formulations or other, as the process may require. The temperature of moving parts can be accurately measured by scanning with a radiation pyrometer to provide indication or full automatic control of the heating cycle. Where small variations in temperature can still meet quality standards desired, an initial installation test may be made with portable instrumentation and thereafter, the cycle will repeat itself with a degree of reliability consistent with the power supply voltage, thus avoiding need for the usual controls required for other types of process heating.

Spot heating of a portion of an object can eliminate the need for energy formerly required to preheat the whole object. Appropriate applications of infrared heating can lead to more efficient use of all energy forms in today's modern production facility.

Comfort Heating with Infrared

In recent years, the use of infrared radiation for personnel heating in commercial and industrial areas has become quite popular. The T-3 quartz lamp as a semi-luminous infrared source has distinguished itself for a wide variety of applications in commercial buildings, marquee areas, industrial plants, warehouses, stadiums, pavilions and other public areas. Units are usually of the pendant or recessed type, with reflector control for the combined visible and infrared radiation, using the quartz tube and quartz lamp sources listed in 1981 Reference Volume, Section 8, Fig. 8-110. In contrast, residential use except for bathroom areas is mostly confined to low temperature sources such as electric base boards and plastered radiant ceilings.

By supplying heat only when and where needed, thermostats of the conventional heating systems can be lowered while comfort is maintained locally and energy conserved over-all.

Applications. Radiant comfort heating applications fall into two broad classifications—general heating and spot heating.[103] General heating installations irradiate complete room areas. Because of economics, this type of heating is most common in areas where fossil fuels are not available or are very limited and quite expensive. High levels of building insulation are generally recommended. The installation in this case often consists of a uniform radiant wattage density in the range of 100 to 320 watts per square meter (10 to 30 watts per square foot) incident to the floor surface. However, some system designers prefer equipment layouts using asymmetric units to provide a somewhat higher density in the areas adjacent to outside walls to help off-set the wall thermal loss. To date, like convection heating systems, over-all radiant systems have an installed capacity sufficient to hold the desired indoor temperature and overcome the building heat loss at the specified outdoor design temperature; but performance data on some installations indicate a capacity 70 to 90 per cent of the building thermal loss is adequate.[104] This reduction is probably due to the direct personnel heating and an improved mean radiant temperature in the space.

Infrared energy passes through air with little absorption and this is particularly true of near infrared. Therefore, installations involving quartz infrared lamps may be mounted at much greater heights than those with far infrared sources. By selecting equipment of narrower beam spread so the radiation can be confined primarily to the floor where it will be most beneficial with limited losses through the walls, the mounting height may be increased without requiring a greater installed capacity. It is good practice to keep the radiation from striking the walls at heights more than 2.4 meters (8 feet) above the floor to limit wall losses.

Although infrared heating of the air is minimal, air in radiation heated areas is warmed from energy absorption by the floor and other solid surfaces. This causes upward gravitational movement, permitting conventional control with air thermostats[104] which are shielded from the infrared sources.

In the case of quartz tubes, metal sheath heaters, and gas-fired infrared units, on-off cycling of the equipment is permissible. When lamps are used, unless the visible energy (about 7 to 8 lumens per watt) is reduced by filtering, the cycling should be from full to half-voltage to prevent a severe change in illumination. Operation of this sort not only gives some lighting but also provides some amount of radiant heat at all times.

Infrared heating systems have an advantage over convection air heating systems for spaces that are subject to high air changes per hour (for example, where overhead doors are opened frequently). In these areas, the warm air is lost immediately and air temperature recovery can be lengthy with convection heating. With a radiant system, most objects are warmer than the air so the air in the space recovers temperature faster.

Spot Heating. The greatest potential use for high intensity radiant heating lies in spot or zone heating in exposed areas where conventional heating is impractical such as marquees, waiting platforms, loading docks; and in infrequently used areas such as stadiums, arenas, viewing stands, churches and assembly halls.

The radiation intensity needed for spot heating varies with a number of factors. The major ones are:

1. The degree of body activity as dictated by the task. The more physical effort expended by the worker, the lower the temperature at which heat is needed. Type of clothing also influences this temperature.
2. The minimum temperature which is apt to exist in the space (or the lowest temperature at which the owner wants to provide comfort).
3. The amount of air movement at the location. Indoor drafts and slight air movements outdoors can be overcome by higher energy densities, but compensation for wind velocities of more than

2.2 to 4.5 meters per second (5 to 10 miles per hour) at temperatures below −1 °C (30 °F) wili not be sufficiently rewarding. Wind screens are far more beneficial than increased radiation levels.

For spot heating, units should be positioned to supply radiation from at least two directions,[103] preferably above and to the side of the area to be heated. Care should be taken to avoid locating equipment directly over a person's head.[104] In practice, levels for spot heating vary from 100 watts per square meter (10 watts per square foot) (at waist level) for an indoor installation supplementing an inadequate convection system to more than 1 kilowatt per square meter (100 watts per square foot) for a marquee or sidewalk people heating system.

At the higher radiation levels, ice and snow are melted[105] and water on the floor is evaporated. This can reduce the safety hazard of a slippery floor and improve housekeeping by minimizing the tracking in of snow and water in inclement weather. Where snow melting is desirable, the heating units should be energized as soon as snow starts to fall to avoid any accumulation and the consequent high reflection of infrared energy.

Infrared heating installations in infrequently used areas can often be turned on before an event to preheat the room surfaces, then turned off before the event is over, with the heat stored in the surfaces and body heat maintaining the comfort level.

REFERENCES

Effects of Radiation on the Eye

1. Ham, W. T., Geeraets, W. J., Williams, R. C., Guerry, D., and Mueller, H. A.: "Laser Radiation Protection", *Proc. First Int. Congr. of Radiat. Prot.*, pp. 933-943, Pergamon Press, New York, 1968.
2. Ham, W. T., Jr., Clarke, A. M., Geeraets, W. J., Cleary, S. F., Mueller, H. A., and Williams, R. C.: "The Eye Problem in Laser Safety", *Arch. Environ. Health*, Vol. 20, pp. 156-160, February 1970.
3. Ham, W. T., Jr., Williams, R. C., Mueller, H. A., Guerry, D., Clarke, A. M., and Geeraets, W. J.: "Effects of Laser Radiation on the Mammalian Eye", *Trans. N.Y. Acad. Sci.*, Vol. 28, pp. 517-526, February, 1966.
4. Ham, W. T., Jr., Mueller, H. A., Goldman, A. I., Newman, B. E., Holland, L. M., and Kuwabara, T.: "Ocular Hazard from Picosecond Pulses of Nd:YAG Laser Radiation", *Science*, Vol. 185, pp. 362-363, July 26, 1974.
5. Ham, W. T., Mueller, H. A., Williams, R. C., and Geeraets, W. J.: "Ocular Hazard from Viewing the Sun Unprotected and Through Various Windows and Filters," *Appl. Opt.*, Vol. 12, No. 9, pp. 2122-2129, September, 1973.
6. Ham, W. T., Jr., Mueller, H. A., and Sliney, D. H.: "Retinal Sensitivity to Damage from Short Wavelength Light", *Nature*, Vol. 260, No. 5547, pp. 153-155, March 11, 1976.
7. Goldman, A. I., Ham, W. T., Jr., and Mueller, H. A.: "Mechanisms of Retinal Damage Resulting from the Exposure of Rhesus Monkeys to Ultrashort Laser Pulses", *Exp. Eye Res.*, Vol. 21, No. 5, pp. 457-469, 1975.
8. Goldman, A. I., Ham, W. T., Jr., and Mueller, H. A.: "Ocular Damage Thresholds and Mechanisms for Ultrashort Pulses of Both Visible and Infrared Laser Radiation in the Rhesus Monkey", *Exp. Eye Res.*, Vol. 24, pp. 45-46, 1977.
9. Matelsky, I.: "The Non-Ionizing Radiations", *Industrial Hygiene Highlights*, (L. V. Cralley, Editor), Industrial Hygiene Foundation of America, Pittsburgh, pp. 140-178, 1968.
10. Sliney, D. H.: "Non-Ionizing Radiation in Industrial Environmental Health", *The Worker and the Community*, (Cralley, L. V., Editor), Vol. 1, pp. 171-241, Academic Press, New York, 1972.
11. Sliney, D. H. and Freasier, B. C.: "The Evaluation of Optical Radiation Hazards", *Appl. Opt.*, Vol. 12, pp. 1-22, January, 1973.
12. Sliney, D. H.: "The Merits of an Envelope Action Spectrum for Ultraviolet Exposure Criteria", *Am. Ind. Hyg. Assoc. J.*, Vol. 33, pp. 644-653, October, 1972.
13. Sliney, D. H. and Wolbarsht, M. L.: *Safety with Lasers and Other Optical Sources*, Plenum Publishing Co., New York, 1980.
14. Zigman, S., Datiles, M. and Torczynski, E.: "Sunlight and Human Cataracts", *Invest. Opthal.*, Vol. 18, No. 5, pp. 462-467.

Skin Effects of Radiation

15. Bachem, A. and Reed, C. I.: "The Penetration of Ultraviolet Light Through the Human Skin", *Arch. Phys. Ther.*, Vol. 11, pp. 49-56, 1930.
16. Sams, W. M.: "Inflammatory Mediators in Ultraviolet Erythema," *Sunlight and Man*, (T. B. Fitzpatrick, Editor), p. 143, University of Tokyo Press, Tokyo, 1974.
17. Daniels, F. Jr. and Johnson, B. E.: "Normal, Physiologic and Pathologic Effects of Solar Radiation on the Skin," *Sunlight and Man*, (T. B. Fitzpatrick, Editor), University of Tokyo Press, Tokyo, 1974.
18. Quevedo, W. Jr. et al.: "Light and Skin Color", *Sunlight and Man*, (T. B. Fitzpatrick, Editor), University of Tokyo Press, Tokyo, 1974.
19. Everett, M. A. et al.: "Physiologic Response of Human Skin to UV Light", *Biologic Effects of Ultraviolet Radiation*, (F. Urbach, Editor), p. 181, Pergamon Press, Oxford, 1969.
20. Freeman, R. et al.: "Requirements for an Erythemal Response", *J. Invest. Dermatol.*, Vol. 47, p. 586, 1966.
21. Berger, D. et al: "Action Spectrum of Erythema", *13th Int. Dermatol. Cong.*, Munich, 1967.
22. Cutchis, P.: "On the Linkage of Solar Ultraviolet Radiation to Skin Cancer," *F.A.A., EQ-78-19 IDA*, paper p-1342, September, 1978, Unclassified.
23. Pathak, M. and K. Stratton.: "Effects of Ultraviolet and Visible Radiation and the Production of Free Radicals in Skin", *Biological Effects of Ultraviolet Radiation*, (F. Urbach, Editor), P. 207, Pergamon Press, Oxford, 1969.

Vitamin D and Calcium Metabolism

24. Neer, R., Clark, M., Friedman, V., Belsey, R., Sweeney, M., Buoncristiani, J., and Potts, J. T. Jr.: "Environmental and Nutritional Influences on Plasma 25-Hydroxyvitamin D Concentration and Calcium Metabolism in Man," *Vitamin D: Biochemical, Chemical and Clinical Aspects Related to Calcium Metabolism, Proc. 3rd Workshop on Vitamin D*, (Norman A. W., Schaefer, K., Coburn, J. W., DeLuca, H. F., Fraser, D., Grigoleit, H. G., and Herrath, D. V., Editors), pp. 595-606, Walter de Gruyter, Berlin, N. Y., 1977.
25. Haussler, M. R. and McCain, T. A.: "Basic and Clinical Concepts Related to Vitamin D Metabolism and Action", *New Engl. J. Med.* Vol. 297, pp. 974-983, 1977.

26. Haussler, M. R.: "Vitamin D", *Nutrition Reviews' Present Knowledge in Nutrition*, 4th Ed., pp. 82-97, The Nutrition Foundation, Inc., New York, Washington, 1976.
27. *The Vitamins.* (Serbrell, W. H. and Harris, R. S., Editors), Vol. 3, pp. 156-290, Academic Press, New York, 1971.
28. Committee on Nutrition, American Academy of Pediatrics: "The Prophylactic Requirement and the Toxicity of Vitamin D", *Pediatrics*, Vol. 31, pp. 513-525, 1963.

Biological Rhythms

29. Brown, F. A., Jr., Hastings, J. W., and Palmer, J. D.: *The Biological Clock*, Academic Press, New York, 1970.
30. Luce, G. G.: *Body Time*, Granada Publishing Ltd., Frogmore, St. Albans, Herts, Great Britian, 1973 (reprinted 1977).
31. Wurtman, R. J.: The Effects of Light on the Human Body", *Sci. Am.*, pp. 69-77, July, 1975.
32. Bünning, E.: *The Physiological Clock*, Revised 2nd Ed., Springer-Verlag, New York, 1967.
33. McGuire, R. A., Rand, W. M., and Wurtman, R. J.: "Entrainment of the Body Temperature Rhythm in Rats: Effect of Color and Intensity of Environmental Light", *Science*, Vol. 182, pp. 956-957, 1973.
34. Zacharias, L. and Wurtman, R. J.: "Blindness: Its Relation to Age of Menarche", *Science* Vol. 144, pp. 1154-1155, 1964.
35. Lynch, H. J., Wurtman, R. J., Moskowitz, M. A., Archer, M. C. and Ho, M. H.: "Daily Rhythm in Human Urinary Melatonin", *Science*, Vol. 187, pp. 169-171, 1975.
36. Binkley, S.: "A Timekeeping Enzyme in the Pineal Gland", *Sci. Am.*, pp. 66-71, April, 1979.
37. Owen, V. M.: "Lighting and Livestock", *Light and Lighting*, pp. 328-329, 1968.
38. Wilson, W. O.: "Lighting Programs for Poultry", *Feedstuffs*, Vol. 48, pp. 39-40, 1977.

Phototherapy of Neonatal Hyperbilirubinemia

39. Sisson, T. R. C.: "Visible Light Therapy of Neonatal Hyperbilirubinemia", (K. C. Smith, Editor), *Photochem. Photobiol. Rev.*, Vol. 1, pp. 241-263, Plenum Press, New York, 1976.

Photochemotherapy

40. Parrish, J. A., Fitzpatrick, T. B., Tanenbaum, L., and Pathak, M. A.: "Photochemotherapy of Psoriasis with Oral Methoxsalen and Longwave Ultraviolet Light", *New Eng. J. Med.*, Vol. 291, pp. 1207-1212, 1974.
41. Parrish, J. A., Fitzpatrick, T. B., Shea, C., and Pathak, M. A.: "Photochemotherapy of Vitiligo with Oral Psoralen and a New High-Intensity Longwave Ultraviolet Light (UV-A) System", *Arch. Dermatol.*, Vol. 112, pp. 1531-1534, 1976.
42. Gilchrest, B. A., Parrish, J. A., Tanenbaum, L., Haynes, H. A., and Fitzpatrick, T. B.: "Oral Methoxsalen Photochemotherapy of Mycosis Fungoides," *Cancer*, Vol. 38, pp. 683-689, 1976.
43. Morison, W. L., Parrish, J. A., and Fitzpatrick, T. B.: "Oral Psoralen Photochemotherapy of Atopic Eczema," *Br. J. Dermatol.*, Vol. 98, pp. 25-30, 1978.
44. Parrish, J. A., LeVine, M. J., Morison, W. L., Gonzalez, E., and Fitzpatrick, T. B.: "Comparison of PUVA and Beta Carotene in the Treatment of Polymorphous Light Eruption", *Br. J. Dermatol.*, pp. 187-191, February, 1970.
45. Pathak, M. A., Kramer, D. M., and Fitzpatrick, T. B.: "Photobiology and Photochemistry of Furocoumarins (Psoralens)", *Sunlight and Man: Normal and Abnormal Photobiologic Responses*, (Pathak, M. A., Harber, L. C., Seiji, M., and Kukita, A., Editors; Fitzpatrick, T. B., Consulting Editor), pp. 335-368, University of Tokyo Press, Tokyo, 1974.
46. Lerman, S.: "A Method for Detecting 8-Methoxypsoralen in the Ocular Lens," *Science*, Vol. 197, pp. 1287-1288, 1977.
47. Lerman, S., Jocoy, M., and Borkman, R. D.: "Photosensitization of the Lens by 8-Methoxypsoralen," *Invest. Ophthalmol. Visual Sci.*, Vol. 16, pp. 1065-1068, 1977.
48. Stern, R. S., Thibodeau, L. A., Kleinerman, R. A., Parrish, J. A., Fitzpatrick, T. B., and 22 participating investigators: "Risk Factors and Increased Incidence of Cutaneous Carcinoma in Patients Treated with Oral Methoxsalen Photochemotherapy for Psoriasis", *New Eng. J. Med.*, April 12, 1979.
49. Photobiology Committee of the IES: "Risks Associated with the Use of UV-A Irradiators Being Used in Treating Psoriasis and Other Conditions", *Light. Des. & Appl.*, pp. 56-60, March, 1979.

Environmental Lighting Safety Criteria

50. Sliney, D. H.: "The Merits of an Envelope Action Spectrum for Ultraviolet Exposure Criteria", *Am. Ind. Hyg. Assn. J.*, Vol. 33, pp. 644-653, October, 1972.
51. Bickford, E. D., Clark, G. W., and Spears, G. R.: "Measurement of Ultraviolet Irradiance from Illuminants in Terms of Proposed Public Health Standards", *J. Illum. Eng. Soc.*, Vol. 4, p. 43 October, 1974.
52. *Threshold Limit Values for Physical Agents*, American Conference of Governmental Industrial Hygienists, Cincinnati, Ohio, 1979.
53. "Safe Use of Lasers," *ANSI Z136.1-1976*, American National Standards Institute, New York, 1976.
54. Sliney, D. H., et al.: "Laser Hazards Bibliography," U. S. Army Environmental Hygiene Agency, Aberdeen Proving Ground, Md., October, 1979.
55. Sliney, D. H. and Wolbarsht, M. L.: *Safety with Lasers and Other Optical Sources*, Plenum Publishing Corp., New York, 1980.

Germicidal (Bactericidal) Ultraviolet

56. Friedberg, E. C., Cook, K. H., Duncan, J., and Mortelmans: "DNA Repair Enzyme in Mammalian Cells," *Photochem. and Photobiol. Rev.*, (Smith, K. C., Editor), Vol. 2, pp. 263-322, Plenum Press, New York, 1977.
57. Harm, W., Rupert, C. S., and Harm, H.: "The Study of Photoenzymatic Repair of UV Lesions in DNA by Flash Photolysis," *Photophysiology*, (Giese, A. C., Editor), Vol. 6, pp. 279-324, Adademic Press, New York, 1971.
58. Kelner, A.: "Effect of Visibility on the Recovery of Streptomyces Griseus Conidia from Ultraviolet Irradiation Injury," *Proc. Natl. Acad. Sci. U.S.*, Vol. 35, pp. 73-79, 1949.
59. Setlow, J. K.: "The Molecular Bases of Biological Effects of Ultraviolet Radiation and Photoreactivation," *Curr. Top. Radiat. Res.*, Vol. 2, pp. 195-248, 1966.
60. Smith, K. C.: "Multiple Pathways of DNA Repair in Bacteria and their Roles in Mutagenesis," *Photochem. Photobiol.*, Vol. 28, No. 2, pp. 121-129, 1978.
61. Snapka, R. M. and Fuselier, C. O.: "Photoreactivating Enzyme from Escherichia Coli," *Photochem. Photobiol.*, Vol. 25, No. 6, pp. 415-420, 1977.

Insect Responses

62. Goldsmith, T. H.: "The Color Vision of Insects," *Light and Life*, (McElroy, W. D., and Glass, B., Editors), Johns Hopkins University Press, Baltimore, MD, 1961.
63. Baker, H. and Hienton, T. E.: "Traps Have Some Value," *Yearbook of Agriculture*, pp. 406-411, US Government Printing Office, Washington, DC, 1952.
64. "Response of Insects to Induced Light", *USDA Agricultural Research Service*, (ARS 20-10), July, 1961.
65. Hollingsworth, J. P., Hartsock, J. G., and Stanley, J. M.: "Electric Insect Traps for Survey Purposes," *USDA, Agricultural Research Service*, (42-3-1), January, 1963.

66. Hollingsworth, J. P. and Hartstack, A. W. Jr.: "Effect of Components On Insect Light Trap Performance," American Society of Agriculture Engineering, paper 71-803, December, 1971.

67. Barrett, J. R., Jr., Killough, R. A., and Hartsock, J. G.: "Reducing Insect Problems in Lighted Areas," *Trans., Am. Soc. Agric. Eng.*, Vol. 17, No. 2, pp. 329, 330, 338, 1974.

68. Barrett, J. R., Jr., Huber, R. T., and Harwood, F. W.: "Selection of Lamps for Minimal Insect Attraction," *Trans. Am. Soc. Agric. Eng.*, Vol. 17, No. 4, pp. 710–711, 1973.

Plant Growth

69. Bickford, E. D. and Dunn, S.: *Lighting For Plant Growth*, Kent State University Press, Kent, Ohio, 1972.

70. Bickford, E. D.: "Interiorscape Lighting," *Light. Des. Appl.*, Vol. 7, No. 10, pp. 22–25, 1977.

71. Langhans, R. W. (Editor): *A Growth Chamber Manual*, Cornell University Press, Ithaca, New York, 1978.

72. Cathey, H. M.: "Guidelines for the Germination of Annual, Pot Plant and Ornamental Herb Seeds", *Florists' Rev.*, pp. 26–29, 75–77, September 4, 1969.

73. Norton, R. A.: "Commercial Lighting of Bedding Plants" *Florists'* Rev., January 4, 1979.

74. Poole, R. T. and Conover, C. A.: "Light Requirements for Foliage Plants," *Florists' Rev.*, pp. 44–45, 94–99, January 16, 1975.

75. Cathey, H. M. and Campbell, L. E.: "Lamps and Lighting—a Horticultural View," *Light Des. Appl.*, Vol. 4, pp. 41–52, November, 1974.

76. Downs, R. J.: *Controlled Environments for Plant Research*, Columbia University Press, New York, 1975.

77. McCree, K. J.: "Test of Current Definitions of Photosynthetically Active Radiation Against Leaf Photosynthesis Data," *Agric., Meterol.*, Vol. 10, pp. 443–453, 1972.

78. Cathey, H. M., Campbell, L. E., and Thimijan, R. W.: "Plant Growth Under Fluorecent Lamps: Comparative Development of 11 Species," *Florists' Rev.*, pp. 26–29, 67–69, August 31, 1978.

79. Elbert, G. and Elbert, V. F.: *Plants That Really Bloom Indoors*, Simon and Schuster, New York, 1974.

80. Fitch, C. M.: *The Complete Book of House Plants*, Hawthorn, New York, 1972.

81. Gaines, R. L.: *Interior Plantscaping: Building Design for Interior Foliage Plants*, Architectural Record Books, 1st edition, New York, 1977.

82. *Guide to Specifications for Interior Landscaping*, Associated Landscape Contractors of America, 1750 Old Meadow Road, McLean, Virginia, 1979.

83. Kranz, F. H. and Kranz, J. L.: *Gardening Indoors Under Lights*, Viking, New York, 1971.

84. McDonald, E.: *The Complete Book of Gardening Under Lights*, Popular Library, New York, 1973.

85. Orans, M.: *Houseplants and Indoor Landscaping*: A. B. Morse Company, Barrington, Illinois, 1977.

86. *A Guide to Interior Planting*, Everett Conklin & Co., Inc., Montvale, New Jersey, 1973.

87. Withrow, R. B. (Editor): "Photoperiodism and Related Phenomena in Plants and Animals," *Am. Assoc. Adv. Sci.*, Publ. 55, Washington, DC, 1959.

88. Rabinowitch, E. and Govinjee, *Photosynthesis*, John Wiley & Sons, Inc., New York, 1969.

89. Hendricks, S. B.: "How Light Interacts with Living Matter," *Sci. Am.*, Vol. 219, p. 174, September, 1968.

90. Shibles, R.: "Committee Report: Terminology Pertaining to Photosynthesis," *Crop Sci.*, Vol. 16, pp. 437–439, 1976.

Fading and Bleaching

91. Giles, C. A. and McKay, R. B.: "The Light-Fastness of Dyes, a Review," *Text. Res. J.*, p. 528, July, 1963.

92. Luckiesh, M. and Taylor, A. H.: "Fading of Dyed Textiles by Radiant Energy," *Am. Dyest. Rep.*, October 14, 1940.

93. Taylor, A. H.: "Fading of Colored Textiles," *Illum. Eng.*, Vol. XLI, p. 35, January, 1946.

94. Pracejus, W. G. and Taylor, A. H.: "Fading of Colored Materials by Light and Radiant Energy," *Illum. Eng.*, Vol. XLV, p. 149, March, 1950.

95. Luckiesh, M. and Taylor, A. H.: "Fading of Colored Materials by Daylight and Artificial Light," *Trans. Illum. Eng. Soc.*, Vol. XX, p. 1078, December, 1925.

96. AATCC Committee on Color-Fastness to Light: "A Study of the Variables in Natural Light Fading," *Am. Dyest, Rep.*, p. 861, November 18, 1957.

97. Delaney, W. B. and Makulec, A.: "A Review of the Fading Effects of Modern Light Sources on Modern Fabrics," *Illum. Eng.*, Vol. LVIII, p. 676, November, 1963.

98. Hansen, L. J. and Sereika, H. E.: "Factors Affecting Color Stability of Prepackaged Frozen Beef in Display Cases," *Illum. Eng.*, Vol. 64, p. 620, October, 1969.

99. Little, A. H.: "The Effect of Light on Textiles," *J. Soc. Dyers Colour*, Vol. 80, pp. 527–534, 1964.

Infrared Energy

100. Hall, J. D.: *Industrial Applications of Infrared*, McGraw-Hill Book Company, New York, 1947.

101. Tiller, F. M. and Garber, H. J.: "Infrared Radiant Heating," *Ind. Eng. Chem.*, July, 1942, and March, 1950.

102. "Standards for Class A Ovens and Furnaces (Including Industrial Infrared Heating Systems)," *Bulletin No. 86A*, National Fire Protection Association, 60 Batterymarch Street, Boston, Mass., 1969.

103. Frier, J. P. and Stephens, W. R.: "Design Fundamentals for Space Heating with Infrared Lamps," *Illum. Eng.*, Vol. LVII, p. 779, December, 1962.

104. "Heating with Infrared," *Elec. Constr. and Maint.*, Vol. 61, pp. 92, 133, August and October, 1962.

105. Frier, J. P.: "Design Requirements for Infrared Snow Melting Systems," *Illum. Eng.*, Vol. LIX, p. 686, October, 1964.

106. Goodell, P. H.: "Radiant Heating - A Full-Fledged Industrial Tool," *Trans. Am. Inst. of Elec. Eng.*, Vol. 60, p. 464, January, 1941.

107. Bennett, H. J. and Haynes, H.: "Paint Baking with Near Infra-Red," *Chem. Metall. Eng.*, Vol. 47, No. 2, p. 106, February, 1940.

108. Haynes, H.: "The Use of Radiant Energy for the Application of Heat," *Illum. Eng.*, Vol. 36, p. 61, January, 1941.

109. Haynes, H.: "Lamps as Radiant Space Heaters," *The Magazine of Light*, Vol. 12, No. 2, p. 26, March, 1943.

Searchlights

SECTION
20

Prior to the 1940's, searchlights were in general use in military operations primarily for coastal defense, anti-aircraft operations and miscellaneous signalling purposes. During the 1940's and thereafter, development concentrated on searchlights for new tactical operations such as airborne searchlights for antisubmarine warfare, tank-mounted searchlights for combat use, and large searchlights for battlefield illumination. The use of searchlights for such purposes has been, or is being, phased out in the United States (US) because of the rapid development of such replacements as image intensifiers, heat detectors, radar and other electronic aids. At the same time civil uses of ground-based and airborne and marine searchlights has greatly increased.

The type A/N TVS-3, originally intended for battlefield illumination is now being used as a mobile general-purpose unit. See Fig. 20-1. This unit has a 20-kilowatt liquid-cooled xenon short-arc lamp and has a maximum beam intensity of 800 megacandelas. An illuminance of approximately 1350 lux (125 footcandles) has been achieved with multiple units to provide camera and television coverage of the various preflight operations for Apollo launches[1]. A battery of up to 50 has been used, beginning with Apollo 8 (see Fig. 20-2). These units operate continuously for periods up to 90 hours unattended.

Searchlights utilized in helicopter operations range in size from a 450-watt, 0.4 megacandela unit using a type PAR-46 incandescent lamp (see Fig. 20-3) up to a 30-kilowatt, xenon-lamped unit with an elliptical reflector producing a 35-degree beam with a peak intensity of one megacandela (see Fig. 20-4).

These searchlights are utilized in such night operations as search-and-rescue, police surveillance and disaster-area lighting.

Marine Searchlights. Shipboard searchlights are used extensively by vessels travelling the inland waterways of the US to illuminate retroreflectorized aids to navigation indicating the navigable channels in addition to search and rescue operations and general purpose illumination. Typical maritime searchlights range from 30 to 60 centimeters (12 to 24 inches) in diameter and use incandescent lamps, carbon arcs and xenon lamps as sources. See Figs. 20-5 and 20-6.

Searchlights Characteristics

Because of the high luminances of the carbon arc, it has been used extensively where maximum beam intensity is required (mainly in high intensity searchlights and motion picture equipment). The carbon arc lamp mechanism is complex and difficult to maintain and the carbon electrodes are consumed rapidly. Automatic magazine fed lamp mechanisms have been developed to provide operation over longer periods of time; however, the total unattended operating time is still relatively short.

Other high intensity sources such as mercury, mercury-xenon, and xenon short-arc lamps are being used increasingly in a wide variety of searchlight applications. These sources have become available in ratings up to 30 kilowatts. Lamps of this type can operate unattended for long periods of time and do not require the complex feed mechanisms common to carbon arcs; however, they have disadvantages, such as high voltage ignition circuitry for starting (up to 50 kilovolts), and average luminances less than those of the carbon arc.

Searchlights may be defocused to increase the beam spread (divergence), but at considerable sacrifice of intensity. Fig. 20-7 shows the effect of defocusing in a typical searchlight. Fig. 20-8 gives important characteristics of several representative searchlights.

Searchlight Calculations

Useful Range of Searchlights. Searchlight applications fall into two general categories: signal lights which are to be seen, and sources of illumination by which distant targets are to be seen. The range of lights which are themselves to be seen is discussed in Section 3 of the 1981

NOTE: References are listed at the end of each section.

Reference Volume under Visual Range of Luminous Signals, starting on page 3-21.

For the second category, range is a highly complex function of numerous variables which are not independent in their effect. These variables include:

1. The peak intensity of the searchlight.
2. The relative intensity distribution of the searchlight.
3. The atmospheric transmittance.
4. The polar scattering function of the atmosphere.
5. The size and shape of the target.
6. The reflectance of the target.

Fig. 20-1. Army A/N TVS-3 20-kilowatt searchlight with liquid cooled xenon short-arc lamps, with spread lens in the open position. U. S. Army photograph (H-06618).

Fig. 20-2. A/N TVS-3 searchlights at Kennedy Space Center illuminate Apollo-8 space vehicle for preflight checkout.

Fig. 20-3. Aircraft search-landing light, retractable and designed for rotational capability in the extended position, uses a 450-watt, PAR-46 lamp.

7. The location of the target within the beam of the searchlight; that is, whether it is near the beam axis, the near edge of the beam or the far edge.
8. The distance between the target and its background.
9. The reflectance of the background.
10. The lateral offset distance of the observer from the searchlight.
11. The luminance of the target and its background with the searchlight off.
12. The applicable contrast or illuminance threshold of the observer.
13. The effect of binoculars, if used.

The range of a target in a searchlight beam has been studied by several researchers, each of whom restricted his study to a few selected conditions[3-8]. An excellent summary of the analyses of references 4, 5, 6 and 7 has been prepared by Middleton[9].

It is possible, however, to develop equations based upon the referenced studies which, when used in conjunction with data on contrast thresholds and on threshold illuminances (see Section 3 of the 1981 Reference Volume) facilitate the computation of the visual range of targets illuminated by the beams of searchlights.

Fig. 20-4. Airborne application of gimbal mounted 30-kilowatt xenon searchlight. Provides 1 million candelas and 35-degree beam spread.

Fig. 20-6. Searchlight with remote electric control using a 1000-watt xenon arc lamp. Spread of the beam can be changed from the operator's position.

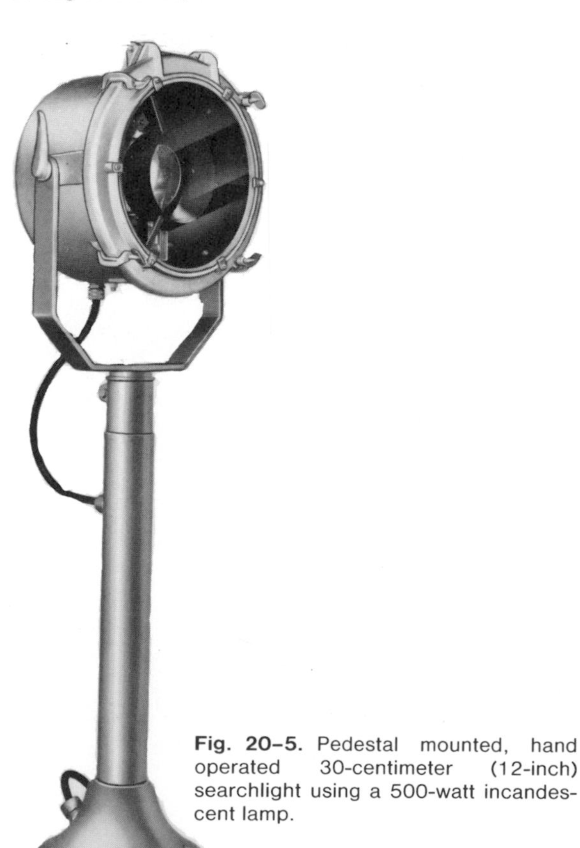

Fig. 20-5. Pedestal mounted, hand operated 30-centimeter (12-inch) searchlight using a 500-watt incandescent lamp.

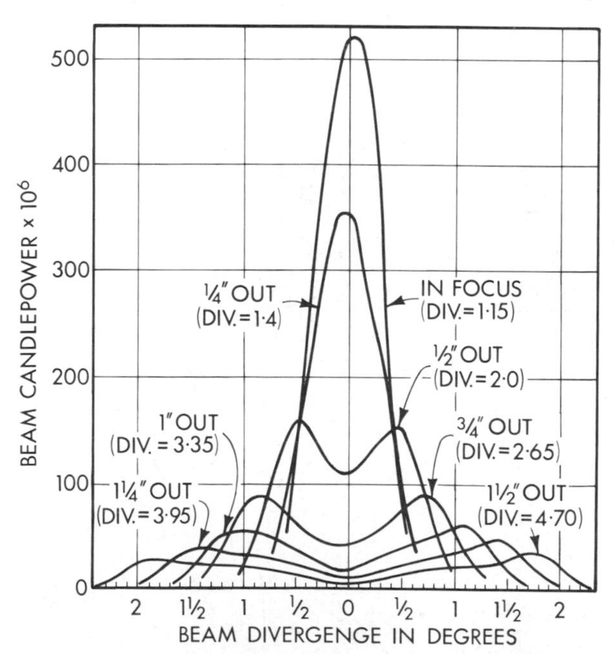

Fig. 20-7. Candlepower distributions of 152-centimeter (60-inch) searchlight with the lamp in focus and out of focus by varying amounts. The lamp is a carbon arc with a 16-millimeter positive operated at 150 amps. The focal length of the reflector is 660 millimeters (26 inches).

Let the arrangement of searchlight, target, background and observer be that shown in Fig. 20-9, where P is a searchlight having an axial intensity I_0 (in candelas) and an intensity distribution which can be approximated by[5, 8]:

$$I_\varphi = I_0 e^{-K\varphi^2} \quad (1)$$

where

K is a constant for a particular beam chosen so that:

$$K = -(\ln 0.01)/\varphi_0{}^2 \quad (2)$$

φ_0 being the half-angle (in radians) of the beam at which the intensity has fallen to one-tenth I_0.

O is the position of the observer, located a distance d from the searchlight.

T is a target, located on the axis of the searchlight beam, a distance R from the searchlight and having a reflectance ρ_T in the direction of point O.

B is the background of the target, located a distance r behind the target, and having a reflectance ρ_B in the direction of point O.

If the apparent contrast C' between the target T and its background B is equal to ϵ, the contrast threshold, then the distance R' is the visual range of the target from the position O. The apparent contrast is given by

$$C' = (L'_T - L'_B)/L'_B \quad (3)$$

where L'_T and L'_B are the apparent luminances of the target and its background at the position O.

If σ is the atmospheric attenuation coefficient (per unit length) and is constant throughout the region of interest, then:

$$L'_T = L_T e^{-\sigma R'} + L_R \quad (4)$$

where

L_T is the inherent luminance of the target and

L_R is the luminance added by light scattered from the searchlight beam in the

Fig. 20-8. Characteristics of Typical Searchlights

Type	Optic	Light Source	Electrical Characteristics	Peak Intensity (approx. megacandelas)	Beam Divergence (degrees)	
					Horizontal	Vertical
Marine	30-cm (12-in) parabolic reflector	Incandescent filament lamp	500W, 120V	0.8	5	5
	46-cm (18-in) parabolic reflector	Incandescent filament lamp	1000W, 120V	3	4½	4¾
	60-cm (24-in) parabolic reflector	Incandescent filament lamp	1000W, 30V	5	3½	3
	36-cm (14-in) parabolic reflector	Carbon-arc	1.4kW, 25A	6	3½	3½
	48-cm (19-in) parabolic reflector	Carbon-arc	2.5kW, 45A	15	4¼	4¼
	36-cm (14-in) short-focus parabolic reflector	Xenon short-arc	1kW, 45A	43	1½	1½
Airborne	15-cm (5.8-in) PAR-46 bulb	Incandescent filament	450W, 28V	0.4	13	14
	25-cm (10-in) parabolic reflector	Xenon short-arc	1.6kW, 63A	4	14	4
	76-cm (30-in) elliptical reflector	Xenon plasma-arc	30kW, 620A	1	35	35
General purpose and (ex-) military	58- x 36-cm (23- x 14-in) parabolic reflector	Xenon short-arc	2.2kW, 95A	100	¾*	¾*
	60-cm (24-in) parabolic reflector	Xenon short-arc	5kW, 140A	250	1⅓*	1⅓*
	76-cm (30-in) parabolic reflector	Xenon short-arc	20kW, 450A	800	1¾	1¾
	152-cm (60-in) parabolic reflector	Carbon-arc	12kW, 150A	500	1¼	1½
Hand-held	14-cm (5.5-in) parabolic reflector**	Incandescent filament lamp	12.5V, 3A	0.2 (in white)	2	2
	38-cm (15-in) parabolic reflector	Xenon short-arc	150W	1	3	3

* Minimum, adjustable.
** Light-signal gun in airport control towers.

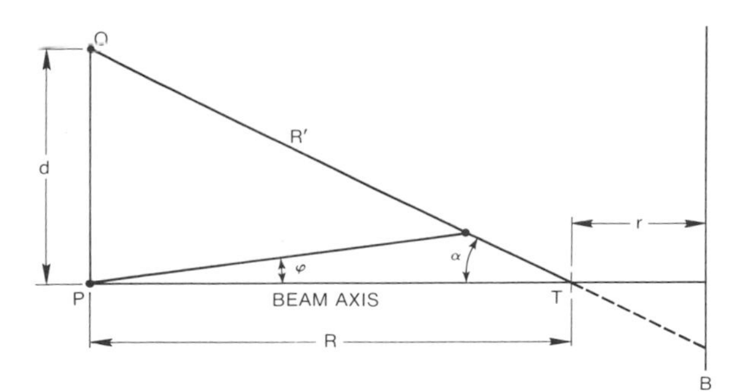

Fig. 20-9. Geometry of target within searchlight beam. *P*, *O*, *T*, and *B* represent the searchlight, observer, target and background locations.

direction of O from points along the line of sight between O and the target. The inherent luminance* of the target is given by:

$$L_T = (\rho_T/\pi)(I_0/R^2 + E_a)e^{-\sigma R} \quad (5)$$

where

E_a is the ambient illuminance on the target,
R is the distance between the searchlight and the target,
ρ is the reflectance of the target, and the target is on the axis of the searchlight.

If the atmospheric attenuation is caused solely by scattering, and R is large in comparison with d, the luminance* added by the light scattered from the searchlight beam is:

$$L_R = (I_0\sigma/\alpha\pi d)\int_0^{\pi/2} S_\varphi d\varphi \quad (6)$$

where

$$S_\varphi = \exp\left[-K\varphi^2 - 2\sigma Rd/(R\varphi + d)\right] \quad (7)$$

Similarly, the apparent luminance of the background, L'_B is :

$$L'_B = L_B e^{-\sigma(R+r)} + L_R + L_r \quad (8)$$

where

L_B is the inherent luminance of the background, L_R and E_a are as previously defined, and L_r is the luminance added by light from the searchlight beam scattered in the direction of O from points along the line of sight between the target and its background. Also:

$$L_B = (\rho_B/\pi)[I_0/(R + r)^2 + E_a]e^{-\sigma(R + r)} \quad (9)$$

* All luminances are in candelas per unit area.

and

$$L_r = (I_0\sigma/8\pi d)\int_\alpha^0 S_\varphi d\varphi \quad (10)$$

where

$$\alpha = rd/R(R + r) \quad (11)$$

Usually the value of E_a is so small in comparison to the values of the other pertinent terms in equations (5) and (8) that it may be neglected. Doing so, combining terms, and simplifying yields:

$$C' = \frac{[(\rho_T/R^2) - (\rho_B/(R + r)^2)e^{-2\sigma r}]e^{-2\sigma R} - (\sigma/8d)\int_\alpha^0 S_\varphi d\varphi}{(\rho_B/(R + r)^2)e^{-2\sigma}(R + r) + (\sigma/8d)\int_\alpha^{\pi/2} S_\varphi d\varphi} \quad (12)$$

If the target is viewed against a sky, or very distant background, r becomes infinite and equation (12) simplifies to:

$$C' = \frac{(\rho_T/R^2)e^{-2\ R} - (\sigma/8d)\int_{\alpha'}^0 S_\varphi d\varphi}{(\sigma/8d\int_{\alpha'}^{\pi/2} S_\varphi d\varphi} \quad (13)$$

where

$$\alpha' = -d/R \quad (14)$$

Note that the apparent contrast is independent of the axial intensity of the searchlight, but is a function of its intensity distribution.

The contrast threshold, ϵ^*, applicable to the viewing situation is a function of the apparent background luminance, L'_B and the angular size of the target. As stated earlier, if C' is equal to ϵ, the distance R is the visual range of the target.

When the target is viewed against a terrestrial background,

$$L'_B = (I_0/\pi)[(\rho_B/(R + r)^2)e^{-2\sigma(R + r)} + (\sigma/8)\int_{\alpha'}^{\pi/2} S_\varphi d\varphi] \quad (15a)$$

and when the target is viewed against a sky background,

$$L'_B = (I_0\sigma/8\pi)\int_{\alpha'}^{\pi/2} S_\varphi d\varphi \quad (15b)$$

* See Fig. 3-31, 1981 Reference Volume.

If the target appears as a point source, instead of an object, it is usually convenient to compute the range R from Allard's Law (see page 3-22 in the 1981 Reference Volume). Under this condition, equation (12) is replaced by:

$$E_m = [I_0 A/\pi][(\rho_T e^{-\sigma R}/R^2) + (\sigma/8d)\left(\int_0^{\pi/2} S_\varphi d\varphi - \int_\alpha^0 S_\varphi d\varphi\right) - \rho_B e^{-\sigma(R+r)}/(R+r)^2](e^{-\sigma R})/R^2 \quad (16)$$

where

E_m is the applicable illuminance threshold.
A is area of the target.

E_m may be determined from Fig. 3-46 by applying the factors discussed on pages 3-23 and 3-24 in the 1981 Reference Volume. The applicable apparent luminance of the background is given by equation 15a or 15b, as applicable.

Under many viewing conditions (for example: targets with a very high reflectance, distant or sky backgrounds, low values of the extinction coefficient), equation (16) may be simplified to:

$$E_m = I_0 \rho_T A e^{-2\sigma R}/\pi R^4 \quad (17)$$

However, equation (17) should be used with caution.

The visual range of a target is affected by its position within the searchlight beam. The magnitude of the effect is a function of most of the parameters listed above. Although quantitative generalizations are not possible, the following qualitative generalizations may be made. Illuminating the target by the "far" side of the beam may lead to losses in range as great as 40 per cent. Small offsets of the target into the "near" side of the beam produce marked increases in range when the observer is close to the searchlight. These increases diminish and finally become decreases as the offset of the target from the beam axis and the displacement of the observer from the searchlight increase[6-8].

The visual range of the target is highly dependent upon the displacement of the observer from the searchlight. For example, with a small target on the axis of the beam of a typical searchlight viewed against the sky in a clear atmosphere, the range increases about 25 per cent as the observer moves from a position 9 meters (30 feet) from the searchlight to a distance of 30 meters (100 feet) from the searchlight, and about 50 and 75 per cent when the displacement is 60 and 150 meters (200 and 500 feet) respectively. The increase is even greater in hazy atmospheres.[6, 7]

For small targets in a clear atmosphere, the range increases roughly as the fourth root of the area and of the reflectance of the target and with the eighth root of the peak intensity of the searchlight, and in haze the increase in range is even less rapid.[5, 6]

The effects of intensity and atmospheric clarity on range are illustrated in Fig. 20-10 presumably based upon the work of Rocard and Reviere.[4] The figure should be considered only as an illustrative example because of the marked effect on range of the parameters discussed above.

Range is also affected by the beam spread of the searchlight, especially in hazy weather with targets of low reflectance.[7, 8] For example, a decrease from a beam spread of 9.6 to 0.6 degrees can increase the range by a factor of three.[7]

The use of binoculars will increase the range of small targets by decreasing the illuminance threshold. For properly designed binoculars the threshold illuminance is inversely proportional to the square of the magnification.[11]

Peak Intensity of Projectors. A useful rough approximation of the peak intensity of a projector may be obtained from the formula:

$$I = L\rho AF$$

where

I = the peak intensity.
L = the average peak luminance of the source within the collecting angle of the optic.
ρ = the reflectance of the reflector or transmittance of the refractor.
A = the area of the searchlight.
F = the ratio of the flashed area to the total area of the aperture.

The factor F is dependent not only upon the proportion of the aperture obscured by the lamp and its supports, but also upon the size of the source and the accuracy of the optics. For projectors producing collimated beams, this factor typically varies from about 0.9 for projectors with precision optics and large sources to about 0.3 for projectors with non-precision optics and small sources. If the beam is spread by use of spreader lenses or by moving the source away from the focus, F is given roughly by the relation:

$$F = F_0 \theta_0 \phi_0 / \theta \phi$$

where

F_0 is the factor which would be applicable if the beam were made collimated by removing the spread

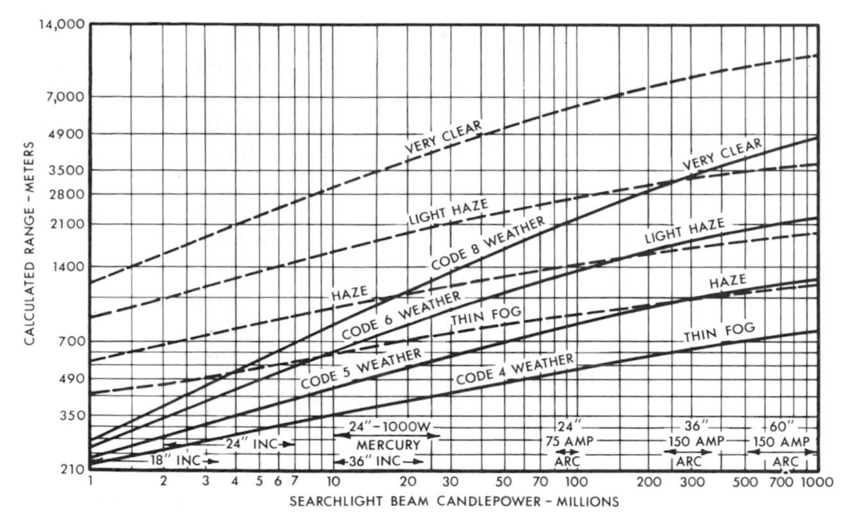

Code Number	Weather	Maximum Daylight Visibility
0	dense fog	50 meters
1	thick fog	200 meters
2	mod. fog	500 meters
3	light fog	1000 meters
4	thin fog	2 km
5	haze	4 km
6	light haze	10 km
7	clear	20 km
8	very clear	50 km
9	except. clear	over 50 km

Fig. 20-10. The effect of atmospheric clarity on the useful range of targets in searchlight beams. Solid lines—Medium target such as a group of men or a dark-colored car against a slightly lighter background such as a field or building. Broken lines—Large target such as a large aircraft or battleship, or a church steeple against the sky. (See page 3-23 in the 1981 Reference Volume for explanation of terms describing atmospheric clarity.)

lens or by placing the source at the focus; and

θ_0 and ϕ_0 are the horizontal and vertical beam spreads which would be obtained if the beam were collimated; and θ and ϕ are the horizontal and vertical beam spreads of the actual beam.

Benford[12] has made a very comprehensive study of the intensity of light and its distribution in a searchlight beam.

Beam Spread (Divergence) of Projectors. To a first approximation, the beam spread of a projector producing a collimated beam can be obtained from the relation:

$$\theta = \tan^{-1}(l/f)$$

where

θ = the horizontal or vertical beam spread.
l = the width or height of the source, as applicable.
f = the focal length of the optic.

Aberrations, imperfections and incorrect adjustments tend to increase the beam spread.

If a collimated beam is made more divergent by the addition of a spreader lens, then a rough approximation of the beam spread may be obtained from the relation:

$$\theta' = \theta + \tan^{-1}(w/f')$$

where

θ' = the spread of the divergent beam.
w = the width of a single flute of the spread lens.
f' = the focal length of the flute in the plane perpendicular to its axis.

If the additional spread is obtained by moving the source away from the focus, the following relation is applicable:

$$\theta' = \tan^{-1}(l/u)$$

where

l = the width or height of the source, as applicable.
u = the distance between the optic and the source.

See also references 10 and 12.

REFERENCES

1. Freeman, R. S. and Ayling, R. J.: "Xenon Arc Searchlight Illumination of the Apollo 8 Launch Area," *J. Soc. Motion Pict. Telev. Eng.*, Vol. 79, p. 313, April, 1970.
2. Chivers, E. W. and Jones, D. E. H.: "The Function and Design of Army Searchlights," *Symposium on Searchlights*, p. 7, Illuminating Engineering Society (London), 1948.
3. Blondel, A.: "A Method for Determining the Visual Range of Searchlights," *The Illuminating Engineer (London)*, Vol. 8, p. 85, February, and p. 153, April, 1915.
4. Rocard, Y.: "Visibilité des Buts Eclaire's par un Projector, Premier Partie," *Revue d'Optique*, Premier Partie, Vol. 11, p. 193, May and p. 257 June-July, 1932; Deuxième Partie, Vol. 11, p. 439, October, 1932; Troisième Partie (by Riviere, J. and Rocard, Y.), Vol. 13, p. 160, April, and 204, May, 1933.
5. Hampton, W. M.: "The Visibility of Objects in a Searchlight Beam," *Proc. Physical Society (London)*, Vol. 45, p. 663, Septem-

ber, 1933. "Discussion," *Symposium on Searchlights*, p. 133, Illuminating Engineering Society (London), 1948.

6. Hulburt, E. O.: "Optics of Searchlight Illumination," *J. Opt. Soc. Amer.*, Vol. 36, p. 483, August, 1946.
7. Chesterman, W. D. and Stiles, W. S.: "The Visibility of Targets in a Naval Searchlight Beam," *Symposium on Searchlights*, p. 7, Illuminating Engineering Society (London), 1948. Biggs, S. S. and Waldram, J. M.: "Some Visibility Problems Associated with Anti-Aircraft Searchlight Beams," ibid, p. 103.
8. Blackwell, H. R., Duntley, S. Q., and Kincaid, W. M.: "Characteristics of Tank-Mounted Searchlights for Detection of Ground Targets," Armed Forces—National Research Council Vision Committee, March 1953, Washington, D.C.
9. Middleton, W. E. K.: "Vision Through the Atmosphere," Section 7.2, University of Toronto Press, 1952.
10. Benford, F. and Bock, J. E.: "The Normal Probability Curve as an Approximation to the Distribution Curve of High Intensity Searchlights," *J. Opt. Soc. Amer.*, Vol. 38, p. 527, June, 1948.
11. Tousey, R. and Hulburt, E. O.: "Visibility of Stars in the Daylight Sky," *J. Opt. Soc. Amer.*, Vol. 38, p. 886, October, 1948. Tousey, R. and Koomen, M. J.: "Visibility of Stars and Planets during Twilight," *J. Opt. Soc. Amer.*, Vol. 43, p. 177, March, 1953.
12. Benford, F.: "The Projection of Light," *J. Opt. Soc. Amer.*, Vol. 35, p. 149, February 1945; and "Studies on the Projection of Light," *General Electric Review*, Vol. 26, p. 75, February, 1923 to Vol. 27, p. 252, April, 1924.

Credits for Illustrations and Tables

The Illuminating Engineering Society is indebted to the many individuals, committees and organizations which contributed the multitude of illustrations and tables published in this Handbook. Many of the illustrations and tables omitted from the following listing appeared in previous publications issued by the Society or were supplied by committees of the Society especially for use in this volume.

Contributors

1. American Society of Heating, Refrigerating and Air Conditioning Engineers, Inc., New York, NY
2. Association of American Railroads, New York, NY
3. Benzio, V., Milan, Italy
4. Carlisle & Finch Co., Cincinnati, OH
5. Crouse-Hinds Company, Syracuse, NY
6. Duro-Test Corporation, North Bergen, NJ
7. EG&G Environmental Equipment Division, Waltham, MA
8. Fairweather, J. R., Lansdale, PA
9. General Electric Company, Nela Park, Cleveland, OH
10. General Railway Signal Company, Rochester, NY
11. Grimes Manufacturing Co., Urbana, OH
12. GTE Products Corporation, Danvers, MA
13. Harmon, D. B., *The Co-ordinated Classroom*, American Seating Co., Grand Rapids, MI and F. W. Wakefield Brass Co., Vermilion, OH
14. Harrison, L. S., Coral Gables, FL
15. Herbert Levine Associates, New York, NY
16. Indiana University, Bloomington, IN
17. International Light, Inc., Newburyport, MA
18. *Lighting & Lamps*, New York, NY
19. Lincoln Center for the Performing Arts, Inc., New York, NY
20. Matsushita Electric Industrial Co., Ltd., Japan
21. The Metropolitan Museum of Art, New York, NY
22. Newport News Shipbuilding and Dry Dock Co., Newport News, VA
23. *Philips Technical Review*, Eindhoven, Holland
24. Prismo Safety Corp., Huntingdon, PA
25. Projector, T. H., Lewes, DE
26. Public Service Company of Indiana, Plainfield, IN
27. Pullman-Standard, Chicago, IL
28. Ripman Lighting Consultants, Belmont, MA
29. Sampson, F. K., Los Angeles, CA
30. Smith, Hinchman & Grylls Associates, Inc., Detroit, MI
31. Society of Motion Picture and Television Engineers, Scarsdale, NY
32. Specter, D. K., New York, NY
33. Steelcase Manufacturing Company, Grand Rapids, MI
34. Stimsonite Plastics, Chicago, IL
35. Tao, W. K. Y., St. Louis, MO
36. Uitg. Stichting "Prometheus," N.A. Voorburgwal 27 1/3, Amsterdam, Netherlands
37. Union Switch & Signal Division of Westinghouse Air Brake Co., Pittsburgh, PA
38. U. S. Army, Fort Belvoir, VA
39. U. S. Army, Washington, DC
40. D. Van Nostrand Co., Inc., New York, NY, *Light, Vision & Seeing*, Luckiesh.
41. Western Cataphote Corporation, Toledo, OH
42. Westinghouse Electric Corporation, Bloomfield, NJ
43. Zahour, R. L., Bloomfield, NJ

Credits

Section 1.

1-1: **9.** 1-3: **9.** 1-4: **9.** 1-5: **9.** 1-6: **9.** 1-7: **9.** 1-8: **18.** 1-9: **33.**

Section 2.

2-6: **29.** 2-7: **29.** 2-11: **23.** 2-12: **13.** 2-13: **40.** 2-17: **1.**

Section 4.

4-5: **35.**

Section 5.

5-2: **29.** 5-3: **42.** 5-4: **42.** 5-6: **33.** 5-7: **33.** 5-8: **30.**

Section 7.

7-8: **26.** 7-15: **28.** 7-16: **28.** 7-17: **28.** 7-18: **28.** 7-19: **28.** 7-21: **28.** 7-22: **28.** 7-26a: **14.** 7-26b: **21.** 7-26c: **21.** 7-26d: **21.** 7-27: **21.** 7-28: **14.**

Section 8.

8-10: **32.** 8-11: **9.**

Section 10.

10-21: **9.**

Section 11.

11-2: **31.** 11-3: **31.** 11-6 upper: **19.** 11-7: **31.** 11-13 lower: **16.** 11-16: **9.** 11-17: **42, 9.** 11-18: **31.** 11-19: **31.** 11-20: **31.** 11-21: **31.** 11-22: **31.** 11-23: **31.** 11-24: **31.**

Section 12.

12-2: **9.** 12-3: **9.** 12-4: **9.** 12-5: **9.** 12-6: **9.** 12-7: **9.** 12-10: **3.** 12-11: **20.** 12-12: **20.** 12-13: **9.** 12-14: **9.** 12-15: **9.** 12-16: **8.**

Section 13.

13-1: **42.** 13-7: **42.** 13-26: **42.**

Section 15.

15–4: **11.**

Section 16.

16–2: **24.** 16–3: **34.** 16–4: **41.** 16–9: **27.** 16–15: **22.** 16–16: **9.** 16–17: **22.** 16–18: **9.** 16–19: **9.** 16–20: **36.** 16–21: **10.** 16–23: **2.** 16–24: **37.**

Section 17.

17–1: **6.** 17–5: **9.** 17–7: **15.** 17–20: **9.**

Section 18.

18–9a: **7.** 18–9c: **7.**

Section 19.

19–1: **9.** 19–13: **42.** 19–14: **9.** 19–16: **9.** 19–17: **9.** 19–21: **17.** 19–23: **12.** 19–24: **43.** 19–25: **12.** 19–26: **9.** 19–27: **9.** 19–28: **12.** 19–29: **43.** 19–30: **12.** 19–32: **9.** 19–33: **9.**

Section 20.

20–1: **38.** 20–2: **6.** 20–3: **11.** 20–4: **39.** 20–5: **5.** 20–6: **4.** 20–7: **25.** 20–10: **9.**

Index*

Pages are numbered consecutively within each section

* Page numbers in **bold face** are in this Handbook. Pages in light face are in the *IES Lighting Handbook-1981 Reference Volume.* For terms not listed, see Section 1 in the 1981 Reference Volume.

* Page numbers in **bold face** are in this Handbook. Pages in light face are in the *IES Lighting Handbook-1981 Reference Volume*. For terms not listed, see Section 1 in the 1981 Reference Volume.

* Page numbers in **bold face** are in this Handbook. Pages in light face are in the *IES Lighting Handbook-1981 Reference Volume.* For terms not listed, see Section 1 in the 1981 Reference Volume.

* Page numbers in **bold face** are in this Handbook. Pages in light face are in the *IES Lighting Handbook-1981 Reference Volume.* For terms not listed, see Section 1 in the 1981 Reference Volume.

* Page numbers in **bold face** are in this Handbook. Pages in light face are in the *IES Lighting Handbook-1981 Reference Volume*. For terms not listed, see Section 1 in the 1981 Reference Volume.

* Page numbers in **bold face** are in this Handbook. Pages in light face are in the *IES Lighting Handbook-1981 Reference Volume*. For terms not listed, see Section 1 in the 1981 Reference Volume.

* Page numbers in **bold face** are in this Handbook. Pages in light face are in the *IES Lighting Handbook-1981 Reference Volume.* For terms not listed, see Section 1 in the 1981 Reference Volume.

* Page numbers in **bold face** are in this Handbook. Pages in light face are in the *IES Lighting Handbook-1981 Reference Volume.* For terms not listed, see Section 1 in the 1981 Reference Volume.

* Page numbers in **bold face** are in this Handbook. Pages in light face are in the *IES Lighting Handbook-1981 Reference Volume*. For terms not listed, see Section 1 in the 1981 Reference Volume.

* Page numbers in **bold face** are in this Handbook. Pages in light face are in the *IES Lighting Handbook-1981 Reference Volume.* For terms not listed, see Section 1 in the 1981 Reference Volume.

* Page numbers in **bold face** are in this Handbook. Pages in light face are in the *IES Lighting Handbook-1981 Reference Volume.* For terms not listed, see Section 1 in the 1981 Reference Volume.

* Page numbers in **bold face** are in this Handbook. Pages in light face are in the *IES Lighting Handbook-1981 Reference Volume.* For terms not listed, see Section 1 in the 1981 Reference Volume.

* Page numbers in **bold face** are in this Handbook. Pages in light face are in the *IES Lighting Handbook-1981 Reference Volume.* For terms not listed, see Section 1 in the 1981 Reference Volume.

* Page numbers in **bold face** are in this Handbook. Pages in light face are in the *IES Lighting Handbook-1981 Reference Volume.* For terms not listed, see Section 1 in the 1981 Reference Volume.

* Page numbers in **bold face** are in this Handbook. Pages in light face are in the *IES Lighting Handbook-1981 Reference Volume.* For terms not listed, see Section 1 in the 1981 Reference Volume.

* Page numbers in **bold face** are in this Handbook. Pages in light face are in the *IES Lighting Handbook-1981 Reference Volume.* For terms not listed, see Section 1 in the 1981 Reference Volume.

* Page numbers in **bold face** are in this Handbook. Pages in light face are in the *IES Lighting Handbook-1981 Reference Volume.* For terms not listed, see Section 1 in the 1981 Reference Volume.

* Page numbers in **bold face** are in this Handbook. Pages in light face are in the *IES Lighting Handbook-1981 Reference Volume.* For terms not listed, see Section 1 in the 1981 Reference Volume.

* Page numbers in **bold face** are in this Handbook. Pages in light face are in the *IES Lighting Handbook-1981 Reference Volume*. For terms not listed, see Section 1 in the 1981 Reference Volume.

* Page numbers in **bold face** are in this Handbook. Pages in light face are in the *IES Lighting Handbook-1981 Reference Volume.* For terms not listed, see Section 1 in the 1981 Reference Volume.

* Page numbers in **bold face** are in this Handbook. Pages in light face are in the *IES Lighting Handbook-1981 Reference Volume.* For terms not listed, see Section 1 in the 1981 Reference Volume.

V

W

---

* Page numbers in **bold face** are in this Handbook. Pages in light face are in the *IES Lighting Handbook-1981 Reference Volume.* For terms not listed, see Section 1 in the 1981 Reference Volume.

* Page numbers in **bold face** are in this Handbook. Pages in light face are in the *IES Lighting Handbook-1981 Reference Volume*. For terms not listed, see Section 1 in the 1981 Reference Volume.

Illuminating Engineering Society of North America

IES of North America is a recognized technical authority for the illumination field. For 75 years its objective has been to communicate information about all aspects of good lighting practice to individual members, the lighting industry and consumers through a variety of programs, publications and services. The strength of IES is in its diversified membership: engineers, architects, designers, educators, students, contractors, distributors, utility personnel, manufacturers and scientists, all *contributing to* and *benefiting from* the Society.

PROGRAMS

IES sponsored local, regional and transnational meetings, and conferences, symposiums, seminars, designers' forums, roadway forums, workshops and lighting exhibitions provide an access to the latest developments in the field through audio-visual presentations and expert speakers. Basic and advanced IES lighting courses are offered by local IES Sections and in cooperation with other organizations. Other Society programs include liaison with school and colleges and career information for students and counselors.

PUBLICATIONS

Lighting Design & Application (LD&A), and the *Journal of the Illuminating Engineering Society* are the official magazines of the Society. *LD&A* is a popular application-oriented monthly magazine. Every issue contains special feature articles and news of practical and innovative lighting layouts, systems, equipment and economics, and news of the industry and its people. The *Journal*, a technical quarterly, contains official transactions: American National Standards, IES recommended practices, technical committee reports, conference papers and research reports; and other technically-oriented materials.

In addition to Handbooks, the Society offers nearly 100 varied publications including: education courses; IES technical committee reports covering many specific lighting applications; forms and guides used for measuring and reporting lighting values, lighting calculations, performance of light sources and luminaires, energy management, etc. Also, the *IES Lighting Library* provides a complete reference package in an oversized loose-leaf binder. The Library, encompassing all essential IES documents, is available with a yearly maintenance service to keep it up-to-date.

Complete lists of current and available IES publications are published periodically in its official magazines or may be obtained by writing to the Publications Office of the Illuminating Engineering Society.

SERVICES

IES provides professional staff assistance with technical problems, reference help and interprofessional liaison with AIA, AID, IEEE, NAED, NECA, NEMA, NSID and other groups. IES is a forum for exchange, professional development and recognition. It correlates the vast amount of research, investigation and discussion through hundreds of qualified members of its technical committees to guide lighting experts and laymen on research-based lighting recommendations.

The Society has two types of membership: individual and sustaining. Applications and current dues schedules are available upon request from the Membership Department of the Illuminating Engineering Society.